T0344918

DESIGN OF ULTRA WIDEBAND POWER TRANSFER NETWORKS

DESIGN OF ULTRA WIDEBAND POWER TRANSFER NETWORKS

Binboga Siddik Yarman

College of Engineering,
Department of Electrical-Electronics Engineering,
Istanbul University, 34320 Avcilar, Istanbul, Turkey.

A John Wiley and Sons, Ltd., Publication

This edition first published 2010
© 2010 John Wiley & Sons, Ltd

Registered office
John Wiley & Sons Ltd, The Atrium, Southern Gate, Chichester, West Sussex, PO19 8SQ, United Kingdom

For details of our global editorial offices, for customer services and for information about how to apply for permission
to reuse the copyright material in this book please see our website at www.wiley.com.

Library of Congress Cataloging-in-Publication Data

Yarman, Binboga Siddik.
  Design of ultra wideband power transfer networks / by Binboga Siddik Yarman.
     p. cm.
  Includes bibliographical references and index.
  ISBN 978-0-470-31989-5 (cloth)
  1. Ultra-wideband antennas—Design and construction.   2. Broadband communication
systems—Power supply—Design and construction.   3. Telecommunication lines—Design and
construction.   4. Broadband amplifiers—Design and construction.   5. Impedance matching.
  6. Electric power transmission.   I. Title.
  TK7871.67.U45Y37 2010
  621.384′135—dc22

                                                                              2009047980

A catalogue record for this book is available from the British Library.

ISBN 978-0-470-31989-5

Set in 9/11pt Times by Integra Software Services Pvt. Ltd, Pondicherry, India
Printed in Singapore by Markono Print Media Pte Ltd.

This book is dedicated to my wife, Prof. Dr Md. Sema Yarman, and my son, Dr Can Evren Yarman, for their continuous support, endless patience and love.

Contents

About the Author

Professor Dr B. S. Yarman received his BSc, MEE and PhD degrees from Istanbul Technical University; Stevens Institute of Technology, Hoboken, NJ; and Cornell University, Ithaca, NY, respectively. He was a Member of Technical Staff at RCA David Sarnoff Research Center, Princeton, NJ, where he was responsible for designing various broadband microwave and satellite communication systems for commercial and military use.

Professor Yarman has taught and done research at Anadolu University, Middle East Technical University, Istanbul University and Istanbul Technical University in Turkey; Stevens Institute of Technology and Cornell University in the USA; Ruhr University in Germany; and Tokyo Institute of Technology in Japan.

He is currently the Chairman of the Department of Electrical–Electronics Engineering and the Coordinator of Scientific Research Projects of Istanbul University, Turkey.

He was the founding president of Isik University and was one of the founders of International Education Research and Engineering Consulting Inc. in Maryland USA; Savronik Defence Electronics Corp. of Turkey; and ARES Electronic Security Inc., Istanbul, Turkey.

He has served as a consultant on the design of various broadband matching networks and microwave amplifiers for many commercial and military agencies in the USA, Europe and Asia Pacific, as well as in Turkey.

He has published more than 300 journal and conference papers as well as technical reports in the area of broadband matching networks, microwave amplifiers, digital phase shifters, speech and biomedical signal processing (ECG, EEC, EMG, etc.) and decision making. He is the author of the books '*Design of Multistage Microwave Amplifiers via Simplified Real Frequency Technique*' published by Scientific Research and Technology Council of Turkey, 1986; '*Design of Ultra Wideband Antenna Matching Networks*' by Springer-Verlag' 2008; and '*Design of Ultra Wideband Power Transfer Networks*' by John Wiley & Sons, Ltd, 2009. He also holds four US patents as assigned to the US Air Force.

Professor Yarman is the recipient of the Young Turkish Scientist Award, Technology Award of National Science-Technology and Research Counsel of Turkey. He is a Fellow of the Alexander Von Humboldt Foundation, Germany; a Member of the New York Academy of Science; 'Man of the Year in Science and Technology' in 1998 of Cambridge Biography Centre, UK; and IEEE Fellow for his contribution to 'Computer Aided Design of Broadband Amplifiers'.

He is married to Prof. Dr Md. Sema Yarman of Istanbul University and is the father of Dr Can Evren Yarman of Schlumberger Houston, Texas.

Preface

Power transfer networks (PTNs) are essential units of communications systems. For example, if the system is a transmitter, a PTN must be placed between the output of the power amplifier and the antenna. If the system is a receiver, the PTN is placed between the antenna and the low-noise amplifier. Any interface or interstage connection must be made over a PTN.

In general, PTNs are lossless two-ports. They transfer the frequency-dependent power between ports over a prescribed frequency band. Depending on the application, they are referred to as filters, matching networks or equalizers.

From the circuit theory point of view, a port may be modeled as a simple resistor or as a complex impedance. In this regard, the power transfer problem is defined as the 'construction of a lossless two-port between the given terminations over a specified frequency band'.

In the course of the PTN design process, power transfer is maximized from the source to the receiving end over the band of interest.

From the physical nature of the problem, we can only transfer a fraction of the available power of the generator. In this case, our concern is with the power transfer ratio, which is defined as the power delivered to the load in relation to the available power of the generator. This ratio is called transducer power gain (TPG).

In practice, our desire is to make the power transfer as flat and as high as possible over the passband.

It is well established that flat TPG level is restricted by the complex terminations. This is called the gain–bandwidth limit of the power transfer problem under consideration. In the classical literature, gain–bandwidth problems are known as broadband matching problems. They may be classified as follows:

- Filter or insertion loss problem: In this problem, a lossless two-port is constructed between the resistive terminations over the specified passband. In other words, the goal of the filter problem is to restrict the power transfer to a selected frequency band. In this case, ideally, the flat TPG level can be unity if an infinite number of reactive elements are used in the lossless two-port.
- Single matching problem: In this problem, a lossless two-port is constructed between a resistive generator and a complex load. It has been shown that the ideal flat TPG level is dictated by the complex load and is less than unity.
- Double matching problem: This is the generalized version of the single matching problem where both the generator and load networks are complex impedances. Therefore, the flat TPG level is even further reduced than those of the single matching problems defined by either the generator or load impedance.
- Active matching problem: In this problem an active device is matched to a complex generator at the frontend and also simultaneously matched to a complex load at the backend.
- Equalization problem: In this problem, a lossless two-port is constructed between resistive terminations which approximate predefined arbitrary TPG shape over a prescribed frequency band.

This book covers all the power transfer problems comprehensively. Solutions to many practical problems are provided with design software (S/W) packages developed on MATLAB®. In this regard, the book is unique.

In order to tackle power transfer problems thoroughly, an understanding of circuit theory with lumped and distributed elements is essential. Furthermore, the practical implementation of PTNs requires a straightforward application of electromagnetic field theory. Hence, the book is organized as follows.

Chapter 1 covers the basic ingredients of circuit theory from the power transfer point of view. In this chapter, it is emphasized that lumped elements are dimensionless. Furthermore, they 'do not care' about the velocity of power transfer. Therefore, they are ideal and excellent tools for designing PTNs. However, in practice we need more.

Chapter 2 is devoted to electromagnetic fields and waves, where we define all the passive lumped circuit components from the field theory perspective by introducing material properties and geometric layouts. Moreover, major properties of ideal transmission lines are derived by employing electromagnetic field theory, which makes power transfer issues physically understandable.

In Chapter 3, transmission lines are introduced as viable practical circuit components having geometric dimensions. From a practical implementation point of view, it is shown that a short transmission line may act like an inductor in series configuration, or like a capacitor in shunt configuration. It may even behave like a transformer or resonance circuit if its length or operating frequency is adjusted properly. In this chapter, we also introduce a complex variable denoted by $\lambda = \Sigma + j\Omega$ which is called the Richard variable. It is shown that lossless networks constructed with equal length or commensurate transmission lines can be described by means of classical network functions such as impedance and admittance functions by $\lambda = \tanh(p\tau)$. In this representation, $p = \sigma + j\omega$ is the classical complex domain variable which is used to describe the functions derived from the networks constructed with lumped elements and τ is the constant delay of the commensurate transmission lines.

In Chapter 4, the concept of unit element (UE) is introduced and the properties of circuits constructed with commensurate transmission lines or UEs are presented. Natural definitions of incident and reflected waves are given and then a scattering description of lossless two-ports is introduced in the Richard domain. The power transfer issue is studied by means of the scattering parameters. Features of filters designed with commensurate transmission lines are provided from experiments run on MATLAB employing the design tools developed for this chapter. Many practical examples are presented to demonstrate the utilization of the design packages.

In Chapter 5, the general equalization problem is solved by employing UEs via the scattering approach, which is known as the simplified real frequency technique (SRFT). Examples and design S/W are provided.

In Chapter 6, the properties of lossless two-ports constructed with lumped circuit elements are investigated by means of scattering parameters. A formal definition of power transfer is introduced and a definition of transfer scattering parameters is given. The power transfer properties of cascade connections of two-ports are ther derived. Examples are presented to help the reader understand the properties of lossless two-ports from a design point of view.

As far as computer-aided design of PTNs is concerned, descriptions of lossless two-ports in terms of 'easy to use' parameters are crucial. For example, a lossless two-port can be described by means of the component values of a chosen circuit topology. In this regard, TPG is expressed as a function of the unknown element values. Then it is optimized to satisfy the design specifications, which in turn yield the element values of the chosen topology. Unfortunately, this task is very difficult since TPG is highly nonlinear in terms of the unknown element values. On the other hand, it is always desirable to deal with quadratic objective functions in optimization problems.

In 1977, Professor H. J. Carlin of Cornell University proposed a design method for the solution of single matching problems, which deals with quadratic objective functions. The new method of broadband matching is called the 'real frequency line segment technique (RFLT)' and it is based on the famous theorem of Darlington, who proved that any positive real (PR) immittance function can be realized as a lossless two-port in resistive termination. In other words, TPG of the power transfer problem can be expressed in terms of the driving point input immittance of the lossless two-port. Then, the PR immittance function is determined in such a way that TPG is optimized. In Carlin's approach, over the real frequencies, the real part of a PR driving point function is described by the mean of line segments; it is shown that TPG is quadratic in terms of the selected points of line segments. Therefore, in the optimization process, we are able to deal with a quadratic objective function which makes the optimization almost always convergent.

Based on the above explanations, in designing PTNs, generation of PR immittance is quite important. Therefore, Chapters 7 and 8 are devoted to generating and modeling the realizable PR driving point functions using a parametric approach and the Gewertz procedure respectively. In this regard, MATLAB S/W tools are developed to solve many practical problems. Examples are presented to show the utilization of S/W tools.

Chapter 9 deals with Darlington synthesis of a PR immittance function, which is essential for the construction of lossless two-ports for real frequency techniques.

In order to understand the nature of the power transfer problem, the analytic theory of broadband matching is indispensable. Therefore, Chapter 10 is devoted to the analytic theory of broadband matching. It is shown that, beyond simple problems, the theory is inaccessible. Nevertheless, it is shown that filter theory can be expanded to solve simple single and double matching problems analytically. Hence, in this chapter, programs are developed to solve practical matching problems by utilizing the modified filter theory. Several examples are presented to show the utilization of the S/W.

The early 1980s witnessed the derivation of the analytic theory of double matching and also the expanded RFLT concept to design complicated single and double matching networks as well as microwave amplifiers. The new techniques are called the 'real frequency direct computational technique or RFDT', the 'RF parametric approach' and the 'simplified real frequency technique or SRFT'. Thus, Chapter 11 deals with all the versions of real frequency techniques. Many complicated real-life problems are solved using the RF design tools developed with MATLAB. Examples include the design of matching networks for a complicated monopole antenna, for a helix GPS antenna and for an ultrasonic piezoelectric transducer. Furthermore, the design of ultra wideband microwave amplifiers using lumped and distributed elements is also presented. Obviously, the reader of this book can utilize the S/W tools provided to solve many crucial matching problems.

In many engineering applications, modeling of the measured immittance data is inevitable. For example, in RFLT, the driving point immittance of the lossless equalizer is generated point by point to optimize TPG. At the end of the design process, computer-generated immittance data must be modeled as a realizable positive real function so that it can be synthesized as a lossless two-port in resistive termination, yielding the desired matching network. Similarly, the analytic theory of matching requires models for both measured generator and load impedances. Therefore, in Chapter 12, we introduce our modeling tools via linear interpolation techniques.

The practical design of broadband matching networks must include both lumped and distributed elements, where all the geometric sizes and related parasitic elements are imbedded into the design. This is a very difficult task from a circuit theory point of view. However, our continuing efforts in the field recently matured in the design of broadband matching networks with mixed lumped and distributed elements. Chapter 13 covers this design. The MATLAB codes provided with this book can be found at http://www.wiley.com/go/yarman_wideband.

Acknowledgments

I should mention that all the design S/W provided with this book has been developed in the scientific sprit of sharing our knowledge, accumulated over the last 30 years. Including myself, the S/W reflects the blessed labor of many outstanding researchers, namely S. Darlington, H. W. Bode, C. Gewertz, R. M.

Fano, V. Belevitch, D. C. Youla, R. Levy, W. K. Chen, H. J. Carlin and A. Fettweis. The S/W is by no means professional and may include some bugs. Nevertheless, it provides solutions to all the worked examples in this book.

On the other hand, our design of broadband matching networks has been developed with the programs provided in this book. As the input, we feed in measured data; as the output, we automatically receive the optimum circuit topology with element values which optimize TPG as desired. This is nice, despite the bugs. At this point, I should mention that lumped element Darlington synthesis of positive real functions is essential to obtain lossless equalizers. The synthesis program in this book was developed by Dr Ali Kilinc of Okan University, Istanbul, Turkey. In this regard, his continuous support is gratefully acknowledged.

It is my hope that, having this book as a base, the readers, namely researchers and professional engineers in the field, will develop outstanding user-friendly design tools to construct optimum matching networks for various kinds of commercial and military applications. Therefore, they should feel free to contact me at yarman@istanbul.edu.tr in case help is needed.

Finally, I would like to take this opportunity to thank to my dear friends Mrs Asli Divris and Dr Birep Aygun for their careful reading of and corrections to the manuscript. I should also acknowledge the constructive guidance of Miss Skinner of Wiley in the course of completing the book.

Binboga Siddik Yarman
yarman@istanbul.edu.tr
Farilya, Bodrum, Turkey
August 2009

In Memory of H. J. Carlin

It would not have been possible to complete this book without the spiritual guidance of Professor H. J. Carlin, who passed away on February 9, 2009. He was the initiator of the real frequency techniques which facilitated the design and implementation process of all kinds of power transfer networks.

1

Circuit Theory for Power Transfer Networks

1.1 Introduction

In circuit theory, a power transfer network is known as a lossless two-port which matches a given voltage generator with internal impedance Z_G to a load Z_L. The lossless two-port consists of lossless circuit elements such as capacitors, inductors, coupled coils, transmission lines and transformers.

In practice, the complex impedances Z_G and Z_L are measured and modeled using idealized lossy and reactive circuit elements. In circuit theory, losses are associated with resistors. Reactive elements can be considered as capacitors, inductors, transmission lines or a combination of these.

It is well known that passive or lossy impedances consume energy. This is also known as power dissipation (i.e. energy consumption per unit time).

For given design specifications, such as the frequency band of operations and a desirable minimum flat gain level, the design problem of a power transfer network involves fundamental concepts of circuit theory. On the other hand, the fundamentals of circuit theory stem from electromagnetic fields. Especially at high frequencies, where the size of the circuit components is comparable to the wavelength of operational signals, use of electromagnetic field theory becomes inevitable for assessing the performance of the circuits. Therefore, at high frequencies, circuit design procedures must include electromagnetic field-dependent behavior of circuit components to produce actual reliable electrical performance.

In designing power transfer networks, we usually deal with mathematical functions employed in classical circuit theory.[1] These functions are determined directly from the given design specifications by means of optimization. Eventually, they are synthesized at the component level, yielding the desired power transfer network. Therefore, a formal understanding of circuit functions and their electromagnetic field assessments is essential for dealing with design problems.

---

[1] Circuit functions may be described as positive real driving point impedance or admittance functions or corresponding bounded real input reflection coefficients. The mathematical properties of these functions will be given in the following chapters.

Design of Ultra Wideband Power Transfer Networks   Binboga Siddik Yarman
© 2010 John Wiley & Sons, Ltd

As mentioned above, power transfer networks are designed as lossless two-ports which may include only reactive lumped elements,[2] or only distributed elements, or a combination of both; that is, lumped and distributed elements. Usually, distributed elements are considered as ideal transmission lines.[3]

In Figure 1.1, a conceptual power transfer network is shown. The input port may be driven by an amplifier which is modeled as a Thévenin voltage source with complex internal impedance Z_G. The output port may be terminated by an antenna which is considered as a complex passive impedance Z_L.

At this point, it may be appropriate to give the formal definitions of ideal circuit components so that we can build some concrete properties of network functions.

Figure 1.1   Conceptual power transfer network

1.2 Ideal Circuit Elements

In classical circuit theory, circuit elements may be described in terms of their terminal or port-related quantities such as voltage and current or incident and reflected wave relations.

In essence, descriptive port quantities are related to power delivered to that port. Referring to Figure 1.2, multiplication of port voltage $v(t)$ by port current $i(t)$ yields the power delivered to that port at any time t.

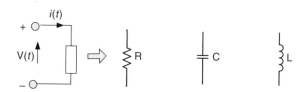

Figure 1.2   Ideal one-port circuit components

For a dissipative or lossy one-port the delivered power

$$P(t) = v(t).i(t)$$

(1.1)

---

[2] Reactive elements are also known as lossless circuit components such as capacitors and inductors.

[3] An ideal transmission line is lossless and propagates uniform transverse electromagnetic waves. These waves are called uniform plane waves.

must be positive. Consequently, the total energy consumed by that 'one-port' is given as the integral of the delivered power such that

$$W = \int_{-\infty}^{+\infty} P(t)dt = \int_{-\infty}^{+\infty} v(t).i(t)dt < 0 \qquad (1.2)$$

Specifically, for a lossless one-port, $W = 0$ since there is no power consumption on it.

Now let us elaborate the concept of power by means of the following examples.

Example 1.1: Let the applied voltage to a port be $v(t) = 3$ volts (or V) (DC) and the corresponding current response be $i(t) = 1$ ampere (or A) (DC) over the entire time domain. Find the power dissipation of the one-port under consideration.

Solution: Power delivered to the port is given by Equation (1.1). Thus, $P(t) = v(t)i(t) = 3\,V \times 1\,A = 3$ watts (or W).

Example 1.2: Let the applied voltage to a port be $v(t) = 3\sin(2\pi \times 50t)$ volts (50 Hz AC) and the corresponding current response be $i(t) = 1\sin(2\pi \times 50t)$ amps (50 Hz AC) over the time domain $t \geq 0$. Find the power dissipation of the one-port at time $t = 5$ milliseconds.

Solution: Instantaneous power dissipation at any time $t \geq 0$ is given by

$$P(t) = v(t) \times i(t) = 3\sin(2\pi \times 50t)\text{volts} \times 1\sin(2\pi \times 50t)\text{amps}$$
$$= 3\sin^2(2\pi \times 50t)\text{watts}$$

Hence, $P(t = 50\,\text{ms}) = 3\,\text{W}$.

Note that, in this problem, the 'voltage and current' pair is sinusoidal with a frequency of $f = 50\,\text{Hz}$; or equivalently with the time period of $T = \frac{1}{50} = 20$ ms. In practice, however, we are interested in average power dissipation over a period. Now let us define the average power dissipation as follows.

1.3 Average Power Dissipation and Effective Voltage and Current

For a one-port, let the port voltage and current pair be specified as

$$v(t) = V_m \sin(\omega_0 t - \varphi_v)$$
$$i(t) = I_m \sin(\omega_0 t - \varphi_i) \qquad (1.3)$$

where

$$\omega_0 = 2\pi f_0 = \frac{2\pi}{T} \qquad (1.4)$$

is the angular frequency with frequency f_0 and the period

$$T = \frac{1}{f_0} \qquad (1.5)$$

Then, for a periodic voltage and current pair, the average power dissipation over a period T is defined as

$$
\begin{aligned}
P_{av} &= \frac{1}{T} \int_0^T v(t)i(t)dt \\
&= \frac{V_m I_m}{T} \int_0^T \sin\left(\frac{2\pi}{T}t - \varphi_v\right) . \sin\left(\frac{2\pi}{T}t - \varphi_i\right) dt
\end{aligned}
\tag{1.6}
$$

Note that

$$
\sin(\alpha) . \sin(\beta) = \frac{1}{2}\cos(\alpha - \beta) - \frac{1}{2}\cos(\alpha + \beta)
$$

Furthermore,

$$
\cos(A) = \cos(-A)
$$

In the above trigonometric equalities, by replacing α by $(2\pi/T)t - \phi_v$ and β by $(2\pi/T)t - \phi_i$, one obtains

$$
P_{av} = \frac{V_m I_m}{2T}\cos(\varphi_i - \varphi_v) \int_0^T dt - \frac{V_m I_m}{2T} \int_0^T \cos\left(\frac{4\pi}{T}t - \varphi_v - \varphi_i\right) dt
\tag{1.7}
$$

Note that the second integral is zero since the area under the cosine function is zero over a full period T. Hence, we have

$$
\begin{aligned}
&P_{av} = \frac{1}{2}V_m I_m \cos(\varphi_v - \varphi_i) \\
&\text{or} \\
&P_{av} = \left[\frac{V_m}{\sqrt{2}}\right]\left[\frac{I_m}{\sqrt{2}}\right]\cos(\varphi_v - \varphi_i)
\end{aligned}
\tag{1.8}
$$

In the above form, the quantities

$$
\begin{aligned}
&V_{eff} = \frac{V_m}{\sqrt{2}} \\
&\text{and} \\
&I_{eff} = \frac{I_m}{\sqrt{2}}
\end{aligned}
\tag{1.9}
$$

are called the effective values of the peak voltage V_m and the peak current I_m respectively.

1.4 Definitions of Voltage and Current Phasors

In the classical circuit theory literature, complex quantities can be expressed in terms of the Euler formula. For example,

$$e^{j\varphi} = \cos(\varphi) + j\sin(\varphi) \tag{1.10}$$

Furthermore, sinusoidal time domain signals can be expressed using the Euler formula such that

$$v(t) = V_m\cos(\omega t - \varphi_v) = \text{real}\{e^{j\omega t}[V_m e^{-j\varphi_v}]\} \tag{1.11}$$

In Equation (1.11) the quantity

$$\mathbf{V} \triangleq [V_m e^{-j\varphi_v}] \tag{1.12}$$

is called the voltage phasor. Similarly, the current phasor is defined as

$$I = [I_m e^{-j\varphi i}]$$

In terms of the current phasor, the actual current is given by $\qquad$ (1.13)

$$i(t) = \text{real}\{I e^{j\omega t}\} = I_m\cos(\omega t - \varphi_i)$$

By means of voltage and current phasors, average power can be expressed as

$$P_{av} = \text{real}\{VI^*\} = \text{real}\{V^*I\} = \frac{1}{2}V_m I_m\cos(\varphi_v - \varphi_i) = V_{eff}I_{eff}\cos(\varphi_v - \varphi_i)$$

Example 1.3: Let $v(t) = 10\cos(\omega t - 10°)$ and $i(t) = 20\cos(\omega t - 40°)$.

(a) Find the voltage and current phasors.
(b) Find the average power dissipated over a period T.

Solution:
(a) By definition, voltage phasor is $V = 10.e^{j10°}$. Similarly, the current phasor is given by $I = 20.e^{j40°}$.
(b) The average power is $P_{av} = \frac{1}{2}10 \times 20.\cos(10° - 40°) = 100.\cos(30°) = 86.6\,\text{W}$.

Example 1.4: Let the voltage phasor be $V = 1.e^{j60°}$. Find the steady state time domain form of the voltage at $\omega = 10\,\text{rad/s}$.

Solution: By formal definition of phasor within this book, we can write $v(t) = \text{real}\{V e^{j10t}\} = \cos(10t - 60°)$.

For the sake of completeness, it should be noted that the steady state voltage $v(t)$ may also be defined as the imaginary part of $\{V e^{j\omega t}\}$ if the input drive is $v_{in}(t) = \sin(\omega t)$.

In general, usage of phasor notation facilitates the sinusoidal steady state analysis of a circuit in the time domain. In principle, network equations (more specifically, equations originating from Kirchhoff's voltage and current laws) are written using voltage and current phasors. Eventually, time domain expressions can easily be obtained by Equation (1.11), like mappings.[4]

---

[4] Here, what we mean is that any steady state time domain expression of a phasor $\dot{A} = \dot{A}_m e^{j\phi_A}\dot{A} = \dot{A}_m e^{j\phi_A}$ can be obtained as $A(t) = \text{real}\{\dot{A} e^{j\omega t}\}$. In this representation $A(t)$ may designate any node or mesh voltage and current in a network.

1.5 Definitions of Active, Passive and Lossless One-ports

Referring to Figure 1.2, let $v(t)$ and $i(t)$ be the voltage and current pair with designated polarity and direction of an ideal circuit component. We assume that these quantities are given as a function of time t. Based on the given polarity and directions:

- A one-port is called passive if $W = \int_{-\infty}^{+\infty} P(t)dt = v(t).i(t) < 0$.
- A one port is called lossless if $W = \int_{-\infty}^{+\infty} P(t)dt = v(t).i(t) = 0$.
- On the other hand, if $W = \int_{-\infty}^{+\infty} P(t)dt = v(t).i(t) < 0$, then the one-port is called active. Obviously, a conventional voltage or current source is an active one-port.

In the following section, we will present elementary definitions of passive and lossless circuit components based on their port voltages and currents.

An ideal circuit component may be a resistor R, a capacitor C or an inductor L. Formal circuit theory definitions of these components are given next.

1.6 Definition of Resistor

A resistor R is a lumped one-port circuit element which is defined by means of Ohm's law:[5]

$$
\begin{aligned}
&v_R(t) \triangleq Ri_R(t) \\
&\text{or} \\
&i_R = \frac{v_R}{R} \triangleq Gv_R \ G = \frac{1}{R}
\end{aligned}
\tag{1.14}
$$

where R is called the resistance and it is measured by means of the ratio of port voltage to port current. The symbol '$\triangleq$' refers to equality by definition.

The units of voltage $v(t)$ and current $i(t)$ are volt (V) and ampere (A) respectively. The unit of resistance R is given by V/A, which is called the ohm and designated by Ω. G is called conductance and it is measured in siemens or Ω^{-1}.[6] The power dissipated on a resistor is given by multiplication of its port voltage and current such that

$$
P_R(t) = v_R(t)i_R = Ri_R^2(t) = Ri_R^2(t) \geq 0
\tag{1.15}
$$

Dissipated power is always non-negative.[7] Therefore, the value of resistance must always be non-negative (i.e. $R \geq 0$).[8]

---

[5] A one-port circuit element is placed between two nodes and described in terms of its port quantities such as voltage and current pairs. These nodes are referred to as terminals of the one-port.

[6] Ω is a Greek letter read as omega.

[7] That is, $P_R(t) \geq 0, \forall t$.

[8] Here, it should be noted that for a real physical system, time is measured as a real number; voltage and current in the time domain are measured as real numbers with respect to selected references. Therefore, energy and power quantities are also measured as real numbers which in turn yield a non-negative real resistance value for the port under consideration.

The unit of power is volt $\times$ ampere which is called the watt and designated by W; 1 watt describes 1 joule of energy (1 J) dissipated per second (s).[9]

1.7 Definition of Capacitor

In electromagnetic field theory, we talk about energy stored both in electric and magnetic fields which produce actual work when applied to a moving electric charge. With this understanding, electric energy is stored on a circuit element called a capacitor and is usually designated by the letter C. As an ideal lumped circuit element, a capacitor C is described in terms of its port voltage v_C and port current i_C as

$$i_C(t) \triangleq C \frac{dv_C(t)}{dt} \qquad (1.16)$$

where C is the capacitance and its unit is the farad (F).[10]

Total electric energy stored in a capacitor C is given in terms of the time integral of the power flow $P_C(t) = v_c(t).i_c(t)$ by

$$W_C = \int_{-\infty}^{t} v_C(\tau)i_C(\tau)d\tau = C \int_{-\infty}^{t} v_C(\tau)dv_C = \frac{1}{2}Cv_C^2$$

provided that initially the capacitor is empty,

i.e. $v(-\infty) = 0.$

(1.17)

Since the stored electric energy W_C must be non-negative (or positive), then capacitance C must always be non-negative (or positive) (i.e. $C \geq 0$). At this point we should mention that this is potential electric energy. It is not dissipation. In other words, it is not consumed by the capacitor; rather it is stored. However, it may generate work or, equivalently, it can be transformed into kinetic energy when it is applied to a moving charge.

In practice, a capacitor is charged with a constant voltage source E_G, say a simple battery which has a series internal resistance R_G. When the charging process is completed within T_C seconds, the capacitor is said to be full and passes no current (i.e. $i_c(T_C) = 0$). The voltage $v_C(T_C)$ across its plates becomes constant and is equal to E_G. In this case, the total stored electric energy is given by $W_C = \frac{1}{2}CE_G^2$. However, consumed energy will be zero since $i_C(T_C) = 0$. In this explanation, any transient process is ignored and the charging time period $T_C = 0^+$ seconds is assumed. This means that the capacitor is immediately charged having $v_C(t < 0^+) = E_G$ and $i_C(t < 0^+) = 0$, yielding no power dissipation (i.e. $P(t) = 0$) or equivalently total energy consumption $W = 0$ (Figure 1.3).

---

[9] That is, 1 W $= 1$ J/s.

[10] In this book all the units are given in the International Standard Unit (ISU) system. Basic units of ISU are the meter, kilogram and second (MKS). Therefore, ISU is also known as the MKS unit system. In MKS, voltage and current units are given as volt (V) and ampere (A).

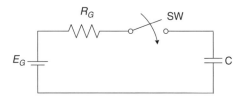

Figure 1.3   Electric energy storage element: capacitor (C)

1.8 Definition of Inductor

An inductor L is an ideal lumped circuit element. It stores magnetic energy. Its formal definition is given in terms of its port voltage $v_L(t)$ and port current $i_L(t)$ as

$$v_L(t) \triangleq L \frac{di_L(t)}{dt} \tag{1.18}$$

where L called inductance and its unit is is given by the henry (H).

Total magnetic energy W_L stored in an inductor L over an interval of time $(-\infty, t]$ is given by

$$W_L = \int_{-\infty}^{t} v_L(\tau).i_L(\tau)d\tau = \int_{-\infty}^{t} Li_L(\tau)di = \frac{1}{2}Li_L^2 \tag{1.19}$$

Since the stored magnetic field energy must be non-negative (or positive), then inductance L must be non-negative (or positive) (i.e.)$L \geq 0$).[11]

In a similar manner to that of a capacitor, as an ideal lumped circuit element, an inductor L is lossless. This means that it does not dissipate power but rather holds magnetic energy over a specified period of time unless it is emptied. When an inductor is connected to an excitation, say to a constant current source I_G with an internal shunt resistance R_G, at time $t=0$ seconds, a constant current $I_L = I_G = i_L(t < 0^+)$ immediately builds up over a very short period of time ending at $t=0^+$ seconds. Then, this current circulates indefinitely within the circuit. Let the voltage drop on R_G be equal to $R_G I_G$ at time $t=0^-$ seconds. Roughly speaking, when the inductor L is connected to the current source I_G, this voltage immediately appears on the inductor $v_L(t=0) = R_G I_G$ and rapidly reduces to zero within $T_L = 0^+$ seconds while the inductor current i_L rises to the level of I_G, yielding zero power transfer. During this process, as i_L increases, current through the shunt resistance R_G goes to zero resulting in zero voltage across inductor L (see Figure 1.4).

We should emphasize that this is a macroscopic explanation. Details are skipped here.

Just to summarize the above discussions based on the definitions, as ideal circuit elements a capacitor or an inductor is a lossless one-port, and it can only store energy. On the other hand, a resistor is a lossy circuit element which dissipates or consumes energy by heating itself. In practice, however, there is no ideal circuit element; one can always associate a real dissipation perhaps in series with an ideal inductance L, say r_L, or in parallel with an ideal capacitance C which may be designated as conductance G_C as shown in Figure 1.5.

---

[11] It should be noted that in Equation (1.16) and (1.19), initially capacitor C and inductor L were assumed to be empty. Therefore, in these equations the integration constant is set to zero.

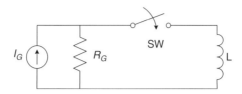

Figure 1.4   Magnetic energy storage element: inductor (L)

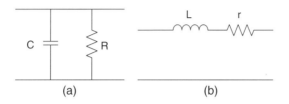

Figure 1.5   A practical capacitor and a practical inductor in shunt and series configurations respectively

Example 1.5: Let a resistive one-port with an input impedance of $10\,\Omega$ be derived by a voltage excitation of $v(t) = 10\cos(5t)$ V.

(a) Find the current through the resistive input.
(b) Find the average power dissipated on the resistive one-port.
(c) Find the energy consumed on the resistive one-port over a period.
(d) Find the total energy consumed on the one-port over a time interval of $[0, +\infty)$.

Solution:

(a) The current of the resistive input is given by $i(t) = 1/Rv(t) = 1.\cos(5t)$A.
(b) $P_{av} = \frac{1}{2}V_m I_m = \frac{1}{2}10 \times 1 = 5\,\text{W}$.
(c) $T = 2\pi/\omega = 2\pi/5$ and $W_T = \int_0^T R.i^2(t)dt = 10\int_0^{2\pi/5}\cos^2(5t)dt$. Note that $\cos^2 5t = \frac{1}{2} + \cos(10t)$.
Then $W_T = 10\left[\frac{1}{2}\frac{2\pi}{5} + \frac{1}{10}\sin\left(10.\frac{2\pi}{5}\right)\right] = \frac{10\pi}{5} = 2\pi \cong 6.28$ J
(d) $W_{[0, +\infty)} = \int_0^{+\infty} R\cos^2(5t)dt = \infty$J. This means that the resistive input consumes energy as time approaches to infinity.

Obviously, a resistor must consume energy as long as it is connected to a power supply. Therefore, one must be very careful to save power when an iron type of house appliance is operated and not let it stand if not in use.

Example 1.6: Let a one-port consist of a single capacitor of $C = 10$ F which is driven by an excitation of $v(t) = 1.\sin(5t)$ V.

(a) Find the current through the capacitor C.
(b) Find the average power dissipated on the capacitor over a period T.
(c) Find the maximum electric field energy stored on the capacitor.

Solution:

(a) By definition, the current of a capacitor is given by $i(t) = C\,dv/dt = 10 \times 5\cos(5t) = 50\cos(5t)$A.

(b) In this problem, the period T is specified as in the above example. Hence, $T = 2\pi/5$ The average power dissipated on the capacitor over a period is specified as

$$
\begin{aligned}
P_{av} &= \frac{1}{T} \int_0^T v(t)i(t)dt \\
&= \frac{50}{T} \int_0^t \sin(5t)\cos(5t)dt \\
&= \frac{50}{T} \frac{1}{2} \int_0^T \sin(10t)dt = \frac{25}{4\pi}[-\cos(4\pi) + \cos(0)] = \frac{25}{4\pi}[-1+1] \\
&= 0W^{12}
\end{aligned}
$$

as it should be, since a capacitor is a lossless lumped circuit element.[13] However, one should keep in mind that instantaneous power may not be zero. It depends on the instantaneous values of the applied voltage and current.

(c) Obviously, a capacitor charges and discharges as the applied voltage across it varies. Then, the maximum stored energy of the capacitor is found at the peak value of the applied voltage. Hence,

$$
W_{max} = \max\left\{\frac{1}{2}Cv^2(t)\right\} = \frac{1}{2}C \times \max\{v^2(t)\}
$$

$$
= \frac{1}{2}CV_m^2 = \frac{1}{2} \times 10 \times 1 = 5\,J
$$

Example 1.7: Let a one-port consist of a single inductor of $L = 1$ H which is driven by an excitation of $i(t) = 1.\sin(5t)$ A.

(a) Find the voltage through the inductor capacitor L.
(b) Find the average power dissipated on the inductor over a period T.
(c) Find the maximum magnetic energy stored on the inductor.

Solution:

(a) By definition, the voltage of an inductor is given by $v(t) = L\,di/dt = 1 \times 5\cos(5t) = 5\cos(5t)$V.
(b) As in the previous problem, it is straightforward to show that

$$
P_{av} = \frac{1}{T} \int_0^T v(t)\,i(t)dt = \frac{5}{T} \int_0^T \sin(5t)\cos(5t)\,dt = 0\,W
$$

as it should be, since an inductor is a lossless lumped circuit element. However, one should keep in mind that instantaneous power may not be zero. It depends on the instantaneous values of voltage and current.

(c) Obviously, an inductor charges and discharges as the current through it varies. Then, the maximum stored magnetic energy is found at the peak value of the current. Hence,

$$
W_{max} = \max\left\{\frac{1}{2}Li^2(t)\right\} = \frac{1}{2}L \times \max\{i^2(t)\} = \frac{1}{2}LI_m^2 = \frac{1}{2} \times 1 \times 1 = 0.5\,J
$$

---

[12] Here, $T = 2\pi/\omega = 2\pi/5$.
[13] Note that $\sin(\alpha).\cos(\alpha) = \frac{1}{2}\sin(2\alpha)$ Furthermore,

$$
\int_0^T \sin(10t)dt = -\frac{1}{10}\cos(10t)|_0^{2\pi/5} = \frac{1}{10} \times 0 = 0
$$

1.9 Definition of an Ideal Transformer

A transformer is an ideal two-port circuit element which consists of two magnetically perfectly coupled coils as shown in Figure 1.6. The coil on the left is called the primary coil with n_1 turns and constitutes port 1 with port voltage $v_1(t)$ and port current $i_1(t)$. The coil on the right is called the secondary coil with n_2 turns and constitutes port 2 with port voltage $v_2(t)$ and port current $i_2(t)$ satisfying the following relations:

$$
\boxed{
\begin{aligned}
&v_2(t) \triangleq \frac{n_2}{n_1} v_1(t) = n v_1(t) \\
&\text{and} \\
&i_2(t) \triangleq -\frac{1}{n} i_1(t)
\end{aligned}
}
\tag{1.20}
$$

where $n = n_2/n_1$ and it is called the turns ratio of the transformer.

From the above equations, it is clear that total power delivered to the transformer is given by

$$
\boxed{P_T = v_1(t). \, i_1(t) + v_2(t). \, i_2(t) = 0}
\tag{1.21}
$$

This means that no power is dissipated on the transformer. Hence, by definition, a transformer is a lossless two-port. The above equations are only valid for time-varying voltage and current. Otherwise, no transformation can occur.[14]

Referring to Figure 1.6, let the impedance seen from port 2 be Z_2. Let port 1 be terminated in impedance Z_1.

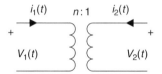

$i_1(t)$ $n:1$ $i_2(t)$

$+$ $+$

$V_1(t)$ $V_2(t)$

Figure 1.6   An ideal transformer

Then, Equation (1.20) reveals that

$$
\boxed{Z_2 = \frac{v_2}{i_2} = -n^2 \frac{v_1}{i_1} = n^2 \left[\frac{-v_1}{i_1} \right] = n^2 Z_1}
\tag{1.22}
$$

where $Z_1 = v_{Z1}/i_{Z1} = -v_1/i_1$. Thus, we see that an ideal transformer also transforms the terminating impedances[15] in a flat manner[16] over the entire frequency band.

---

[14] That is, transformation neither from the primary to the secondary nor from the secondary to the primary coils can occur.

[15] Loosely speaking, the voltage to current ratio is defined as the impedance and its inverse is called the admittance. However, their formal definitions will be given in the following sections.

[16] Here, we use the term 'flat' to indicate the entire frequency band $(-\infty < \omega < \infty)$.

1.10 Coupled Coils

Referring to Figure 1.7, inductors L_1 and L_2 are said to constitute a coupled structure with a coupling coefficient M if the following equations are satisfied:

$$
\begin{aligned}
v_1 &\triangleq L\frac{di_1}{dt} + M\frac{di_2}{dt} \\
v_2 &\triangleq M\frac{di_1}{dt} + L_2\frac{di_2}{dt}
\end{aligned}
\tag{1.23}
$$

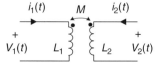

Figure 1.7   Coupled coils

Here, we should note that the position of the dots in Figure 1.7 defines the sign of the coupling coefficient M. This coefficient is positive if the dots are on the same side (i.e. they are either at the top or at the bottom) and negative if the dots are on opposite sites (i.e. one of the dots is at the top and the other is at the bottom).[17]

1.11 Definitions: Laplace and Fourier Transformations of a Time Domain Function $f(t)$

Two-sided or bilateral Laplace transformation of a continuous function $f(t)$ over the entire time domain is defined as

$$
F(p) \triangleq \mathcal{L}\{f(t)\} = \int_{-\infty}^{\infty} f(t).e^{-pt}dt
\tag{1.24}
$$

where $p = \sigma + j\omega$ is a complex variable and known as the complex frequency.

We may define a similar transformation for complex variable $p = j\omega$ by setting $\sigma = 0$. In this case, Equation (1.24) can be written as

$$
F(p) \triangleq \mathcal{F}\{f(t)\} = \int_{-\infty}^{+\infty} f(t)e^{-j\omega t}dt
\tag{1.25}
$$

---

[17] It should be noted that one may also define coupled capacitors in a similar manner such that $i_1(t) = C_1\,dv_1/dt + M_C\,dv_2/dt$ and $i_2(t) = M_C\,dv_1/dt + C_2\,dv_2/dt$ where C_1 and C_2 are coupled capacitors and M_c is the capacitive coupling coefficient. However, for most practical applications M_c becomes negligibly small and therefore is set to zero.

Equation (1.25) is called the Fourier transformation of a continuous time domain function $f(t)$.[18] In this equation, ω is called the real angular frequency or, in short, the real frequency, and it is given by

$$\omega = 2\pi f \tag{1.26}$$

where f is called the frequency or the actual frequency.[19]

In everyday problems, we usually deal with real time functions (or signals) which are bounded and, perhaps, we know 'how and when' they are initiated. The initiation point of these functions can be fixed at $t = 0$ seconds. In this case, we can say that these functions do not exist before they are initiated and therefore they are set to zero before their initiation time. Hence, mathematically, we can state that

$$f(t) = \begin{cases} 0 \text{ for } -\infty < t < 0 \\ \neq 0 \text{ and bounded for } 0 \leq t < +\infty \end{cases} \tag{1.27}$$

Those functions which satisfy Equation (1.27) are called causal functions.

Strictly speaking, for the existence of Laplace and Fourier transforms, $f(t)$ must be a member of the class of functions which are integrable. For example, a class of functions is called L^1 if $f(t)$, as a member of the class, satisfies the inequality of $\int_0^\infty |f(t)| \, dt < M < \infty$. $f(t)$ is called L^2 if it belongs to a class of functions which satisfies the inequality $\int_0^\infty |f(t)|^2 dt < M < \infty$.

In this context, the causal functions that we deal with must belong either to L^1 or L^2. Therefore, their Laplace and Fourier transform integrals start from $t = 0$ seconds. In practice, however, we employ L^2 functions. Thus, for a bounded causal (perhaps L^2) real time domain signal $f(t)$, the 'one-sided Laplace transform'

$$F(p) = \int_0^\infty f(t)e^{-pt}dt = \int_0^\infty \left[f(t)e^{-\sigma t} \right] e^{-j\omega t}dt \tag{1.28}$$

exists if and only if the integrand $f(t)e^{-\sigma t}$ is bounded, i.e. $|f(t)e^{-\sigma t}| < \infty$.

This is possible if σ is non-negative. Therefore, for a bounded causal signal, the 'one-sided Laplace transformation' exists for the values of $\sigma \geq 0$. So, roughly speaking, we say that $F(p)$ is analytic (or differentiable everywhere) in the right half plane of $p = \sigma + j\omega$ (or simply in the RHP).[20]

By setting $\sigma = 0$, the Laplace transform of Equation (1.27) becomes the one-sided Fourier transform and for bounded causal signals it is given by

$$F(j\omega) = \int_0^\infty f(t)e^{-j\omega t}dt \tag{1.29}$$

---

[18] In circuit theory, $f(t)$ may refer to a time domain signal such as voltage or current. Then, $F(j\omega)$ is called the Fourier transform of a time domain signal $f(t)$.

[19] In the classical literature of power transfer networks or broadband matching theory, for the sake of simplicity, angular frequency ω and the actual frequency f are both referred to as the *real frequency* since they carry the same information within a constant 2π.

[20] A rigorous discussion of Fourier and Laplace transforms of causal signals can be found in Chapter 3 of the book '*Wideband Circuit Design*' by H. J. Carlin and P. C. Civalleri, CRC Press, 1998.

For bounded causal signals, if the Fourier transform $F(j\omega)$ is known, the original time domain signal $f(t)$ can be uniquely determined from $F(j\omega)$ by means of the inverse Fourier transform as

$$f(t) = \mathcal{F}^{-1}\{F(j\omega)\} = \frac{1}{2\pi} \int_{-\infty}^{\infty} F(j\omega)e^{j\omega t} d\omega \tag{1.30}$$

Or, replacing $j\omega$ by p, the inverse Laplace transform is given by

$$f(t) = \mathcal{L}^{-1}\{F(p)\} = \frac{1}{2\pi j} \int_{-j\infty}^{j\infty} F(p)e^{pt} dp \tag{1.31}$$

Fourier and Laplace transformations are handy tools for solving integral and differential equations, as demonstrated below.

In Equation (1.31), using the 'integration by parts' rule, it is straightforward to show that

$$\mathcal{L}\left\{\frac{df}{dt}\right\} = pF(p)$$

and

$$\mathcal{L}\left\{\int f(t)dt\right\} = \frac{1}{p}F(p) \tag{1.32}$$

1.12 Useful Mathematical Properties of Laplace and Fourier Transforms of a Causal Function[21]

In the following subsection useful properties of the Laplace and Fourier transforms are introduced from the design perspective of power transfer networks.

Even and Odd Parts

Assume that complex function $F(p)$ is defined as the one-sided Laplace transform of a causal, continuous, bounded (CCB) function. As was discussed earlier, $F(p)$ exists for all values of $t \geq 0$ if and only if real$\{p\} = \sigma \geq 0$. This condition immediately yields that $F(p)$ must be analytic in the closed RHP. Let us comment on this statement for some practical situations.

For example, let $v(t)$ and $i(t)$ be the CCB voltage and current response of a one-port which consists of various combinations of connections of R, L and C elements in a sequential manner.[22] In this case, it can be shown that Laplace transforms $V(p) = \mathcal{L}\{v(t)\}$ and $I(p) = \mathcal{L}\{i(t)\}$ are rational functions. Obviously, rational functions should have their singularities at the poles. Therefore, $V(p)$ and $I(p)$ must be free of closed RHP poles.

Generically speaking, a complex variable function $F(p)$ which is free of open RHP zeros and closed RHP poles is called a minimum function.[23]

---

[21] Formal discussions of the subject can be found in Chapter 3 of the book '*Wideband Circuit Design*' by H. J. Carlin and P. P. Civalleri, CRC Press, 1998, and also in Chapters 1–4 of the book '*Signal Processing and Linear Systems*' by B. P. Lathi, Cambridge University Press, 1998.

[22] Such as a cascaded connection of parallel and series circuit elements.

[23] Open RHP excludes the entire $j\omega$ axis whereas closed RHP includes the $j\omega$ axis. It should be noted that the origin ($\omega = 0$) is also included on the $j\omega$ axis.

The Fourier transform of a minimum function can be uniquely determined from its Laplace transform $F(p)$ by setting $\sigma = 0$ in $p = \sigma + j\omega$ or equivalently by setting $p = \sigma + j\omega$. Let $F(p)$ be a rational function. Then,

$$F_e(p^2) = \frac{1}{2}[F(p) + F(-p)] \qquad (1.33)$$

and

$$F_o(p) = \frac{1}{2}[F(p) - F(-p)] \qquad (1.34)$$

represent its even and the odd parts respectively. In this case, Fourier transform $F(j\omega)$ can be written as

$$F(j\omega) = A(\omega^2) + jB(\omega) \qquad (1.35)$$

where $A(\omega^2) = F_e(p^2 = -\omega^2)$ is an even function in ω^2 and $jB(\omega) = F_0(j\omega)$ or $B(\omega) = -jF_0(j\omega)$ is an odd function of ω.

Similarly,

$$F(j\omega) = \rho(\omega) e^{j\varphi(\omega)} \qquad (1.36)$$

where $\rho^2(\omega) = A^2 + B^2$ and $\phi(\omega) = \tan^{-1}(B/A)$ are even and odd functions of ω respectively.

Example 1.8: Let $F(p) = 1/(1 + p)$.

(a) Find the even part of $F(p)$.
(b) Find the odd part of $F(p)$.
(c) Find the frequency representation of even and odd parts.

Solution:

(a) By Equation(1.33),

$$F_e(p) = \frac{1}{2}[F(p) + F(-p)] = \frac{1}{2}\left[\frac{1}{1+p} + \frac{1}{1-p}\right] = \frac{1}{1-p^2}$$

(b) By Equation (1.34),

$$F_o(p) = \frac{1}{2}\left[\frac{1}{1+p} - \frac{1}{1-p}\right] = -\frac{p}{1-p^2}$$

(c) Replacing p by $j\omega$, we have

$$F_e(j\omega) = A(\omega) = \frac{1}{1+\omega^2}$$

$$F_o(j\omega) = -j\,\frac{\omega}{1+\omega^2}$$

which yields

$$B(\omega) = -\frac{\omega}{1+\omega^2}$$

Example 1.9: Let

$$F(p) = \frac{2p^2 + 2p + 1}{1 + p}$$

(a) Find the even part of $F(p)$.
(b) Find the odd part of $F(p)$.
(c) Find the frequency domain representation of the even part.
(d) Find the frequency domain representation of the odd parts.

Solution:

(a) By Equation (1.33),

$$F_e(p) = \frac{1}{2}[F(p) + F(-p)] = \frac{1}{2}\left[\frac{2p^2 + 2p + 1}{1 + p} + \frac{2p^2 - p + 1}{1 - p}\right] = \frac{1}{1 - p^2}$$

(b) By Equation (1.34),

$$F_o(p) = \frac{1}{2}[F(p) - F(-p)] = \frac{1}{2}\left[\frac{2p^2 + 2p + 1}{1 + p} - \frac{2p^2 - p + 1}{1 - p}\right] = \frac{-2p^3 + p}{1 - p^2}$$

(c) In this case, $F_e(j\omega) = A(\omega) = \frac{1}{\omega^2 + 1}$
(d) Here,

$$F_o(j\omega) = \frac{j\omega^3 + j\omega}{\omega^2 + 1} = j\frac{\omega^3 + \omega}{\omega^2 + 1} = jB(\omega)$$

which yields[24]

$$B(\omega) = \frac{2\omega^3 + \omega}{\omega^2 + 1} = \frac{\omega}{\omega^2 + 1} + \frac{2\omega^3}{\omega^2 + 1}$$

It is interesting to note that in this example:

- The real part $A(\omega)$ of $F(j\omega)$ is the same as the one determined in Example 1.8 even though $F(j\omega)$ is different.
- On the other hand, the odd term $B(\omega)$ differs from that of Example 1.8 by a single additive term of $2\omega^3/(\omega^2 + 1)$.
- The above points can easily be verified by writing

$$F(p) = \frac{2p^2 + 2p + 1}{p + 1} = 2p + \frac{1}{p + 1}$$

  which only differs from that of $F(p)$ of Example 1.8 by a *simple* term $2p$.
- It is straightforward to show that $F(p)$ of Example 1.8 is a minimum function whereas $F(p)$ of Example 1.9 is not, since it has a pole on the $j\omega$ axis at $\omega = \infty$ (or $p = \infty$).

Analytic Continuation of F(jω)

Let $f(t)$ be a 'CCB' real-valued function in real-time variable $t \geq 0$. Let $F(j\omega)$ be its Fourier transform. Then, its one-sided Laplace transform $F(p)$ can be uniquely determined from $F(j\omega)$, replacing $j\omega$ by

---

[24] Note that $(j\omega)^3 = -j\omega^3$. Therefore, $-p^3$ yields $+j\omega^3$. Then, the above results follow.

$p = \sigma + j\omega$ for all $\sigma \geq 0$ since $F(p)$ is analytic in the closed RHP. This process is called the analytic continuation of $F(j\omega)$ in the closed RHP.

Hilbert Transformation

Let $F(p) = N(p)/D(p)$ be a minimum rational function which is free of closed RHP poles as defined above. Let $F(j\omega) = A(\omega) + jB(\omega)$ be its Fourier transform. Then, $F(j\omega)$ can uniquely be determined from the given real part $A(\omega)$ by

$$B(\omega) = A_\infty + \frac{1}{\pi} \int_{-\infty}^{+\infty} \frac{A(\Omega)}{\omega - \Omega} d\Omega$$

where the constant term A_∞ is the value of $A(\omega)$ at infinity.

Since $A(\omega)$ is an even function,

$$B(\omega) = A_\infty + \frac{2\omega}{\pi} \int_{0}^{+\infty} \frac{A(\Omega)}{\Omega^2 - \omega^2} d\Omega$$

Using integration by parts, it is straightforward to show that

$$B(\omega) = A_\infty + \frac{1}{\pi} \int_{0}^{+\infty} \left[\frac{dA(\Omega)}{d\Omega} \right] \ln \left| \frac{\Omega + \omega}{\Omega - \omega} \right| d\Omega$$

(1.37)

or in compact form

$$F(p) = A_\infty + \frac{2p}{\pi} \int_{0}^{+\infty} \frac{A(\Omega)}{p^2 + \Omega^2} d\Omega$$

Similarly, $A(\omega)$ is given in terms of $B(\omega)$ as

$$A(\omega) = A_\infty - \frac{2\omega}{\pi} \int_{0}^{\infty} \frac{B(\Omega)}{\Omega^2 - \omega^2} d\Omega$$

Equation (1.37) is known as the Hilbert transformation relations.[25]

The above $\{A(\omega)$ and $B(\omega)\}$ Hilbert transform pairs are useful tools for generating minimum functions from the real part $A(\omega)$. On the other hand, depending on the form of the even part, evaluation of these integrals may be troublesome. Therefore, we prefer to work with various numerical or algebraic techniques to generate the minimum functions from their even parts without evaluating the integral. Especially in designing ultra wideband power transfer networks, 'non-integral techniques' are employed. In the following chapters details of these techniques will be presented. Nevertheless, for the sake of completeness, let us look at a simple example to show how complicated it is to evaluate the Hilbert integrals.

Example 1.10: Let $A(\omega^2) = 1/(1 + \omega^2)$ be the real part of a minimum function $F(p)$ on the real frequency axis $j\omega$ as in Example 1.8. Find the analytic form of the $F(p)$ using the Hilbert transformation.

Solution: From Example 1.8 we know that $F(p) = 1/(p + 1)$. However, in a more formal way, by Equation (1.37),

---

[25] In many practical situations, the constant term A_∞ approaches zero. Therefore, it may be neglected.

$$F(p) = A_\infty + \frac{2p}{\pi} \int_0^{+\infty} \frac{A(\Omega)}{p^2 + \Omega^2} d\Omega = A_\infty + \frac{p}{\pi} \int_{-\infty}^{+\infty} \frac{A(\Omega)}{p^2 + \Omega^2} d\Omega$$

Close examination of $A(\omega^2) = 1/(1 + \omega^2)$ reveals that $A_\infty = 0$.

This integral may be evaluated by means of a complex contour integral technique employing the Cauchy residue formula. In short, it is known that for any function $\alpha(\xi)$ which is analytic inside a closed region C,

$$(2\pi j).\alpha(p) = \oint \frac{\alpha(\xi)}{\xi - p} d\xi = \int_{-\infty}^{+\infty} \frac{\alpha(\xi)}{\xi - p} d\xi$$

provided that $\lim_{R \to \infty} \alpha(\xi)/(\xi - p) = 0$ where R is the radius of the closed contour C.[26] Now, to go back to our problem, let $\xi = j\Omega$ and $\alpha(\xi) = 1/(1 + \xi)(p + \xi)$. Then,

$$\Omega^2 = -\xi^2 \quad d\Omega = \frac{1}{j} d\xi \quad A(\xi) = \frac{1}{1 - \xi^2} = \frac{1}{(1 - \xi)(1 - \xi)}$$

and $p^2 + \Omega^2 = p^2 - \xi^2 = (p + \xi)(p - \xi)$.

Thus, the original Hilbert transformation becomes

$$\frac{(j\pi)F(p)}{p} = \int_{-\infty}^{+\infty} \frac{A(\Omega)}{p^2 + \Omega^2} d\Omega = \int_{-\infty}^{+\infty} \frac{\alpha(\xi)}{(1 - \xi)(p - \xi)} d\xi$$

$$= \int_{-\infty}^{+\infty} \left[\frac{\mu(1)}{1 - \xi} + \frac{\eta(p)}{p - \xi} \right] d\xi = (2\pi j)[\mu(1) + \eta(p)]$$

where

$$\mu(\xi) = \frac{\alpha(\xi)}{p - \xi}$$

and

$$\eta(\xi) = \frac{\alpha(\xi)}{1 - \xi}$$

Note that

$$\alpha(1) = \frac{1}{2(p + 1)}$$

$$\alpha(p) = \frac{1}{2p(1 + p)}$$

---

[26] For this problem it is assumed that the closed countour is the closed right half of the complex p plane.

Then,

$$[\mu(1) + \eta(p)] = \frac{1}{2p(p+1)}$$

Thus,

$$F(p) = \frac{1}{1+p}$$

confirms the original function specified by Example 1.8.

There are some other methods for evaluating the above Hilbert integral but they are all tedious. In designing power transfer networks, that is why we wish to work with numerical or algebraic methods to

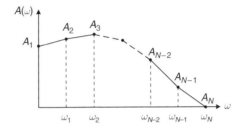

Figure 1.8 Piecewise linearization of real part $A(\omega)$

generate a minimum function from its even part.

As will be shown in the following section, $A(\omega)$ is always non-negative, and a smooth function of angular frequency or real frequency ω in many practical situations eventually reaches zero as frequency goes to infinity. Therefore, its piece wise linear approximation is possible as shown in Figure 1.8.

Let us consider that $A(\omega)$ as expressed in terms of N sampled point pairs $\{(A_k, \omega_k); k = 1, 2, \ldots, N\}$ or equivalently total $(N-1)$ pieces of line segment such that

$$\left.\begin{array}{l} A(\omega) = \begin{cases} a_j\omega + b_j & \text{for } \omega_j \leq \omega \leq \omega_{j+1}, \quad j = 1, 2, \ldots, (N-1) \\ 0 & \omega \geq \omega_N \end{cases} \\[2em] \text{where} \\[1em] \qquad a_j = \dfrac{A_j - A_{j+1}}{\omega_j - \omega_{j+1}} = \dfrac{\Delta A_j}{\omega_{j+1} - \omega_j} \\[2em] \text{and} \\[1em] \qquad b_j = \dfrac{(A_{j+1})\omega_j - (A_j)\omega_{j+1}}{\omega_j - \omega_{j+1}} \\[2em] \text{where} \quad \Delta A_j = A_{j+1} - A_j \end{array}\right\} \tag{1.38}$$

The above form of $A(\omega)$ is suitable for numerical evaluation of $B(\omega)$ by employing the Hilbert transformation as illustrated in the following section.

1.13 Numerical Evaluation of Hilbert Transform

Referring to Equation (1.38), let

$$
\hat{B}_j(\omega) = \frac{1}{\pi} \int\limits_{\omega_j}^{\omega_{j+1}} \left[\frac{dA(\Omega)}{d\Omega} \right] \ln \left| \frac{\Omega + \omega}{\Omega - \omega} \right| d\Omega
$$

$$
= \frac{a_j}{\pi} \int\limits_{\omega_j}^{\omega_{j+1}} \ln \left| \frac{\Omega + \omega}{\Omega - \omega} \right| d\Omega
$$

or   (1.39)

$$
\hat{B}_j(\omega) = \frac{A_j - A_{j+1}}{\pi(\omega_j - \omega_{j+1})} \int\limits_{\omega_j}^{\omega_{j+1}} \ln \left[\frac{\Omega + \omega}{\Omega - \omega} \right] d\Omega
$$

Then,

$$
B(\omega) = \sum_{k=1}^{N} \hat{B}_k
$$

For numerical reasons, it may be appropriate to express Equation (1.39) in terms of excursions of $\Delta A_j = A_{j+1} - A_j$ and its multiplying coefficients

$$
B_j(\omega) = \frac{1}{\pi(\omega_j - \omega_{j+1})} \int\limits_{\omega_j}^{\omega_{j+1}} \ln \left| \frac{\Omega + \omega}{\Omega - \omega} \right| d\Omega
$$ (1.40)

In this regard, let

$$
F_j(\omega) = (\omega + \omega_j)\ln(|\omega + \omega_j|) + (\omega - \omega_j)\ln(|\omega - \omega_j|)
$$ (1.41)

Then, it is straightforward to show that

$$
B_j(\omega) = \frac{1}{\pi(\omega_j - \omega_{j+1})} [F_{j+1}(\omega) - F_j(\omega)]
$$ (1.42)

Thus,

$$
B(\omega) = \sum_{j=1}^{N-1} B_j(\omega)\Delta A_j
$$ (1.43)

This equation clearly indicates that the imaginary part $B(\omega)$ of a minimum function $F(j\omega)$ can also be expressed in terms of a linear combination of the same break points $\{A_k, \omega_k;\ K = 1, 2, \ldots, N\}$.

Details will be presented in Chapter 11.

1.14 Convolution

Let $f(t)$ and $h(t)$ be CCB signals. Their convolution is defined as

$$
\begin{aligned}
y(t) &= f(t) * h(t) \\
&= \int_{-\infty}^{+\infty} f(\tau)h(t-\tau)d\tau \\
&= \int_{-\infty}^{+\infty} f(t-\tau)h(\tau)d\tau, t \geq 0
\end{aligned}
\tag{1.44}
$$

Let $Y(j\omega) = \mathcal{F}\{y(t)\}$, $f(j\omega) = \mathcal{F}\{f(t)\}$ and $H(j\omega) = \mathcal{F}\{h(t)\}$. Then, it can be shown that

$$
\begin{aligned}
&Y(j\omega) = F(j\omega)\ H(j\omega) \\
&\text{or, replacing } j\omega \text{ by } p, \text{ we can write} \\
&Y(p) = F(p)H(p)
\end{aligned}
\tag{1.45}
$$

1.15 Signal Energy

Let $v(t)$ and $i(t)$ be the port voltage and current of a circuit component. Let $f(t)$ designate either $v(t)$ or $i(t)$. In general these quantities can be complex. In this case, it can be shown that total energy W delivered to a component or to a one-port is proportional to $W \sim \int_{-\infty}^{+\infty} |f(t)|^2 |dt$.

It must be clear that for a passive one port W must be non-negative. For example, for a resistor R[27]

$$
\begin{aligned}
W_R &= \int_{-\infty}^{+\infty} v(t)i^*(t)dt \\
&= \int_{-\infty}^{+\infty} R|i(t)|^2 dt = \int_{-\infty}^{+\infty} \frac{1}{G}|v(t)|^2 dt \\
&\geq 0
\end{aligned}
\tag{1.46}
$$

For real signals we can simply write

$$
w = \int_{-\infty}^{+\infty} v(t)i(t)dt \geq 0
$$

where the superscript $*$ designates the complex conjugate of the signal.

---

[27] Strictly speaking, the resistor R may be a complex quantity. However, for a real passive physical system, R must be real.

In general,

$$W_f \triangleq \int_{-\infty}^{+\infty} |f|^2 dt \geq 0 \tag{1.47}$$

is called the signal energy.

Obviously, for physical passive one-ports which are defined in terms of real parameters, W_f must be non-negative, finite and L^2. Therefore, we can comfortably state that for any real, passive one-port, the port descriptive functions $v(t)$ and $i(t)$ must be real, CCB and L^2. Let $V(j\omega)$ and $I(j\omega)$ be the Fourier transforms of $v(t)$ and $i(t)$ respectively. Using the inverse Fourier transform, we can write $v(t) = \frac{1}{2\pi} \int_{-\infty}^{+\infty} V(j\omega) e^{j\omega t} d\omega$.

Then, for any real, passive one port, Equation (1.46) becomes

$$W = \frac{1}{2\pi} \int_{-\infty}^{+\infty} V(j\omega) \left[\int_{-\infty}^{+\infty} i(t) e^{-j\omega t} dt \right]^* d\omega \tag{1.48}$$

or

$$W = \int_{-\infty}^{+\infty} v(t) i(t) dt = \frac{1}{2\pi} \int_{-\infty}^{+\infty} V(j\omega) I^*(j\omega) d\omega \geq 0 \tag{1.49}$$

1.16 Definition of Impedance and Admittance

Equation (1.49) is very useful for expressing definitions of circuit components in terms of Laplace transforms since the derivatives of voltage and current are removed from the mathematical expressions. Thus, employing the Laplace-transformed pairs for voltage and current as $v \leftrightarrow V(p)$ and $i \leftrightarrow I(p)$; passive ideal resistor R, capacitor C and inductor L definitions can be given in a straightforward manner as

$$\begin{aligned} V_R(p) &\triangleq R I_R(p) \\ I_c(p) &\triangleq [Cp] V_c(p) = Y_c(p) V_c(p) \\ V_c(p) &\triangleq [CL] I_c(p) = Z_c(p) I_L(p) \end{aligned} \tag{1.50}$$

where $Z_L(p) = pL = V_L(p)/I_L(p)$ is called the impedance of an inductor and $Y_L(p) = pC = I_C(p)/V_C(p)$ is called the admittance of a capacitor.

Similarly the coupled coil definition is given by

$$\begin{aligned} V_1(p) &= pL_1 I_1(p) + pM\, I_2(p) \\ V_2(p) &= pM_1(p) + pL_2\, I_2(p) \end{aligned} \tag{1.51}$$

or

$$\begin{bmatrix} V_1 \\ V_2 \end{bmatrix} = \begin{bmatrix} pL_1 & pM \\ pM & pL_2 \end{bmatrix} \begin{bmatrix} I_1 \\ I_2 \end{bmatrix} = Z_M(p) \begin{bmatrix} I_1 \\ I_2 \end{bmatrix} \tag{1.52}$$

where $Z_M(p)$ is called the impedance matrix of the coupled coils.

1.17 Immittance of One-port Networks

Referring to Figure 1.9, consider a one-port network which consists of passive lumped R, L and C elements and is driven by a bounded causal excitation.[28]

Let $v_{in} \leftrightarrow V_{in}(p)$ and $i_{in} \leftrightarrow I_{in}(p)$ be the time and Laplace transform pairs of the one-port. Then, the driving point impedance function $Z_{in}(p)$ of the one-port is defined as

$$Z_{in}(p) \triangleq \frac{V_{in}(p)}{I_{in}(p)} \tag{1.53}$$

Similarly,

$$Y_{in}(p) \triangleq \frac{I_{in}(p)}{V_{in}(p)} \tag{1.54}$$

is called the driving point admittance function. In short, if we drop the subscripts *in* then, for any one port network, $Z(p) = V(p)/I(p)$ and $Y(p) = I(p)/V(p)$ are called the impedance and admittance functions respectively. By nature, impedance and admittance functions possess the same mathematical properties and are referred to as immittance functions. It is straightforward to show that, for a linear passive lumped element circuit, the 'immittance' function is rational with real numerator $N(p)$ and real denominator $D(p)$ polynomials. Hence,

$$F(p) = \frac{N(p)}{D(p)} = \frac{a_n p^n + a_{n-1} p^{n-1} + \ldots + a_1 p + a_0}{b_m p^m + b_{m-1} p^{m-1} + \ldots + b_1 p + b_0} \tag{1.55}$$

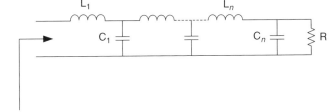

One-port consists of R, L, C components

Figure 1.9   One-port network consisting of passive elements

---

[28] That means port voltage $v(t)$ and port current $i(t)$ are bounded and causal signals.

Clearly, an immittance function $F(p)$ is a complex function in $p = \sigma + j\omega$. In addition, voltage $V(p)$ and current $I(p)$ can be expressed in terms of impedance and admittance functions as

$$
\begin{aligned}
V(p) &= Z(p)I(I) \\
I(p) &= Y(p)V(p)
\end{aligned}
\tag{1.56}
$$

Close examination of Equation (1.55) reveals that $Z(p)$ and $Y(p)$ must be regular or pole free in the RHP to make $v(t)$ and $i(t)$ causal.[29] Furthermore, if they have poles on $j\omega$ the axis, these poles must be simple to make $v(t)$ and $i(t)$ bounded.[30]

Using classical circuit nomenclature, the transformed functions in $p = j\omega$[31] can be expressed as follows:

$$
\begin{aligned}
V(j\omega) &= |V(j\omega)|.e^{i\varphi_v(\omega)} \\
I(j\omega) &= |I(j\omega)|.e^{i\varphi_v(\omega)} \\
Z(j\omega) &= R(\omega) + jX(\omega) \\
Y(j\omega) &= G(\omega) + jB(\omega) \\
F(j\omega) &= A(\omega) + jB(\omega)
\end{aligned}
\tag{1.57}
$$

Based on the description of immittance functions, Equations (1.48–1.49) can be expressed as

$$
\begin{aligned}
W &= \int_{-\infty}^{+\infty} v(t)i(t)dt \\
&= \frac{1}{2\pi} \int_{-\infty}^{+\infty} V(j\omega)I(-j\omega)d\omega \\
&= \frac{1}{2\pi} \left[\int_{-\infty}^{+\infty} R(-\omega^2)|I(j\omega)|^2 d\omega + j\int_{-\infty}^{+\infty} X(\omega)|I(j\omega)|^2 d\omega \right] \geq 0
\end{aligned}
\tag{1.58}
$$

In Equation (1.57) $\int_{-\infty}^{+\infty} X(\omega)|I(j\omega)|^2 d\omega = 0$ since $X(\omega)$ is an odd function of ω and integral $\int_{-\infty}^{+\infty} R(-\omega^2)|I(j\omega)|^2 d\omega = 2\int_{0}^{+\infty} R(\omega)|I(j\omega)|^2 d\omega \geq 0$ since $R(-\omega^2)$ is an even function of ω. Thus, the total energy delivered to the one-port is given by

$$
W_p = \frac{1}{\pi} \int_{0}^{+\infty} R(\omega)|I(j\omega)|^2 d\omega \geq 0
\tag{1.59}
$$

---

[29] A complex function is said to be regular in a domain D if it is analytic (differentiable everywhere) in D. Note that inverse Laplace transformation of a single pole term $K_\sigma/(p - \sigma_k)$ is equal to $K_\sigma.e^{\sigma_k t}u(t)$, which is not bounded for $\sigma_k < 0$. Therefore, an immittance function $F(p)$ must be free of RHP poles. In this representation, $u(t)$ designates the unit step function.

[30] Suppose that $F(p)$ has simple complex conjugate poles as in $1/(p^2 + \omega_r^2)$. Then, considering the inverse Laplace transformation of $1/(p^2 + \omega_r^2)$, a term $\mathcal{L}^{-1}\{F(p)\} = (1/\omega_r)\sin(\omega_r t) u(t)$ must appear either on $v(t)$ or on $i(t)$, or on both, which is causal. Therefore, simple $j\omega$ poles are acceptable in $F(p)$. On the other hand, multiple $j\omega$ poles are not acceptable since they yield non-dissipative immittance functions. Details of this issue are skipped here.

[31] That is, by choosing $\sigma = 0$, we set $p = j\omega$.

yielding $R(\omega) \geq 0$ for all ω; equivalently, $R(p^2) \geq 0$ for all real $\{p\} \geq 0$. Similarly, one can write that

$$W_p = \frac{1}{\pi} \int_0^{+\infty} G(-\omega^2)|V(j\omega)|^2 d\omega \geq 0 \tag{1.60}$$

yielding $G(-\omega^2) \geq 0$ for all ω; equivalently, $G(p^2) \geq 0$ for all real $\{p\} \geq 0$.

1.18 Definition: 'Positive Real Functions'

The above derivations lead us to define a positive real function.

A complex variable function F(p) which designates either the impedance or the admittance function of a one-port is said to be 'positive real' (or in short PR) if:

1. $F(p)$ is real when p is real.
2. $A(p^2) = \text{real}\{F(p)\} \geq 0$ and bounded when $\sigma = \text{real } \{p\} \geq 0$.

Condition 1 implies that all the parameters in $F(p)$ other than p (such as coefficients, exponents, etc.) must be real. In condition 2, the two equality signs do not hold simultaneously unless they hold identically.

Clearly, for a passive one-port network described by Equations (1.58–1.59), the driving point immittance function $F(p)$ must be positive real. Furthermore, if the one-port consists of lumped circuit elements such as resistors, inductors, capacitors and transformers, then $F(p)$ must be a rational PR function.

As verified from the above discussions, a rational function

$$F(p) = \frac{N(p)}{D(p)} = \frac{a_m p^m + a_{m-1}p^{m-1} + \ldots + a_1 p + a_0}{b_n p^n + b_{n-1}p^{n-1} + \ldots + b_1 p + b_0}$$

is PR if and only if:

- Coefficients $\{a_i, b_j\}$ are real.
- It is free of RHP zeros and poles. In other words, both itself and its inverse are analytic in RHP.
- Its $j\omega$ axis zeros and poles are simple, which implies that
- $|m - n| \leq 1$, meaning either $m = n$ or $|m - n| = 1$.

It can be shown that any rational PR impedance function Z(p) can be represented as

$$F(p) = \left[K_\infty p + \frac{k_0}{p} + \sum_{j=1}^{N_\omega} \frac{k_{\omega j}p}{p^2 + \omega_j^2} \right] + \left[K + \sum_{j=1}^{n} \frac{k_j}{p - p_j} \right] \tag{1.61}$$

or

$$Z(p) = Z_F(p) + Z_{min}(p) \tag{1.62}$$

where

$$Z_F(p) = \left[k_\infty p + \frac{k_0}{p} + \sum_{j=1}^{N_\omega} \frac{k_{\omega j}p}{p^2 + \omega_j^2} \right] \tag{1.63}$$

is an odd function with all its poles on the $j\omega$ axis; more specifically, at infinity, at zero and at finite frequencies ω_j with real non-negative residues k_∞, k_0 and $k_{\omega j}$ respectively. In classical circuit theory it is known as the Foster function. Also,

$$Z_{min}(p) = \left[K + \sum_{i=1}^{N_p} \frac{k_i}{p - p_i} \right] \tag{1.64}$$

is the $j\omega$ pole-free part of the impedance $Z(p)$ and is called the minimum reactance function. The complex residue k_i of the poles p_i must have a negative real part. Obviously, a pole $p_i = -\alpha_i + j\beta_i$ must be placed in the open left hand plane (LHP) and therefore α_i must be positive. Furthermore, if $\beta \neq 0$, then it must be accompanied with its conjugate to make the impedance function real.

It is interesting to note that the even part of $Z(p)$ is equal to the even part of $Z_{min}(p)$ and it is given by

$$
\begin{aligned}
R(p^2) &= \text{even}\{Z(p)\} = \text{even}\{Z_{min}(p)\} \\
&= \frac{1}{2} \left[2K + \sum_{i=1}^{n} \left(\frac{k_i}{p - p_i} + \frac{k_i}{-p - p_i} \right) \right] \\
&= \left[K + \sum_{i=1}^{n} \frac{k_i p_i}{(p^2 - p_i^2)} \right]
\end{aligned}
\tag{1.65}
$$

In this case, for a pole p_j, the complex residues k_j can be directly computed from Equation (1.65) as

$$
k_j = (p^2 - p_j^2) \frac{R(p^2)}{p_j} \Big|_{p = p_j}
$$
and
$$
K = \lim_{p \to \infty} R(p^2)
\tag{1.66}
$$

Equation (1.66) indicates that a minimum reactance function can be uniquely generated from its real part in a similar manner to that of Equation (1.39). This greatly facilitates the generation of PR functions to construct lossless power transfer networks. By Darlington's theorem it is also well known that any PR function can be represented as a lossless two-port in resistive termination. In this case, The PR function can be utilized to describe lossless power transfer networks. Furthermore, if we deal with minimum reactance or minimum susceptance functions, then the lossless power transfer network can be uniquely described in terms of the real part of the driving point input immittance of the power transfer network when it is terminated in a resistance. Details of this issue will be covered in the following chapters.

Based on Darlington's theorem, we can also consider the generation of PR functions by means of lossless two-ports when they are terminated in a resistance R (or equivalently a conductance $G = 1/R$).

In practice, lossless filters and matching networks or, in short, power transfer networks, are constructed using ladder networks. Therefore, the following chapters are devoted to generating PR functions from ladder structures.

Below we give an example to test a rational function if it is PR.

Example 1.11: Let

$$F(p) = \frac{N(p)}{D(p)} = \frac{30p^5 + 30p^4 + 23p^3 + 17p^2 + 4p + 1}{24p^6 + 24p^5 + 68p^4 + 44p^3 + 23p^2 + 3p + 1}$$

Find:

(a) If $F(p)$ is PR.
(b) If it is a minimum function.
(c) The analytic form of the even part of $F(p)$.
(d) The real part $R(\omega) = \mathrm{re}\{F(j\omega)\}$ and $X(\omega) = \mathrm{im}\{F(j\omega)\}$ and plot them vs. angular frequency ω.
(e) The imaginary part $X(\omega) = \mathrm{Imag}\{F(j\omega)\}$ using the numerical Hilbert transformation of Equations (1.40–1.43).

Solution:

(a) The above function is rational. If it is PR, then it must have all its zeros and poles in the LHP of the complex p domain. Furthermore, if there exist $j\omega$ poles, they must be simple.

Using MATLAB®, we can easily compute the zeros and the poles and verify if $F(p)$ is PR.

Let $N = [30\ 30\ 23\ 17\ 4\ 1]$ be the vector which contains all the coefficients of the numerator polynomial $N(p)$ in descending order. Then, using MATLAB command '>>roots(N)', we can compute the zeros of $F(p)$.

Thus, we have the following zeros (Table 1.1):

Table 1.1   Zeros of $F(p)$

Zero 1	-0.7654
Zero 2	$-0.0000 + 0.7071i$
Zero 3	$-0.0000 - 0.7071i$
Zero 4	$-0.1173 + 0.2708i$
Zero 5	$-0.1173 - 0.2708i$

As can be seen from Table 1.1, all the zeros of $F(p)$ are in the LHP, which is acceptable.

Similarly, we can compute the poles of $F(p)$. Let $D = [24\ 24\ 68\ 44\ 23\ 3\ 1]$ be the vector which includes all the coefficients of the denominator polynomial in descending order. Then, the poles are found using the MATLAB command '>>roots(D)' as in Table 1.2:

Table 1.2   Poles of $F(p)$

Pole 1	$-0.1244 + 1.4886i$
Pole 2	$-0.1244 - 1.4886i$
Pole 3	$-0.3548 + 0.4505i$
Pole 4	$-0.3548 - 0.4505i$
Pole 5	$-0.0207 + 0.2374i$
Pole 6	$-0.0207 - 0.2374i$

These poles are also in the LHP. Thus $F(p)$ is PR.

(b) Obviously, $F(p)$ is minimum since it has all its zeros and poles in the LHP. That is, it is free of $j\omega$ poles.

(c) The even part of $F(p)$ is given by

$$R(p) = \frac{1}{2} [F(p) + F(-p)]$$

$$= \frac{576p^8 + 672p^6 + 244p^4 + 28p^2 + 1}{576p^{12} + 2688p^{10} + 3616p^8 + 1096p^6 + 401p^4 + 37p^2 + 1}$$

(d) Setting $p = j\omega$, we can plot the real part $R(\omega) = \text{real}\{F(j\omega)\}$ as well as $X(\omega) = \text{imaginary}\{F(j\omega)\}$ as shown in Figure 1.10 and 1.11 using MATLAB.

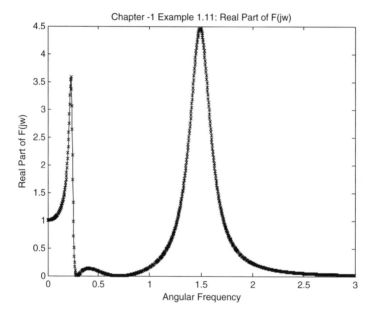

Figure 1.10   Plot of the real part of $F(j\omega)$

(e) In order to evaluate $X(\omega)$ by means of the numerical Hilbert transformation, firstly, we have to sample the real part $R(\omega)$ to specify the break points $\{R_k, k = 1, 2, \ldots, NB\}$ and the break frequencies $\omega_k, k = 1, 2, \ldots, NB$. Obviously, the quality of the computations depends on the total number of sampling points. The more the number of break points, the better the quality in computations. For our computational purposes, we developed a MATLAB program 'Chapter 1, Example 1.11, which is listed at the end of this chapter.

Close examination of Figure 1.10 reveals that, after $\omega \geq 3$, $R(\omega)$ practically approaches zero. Therefore, break frequencies can be uniformly distributed over the angular frequency interval $0 \leq \omega \leq 3$. In this case, choosing $NB = 10$ break point (Table 1.3), we end up with the break point pairs shown in Figure 1.12.

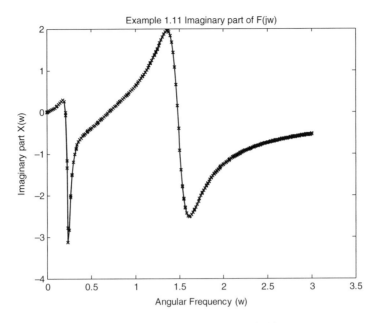

Figure 1.11   Plot of $X = \text{im}\{F(j\omega)\}$

Table 1.3   Selection of break points $NB = 10$ uniform samples for Example 1.11

Angular break frequencies, ω_k	Corresponding break points, R_k
0	1.0000
0.3333	0.0798
0.6667	0.0035
1.0000	0.1565
1.3333	1.6720
1.6667	1.4607
2.0000	0.2310
2.3333	0.0785
2.6667	0.0361
3.0000	0.0194

Employing Equation (1.43) we have

$$X_H(\omega) = \sum_{k=1}^{NB-1} B_k(\omega) \Delta R_k \qquad (1.67)$$

where the excursions ΔR_k and the coefficients B_k and $X_H(\omega_k)$ are given in Table 1.4.

Table 1.4 is illustrated in Figure 1.13.

As can be seen from Figure 1.13 $NB = 10$ points results in poor estimation of the imaginary part of $F(j\omega)$. Nevertheless, resolution can be improved by increasing sample points on $R(\omega)$.

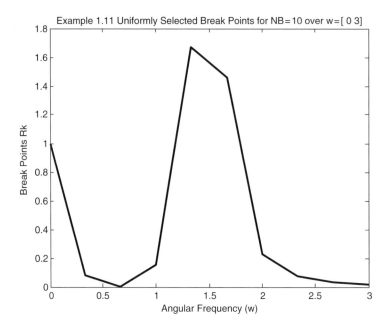

Figure 1.12   Break point distribution of Example 1.11

Table 1.4   Evaluation of numerical Hilbert transform of $R(\omega)$ for $NB = 10$ break points

	Break frequencies, ω_k	$\Delta\omega_k = \omega_k - \omega_{k-1}$	ΔR_k	B_k	Numerically computed X_H
1	0	—	—	—	−0.0000
2	0.3333	0.3333	−0.9202	0.0354	−0.3194
3	0.6667	0.3333	−0.0763	0.1072	0.1444
4	1.0000	0.3333	0.1530	0.1818	0.7350
5	1.3333	0.3333	1.5155	0.2617	0.4005
6	1.6667	0.3333	−0.2113	0.3503	−0.8337
7	2.0000	0.3333	−1.2297	0.4535	−1.0609
8	2.3333	0.3333	−0.1525	0.5829	−0.6620
9	2.6667	0.3333	−0.0424	0.7693	−0.4911
10	3.0000	0.3333	−0.0166	1.2293	−0.3900

For example, $NB = 30$ yields uniformly distributed sample points and one can obtain much better estimation of $X_H(\omega)$ as shown in Figure 1.14.

Finally, note that an almost original approximation can be obtained with $NB = 50$ break points. This is left as an excercise to the reader.

Details of the Hilbert transformation will be given in the following chapters.

Finally, we present the MATLAB programs developed for this chapter to enable the reader to verify the above results.

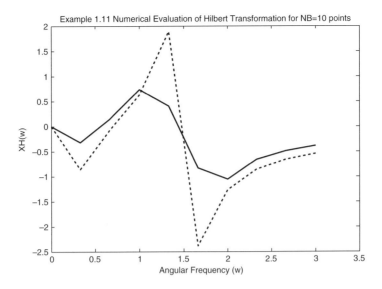

Figure 1.13   Approximation of the imaginary part via Hilbert transformation

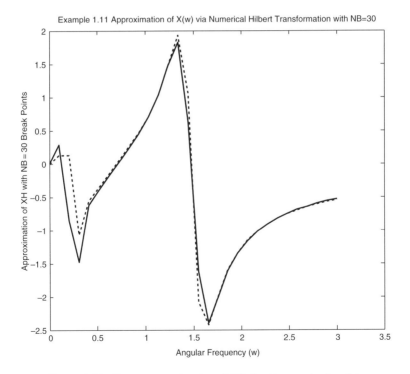

Figure 1.14   Better approximation of $X(j\omega)$ with more beak points

Program Listing 1.1   Chapter 1, Example 1.11, main program listing

```
% Program: Chapter 1; Example 1.11B
clear
% Given  F(p) = N/D  or  F(p) = a(p)/b(p)
% Given  N = [30 30 23 17 4 1]  in descending order
% Given  D = [24 24 68 44 23 3 1]in descending order
% Computation of R(w) and X(w)
N = [30 30 23 17 4 1];% Input
D = [24 24 68 44 23 3 1]; %Input

wstart = 0.0; % Beginning of angular frequency
wend = 3.0;   % End of angular frequency
NB = 30;      % Total number of sampling frequencies
dw = (wend-wstart)/(NB-1);% sweeping step size
%
%Computation of Break frequencies WB and Break Points RB
w = wstart;
j = sqrt(-1);
for i = 1:NB
    p = j*w;                % Complex variable on jw axis
    Num = polyval(N,p);     % Complex value of the numerator at w
    Denom = polyval(D,p);   % Complex value of the denominator at w
    F = Num/Denom;          % Complex value of the given function at w
    RB(i) = real(F);        % Break Points
    X(i) = imag(F);         % Original Imaginary part
    WB(i) = w;              % Break Frequencies
    w = w + dw;             % Sweeping frequency
end

% Numerical Hilbert Transformation
wstart = 0.0 + 1e-6; % Beginning of angular frequency
w = wstart;
%
for i = 1:NB
    [DW,DR,B,XA] = Num_HilbertB(w,WB,RB);
    XH(i) = XA;
    W(i) = w;
    w = w + dw;
end
figure (1)
plot (WB,RB)
xlabel('Angular Frequency (w)')
ylabel(' Break Points Rk')
title(' Example 1.11 Break Points Rk NB points')
figure (2)
plot(W,XH,WB,X)
xlabel('Angular Frequency (w)')
```

```
ylabel (' Approximation of XH with NB points')
title (' Example 1.11 Approximation of X(w) via Numerical Hilbert
Transformation NB points')

% Numerical Hilbert Transformation
wstart = 0.0 + 1e-6; % Beginning of angular frequency
w = wstart;
%
for i = 1:NB
    [DW,DR,B,XA] = Num_HilbertB(w,WB,RB);
    XH(i) = XA;
    W(i) = w;
    w = w + dw;
end
figure
plot(WB,RB)
figure
plot(W,XH,WB,X)
```

Program Listing 1.2   Evaluation of numerical Hilbert transformation for Example 1.11

```
function [DW,DR,B,XA] = Num_HilbertB(w,W,R)
%Numerical Hilbert Transformation
N = length(W);
%
for k = 2:N
    DW(k) = W(k)-W(k-1);
    DR(k) = R(k)-R(k-1);
    M(k-1) = DR(k)/DW(k);
end
    for k = 1:N
    F(k) = (w+W(k))*log(abs(w+W(k))) + (w-W(k))*log(abs(w-W(k)));
    end
for k = 2:N
    DF = (1/pi)*(F(k)-F(k-1));
    B(k) = (1/pi)*(F(k)-F(k-1))/DW(k);
    X(k) = B(k)*DR(k);
  end
  %
      XA = 0;
      for k = 2:N
          XA = XA + X(k);
      end
```

2

Electromagnetic Field Theory for Power Transfer Networks: Fields, Waves and Lumped Circuit Models

2.1 Introduction

The fundamentals of circuit theory stem from electromagnetic fields. Especially at high frequencies, where the size of the circuit components is comparable to the wavelength of operational signals, use of electromagnetic field theory becomes essential for assessing the performance of the circuits.

On the other hand, in order to obtain dependable circuit performance, design procedures must include the electromagnetic behavior of circuit components as well as interconnectors.

In designing power transfer networks, we usually deal with mathematical functions employed in classical circuit theory. These functions are derived directly from the given design specifications such as termination impedances to be matched, frequency bandwidth of the operation, desirable flat gain level, etc.

On the other hand, a realizable circuit function can be generated to optimize the electrical performance of the system under consideration. In this case, it must be synthesized at the component level yielding the desired power transfer network. Therefore, a formal understanding of circuit functions is essential for dealing with design problems.

Obviously, circuits, or more specifically two-ports, which are used to transfer power, may consist of only lumped elements, or distributed elements, or combinations of both; that is, lumped and distributed elements.

Generally, a power transfer network is considered as a lossless two-port. It is derived by a Thévenin voltage source at one end, say at the input port (or at port 1), and terminated with a passive load at the other end, say at the output port (or port 2) as shown in Figure 2.1; it contains lossless reactive elements such as ideal inductors, ideal capacitors or ideal transformers.

From a theoretical point of view, we may also consider mutual coupling between inductors and capacitors. However, in practice, we generally avoid using mutually coupled elements as much as we can.

Design of Ultra Wideband Power Transfer Networks   Binboga Siddik Yarman
© 2010 John Wiley & Sons, Ltd

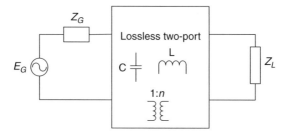

Figure 2.1   Lossless power transfer two-port consists of ideal lossless elements

At this point, it may be appropriate to give the formal definitions of circuit components so that we can construct some concrete properties of network functions.

In classical circuit theory, circuit elements may be described in terms of their terminal or port-related voltage and current relations. In designing power transfer networks, we may employ lumped inductors, lumped capacitors, transformers and transmission lines. The roots of all the circuit components, as well as power transfer networks, go back to electromagnetic field theory. Therefore, let us review electromagnetic field theory from the circuit theory perspective, then state Kirchhoff's law and give the formal definitions of circuit components.

2.2 Coulomb's Law and Electric Fields

From classical Newtonion physics we know that masses attract each other as

$$F = K_N \frac{m_1 m_2}{R^2}$$

(2.1)

where m_1 and m_2 represent masses; R is the distance between them; K_N is the gravitational constant which is specified as $K_N = 6.673 \times 10^{-11} \, \text{N} \, \text{m}^2/\text{kg}^2$; and F is the force of attraction Equation (2.1) is known as Newton's law.

A similar force equation was written for electric charges by Coulomb. Let Q_1 and Q_2 be two distinct electric charges measured in coulombs. These charges could be either in the same or opposite polarities. It is well established that if they have opposite polarities, they attract each other, otherwise they repel. The attractive or repulsive force between Q_1 and Q_2 is measured by

$$F = K_E \frac{Q_1 Q_2}{R^2}$$

(2.2)

where K_E is a constant which depends on the medium.

For example, in free space K_E is specified by $K_E = 1/4\pi\varepsilon_0$ where ε_0 is called the permittivity of free space and approximated as $(1/36\pi)10^{-9} \text{F/m}$.

Note that Equation (2.2) describes a macroscopic physical event for which R cannot be zero. In essence, the force F can never be infinity; therefore, R is limited by a minimum quantum distance R_{min} which is never zero. Details of this issue are beyond the scope of this book and therefore it is omitted here. Equation (2.2) is known as Coulomb's law.

2.3 Definition of Electric Field

Force is an observable quantity. Obviously we can measure it. It also has a direction. In mathematics, directional quantities are represented by vectors. Therefore, we can say that electric force is a vector quantity and it is a function of Q_1, Q_2 and the distance R. Thus, we can write

$$F = K_E \frac{Q_1 Q_2}{R^2} u_R \qquad (2.3)$$

where u_R is the unit vector in the direction of vector R.

Let us assume that a positive charge Q_1 is fixed at a point P_1 and a negative charge Q_2 is free to move starting from the point P_2. By Coulomb's law, it is observed that Q_2 moves toward Q_1 over the vector distance R. In this case, F is the attractive force directed from P_2 to P_1 or on the vector R. The unit vector u_R is then given by $u_R = R/R$.

Hence, the force vector is expressed as

$$F = K_E \frac{Q_1 Q_2}{R^3} R \qquad (2.4)$$

Based on the above discussion, Equation (2.2) specifies the intensity of the attractive force developed between Q_1 and Q_2 and Equation (2.4) indicates the direction of the force as shown in Figure 2.2.

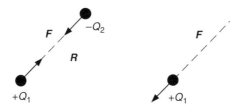

Figure 2.2   The Coulomb force

Literally, the electric field vector is defined as the electric force per positive unit charge. In order to make a formal definition, let the infinitesimally small charge $Q_{test} = Q_2$ and $Q = Q_1$. Then, the electric field vector E which is generated by a static charge Q is defined as

$$E = \frac{1}{Q_{test}} F = K_E \frac{Q}{R^2} u_R = E u_R \qquad (2.5)$$

where $E = K_E Q/R^2$ is the intensity of the electric field vector E, $F = K_E QQ_{test}/R^2$ is the intensity of the Coulomb force developed between the charges Q_{test} and Q. Equivalently, the Coulomb force F imposed on a test charge Q_{test} can be given in terms of the electric field E generated by a static charge Q as follows:

$$F = Q_{test} E \qquad (2.6)$$

Equivalently, the electric field intensity can be given in terms of the intensity of the Coulomb force F such that

$$E = \frac{F}{Q_{test}}$$
(2.7)

Based on the labeling convention introduce above, we say that electric field vectors are directed from negative charges to positive charges, or we can simply say that electric field lines are initiated from positive charges and end at negative charges as shown in Figure 2.3.

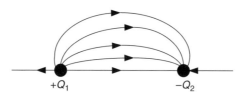

Figure 2.3 Direction of electric field lines

2.4 Definition of Electric Potential

Referring to Figure 2.4, let Q_1 be a positive electric charge generating an electric field E which is radial and outward from the charge. Let Q_2 be a negative electric charge at infinity. In this situation, the attractive force $F = K_E Q_1 Q_2 / r^2$ between Q_1 and Q_2 must be zero since the distance r is infinity. Now, let us drag Q_2 from infinity to a point P_B on the 'ray r' by means of an external force. Obviously, we do work by bringing in Q_2 which is expressed by[1]

$$W \triangleq \int_{\substack{\text{over an} \\ \text{arbitrary path C}}} F.dc = \int_{\infty}^{P_B} F.dr$$
(2.8)

---

[1] It should be noted that the dot product $F.dc$ is equal to $|F||dc|\cos(\theta)$ where θ is the angle between F and dc and $dr = |dc|\cos(\theta)$ is the projection of dc on the vector F and therefore it describes the shortest distance subject to evaluation of the work. Hence, evaluation of work or equivalently the potential integral between two points, say A and B, over any arbitrary path C, must be equal to the minimum of $\left\{ \int_{\text{over } c} F.dc \right\} = \int_A^B F.dr = \pm \int_A^B F(r)dr$ where dr is an infinitesimally small distance along the line (or ray) or the shortest distance between A and B aligned with the direction of dc. The $\pm$ sign is determined based on the directions of F and dr. Obviously, if they are in the same direction the plus sign $(+)$ is selected, otherwise it is the minus $(-)$.

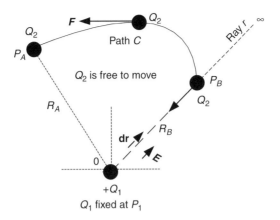

Figure 2.4 Electric potential between two points

In this case, we say that the potential energy W_P of point P_B is equal to the negative of the kinetic energy or the work done. Hence, by definition,

$$W_p \triangleq -W = -\int_{\infty}^{P_B} \boldsymbol{F}.d\boldsymbol{c} = \int_{P_B}^{\infty} \boldsymbol{F}.d\boldsymbol{c}$$

Or, by taking $\boldsymbol{F} = -$Coulomb force $\boldsymbol{F}_C =$

$-\left[K_E\left(Q_1 Q_2/r^2 \right)\boldsymbol{u_R}\right]$ and the direction of $d\boldsymbol{c}'$ along with $\qquad\qquad$ (2.9)

the Coulomb force then, we have $\boldsymbol{F}.d\boldsymbol{c} = \boldsymbol{F}_C.d\boldsymbol{c}'$

$$W_p = Q_2 \int_{\text{over a path } C} \boldsymbol{E}.d\boldsymbol{c}' = Q_2 \int_{P_B}^{\infty} \boldsymbol{E}.d\boldsymbol{r}$$

Clearly, by sending Q_2 to infinity, this potential energy can always be spent or converted to kinetic energy $W_k = W$. Based on the above definition, at any point P_B, the sum of the potential and kinetic energies must be zero. In other words,

$$W_p + W_k = 0 \qquad\qquad (2.10)$$

Therefore, we can state that

$$W_p + W = 0 \quad \text{or} \quad W = -W_P \qquad\qquad (2.11)$$

Literally, one can define the potential difference between the points P_B and infinity as

$$V \triangleq V_B - V_\infty \triangleq \frac{W_p}{Q_2} = -\int_{\infty}^{R_B} \boldsymbol{E}.d\boldsymbol{r} = \int_{R_B}^{\infty} \boldsymbol{E}.d\boldsymbol{r} \qquad\qquad (2.12)$$

where the intensity of the electric field vector is measured by

$$E = \frac{F_C}{Q_2} = K_E \frac{Q_1}{r^2}$$

In the above equation, V_B is read as the potential of point P_B with respect to infinity or with respect to a reference point.

Obviously, Equation (2.12) can be evaluated along any path (or simply on the ray r) connecting the points P_B and infinity. The electrostatic field is conservative; hence, under any choice, the work completed or 'done' between two points must be the same.

Note that for the present choice $\mathbf{E}.\mathbf{dr} = E.dr$. Therefore,

$$
\begin{aligned}
V = V_B - V_\infty &\triangleq K_E Q_1 \int_{R_B}^{\infty} \frac{dr}{r^2} = -K_E \left(\frac{1}{r} \right) \Big|_{R_B}^{\infty} \\
&= -\lim_{r \to \infty} K_E \frac{Q_1}{r} + K_E \frac{Q_1}{R_B}
\end{aligned}
\tag{2.13}
$$
or
$$V = V_B - V_\infty = K_E \frac{Q_1}{R_B} - [0] = K_E \frac{Q_1}{R_B}$$

The above result suggests the following:

- The potential V_B of point P_B is $V_B = K_E Q_1 / R_B$.
- The potential V_∞ of infinity is $V_\infty = \lim_{r \to \infty} K_E Q/r = 0$, which is also called the reference potential.
- For any arbitrary point A, potential $V_A = K_E Q_1/R_A$. Clearly, this measurement or reading is made with respect to infinity.
- The potential of a point is a scalar quantity and it is expressed with respect to a reference point.
- In general, we talk about the potential difference between two points, say $V_{AB} = V_A - V_B$, which corresponds to 'work done per unit charge' when we drag charge Q_2 from point P_B to point P_A along a *path C*. Then, Equation (2.13) takes the following form:

$$V_{AB} = V_A - V_B = -\int_{P_B}^{P_A} \mathbf{E}.d\mathbf{c} = -K_E \int_{R_B}^{R_A} \frac{Q_1}{r^2} dr = K_E \frac{Q_1}{R_A} - K_E \frac{Q_1}{R_B}$$

yielding

$$V_A = K_E \frac{Q_1}{R_A} \quad \text{and} \quad V_B = K_E \frac{Q_1}{R_B}$$

It should be clear for the reader that the potentials of points A and B are specified with respect to infinity.

Based on the above definitions, it is interesting to note the following remarks.

Remark 1: When moving a charge Q_{Test} over a path C from point B to point A under a uniform electric field $\mathbf{E}$, the potential difference between point A and point B is given by Equation (2.8) as

$$V_{AB} = V_A - V_B = \int_A^B \mathbf{E}.d\mathbf{c}$$

Remark 2: If the path is a straight line, say along the x axis, and if the uniform field is in the same direction of the x axis and assumed to be constant or space independent, then the potential difference between two close points X_A and X_B is expressed as

$$V_{AB} = \int_{X_A}^{X_B} \boldsymbol{E}.d\boldsymbol{x} = E[X_B - X_A] = Ed$$

where E, the intensity of the constant electric field, is a vector and d is the distance measured as $d = X_B - X_A$.

Remark 3: Integration of the electric field over a closed path must be zero since there is no work done by dragging a charge starting at a point and ending with the same point under a static electric field. Therefore we state that

$$\oint \boldsymbol{E}.d\boldsymbol{c} = 0$$
over a closed path C

2.5 Units of Force, Energy and Potential

In the International Standard Unit system (ISU system), which is also known as the meter, kilogram, second system (MKS system), the force is measured in newtons, no matter if it is mechanical, electrical or magnetic force. Since the Newton force is expressed as the multiplication of mass by acceleration (i.e. $F = m\gamma = m\,dv/dt$, where $\gamma = dv/dt$ is the acceleration and $v = v(t)$ is the velocity specified as a function of time in meters per second), then

$$1 \text{ newton} = 1 \text{ kg} \times \frac{1 \text{ meter}}{(1 \text{ second})^2}$$
$$\text{or}$$
$$1 \text{ N} = 1 \text{ kg m/s}^2 \tag{2.14}$$

Similarly, the electric force unit of newtons can also be given in terms of the units of electric charges (coulomb), distance (meter) and the electric medium constant K_E which has the dimensions of m/F. In this case,

$$1 \text{ N} = \left(\frac{\text{m}}{\text{F}}\right) \times \frac{\text{C}^2}{\text{m}^2} = 1 \text{ C}^2/(\text{F} \times \text{m}) \tag{2.15}$$

The work done by 1 N force over a 1 m distance is called 1 joule of energy.

In other words, in MKS, the unit of energy is measured in joules and given by

$$1 \text{ J} = 1 \text{ N} \times 1 \text{ m} \tag{2.16}$$

As stated by Remark 2 of Section 2.4, the potential difference between any points A and B is given by

$$V = V_{AB} = V_A - V_B \triangleq \int_A^B \boldsymbol{E}.d\boldsymbol{l} \tag{2.17}$$

and it is measured in volts, which is equivalent to '1 J of work done to drag 1 C of electric charge'. Thus,

$$
\begin{aligned}
1\,\mathrm{V} &= 1\,\mathrm{J}/1\,\mathrm{C} \\
\text{or in detail} \\
&= 1\,\mathrm{N} \times 1\,\mathrm{m}/1\,\mathrm{C}
\end{aligned} \tag{2.18}
$$

2.6 Uniform Electric Field

Referring to Figure 2.5, let us consider two flat plates 'd' meters apart from each other and loaded with opposite static electric charges.

Plate 1 is loaded with total positive charges of '$+Q$' coulombs over the surface S. Similarly, mirror image plate 2 is loaded with total negative charges of '$-Q$'. In this case, electric field vectors are drawn as straight lines from positively charged plate 1 down to negatively charged plate 2. Thus, the electric field between the plates is uniform and assumed to be constant. Therefore, $E.dl = -E.dl$. Then, by Remark 2 of Section 2.4, the potential difference between P_1 and P_2 is given by[2]

$$
V = \int_{P_1}^{P_2} E.\,dl = V_1 - V_2 = -\int_{d}^{0} E.\,dl = E.d \tag{2.19}
$$

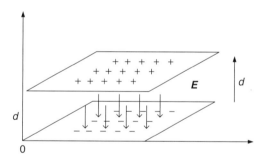

Figure 2.5 Generation of constant uniform electric field by means of charge-loaded, closely placed, parallel plates

An alternative way of defining the potential difference between the charged plates may be as follows.

Assume that the plates are idealized as two opposite Coulomb charges $+Q$ and $-Q$ which are placed d meters apart. In this case, the intensity of the Coulomb force is given by $F_C = K_E\, Q^2/d^2$ or equivalently the intensity of the uniform electric field between the plates is defined as $E \triangleq F_C/-Q = K_E\, Q/d^2$, which is directed from the positive plate down to the negative plate. This uniform Coulomb force must

---

[2] Here, the uniform and space-independent electric field can be generated by applying an external voltage source across the plates of the capacitor, which imposes the boundary conditions on the plates.

generate the potential energy $W_p \triangleq \int_d^0 F_C dx = F_C d = Q|E|d$. Hence, by definition, the potential difference between plates 1 and 2 is given as $V_{12} \triangleq W_p / Q = |E|d$ or simply $V = Ed$, which verifies Equation (2.19).

2.7 Units of Electric Field

Equation (2.19) indicates that one can readily determine the intensity of the uniform electric field E developed between charged plates as

$$\boxed{E = \frac{V}{d}}$$

(2.20)

where $V = V_1 - V_2$ is the potential difference between plate 1 and plate 2 and d is the distance between the parallel plates. In this case, the electric field intensity is either measured by Equation (2.20) as volts per meter or it can be measured by Equation (2.7) as newtons per coulomb (in short, N/C).

2.8 Definition of Displacement Vector or 'Electric Flux Density Vector' D

In Equation (2.2), Coulomb's law or the electric force between charges was expressed in terms of a constant

$$\boxed{K_E = \frac{1}{4\pi\varepsilon}}$$

(2.21)

In this equation ε is called the permittivity and it is considered as a measure of the electrical property of the medium where the electric forces are exercised.

In practice, the electrical property or permittivity of a medium is measured by means of the permittivity of free space ε_0 within a constant ε_r, called relative permittivity, as $\varepsilon = \varepsilon_r \varepsilon_0$.

For example, in a medium with permittivity ε, the electric field generated by a concentrated charge Q, can be measured by means of the force applied on an infinitesimally small test charge Q_{Test} which is placed R meters away from Q. In this case, the electric field intensity E is measured from the center of Q within a radial distance R or on the surface of a sphere with radius R as

$$\boxed{E = \frac{F_C}{Q_{test}} = \frac{Q}{4\pi\varepsilon R^2} = \left[\frac{1}{\varepsilon}\right]\left[\frac{Q}{4\pi R^2}\right]}$$

(2.22)

Close examination of Equation (2.22) reveals that one can define a quantity D which has similar mathematical features to E but is free of the electrical properties of the medium such that

$$\boxed{D = \varepsilon E = \frac{Q}{4\pi R^2} = \frac{Q}{\text{surface of a sphere of radius } R}}$$

(2.23)

D is called the displacement and it can be considered as the intensity of a vector D which is also expressed in terms of the electric field vector E. Thus,

$$D = \varepsilon E = \frac{Q}{4\pi R^2} u_R \ (\mathrm{C/m}^2).$$ (2.24)

Note that, in Equation (2.24), the unit of the displacement vector is coulombs per meter squared due to the term $S = 4\pi R^2$ which is the surface of a sphere of radius R centered at the point charge Q. This charge generates the electric field E, perpendicular to the surface $S = 4\pi R^2$.

Hence, one can state that

$$D.S.u_R = \varepsilon E.Su_R = Q$$ (2.25)

Equation (2.25) may be written in integral form as follows. Let dS be an infinitesimally small surface on S. Since u_R is perpendicular to this surface, then we can define an infinitesimally small surface vector $dS = dS.u_R$ so that

$$\oiint_S D.dS = \oiint_S \varepsilon E.dS = \oiint_S \varepsilon E.u_R dS = Q$$ (2.26)

In Equation (2.26), the displacement vector D can be considered as a field or flux density vector which is free of electrical properties of the medium and the total charge Q may be regarded as the net electric flux ψ that flows out of surface S. Thus, we can write that

$$\psi = \oiint_S D.dS = \oiint_S \varepsilon E.dS = Q$$ (2.27)

This integral indicates that total charge Q enclosed by any closed surface S describes the flux ψ and is equal to the surface integral of the displacement vector D. Equation (2.27) is known as Gauss's law.

We should note that the enclosure or the surface S which contains electric charges obviously closes a volume V as shown in Figure 2.6.

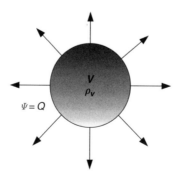

Figure 2.6 S encloses a volume V with charge density ρ_v

Generally speaking, the charges in this volume may be continuous, with charge density ρ_v yielding the total flux or charge as

$$Q = \psi = \iiint_V \rho_v d_v \tag{2.28}$$

On the other hand, by the divergence theorem of calculus, it is well established that

$$\oiint_S \boldsymbol{D}.d\boldsymbol{S} = \iiint_V \operatorname{div}(\boldsymbol{D})dV = \iiint_V \boldsymbol{\nabla}.\boldsymbol{D}dV \tag{2.29}$$

By equating Equations (2.28) and (2.29) we have

$$\operatorname{div}(\boldsymbol{D}) = \boldsymbol{\nabla}.\boldsymbol{D} = \rho_v \tag{2.30}$$

where $\boldsymbol{\nabla}$ is called the del operator and in the Cartesian coordinate system (CCS) it is given by

$$\boldsymbol{\nabla} = \frac{\partial}{\partial x}\boldsymbol{u}_x + \frac{\partial}{\partial y}\boldsymbol{u}_y + \frac{\partial}{\partial z}\boldsymbol{u}_z \tag{2.31}$$

where $\boldsymbol{u}_x$, $\boldsymbol{u}_y$ and $\boldsymbol{u}_z$ are the unit vectors along the x, y, z axes of CCS.

Obviously, the del operator is a vector, so it is expressed here in a bold font in plain form as $\boldsymbol{\nabla}$.

In CCS, the displacement of flux density vector $\boldsymbol{D}$ is given in terms of its components D_x, D_y, D_z as

$$\boldsymbol{D} = D_x\boldsymbol{u}_x + D_y\boldsymbol{u}_y + D_z\boldsymbol{u}_z \tag{2.32}$$

and $\boldsymbol{\nabla}.D$, which is read as divergence D, is given by

$$\boldsymbol{\nabla}.\boldsymbol{D} = \frac{\partial D_x}{\partial x} + \frac{\partial D_y}{\partial y} + \frac{\partial D_z}{\partial z} \tag{2.33}$$

Thus,

$$\boldsymbol{\nabla}.\boldsymbol{D} = \frac{\partial D_x}{\partial x} + \frac{\partial D_y}{\partial y} + \frac{\partial D_z}{\partial z} = \rho_v \tag{2.34}$$

Equation (2.34) is known as the differential form of Gauss's law.

2.9 Boundary Conditions in an Electric Field

Referring to Figure 2.7, Equation (2.27) yields the boundary conditions in an electric field when passing from medium 1 to medium 2.

Let us consider an infinitesimally small closed path '*abcd*' which starts in medium 1, goes along the path '*ab*' and then into medium 2 along '*bc*', and continues in medium 2 along '*cd*' and finally comes back to medium 1 over '*da*'. In this case,

$$\oint_{abcd} E.\,dl = \int_a^b E.\,dl + \lim_{bc \to 0} \int_b^c E.\,dl + \int_c^d E.\,dl + \lim_{da \to 0} \int_d^a E.\,dl$$

Let the tangential components of the electric field be E_{t1} in medium 1 and E_{t2} in medium 2. Then,

$$(E_{t1})(ab) + (E_{t2})(cd) = 0$$

or

$$E_{t1} - E_{t2} = 0$$

Thus,

$$\boxed{E_{t1} = E_{t2}} \tag{2.35}$$

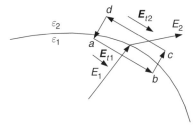

Figure 2.7   Boundary conditions in an electric field

In order to find the boundary conditions in the normal direction, we consider an infinitesimally small cylinder with cross-sectional area A (see Figure 2.8). Then, Gauss's law yields

$$\psi = \oiint_S D.\,dS = Q = \lim_{side \to 0} \oiint_{side} D.\,dS + \iint_{A(\text{bottom})} D_1 n_1 dS + \iint_{A(\text{top})} D_2 n_2 dS$$

Let D_{n1} and D_{n2} be the normal components of the displacement vector in medium 1 and medium 2 respectively. Then,

$$(D_{n1} - D_{n2})A = Q_A$$

Further, let us designate the surface charge density by ρ_S so that

$$Q_A = \rho_S A$$

Then, the boundary condition along the normal components yields

$$\boxed{D_{n1} - D_{n2} = \rho_S}$$

(2.36)

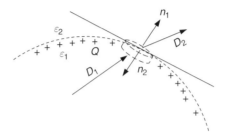

Figure 2.8   Boundary condition along the normal components

2.10 Differential Relation between the Potential and the Electric Field

In Equation (2.17), the potential is a scalar function of geometry as well as the electrical property of the medium. It may be written as $V = V(x, y, z)$ and the electric field vector can be expressed in terms of it is (X, Y, Z) components as $E = E_x \mathbf{i} + E_y \mathbf{j} + E_z \mathbf{k}$. In addition, the differential form of Equation (2.17) is $dV = -\mathbf{E} \cdot \mathbf{dl}$ where dl is an infinitesimally small tangential vector along a path C and may be expressed as $dl = dx\mathbf{i} + dy\mathbf{j} + dz\mathbf{k}$. In this case, the total differential dV is written as

$$dV = \frac{\partial V}{\partial x}dx + \frac{\partial V}{\partial y}dy + \frac{\partial V}{\partial z}dz = -[E_x dx + E_y dy + E_z dz]$$

Hence, the above equation reveals that

$$E_x = -\frac{\partial V}{\partial x} \quad E_y = -\frac{\partial V}{\partial y} \quad E_z = -\frac{\partial V}{\partial z}$$

or simply,

$$\boxed{E = -\nabla V}$$

(2.37a)

By Equation (2.34) we can write that

$$\boxed{\nabla \cdot (E) = \frac{\rho}{\varepsilon}}$$

(2.37b)

or

$$\boxed{\nabla^2 V(x,y,z) = -\frac{\rho(x,y,z)}{\varepsilon}}$$

(2.37c)

The above equation is known as the Poisson equation and its solution is given by

$$\boxed{V(x,y,z) = \frac{1}{4\pi\varepsilon} \iiint_{V'} \frac{\rho(r')dv'}{R}}$$

(2.37d)

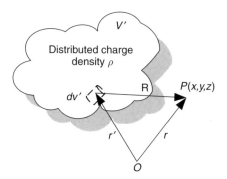

Referring to the above figure, in Equation (2.37d) R is the distance measured from an infinitesimally small volume element $dv\prime$ to a point $p(x,y,z)$. Let r be the distance measured from the origin O to point P and r' be the distance defined from the origin O to an infinitesimally small volume $dv\prime$. Then, $r = |OP| = \sqrt{x^2 + y^2 + z^2}$ and $R = r - r'$.

Here the del operator is specified as

$$\nabla = \frac{\partial}{\partial_x}i + \frac{\partial}{\partial y}j + \frac{\partial}{\partial z}k$$

and

$$\nabla \cdot (\nabla) = \nabla^2 = \frac{\partial^2}{\partial x^2} + \frac{\partial^2}{\partial y^2} + \frac{\partial^2}{\partial z^2}$$

which is a scalar operator.

Equations (2.37) say that the electric field vector of a medium can simply be generated by taking the gradient of the measured potential at each point. Furthermore, if the charge density $\rho(x, y, z)$ is known, one can immediately compute the potential $V(x, y, z)$ of a point $P(x, y, z)$ by integrating the Poisson equation.

2.11 Parallel Plate Capacitor

In general, wherever we see a charge concentration Q, we can always define an electric field E at a distance R meters away from it.[3]

If Q is a point charge, then the field lines are considered as radial lines, initiating from the charge source Q.

Furthermore, we can measure the intensity of the electric field at any point P with the help of a test charge. $Q_{test} = 1$ C.

Whenever we test the existence of an electric field with a test charge (or simply with a charge), then a potential difference between two points can be measured.

When it is charged by a voltage source, a practical parallel plate capacitor can be regarded as a medium which holds a space-independent uniform electric field confined within the plates. In this case, two opposite point charges $+Q$ and $-Q$ are developed on the plates which are placed 'd' meters apart from each other, as shown in Figure 2.9.

This idealized model is called an electric dipole. In this case, the point charge $+Q$ can be regarded as a

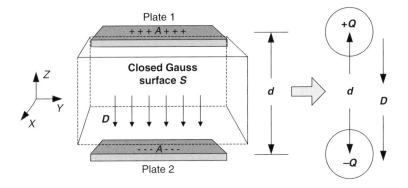

Figure 2.9 A parallel plate capacitor

charge source which creates a uniform electric field from $+Q$ down to $-Q$ along the attractive force $\boldsymbol{F}$. In this regard, concentrated charge $-Q$ can be considered as the test charge to measure the electric field intensity. Thus, due to the applied external source V, the potential difference between plates 1 and 2 is specified as

$$V = V_1 - V_2 = \int_{0(\text{from the level of plate 1})}^{d(\text{to plate 2})} \boldsymbol{E}.d\boldsymbol{l} = Ed \tag{2.38}$$

---

[3] It should be noted that electric field E is a force described as $F = Q \times 1/4\pi r^2$ per unit charge. Therefore, whenever we see a charge Q, it is assumed that the electric field is created by bringing a test charge of 1 C to a distance r meters measured from Q.

where the electric field vector E is space independent.

Let ε be the permittivity of the material which fills the region between the plates. Then, by writing $D = \varepsilon E$, we find

$$V = \frac{D}{\varepsilon} d \tag{2.39}$$

Now let us also determine the displacement vector D by means of Gauss's law. Obviously, the displacement vector D is directed from the positively charged plate down to the negatively charged plate and is only confined within these plates. Otherwise, it is zero. Let the plate surfaces of the capacitor be A m^2.

In order to determine the electric filed at a point in the given medium, one must properly choose a closed surface to apply Gauss's law. Let us refer to the chosen surface as the Gauss surface or GS. In this case, we have the following possibilities:

1. If GS is chosen outside the capacitor, {net charge of GS} $= \psi = \oint D . dS = 0$, since there is no charge within the enclosure of the selected region. Therefore, we conclude that there exists no electric field outside the plates of the capacitor.
2. If GS is chosen to include both plates of the capacitor, net electric charge within the selected enclosure is zero. Therefore, the electric field outward of GS is still zero.
3. If we select GS to include the closed dielectric medium confined within the capacitor plates, including the upper plate (but excluding the lower plate), at any perpendicular distance measured from the upper plate downward to the lower plate, Gauss's law reveals that

$$+Q = \oint_{GS} D . dS = \oint_A D . dS + \oint_A (D . n dS)$$

where n is the normal vector of GS and $\bar{A}$ is the rest of the area beyond A.

At any cross-section of GS, the electric field (or displacement vector) should pass through a constant area A. The normal vector n is always perpendicular to the surface A and is aligned with the displacement vector $D = D u_z = D = \varepsilon E u_z$.

On the other hand, along the X and Y directions, there is no displacement vector. Therefore,

$$\oiint_{\text{rest of } GS \text{ beyond } A} D . dS$$

must be zero.

Hence, for a parallel plate capacitor, Gauss's law reveals that

$$Q = \oint_{GS} D . n dS = \oint_A D . dS = DA \tag{2.40}$$

The capacitance C of a capacitor is defined as

$$C \triangleq \frac{Q}{V}$$
$$\text{provided that } V \neq 0. \tag{2.41}$$

Then, for a parallel plate capacitor, the capacitance is found as

$$C = \frac{DA}{V} = \frac{\varepsilon A}{d} \text{ farads} \quad d \neq 0 \tag{2.42}$$

As the result of Gauss's law and by Equation (2.41), we can state that 'for an enclosure (GS) which includes a net charge Q, one can always associate a capacitance C between two points where a voltage difference is measured'.

In designing power transfer networks, capacitors and inductors are essential circuit components. Therefore, one must practically understand their operation and physical implementation. Now, let us consider a simple example to compute the value of capacitance of a 1 mm^3 cube made of different materials.

Example 2.1: Let us consider a cube of $V = 1 \times 1 \times 1 \text{ mm}^3$ volume made from quartz, mica, impregnated paper, barium titanate and barium strontium titanate. Assume that, by metal plating the top and the bottom surfaces, we form five capacitors with these cubes:

(a) Find the capacitances of the cubes.
(b) Comment on the results.

Solution:

(a) For free space, the permittivity is given as

$$\varepsilon_0 = 8.854 \times 10^{-12} \cong \frac{1}{36\pi} 10^{-9} \text{ F/m}$$

The relative dielectric constants (or permittivity) ε_r for quartz, mica, paper, barium titanate and barium strontium titanate are given in Table 2.1.

Table 2.1 Relative permittivity of some materials

Material	Permittivity, ε_r
Quartz	5
Mica	6
Paper (impregnated)	3
Barium titanate	1200
Barium strontium titanate	10 000

Then, employing Equation (2.42), the capacitance of each cube is specified as $C = \varepsilon A/d = 10^{-3} \varepsilon_0 \varepsilon_r$ farads. Thus, employing the relative permittivity values given in Table 2.1, the computed capacitances are as given in Table 2.2.

(b) The permittivity or dielectric constant is the measure of the charge handling capacity of the medium. As can be seen from Table 2.2 'Barium strontiumtitanate' can handle 2000 times as much charge than quartz. In designing capacitors, the choice of material is determined by:

(i) the value of the capacitance;
(ii) the required size of the application; and
(iii) the cost of production.

For ordinary discrete applications, quartz, mica and paper, perhaps, are cheap enough to be utilized to manufacture reasonable-size capacitors. On the other hand, if the application demands a high value of capacitance, then barium alloys may be considered despite their high costs.

Table 2.2   Relative permittivity of materials

Material	Permittivity, ε_r	Capacitance (F) $C = 10^{-3}\varepsilon_0\varepsilon_r$	Capacitance
Quartz	5	44.3×10^{-15}	44.3 fF
Mica	6	53.1×10^{-15}	53.1 fF
Paper (impregnated)	3	26.6×10^{-15}	26.6 fF
Barium titanate		10.6×10^{-12}	10.6 pF
Barium strontium titane	10 000	88.5×10^{-12}	88.5 pF

We should also mention that capacitors are considered as lumped circuit elements. Therefore, the size of a capacitor must be much less than that of the wavelength of the operating frequencies in the circuit under consideration. From a practical point of view, the largest size of the capacitor geometry can be 100 times smaller than the wavelength. In this regard, the designed capacitors may safely be used in circuits where the operating electromagnetic wavelength is greater than 100 mm which corresponds to an operating frequency of $f \leq C/100$ mm $= 3 \times 10^8/10^{-1} = 3$ GHz. In other words, the above capacitors can safely be used up to 3 GHz operating frequency.

2.12 Capacitance of a Transmission Line

Referring to Figure 2.10, let us consider two oppositely charged parallel cylindrical conductors with diameter a and $L = 1$ m. This is a typical 'uniform–two–pair wire' transmission line structure with unit length.

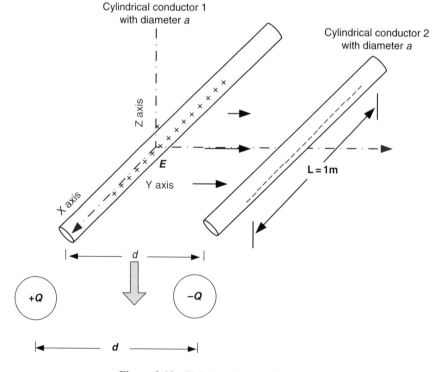

Figure 2.10   Unit length transmission line

These lines can be considered as a dipole with uniform charge Q. Let the volume charge density of these lines be designated as ρ_V. Then the total line charge Q is given by

$$Q = \rho_V(\pi a^2).L \tag{2.43}$$

In order to determine the electric field developed between the lines at a distance of radius r, we should choose a GS which completely encloses the total charge of conductor 1. For this case, electric field lines must be perpendicular to the side surface $A_r = (2\pi r) \times L$ of the cylinder and there should be no field component normal to the front and back cross-sections of the charged line. In this case, Gauss's law is expressed as

$$Q = \rho_V \pi a^2 . L = D_r . A_r = D_r(2\pi r)L$$

or

$$D_r = \frac{\rho_V a^2}{2r}$$

Hence, the electric field along the radial direction is given by

$$E_r = \frac{\rho_V a^2}{2\varepsilon r} u_r \tag{2.44}$$

On the other hand, the potential differences between the lines are given by

$$V_{12} = \int_a^d E_r d\mathbf{r} = \int_a^d \frac{\rho_V a^2}{2\varepsilon r} dr = \frac{\rho_V a^2}{2\varepsilon} \ln\left[\frac{d}{a}\right] \tag{2.45}$$

Hence, the capacitances between the lines are given by

$$\hat{C} = \frac{Q}{V_{12}} = 2\varepsilon \frac{\rho_V \pi a^2 L}{\rho_V a^2} \cdot \frac{1}{\ln\left[\dfrac{d}{a}\right]} = \frac{2\pi\varepsilon L}{\ln\left[\dfrac{d}{a}\right]} F \tag{2.46}$$

or

$$C = \frac{\hat{C}}{L} = \frac{2\pi\varepsilon}{\ln\left[\dfrac{d}{a}\right]} \text{ F/M} \tag{2.47}$$

It should be noted that ideally, for a long uniform transmission line with length L, electric charges should be accumulated on the conductor surfaces facing each other like sequential dipoles, along the longitudinal axis $[L]$. On the other hand, the electric field vectors must be confined within the area formed tidily between the conductors. This area may be approximated by $S_E = [L \times d]$ m^2. Here, the assumption is that there is no external electric disturbance imposed on the lines. However, in real life, electric disturbances are always present but their effects can be eliminated or drastically reduced by using shielded cables. Thus, especially for high-frequency or microwave applications, the use of coaxial cables becomes essential. Therefore, in the following section we will calculate the capacitance of coaxial cables.

2.13 Capacitance of Coaxial Cable

A shielded coaxial cable or, in short, a coax cable consists of two cylindrical conductors with radii 'a' and 'b' placed on the same axis as shown in Figure 2.11.

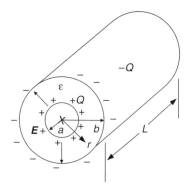

Figure 2.11   A coax cable structure with length L

The medium between the conductors is filled with a dielectric material like a polymer with a dielectric constant ε and the outer conductor is covered with a high-conductive metallic mesh like a shield.

Assume that the inner and outer conductors are loaded with positive and negative charges respectively. For this geometric formation, the radial electric field vectors must be confined within the conductors from 'a' to 'b' as shown in Figure 2.11. Hence, Gauss's law can be written on a Gauss surface at any point r as $D_r S_r = \varepsilon E_r [2\pi r \times L] = Q$, or the intensity of the radial electric field vector is given by $E_r = Q/2\pi\varepsilon r L$ and the voltage difference between the conductors 'a' and 'b' is calculated by

$$V_{ab} = \int_a^b E_r dr = \frac{Q}{2\pi\varepsilon L} \int_a^b \frac{dr}{r} = \frac{Q}{2\pi\varepsilon L} \ln\left[\frac{b}{a}\right]$$

Thus, the capacitance of the coax cable per unit length is calculated as

$$C = \frac{\hat{C}}{L} = \left(\frac{1}{L}\right) \cdot \frac{Q}{V_{ab}} = \frac{2\pi\varepsilon}{\ln\left[\dfrac{b}{a}\right]} \qquad (2.48)$$

where $\hat{C}$ is the total capacitance of the coax cable of length L.

2.14 Resistance of a Conductor of Length L: Ohm's Law

Let us consider a conductor of length L, with a cross-sectional area S, connected to a DC power supply with voltage V as shown in Figure 2.12.

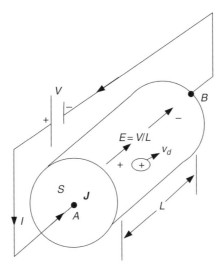

Figure 2.12   A conductor of length L connected to a DC power supply

For the given geometry, the DC power supply V develops an electric field E directed from point A to point B. The intensity of this field is given by $E = V/L$. The field must drag the free electric charges with a constant speed v_d which is called the drift velocity. It is a physical phenomenon that, for an isotropic material, v_d is proportional to the applied uniform electric field and given by

$$v_d = \mu E \tag{2.49}$$

In Equation (2.49), the constant μ is called the mobility of the majority carriers (i.e. free-traveling electric charges), and it is a macroscopic physical property of the selected material.[4]

The drift velocity is measured as a small distance ΔL over which the free electric charges are dragged for a short time Δt. Hence, we can write

$$v_d = \lim_{\Delta t \to 0} \frac{\Delta L}{\Delta t} \tag{2.50}$$

This depends linearly on the applied uniform electric field.

---

[4] Units of mobility are specified as m²/Vs.

On the other hand, the current I through the cross-section S is defined as the rate of change of the charges within the infinitesimally small volume specified by $\Delta V = S \times \Delta L$, and it is given by

$$I = \lim_{\Delta t \to 0} \frac{\Delta Q}{\Delta t} \qquad (2.51)$$

By definition, the units of current are C/s, or the ampere (A). That is to say, $1\,A = 1\,C$ charge flow per second.

Here, one can always define a current density J on the cross-section S such that

$$J = \frac{I}{S}\ A/m^2 \qquad (2.52)$$

Then,

$$J = \lim_{\Delta t \to 0} \frac{\Delta Q}{\Delta t}\frac{1}{S} \qquad (2.53)$$

Let ρ_v be the charge density in $\Delta V = \Delta L \times S$. Then, the total charge in ΔV is given by $\Delta Q = \rho_v \Delta V$. In this case, Equation (2.50) can be rearranged as

$$J = \rho_v \lim_{\Delta t \to 0} \frac{\Delta L}{\Delta t} \times \frac{S}{S} \qquad (2.54)$$

or

$$J = \rho_v v_d \qquad (2.55)$$

Substituting Equation (2.49) into Equation (2.55) we obtain

$$J = [\rho_v \mu]E \qquad (2.56)$$

From the practical point of view, we can comfortably state that the mobility μ and the charge density ρ_v are constant; they are dictated by the physical features of the material under consideration. Therefore, for any material, multiplication of these quantities is also constant. The quantity $\rho_v \mu$ is known as the 'conductivity' and it is designated by σ. The units of σ are $c/m^3 \times m^2/V\,s = A/V\,m$. In other words, conductivity is defined by

$$\sigma = \rho_v \mu \qquad (2.57)$$

which reflects a macroscopic physical property of the material under consideration; its units are A/V m. It should be noted that the ratio of 1 V/A is known as 1 ohm resistance and is designated by $1\,\Omega$. The inverse of $1\,\Omega$ is known as 1 Siemens and is designated by $1\,\Omega^{-1}$ or $1\,\mho$. The ratio A/V is named conductance. Therefore, we can say that the conductivity is the conductance per unit meter.

Now let us list the measured values of conductivity for the selected materials given below at room temperature (300 kelvin):

- For aluminum, conductivity is measured as $\sigma_{Al} = 3.72 \times 10^7\,\mho/m$.
- For copper, $\sigma_{Cu} = 5.80 \times 10^7\,\mho/m$.
- For silver, $\sigma_{Ag} = 6.17 \times 10^7\,\mho/m$.
- For brass, $\sigma_{brass} = 1.17 \times 10^7\,\mho/m$.

The above materials are known as conductors.

For germanium, $\sigma_{Ge} = 2.3$, whereas for silicon $\sigma_{Si} = 3.9 \times 10^{-4}$. It can be shown that as we heat up germanium and silicon, the conductivity of these materials increases sharply to a level of 10^7, similar to that of a conductor. Therefore, they are called semiconductors.

On the other hand, glass has a conductivity of $\sigma_{glass} = 10^{-12}$, and rubber's conductivity is about 10^{-15}. These types of materials are called insulators.

Finally, substituting Equation (2.57) into Equation (2.56) we obtain

$$\boxed{J = \sigma E} \qquad (2.58)$$

Equation (2.58) known as the microscopic Ohm's law.

Referring to Figure 2.12, for an isotropic material which is placed in a uniform electric field E, the total current I can be expressed as

$$I = \iint_S J.dS = J.S$$

On the other hand, the uniform field E between the terminals A and B is specified in terms of the applied DC voltage V as. $E = V/L$. Hence,

$$\boxed{J = \sigma E = \frac{I}{S} = \sigma\frac{V}{L}} \qquad (2.59)$$

or

$$\boxed{V = RI} \qquad (2.60)$$

with

$$\boxed{R = \frac{L}{\sigma S} = \rho_l\frac{L}{S} \quad \rho_l = \sigma^{-1}} \qquad (2.61)$$

Equation (2.60) is known as Ohm's law. In this equation, $R = L/(\sigma S)$ is the resistance of the conductor with specified geometry and physical properties.

Metallic conductors obey Ohm's law quite faithfully. In superconductors, σ approaches infinity at very low temperatures. For example, aluminum acts like a superconductor at temperatures below 1.14 K.

Example 2.2: Let us compute the resistance of a copper #16 wire with $L=1$ mile (1.608 km).

Solution: For copper #16 wire, the cross-sectional radius is $2r=0.0508$ in $=1.291$ mm and the conductivity is 5.8×10^7 S/m. Then,

$$S = \pi r^2 = (3.141\,59) \times \left(\frac{1.291 \times 10^{-3}}{2} \right)^2 = 1.308 \times 10^{-6}\,\text{m}^2$$

Now $L=1$ mile $= 1608$ m. Hence, the resistance R of the wire is computed as

$$R = \frac{L}{\sigma S} = \frac{1609}{(5.8 \times 10^7) \times (1.308 \times 10^{-6})} = 21.2\,\Omega$$

Example 2.3: It is tested that copper #16 wire can safely handle a current density of $J=7.65$ A/mm^2. Find a safe value of the current which can be handled with this cable.

Solution:

$$I = J \times S = 7.65 \times 1.308 = 10\,\text{A}$$

Example 2.4:

(a) For copper #16 cable, the electron mobility is $\mu_e = 3.2 \times 10^{-3}\,\text{m}^2/\text{V s}$. Find the drift velocity at $I=10$ A.
(b) Find the free electron charge density ρ_v.

Solution:

(a) The voltage drop on the cable is given by $V = R \times I = 21.2 \times 10 = 212$ V. Then the intensity of the electric field $E = V/L = 212/1609 = 0.312$ V/m. Hence, $v_d = \mu_e E = 3.2 \times 10^{-3} \times 0.312 = 3.98 \times 10^{-4}$ m/s.
(b) $\rho_v = \sigma/\mu_e = 5.8 \times 10^7/3.2 \times 10^{-3} = 1.31 \times 10^{10}$ C/m^3.

Example 2.5: The purpose of this example is to develop a practical model for a parallel plate capacitor in shunt configuration using the plates and dielectric losses.

Let us design a capacitor of $V = 1 \times 1 \times 1$ mm^3 volume employing high-resistivity silicon dielectric material which has a conductivity of $\sigma = 3.9 \times 10^{-4}\,\mho/\text{m}$ and a permittivity $\varepsilon_r = 11.7$. Let us gold plate the top and bottom surfaces to a $10\,\mu$m thickness. For the gold plating process, the conductivity is measured as $\sigma_{Au} = 4.1 \times 10^7\,\mho/\text{m}$.

(a) Find the value of the capacitance C.
(b) Find the resistance of the top and bottom plates.
(c) Find the resistance of the dielectric medium.
(d) Develop an approximate model for the lumped capacitor under consideration.

Solution:

(a) For high-resistivity silicon, $\varepsilon_r = 11.7$. Then,

$$C = \varepsilon_0 \varepsilon_r \frac{A}{d} = 8.854 \times 10^{-12} \times 11.7 \frac{10^{-6}}{10^{-3}} = 1.035\,92 \times 10^{-13}\,\text{F} = 103.6\,\text{fF}$$

(b) Both for the top and bottom plates, the resistance is given by

$$R_{top} = R_{bottom} = \frac{L}{\sigma_{Au}S_{plate}} = \frac{1 \times 10^{-3}}{4.1 \times 10^7 \times 1 \times 10^{-3} \times 10 \times 10^{-6}} = \frac{1}{410}$$
$$= 0.002\,439\,024\,\Omega = 2.439 \times 10^{-3}\,\Omega \cong 2.5\,\text{m}\Omega$$

(c) Here,

$$R_{Si} = \frac{L}{\sigma_{Si}S_{Si}} = \frac{1 \times 10^{-3}}{3.9 \times 10^{-4} \times 1 \times 10^{-3} \times 1 \times 10^{-3}} = \frac{10^7}{3.9} = 0.256\,410\,256 \times 10^7$$
$$\cong 2.56 \times 10^6\,\Omega = 2.56\,\text{M}\Omega$$

(d) Referring to Figure 2.13a, a floating capacitor may be regarded as a one-port physical device. In this case, one should consider the losses of both the top and bottom plates as computed: $R_{top} = R_{bottom} \cong 2.5\,\text{m}\Omega$. The bulk resistance R_{Si} appear between the plates, which is connected to the capacitor C in a shunt configuration. Hence, we end up with the simplified model shown in Figure 2.13(a).

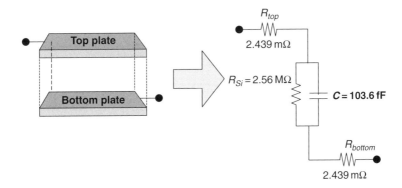

Figure 2.13(a)   Simplified model for a parallel plate capacitor in shunt configuration

On the other hand, we may also consider the capacitor as grounded, as shown in Figure 2.13(b). In this case, the bottom plate loss should disappear since it is thoroughly grounded.

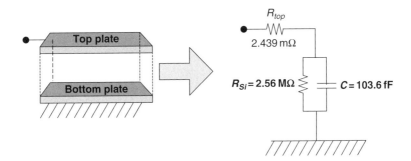

Figure 2.13(b)   Grounded capacitor model in shunt configuration

2.15 Principle of Charge Conservation and the Continuity Equation

One of the fundamental principles of electric field theory is that electric charges are conserved. In this regard, conservation means that, in a given region, moving charges can never disappear. We can keep track of their movements and count them one by one. For example, consider a rectangular volume of moving charges as shown in Figure 2.14. Here, if there is a net current flow I out of the rectangular box, this current must be equal to the rate of decrease of positive charge. In this case, the rate of decrease of charge can be expressed as $-dQ/dt$. Hence, the complete mathematical expression for this statement becomes

$$I = \text{net current flow out of the rectangular box} = -\frac{\partial Q(t,x,y,z)}{\partial t}$$

In the above equation we prefer to use partial differentiation just to keep the generality of the current equation. Therefore, this equation can simply be written as $I - \partial Q/\partial t$. Furthermore, since we are talking about a volume charge, say of length L and cross-sectional area dS, then we can express the net current flow in terms of the surface current density J such that

$$I = \oiint_s \mathbf{J}.\,dS = \iiint_V \nabla.\mathbf{J}\,dV$$

On the other hand, let the charge density in that volume be ρ_v; then,

$$Q = \iiint_v \rho_v dV \quad \text{or} \quad \iiint_V \nabla.\mathbf{J}\,dV = -\frac{\partial}{\partial t} \iiint_v \rho_v dV$$

or simply,

$$\boxed{\nabla.\mathbf{J} = -\frac{\partial \rho_v}{\partial t}} \tag{2.62}$$

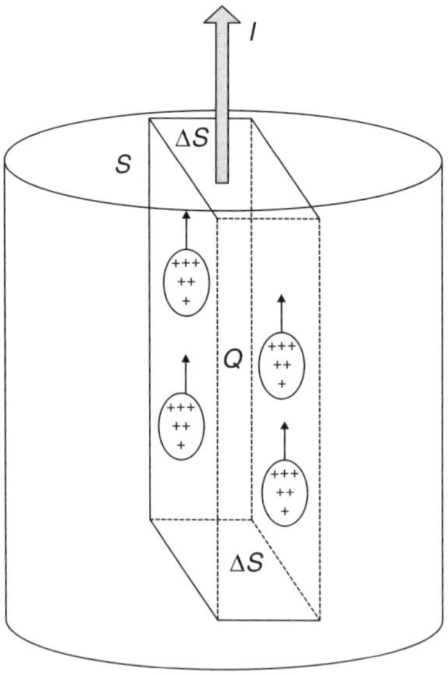

Figure 2.14 Moving charges creating a current I in a dielectric medium

Equation (2.62) is known as the continuity equation.

It is interesting to note that if the charge density is constant then the divergence of the current is zero. However, if the charge density or equivalently the current varies over time, the divergence of the current density must be different to zero. This statement has a significant impact on the electromagnetic field theory as will be shown later.

2.16 Energy Density in an Electric Field

Electric field energy is stored in a capacitance. By definition, capacitance is defined as $C = Q/V$ or $Q = CV$. For a linear, isotropic, homogeneous material the capacitance must be constant.

In this case, when we apply a time-varying voltage to a capacitor, then the charge Q accumulated on the capacitor must also change with respect to time. In this case, we can write

$$\frac{dQ}{dt} = i(t) = C\frac{dv}{dt}$$

which is in harmony with the definition of a capacitor given in Chapter 1.

Then, as shown, the total electric field or energy stored in a capacitor is

$$W_E = \int_0^t v(\tau)i(\tau)d\tau = \frac{1}{2}Cv^2$$

Thus, the energy density per volume is defined as

$$w_E = \frac{W_E}{Vol} \tag{2.63}$$

Now, let us consider a parallel plate capacitor C. By Equation (2.39), $C = \varepsilon A/d$. In this case,

$$w_E = \frac{1}{2}\frac{1}{A.d}\frac{\varepsilon A}{d}[E.d]^2$$

or

$$w_E = \frac{1}{2}\varepsilon E^2 \tag{2.64}$$

2.17 The Magnetic Field

'Magnetic Force'

Let us consider a moving charge Q_j with a constant velocity v_{Qj} between the north (N) and the south (S) poles of a permanent magnet. It is observed that a force F_j is exerted on Q_j which deflects the direction of the moving charge. It is postulated that the observed force is due to the magnetic field flux density which runs from the north pole to the south pole over the cross-sectional area $A(\mathrm{m}^2)$ of the magnet as shown in Figure 2.15. Hence, the magnetic flux density is a vector quantity $\mathbf{B}$. It has intensity and direction. Its intensity is designated by B. The magnetic flux density is associated with magnetic flux ϕ which is simply given by $\phi = B.A$ for the case under consideration.

The force $\mathbf{F}_j$ is measured as a perpendicular vector, upward on the plane formed by the vectors $\mathbf{v}_Q$ and $\mathbf{B}$ as in Figure 2.15.

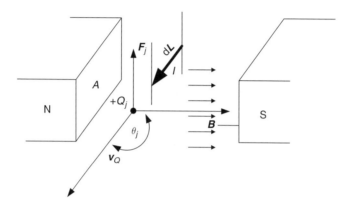

Figure 2.15   Force exerted on a moving charge in a magnetic field

The moving charge Q_j may be considered as created by means of the current I which flows in an infinitesimally small piece of wire dL, called the current element and designated by IdL. For a current element, the direction of dL is the direction of the current I. Let θ_j be the angle measured between $\mathbf{v}_{Qj}$ and vector $\mathbf{B}$. Then it can be found that the ratio $B = F_j / [Q_j v_{Qj} \sin(\theta_j)]$ is invariant for all $\{Q_j v_{Qj}, \theta_j\}$ triplets. Therefore, we conclude that the measured force F_j is the result of the applied magnetic field which is created due to the permanent magnet. The magnetic field vector is designated by $\mathbf{H}$ and it is linearly related to the magnetic field density vector $\mathbf{B}$ with a constant $1/\mu$, where μ is known as the permeability and it reflects the magnetic property of the medium.

Hence,

$$\begin{aligned} \mathbf{H} &= \frac{1}{\mu}\mathbf{B} \\ \mathbf{B} &= \mu\mathbf{H} \end{aligned}$$

(2.65)

and the constant intensity of the vector $\mathbf{B}$ is then specified in terms of the measured force and the given triplet $\{Q_j, v_{Qj}, \theta_j\}$ by

$$B = \frac{F_j}{Q_j.v_{Q_j}.\sin(\theta_j)} \text{ for all } j$$
or the force is given by
$$F_j = Q_j.v_{Q_j}.B.\sin(\theta_j)$$

(2.66)

Hence, once **B** is determined, then for any arbitrary moving charge Q with constant velocity v_Q the magnetic force **F** is given by[5]

$$\boxed{F = Q[v_Q \times B]} \qquad (2.67)$$

The magnetic flux density **B** is also called the magnetic induction.

On a current element, the differential form of Equation (2.67) may be expressed as

$$
\begin{array}{l}
dF = dQ \dfrac{dL}{dt} \times B \\[2mm]
\text{or} \\[1mm]
dF = \dfrac{dQ}{dt} dL \times B \\[2mm]
\text{or} \\[1mm]
dF = IdL \times B \\[2mm]
\text{or} \\[1mm]
F = \oint IdL \times B = \oint_L I \times B\, dL
\end{array}
\qquad (2.68)
$$

Hence, the integral specified by Equation (2.68) can be evaluated over a closed current loop as shown in Figure 2.16, which constitutes the principle of an electric machine.

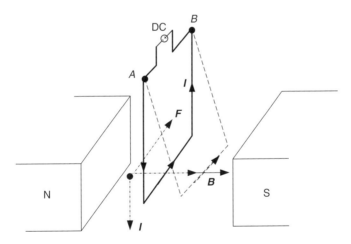

Figure 2.16   A closed current loop placed in a magnetic field produced by a permanent magnet

---

[5] Note that the quantity $F/Q = v_Q \times B$ has the dimensions of an electric field. Therefore, it may be considered as the induced electric field due to moving charges in a magnetic field.

2.18 Generation of Magnetic Fields: Biot–Savart and Ampère's Law

Referring to Figure 2.17, it is well established that a current-carrying wire produces a magnetic field. The existence of this field can be observed by measuring the force exercised on a moving charge which is brought close to the wire under consideration. Here, the current-carrying wire can be considered as the imitation of a permanent magnet. In this case, the magnetic force dF is specified by $dF = dQ\boldsymbol{v}_q \times \boldsymbol{B}$. On the other hand, it is known that the magnetic induction $\boldsymbol{B}$ is generated by the current I flowing in the wire L and it is given by

$$dB = \frac{\mu}{4\pi} \frac{I \times \mu_R}{R^2} dL \qquad (2.69)$$

where $\boldsymbol{u}_R = \boldsymbol{R}/R$ is the unit vector along the direction of the diameter $\boldsymbol{R}$ as shown in Figure 2.17 and μ is the permeability, the measure of magnetization of the medium.

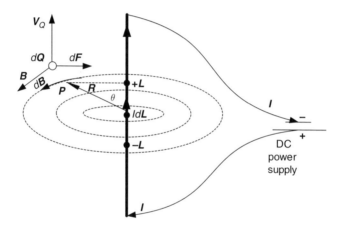

Figure 2.17   Production of a magnetic field by a current-carrying wire

It is postulated that the intensity dB of the magnetic induction vector satisfies the following equation:

$$dB = \frac{\mu}{4\pi} \frac{I \cdot \sin(\theta)}{R^2} dL$$
or, equivalently,
$$dH = \frac{dB}{\mu} = \frac{I}{4\pi} \cdot \frac{\sin(\theta)}{R^2} dL \qquad (2.70)$$

where θ is the angle between $\boldsymbol{I}$ or $d\boldsymbol{L}$ and $\boldsymbol{R}$.

Equation 2.70 was first proposed by Biot and Savart and is known as the Biot–Savart law.

Let us now investigate the intensity of the magnetic field locus on a circular plane formed by a circle of fixed diameter R which is perpendicular to the infinitely long current-carrying wire, as shown in Figure 2.18.

For ease of computation, let us assume that the length of the wire is L meters. Equation (2.70) reveals that

$$H = \lim_{L \to \infty} \int_{-L}^{+L} \frac{I}{4\pi} \frac{\sin(\theta)}{r^2} dL \qquad (2.71)$$

In order to evaluate the above integral, we need to find a relationship between $\{R, L, r, \theta\}$. Close examination of Figure 2.18 indicates that

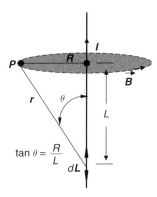

Figure 2.18   Magnetic field around the current-carrying wire

$$\tan(\theta) = \frac{R}{L}$$

or

$$L\tan(\theta) = R = \text{fixed diameter}$$

Differentiating totally we find that

$$dL\frac{\sin(\theta)}{\cos(\theta)} + L\frac{d\theta}{\cos^2(\theta)} = 0$$

or

$$\cos(\theta)\sin(\theta)dL = -Ld\theta \qquad (2.72)$$

On the other hand,

$$r.\sin(\theta) = R$$

and

$$r.\cos(\theta) = L$$

In order to find the corresponding limits of θ with respect to L, we use the relation $L = R/\tan(\theta)$. For $L = -\infty$, $\tan(\theta) - 0$ or $\theta = \pm\pi$. Similarly for $L = +\infty$, $\tan(\theta) = +0$ or $\theta = 0$.

Using Equation (2.27) in Equation (2.71), we get

$$
\frac{\sin(\theta)\,dL}{r^2} = -\frac{L\,d\theta}{\cos(\theta)r^2} = \frac{L\,d\theta}{\frac{L}{r}r^2} = \frac{d\theta}{\frac{R}{\sin(\theta)}}
$$

$$
= \frac{\sin(\theta)\,d\theta}{R}
$$

Hence,   (2.73)

$$
H = \frac{I}{4\pi R}\int_0^\pi \sin(\theta)\,d\theta = \frac{I}{2\pi R}
$$

Note that the circumference of the circle with fixed diameter R is $C = 2\pi R$ (meters). The magnetic field vector $\boldsymbol{H}$ is always a tangent to the circumference which is considered as a closed path $\boldsymbol{C}$ following the direction of $\boldsymbol{H}$. Therefore, we can safely state that

$$
HC = \boldsymbol{H}.\boldsymbol{C} = H.C.\cos(\varphi) = (2\pi R)H = I
\tag{2.74}
$$

where φ is the angle between the vectors $\boldsymbol{H}$ and $\boldsymbol{C}$.

Literally, the vector $\boldsymbol{C}$ is the tangential component of the circumference $\boldsymbol{C}$ at each point. For the case under consideration, obviously $\varphi = 0$ Therefore, in general it is wise to express Equation (2.74) as the integral of the differential form of $\boldsymbol{H}.\,d\boldsymbol{c} = dI$.

Hence, referring to Figure 2.19, we have

$$
\oint_c \boldsymbol{H}.d\boldsymbol{c} = I
\tag{2.75}
$$

Equation (2.75) indicates that the integral of the magnetic field vector over a closed path is equal to total net current enclosed within that path. The above form of the Biot–Savart law is known as Ampère's law.

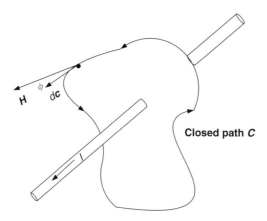

Figure 2.19   Ampères law: The integral of a magnetic field over a closed path

Ampère's or the Biot–Savart law provides a practical means to determine the magnetic field due to the net current enclosed by that closed path.

2.19 Direction of Magnetic Field: Right Hand Rule

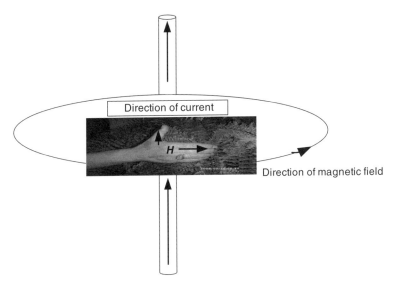

Figure 2.20   Right hand rule to determine the direction of the magnetic field arising from the current I

Referring to Figure 2.20, the direction of the magnetic field which arises from the applied current over an infinite length wire can be simply determined by what we call the right hand rule (RHR). In the RHR, the thumb of the right hand is aligned with the direction of current as shown in Figure 2.20. Then, the rest of the fingers which cover the circular path of fixed diameter R point to the direction of the magnetic field H.

2.20 Unit of Magnetic Field: Related Quantities

Unit of Magnetic Field H
In ISU, the unit of the magnetic field is ampere/meter (see Equations (2.74–2.75).

Example 2.6: Let the current in an infinitely long wire be 10 amperes. Find the magnetic field due to this current 5 meters away from the wire.

Solution: By Equation (2.74) we have

$$H = \frac{I}{2\pi R} = \frac{10\,\text{A}}{2\pi 5m} = \frac{1}{\pi} = 0.3183\ \text{A/m}$$

2.21 Unit of Magnetic Flux Density B

Magnetic force is given in terms of newtons. Equation (2.66) indicates that

$$B = \frac{F_j}{Q_j \cdot v_{Q_j} \cdot \sin(\theta_j)} \rightarrow \frac{newton}{coulomb \times \dfrac{m}{s}}$$

$$\rightarrow \frac{N}{Ampere \times meter}, \text{ i.e. the tesla (T)}$$

It is recognized that, in MKS, the tesla is relatively large. Therefore, the unit of the magnetic flux is generally expressed in the centimeter, gram, second system. In the cgs system, the unit of magnetic flux is called the Gauss:

$$1\,tesla(T) = 10^4 gauss$$

Just to give a few examples, the Earth's magnetic flux density is 0.5 gauss (G); that of a small permanent magnet B is in the range of 20 000 gauss or 2 tesla; for a large particle accelerator, it is in the range of 60 000 gauss.

2.22 Unit of Magnetic Flux ϕ

In MKS, the unit of magnetic flux is the weber (Wb). Actually $1\,weber(Wb) = 1\,tesla \times 1\,meter^2$. In this regard, the unit of the flux density is usually expressed in terms of the unit of the magnetic flux. Hence, the unit of B is also known as the weber/meter^2 or Wb/m^2.

In the cgs system, the unit of magnetic flux is known as the maxwell or Mx, and $1\,Wb = 10^8\,Mx$.

Example 2.7: Referring to Figure 2.9, it is interesting to compute the flux of an ordinary magnet of 20 000 gauss over a cross-sectional area of $1 \times 4\,cm^2$.

Solution: For this example, the cross-sectional area of the magnet is given as $A = 4\,cm^2$. In this case, the flux from the cross-section is given by

$$\phi = B.A = 2 \times 4 \times 10^{-4} = 8 \times 10^{-4}\,Wb = 8 \times 10^4\,Mx.$$

2.23 Definition of Inductance L

In magnetic fields an important concept is the definition of an inductance L, which is described over a cross-sectional area A (m^2). Let the current I (A) generate a magnetic flux density B. Then, the magnetic flux ϕ is given in integral form as

$$\boxed{\phi = \int_A B.dA \,(webers)} \tag{2.76}$$

where dA is the vector perpendicular to the infinitesimally small area dA and given by $dA = dAn$ such that n is the unit vector perpendicular to the surface dA.

Then, inductance L is defined as the net flux over the cross-sectional area A per unit current. Thus,

$$L \triangleq \frac{\phi}{I}$$
Which has the unit of wb/A.

(2.77)

In MKS, the unit of inductance is the henry, where 1 henry (H) = 1 Wb/A.

In a similar manner to that of a capacitor, inductance is the measure of the magnetic energy storage capacity of the medium. As shown in Chapter 1, the stored magnetic energy can be expressed as

$$W_m = \frac{1}{2}LI^2 \text{ J}$$

(2.78)

2.24 Permeability μ and its Unit

The magnetic property of a medium is characterized in terms of the magnetic flux density per unit current over a distance of unity. In other words,

$$\mu = \frac{B}{H}$$

(2.79)

In MKS, the unit of permeability is given by

$$\mu \rightarrow \frac{\text{Wb/m}^2}{\text{A/m}} \rightarrow \frac{\text{Wb}}{\text{A} \times \text{m}} \rightarrow \frac{\text{henry}}{\text{m}} \rightarrow \text{H/m}$$

Hence, we can say that the permeability of a medium is specified as the inductance per unit length.

For free space the permeability $\mu_0 = 4\pi \times 10^{-7} \text{ H/m}$.

We should emphasize that the magnetic field vector H is a function of the environmental geometry and the current that generates the magnetic field, whereas the magnetic flux density depends on the magnetic property of the medium and is specified by permeability. Therefore, the magnetic energy produced or stored in an environment depends on the quality of the material. In general, the quality of the magnetic material which is measured by μ is expressed in terms of the relative permeability μ_r and the permeability of free space as

$$\mu = \mu_r \mu_0$$

(2.80)

In Table 2.3 we present the relative permeability values of some materials. It is striking to note that most of the conductive materials such as gold, silver, lead, copper, aluminum, etc., present almost equal permeability to that of free space. On the other hand, materials such as cobalt, nickel, iron, etc., possess quite high permeability compared to free space. In this regard, materials are classified under three major categories;

1. Diamagnetic materials, which possess negative or very small permeability with $\mu_r \ll 1$, e.g. bismuth, mercury, gold, silver.
2. Paramagnetic materials, which have little permeability equal to or slightly higher than 1, i.e. $\mu_r \geq 1$, such as vacuum, air, aluminum.
3. Ferro magnetic materials, which possess very high relative permeability, i.e. $\mu_r \gg 1$, e.g. cobalt, nickel, iron.

It should also be mentioned that diamagnetic and paramagnetic materials have no practical use in magnetic field applications, whereas ferromagnetic materials are excellent for generating magnetic forces which are used in electric motors, generators, accelerators, medical instruments, etc.

Table 2.3 Relative permeability of some materials

Material	Relative permeability, μ_r	Classification
Bismuth	0.999 83	Diamagnetic
Mercury	0.999 968	Diamagnetic
Gold	0.999 964	Diamagnetic
Silver	0.999 99 83	Diamagnetic
Vacuum	1	Paramagnetic
Air	1.000 000 36	Paramagnetic
Aluminum	1.000 021	Paramagnetic
Palladium	1.000 82	Paramagnetic
Cobalt	250	Ferromagnetic
Nickel	600	Ferromagnetic
Commercial iron	6000	Ferromagnetic
High-purity iron	2×10^5	Ferromagnetic
Supermalloy (79% Ni, 5% Mo)	1×10^6	Ferromagnetic

2.25 Magnetic Force between Two Parallel Wires

Let us consider two infinite length wires which are parallel and R meters distant from each other as shown in Figure 2.21.

Clearly, each wire will find itself in a magnetic field generated by the other wire. Assume that the wires are identical and produce the same amount of current with similar or opposite charges. In this case, by Ampère's law, the intensity of the magnetic field density can be expressed as

$$B = \frac{\mu_0 I}{2\pi R}$$

For example, to find the force acting on the second wire L_2 per unit meter, one can write[6]

$$\boxed{\frac{F}{L} = I_2 B = I_2 \frac{\mu_0 I_1}{2\pi R}} \tag{2.81}$$

---

[6] Noted that **I**, **B**, **F** are at right angles to each other; therefore $\theta = \pi/2$ which is the angle between **B** and **R**, yielding $\sin(\theta) = 1$.

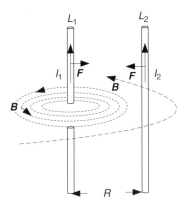

Figure 2.21   Magnetic force between two parallel wires

Since the geometry is symmetric for both wires, the same force also applies on wire 1. Two wires attract each other if the currents I_1 and I_2 are in the same direction, otherwise they repel.

Example 2.8: Let us increase the force between the wires of Figure 2.21 by 100 A. Let the separation between the wires be $R = 0.5$ cm. Find the attractive force between the wires per unit distance.

Solution: Employing Equation (2.81), we have

$$\frac{F}{L} = 100 \frac{4\pi \times 10^{-7}}{0.25 \times 10^{-4}} \times (100) \cong 503 \, \text{N/m}$$

In order to assess the level of the strength of this force let us consider a mass of 50 000 g or $m = 50$ kg. This mass will be attracted by a gravitational force of $F_g = mg \cong 50 \, \text{kg} \times 10 \, \text{m/s}^2 = 500$ N. This is the amount of force exerted on the wires, which is something to consider when designing overhead power lines.

2.26 Magnetic Field Generated by a Circular Current-Carrying Wire

A circular current-carrying wire is an important geometry for many applications where the magnetic field is concentrated on the central axis, which is perpendicular to the plane formed by the circle as shown in Figure 2.22.

Let us calculate the magnetic field generated by the current elements Idl and Idl'. Due to the circular symmetry, the magnetic field components along the x and y directions are cancelled. However, the field component on the z axis adds up. Employing the Biot–Savart law, the intensity of the magnetic flux density vector $\mathbf{B}$ is given by

$$d\mathbf{B} = \frac{\mu_0}{4\pi} \frac{Idl \times \mathbf{u}_r}{r^2}$$

Note that the vector $\mathbf{r}$ is always perpendicular to current element IdL. Therefore,

$$dB = \frac{\mu_0}{4\pi} \frac{Idl}{r^2}$$

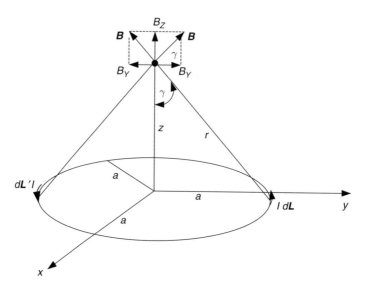

Figure 2.22   Magnetic field of current-carrying circular loop

The z component of dB is given by

$$dB_z = dB\sin(\gamma) = \frac{\mu_0}{4\pi}\sin(\gamma)\frac{Idl}{r^2}$$

with

$$\sin(\gamma) = \frac{a}{r} \qquad r = (a^2 + z^2)^{1/2}$$

Here, $\{\gamma, r, z, I\}$ are independent of the integration variable dL. Furthermore, integration of dL over the closed loop yields the circumference of the circle, which is $2\pi a$. Thus,

$$B_z = \frac{\mu_0}{4\pi}\sin(\gamma)\frac{I}{r^2} \oint dL = \frac{\mu_0}{4\pi}\sin(\gamma)\frac{I2\pi a}{r^2}$$

or

$$\boxed{B_z = \frac{\mu_0 Ia}{2(a^2 + z^2)^{3/2}}\ \text{Wb/m}^2} \tag{2.82}$$

At the center of the circle, the field intensity is found by setting $z = 0$. Hence,

$$\boxed{B_0 = \frac{\mu_0}{2a}I} \tag{2.83}$$

2.27 Magnetic Field of a Tidily Wired Solenoid of *N* Turns

Referring to Figure 2.23, let us consider a solenoid consisting of tidily wired *N* current loops concentrated within a small length compared to the diameter a. In other words, length z of the solenoid is negligibly small compared to a. In this case, we can assume that the magnetic field components for each loop are accumulated at the center of the circle yielding a total field of

$$B = NB_0 = \frac{\mu_0}{2a}NI \text{ Wb/m}^2, \quad z \ll a \tag{2.84}$$

On the other hand, consider a solenoid of length L which is much longer than the diameter a. In this case we can consider a small section dz along the z axis. This small section contributes to the magnetic field component on the z axis in the form

$$dB_z = \frac{\mu_0 a^2}{2(a^2 + z^2)^{3/2}} \left[\frac{NI}{L} \right] dz$$

or

$$B_z = \int_{-L/2}^{+L/2} \frac{\mu_0 a^2}{2(a^2 + z^2)^{3/2}} \left[\frac{NI}{L} \right] dz = \frac{\mu_0 NI}{(4a^2 + L^2)^{1/2}}$$

If $a \ll L$, then

$$B_z = \frac{\mu_0}{L}NI, \quad L \gg a \tag{2.85}$$

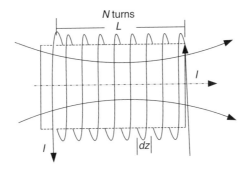

Figure 2.23 Tidily wired *N* current loops in a solenoid

2.28 The Toroid

A toroid of *N* turns is shown in Figure 2.24. This is also a useful geometry for storing magnetic energy. The toroid is formed by tidily winding closely spaced *N* turns of current-carrying wires on a ring-type

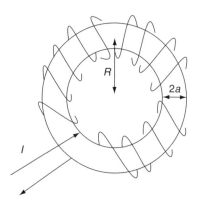

Figure 2.24   Toroid of N turns

ferrite of diameter R. The circular current loops have a diameter of a. For this geometry, the magnetic field is confined within the ferrite along the circular axis.

One can consider this structure as a solenoid which is circularly bent and closed at both ends. In this case, when the toroid is cut and opened out the co-centric circular axis becomes the z direction and the length of the solenoid is considered as $L = 2\pi R$. Hence, the magnetic field density is given by

$$B = \mu \frac{NI}{2\pi R} \text{ Wb/m}^2$$

(2.86)

2.29 Inductance of N-Turn Wire Loops

For a single current loop we talk about magnetic flux generated by the applied current. When closely spaced wire loops of N turns are considered, the magnetic flux of each loop is linked with another which in turn creates a multiplication effect. Hence, the total flux for an N-turn current loop is then given by

$$\psi = N\phi$$

(2.87)

where $\phi = \iint_A B.dA$ is the flux due to the combined magnetic flux density B. For an N-turn solenoid of length l, B is approximately as in Equation (2.75) such that $B = (\mu/l)NI$.[7]

In this case, the inductance is described in terms of the total flux linkage ψ as

$$L = \frac{\psi}{I} = \frac{\mu N^2 A}{l}$$

(2.88)

---

[7] Here, it is assumed that the total magnetic flux density is concentrated at the center of the solenoid.

For a short solenoid Equation (2.74) yields

$$L = \frac{\mu}{2a} N^2 (\pi a^2) = \mu N^2 \frac{\pi a}{2} \qquad (2.89)$$

For power transfer networks, practical implementations of inductors are quite important since they occupy a significant amount of circuit/chip area. The above expressions give a rough idea to the reader of realizing inductors. Now, let us looks at an example.

Example 2.9: Consider a spiral inductor of 0.5 nH printed in a series configuration on a silicon substrate. Let the diameter of the spiral be 200 μm. Find the total number of turns N.

Solution:

Equation (2.89) may be employed to estimate the total number of turns. In this case,

$$N = \sqrt{\frac{2 \times L}{\mu \times \pi \times a}}$$

For silicon, $\mu \approx \mu_0 = 4 \times \pi \times 10^{-7}$. Then,

$$N = \sqrt{\frac{2 \times 0.5 \times 10^{-9}}{4 \times \pi \times 10^{-7} \times \pi \times 200 \times 10^{-6}}} \approx \frac{1}{\pi} \sqrt{\frac{1 \times 10^4}{800}} = 1.254 \text{ turns}$$

It should be mentioned that when we realize an inductor, chip losses must be considered. These losses may be classified as sheet metal resistance, eddy current resistance, capacitive and magnetic losses of the substrate. Obviously, losses affect the frequency range of operation. Details are omitted here. However, there are commercially available layout programs which immediately yield practical realizations of inductors with associated losses.

Now we present an example to practically build a simplified model of a spiral inductor printed on high-resistivity silicon. For many applications, eddy current and magnetic losses are neglected. Therefore, in the following examples, we have omitted these quantities.

Example 2.10: Referring to the previous example, assume the printed spiral inductor is made of gold with width $W = 10$ μm and thickness $d = 5$ μm.

(a) Find the conduction loss of the printed sheet metal.
(b) Find the quality factor of the inductor at $f = 1$ GHz
(c) Let the printed circuit board thickness be 6 μm. Find the substrate capacitance from the inductor down to the ground plane.
(d) Build a simplified model for the spiral inductor using the above calculated loss and substrate capacitance.
(e) Find the resonance frequency of the inductor model.

Note that for the gold plating process, the conductivity is measured as $\sigma_{Au} = 4.1 \times 10^7$ ℧/m and for high-resistivity silicon $\varepsilon_r = 11.7$; for free space $\varepsilon_0 = 8.854 \times 10^{-12}$ F/m.

Solution:

(a) Let us find the required length of the inductor. For one turn $L_1 = 2 \times \pi \times a = 2 \times \pi \times 200 \times 10^{-6}$. For the fractional turn of 0.254 the additional length required is $L_2 = 2 \times \pi \times a \times 0.254$. Thus the total length is $L = L_1 + L_2 = 2 \times \pi \times 200 \times 10^{-6} \times 1.254 = 0.0014$ m.

Hence, the sheet metal loss which may be considered in series with the inductor is given by

$$r_L = \frac{L}{\sigma_{Au}[W \times d]} = \frac{0.0014}{4.1 \times 10^7 \times 10 \times 10^{-6} \times 5 \times 10^{-6}} = 0.682 \, \Omega$$

(b) The quality factor Q_L of the inductor is given by

$$Q_L = \frac{L\omega}{r_L} = \frac{0.5 \times 10^{-9} \times 2\pi \times 1 \times 10^9}{0.6829} = 4.6004$$

Actually, for practical applications this quality factor may not be sufficient. However, it can be improved by means of several techniques as discussed in the literature.[8]

(c) The substrate capacitor may be approximated by

$$C_P = \varepsilon_{si} \frac{A}{d} = 11.7 \times 8.854 \times 10^{-12} \frac{10 \times 10^{-6} \times 0.0014}{6 \times 10^{-6}} = 2.417 \times 10^{-13} \approx 0.242 \text{ pF}$$

(d) In order to build a simplified model for the spiral inductor in a series configuration, the substrate capacitor C_P may evenly be distributed to both terminals; similarly, series loss r_L can be separated to both ends as shown in Figure 2.25.

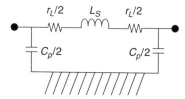

Figure 2.25   Simplified model for a spiral inductor printed on a high-resistivity silicon chip

(e) The resonance frequency of the model circuit is specified by

$$f_{ReS} = \frac{1}{2\pi\sqrt{L_S C_P}} = \frac{1}{2\pi\sqrt{0.5 \times 10^{-9} \times 0.242 \times 10^{-12}}} = 1.4469 \times 10^{10} \text{ Hz} = 14.69 \text{ GHz}$$

Here, we should mention that these types of inductors can be safely used up to frequencies which are less than $f_{Res}/10$. In other words, this inductor can be utilized up to 1.4 GHz as a linear lumped circuit element.

2.30 Inductance of a Coaxial Transmission Line

Consider a long coaxial cable with a cross-section as shown in Figure 2.26.

---

[8] For example, Inder Bahl, *Lumped Elements for RF and Microwave Circuits*, Artech House, 2003.

In this geometry, it is assumed that the current flow is in the center conductor and returns to the outer conductor. Therefore, the magnetic flux is confined between the inner and outer conductors, otherwise it is zero. Hence, applying Ampère's law, the magnetic flux density is given by

$$B = \frac{\mu I}{2\pi r}, \quad a < r < b - t \tag{2.90}$$

where t is the thickness of the outer conductor which may be negligible compared to b.

Over the length L, we can define an infinitesimally small cross-sectional area $dA = Ldr$ such that the magnetic flux $d\psi$ is generated and given by $d\psi = B.Ldr$. Then, the total flux per unit length is specified as

$$\frac{\psi}{L} = \int_a^{b-t} \frac{\mu I}{2\pi r} dr = I \frac{\mu}{2\pi} \ln\left(\frac{b-t}{a}\right) \tag{2.91}$$

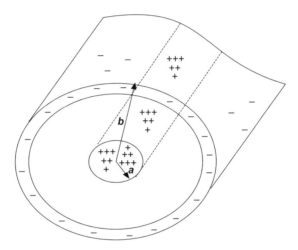

Figure 2.26 Coax cable cross-section with inner and outer diameters a and b respectively

In this case, the inductance per unit length is given as

$$L_{coax} = \frac{\psi}{LI} = \frac{\mu}{2\pi} \ln\left(\frac{b-t}{a}\right) \text{ H/m} \tag{2.92}$$

Now let us consider an example to quantify the inductance given for a coax cable per unit length.

Example 2.11: Let the co-centric radii of a coaxial cable be $a = 1$ mm and $b = 10$ mm.

(a) Find the inductance per unit meter for this cable.
(b) Find the capacitance of this line per unit meter.
(c) Find the ideal (i.e. loss-free) equivalent circuit of the coax cable per unit length.

Notes:

- The thickness of the outer conductor is about 1 mm.
- The filling material of the coax cable is polystyrene.
- Its permittivity is $\varepsilon = \varepsilon_0\varepsilon_r = 10^{-9}/36\pi$ F/m and its permeability is almost equal to the permeability of free space, which is $\mu = \mu_0 = 4\pi \times 10^{-7}$ H/m.

Solution:

(a) Straightforward manipulation of Equation (2.92) reveals that

$$L_{coax} = \frac{\mu}{2\pi} \ln\left(\frac{b-t}{a}\right) = \frac{4\pi \times 10^{-7}}{2\pi} \ln\left(\frac{10-1}{1}\right) = 2 \times 10^{-7}\ln(9) = 2 \times 2.1972 \times 0.1 = 0.4394 \ \mu H/m$$

(b) From Equation (2.48) we have

$$C_{coax} = \frac{2\pi\varepsilon}{\ln\left(\dfrac{b-t}{a}\right)} = \frac{2\pi\dfrac{10^{-9}}{36\pi}}{\ln(9)}$$

or

$$C_{coax} = 2.4955 \times 10^{-10} \ \mathrm{F/m} \cong 24.5 \ \mathrm{nF/m}$$

(c) The ideal (loss-free) equivalent circuit per unit length of the coax cable may be given as symmetric LC low-pass ladder sections. The sections can be organized either in T or in pi configurations as shown in Figure 2.27.

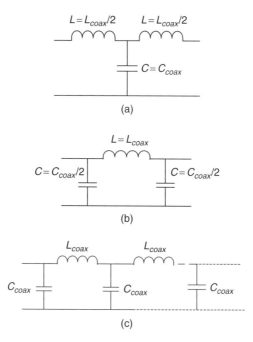

Figure 2.27 Loss-free models of coax cable with low-pass LC sections: (a) T model; (b) pi model; (c) a transmission line

We should note that in the above models one can always include small-resistor values r_L in series with inductors which reflect metal losses. Similarly, high values of shunt resistors $R_c = 1/G_c$ may be attached to capacitors which are considered as dielectric losses of the filling material. Depending on the frequency band of operations, these losses may be neglected. Let us, next, consider an example to see the effect of the loss resistors.

Eventually, a transmission line can be modeled by means of cascaded connections of infinitely many LC T or pi sections with unit inductance L_{coax} and capacitance C_{coax}, as shown in Figure 2.27(c).

Example 2.12: In Table 2.4, open air TV broadcast frequencies are given. The coax cable of Example 2.11 will be used to connect the external VHF and UHF antennas to the TV set. The conducting material of the cable is made of copper. The outer conductor of the cable is shielded (i.e. grounded).

(a) Calculate the inner conductor loss per unit meter.
(b) Calculate the dielectric loss of the filling material.
(c) Compare the calculated losses at the operating frequencies and see if they are negligible.

Solution:

(a) Assume that when we cut and open out both the inner and outer conductors about their circumferences we obtain the layout as shown in Figure 2.28. In this case, we need to calculate the concentrated loss of the dielectric new geometry of unit length L with a uniform cross-section of $S_{poly} = 2\pi(b - t - a) \times (L)$ and length $h = b - t - a = 8\,\text{mm}$. In this case, the conductance G_{poly} is given by

$$G_{poly} = \sigma_{poly} S_{poly}/H$$

Hence the conductance per unit length is then given by

$$G = \frac{G_{poly}}{L} = 2\pi\sigma_{poly}$$

or

$$G = 6.2832 \times 10^{-12}\ \mho/\text{m}$$

Table 2.4 General TV broadcast frequencies[9]

Sub CATV band, T7–T13	7–58 MHz
VHF band, Ch. 2–13	54–216 MHz
Low band, VHF Ch. 2–6	59–88 MHz
Mid band, UHF Ch. 14–22 UHF Ch. 95–99	121–174 MHz 91–120 MHz
High band, VHF Ch. 7–13	175–216 MHz
Super band, CATV Ch. 23–36	216–300 MHz
Hyper band, CATV Ch. 37–62	300–456 MHz
Ultra band, CATV Ch. 63–158	457–1002 MHz
UHF band, Ch.14–83 CATV Ch. 63–158	70–1002 MHz

---

[9] This table is the frequency chart for television channels in the USA. The abbreviation 'Ch.' stands for TV channel.

(b) Referring to Table 2.4, at the lower end of the VHF band the operating frequency is about 54 MHz. At this frequency, the admittance of the capacitor per unit length is specified as

$$Y_c = j\omega C = j2 \times \pi \times 54 \times 10^6 \times 2.4955 \times 10^{-10} = j0.084 \qquad \mho/m$$

On the other hand, the shunt conductance per unit length is given by

$$G = 6.2832 \times 10^{-12} \qquad \mho/m$$

Then, total shunt admittance per unit length will be as follows:

$$Y = G + j\omega C = 6.283 \times 10^{-12} + j0.084$$

From a practical point of view, the real part of the admittance is negligible compared to the imaginary part.

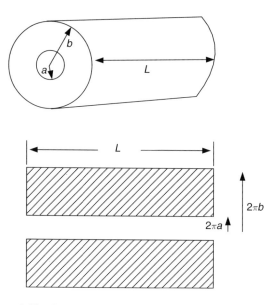

Figure 2.28   Opened form of the dielectric fill of the coax cable

Similarly, in the series arms, total impedance per unit length is given by

$$z = r_L + j\omega L_m = 5.6 \times 10^{-3} + j \times 2 \times \pi \times 54 \times 10^6 \times 0.4394 \times 10^{-6} = 0.0056 + j148.9 \ \Omega$$

(c) Clearly, the imaginary part of the impedance per unit length is much greater than that of the real part. Therefore, at the lower end of the VHF band, this coax cable can safely be used as a lossless connecting line between the antenna and the TV set.

Obviously, as the operating frequencies increase toward the UHF band (1000 MHz, see Table 2.4), the contribution of the imaginary part to the series arm impedance and the shunt arm admittance becomes more apparent.

Finally, for a coax cable, a simplified lumped circuit model per unit length is depicted in Figure 2.29. Here, note that only conduction losses are included in the model. However, one may develop more sophisticated loss models

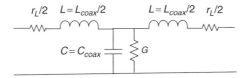

Figure 2.29   A simplified lumped element circuit model per unit length for a coaxial transmission line

2.31 Parallel Wire Transmission Line

In Section 2.10, the capacitance per unit length was derived for an infinitely long, parallel wire transmission line, Here, in this section, using the same geometry, the inductance per unit length will be obtained.

Referring to Figure 2.30, assume that line L_1 carries current I_1 in one direction, whereas in line L_2, the same current returns in the opposite direction. In this case, each line produces a magnetic flux density B with the same sense, producing a total magnetic flux density of $2B$ between the lines. It must be noted that, outside the conducting lines, the flux density must be zeros by Ampère's law since $\oint_c \boldsymbol{B} \cdot dc = 1 - 1 = 0$, where closed C contour includes both lines L_1 and L_2.

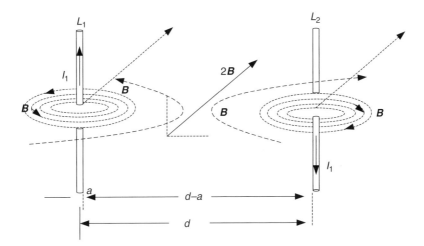

Figure 2.30   Buildup of magnetic flux density between parallel lines carrying currents of opposite direction

By Equation (2.82) we can determine the magnetic flux density B for each line or total fulx density:

$$B_T = 2B = 2\frac{\mu I}{2\pi r} = \frac{\mu I}{\pi r}, \quad a < r < d - a, \, d \gg a \tag{2.93}$$

Considering the single wire, say, on the left of Figure 2.30, the magnetic flux linkage is given by

$$\psi = \int_A B dA = \int_a^{d-a} B(L dr) \cong L \int_a^d \frac{\mu I}{2\pi r} dr = L\frac{\mu I}{2\pi} \ln\left[\frac{d}{a}\right] \tag{2.94}$$

Hence, by definition, the inductance per unit length is given by

$$L_r = \frac{\psi/I}{L} = \frac{\mu}{2\pi} \ln\left[\frac{d}{a}\right] \, \text{H/m}$$

(2.95)

Recall that, in Equation (2.47), the equivalent shunt capacitance per unit length of a parallel wire transmission line was determined as

$$C_r = \frac{2\pi\varepsilon}{\ln\left[\dfrac{d}{a}\right]}$$

Hence, in a similar manner to that of Figures 2.27 and 2.29, one can describe an ideal parallel transmission line as an infinitely many cascaded connection of T sections, i.e. $L_T/2 - C_T - L_T/2$, as depicted in Figure 2.31. Obviously, as in the coaxial cable case, one can associate losses with series inductors L_T and shunt capacitors C_T designated by r_s and G respectively.[10]

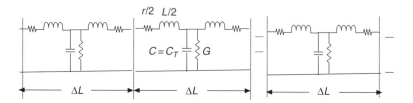

Figure 2.31 Equivalent circuit of a parallel wire transmission line consisting of infinitely many cascaded connections of T sections specified over an infinitesimally small length ΔL by means of L_T and C_T and their associated losses r_S and G respectively

2.32 Faraday's Law

Faraday showed that a changing magnetic flux generates an electric field which may be measured in the form of an induced voltage. In this manner, his classical experiment may be carried out by moving a magnet inside a solenoid as shown in Figure 2.32.

Movement of the magnet changes the magnetic flux with respect to time which in turn produces an induced voltage measured at the terminals of the solenoid. In this case, the induced voltage is given by

$$\nu_L = \frac{d\psi}{dt} = -\frac{d}{dt}\left[LI(t)\right]$$

(2.96)

where $I(t)$ is the current built in the circuit and L is the inductance of the solenoid.

---

[10] If the transmission line is lossless, which may be a reasonable assumption at high frequencies, then we set $r_S = 0$ and $G = \infty$.

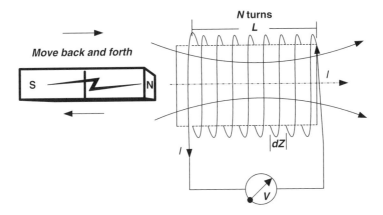

Figure 2.32   Faraday's experiment

Equation (2.96) is known as Faraday's law.

2.33 Energy Stored in a Magnetic Field

As discussed in the previous sections, coil, solenoid and toroid-like geometries store magnetic energy. In fact, the inductance of these physical structures reflects the measure of stored magnetic energy.

For a given two-terminal structure such as a solenoid, the stored magnetic energy may be expressed as the integral of the power such that $W_m = \int_0^t \nu(\tau)i(\tau)dr$ where $v(t)$ and $i(t)$ are the induced voltage and current measured on the device. Hence, using Faraday's law, the stored energy can be written as

$$W_m = \int_0^t Li(\tau)di = \frac{1}{2}L^2 i(t) \tag{2.97}$$

which verifies the result obtained in Chapter 1.

2.34 Magnetic Energy Density in a Given Volume

Assume that the magnetic energy is confined to or stored in a closed volume V. Then, the energy density at each point within the volume is defined by

$$w_m = \frac{W_m}{V} \tag{2.98}$$

In order to obtain the open form for the energy density in a magnetic field, let us consider a solenoid of length L, with a cross-sectional area A and N turns. Then, by Equation (2.89), the inductance L_S of the solenoid is given as $L_s = \mu N^2 A/L$.

Furthermore, by Ampère's law, it can be shown that $H.L = NI$ or $I = H.L/N$. The volume $V = A.L$. Hence, using all these relations in Equation (2.97), it is found that

$$w_m = \frac{1}{2}\frac{1}{AL}\left[\frac{\mu N^2 A}{L}\right]\left[\frac{HL}{N}\right]^2 = \frac{1}{2}\mu H^2 \tag{2.99}$$

The importance of Equation (2.99) is that, once the magnetic field H is known, the magnetic energy density within a given volume can be easily computed. Usually, the magnetic field is generated by means of an applied current. Thus, applying Ampère's law, the magnetic field can then be computed. Eventually, the stored energy in the field can be computed by integration such that

$$W_m = \iiint_V w_m dV \tag{2.100}$$

2.35 Transformer

The Voltage Ratio

A representation of a physical transformer is shown in Figure 2.33. Primary windings of N_1 turns and secondary windings of N_2 turns are wound on a high-quality ferromagnetic material of permeability μ. When a time-varying voltage excitation $V_1(t)$ is applied on the primary site, it generates a current $I_1(t)$ and a magnetic flux ϕ which is practically confined within the ferromagnetic material. For example, if the applied voltage variation is a sinusoidal waveform with an angular frequency of $\omega_0 = 2\pi f_0$, then, by linearity, the magnetic flux is also sinusoidal. If $\phi(t) = \phi_m \sin(2\pi f_0 t)$, by Faraday's law it follows that

$$V_{induced} = [-V_1(t)] = -\frac{d\psi_1}{dt} = -N_1\frac{d\phi(t)}{dt} \tag{2.101}$$

or

$$V_1(t) = (2\pi f_0)N_1\phi_m\cos(2\pi f_0 t) = V_m\sin\left(2\pi f_0 t - \frac{\pi}{2}\right) \tag{2.102}$$

where ψ_1 is the total magnetic flux generated on the primary windings, ϕ_m is the amplitude or maximum allowable value of the sinusoidal flux variation, and, $V_m = 2\pi f_0 \phi_m$.

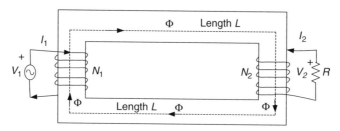

Figure 2.33 A practical transformer constructed on a ferromagnetic core with primary windings of N_1 turns, secondary windings of N_2 turns, cross-sectional area of A (m^2) and length L (m)

Similarly, the induced voltage on the secondary windings is expressed as

$$[-V_2(t)] = V_R = -\frac{d\psi_2}{dt} = -N_2\frac{d\phi(t)}{dt}$$

Hence, the voltage ratio of the transformer is found to be

$$\frac{V_1}{V_2} = \frac{N_1}{N_2} \qquad\qquad (2.103)$$

The Current Ratio

In order to find the current ratio I_1/I_2, let us terminate the secondary windings with a resistor R. The magnetic flux $\phi(t)$ confined within the magnetic circuit is given by

$$\phi = BS = \mu H$$

However, by Ampère's law

$$H.L = N_1 I_1$$

Therefore,

$$\phi(t) = \mu\frac{N_1 I_1}{L}S = \frac{N_1 I_1}{\mathcal{R}} \qquad\qquad (2.104)$$

where $\mathcal{R}$ is called the reluctance of the magnetic circuit and defined as

$$\mathcal{R} = \frac{L}{\mu S} \qquad\qquad (2.105)$$

On the other hand, when we apply Ampère's law on the secondary windings, we find that[11]

$$\phi(t) = \frac{N_2(-I_2)}{\mathcal{R}} \qquad\qquad (2.106)$$

Hence, Equations (2.104) and (2.106) yield that

$$\frac{I_1}{I_2} = -\frac{N_2}{N_1} \qquad\qquad (2.107)$$

---

[11] Note that currents I_1 and I_2 are in opposite directions. Therefore, if the direction of I_1 is designated as positive then I_2 must be negative.

Power Dissipation on the Transformer

As was shown in Chapter 1, it is straightforward to conclude that [power delivered to the input port of the transformer $P_1(t) = V_1(t).I_1.(t)$] + [power delivered to the output port $P_2(t) = V_2(t).I_2(t)$] = 0. This representation directly follows from multiplications of Equations (2.103) and (2.107).

Hence, we say that the transformer is a lossless two-port.

Self-inductances of Primary and Secondary Coils of a Transformer

In Equation (2.101), the magnetic flux may also be expressed in terms of the primary current I_1 as follows:

$$\phi = B.S$$

Where S is the cross-sectional area of the iron core.

In the linear operating region of the transformer, $B = \mu H$. On the other hand, by Ampère's law, $H.L = N_1 I_1$. Then, Equation (2.102) becomes

$$V_1(t) = \mu N_1^2 \frac{S}{L} \frac{dI_1}{dt}$$

Let

$$\boxed{\text{or} \quad \begin{aligned} L_{11} &= \mu N_1^2 \frac{S}{L} \\ L_{11} &= \frac{N_1^2}{\mathcal{R}} \end{aligned}}$$

(2.108)

be the self-inductance of the primary coil of the transformer. Then,

$$\boxed{V_1(t) = L_{11} \frac{dI_1}{dt}}$$

(2.109)

complies with the formal circuit theory definition of an inductor.

The above derivations, completed for the primary coil, can be identically carried out for the secondary coil. In this case, the self-inductance L_{22} of the secondary coil is defined as

$$\boxed{\text{or} \quad \begin{aligned} L_{22} &= \mu N_2^2 \frac{S}{L} \\ L_{22} &= \frac{N_2^2}{\mathcal{R}} \end{aligned}}$$

(2.110)

and it is straightforward to show that

$$\boxed{V_2(t) = L_{22} \frac{dI_2}{dt}}$$

(2.111)

Nonlinear Behavior of a Transformer

As was shown by Equation (2.102), it is important to note that the voltage $V_1(t)$ is a function of the operating frequency f_0. On the other hand, it is well known that ferromagnetic materials exhibit a hysteresis type of nonlinearity in the B versus H relationship, or, equivalently, the ϕ versus I relationship, as shown in Figure 2.34.[12] Thus, the linear operating region $0 \leq \phi_L < \phi_m$ of the ferrite core is bounded by ϕ_m, which in turn limits the upper bound $V_m = 2\pi f_0 \phi_m$ of the applied voltage as a function of frequency. Therefore, in designing a transformer, the choice of material and operating frequency is quite important. They determine the allowable limits of the input and output voltages. Details are beyond the scope of this book.

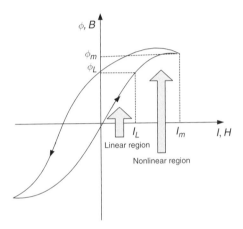

Figure 2.34   Nonlinear hysteresis type of variation of ϕ versus I

2.36 Mutual Inductance

Consider a toroid made of a ferrite core with

- tidily wound N_1 turns, for the primary coil;
- a cross-sectional area of $S \, (\mathrm{m}^2)$; and
- average circular length $L = 2\pi R \, (\mathrm{m})$ where R is the diameter of the toroid measured from the center to the coverage circumference of the ferrite core as shown in Figure 2.35.

Let us use a second winding, for the secondary coil, with N_2 turns on the primary.

---

[12] Note that for a uniform magnetic field density B over a constant cross-sectional area S, the magnetic flux is given by $\phi \equiv B.S$. Furthermore, in the linear operating region of the transformer, $B = \mu H$. On the other hand, by Ampère's law the magnetic field H is related to the current I_1 such that $N_1 I_1 = H.L$ where L is the average length of the ferrite core of the transformer and describes a closed path.

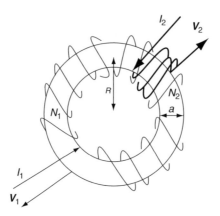

Figure 2.35   Mutual inductance between two windings on a toroid

In this case, the self-inductance L_{11} of the primary coil is given as in the transformer case. That is, when $I_2 = 0$, L_{11} is defined as the ratio of the flux linkage ϕ on N_1 turns to the current I_1. Thus,

$$L_{11} = N_1 \frac{\phi}{I_1} = \frac{N_1^2}{\mathcal{R}}$$

where $\phi = B.S$, $B = \mu H = \mu N_1 I_1 / L$ and $\mathcal{R} = L/\mu S$ wHen $I_2 = 0$

$$(2.112)$$

Similarly, the self-inductance L_{22} of the secondary coil is given by

$$L_{22} = N_2 \frac{\phi}{I_2} = \frac{N_2^2}{\mathcal{R}}$$

where $\phi = B.S$, $B = \mu H = \mu N_2 I_2 / L$ and $\mathcal{R} = L/\mu S$ wHen $I_1 = 0$

$$(2.113)$$

The mutual inductance M_{21} is defined as the ratio of the flux linkage ϕ on N_2 turns to the current I_1 when $I_2 = 0$. Hence,

$$M_{21} = N_2 \frac{\phi}{I_1} = N_2 \frac{B.S}{I_1} = N_2 \frac{\mu H S}{I_1} = N_1 N_2 \frac{\mu S}{L} = \frac{N_1 N_2}{\mathcal{R}}$$

$$(2.114)$$

In other words, the current I_1 or equivalently the magnetic flux $\phi(t)$ within the iron core induces a voltage $V_2(t)$ on the secondary coil of N_2 turns such that

$$V_2(t) = N_2 \frac{d\phi}{dt} = M_{21} \frac{dI_1}{dt}$$

when $I_2 = 0$.

Here, it is crucial to mention that the flux ϕ linked with the secondary coil over N_2 turns is the same as the one linked with the primary coil over N_1 turns. In other words, there is no loss of flux linkage out of the ferrite core. This is called perfect coupling. In practice, however, there are always fringing effects which cause a loss of flux on the secondary coil. In this case, we can say that the flux linked with the secondary coil is the fraction of ϕ which may be designated by $\phi_2 = k\phi$, such that $0 < k < 1$. The constant coefficient is called the coupling coefficient. Obviously, for perfectly coupled coils, $k = 1$. Considering the fringing effects, the mutual inductance M_{21} is revised as

$$M_{21} = N_2 \frac{k\phi}{I_1} = k\frac{N_1 N_2}{\mathcal{R}}$$

(2.115)

Similarly, the mutual inductance M_{21} is defined as the ratio of the flux linkage ϕ on the primary coil of N_1 turns to the current I_2 when $I_1 = 0$. In this case,

$$M_{12} = N_1 \frac{k\phi}{I_1} = N_1 k \frac{B.S}{I_2} = N_1 k \frac{\mu HS}{I_2} = N_1 N_2 k \frac{\mu S}{L}$$
$$= k\frac{N_1 N_2}{\mathcal{R}}$$

(2.116)

Comparison of Equations (2.115) and (2.116) yields that $M_{12} = M_{21}$, which must be expected due to reciprocity.

Regarding the fringing effects, it should be noted that Equations (2.112–2.116) reveal that

$$M = M_{12} = M_{21} = k\sqrt{L_{11} L_{22}} = \frac{N_1 N_2}{\mathcal{R}}$$

However, for perfect coupling, $k = 1$.

(2.117)

2.37 Boundary Conditions and Maxwell Equations

Equation (2.96) describes a significant result postulated by Faraday. It states that a time-varying magnetic field results in an electric potential which in turn is associated with an electric field. Certainly this result is bilateral, i.e. a time-varying electric field produces a magnetic field.

The interesting aspect of time variation is that when an electric field is linked to a magnetic field via time variation, the solution of the field equation results in a wave equation; hence, we obtain the propagation of electro magnetic field energy in a medium. Let us see how the mathematics works on this issue.

Fact 1: Differential Form of Faraday's Law

Induced voltage via the time variation of a magnetic field has the potential to do work in an electric field. In this case, Equation (2.96) can be expressed in terms of the induced electric field as

$$V = \oint_C E. \, dl = -\frac{\partial}{\partial t} \iint_S B. \, dS$$

where C is the closed path encircling cross-sectional area S.

We know from the mathematics that

$$\oint_C E. \, dl = \iint_S \nabla \times E. \, dS$$

Hence,

$$\text{curl } \boldsymbol{E} = \boldsymbol{\nabla} \times \boldsymbol{E} = -\frac{\partial \boldsymbol{B}}{\partial t} \qquad (2.118)$$

This equation is the differential form of Faraday's law and is also known as one of the four Maxwell equations.

Fact 2: Differential Form of Gauss's Law

Previously, the differential form of Gauss's law was given as

$$\text{div } \boldsymbol{D} = \rho_v \qquad (2.119)$$

Equation (2.101) is recognized as of one the Maxwell equations as well.

Fact 3: Magnetic Flux Theorems

Remember that Gauss's Law is a straightforward expression of the flux in an electric field and was originally written as $\psi_E = \displaystyle\oiint_S \boldsymbol{D}.\,d\boldsymbol{S}$. where S is a closed surface which encloses a volume V. In an electric field, if there is no charge in the closed volume then $\psi_E = 0$, meaning that 'whatever goes in, completely comes out'. However, if a net charge Q is enclosed, then the electric field flux must be equal to it. That is to say,

$$Q = \psi_E = \oiint_S \boldsymbol{D}.\,d\boldsymbol{S}$$

In a magnetic field, however, when we consider magnetic flux lines entering a closed volume, then they leave in the way they enter as shown in Figure 2.36.

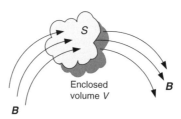

Figure 2.36   Magnetic flux over a closed surface S which encloses a volume V

Hence, the net flux captured within that region must be zero. Therefore, the net flux ψ in a closed region S is

$$\psi = \oiint_S \boldsymbol{B}.\,d\boldsymbol{S} = 0$$

By the Stokes–Green theorem,

$$\oiint_S \boldsymbol{B}.\,dS = \iiint_V \boldsymbol{\nabla}.\,\boldsymbol{B}dV = 0$$

Then, we can state that

$$\boxed{\text{div } \boldsymbol{B} = \boldsymbol{\nabla}.\,\boldsymbol{B} = 0} \qquad (2.120)$$

This equation is also known as one of the Maxwell Equations.

Note that Equation (2.120) is a straightforward mathematical result stating that the integral of any vector field over a closed surface S must be zero. Therefore, it has nothing to do with magnetic theory. However, magnetic flux density B defines a typical continuous vector field without interruption. Hence, its integral over any surface is zero. Then, Equation (2.120) follows.

Nevertheless, in electromagnetic field theory the above results may be referred to as the magnetic flux theorem (MFT).

Concept of Vector Potential

In mathematics, it is well known that the divergence of any curl vector, i.e. $\boldsymbol{\nabla}.\,(\boldsymbol{\nabla}\times A)$, is zero since the vector $\boldsymbol{B} = \boldsymbol{\nabla}\times A$ is always perpendicular to the vector nabla $\boldsymbol{\nabla}$. Therefore, Equation (2.120) suggests that the magnetic flux density vector $\boldsymbol{B}$ is specified as

$$\boxed{\boldsymbol{B} = \boldsymbol{\nabla}\times A} \qquad (2.121)$$

In this presentation, the fictitious vector A is called the vector potential; it facilitates computation of the magnetic field vector using Ampère's law as follows.

Ampère's law states that over a closed path C

$$\oint \boldsymbol{H}.\,dc = I_T = \iint_S \boldsymbol{J}_T.\,dS$$

where I_T is the net current encircled by the closed path C, S is the cross-sectional area formed by the closed path C and J_T is the current density over the surface S. Or

$$\oint \boldsymbol{B}.\,dc = \mu I_T = \mu \iint_S \boldsymbol{J}_T.\,dS$$

since $B = \mu H$.

By Stoke's theorem

$$\oint \boldsymbol{B}.\,dc = \iint_S (\boldsymbol{\nabla}\times\boldsymbol{B}).\,dS$$

Hence,

$$\iint_S \mathbf{\nabla} \times \mathbf{\nabla} \times A \, . \, dS = \mu \iint_S J_T \, . \, dS$$

or

$$\boxed{\mathbf{\nabla} \times \mathbf{\nabla} \times A = \mu J_T} \tag{2.122}$$

Note that the fictitious vector A is arbitrary and only defined to yield $B = \mathbf{\nabla} \times A$. However, we can freely impose a condition on A such that its divergence is zero. For instance, let us select a vector A yielding $\mathbf{\nabla}.A = 0$.[13]

Mathematically, it is straightforward to show that

$$\mathbf{\nabla} \times \mathbf{\nabla} \times A = -\mathbf{\nabla}^2 A + \mathbf{\nabla}(\mathbf{\nabla}.A)$$

Hence,

$$\boxed{\mathbf{\nabla}^2 A = -\mu J_T} \tag{2.123}$$

This equation very much resembles that of Equation (2.37c). Therefore its solution is written as in

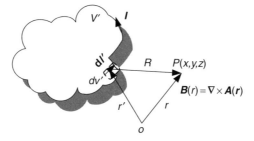

Figure 2.37 Determination of magnetic flux density generated by an arbitrary current loop enclosed by a volume V'.

Equation (2.37d). Thus, referring to Figure 2.37,

$$\boxed{A(r) = \frac{\mu}{4\pi} \iiint_{V'} \frac{J(r')}{R} \, dv'} \tag{2.124a}$$

---

[13] The condition $\mathbf{\nabla}.A = 0$ is known as the gauge condition.

> Or, if the current source is an arbitrary loop with cross-sectional
> area dS, then $dv' = dS.dl'$ and $[J(r')dS]dl' = I(r')dl'$. Hence, the magnetic
> potential vector A is then given by
>
> $$A(r) = \frac{\mu}{4\pi} \oint \frac{I(r')}{R} dl'$$

(2.124b)

Use of the vector potential concept and Equations (2.124) facilitates numerical computation of the magnetic field stemming from a current source. Especially, Equation (2.124) is useful for antenna pattern computations.

Boundary Conditions in Magnetic Fields

As in an electric field, one can determine the boundary conditions in a magnetic field between two media with permeabilities of μ_1 and μ_2.

Continuity of Normal Components

The boundary conditions of a magnetic field along the normal components may be derived by means of the MFT as $\psi = \oiint_S B.\, dS = 0.$

Referring to Figure 2.38, let us consider an infinitesimally small cylinder with a total surface S and cross-section of ΔS. In this case, the total magnetic flux is given by

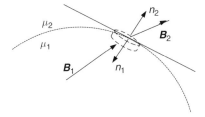

Figure 2.38 Boundary conditions along normal components of the magnetic field

$$\psi = \oiint_S B.\, dS = 0 = B.\, S_{side\ surface} + B_{n1}.\, n_1.\Delta S + B_{n2}.\, n_2.\Delta S$$

On the side surface, the dot product of $B.\, S_{side\ surface}$ is zero. Therefore, on the bottom and the top cross sections, dot products must yield

$$B_{n1}.\, n_1.\, \Delta S + B_{n2}.\, n_2.\, \Delta S = -B_{n1}.\, \Delta S + B_{n2}\Delta S = 0$$

Hence,

$$\boxed{B_{n1} = B_{n2}}$$

(2.125)

Since $B = \mu H$, then $\mu_1 H_{n1} = \mu_2 H_{n2}$ or

$$\boxed{\frac{H_{n1}}{H_{n2}} = \frac{\mu_2}{\mu_1}} \tag{2.126}$$

Continuity of Tangential Components

Referring to Figure 2.39, let us consider two magnetic materials with permeabilities μ_1 and μ_2 placed side by side. Let us further assume that medium 1 is a perfect conductor and medium 2 is an ideal isolator. Further, there is a current flow I (A) on the border of medium 1 due to a current density $J = 1/\Delta A = 1/\Delta L \times \Delta Y = $ constant. Then, the boundary condition of the magnetic field along the tangential axis can be determined by means of Ampère's law: $\oint_C H . dl = I$. In this case, one can consider an 'almost zero-height rectangular path of length ΔL'. Designating the tangential magnetic field components as H_{t1} and H_{t2} we can write

$$\oint H . dl = H_{t1} . \Delta L_{t1} + H_{t2} . \Delta L_{t2} = H_{t1}\Delta L - H_{t2}\Delta L = I = J(\Delta L \times \Delta Y)$$

where ΔY is a small distance perpendicular to ΔL. Hence,

$$\boxed{H_{t1} - H_{t2} = J\Delta Y} \tag{2.127}$$

Clearly, for current free media, Equation (2.127) becomes zero yielding

$$H_{t1} = H_{t2}$$

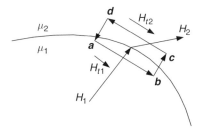

Figure 2.39   Tangential boundary condition in a magnetic field, $\Delta L = |ab| = |cd|$

Fact 4: Differential Form of Ampère's Law with Maxwell's Revision

As discussed above, Ampère's law asserts that the integral of the magnetic field over a closed path is equal to the net current encircled by that path. In other words,

$$\oint_C H . dl = I$$

The current I can be expressed in terms of the current density J over the cross-sectional area S encircled by the closed path C such that

$$\oint_C \boldsymbol{H}.\, dl = I = \iint_S \boldsymbol{J}.\, d\boldsymbol{S}$$

By Stokes's theorem,

$$\oint_C \boldsymbol{H}.\, dl = \iint_S (\boldsymbol{\nabla} \times \boldsymbol{H}).\, d\boldsymbol{S} = \iint_S \boldsymbol{J}.\, d\boldsymbol{S}$$

Hence,

$$\boxed{\text{curl}\, \boldsymbol{H} = \boldsymbol{\nabla} \times H = \boldsymbol{J}} \qquad (2.128)$$

where J is the current density.

Now, taking the divergence of both sides of Equation (2.128),

$$\boxed{\boldsymbol{\nabla}.\, (\boldsymbol{\nabla} \times \boldsymbol{H}) = \boldsymbol{\nabla}.\, \boldsymbol{J}} \qquad (2.129)$$

From the mathematics, the divergence of any curl vector X must be zero, i.e. $\boldsymbol{\nabla}.\, (\boldsymbol{\nabla} \times X) = 0$. Therefore, Equation (2.129) requires that $\boldsymbol{\nabla}.\, \boldsymbol{J} = 0$.[14] Actually, this is true for only static electric fields where the charge density is constant. In general, however, the principle of charge conservation indicates that $\boldsymbol{\nabla}.\, \boldsymbol{J} = -\partial \rho_v / \partial t = (\partial / \partial t)(\boldsymbol{\nabla}.\, \boldsymbol{D})$. In this case we see that Equation (2.129) is incomplete; it does not reflect the principle of charge conservation. This gap was filled by Maxwell. He asserted that the current density has two components. One is due to the conduction current, which is the current that we observe on the conductors as its density is given by $J_c = \sigma E$; and the other one is what he calls the displacement current created by the time variation of the electric field. In this case, considering the current continuity equation of Equation (2.62), he proposed that the divergence of the displacement current density, which is represented by $\boldsymbol{\nabla}.\, \boldsymbol{J}_D$, be equal to $-\partial \rho_v / \partial t$. On the other hand, by Gauss's law, $\text{div}\, \boldsymbol{D} = \boldsymbol{\nabla}.\, \boldsymbol{D} = \rho_v$. Then,

$$\boxed{\boldsymbol{J}_D = -\frac{\partial \boldsymbol{D}}{\partial t}} \qquad (2.130)$$

---

[14] It is a standard practice for a del operator $\boldsymbol{\nabla}$ to be considered as an ordinary vector. The cross-product of $\boldsymbol{\nabla}$ and $\boldsymbol{H}$ is a vector, say Ξ, which is perpendicular to the surface formed by $\boldsymbol{\nabla}$ and $\boldsymbol{H}$. Therefore, vector $\boldsymbol{\nabla}$ must be perpendicular to the vector $\Xi = \boldsymbol{\nabla} \times \boldsymbol{H}$. Hence, their dot product $\boldsymbol{\nabla}.\, \Xi = \boldsymbol{\nabla}.\, (\boldsymbol{\nabla} \times \boldsymbol{H})$ must be zero.

where the subscript D indicates the current density produced due to the rate of displacement vector. Eventually, he expressed the continuity equation as

$$\nabla . J_C = -\frac{\partial \rho_v}{\partial t} = -\nabla . J_D$$

or

$$\nabla . (J_C + J_D) = 0$$

Setting

$$J = J_C + J_D$$

Maxwell revised Ampère's law in the following differential form:

$$\boxed{\nabla \times H = J_C + \frac{\partial D}{\partial t} = \sigma E + \varepsilon \frac{\partial E}{\partial t}} \qquad (2.131)$$

This equation is known as the fourth Maxwell equation.

Maxwell simply says that time variation of an electric field, or, more specifically, time variation of charge density, produces a magnetic field. This may be regarded as Maxwell's law, which reciprocates Faraday's law in electric fields.

Hence, by Faraday's and Maxwell's laws, we can say that a changing magnetic field in time produces an electric field. Similarly, the time variation of an electric field produces a magnetic field.

In this case, Maxwell's version of Ampère's law is written as,

$$\boxed{\oint_{C(S)} H . dl = \iint_S J_C . dS + \iint_S \frac{\partial D}{\partial t} . dS} \qquad (2.132)$$

2.38 Summary of Maxwell Equations and Electromagnetic Wave Propagation

All the details above can be summarized in four major equations, referred to as the Maxwell equations, which lead to electromagnetic wave propagation, as follows:

Maxwell Equations:

Faraday's law

$$\boxed{\nabla \times E = -\frac{\partial B}{\partial t} = -\mu \frac{\partial H}{\partial t}} \qquad (2.133)$$

Revised Ampère's law

$$\nabla \times H = J_C + \frac{\partial D}{\partial t} = J_C + \varepsilon \frac{\partial E}{\partial t}$$

(2.134)

Gauss's law

$$\nabla \cdot D = \rho$$

(2.135)

Flux law of magnetic field

$$\nabla \cdot B = 0$$

(2.136)

In these equations the nomenclature used is the same as in previous sections.

Note that the Maxwell equations are excellent mathematical models based on the observed quantities of electric and magnetic forces. Therefore, we can say that if there is no electric force observed, then there is no assumption of an electric field or vice versa. The reason for this is that the electric field is defined as the electric force applied on a test charge. One can make a similar statement for a magnetic field. In fact, as far as a charge q is concerned, moving with velocity $v(t)$, the forces observed will be a combination of both the electric and magnetic forces and specified as

$$F = qE + qv(t) \times B$$

(2.137)

This force is known as the 'Lorenz force'. Obviously, in order to observe the Lorenz force, first of all we must have a test charge Q in the medium of measurement. Then, the moving charge q with velocity $v(t)$ must be brought from infinity to the vicinity of charge Q, As q moves, the electric force

$$F_E = \frac{1}{4\pi\varepsilon} \frac{qQ}{R^2}$$

varies as the distance $R(t) = v(t).t$ between q and Q varies. Hence, the electric field

$$E = \frac{F_E}{q} = \frac{1}{4\pi\varepsilon} \frac{Q}{[v(t).t]^2} u_R$$

varies, which in turn generates a time-varying magnetic flux density

$$B = \mu H = \int_t [\nabla \times E(\tau)] d\tau$$

as specified by Faraday's law in Equation (2.133). Or, equivalently, a magnetic field is generated as

$$H = -\frac{1}{\mu} \int_t \mathbf{\nabla} \times E(\tau) d\tau \qquad (2.138)$$

On the other hand, a time-varying magnetic field generates a displacement current density $\partial \vec{D}/\partial t$ by the principle of charge conservation and Ampère's law, or, in short, by the revised Ampère's law, meaning that it generates a time-varying electric field

$$E = \frac{1}{\varepsilon} \int_t [\mathbf{\nabla} \times H(\tau) - J_C] d\tau \qquad (2.139)$$

as specified by Equation (2.134).

As can be seen from the above mathematical models, a time-varying electric field generates a time-varying magnetic field and vice versa. So, the process builds on itself. Thus, intuitively, we see that wave-like motion dynamics is created.

Now let us consider mathematically 'how this motion is simply generated'.

Assume that we have a time-varying uniform electric field just like that within the plates of a capacitor. Here, the direction of the electric field is along the x axis. Similarly, assume that the magnetic field component only exists along the y axis. In this case, we have only E_x and H_y components of electric and magnetic fields respectively. Further assume that we investigate the variation of the electric and magnetic fields far away from a source which is a moving charge density within a confined region. In this case, the conduction current at a far distance away from the moving charge region must be zero. Thus, Equations (2.133) and (2.134) can be simplified as

$$-\frac{\partial E_x}{\partial z} = -\mu \frac{\partial H_y}{\partial t} \qquad (2.140)$$

and

$$-\frac{\partial H_y}{\partial z} = \varepsilon \frac{\partial E_x}{\partial t} \qquad (2.141)$$

Differentiating Equation (2.140) along the z axis,

$$-\frac{\partial E_x}{\partial z^2} = -\mu \frac{\partial^2 H_y}{\partial z \partial t} \qquad (2.42)$$

Now, differentiating Equation (2.141) with respect to time, we have

$$-\frac{\partial^2 H_y}{\partial z \partial t} = \varepsilon \frac{\partial^2 E_x}{\partial z^2} \qquad (2.143)$$

Hence, comparing Equation (2.142) to Equation (2.143), we find that

$$\frac{\partial^2 E_x}{\partial^2 z} = \mu\varepsilon \frac{\partial^2 E_x}{\partial t^2} \qquad (2.144)$$

Similarly, we can derive a second-order partial differential equation in a magnetic field H_y as

$$\frac{\partial^2 H_y}{\partial^2 z} = \mu\varepsilon \frac{\partial^2 H_y}{\partial t^2} \qquad (2.145)$$

In mathematics, the generic form of a second-order time-dependent partial differential equation given as

$$\frac{\partial^2 \psi}{\partial z^2} = \frac{1}{v^2} \frac{\partial^2 \psi}{\partial t^2} \qquad (2.146)$$

is known as the one-dimensional wave equation.

It is straightforward to verify that its generic solution is given by

$$\psi(z, t) = \psi_i\left(t - \frac{z}{v}\right) + \psi_r\left(t + \frac{z}{v}\right) \qquad (2.147)$$

where v is known as the velocity of propagation and its direction is aligned with the z axis, $\psi_i(t - z/v)$ is known as the incident wave, and $\psi_r(t + z/v)$ is known as the reflected wave.

Hence, comparing Equations (2.144) and (2.145) to (1.146), we see that electric field $\boldsymbol{E}_x$ and magnetic field $\boldsymbol{H}_y$ are obtained as the solution of a one-dimensional wave equation. In other words, they are in fact waves. The so-called electromagnetic waves propagate together with velocity v specified as

$$v = \frac{1}{\sqrt{\varepsilon\mu}} \qquad (2.148)$$

In a vacuum, the propagation velocity is specified by

$$v = \frac{1}{\sqrt{\varepsilon_0 \mu_0}} = \frac{1}{\sqrt{\left[4 \times \pi \times 10^{-7}\right] \times \left[\dfrac{10^{-9}}{36 \times \pi}\right]}} \\ = 3 \times 10^8 \mathrm{m/s} \qquad (2.149)$$

We can see that, in a vacuum, the propagation velocity is equal to the velocity of light. Therefore, we can state that the velocity of light is related to the electro magnetic properties of the propagation medium such that

$$c = \frac{1}{\sqrt{\varepsilon_0 \mu_0}}$$

(2.150)

In general, in a medium specified by permeability $\mu = \mu_r \mu_0$ and permittivity $\varepsilon = \varepsilon_r \varepsilon_0$, the electromagnetic wave propagation velocity is given by

or equivalently

$$v = \frac{c}{\sqrt{\varepsilon_r \mu_r}}$$

$$\frac{c}{v} = \sqrt{\varepsilon_r \mu_r}$$

(2.151)

Obviously, the direction of energy propagation must be the direction of propagation. For the electromagnetic wave propagation, described above along the x axis we have the electric field component $\boldsymbol{E}_x$, along the y axis we have the magnetic field component $\boldsymbol{H}_y$ and finally the direction of propagation is along the z axis, as shown in Figure 2.40. The electromagnetic waves depicted in Figure 2.40 are called uniform plane waves since $\boldsymbol{E}_x$ and $\boldsymbol{H}_y$ describe a flat plane while moving along the z axis.

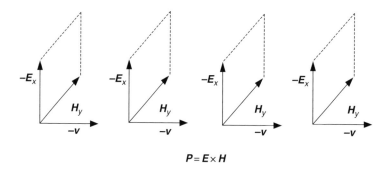

$$P = E \times H$$

Figure 2.40   Electromagnetic wave propagation: plane waves

It is interesting to note that direction of propagation v coincides with vector $\boldsymbol{P} = \boldsymbol{E}_x \times \boldsymbol{H}_y$. The vector $\boldsymbol{P}$ is called the Poynting vector and its intensity is given by $P = E_x H_Y \sin(90°) = E_x H_y$. Its dimension is $V/m \times A/m = W/m^2$. Hence, the Poynting vector describes power density over a surface.

2.39 Power Flow in Electromagnetic Fields: Poynting's Theorem

In fact, straightforward manipulation of the Maxwell equations yields[15]

$$-\iint_S (\boldsymbol{E} \times \boldsymbol{H}).d\boldsymbol{S} = \iiint_V \left[\frac{\partial}{\partial t}\left(\frac{\boldsymbol{B}.\boldsymbol{H}}{2}\right) + \frac{\partial}{\partial t}\left(\frac{\boldsymbol{D}.\boldsymbol{E}}{2}\right) + \boldsymbol{E}.\boldsymbol{J}\right]dV \qquad (2.152)$$

In the above equation, the first term $\iiint_V (\partial/\partial t)(\boldsymbol{B}.\boldsymbol{H}/2)\,dV$ is the energy stored in the magnetic field per unit time in volume V. Therefore, $(\boldsymbol{B}.\boldsymbol{H}/2)$ is the power density of the magnetic field. The second term $\iiint_V (\partial/\partial t)(\boldsymbol{D}.\boldsymbol{E}/2)\,dV$ refers to the energy stored in the electric field per unit time in volume V. Hence, the term $(\boldsymbol{D}.\boldsymbol{E}/2)$ refers to the power density of the electric field. The last term $\iiint_V \boldsymbol{E}.\boldsymbol{J}\,dV$ is the power loss of the medium.

The left hand side of Equation (2.152) is the integral of the Poynting vector over the closed surface S of the volume V. Therefore, $\boldsymbol{P}$ is related to power density. The negative sign on the right hand side of Equation (2.152) indicates the outward power flow toward the direction of propagation.

In electromagnetic field theory, Equation (2.152) is known as Poynting's theorem. Details of the proof can be found in any electromagnetic textbook.[16]

2.40 General Form of Electromagnetic Wave Equation

Now, consider Equation (2.133): $\boldsymbol{\nabla} \times \boldsymbol{E} = -\partial \boldsymbol{B}/\partial t = -\mu\,\partial \boldsymbol{H}/\partial t$. Taking the curl of both sides we can write

$$\boldsymbol{\nabla} \times \boldsymbol{\nabla} \times \boldsymbol{E} = -\frac{\partial(\boldsymbol{\nabla} \times \boldsymbol{B})}{\partial t} = -\mu\frac{\partial(\boldsymbol{\nabla} \times \boldsymbol{H})}{\partial t}$$

Note that from Equation (2.134)

$$\boldsymbol{\nabla} \times \boldsymbol{H} = J_C + \frac{\partial \boldsymbol{D}}{\partial t} = \varepsilon\frac{\partial \boldsymbol{E}}{\partial t}$$

if the medium is free of a current source, i.e. $J_C = 0$. On the other hand, by the vector multiplication identity

$$-\boldsymbol{\nabla}^2 \boldsymbol{E} + \boldsymbol{\nabla}(\boldsymbol{\nabla}.\boldsymbol{E}) = -\mu\varepsilon\frac{\partial^2 \boldsymbol{E}}{\partial t^2}$$

---

[15] Note that by mathematics $\boldsymbol{\nabla}.(\boldsymbol{E} \times \boldsymbol{H}) = \boldsymbol{H}.(\boldsymbol{\nabla} \times \boldsymbol{E}) - \boldsymbol{E}.(\boldsymbol{\nabla} \times \boldsymbol{H})$. On the other hand, by the Maxwell equation, $\boldsymbol{\nabla} \times \boldsymbol{E} = -\partial \boldsymbol{B}/\partial t = -\mu\,\partial \boldsymbol{H}/\partial t$ and $\boldsymbol{\nabla} \times \boldsymbol{H} = \boldsymbol{J} + \partial \boldsymbol{D}/\partial t = \boldsymbol{J} + \varepsilon\,\partial \boldsymbol{E}/\partial t$. Then, $\iiint_v \boldsymbol{\nabla}.(\boldsymbol{E} \times \boldsymbol{H}) = \iint_S (\boldsymbol{E} \times \boldsymbol{H})d\boldsymbol{S} = -\left\{\iiint [\boldsymbol{B}.\partial \boldsymbol{H}/\partial t + \boldsymbol{D}\,\partial \boldsymbol{E}/\partial t + \boldsymbol{J}.\boldsymbol{E}]dv\right\}$. By straightforward algebraic manipulations, it can be shown that

$$\frac{\partial}{\partial t}\left[\frac{1}{2}(\boldsymbol{B}.\boldsymbol{H})\right] = \boldsymbol{B}.\frac{\partial \boldsymbol{H}}{\partial t^2} \quad \text{and} \quad \frac{\partial}{\partial t}\left[\frac{1}{2}\boldsymbol{D}.\boldsymbol{E}\right] = \boldsymbol{D}.\frac{\partial \boldsymbol{E}}{\partial \boldsymbol{E}}$$

Hence, the result of Equation (2.152) follows.

[16] For example, see pp. 139–140 of *Fields and Waves in Communication Electronics*, 3rd edition, by Simon Ramo, John R. Whinnery and Theodore Van Duzer, John Wiley & Sons, Inc., 1994.

If the medium of propagation is free of electric charges then, by Equation (2.135), $\mathbf{\nabla}.\mathbf{D} = \varepsilon\mathbf{\nabla}.\mathbf{E} = \rho = 0$. Thus, the general form of the electric field equation is simplified as

$$\nabla^2 E = \mu\varepsilon\frac{\partial^2 E}{\partial t^2} \tag{2.153}$$

The solution of this is given by

$$E(x,y,z,t) = E_i\left(t - \frac{d}{v}\right) + E_r\left(t + \frac{d}{v}\right)$$

where $E_i(t - d/v)$ is the incident electric field wave, $E_r(t + d/v)$ is the reflected electric field wave, $v = 1/(\mu\varepsilon)^{1/2}$ is the velocity of propagation and $d = d(x,y,z)$ is the distance (vector) which is measured from the source (origin) to an arbitrary point $G(x,y,z) : d = OG = (x^2 + y^2 + z^2)^{1/2}$ in the direction of propagation. In this case, the electric field vector is measured at the point $G(x,y,z)$.

Similarly, the general form of the magnetic field equation is given by

$$\nabla^2 H = \mu\varepsilon\frac{\partial^2 H}{\partial t^2} \tag{2.154}$$

In a similar manner to that of Equation (2.153), solution of Equation (2.154) reveals that

$$H(x,y,z,t) = H_i\left(t - \frac{d}{v}\right) + H_r\left(t + \frac{d}{v}\right) \tag{2.155}$$

where $H_i(t - d/v)$ is the incident magnetic field wave and $H_r(t + d/v)$ is the reflected magnetic field wave.

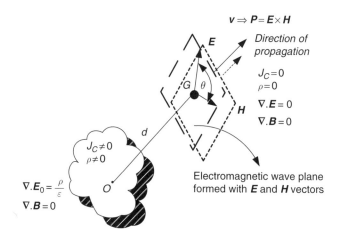

Figure 2.41 Generation of electromagnetic waves from an arbitrary source with charge density ρ

At this point we must clarify the following issues:

- *Any time-varying source*, with conduction current density J_C and moving charge density ρ, *radiates energy*. This form of radiated energy is called an '*electromagnetic wave*'. In other words, an electromagnetic wave is a form of traveling or radiated energy. In this regard, light is a specific form of electromagnetic wave or radiated energy. Such radiated energy travels and consists of infinitesimally small energy packages referred to as photons in quantum physics.
- It is presumed that electromagnetic waves consist of electric and magnetic field components.
- The electric and magnetic field components of the radiated energy must co exist. We cannot have one of them without the other.
- The propagation velocity of the radiated energy or electromagnetic wave is specified by the electric and magnetic properties of the transmission medium as $v = 1/(\mu\varepsilon)^{1/2}$. In free space or in vacuum, $v = 3 \times 10^8 \, \text{m/s}$.
- The direction of propagation is the negative direction of the Poynting vector $P = E \times H$ which is perpendicular to the plane formed with the electric E and magnetic field H vectors as shown in Figure 2.41.
- On the source side, the electric field is specified in terms of the charge density by means of Gauss's law: $\nabla . E(t,d) = -\rho(t)/_\varepsilon$. In return, the magnetic field is specified either by the revised Ampère's law (i.e. $\nabla \times H = J_c + \varepsilon \, \partial E/\partial t$) or by Faraday's law (i.e. $\nabla \times E = -\mu \, \partial H/\partial t$ whichever is appropriate).
- Consider an antenna which is driven by a time-varying voltage source $v_s(t)$. Let the current of the antenna be designated by $i_s(t)$. In this case, average power delivered to the antenna must be specified by $P_T = v_s(t).i_s(t) = V_s I_s$ where V_s and I_s are the effective values of voltage and current respectively. Here, it is interesting to observe that some of the power pumped into the antenna will be dissipated on the antenna itself, which may be a fraction of P_T, and the rest will be radiated. For example, for a typical cellular handy, the antenna efficiency could be 50%. If the total power pumped into the antenna is about $P_T = 1W$ then $P_{Loss} = 0.5W$ is dissipated on the antenna itself and the rest must be radiated, which is $P_r = 0.5W$. Generally speaking, the loss of an antenna is expressed in terms of its ohmic resistance R_{loss}. That is, $P_{loss} = R_{loss} I_S^2$ Similarly, the radiated power can be expressed in terms of a virtual quantity called the radiation resistance R_r, such that $P_r = R_r I_s^2$. Nevertheless, the radiation pattern of the antenna is obtained by solving the Maxwell equation for the antenna under consideration.
- From a physical point of view, in the wave solutions, the terms $E_r(t + d/v)$ and $H_r(t + d/v)$ are obtained as a result of reflection of the incident waves from an obstacle which is placed in the transmission medium. Therefore, in a loss and obstacle-free environment, ideal solutions of Equations (2.153) and (2.154) are simplified as

$$
\begin{aligned}
E(x,y,z,t) &= E(d,t) = E_i\left(t - \frac{d}{v}\right) \\
H(x,y,z,t) &= H(d,t) = H_i\left(t - \frac{d}{v}\right)
\end{aligned}
\tag{2.156}
$$

2.41 Solutions of Maxwell Equations Using Complex Phasors

Assume that at the source side, electric and magnetic fields are generated using a sinusoidal voltage or current source of angular frequency $\omega = 2\pi f$. In this case, the steady state solution of the Maxwell equations can be obtained using complex phasor notation.

In this approach, regarding the solutions of Equations. (2.144) and (2.145), the electric field is represented as $E(x,t) = \text{real}\{E_x(z)e^{j\omega t}\}$ where $E = E_x(z)e^{j\omega t}$ is called the complex phasor of the **E** field. Similarly we can define the complex phasor of the **H** field as $H = H_y(z)e^{j\omega t}$.

Then, along the direction of propagation, Equation (2.144) is written as

$$\frac{d^2 E_x(Z)}{dz^2} = -\omega^2 \mu \varepsilon E x(z) \qquad (2.157)$$

which is known as the 'one-dimensional Helmholtz equation' and its solution is given in terms of complex exponentials

$$E_x(z) = A e^{-j\beta z} + B e^{+j\beta z} \qquad (2.158)$$

where

$$\beta = \omega \sqrt{\mu \varepsilon} = \frac{\omega}{v} \qquad (2.159)$$

In the above equations, β is known as the phase constant or wave number. In the classical electromagnetic literature[17] it is also designated by the letter 'k'. In Equation (2.158) the terms $E_A = A e^{-j\beta z}$ and $E_B = B e^{+j\beta z}$ are known as the complex incident and reflected wave phasors of the **E** Field.

Similarly, using Equation (2.141) one can obtain the phasor form of the magnetic field:

$$H_y(z) = -\frac{1}{j\omega\mu} \frac{dE_x}{dz} = \frac{\beta}{\omega\mu} \left(A e^{j\beta z} - B e^{j\beta z} \right)$$
or
$$H_y(z) = -\sqrt{\frac{\varepsilon}{\mu}} \left(A e^{j\beta z} - B e^{j\beta z} \right)$$
or
$$H_y(z) = \frac{1}{Z_0} \left(A e^{j\beta z} - B e^{j\beta z} \right) \qquad (2.160)$$
where
$$Z_0 = \sqrt{\frac{\mu_0}{\varepsilon_0}}$$
is known as the characteristic impedance of the transmission medium.

Eventually, sinusoidal steady state time domain solutions of the **E** and **H** Fields are obtained as

$$E(x,t) = \text{real}\{E_x(z)e^{j\omega t}\} = A\cos(\omega t - \beta z) + B\cos(\omega t + \beta z)$$
or
$$E(x,t) = A\cos\left[\omega\left(t - \frac{z}{v}\right)\right] + B\cos\left[\omega\left(t + \frac{z}{v}\right)\right]$$
and similarly
$$H(y,t) = \frac{1}{Z_0}\left\{A\cos\left[\omega\left(t - \frac{z}{v}\right)\right] - B\cos\left[\omega\left(t + \frac{z}{v}\right)\right]\right\} \qquad (2.161)$$

---

[17] Carl T. A. Jonk, '*Engineering Electromagnetic Fields and Waves*', John Wiley & Sons, Inc., 1988. Simon Ramo, John R. Whinnery and Theodore Van Duzer, '*Fields and Waves in Communication Electronics*', 3rd edition, John Wiley & Sons, Inc., 1994.

Note that the characteristic impedance of free space or a vacuum is given as

$$Z_0 = \sqrt{\frac{4\pi \times 10^{-7}}{10^{-9}/36 \times \pi}} = 120\pi \cong 377 \ \Omega$$

2.42 Determination of the Electromagnetic Field Distribution of a Short Current Element: Hertzian Dipole Problem

While the size of very large-scale integrated (VLSI) circuits shrinks, the frequency of operation increases tremendously, forcing all circuit components to act like antennas radiating electromagnetic energy with wavelength $\lambda = v/f$. This physical phenomenon is due to the comparable size of lumped and distributed circuit elements as well as their interconnections placed on the layout with wavelength λ. Therefore, it is essential to understand the electromagnetic wave behavior of the designed circuits to assess the electrical performance. In order to do so, it may be helpful to study the Hertzian dipole field distribution, as short current elements resemble the electromagnetic behavior of lumped circuit elements.

Hertzian Dipole

Referring to Figure 2.42, and using the spherical coordinate system with independent variables $\{\phi, r, \theta\}$, for a short current element of length 'l' placed at the center of the coordinate system along the z axis, phasor form of the vector potential A is given by

$$\begin{aligned} A(r) = A_z &= \left[\frac{\mu}{4\pi}\right] l . \frac{I_m}{r} e^{-j\beta r}, \quad \beta = \frac{\omega}{v} \\ A_r &= A_z \cos(\theta) \\ A_\theta &= -A_z \sin(\theta) \\ A_\phi &= 0 \end{aligned} \tag{2.162}$$

where I_m is the peak value of the current on the antenna.

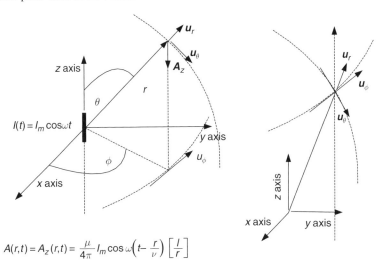

$$A(r,t) = A_z(r,t) = \frac{\mu}{4\pi} I_m \cos \omega \left(t - \frac{r}{v}\right) \left[\frac{l}{r}\right]$$

Figure 2.42 Spherical coordinate representation of a Hertzian dipole

By definition, vector potential A is related to the magnetic field vector as $\boldsymbol{B} = \boldsymbol{\nabla} \times \boldsymbol{A}$ or $\boldsymbol{H} = (1/\mu)\boldsymbol{\nabla} \times \boldsymbol{A}$. In this regard, it can be deduced that[18]

$$
\begin{aligned}
H_r &= 0 \\
H_\theta &= 0 \\
H_\phi &= \frac{I_m}{4\pi}.l.e^{-j\beta r}\left[\frac{j\beta}{r} + \frac{1}{r^2}\right]\sin(\theta)
\end{aligned}
\tag{2.163}
$$

Once the magnetic field is found, the electric field can be determined from the phasor form of the Maxwell equation of $\boldsymbol{\nabla} \times \boldsymbol{H} = J + j\omega\varepsilon\boldsymbol{E}$ for which the conduction current density J is set to zero for free space.

Hence, the spherical form of the electric field is expressed as follows:

$$
\begin{aligned}
E_\phi &= 0 \\
E_r &= \frac{I_m}{4\pi}.l.e^{-j\beta r}\left[\frac{2\eta}{r^2} + \frac{2}{j\omega r^3}\right]\cos(\theta) \\
E_\theta &= \frac{I_m}{4\pi}.l.e^{-j\beta r}\left[\frac{j\omega\mu}{r} + \frac{1}{j\omega\varepsilon r^3} + \frac{\eta}{r^2}\right]\sin(\theta)
\end{aligned}
\tag{2.164}
$$

It is interesting to note that, in generating the E and H fields, for the near-field computations, the terms which are associated with $1/r^2$ and $1/r^3$ will be more effective compared to terms attached to $1/r$. On the other hand, for far-field computations the terms associated with $1/r$ will be more effective than those terms in $1/r^2$ and $1/r^3$. Thus for far-field computations Equations (2.163) and (2.164) will be approximated as

$$
\begin{aligned}
H_\phi &= \frac{I_m}{4\pi}.l.e^{-j\beta r}\left[\frac{j\beta}{r}\right]\sin(\theta) \\
E_\theta &= \frac{I_m}{4\pi}.l.e^{-j\beta r}\left[\frac{j\omega\mu}{r}\right]\sin(\theta)
\end{aligned}
\tag{2.165}
$$

yielding

$$
E_\theta = \eta H_\phi
$$

The above equation describes a plane wave.

At this point it is meaningful to compute the radiated power using Poynting's theorem. In this case, the radiated time average of the power density P_r can be determined as

$$
P_r = \frac{1}{2}\text{real}\{E_\theta H_\phi^*\} = \frac{\eta\beta^2 I_m^2 l^2}{32\pi^2 r^2}\sin^2(\theta) \ \text{W/m}^2
\tag{2.166}
$$

---

[18] In the spherical coordinate system, for any vector A curl A is given by

$$
\boldsymbol{\nabla} \times \boldsymbol{A} = \left\{\frac{\partial}{\partial\theta}\left[A_\phi\sin(\theta)\right] - \frac{\partial A_\theta}{\partial\phi}\right\}\frac{\boldsymbol{u}_r}{r\sin(\theta)} + \left[\frac{1}{\sin(\theta)}\frac{\partial A_r}{\partial\phi} - \frac{\partial}{\partial r}(rA_\phi)\right]\frac{\boldsymbol{u}_\theta}{r} + \left[\frac{\partial}{\partial r}(rA_\theta) - \frac{\partial A_r}{\partial\theta}\right]\frac{\boldsymbol{u}_\phi}{r}
$$

Total power can be determined by integrating the above equation:

$$W = \oint\limits_{S} \left[\boldsymbol{E}_\theta \times \boldsymbol{H}_\phi \right] d\boldsymbol{S}$$

For the sake of simplicity, this integral can be taken over the surface of a sphere of radius r. In this case, for a spherical coordinate system $dS = 2\pi r^2 \sin(\theta) d\theta$. Therefore,

$$W = \int_0^\pi P_r 2\pi r^2 \sin(\theta) d\theta$$

or

$$W = \frac{\eta \beta^2 I_m^2 l^2}{16\pi} \int_0^\pi \sin^3(\theta) d\theta$$

where

$$\beta = \frac{\omega}{v} = \frac{2\pi}{T \times \left(\dfrac{\lambda}{T} \right)} = \frac{2\pi}{\lambda}$$

Therefore,

$$\boxed{ W = \frac{\eta \pi I_m^2}{3} \left[\frac{l}{\lambda} \right]^2 \cong 40\pi^2 I_m^2 \left[\frac{l}{\lambda} \right]^2 \mathrm{W} } \qquad (2.167)$$

It is very important to note that one can always associate a radiation resistance R_r with the above equation such that

$$W = \frac{1}{2} R_r I_m^2$$

Or we can define the radiation resistance for a Hertzian dipole as

$$\boxed{\begin{array}{l} R_r \triangleq \dfrac{2W}{I_m^2} = 80\pi^2 \left[\dfrac{l}{\lambda} \right]^2 \\[2mm] \text{or setting} \\[2mm] \lambda = \dfrac{v}{f} = \dfrac{2\pi v}{2\pi f} = \dfrac{2\pi v}{\omega} \\[2mm] \text{then} \\[2mm] R_r = 20 \left[\dfrac{l}{v} \right]^2 \omega^2 = 20[\omega\tau]^2 \\[2mm] \text{where } \tau = 1/v \text{ is called the delay length of the antenna,} \\ \text{which is measured in seconds.} \end{array}} \qquad (2.168)$$

Equations (2.167) and (2.168) clearly indicate that in order to increase the radiation power, the normalized or relative antenna length $[l/\lambda]$ of the antenna must be increased. On the other hand, antenna radiation resistance is an even function of the angular frequency ω, as it should be.

Example 2.13: For a Hertzian dipole, the relative length of the antenna is given by $l/\lambda = 0.1$. Compute the radiation resistance R_r.

Solution: By Equation (2.168)

$$R_r \triangleq \frac{2W}{I_m^2} = 80\pi^2 \left[\frac{l}{\lambda}\right]^2$$

Then, $R_r = R_r \triangleq 2W/I_m^2 = 80\pi^2 \times (0.01) \cong 7.8957\,\Omega$.

2.43 Antenna Loss

One can always associate a loss to a wire of length l. In this case, the loss resistance R_L of a Hertzian dipole is given by

$$R_L = \frac{l}{\sigma S}$$

where S is the cross-sectional area of the wire of length 'l' and σ is the conductivity of the wire.

Example 2.14: For an aluminum Hertzian dipole the length is given as $[l/\lambda]$ with cross-sectional area $S = 1\,\text{mm}^2$. Let the conductivity of the wire be $\sigma = 3.5 \times 10^7\,\text{S/m}$. Find the loss of the antenna.

Solution: Here

$$\lambda = \frac{c}{f} = \frac{3 \times 10^8}{3 \times 10^9} = 0.1\,\text{m}$$

Then,

$$l = 0.1 \times 0.1 = 0.01\,\text{m} = 1\,\text{cm}$$

Hence,

$$R_L = \frac{10^{-2}}{3.5 \times 10^7 \times 1 \times 10^{-6}} = 0.2857\,\text{m}\Omega$$

2.44 Magnetic Dipole

A magnetic dipole is a small current loop antenna and its electromagnetic field radiation can be derived using the solution obtained for the Hertzian dipole by means of the duality principle. In this principle, the magnetic dipole is defined as $m = I \times (\pi a^2)$ where a is the diameter of the loop. Similarly, the electric dipole is given by $p = ql$. The current of the antenna is expressed as $I = dq/dt$ or, using phasor notation, $I = j\omega q$ or $q = I/j\omega$. Then, the electric dipole $p = I/j\omega l$. According to the duality principle, the radiation solution for the magnetic dipole is obtained directly from the Hertzian dipole by replacing: p by μm; H by E; $-E$ by H; and ε by μ. Thus, the complete electric and magnetic field solutions are given by

$$E_r = 0$$
$$E_\theta = 0$$
$$E_\phi = -j\omega\mu \frac{I_m a^2}{4} . e^{-j\beta r} \left[\frac{j\beta}{r} + \frac{1}{r^2} \right] \sin(\theta) \qquad (2.169)$$

and

$$H_\phi = 0$$
$$H_r = j\omega\mu \frac{I_m a^2}{4} . e^{-j\beta r} \left[\frac{2}{\eta r^2} + \frac{2}{j\omega\mu r^3} \right] \cos(\theta)$$
$$H_\theta = j\omega\mu \frac{I_m a^2}{4} . e^{-j\beta r} \left[\frac{j\omega\varepsilon}{r} + \frac{1}{j\omega\mu r^3} + \frac{1}{\eta r^2} \right] \sin(\theta) \qquad (2.170)$$

For the far field,

$$E_\phi = \omega\mu \frac{I_m a^2}{4} . e^{-j\beta r} \left[\frac{\beta}{r} \right] \sin(\theta)$$
$$H_\theta = -\omega\mu \frac{I_m a^2}{4} . e^{-j\beta r} \left[\frac{\omega\varepsilon}{r} \right] \sin(\theta) \qquad (2.171)$$

yielding

$$E_\phi = -\eta H_\theta$$

2.45 Long Straight Wire Antenna: Half-Wave Dipole

Referring to Figure 2.43, for a long straight antenna of arbitrary length l, the radiation pattern can be driven by means of the integral of the radiation pattern of the Hertzian dipole. Detailed computations are beyond the scope of this book. Nevertheless, we will summarize the mathematical expression for the far field as follows:[19]

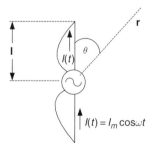

Figure 2.43 Dipole antenna of arbitrary length l.

---

[19] Simon Ramo, John R. Whinnery and Theodore Van Duzer, *Fields and Waves in Communication Electronics*, 3rd edition, John Wiley & Sons, Inc., 1994, Chapter 12. R. S. Elliot, *Antenna Theory and Design*, Prentice Hall, 1981.

$$E_\theta = j \frac{\eta I_m}{2\pi r} e^{-j\beta r} \left[\frac{\cos(\beta l.\cos\theta) - \cos(\beta l)}{\sin(\theta)} \right]$$

$$H_\phi = \frac{E_\theta}{\eta}$$

The radiated power is given by

$$W = \frac{\eta I_m^2}{4\pi} \int_0^\pi \frac{[\cos(\beta l.\cos\theta) - \cos(\beta l)]^2}{\sin(\theta)} d\theta$$

(2.172)

From the radiated power we can always generate the

antenna radiation resistance and the antenna loss as

$$R_r \triangleq \frac{2W}{I_m^2}$$

and

$$R_L = \frac{l}{\sigma S}$$

Half-Wave Dipole

Field computations of the half-wave dipole can be obtained by setting $2l = \lambda/2$ *or* $l = \lambda/4$. Hence, we have

$$|E_\theta| = \frac{60 I_m}{r} \left| \frac{\cos\left[\left(\frac{\pi}{2}\right)\cos(\theta)\right]}{\sin(\theta)} \right|$$

(2.173)

2.46 Fourier Transform of Maxwell Equations: Phasor Representation

By taking the Fourier transform (FT), one can simply get rid of the time dependence of the Maxwell equations by replacing the $\partial/\partial t$ terms with $j\omega$.

For harmonic excitations, this method greatly facilitates the solution of Equation (2.133–2.136). In this case, using the phasor nations for the field variables E and H, we obtain[20]

Faraday's law

$$\nabla \times E = -j\omega B = -j\omega\mu H$$

(2.174)

---

[20] Note that during the solution process of the Maxwell equations, the conductivity σ and the charge density ρ must be considered on the geometry of the excitation, which is separated from the free medium of propagation. On the other hand, away from the excitation, these quantities must be zero since we assume no free charge in the medium of propagation (i.e. $\sigma = 0$ and $\rho = 0$ in the propagation medium).

Revised Ampère's law

$$
\begin{aligned}
\boldsymbol{\nabla} \times \boldsymbol{H} &= J_c + j\omega\boldsymbol{D} \\
&= \sigma\boldsymbol{E} + j\omega\varepsilon\boldsymbol{E} \\
&= (\sigma + j\omega\varepsilon)\boldsymbol{E}
\end{aligned}
\tag{2.175}
$$

Gauss's law

$$
\boldsymbol{\nabla} . \boldsymbol{D} = \rho
\tag{2.176}
$$

Flux law of magnetic field

$$
\boldsymbol{\nabla} . \boldsymbol{B} = 0
\tag{2.177}
$$

In the above equations, all the variables are specified in phasor form.

For isotropic media, permittivity ε and permeability μ are real quantities. Therefore, the complex quantity $j\omega\mu\boldsymbol{H}$ of Equation (2.174) is associated with the magnetic field energy stored in the medium of propagation, which is considered a loss-free environment. Similarly, in Equation (2.175), the complex quantity $j\omega\varepsilon\boldsymbol{E}$ refers to the stored electric field energy in the same region. On the other hand, the real quantity $\sigma\boldsymbol{E}$ is a loss term which is due to the conduction current generated by the excitation under consideration. Obviously, this loss must be measured on the lossy medium of excitation rather than on the loss-free environment of propagation.

If the medium of propagation is not homogeneous and isotropic, then the permittivity and permeability of the medium become complex quantities such that

$$
\begin{aligned}
\varepsilon &= \varepsilon' - j\varepsilon'' \\
\mu &= \mu' - j\mu''
\end{aligned}
\tag{2.178}
$$

This form of permittivity and permeability introduces some sort of fictitious losses called hysteresis of dielectric and magnetic materials in such a way that Equations (2.174) and (2.175) take the following form:

$$
\begin{aligned}
\boldsymbol{\nabla} \times \boldsymbol{E} &= -j\omega\boldsymbol{B} = -j\omega[\mu' - j\mu'']\boldsymbol{H} = \omega\mu''\boldsymbol{H} + j\omega\mu'\boldsymbol{H} \\
\boldsymbol{\nabla} \times \boldsymbol{H} &= (\sigma + j\omega\varepsilon)\boldsymbol{E} = [\sigma + \omega\varepsilon']\boldsymbol{E} + j\omega\varepsilon''\boldsymbol{E} \\
&= (\sigma' + j\omega\varepsilon'')\boldsymbol{E}
\end{aligned}
$$

where

$$
\sigma' = \sigma + \omega\varepsilon''
$$
$$\tag{2.179}$$

In Equation (2.179) the term $\omega\varepsilon''$ acts like a fictitious conductivity which increases the actual conductivity σ as a function of angular frequency ω. However, for good conductors, if $\sigma \gg \omega\varepsilon''$ then the fictitious conductivity can be neglected.

After some manipulation and elimination, Equations (2.174) and (2.175) can be written in phasor form as

$$
\boldsymbol{\nabla}^2\boldsymbol{E} + \gamma^2\boldsymbol{E}^2 = 0
$$

and

$$\nabla^2 H + \gamma^2 H^2 = 0$$

where the wave number or the propagation constant is still given as

$$\gamma = \frac{1}{\sqrt{\varepsilon \mu}} = \alpha + j\beta$$

and is a complex number since permittivity and permeability are complex quantities. The frequency-dependent quantity $\alpha = \alpha(\omega)$ is called the attenuation and the frequency-dependent quantity $\beta = \beta(\omega)$ is called the phase constant.

In this case, wave solutions are given as

$$E(r) = E_+ e^{-\gamma r} + E_- e^{+\gamma r}$$

$$H(r) = H_+ e^{-\gamma r} + H_- e^{+\gamma r}$$

or

$$H = \frac{1}{\eta} \left[E_+ e^{-\gamma r} - E_- e^{+\gamma r} \right]$$

where the terms $E_+ e^{-\gamma r}$ and $H_+ e^{-\gamma r}$ represent the incident waves and the terms $E_- e^{+\gamma r}$ and $H_- e^{+\gamma r}$ designate the reflected terms. For media with no reflections $E = E_+ = \eta H = \eta H_+$.

Phasor Form of the Power Density

The radiated power of the electromagnetic field is given by Equation (2.152) as

$$\iiint_v \left[\frac{\partial}{\partial t} \left(\frac{B.H}{2} \right) + \frac{\partial}{\partial t} \left(\frac{D.E}{2} \right) + E.J \right] dV = -\oiint_S (E \times H) dS$$

This equation can be written in the frequency domain as a complex power density such that

$$P_c = \frac{dP_r}{dV} = E.J^* + \frac{1}{2} j\omega B.H^* + \frac{1}{2} j\omega D.E^*$$

or

$$P_D = P_R + jP_X = \sigma |E|^2 + \frac{1}{2} \omega \varepsilon'' |E|^2 + \frac{1}{2} \omega \mu'' |H|^2 + \frac{1}{2} j\omega \left[\frac{1}{2} \varepsilon' |E|^2 + \frac{1}{2} \mu' |H|^2 \right] \qquad (2.180)$$

The real part $P_R = \sigma |E|^2 + \frac{1}{2} \omega \varepsilon'' |E|^2 + \frac{1}{2} \omega \mu'' |H|^2$ of the above power density represents the loss of the radiated power per volume in free space, $\sigma = 0$. Therefore, the ohmic loss term $P_\Omega = \sigma |E|^2$ must be zero. The term $P_E = \frac{1}{2} \omega \varepsilon'' |E|^2$ represents the power-loss density due to the electric field. Similarly, the term $P_M = \frac{1}{2} \omega \mu'' |H|^2$ is the power-loss density due to the magnetic field.

Phase Velocity and Refractive Index

In a lossless medium the propagation constant is real and the phase velocity v_ϕ is defined as $v_\phi = [d\beta/d\omega]^{-1} = 1/(\varepsilon\mu)^{1/2}$. Let the speed of light be $c = 1/(\varepsilon_0\mu_0)^{1/2}$; then the refractive index n is defined as $n = c/v_\phi$.

Loss Tangent

For an anisotropic medium the dielectric constant was described as

$$\varepsilon = \varepsilon' - j\varepsilon'' = |\varepsilon|e^{-j\delta}$$

In the above equation the quantity $\tan(\delta) = \varepsilon''/\varepsilon'$ is called the loss tangent.

Penetration Depth δ and the Internal Impedance Z_s of a Conductor

For a good conductor, electric charges accumulate mostly on the surface which in turn reduces the time and space variation of the electric field significantly. Therefore, the displacement current density $J_D = \partial D/\partial t$ or, in phasor form, $J_D = j\omega D = j\omega\varepsilon E$, becomes negligible compared to the conduction current density J_C. In this case, the phasor forms of the Maxwell equations can be simplified as

$$\boxed{\begin{aligned} \nabla \times H = J_T = J_c + J_D = \sigma E + j\omega\varepsilon E \approx \sigma E \\ \nabla \times E = j\omega\mu H \end{aligned}}$$

(2.181)

By taking the curl of both equations, we end up with

$$\nabla \times (\nabla \times E) = \nabla(\nabla . E) - \nabla^2 E = -j\omega\mu\nabla \times H = j\omega\mu\sigma E$$

As expressed above, inside a good conductor the charge density ρ is negligible. Therefore, $\nabla . D = \varepsilon\nabla . E = \rho \approx 0$. In this case, we can state that

$$\boxed{\nabla^2 E = j\omega\mu\sigma E}$$

(2.182)

Similarly, one can obtain the following equations:

$$\boxed{\begin{aligned} \nabla^2 H = j\omega\mu\sigma H \\ \nabla^2 J_c = j\omega\mu\sigma J_c \\ \omega = 2\pi f \end{aligned}}$$

where ω is the angular frequency and f is the actual frequency.

(2.183)

As can be seen, the above equations are in the form of the Helmholtz wave equation. By letting $\tau^2 \triangleq j\omega\mu\sigma$ we can state that[21]

$$\tau = (1+j)\sqrt{\pi\mu\sigma} = \frac{1+j}{\delta} = \frac{1}{\delta} + j\frac{1}{\delta}$$

---

[21] Note that $j = e^{j\frac{\pi}{2}}$ and $\sqrt{j} = e^{j\frac{\pi}{4}} = \cos(\pi/4) + j\sin(\pi/4) = (1+j/2)$

where

$$\delta = \frac{1}{\sqrt{\pi f \mu \sigma}} \text{ measured in meter}$$

is called the skin depth or penetration.

(2.184)

In order to simplify the solution of the above differential equations, let us assume that a plane wave penetrates a conductor in the x direction while the electric and magnetic fields propagate in the z and y directions respectively. In this case, the general solution for the electric field can be given as

$$E_z(x) = \left[E_+ e^{-\frac{x}{\delta}}\right] e^{-j\frac{x}{\delta}} + \left[E_- e^{\frac{x}{\delta}}\right] e^{j\frac{x}{\delta}}$$

where $\left[E_+ e^{-\frac{x}{\delta}}\right]$ and $\left[E_- e^{\frac{x}{\delta}}\right]$ designate the incident and reflected waves within the good conductor respectively. Obviously, the reflected wave must diminish since the intensity of the incident wave drastically reduces with the attenuation term of $e^{\frac{x}{\delta}}$, leaving no reflection. On the other hand, when x approaches infinity, this makes the reflected term $\left[E_- e^{\frac{x}{\delta}}\right]$ go to infinity. This is impossible physically. Therefore, the real coefficient E_-(or reflection) must be zero. Hence,

$$E_z(x) = \left[E_+ e^{-\frac{x}{\delta}}\right] e^{-j\frac{x}{\delta}}$$

(2.185)

Similarly,

$$H_y(x) = \left[H_+ e^{-\frac{x}{\delta}}\right] e^{-j\frac{x}{\delta}}$$

$$J_z(x) = \left[J_0 e^{-\frac{x}{\delta}}\right] e^{-j\frac{x}{\delta}}$$

(2.186)

where E_+, H_+ and J_0 are respectively the incident electric field phasor, incident magnetic field phasor and current density component right on the conductor's surface.

Note that when $x = \delta$ the attenuation of the above quantities is given by

$$\left|\frac{E_{z(\delta)}}{E_+}\right| = \left|\frac{H_y(\delta)}{H_+}\right| = \left|\frac{J_z(\delta)}{J_0}\right| = e^{-1} = 36.79\%$$

Obviously, the skin depth is the measure of penetration of the electromagnetic energy into conductors and therefore associated with surface loss as described in the following.

In general, one can determine the total current J_{SZ} flowing into the conductor depth as x approaches infinity from

$$J_{sz} = \int_0^\infty J_z(x)dx = \int_0^\infty J_0 e^{-(1+j)\frac{x}{\delta}} dx = \frac{J_0 \delta}{(1+j)}$$

(2.187)

On the other hand, the current density on the surface is given by $J_0 = \sigma E_+$. Then, we can define the conductor's internal impedance as

$$
Z_s \triangleq \frac{E_+}{J_{sz}} = \frac{1+j}{\sigma\delta} = R_{sz} + j\omega L_{sz} = \frac{1}{\sqrt{2}}\frac{1}{\sigma\delta}e^{j\frac{\pi}{2}}
$$

where the frequency-dependent surface resistance
R_{sz} over a unit length $(z = 1\,\text{m})$ and unit width $(y = 1\,\text{m})$ is given by

(2.188)

$$
R_{sz} = \frac{1}{\sigma\delta} = \sqrt{\frac{\pi f \mu}{\sigma}} = \sqrt{\frac{\omega\mu}{2\sigma}}
$$

and the frequency-dependent inductance L_{sz} is defined by

$$
\omega L_{sz} = \frac{1}{\sigma\delta} = R_{sz}
$$

Here, what we see is that the surface resistance along the z axis per unit length and along the y axis per unit width is simply expressed as the function of skin depth δ, even though the integration bound of Equation (2.187) goes to infinity.

Remember that the loss of a conducting slab of, $l = 1\,\text{m}$, $S_{xy} = \delta \times 1\,\text{m}^2$ and conductivity σ is specified by $R_{loss} = 1/(\sigma S_{xy}) = 1/(\sigma\delta)\,\Omega$. Therefore, the skip depth or penetration δ describes the loss of a conductor in a straightforward manner.

Example 2.15: The conductivity of a copper slab is specified as $\sigma = 5.8 \times 10^7$ at $300\,\text{K}$ The permeability for copper is approximately equal to $\mu_0 = 4\pi \times 10^{-7}\,\text{H/m}$.

(a) Find the surface resistance R_{sz}.
(b) Find the skin depth δ.

Solution:

(a) Using Equation (2.188),

$$
R_{sz} = \frac{1}{\sigma\delta} = \sqrt{\frac{\pi f \mu}{\sigma}} = \sqrt{\frac{\omega\mu}{2\sigma}} = \sqrt{\frac{\pi \times 4 \times \pi \times 10^{-7}}{5.8 \times 10^7}}\left[\sqrt{f}\,\right]
$$

$$
R_{sz} = 2.16 \times 10^{-7}\left[\sqrt{f}\,\right]\Omega
$$

(b) Here

$$
\delta = \frac{1}{\sigma \times R_{sz}} = \frac{1}{5.8 \times 10^7 \times 2.61 \times 10^{-7}}\left[\sqrt{f}\,\right]^{-1} = 0.066\left[\sqrt{f}\,\right]^{-1}\text{m}
$$

Antenna Input Impedance $Z_{in}(j\omega)$ and its Equivalent Circuit

Referring to Figure 2.44, an antenna may be considered as a one-port passive circuit component with input impedance $Z_{in}(j\omega) = R_{in}(\omega^2) + jX_{in}(\omega)$ where the even real part can be expressed in terms of the loss resistance R_L and the radiation resistance $R_r(\omega^2)$ such that

$$R_{in}(\omega^2) = R_L + R_r(\omega^2)$$

The reactive part $X(\omega)$ could be inductive (positive) or capacitive (negative) or it could exhibit positive and negative variation as a function of frequency.

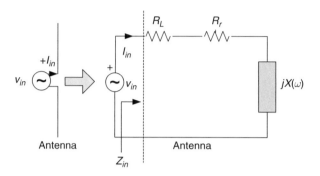

Figure 2.44   Antenna equivalent circuit

The frequency ω_k which makes the reactance $X(\omega_k)$ zero is called the resonance frequency of the antenna. An antenna may resonate at single or multiple frequencies.

In general, antennas may be manufactured by utilizing wires, waveguides, any conducting planar surface like microstrip patches, or by utilizing anisotropic dielectric slabs perhaps having complex permittivity and permeability etc. In this case, the antenna loss resistance R_L must include surface resistance R_{sz} which rises due to skip depth δ and other frequency-dependent terms raised due to power loss in the electric field (which is specified by $P_{E\,loss} = \sigma|\boldsymbol{E}|^2 + \frac{1}{2}\omega\varepsilon''|\boldsymbol{E}|^2$) and power loss raised due to the magnetic field (which is given by $P_{H\,loss} = \frac{1}{2}\omega\mu''|\boldsymbol{H}|^2$). Computation of all of these may be very tedious and troublesome. Therefore, in practice, we always prefer to measure the input impedance of an antenna as a function of frequency, and also to measure the radiation resistance, to end up with the microscopic loss resistance $R_L(\omega)$. From the design perspective of power transfer networks for antennas, it may be useful to model the measured antenna impedance to obtain the gain–bandwidth limitations. In this regard, a useful method for modeling the measured impedance data is presented in a following chapter.

Component or device models greatly facilitate many design problems by passing the solution of the Maxwell equations. Especially at microwave and millimeter wave frequencies, when the operating frequency wavelength $\lambda = v/f$ becomes comparable to the size of the circuit components such as antennas, capacitors and inductors, models and the design of power transfer networks must include the physical sizes which make the circuit components act like transmission lines. Thus, utilization of ideal transmission lines becomes inevitable in both design and model problems. Therefore, in the following chapter, the theory of transmission lines as circuit components is reviewed. Ideal transmission lines are regarded as lossless parallel plates or two-pair wires made of perfect conductors guiding the plane waves, or coaxial cables which are known as transverse electric and magnetic (TEM) field lines.

3

Transmission Lines for Circuit Designers: Transmission Lines as Circuit Elements

3.1 Ideal Transmission Lines

Transmission lines are very simple conducting geometric structures which guide electromagnetic energy. For example, two long parallel conducting plates constitute the simplest version of such structures, yielding voltage versus current or electric field versus magnetic field relationships. Similarly, two conducting parallel wires or coaxial cables are also excellent examples of electromagnetic energy guiding systems.

Let us consider two parallel long plates made of perfect conductors as shown in Figure 3.1. This structure guides the electromagnetic field energy as it propagates from the source end down to the load end in a loss-free medium. Therefore, it is called the ideal transmission line.

Assume that a time-varying current is initiated on the top plate toward the positive z direction and returns on the bottom plate. For this geometry, straightforward application of Gauss's and the revised Ampère's laws reveal that the electromagnetic field must be confined within the plates uniformly over the distance d. The conduction current $I(t, z)$ must be on the surface of the perfect conductors, which in turn yields a uniform magnetic field $\boldsymbol{H}_y$ in y direction; and it is aligned with the width W of the conducting plates. In this case, over a unit length $L = 1$ m, closed path integration of the uniform magnetic field reveals that

$$\oint \boldsymbol{Hdl} = H_y W = I(t, z)$$

or

$$H_y = \frac{I(t, z)}{W}$$

Design of Ultra Wideband Power Transfer Networks Binboga Siddik Yarman
© 2010 John Wiley & Sons, Ltd

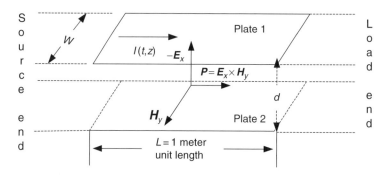

Figure 3.1 A parallel plate transmission line made of perfect conductors

Obviously, this magnetic field generates a uniform magnetic field density vector $B_y = \mu H_y$. Since the magnetic field density vector is outward, the cross-sectional area $A = d\,(\mathrm{m}) \times L\,(\mathrm{m}) = d\,(\mathrm{m}) \times (1\,\mathrm{m}) = d\,\mathrm{m}^2$; then, the magnetic flux is given by

$$\emptyset = \iint_A \boldsymbol{B}.d\boldsymbol{S} = \mu H_y A = \mu H_y d$$

In this case, we can define the inductance L over the unit length such that

$$L = \frac{\emptyset(t, z)}{I(t, z)} = \mu \frac{d}{W}$$

Furthermore, we can assume that the displacement or leakage current due to uniform electric field E_x is generated by the uniform displacement vector $\boldsymbol{D}_x = \varepsilon \boldsymbol{E}_x$. Then, over the surface $A_W = W\,(\mathrm{m}) \times L\,(\mathrm{m}) = W \times 1 = W\,\mathrm{m}^2$, the Gauss's law reveals that

$$\iint_{A_W} \boldsymbol{D}(t, z)\,.\,d\boldsymbol{S} = Q(t, z)$$

or

$$\varepsilon E_x(t, z) W = Q(t, z)$$

On the other hand, at any point z, the potential difference $V(t, z)$ between the palates can be expressed as $V(t, z) = E_x(t, z).d$.

Hence, we can define the capacitance per unit length as

$$C = \frac{Q(t, z)}{V(t, z)} = \varepsilon \frac{W}{d}$$

which depends on the geometry and the electric property of the medium.

In summary, this ideal parallel plate transmission line can be modeled by means of infinitely many cascaded T sections of L/2–C–L/2 or equivalently pi sections of C–L–C per unit length as depicted in

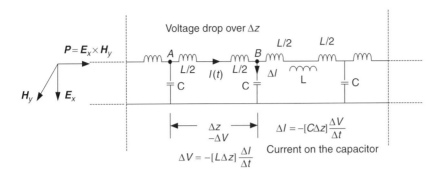

Figure 3.2   T or equivalently pi section model for an ideal transmission line

Figure 3.2.[1] In this configuration, letters L and C denote the inductor and capacitor per unit length respectively.

It must be mentioned that plane waves which have electric and magnetic field components on the same plane, perpendicular to the direction of propagation, just like the ones described above, are called transverse electric and magnetic field waves or, in short, TEM waves.

For this line, we presume that the electric field vector is in the negative x direction (from plate 1 to plate 2), the magnetic field vector is in the positive y direction and the electromagnetic wave energy propagates in the positive z direction.

Certainly, the solution of the Maxwell Equation yields the propagation pattern of the fields.

On the other hand, one should be able to obtain wave equations when the descriptive voltage and current relations for inductors and capacitors are written. In fact, considering the pi section of C–L–C, the voltage drop $(-\Delta V)$ on the inductance $[L\Delta z]$ over an infinitesimally small length Δz is given by

$$-\Delta V = V_L = [L\Delta z]\frac{\Delta I}{\Delta t}$$

or as $\Delta z \to 0$

$$\lim_{\Delta z \to 0} \frac{\Delta V}{\Delta z} = \frac{\partial V}{\partial z} = -L\frac{\partial I}{\partial t} \tag{3.1}$$

which is Faraday's law and must be consistent with

$$\nabla \times E = \frac{\partial E_x}{\partial z} = -\mu\frac{\partial H}{\partial t}$$

---

[1] An ideal transmission line may consist of 'two parallel perfect conductors', 'two parallel wires' made of perfect conductors, or it may be a lossless coaxial cable made of perfect conductors. They all posses the same pi or T section type of low-pass L–C models with different values of inductors and capacitors.

In a similar manner, over an infinitesimally small length Δz, the leakage or the displacement current $-\Delta I = I_C$ through the capacitor $[C\Delta z]$ is expressed as

$$-\Delta I = I_C = [C\Delta z]\frac{\Delta V}{\Delta t}$$

or as $\Delta z \to 0$

$$\lim_{\Delta z \to 0} \frac{\Delta I}{\Delta z} = \frac{\partial I}{\partial z} = -C\frac{\partial V}{\partial t}$$

which must be consistent with the revised version of Ampère's law

$$\nabla \times \boldsymbol{H} = \frac{\partial \boldsymbol{H}_y}{\partial t} = J_c + \varepsilon\frac{\partial E_x}{\partial t}$$

with $J_c = 0$, meaning that the capacitor is free of conduction current.

(3.2)

By taking the derivative' with respect to z, Equation (3.1) can be written as

$$\frac{\partial^2 V}{\partial z^2} = -L\frac{\partial^2 I}{\partial z \partial t}$$

(3.3)

Similarly, by taking the derivative with respect to time t, Equation (3.2) takes the following form:

$$\frac{\partial^2 I}{\partial z \partial t} = -C\frac{\partial^2 V}{\partial t^2}$$

(3.4)

Hence, Equation (3.3) and Equation (3.4) reveal that

$$\frac{\partial^2 V(z,t)}{\partial z^2} = LC\frac{\partial^2 V(z,t)}{\partial t^2}$$

or

$$\frac{\partial^2 V(z,t)}{\partial z^2} = \frac{1}{v^2}\frac{\partial^2 V(z,t)}{\partial t^2}$$

where

$$v = \frac{1}{\sqrt{LC}}$$

is the speed of the voltage wave propagation. Actually, the above equation is equivalent to

$$\frac{\partial^2 E_x(z,t)}{\partial z^2} = \frac{1}{v^2}\frac{\partial^2 E_x(z,t)}{\partial t^2}$$

where

$$v = \frac{1}{\sqrt{\mu\varepsilon}}$$

is the velocity of wave propagation.

(3.5)

In fact, as shown above, for parallel conducting plates, which constitute an ideal TEM line, the capacitance C and the inductance L per unit length are given by $C = \varepsilon W/d$ and $L = \mu d/W$ respectively. Thus, the velocity of wave propagation

$$v = 1/\sqrt{LC} = \frac{1}{\sqrt{\mu\varepsilon}}$$

is verified.

In a manner similar to that of Equation (3.5), the current wave equation $I(z, t)$ and the magnetic field wave equation are written as

$$
\begin{aligned}
\frac{\partial^2 I(z,t)}{\partial z^2} &= \frac{1}{v^2}\frac{\partial^2 I(z,t)}{\partial t^2} \\
\frac{\partial^2 H_y(z,t)}{\partial z^2} &= \frac{1}{v^2}\frac{\partial^2 H_y(z,t)}{\partial t^2} \\
\text{with } & LC = \mu\varepsilon \text{ and } v = 1/(LC)^{1/2}
\end{aligned}
\tag{3.6}
$$

Here, it should be clear that the uniform electric field $E_x(z, t)$ can easily be obtained from the voltage wave $V(z, t)$ by

$$V(z,\ t) = -\int_{line\,2}^{line\,1} E(z,\ t)dx$$

which simply yields

$$V(z,\ t) = E_x(z,t).d$$

or

$$E_x(t,z) = \frac{V(t,\ z)}{d}$$

Furthermore, by employing Ampère's law, the relation between magnetic field and current is given by

$$H_y(z,t) = \frac{1}{w}I(z,t)$$

For historical reasons, Equations (3.1) and (3.2) are called the telegrapher's equations and they are summarized as below.

Telegrapher's Equations

$$
\begin{aligned}
\frac{\partial V}{\partial z} &= -L\frac{\partial I}{\partial t} \\
\frac{\partial I}{\partial z} &= -C\frac{\partial V}{\partial t}
\end{aligned}
\tag{3.7}
$$

3.2 Time Domain Solutions of Voltage and Current Wave Equations

As shown by Equation (2.147) of Section 2.38, the solution of the voltage wave equation is given by

$$
\boxed{
\begin{array}{l}
V(z,t) = V_i\left(t - \dfrac{z}{v}\right) + V_r\left(t + \dfrac{z}{v}\right) \\
\text{where} \\
\qquad V_i\left(t - \dfrac{z}{v}\right) \\
\text{refer to the incident wave and the term} \\
\qquad V_r\left(t + \dfrac{z}{v}\right) \\
\text{is the reflected wave.}
\end{array}
}
\tag{3.8}
$$

The current wave solution can be obtained by taking the derivative of Equation (3.8) with respect to z (i.e $\partial V/\partial z = -L\,\partial I/\partial t$)[2]

$$
-L\frac{\partial I(z,\,t)}{\partial t} = -\frac{1}{v}V_i'\left(t - \frac{z}{v}\right) + \frac{1}{v}V_r'\left(t + \frac{z}{v}\right)
$$

Integrating this equation over time, one obtains the following solution:

$$
\boxed{
\begin{array}{l}
I(z,\,t) = \dfrac{1}{Lv}\left[V_i\left(t - \dfrac{z}{v}\right) - V_r\left(t + \dfrac{z}{v}\right)\right] \\
\text{or, by setting } Z_0 = Lv = \sqrt{L/C}, \\
I(z,\,t) = \dfrac{1}{Z_0}\left[V_i\left(t - \dfrac{z}{v}\right) - V_r\left(t + \dfrac{z}{v}\right)\right] \\
\text{where } Z_0 \text{ is called the characteristic impedance of the transmission line.}
\end{array}
}
\tag{3.9}
$$

3.3 Model for a Two-Pair Wire Transmission Line as an Ideal TEM Line

For a two-pair parallel wire transmission line, Equation (2.47) of Chapter 2 specifies the unit length capacitor as.

$$
C = \frac{2\pi\varepsilon}{\ln\left[\dfrac{d}{a}\right]}
$$

---

[2] In this representation V' denotes the derivatives with respect to time. That is,

$$
V_i'\left(t - \frac{z}{v}\right) = \frac{\partial V_i\left(t - \frac{z}{v}\right)}{\partial t} \quad \text{and} \quad V_r'\left(t + \frac{z}{v}\right) = \frac{\partial V_r\left(t + \frac{z}{v}\right)}{\partial t}
$$

On the other hand, the unit length inductor is given by Equation (2.95) of Chapter 2 as

$$L = \frac{u}{2\pi} \ln\left[\frac{d}{a}\right]$$

In these expressions a is the radius of parallel wires and d is the distance between them. Thus, we see that for this uniform TEM line, $LC = \mu\varepsilon$, which verifies the result obtained for ideal parallel plates. For this case, the characteristic impedance is given by

$$Z_0 = \sqrt{\frac{L}{C}} = \frac{\ln\left[\frac{d}{a}\right]}{2\pi}\sqrt{\frac{\mu}{\varepsilon}} \tag{3.10}$$

3.4 Model for a Coaxial Cable as an Ideal TEM Line

Coaxial cables also guide uniform electric and magnetic fields which in turn yield voltage and current wave equations as above. In this regard, Equation (2.48) and Equation (2.92) specify the unit length capacitance and inductance respectively as

$$C_{coax} = \frac{2\pi\varepsilon}{\ln\left[\frac{b}{a}\right]}$$

$$L_{coax} = \frac{\mu}{2\pi} \ln\left[\frac{b}{a}\right]$$

Hence, the velocity of propagation verifies the above results:

$$v = \frac{1}{\sqrt{L_{coax} . C_{coax}}} = \frac{1}{\sqrt{\mu\varepsilon}}$$

In a similar manner to that of Equation (3.10), the characteristic impedance of the coaxial line is given by

$$Z_0 = \sqrt{\frac{L_{coax}}{C_{coax}}} = \frac{\ln\left[\frac{b}{a}\right]}{2\pi}\sqrt{\frac{\mu}{\varepsilon}} \tag{3.11}$$

3.5 Field Solutions for TEM Lines

For any TEM line, regardless of whether it is a parallel plate, two-pair wire or coax cable, the time domain solutions of the electric and magnetic fields are given by

$$E_x(t, z) = E_{xi}\left(t - \frac{z}{v}\right) + E_{xr}\left(t + \frac{z}{v}\right)$$

$$H_y(t, z) = \frac{1}{Z_0}\left[E_{xi}\left(t - \frac{z}{v}\right) - E_{xr}\left(t + \frac{z}{v}\right)\right]$$

(3.12)

where

$$v = \frac{1}{\sqrt{LC}} \quad \text{and} \quad Z_0 = \sqrt{\frac{L}{C}}$$

Furthermore, uniform electric field vector intensity is specified in terms of the potential difference between lines $V(t,z)$ at any point z as

$$E_x(t, z) = \frac{V(t, z)}{d} \qquad E_{xi} = \frac{V_i\left(t - \frac{z}{v_p}\right)}{d} \qquad E_{xr} = \frac{V_r\left(t + \frac{z}{v_p}\right)}{d}$$

where d is the distance between the lines.
Similarly,

$$H_y(t, z) = \frac{1}{W}I(t, z) = \frac{1}{WZ_0}\left[V_i\left(t - \frac{z}{v_p}\right) - V_r\left(t + \frac{z}{v_p}\right)\right]$$

3.6 Phasor Solutions for Ideal TEM Lines

Phasor forms of the telegrapher's equations can be obtained by replacing $\partial/\partial t$ by $j\omega$ and $\partial^2/\partial t^2$ by $-\omega^2$.

Hence, Equation (3.8) reveals that

$$\frac{\partial V(z, j\omega)}{\partial z} = -j\omega I(z, j\omega)$$

$$\frac{\partial I(z, j\omega)}{\partial z} = -j\omega V(z, j\omega)$$

Or, by taking the derivative of the first equation with respect to z and using it in the second equation, we obtain

(3.13)

$$\frac{\partial^2 V(z, j\omega)}{\partial z^2} = -\omega^2 LCV(z, j\omega) = (j\omega L)(j\omega C)V(z, j\omega)$$

$$\frac{\partial^2 I(z, j\omega)}{\partial z^2} = -\omega^2 LCI(z, j\omega) = (j\omega L)(j\omega C)I(z, j\omega)$$

Let

$$
\boxed{
\begin{aligned}
z(j\omega) &= j\omega L \\
Y(j\omega) &= j\omega C \\
\gamma^2 &= -\omega^2 LC = Z(j\omega)Y(j\omega)
\end{aligned}
}
$$

or

$$
\boxed{
\begin{aligned}
\gamma &= j[\omega LC] = j\left(\frac{\omega}{v}\right) = j\beta(\omega) = \sqrt{Z(j\omega)Y(j\omega)} \\
\beta(\omega) &= \frac{\omega}{v_p} \\
v_p &= \frac{1}{\sqrt{LC}}
\end{aligned}
}
\tag{3.14}
$$

Then, the phasor solution for the voltage and current is given by

$$
\boxed{
\begin{aligned}
V(z,j\omega) &= Ae^{-\gamma z} + Be^{+\gamma z} \\
I(z,j\omega) &= \frac{1}{Z_0}\left[Ae^{-\gamma z} - Be^{+\gamma z}\right]
\end{aligned}
}
$$

where

$$
\gamma = j\beta
$$

is the propagation constant and

$$
Z_0 = \sqrt{\frac{Z(j\omega)}{Y(j\omega)}} = \sqrt{\frac{L}{C}}
$$

is the characteristic impedance of the TEM line.

$$\tag{3.15}$$

3.7 Steady State Time Domain Solutions for Voltage and Current at Any Point z on the TEM Line

Assume that a TEM line is excited by a sinusoidal voltage source

$$
E_G(t) = E_m \cos(\omega t) = \text{real}\{E_m e^{j\omega t}\}
$$

Then, the time domain solution for voltage and current at any point z on the line is given by

$$
\begin{aligned}
V(t,z) &= \text{real}\{V(z,j\omega)e^{j\omega t}\} \\
I(t,z) &= \text{real}\{I(z,j\omega)e^{j\omega t}\}
\end{aligned}
$$

In these equations, the voltage and current phasors are given by

$$
\begin{aligned}
V(z,jw) &= V_m(z,\omega)e^{-j\varphi_v(\omega)} \\
I(z,jw) &= I_m(Z,\omega)e^{-j\varphi_i(\omega)}
\end{aligned}
$$

In this case, at any point z, sinusoidal steady state solutions for voltage and current are given by

$$V(t,z) = V_m(z,\omega)\cos(\omega t - \varphi_v)$$
$$I(t,z) = I_m(Z,\omega)\cos(\omega t - \varphi_i)$$

Some practical definitions and notations for transmission lines are given in the following sections.

3.8 Transmission Lines as Circuit Elements

Propagation Constant γ

In the above, $\gamma(j\omega)$ is called the propagation constant. For ideal lossless TEM lines the propagation constant $\gamma = [Z(j\omega)Y(j\omega)]$ is purely imaginary and specified as $\gamma(j\omega) = j\beta(\omega)$. However, referring to Figure 3.2, for a lossy line, a series loss resistance r and a shunt conductance G are connected with inductance L and capacitance C respectively. In this case, the propagation constant takes the following form:

$$\gamma(j\omega) = \alpha(\omega) + j\beta(\omega) = \sqrt{(r + j\omega L)(G + j\omega C)} \qquad (3.16)$$

Phase Constant β

The quantity imaginary $\{\gamma\} = \beta(\omega) = \omega/v_p$ is called the phase constant. For a lossless TEM line, it is interesting to observe that the phase constant is a linear function of the frequency of the operation, yielding

$$\frac{d\beta}{d\omega} = \frac{1}{v_p}$$

For a fixed operating frequency f, the wavelength λ is given by.

$$\lambda = \frac{v_p}{f} = \frac{2\pi v_p}{2\pi f} = \frac{2\pi v_p}{\omega}$$

In this case, the phase constant β is expressed as

$$\beta = \frac{2\pi}{\lambda} \text{ rad/m}$$

Attenuation Constant α

The real part of γ, which is $\alpha(\omega) = \text{real}\{\gamma(j\omega)\}$, is called the attenuation constant.

Phase Velocity v_p

The quantity $\left[d\beta/d\omega \right]^{-1}$ is called the phase velocity and designated by v_p. For ideal TEM lines,

$$v_p = \frac{1}{\sqrt{\mu\varepsilon}}$$

Delay Length τ

Let us feed a TEM line of length L with a sinusoidal source of frequency $f = 1/T$ period $T = 1/f$ and wavelength $\lambda = v_p/f = Tv_p$. In this case, it takes $\tau = L/v_p$ seconds for a sinusoidal waveform of one period to travel from one end of the line to the other. Hence, τ is called the delay length of the line.

Phase Delay

At any point $Z = l$ on the line, the phase delay is measured by multiplication of $\varphi = \beta l$ which has either degree or radian units. Usually, βl is measured in radians. It is straightforward to see that the phase delay $\varphi = \beta l = \omega l / v_p = \omega \tau$.

Incident Wave A(z, jω)

In Equation (3.15) the term

$$A(z, j\omega) = Ae^{-\gamma z}$$

is called the incident wave at point z.

Reflected Wave B(z, jω)

In Equation (3.15) the term

$$B(Z, j\omega) = Be^{+\gamma z}$$

is called the reflected wave.

Voltage and Current Expressions in Terms of Incident and Reflected Waves

Based on the definitions of incident and reflected waves, voltage and current expressions of Equation (3.15) are written as

$$
\begin{aligned}
V(z, j\omega) &= A(z, j\omega) + B(z, j\omega) \\
I(z, j\omega) &= \frac{1}{Z_0}[A(z, j\omega) - B(z, j\omega)]
\end{aligned}
\tag{3.17}
$$

Reflection Coefficient S(z, jω)

At any point 'z' on the line, the ratio of the reflected wave to the incident wave is called the reflection coefficient and given by

$$S(z, j\omega) = \frac{B}{A}e^{2\gamma z} \tag{3.18}$$

3.9 TEM Lines as Circuit or 'Distributed' Elements

In the classical microwave circuit literature, a typical TEM line is called a distributed circuit element, with physical length L as depicted in Figure 3.3.

In this physical layout, the conducting plates or wires are designated by two long rectangular boxes. At microwave frequencies, these conducting boxes or planes may be realized as microstrip lines or coplanar lines on a selected substrate such as alumina, silicon, etc., using metal deposition techniques.

In Figure 3.3, the line is driven by a voltage source E_G with internal impedance Z_G. At the end, the line is terminated by a complex load impedance called $Z_L(j\omega)$. The line has a physical length of L. For convenience, the origin of the z axis is selected at the far end of the line and geometrically designated by a $z = 0$ plane. The line is physically placed on the negative z axis. Therefore, the distance along the line is measured in negative lengths. With this convention, the voltage source at the input is connected to the line at the $z = -L$ plane.

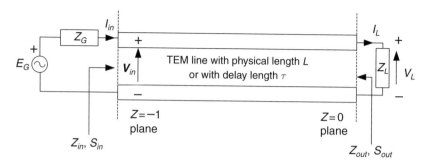

Figure 3.3   A typical TEM transmission line as a 'distributed circuit element'

As a distributed circuit element, a TEM line is considered as a two-port, which introduces a physical length in general purpose designs.

Now let us look at the voltage and current relations at the input (or source end; $z = -L$ plane) and the output (or load end; $z = 0$ plane).

Voltage and Current Expressions at the Load End; Load Reflection Coefficient on the $z = 0$ Plane

First of all, at the far end of the line (on the $z = 0$ plane), the reflection coefficient is given by

$$S(0, j\omega) = S_L = \frac{B}{A}$$

S_L is called the load reflection coefficient.

$$(3.19)$$

Equation (3.20) below reveals a boundary condition such that the voltage across the load must be equal to the line voltage at the $z = 0$ plane. Therefore,

$$V_L = Z_L I_L = A + B = A\left(1 + \frac{B}{A}\right) = A(1 + S_L)$$

On the other hand, the current at the load end or at the $z = 0$ plane is given by

$$I_L = \frac{1}{Z_0}[A - B] = \frac{A}{Z_0}(1 - S_L)$$

Then, by setting $Z_L = V_L/I_L$ it is found that

$$Z_L = Z_0 \frac{1 + S_L}{1 - S_L}$$

or

$$z_L = \frac{Z_L}{Z_0} = \frac{1 + S_L}{1 - S_L}$$

$$(3.20)$$

In Equation (3.20), z_L is called the normalized load impedance with respect to the given standard impedance Z_0. The characteristic impedance Z_0 could be either real or complex impedance. If the line is lossless then, $Z_0 = (L/C)^{1/2}$ is real, otherwise it is a complex quantity.

Utilizing Equation (3.20), the reflection coefficient S_L may also be expressed in terms of the normalized load impedance z_L such that

$$S_L = \frac{\text{reflected wave}}{\text{incident wave}} = \frac{B}{A} = \frac{z_L - 1}{z_L + 1} = \frac{Z_L - Z_0}{Z_L + Z_0} \tag{3.21}$$

Voltage and Current Expressions at the Source end; Input Reflection Coefficient on the $z = -L$ plane

The source end or $z = -L$ plane of the transmission line may be named the input. In this case, at $z = -L$ the reflection coefficient of Equation (3.18) takes the following form:

$$S(-L, j\omega) = S_{in} = \frac{\text{reflected wave}}{\text{incident wave}} = \frac{Be^{-\gamma l}}{Ae^{\gamma L}}$$

$$= S_L e^{-2\gamma L} \tag{3.22}$$

S_{in} is called the input reflection coefficient.

On the other hand, the voltage and current relations of Equation (3.17) become

$$V(-L, j\omega) = V_{in} = Ae^{\gamma L} + Be^{-\gamma L} = Ae^{\gamma L}\left(1 + \frac{B}{A}e^{-2\gamma L}\right)$$

or

$$V_{in} = Ae^{\gamma L}(1 + S_L e^{-2\gamma L})$$

or

$$V_{in} = Ae^{\gamma L}[1 + S_{in}]$$

Similarly,

$$I(-L, j\omega) = I_{in} = \frac{1}{Z_0}Ae^{\gamma L}[1 - S_L e^{-2\gamma L}]$$

or

$$I_{in} = \frac{1}{Z_0}Ae^{\gamma L}[1 - S_{in}]$$

Then, the input impedance $Z_{in} = V_{in}/I_{in}$ is written as

$$Z_{in} = Z_0 \frac{1 + S_{in}}{1 - S_{in}}$$

Let the normalized input impedance be

$$z_{in} = \frac{Z_{in}}{Z_0}$$

$$z_{in} = \frac{1 + S_{in}}{1 - S_{in}}$$

or

$$S_{in} = \frac{z_{in} - 1}{z_{in} + 1} = \frac{Z_{in} - Z_0}{Z_{in} + Z_0}$$

(3.23)

Output Reflection Coefficient at the z = 0 Plane

Note that if the transmission line is driven at the $z = 0$ plane by a voltage source with an internal impedance Z_L, and if it is terminated by an impedance Z_G at the $z = -L$ plane, then one can derive the expressions for the output reflection coefficient S_{out} and output impedance Z_{out} and normalized output impedance z_{out} as above.

Thus, the reflection coefficient for Z_G is given by

$$S_G = \frac{Z_G - Z_0}{Z_G + Z_0} = \frac{z_G - 1}{z_G + 1}$$

where

$$z_G = \frac{Z_G}{Z_0}$$

is the normalized impedance of the actual impedance Z_G.

(3.24)

The output reflection coefficient is given by

$$S_{out} = \frac{text > reflected\,wave}{incident\ wave} = \frac{Z_{out} - Z_0}{Z_{out} + Z_0} = \frac{z_{out} - 1}{z_{out} + 1}$$

$$= S_G e^{-2\gamma L}$$

where

$$Z_{out} = Z_0 \frac{1 + S_{out}}{1 - S_{out}}$$

Or the normalized output impedance is given by

$$z_{out} = \frac{Z_{out}}{Z_0} = \frac{1 + S_{out}}{1 - S_{out}}$$

(3.25)

Voltage Standing Wave Ratio

At any point on the ideal TEM line, the maximum voltage phasor can be expressed as

$$V_{max} = \left| A e^{-j\beta l} \right| + \left| B e^{-j\beta l} \right|$$

or

$$V_{max} = A + B = A\left(1 + \frac{B}{A}\right) = A[1 + |S_L|]$$

Similarly, the minimum voltage is given by

$$V_{min} = \left|Ae^{-j\beta l}\right| - \left|Be^{-j\beta l}\right|$$

or

$$V_{min} = A - B = A[1 - |S_L|]$$

Then, the voltage standing wave ratio (VSWR) of the line is defined as the ratio of the maximum voltage amplitude to the minimum voltage amplitude. Thus,[3]

$$\boxed{\text{VSWR} = \frac{1 + |S_L|}{1 - |S_L|}} \tag{3.26}$$

Open Expressions for Input and Output Reflection Coefficients and Impedances

Equations (3.24) and (3.26) specify the input and output reflection coefficients by means of the term $e^{-2\gamma l}$ as $S_{in} = S_L e^{-2\gamma l}$ and $S_{out} = S_G e^{-2\gamma l}$.

On the other hand, the term $e^{-2\gamma l}$ can be expressed as the tangent hyperbolic function as follows:

$$\tanh(\gamma l) = \frac{e^{\gamma l} - e^{-\gamma l}}{e^{\gamma l} + e^{-\gamma l}} = \frac{e^{\gamma l}(1 - e^{-2\gamma l})}{e^{\gamma l}(1 + e^{-2\gamma l})} = \frac{1 - e^{-2\gamma l}}{1 + e^{-2\gamma l}}$$

or

$$e^{-2\gamma l} = \frac{1 - \tanh(\gamma l)}{1 + \tanh(\gamma l)}$$

Let us define a new complex variable λ, called the Richard variable, as

$$\boxed{\lambda = \tanh(\gamma l) = \sum + j\Omega} \tag{3.27}$$

Then, in terms of the Richard variable,

$$\boxed{e^{-2\gamma l} = \frac{1 - \lambda}{1 + \lambda}} \tag{3.28}$$

---

[3] Note that the VSWR is a positive real number and varies between zero and infinity since the load reflection coefficient is bounded by 1 (i.e. $|S_L| \leq 1$).

Using the above identity, the input and output reflection coefficients are given by

$$
\begin{aligned}
S_{in}(\lambda) &= \frac{1 - \lambda}{1 + \lambda} S_L \\
S_{out}(\lambda) &= \frac{1 - \lambda}{1 + \lambda} S_G
\end{aligned}
$$

(3.29)

Using Equation (3.29) in Equation (3.24) we have

$$
Z_{in}(\lambda) = Z_0 \frac{(1 + \lambda) + (1 - \lambda)S_L}{(1 + \lambda) - (1 - \lambda)S_L}
$$

(3.30)

Or, employing

$$
S_L = \frac{Z_L - Z_0}{Z_L + Z_0}
$$

in the above equation for $Z_{in}(\lambda)$, we find that

$$
Z_{in}(\lambda) = Z_0 \frac{Z_L + Z_0\lambda}{Z_0 + Z_L\lambda}
$$

or in open form

$$
Z_{in} = Z_0 \frac{Z_L + Z_0\tanh(\gamma l)}{Z_0 + Z_L\tanh(\gamma l)}
$$

(3.31)

In a similar manner to that of Equation (3.31), the output impedance Z_{out} is given by

$$
Z_{out}(\lambda) = Z_0 \frac{Z_G + Z_0\lambda}{Z_0 + Z_G\lambda}
$$

or in open form

$$
Z_{out} = Z_0 \frac{Z_G + Z_0\tanh(\gamma l)}{Z_G + Z_0\tanh(\gamma l)}
$$

(3.32)

For an ideal TEM transmission line, the attenuation constant $\alpha(\omega)$ must be zero. Therefore, the propagation constant has only an imaginary part and it is specified as equal to $j\beta$. On the other hand, the characteristic impedance is given as a positive real number such that $z_0 = (L/C)$. In this case, for an ideal TEM line, Equations (3.27), (3.31) and (3.32) are simplified as

$$
\begin{aligned}
\lambda &= j\Omega \\
\Omega &= \tan(\beta l) = \tan(\omega\tau) \\
Z_{in} &= Z_0 \frac{Z_L + jZ_0\tan(\beta l)}{Z_0 + jZ_L\tan(\beta l)} \\
Z_{out} &= Z_0 \frac{Z_G + jZ_0\tan(\beta l)}{Z_0 + jZ_G\tan(\beta l)}
\end{aligned}
$$

(3.33)

Equation (3.33) may also be expressed in terms of admittances. In this case, all impedances which are designated by Z_{xx} are replaced with corresponding admittances which are denoted by Y_{xx}. In this notation, subscripts 'xx' refer to input, output, load and generator, and the characteristic impedances and corresponding admittances respectively:

$$
\boxed{
\begin{aligned}
Y_{in} &= Y_0 \frac{Y_L + jY_0\tan(\beta l)}{Y_0 + jY_L\tan(\beta l)} \\[2mm]
Y_{out} &= Y_0 \frac{Y_G + jY_0\tan(\beta l)}{Y_0 + jY_G\tan(\beta l)}
\end{aligned}
}
\tag{3.34}
$$

Ideal TEM lines may be utilized as distributed capacitors and inductors by properly choosing the real characteristic impedance $Z_0 = R_0 = (L/C)^{(1/2)}$, the line phase length $\varphi_L = \beta L = \omega\tau$ and termination impedances Z_G and Z_L. In the following section we present some examples of distributed one- or two-port components which are useful in the design of power transfer networks.

An Open-End TEM Line as a Capacitor

When the far end of a transmission line is left open, it is considered to be terminated with an infinite impedance load, i.e. $Z_L = \infty$. In this case, the input impedance acts like a capacitace C in the λ domain such that

$$
Z_{in}(\lambda) = \lim_{z_L \to \infty} Z_0 \frac{Z_L + Z_0\lambda}{Z_0 + Z_L\lambda} = \frac{1}{\lambda C}
$$

or equivalently

$$
Y_{in}(\lambda) = \lim_{Y_L \to 0} Y_0 \frac{Y_L + Y_0\lambda}{Y_0 + Y_L\lambda} = \lambda C
$$

where the capacitance is given by

$$
C = \frac{1}{Z_0} = Y_0
$$

$Y_0 = 1/Z_0$ is the characteristic admittance.

The open form of the admittance equation is given by

$$
Y_{in}(\lambda) = \lambda C = j\Omega C = j[Y_0 \tan(\omega\tau)]
$$

This is a typical capacitive admittance defined in the transform frequency domain $\Omega = \tan(\omega\tau)$. However, one should never forget that, here, we are dealing with transmission lines. The frequency Ω is expressed as a tangent function of the actual frequency ω. In this case, if $\omega\tau$ is small enough, then one can approximate $\tan(\omega\tau)$ as $\tan(\omega\tau) \approx \omega\tau$. Hence, a lumped element capacitor C_{lump} may be approximated with a short-length open-end transmission line of characteristic admittance Y_0 such that

$$
j[Y_0 \tan(\omega\tau)] \approx j\omega[Y_0\tau] = j\omega C_{lump}
$$

Hence,

$$
\boxed{C_{lump} \cong \tau Y_0}
\tag{3.35}
$$

By short-length TEM line, what we mean is that the physical length L of the line must be much smaller than that of the wavelength λ of the operating frequency. Therefore, we assume that $\omega\tau = \beta L \ll \beta\lambda$. For many practical situations, it may be sufficient to choose the physical length of the line as

$$L < 0.1\lambda \quad \lambda = \frac{v_p}{f}$$

Certainly, selection of $L \cong 0.01\lambda$ is a good choice.

Now let us consider an example to compute the capacitor of a short line.

Example 3.1: At $f = 1$ GHz operating frequency, compute the capacitance of an open-ended short transmission line of length $L \cong 0.01\lambda$ with characteristic impedance $Z_0 = 20\ \Omega$.

Solution: If $f = 1$ GHz then

$$L = 0.01 \times \frac{3 \times 10^8}{10^9} = 0.003 \text{ m} = 3 \text{ mm}$$

is a good choice for realizing the capacitor as an open-ended TEM line. In this case, the delay length is approximately

$$\frac{3 \times 10^{-3}}{3 \times 10^8} = 1 \times 10^{-11}\text{s} = 10 \text{ picoseconds (PS)}$$

Hence, the lumped equivalent of the capacitance is given by[4]

$$C_{Lump} = 10Y_0 \text{ pF}$$

By setting $Y_0 = 1/20 = 0.05$ we have[5]

$$C_{lump} = 10 \times 0.05 \text{ pF} = 0.5 \text{ pF} \text{ at } 1\text{ GHz}$$

In fact, considering a wide microstrip patch, a short-length open-ended transmission line can be considered as a parallel plate capacitor as shown in Figure 3.4

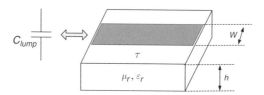

Figure 3.4   An short-length open-ended microstrip line as a capacitor

For the above structure, we can roughly estimate the characteristic admittance Y_0 as follows. The value of the capacitance is given by

---

[4] Note that, in order to get a big value of the capacitance, the characteristic admittance must be big or, equivalently, the characteristic impedance must be as small as possible. For a selected dielectric material, employing microstrip line technology, we can safely achieve a small characteristic impedance of $Z_0 = 20\ \Omega$. Low-value characteristic impedances exhibit a fairly wide patch on microstrips.

[5] Here, as the velocity of propagation we used the free space velocity of light, but it should be modified depending on the selected dielectric material as $v_p = c/(\sqrt{\mu_r \varepsilon_r})$.

$$C_{lump} = \varepsilon \frac{W \times L}{h} \cong \tau Y_0$$

Hence,

$$Y_0 \cong \varepsilon \left(\frac{W}{h}\right)\left(\frac{L}{\tau}\right)$$

where h is the thickness of the substrate. It should be noted that, usually, the permeability μ of a good dielectric material is equal to $\mu_r \mu_0 \cong \mu_0$ (i.e. $\mu_r \cong 1$). On the other hand, the phase velocity is $v_p = L/\tau = 1/\sqrt{\mu\varepsilon}$ and the characteristic impedance of free space is $\eta_0 = (\mu_0/\varepsilon_0 \cong 120\pi)$.

With these figures, the characteristic admittance can be approximated as

$$Y_0 \cong \sqrt{\frac{\varepsilon}{\mu}}\left(\frac{W}{h}\right) = \sqrt{\varepsilon_r \eta_0^{-1}}\left(\frac{W}{h}\right)$$

Or a rough estimate of the characteristic impedance is given by

$$Z_0 \cong \frac{120\pi}{\sqrt{\varepsilon_r}}\left(\frac{h}{W}\right) \qquad (3.36)$$

(for a short-length open-ended TEM line).

Example 3.2: For an aluminum substrate of thickness $h = 0.635$ mm, compute the top-surface gold-conducting patch W width for a characteristic impedance of $Z_0 = 20\,\Omega$.

Solution: For aluminum $\varepsilon_r \cong 9.8$. Then,

$$\left(\frac{h}{W}\right) = \frac{\sqrt{\varepsilon_r}Z_0}{120\pi} = \frac{3.1305 \times 20}{120 \times \pi} = 0.1661$$

or

$$W = \frac{0.635}{0.1661} = 3.8235 \text{ mm}$$

A Shorted TEM Line as an Inductor

A shorted TEM line (i.e. $Z_L = 0$) acts like an inductance L in the λ domain such that

$$\lim_{Z_L \to 0} Z_{in}(\lambda) = Z_0 \frac{Z_L + Z_0\lambda}{Z_0 + Z_L\lambda} = \lambda L$$

where the inductance L is given by

$$L = Z_0$$

As shown in the previous section, we should remember that we are dealing with a transmission line. Therefore, the input impedance of a shorted transmission line is expressed as function of $\Omega = \tan(\omega\tau)$. In this case, a lumped inductance can be realized as the input impedance of a small-length transmission line when it is shorted at the other end.

Hence,

$$Z_{in} = jZ_0 \tan(\omega\tau) \cong j\omega[\tau Z_0] = jL_{lump}$$

or

$$L_{lump} \cong \tau Z_0$$
where τ is the delay length of the line which is
selected as the time delay of the physical length of
$L \cong 0.01\lambda = 0.001v_p/f.$

(3.37)

As can be seen from this equation, a high-value inductance can be realized with a high-value characteristic impedance Z_0 of a shorted small-length line.

In practice, a shorted line can be realized on a dielectric substrate as a very thin-plate conductor of width W as a microstrip conducting patch as shown in Figure 3.5.

$$L_{plate} = \mu\left(\frac{h}{w}\right)L$$

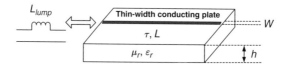

Figure 3.5   Realization of an inductor as a thin-width microstrip conducting patch

The series inductance between conducting plates h = meters apart of length L can roughly be estimated as in Section 3.1 such that

$$L_{plate} = \mu\left(\frac{h}{w}\right)(L)$$

On the other hand, this inductance must be equal to the input impedance of the shorted line. Therefore,

$$L_{lump} = Z_0\tau = L_{plate} = \mu\left(\frac{h}{w}\right)(L)$$

Using $v_p = L/\tau$, and $\eta_0 = \mu_0/\varepsilon_0$, the characteristic impedance of the line is roughly estimated as

$$Z_0 \cong \frac{\eta_0}{\sqrt{\varepsilon_r}}\left(\frac{h}{W}\right) = \frac{120\pi}{\sqrt{\varepsilon_r}}\left(\frac{h}{W}\right)$$

(3.38)

Let us consider a practical example to compute the possible high value of an inductance manufactured on an alumina substrate.

Example 3.3: On alumina, we can safely build a high characteristic impedance of $Z_0 = 100\ \Omega$. For the sake of simplicity, let us choose the relative dielectric constant as $\varepsilon_r \cong 9$ and relative permeability $\mu_r \cong 1$. In this case, the propagation or phase velocity is given by $v_p \approx 3 \times 10^8/\sqrt{9} = 10^8\ \text{m/s}$. At, $f = 1\text{GHz}$ the wavelength is given by $\lambda = v_p/f = 0.1\text{m} = 10\ \text{cm}$. Then, let us choose the physical length as $L = 0.01\lambda = 10^{-3}\ \text{m}$. The time delay of the line is $\tau = L/v_p = 10^{-11}\text{s}$.[6] Thus, the characteristic impedance of $Z_0 = 100\Omega$ yields an equivalent lumped inductance of $L_{lump} = 100 \times 10^{-11} = 10^{-9}\ \text{H} = 1\text{nH}$.

For this example, the estimated (h/W) is computed as

$$\left(\frac{h}{W}\right) = \frac{\sqrt{\varepsilon_r}}{120\pi}Z_0 = \frac{3 \times 100}{120\pi} = 0.7958 \cong 0.8$$

It should be noted that all the above derivations are carried out for ideal parallel plate TEM lines which may approximate the behavior of microstrip lines. However, microstrip lines are not ideal. Regarding the practical implementation of lumped element inductors and capacitors, the reader is encouraged to refer to the books by Inder Bahl and Prakash Bhartia.[7]

A Quarter-Wavelength TEM Line at Resonance Frequency

On an ideal TEM line, when the phase shift is $\varphi = \beta L = \pi/2 = (2\pi/\lambda)L$ or $L = \lambda/4$ then the Richard variable becomes infinite (i.e. $\lambda = j\tan(\beta L) = \infty$). In this case, the input impedance Z_{in} becomes

$$Z_{in}(\lambda) = \lim_{\lambda \to \infty} Z_0 \frac{Z_L + Z_0\lambda}{Z_0 + Z_L\lambda} = \frac{Z_0^2}{Z_L}$$

or

$$\boxed{Z_{in}Z_L = Z_0^2} \qquad (3.39)$$

From the design point of view, Equation (3.39) may be utilized to adjust the input impedance of the line for matching purposes, as in a transformer, by playing with the characteristic impedance Z_0.

So far, we have investigated the behavior of open- and short-ended transmission lines with very small physical lengths compared to the wavelength of operations. However, we can directly investigate the input impedance or equivalently the input admittance of Equation (3.34) by varying the termination immittances (i.e. impedance Z_L or admittance Y_L) starting from small values down to zero. This way, we will have the chance to investigate the limit cases and, perhaps, develop simple lumped circuit models for TEM lines which are basic resonance circuits. These models may also be utilized to understand the basic operation of dipole antennas as open- or short-ended TEM lines.

Let us first work on the limit case where the termination impedance approaches infinity (i.e. $Z_L \to \infty$) or equivalently the termination admittance approaches zero, which is much easier to consider (i.e. $Y_L \to 0$).

---

[6] Note that, when the delay length of the line L is selected as the fraction δ of the wavelength λ, then the time delay is given by $\tau = \delta/f$ which is independent of the propagation velocity. For the example under consideration, $\delta = 0.01$ and $f = 1\ \text{GHZ}$ are selected. Thus, $\tau = 10^{-11}\ \text{s}$ is found.

[7] Inder Bahl, *Lumped Elements for RF and Microwave Circuits*, Artech House, 2003. Inder Bahl and Prakash Bhartia, *Microwave Solid State Circuit Design*, Wiley-Interscience, 2003.

Open-Ended TEM Line of Arbitrary Length

Using normalized admittance values with respect to characteristic admittance Y_0 (i.e. $Y_0 = 1$selected) and setting $Y_L = 0.1$, we can plot the real and imaginary parts of input admittance $Y_{in}(j\omega\tau) = G_{in}(\omega\tau) + jB_{in}(\omega\tau)$ as shown in Figure 3.6 and Figure 3.7 respectively. Close examination of these figures reveals that the operation of a low-admittance loaded or shorted transmission line (i.e. normalized $Y_L = 0.1$) resembles the operation of a series resonance circuit as shown in Figure 3.8. As Y_L approaches zero we reach the ideal case; that is to say, we reach the lossless series resonance circuit configuration. In this case, as determined above, the equivalent lumped element capacitance, inductance and resistance are given by $C_{lump} = \tau Y_0$, $L_{lump} = \tau Z_0$, and a series resistance

$$r = \frac{1}{G_{in}\left(\dfrac{\pi}{2}\right)} = \frac{1}{Y_L}$$

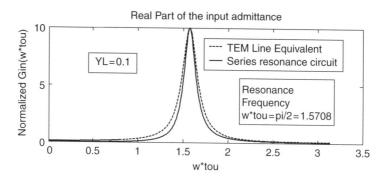

Figure 3.6   Real part of the input admittance

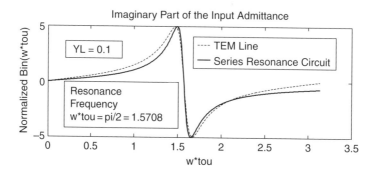

Figure 3.7   Imaginary part of the input admittance

respectively. The resonance frequency of the line is located at $\omega\tau = \pi/2$ or the resonance circuit resonates at $\pi/2\tau$.

Now let us consider a half-wavelength dipole antenna. This structure can be obtained by opening wide the arms of the TEM line as shown in Figure 3.8. This antenna should act like a series resonance circuit and resonate at the resonance frequency such that angular resonance frequency is given by

$$\omega = 2\pi f = \frac{\pi}{2\tau}$$

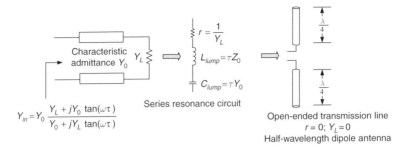

$$Y_{in} = Y_0 \frac{Y_L + jY_0 \tan(\omega\tau)}{Y_0 + jY_L \tan(\omega\tau)}$$

Figure 3.8 Low-admittance loaded transmission line, its lumped element and open-ended transmission line as half-wave length dipole antenna

or the actual resonance frequency is

$$f = \frac{1}{4\tau}$$

At the resonance frequency, obviously we must see a resistive input admittance which is associated with the series loss resistance of the wires specified by $r = Y_0^2/Y_L = 1/Y_L$ (normallized with respect to Y_0). However, for the antenna case, at resonance, the input resistance must also include the radiation resistance of the dipole antenna.

Shorted TEM Line of Arbitrary Length

In a similar manner to that for open-ended lines, it is straightforward to show that a shorted TEM line acts like a parallel resonance circuit at an operating frequency of $f = 1/4\tau$. In this case, it is appropriate to work with the input impedance.

Now, let us normalize all the impedances with respect to the characteristic impedance Z_0. Then, as above, we set $Z_0 = 1$ and the termination resistance to a small value such as $Z_L = 0.1$ to investigate the impedance variation versus frequency. Under these assumptions, we developed a MATLAB® program to analyze the input impedance of

$$Z_{in} = Z_0 \frac{Z_L + jZ_0 \tan(\beta l)}{Z_0 + jZ_L \tan(\beta l)}$$

called TRLINE as listed at the end of this chapter. This program was tested for different values of Z_L. As Z_L approaches zero, the fit between the line impedance and the model impedance improves, as expected.

Hence, we end up with the variation of the real part R_{in} and the imaginary part X_{in} of the input impedance as a function of $\omega\tau$ as depicted in Figures 3.9 and 3.10 respectively. On the same figures, we also plot the parallel resonance circuit model for the line under consideration. As in the previous subsection, the circuit elements of the parallel resonance circuit are given as follows:

$$
\begin{array}{lll}
\text{resonance frequency} & f = \dfrac{1}{4\tau} & \\[2mm]
\text{parallel capacitance} & C_p = \tau Y_0 & \\[2mm]
\text{parallel inductance} & L_p = \tau Z_0 & \\[2mm]
\text{parallel resistance} & R_p = \dfrac{Z_0^2}{Z_L} = \dfrac{1}{Z_L} &
\end{array}
\qquad (3.40)
$$

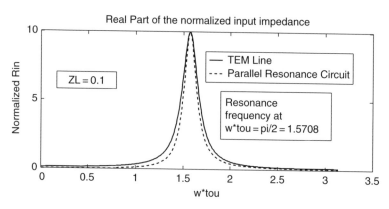

Figure 3.9 Real part of the input impedance when it is terminated in a normalized load impedance $Z_L = 0.1$

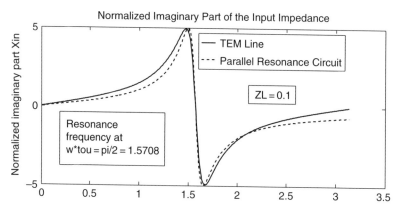

Figure 3.10 Normalized imaginary part of the input impedance of a TEM line which is terminated at $Z_L = 0.1$

We should also mention that the parallel resonance circuit model can also be used to model loop antennas of arbitrary length as shown in Figure 3.11.

The antenna resonates at a frequency of $f = 1/4\tau$ and it sees a resistive input impedance R_{in} which is partially related to the loss of the wires specified by $R_P Z_0^2/Z_L = 1/Z_L$ (normalized with respect to Z_0) and the radiation resistance.

Remarks: tan($\omega\tau$) as a Foster Function

1. We should note that for an open- or short-ended transmission line the input immittance $F_{in} = jX_{in}(\omega) = j\tan(\omega\tau)F_0 = j\Omega F_0$ is a Foster function for all ω since

$$\frac{dX_{in}(\omega)}{d\omega} = \frac{[\tau F_0]}{\cos^2(\omega\tau)} \geq 0, \quad \forall \omega.$$

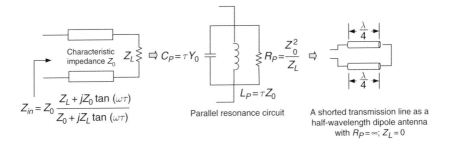

$$Z_{in} = Z_0 \frac{Z_L + jZ_0 \tan(\omega\tau)}{Z_0 + jZ_L \tan(\omega\tau)}$$

Parallel resonance circuit

A shorted transmission line as a half-wavelength dipole antenna with $R_P = \infty$; $Z_L = 0$

Figure 3.11   Shorted transmission line model as a parallel resonance circuit which resembles a shorted dipole antenna

Therefore, it is lossless. In this nomenclature, the symbol F_0 denotes either the characteristic impedance Z_0 or characteristic admittance Y_0.

2. Over the interval $0 \leq \omega\tau < 3\pi/4$, the function $X(\omega) = \tan(\omega\tau)$ exhibits a resonance circuit type of immittance variation with the resonance frequency placed at $\omega\tau = \pi/2$ as shown in Figure 3.12. Therefore, it may be meaningful to build approximate models for open or shorted transmission lines employing resonance circuits as we have shown in the above sections. However, one should bear in mind that, at the operating frequency f, the line delay length τ should not violate the inequality of $\omega\tau < 3\pi/4$. Therefore, it may always be appropriate to choose the fixed length of the transmission line to satisfy the inequality of $\leq \omega\tau < 3\pi/4$.

We would like to draw the reader's attention to Figures 3.7 and 3.10. These are plots of the imaginary parts of the input immittance. Close examination of these figures reveals that, for TEM lines, the zero crossing point is at $\omega\tau = 3\pi/4 \cong 2.356$. On the other hand, for resonance structures, the imaginary parts of the immittance continue to infinity below zero as shown in Figure 3.12.

3. The resonance circuit types of models of transmission lines provide a very rough fit between the model and the actual data. In practice, we prefer to work with precise models which probably include three or more reactive elements. In Chapter 12, we present an elegant procedure to model the measured data as a positive real function which in turn yields the desired circuit.

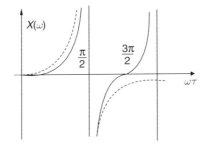

Figure 3.12   Frequency behavior of the $X(\omega) = \tan(\omega\tau)$ function. Dashed lines denote the immittance variation of a resonance circuit

3.10 Ideal TEM Lines with No Reflection: Perfectly Matched and Mismatched Lines

Reflection-free TEM lines have special importance in designing power transfer networks.

When the line is terminated in complex load impedance $Z_L(j\omega)$, the average power delivered to this load is expressed as

$$P_L = \text{real } \{V_L(j\omega)I_L(-j\omega)\}$$

As shown earlier, in terms of the voltage incident and reflected waves, the voltage and current are given by

$$V_L(j\omega) = A(0, j\omega) + B(0, j\omega) = A(0, j\omega)[1 + S_L(j\omega)]$$
$$I_L(j\omega) = \frac{1}{Z_0}[A(0, j\omega) - B(0, j\omega)] = \frac{A(0, j\omega)}{Z_0}[1 - S_L(j\omega)]$$

with

$$S_L(j\omega) = \frac{B(0, j\omega)}{A(0, j\omega)} = \frac{Z_L - Z_0}{Z_L + Z_0} \quad Z_L = \frac{V_L}{I_L} = Z_0\frac{1 + S_L}{1 - S_L}$$

For lossless TEM lines, the characteristic impedance is real. In this case, $Z_0(j\omega) = Z_0(-j\omega)$. and it is a positive real number. Therefore,

$$P_L = \text{real}\{V_L(j\omega)I_L(-j\omega)\} = \frac{|A(0, j\omega)|^2}{Z_0}\left\{\left[1 - \left|\frac{B(0, j\omega)}{A(0, j\omega)}\right|^2\right]\right\}$$

or

$$P_L = \left[\frac{|A(0, j\omega)|^2}{Z_0} - \frac{|B(0, j\omega)|^2}{Z_0}\right]$$
$$P_L = \frac{|A(0, j\omega)|^2}{Z_0}[1 - |S_L|^2]$$

(3.41)

It should be noted that, in Equation (3.41),

$$0 \le |S_L|^2 = \frac{(R_L - Z_0)^2 + (X_L)^2}{(R_L + R_0)^2 + (X_L)^2} \le 1.$$

Therefore, the maximum value of the average load power is obtained when $S_L = 0$, which requires $Z_L = Z_0$. Then, the maximum of the average power delivered to the load is obtained as

$$(P_L)_{max} = \frac{|A(0, j\omega)|^2}{Z_0}$$

(3.42)

Furthermore, at the generator end, if $Z_G = Z_0$ then Equation (3.34) yields

$$Z_{in} = Z_0$$
$$Z_{out} = Z_0$$

(3.43)

Definition 1: An ideal TEM line which satisfies Equation (3.43) is called a perfectly matched line at the input and the output respectively.

For a loss-free line, at the generator end, the amplitude of the incident wave $|A(l, j\omega)|$ must be equal to $|A(0, j\omega)|$. Therefore, at the generator end, the maximum deliverable input power is expressed as

$$(P_{in})_{max} = \frac{|A(l, j\omega)|^2}{Z_0} = \frac{|A(0, j\omega)|^2}{Z_0} \tag{3.44}$$

Definition 2: We should indicate that $(P_{in})_{max}$ is also known as the available power of the generator and denoted by P_A.

For a loss-free and reflection-free TEM line, the above electrical conditions are pictured by means of Thévenin equivalent circuits at the input and output ends of the TEM line as shown in Figure 3.13.

$$(P_{in})_{max} = \frac{|A(j, j\omega)|^2}{Z_0} = \frac{|A(0, j\omega)|^2}{Z_0} = P_A = \frac{|V_G|^2}{4Z_0}$$

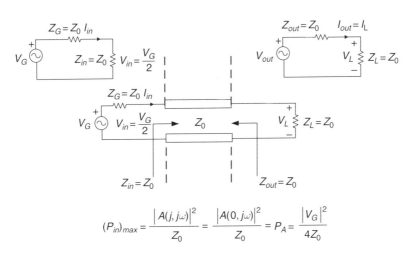

$$(P_{in})_{max} = \frac{|A(j, j\omega)|^2}{Z_0} = \frac{|A(0, j\omega)|^2}{Z_0} = P_A = \frac{|V_G|^2}{4Z_0}$$

Figure 3.13   Thévenin equivalent circuits of the input and output ends of the line

Hence, by Equation (3.41), it is seen that

$$P_L = P_A[1 - |S_L|^2] \tag{3.45}$$

Definition 3: A TEM transmission line can be regarded as a two-port which has a physical length l. In this case, the transducer power gain (TPG) of this two-port can be defined as the ratio of the power delivered to the load (P_L) to the available power of the generator (P_A). Thus, by definition,

$$\text{TPG} \triangleq \frac{P_L}{P_A} = 1 - |S_L|^2 \tag{3.46}$$

Consequently, we can make the following statements and definitions.

Statement 1: When a TEM line is terminated in its characteristic impedance (i.e. $Z_L = Z_0$), then the refection coefficient S_L vanishes (i.e. $S_L = 0$), which makes the line free of reflections. Hence, the reflected voltage wave at the load end $B(0, j\omega)$ becomes zero, which requires no reflection on the line. Therefore, $B(z, j\omega)$ must be zero for all z.

Statement 2: If the line is reflection free and lossless, in other words, if the ideal TEM line is terminated in its characteristic impedance, then the average power delivered to the load is maximum, and by Equation (3.46) it is given as

$$(P_L)_{max} = \frac{|A(0, j\omega)|^2}{Z_0} = P_A, \quad Z_0 = Z_L$$

which yields

$$|A(0, j\omega)|^2 = |V_L(j\omega)|_{max}^2$$

Definition 4: For a perfectly matched line, the input incident wave power $|a_{in}|^2$ is defined as

$$P_A \triangleq |a_{in}|^2 = P_{incident} = \frac{|A(-l, j\omega)|^2}{Z_0} \tag{3.47}$$

If the TEM line is mismatched at the input, i.e. if $Z_{in} \neq Z_0$, then the total power delivered to the input (P_{in}) is given in a similar manner to that of Equation (3.41):

$$
\begin{aligned}
&P_{in} = \text{real}\{V_{in}(j\omega)I_{in}(-j\omega)\} \\
\text{or} \\
&P_{in} = \frac{|V_{in}|^2}{Z_0} \\
\text{or} \\
&P_{in} = \frac{1}{Z_0}[A(-l, j\omega) + B(-l, j\omega)][A(-l, -j\omega) + B(-l, -j\omega)] \\
&P_{in} = \frac{|A(-l, j\omega)|^2}{Z_0} - \frac{|B(-l, j\omega)|^2}{Z_0}
\end{aligned}
\tag{3.48}
$$

Definition 5: Equation (3.48) suggests that the reflected wave power can be defined as

$$P_{reflected} \triangleq |b_{in}|^2 \triangleq \frac{|B(-l, j\omega)|^2}{Z_0} \tag{3.49}$$

Statement 3: Based on Definitions 4 and 5, we can say that the power delivered to the input port of the transmission line is equal to the difference between the incident and reflected powers. That is,

$$P_{in} = |a_{in}|^2 - |b_{in}|^2 \tag{3.50}$$

Based on the definitions of incident and reflected powers, Equation (3.48) may be rewritten as

$$P_{in} = \frac{1}{Z_0} [A(-l,j\omega) + B(-l,j\omega)][A(l,-j\omega) + B(-l,-j\omega)]$$

Or, emphasizing the positive realness of the characteristic impedance Z_0 by setting $R_0 = Z_0$,

$$P_{in} = \text{real}\left\{ \left[\frac{1}{\sqrt{R_0}}A(-l,j\omega) + \frac{1}{\sqrt{R_0}}B(-l,j\omega) \right] \left[\frac{1}{\sqrt{R_0}}A(-l,-j\omega) - \frac{1}{\sqrt{R_0}}B(-l,-j\omega) \right] \right\}$$

or

$$P_{in} = \left[\frac{V_{in}}{\sqrt{R_0}} \right] \left[\frac{V_{in}}{\sqrt{R_0}} \right]^{*}$$

where $*$ denotes the complex conjugate of a complex quantity.

(3.51)

Then, Equation (3.51) suggests the following definition for voltage-based normalized quantities.

Definition 6 (Normalized input voltage v_{in}): Referring to Equation (3.51), the quantity

$$v_{in} = \frac{V_{in}}{\sqrt{R_0}} = \frac{1}{\sqrt{R_0}}A(-l,j\omega) + \frac{1}{\sqrt{R_0}}B(-l,j\omega)$$

(3.52)

is called the normalized input voltage with respect to the normalization resistance or number R_0.

The normalized voltage definition above leads us to define voltage-based normalized incident a_{in} and reflected b_{in} waves as follows.

Definition 7 (Voltage-based incident wave a_{in}): The quantity

$$a_{in} = \frac{A(l,j\omega)}{\sqrt{R_0}}$$

is called the voltage-based normalized incident wave or, in short, the 'voltage-based incident wave'

(3.53)

Definition 8 (Voltage based reflected wave b_{in}): The quantity

$$b_{in} = \frac{B(-l,j\omega)}{\sqrt{R_0}}$$

is called the voltage-based normalized reflected wave or, in short, the 'voltage-based reflected wave'.

(3.54)

These definitions of normalized incident and reflected waves lead us to the following statement.

Statement 4: The normalized input voltage v_{in} is expressed as the sum of incident and reflected waves such that

$$v_{in} = a_{in} + b_{in}$$

(3.55)

The input current I_{in} of the transmission line can also be expressed in terms of the actual input and reflected waves as

$$I_{in} = \frac{1}{R_0} [A(l, j\omega) - B(-l, j\omega)]$$

Multiplying both sides of the above equation by R_0, we obtain

$$\sqrt{R_0} I_{in} = \frac{\sqrt{R_0}}{R_0} [A(l, j\omega) - B(-l, j\omega)]$$
$$= \frac{A(l, j\omega)}{\sqrt{R_0}} - \frac{B(-l, j\omega)}{\sqrt{R_0}}$$

or

$$\sqrt{R_0} I_{in} = a_{in} - b_{in}$$

(3.56)

Definition 9 (Normalized input current i_{in}): Equation (3.56) leads to a definition of the normalized input current. The quantity

$$i_{in} = \sqrt{R_0} I_{in}$$

(3.57)

is called the normalized input current with respect to the normalization number R_0.

Statement 5: The above definition of input current reveals that the normalized input current of a TEM line can be expressed as the difference of the incident and reflected waves such that

$$i_{in} = a_{in} - b_{in}$$

(3.58)

Statement 6: Note that all the above definitions can be made for the output port of a TEM line when it is terminated Z_G at the input end and driven by a voltage source V_L of internal impedance Z_L at the back end, as shown in Figure 3.14.

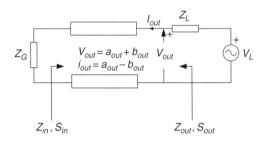

Figure 3.14   TEM line driven on the back end

In this case, normalized output voltage v_{out} and output current i_{out} are described in terms of the voltage-based output incident a_{out} and reflected b_{out} waves as follows:

$$v_{out} = a_{out} + b_{out}$$

(3.59)

and

$$i_{out} = a_{out} - b_{out}$$

(3.60)

Or, equivalently, output incident and reflected waves are given in terms of the measured output voltage and current as

$$a_{out} = \frac{1}{2}\left[\frac{V_{out}}{\sqrt{R_0}} + \sqrt{R_0}I_{out}\right] = \frac{1}{2}[v_{out} + i_{out}]$$

$$b_{out} = \frac{1}{2}\left[\frac{V_{out}}{\sqrt{R_0}} - \sqrt{R_0}I_{out}\right] = \frac{1}{2}[v_{out} - i_{out}]$$

(3.61)

Statement 7: Based on the above definitions and derivations, in terms of voltage-based normalized incident and reflected waves, the actual powers delivered to the input and output ports of a TEM line are expressed as

$$P_{in} = |a_{in}|^2 - |b_{in}|^2 = |a_{in}|^2[1 - |S_{in}|^2]$$

and

$$P_{out} = |a_{out}|^2 - |b_{out}|^2 = |a_{out}|^2[1 - |S_{out}|^2]$$

where

$$S_{in} = \frac{Z_{in} - R_0}{Z_{in} + R_0}$$

and

$$S_{out} = \frac{Z_{out} - R_0}{Z_{out} + R_0}$$

such that by Equations(3.31) and (3.32)

$$Z_{in}(\lambda) = R_0\frac{Z_L + R_0\lambda}{R_0 + Z_L\lambda}$$

$$Z_{out}(\lambda) = R_0\frac{Z_G + R_0\lambda}{R_0 + Z_G\lambda}$$

(3.62)

In this chapter we focused on the extension of electromagnetic wave theory, summarizing the basic properties of transverse electric and magnetic field lines or, in short, TEM lines. The next chapter is devoted to circuits built using equal length ideal TEM lines, or the so-called commensurate lines.

MATLAB Program TRLINE: Program Listing 3.1 Computation of the input immittance of a transmission line of arbitrary length

```
% TEM Line Input impedance computation
%
%                     ZL + jZ0*tan(beta*L)
%          Zin = Z0  ----------------------
%                     Z0 + jZL*tan(beta*L)
%
% Note that phase = beta*L; varies between 0 and pi
```

Program Listing 3.1 (Continued)

```
% or beta*L = (w/vp)*L where vp is the phase velocity
%
% Input
%      ZL: Load termination
%      Z0: Characteristic Impedance
%      phase = beta*L = w*tou; we sweep over w*tou axis
% The phase is being scanned
%
% Output
%      Zin = Rin(w) + jXin(w)
% Program TEM Line: Computation of input impedance
ZL = 0.1    % INPUT: Normalized with respect to 50 ohms
Z0 = 1;     % INPUT: Normalized with respect to 50 ohms
N = 501;    % Sampling number
delta = pi/N; % Step size for sweeping over w*tou axis
phase = 0.0;% phase = w*tou axis

% Quarter wavelength equivalent resonance circuit components
L = 2/pi;  % Normalized inductance with respect to tou
C = L;     % Normalized Capacitance
R = 1/ZL; % Resistance at resonance frequency: w*tou = pi/2
j = sqrt(-1);
for i = 1:N
    Nin = ZL + j*tan(phase); % Numerator of Zin(w*tou)
    Din = Z0 + j*ZL*tan(phase);% Denominator of Zin(w*tou)
    Zin = Nin/Din; % TEM Line input impedance
    Rin(i) = real(Zin);
    Xin(i) = imag(Zin);
    fi(i) = phase;% array for plot purpose
% Equivalent parallel resonance circuit impedance computations
    Yres = (1/L/phase/j) + j*phase*C + ZL;% Admittance
    Zres = 1/Yres;% Impedance of the resonance circuit
    Rres(i) = real(Zres);
    Xres(i) = imag(Zres);
    Yin = 1/Zin;
    Gin(i) = real(Yin);
    Bin(i) = imag(Yin);
    phase = phase + delta;
end
figure
plot(fi,Rin,fi,Rres); % plots of real parts
figure
plot(fi,Xin,fi,Xres); % plots of imaginary parts
```

4

Circuits Constructed with Commensurate Transmission Lines: Properties of Transmission Line Circuits in the Richard Domain

4.1 Ideal TEM Lines as Lossless Two-ports

An ideal TEM line of fixed physical length l with positive real characteristic impedance Z_0 can be considered as a lossless two-port between the input and output measurement planes or, equivalently, ports as depicted in Figure 4.1.[1]

Definition 1 (Unit element (UE)): Referring to Figure 4.1, a single, fixed physical length TEM line is called a unit element (UE) in cascade configuration.

Definition 2 (Commensurate transmission line): Referring to Figure 4.3 below, circuits can be constructed by properly connecting equal length ideal TEM lines. In this context, equal length transmission lines are called commensurate.[2]

Definition 3 (A distributed circuit element): A single transmission line of physical length l and characteristic impedance Z_0 is also called a distributed circuit element. A distributed circuit element could be in cascade configuration as a UE, or it may be either shorted or open ended in series or shunt configuration.

When we deal with distributed circuit elements, impedance normalization simplifies the design and the analysis processes of circuits constructed with Commensurate transmission lines. If one utilizes a single line, then the normalization impedance may be selected as the characteristic impedance Z_0 of that line. Normalized impedances may be denoted by lower case letters and the normalization impedance can be put in parentheses. For example, let Z_{in} denote the actual input impedance of a transmission line. Then the

---

[1] Recall that an ideal TEM line is lossless and has a constant speed of electromagnetic wave propagation or, equivalently, constant phase velocity $v_p = (L/C)^{(1/2)}$ over the entire frequency band.
[2] The dictionary meaning of 'commensurate' is 'equal'.

Design of Ultra Wideband Power Transfer Networks   Binboga Siddik Yarman
© 2010 John Wiley & Sons, Ltd

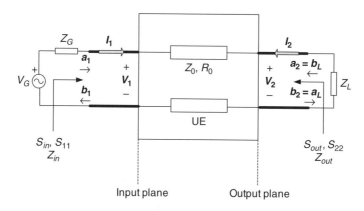

Figure 4.1 Transmission line as a unit element

normalized input impedance is denoted by z_{in} or $z_{in}(Z_0)$. Assume that the line terminations are resistive such that $Z_G = R_G$ and $Z_L = R_L$. In this case, in terms of the normalized termination r_L, the normalized input impedance z_{in} with respect to normalization Z_0 is given by Equation (3.31) as

$$z_{in} = \frac{r_L + \lambda}{1 + r_L \lambda} \tag{4.1}$$

where

$$z_{in} = \frac{Z_{in}}{Z_0}$$

$$r_L = \frac{R_L}{Z_0} \tag{4.2}$$

and Z_{in}, R_L and Z_0 are the actual input, load and characteristic impedances respectively.

The Input reflection coefficient

$$S_{in} = \frac{z_{in} - 1}{z_{in} + 1}$$

reveals that

$$S_{in}(\lambda) = \left[\frac{1 - \lambda}{1 + \lambda} \right] S_L$$

where S_L is the load reflection coefficient and given by

$$S_L = \frac{r_L - 1}{r_L + 1} \tag{4.3}$$

and

$$e^{-2\gamma l} = \frac{1 - \lambda}{1 + \lambda}$$

It is interesting to note that the even part of z_{in} is determined by $r_{in}(\lambda) = \frac{1}{2}\left[z_{in}(\lambda) + z_{in}(-\lambda)\right]$. Thence, it is found as

$$r_{in}(\lambda) = r_L \frac{1 - \lambda^2}{1 - r_L\lambda^2} \tag{4.4}$$

4.2 Scattering Parameters of a TEM Line as a Lossless Two-port

In circuit theory, the description of one-ports (such as a two-terminal circuit component like an inductor, a capacitor or a resistor) or two-ports can be electrically identified in terms of the responses to applied excitations. For example, if an excitation is specified as a voltage source $v(t)$ or $V(j\omega)$, then the response is measured as a current $i(t)$ or $L(j\omega)$ respectively.[3] Similarly, if the excitation is specified as the voltage-based incident wave a, then the response is the reflected wave b.

In this regard, for the transmission line of Figure 4.1, excitation of the input port may be selected as the normalized incident wave $a_1 = a_{in}$ and then the response is specified as the port reflected wave $b_1 = b_{in}$. Similarly, for the selected polarities at the output port, the incident wave is described by a_2 which must be the reflected wave b_L of the load reflection coefficient

$$S_L = \frac{b_L}{a_L} = \frac{Z_L - R_0}{Z_L + R_0}$$

and the reflected wave b_2 must be equal to the incident wave of the load.[4] In other words, at the output port $a_2 = b_L$ and $b_2 = a_L$.

An ideal TEM line is considered as a linear, time-invariant, reciprocal and lossless two-port. Therefore, in the λ domain its input and output responses, $b_1(\lambda)$ and $b_2(\lambda)$, can be expressed in terms of the input and the output excitations $a_1(\lambda)$ and $a_2(\lambda)$ in a linear manner as follows:

$$\begin{aligned} b_1 &= S_{11}a_1 + S_{12}a_2 \\ b_2 &= S_{21}a_1 + S_{22}a_2 \end{aligned} \tag{4.5}$$

In Equation (4.5), the quantities $\{S_{11}, S_{12}, S_{21}, S_{22}\}$ are called the real normalized scattering parameters of the lossless two-port. For the present case, the two-port is the TEM line. Literally speaking, the measured scattering parameters fully describe the linear two-ports under consideration; they are determined by means of the normalized port voltage and current measurements on the specified planes as described in the following sections. Port normalization numbers may be the same for each port, say R_0.

R_0 could be selected as an arbitrary positive number. However, it may also be the characteristic impedance Z_0. In general, one may select independent normalization numbers for each port.

At the input port, the normalization number is represented by R_1, which may be selected as the internal impedance Z_G of the generator. Similarly, at the output port, the normalization number is designated by

---

[3] Note that $V(j\omega)$ and $I(j\omega)$ denote the voltage and current phasors respectively.
[4] See Equation (3.21) of Chapter 3.

R_2 and may be selected as the load termination Z_L. Then, the incident and reflected wave definitions of Chapter 3 are generalized as

$$
\begin{aligned}
b_1 &= \frac{V_1}{\sqrt{R_1}} - \sqrt{R_1}I_1 = v_1 - i_1 \\
a_1 &= \frac{V_1}{\sqrt{R_1}} + \sqrt{R_1}I_1 = v_1 + i_1 \\
b_2 &= \frac{V_2}{\sqrt{R_2}} - \sqrt{R_2}I_2 = v_2 - i_2 \\
a_2 &= \frac{V_2}{\sqrt{R_2}} + \sqrt{R_2}I_2 = v_2 + i_2
\end{aligned}
\tag{4.6}
$$

Similarly, the input and output voltages and currents are given in terms of incident and reflected waves:

$$
\begin{aligned}
v_1 &= \frac{1}{2}[a_1 + b_1] \\
i_1 &= \frac{1}{2}[a_1 - b_1] \\
v_2 &= \frac{1}{2}[a_2 + b_2] \\
i_2 &= \frac{1}{2}[a_2 - b_2]
\end{aligned}
\tag{4.7}
$$

- Thus, the input reflection coefficient S_{11} is measured as $S_{11} = S_{in} = b_1/a_1$ when $a_2 = 0$. This condition implies that

$$
a_2 = b_L = S_L a_L = \frac{Z_L - R_2}{Z_L + R_2} a_L = 0 \quad \text{or} \quad Z_L = R_2
$$

In other words, the Input reflection coefficient $S_{in} = S_{11}$ is measured as the ratio of $(v_1 - i_1)/(v_1 + i_1)$ when the far end of the line is terminated in its normalization number R_2.[5]

- The backward scattering parameter S_{12} is measured as the ratio of

$$
\frac{v_1 - i_1}{v_2 + i_2} = S_{12} = \frac{b_1}{a_2}
$$

when $a_1 = 0$. That is, S_{12} is measured when the line is derived from the backend while the input port is terminated in $R_1 = Z_G$, which makes $S_G = 0$. In other words, S_{12} is measured when

$$
b_G = a_1 = S_G a_G = \frac{Z_G - R_1}{Z_G + R_1} b_1 = 0
$$

This condition implies that the line is terminated at port 1 in its normalization number $R_1 = Z_G$.

---

[5] Noted that if the load Z_L is taken as the reference port, the reflected wave b_L of the load becomes the incident wave of the output port. Similarly, the incident wave a_L of the load becomes the reflected wave of the output port.

- Similarly, the forward scattering parameter $S_{21} = (b_2/a_1)|a_2 = 0$ is measured as the ratio of $(v_2 - i_2)/(v_1 + i_1)$ when the load end is terminated in $Z_L = R_2$. This condition makes the load reflection coefficient S_L zero.
- Finally, the backend reflection coefficient $S_{22} = S_{out}(b_2/a_2)|a_1 = 0$ is measured as the ratio of $(v_2 - i_2)/(v_2 + i_2)$ when the input is in its port normalization impedance R_1 forcing

$$a_1 = S_G b_1 = \frac{Z_G - R_1}{Z_G + R_1} b_1 = 0$$

which implies that $Z_G = R_1$.

4.3 Input Reflection Coefficient under Arbitrary Termination[6]

Literally speaking, under an arbitrary actual termination Z_L, the input reflection coefficient

$$S_{in} = \frac{b_1}{a_1} = \frac{z_{in} - 1}{z_{in} + 1}$$

is specified as

$$\boxed{S_{in} = S_{11} + \frac{a_2}{a_1} S_{12}} \qquad (4.8)$$

On the other hand, the load reflection coefficient is given by

$$S_L = \frac{b_L}{a_L} = \frac{a_2}{b_2} \quad \text{or} \quad b_2 = \frac{a_2}{S_L}$$

Using this form of b_2 in Equation (4.6) it is found that

$$\boxed{\left(\frac{a_2}{a_1}\right) = \frac{S_{21} S_L}{1 - S_{22} S_L}} \qquad (4.9)$$

Hence, using Equation (4.9) in Equation (4.8) it is found that

$$\boxed{S_{in} = S_{11} + \frac{S_{12} S_{21}}{1 - S_{22} S_L} S_L} \qquad (4.10)$$

---

[6] In the classical microwave engineering literature, the term 'reflection coefficient' may be replaced by the term 'reflectance'. Therefore, in this book, these terms are equivalently used with the same meaning.

4.4 Choice of the Port Normalizations

As mentioned above, in defining scattering parameters, the choice of port normalization numbers is just a matter of convenience. They may be carefully selected to facilitate the circuit design and analysis. For many practical cases, however, R_1 and R_2 are chosen to match the line terminations of two-ports. In microwave engineering, most commercial products are designed to yield $50\,\Omega$ input and output impedances. Therefore, for many applications, it may be proper to choose the normalization numbers $R_1 = R_2 = R_0$ to match the standard termination of $R_0 = 50\,\Omega$ If one is dealing with a single TEM line, then R_0 may be selected as the characteristic impedance Z_0 of that line. Let us investigate the effect of the port normalization numbers on actual incident and reflected waves.

4.5 Derivation of the Actual Voltage-Based Input and Output Incident and Reflected Waves

Referring to Figure 4.1, let the input and the output port normalization numbers be $R_1 = Z_G$ and $R_2 = Z_L$. No matter what the terminations are, the actual input voltage V_{in} and the actual input current I_{in} of the TEM line are specified by Equation (3.17) for the line length of $z = -l$ such that

$$\boxed{\begin{aligned} V_{in} &= Ae^{+\gamma l} + Be^{-\gamma l} \\ Z_0 I_{in} &= Ae^{+\gamma l} - Be^{-\gamma l} \end{aligned}} \qquad (4.11)$$

A and B are found from the above equation as follows:

$$A = \frac{\begin{vmatrix} V_{in} & e^{-\gamma l} \\ Z_0 I_{in} & -e^{-\gamma l} \end{vmatrix}}{\begin{vmatrix} e^{+\gamma l} & e^{-\gamma l} \\ e^{+\gamma l} & -e^{-\gamma l} \end{vmatrix}} = \frac{1}{2} e^{-\gamma l} [V_{in} + Z_0 I_{in}]$$

$$B = \frac{\begin{vmatrix} e^{+\gamma l} & V_{in} \\ e^{+\gamma l} & Z_0 I_{in} \end{vmatrix}}{\begin{vmatrix} e^{+\gamma l} & e^{-\gamma l} \\ e^{+\gamma l} & e^{-\gamma l} \end{vmatrix}} = \frac{1}{2} e^{+\gamma l} [V_{in} - Z_0 I_{in}]$$

Note that the actual input voltage is

$$V_{in} = Z_{in} I_{in}$$

Therefore,

$$A = \frac{1}{2} e^{-\gamma l} [Z_{in} + Z_0] I_{in}$$

and

$$B = \frac{1}{2} e^{+\gamma l} [Z_{in} - Z_0] I_{in}$$

where Z_{in} is given by Equation (3.31)

$$Z_{in} = Z_0 \frac{Z_L + Z_0 \lambda}{Z_0 + Z_L \lambda}$$

In this case,

$$[Z_{in} - Z_0] = Z_0(Z_L - Z_0)(1 - \lambda)\left[\frac{1}{Z_0 + Z_L \lambda}\right] \tag{4.12}$$

Similarly,

$$[Z_{in} + Z_0] = Z_0(Z_L + Z_0)(1 + \lambda)\left[\frac{1}{Z_0 + Z_L \lambda}\right] \tag{4.13}$$

Here, note that division of Equation (4.12) by Equation (4.13) reveals that

$$S_{in}(Z_0) = \left(\frac{1 - \lambda}{1 + \lambda}\right) S_L(Z_0)$$

where

$$S_{in}(Z_0) = \frac{Z_{in} - Z_0}{Z_{in} + Z_0}$$

and

$$S_L(Z_0) = \frac{Z_L - Z_0}{Z_L + Z_0} \tag{4.14}$$

are the input and the load reflection coefficients defined with respect to the normalization number Z_0.

The input current I_{in} is determined in terms of the excitation phasor V_G as

$$I_{in} = \frac{V_G}{Z_{in} + Z_G}$$

Using the open form of Z_{in} we obtain

$$I_{in} = V_G \frac{Z_0 + Z_L \lambda}{c + d\lambda}$$

where

$$c = Z_0[Z_G + Z_L]$$

and

$$d = Z_0^2 + Z_G Z_L \tag{4.15}$$

Furthermore, by Equation (3.28),

$$e^{+\gamma l} = \left(\frac{1+\lambda}{1-\lambda} \right)^{1/2}$$

$$e^{-\gamma l} = \left(\frac{1-\lambda}{1+\lambda} \right)^{1/2}$$

Thus, using all the above results it is found that

$$A = \left[\frac{V_G}{2} \right] [Z_0(Z_L + Z_0)] \left[\frac{\sqrt{1-\lambda^2}}{c + d\lambda} \right]$$

or

$$A = \left[\frac{V_G}{2} \right] \left[\frac{\sqrt{1-\lambda^2}}{1 + \left(\dfrac{d}{c} \right) \lambda} \right] \tag{4.16}$$

$$B = \left[\frac{V_G}{2} \right] [Z_0(Z_L - Z_0)] \left[\frac{\sqrt{1-\lambda^2}}{c + d\lambda} \right]$$

with

$$c = Z_0[Z_G + Z_L]$$
$$d = Z_0^2 + Z_G Z_L$$

Note that by Equation (4.16) the ratio of B/A results in the Z_0-based load reflection coefficient S_L such that

$$\frac{B}{A} = S_L(Z_0) = \frac{Z_L - Z_0}{Z_L + Z_0}$$

In this case, in Equation (4.16), the actual reflected voltage wave B is simplified as

$$B = [S_L(Z_0)]A \tag{4.17}$$

which confirms the definition given by Equation (3.21).

Now let us present a simple example to generate the actual voltage-based incident and reflected waves in terms of the applied input excitation phasor V_G.

Example 4.1: Let $V_G = 10 \cos(\omega t)$, $Z_0 = 10\,\Omega$ and the termination $Z_G = Z_L = 50\,\Omega$. Determine the actual voltage-based incident and reflected waves $A(\lambda)$ and $B(\lambda)$ at the load end.

Solution: The applied input excitation phasor is $V_G = 10$ V.

The characteristic impedance Z_0-based load reflection coefficient is given by

$$S_L(Z_0) = \frac{Z_L - Z_0}{Z_L + Z_0} = \frac{50 - 10}{50 + 10} = \frac{40}{60} = \frac{2}{3} = 0.6667$$

Using Equation (4.16), $c = Z_0[Z_G + Z_L] = 10 \times (50 + 50) = 10 \times 100 = 1000$ and $d = Z_0^2 + Z_G Z_L = 10 \times 10 + 50 \times 50 = 100 + 2500 = 2600$, yielding $d/c = 2600/1000 = 2.6$. Then, by Equation (4.16),

$$A = \left[\frac{V_G}{2}\right]\left[\frac{\sqrt{1-\lambda^2}}{1+\left(\dfrac{d}{c}\right)\lambda}\right] = 5\frac{\sqrt{1-\lambda^2}}{1+2.6\lambda}$$

and, by Equation (4.17),

$$B = \frac{2}{3} \times 5 \times \frac{\sqrt{1-\lambda^2}}{1+2.6\lambda} = 3.333\,33\frac{\sqrt{1-\lambda^2}}{1+2.6\lambda}$$

4.6 Incident and Reflected Waves for Arbitrary Normalization Numbers

As mentioned earlier, the normalization numbers of the ports could be different than those of the input and output terminations. In this case, port normalization numbers R_1 and R_2 are regarded as part of the measurement setup for the scattering parameters.

Let R_1 and R_2 be the port normalization numbers of the input and output of the line respectively. Input and output reflection coefficients are measured as $S_{11} = (b_1/a_1)|a_2 = 0$ and $S_{21} = (b_2/a_1)|a_2 = 0$ respectively when the line is driven at the input port by a generator of internal impedance R_1 while the other end is terminated in R_2.

Under the above conditions, the actual input current and voltage are derived as follows:

$$I_1 = \frac{V_G}{Z_{in} + R_1}$$

where

$$Z_{in} = Z_0 \frac{R_2 + Z_0\lambda}{Z_0 + R_2\lambda}$$

and

$$Z_{in} + R_1 = \frac{Z_0 R_2 + Z_0^2\lambda + Z_0 R_1 + R_1 R_2\lambda}{Z_0 + R_2\lambda}$$
$$Z_{in} + R_1 = \frac{Z_0(R_1 + R_2) + (Z_0^2 + R_1 R_2)}{Z_0 + R_2\lambda}$$

Then,

$$I_1 = V_G\left[\frac{Z_0 + R_2\lambda}{C_1 + D_1\lambda}\right]$$

where

$$C_1 = Z_0(R_1 + R_2)$$
$$D_1 = Z_0^2 + R_1 R_2$$

(4.18)

Similarly,

$$V_1 = Z_{in}I_1$$

or

$$V_1 = V_G \left[\frac{Z_0 R_2 + Z_0^2 \lambda}{C_1 + D_1 \lambda} \right] \qquad (4.19)$$

Now, we are ready to generate $a_1 = v_1 + i_1$ and $b_1 = v_1 - i_1$:

$$a_1 = \frac{1}{\sqrt{R_1}} [V_1 + R_1 I_1]$$

Note that

$$V_G = [V_1 + R_1 I_1]$$

and

$$b_1 = \frac{1}{\sqrt{R_1}} [V_1 - R_1 I_1]$$

Thus, we have

$$a_1 = \frac{V_G}{\sqrt{R_1}} \qquad (4.20)$$

and

$$b_1 = \frac{V_G}{\sqrt{R_1}} \left[\frac{Z_0(R_2 - R_1) + (Z_0^2 - R_1 R_2)\lambda}{C_1 + D_1 \lambda} \right] \qquad (4.21)$$

The Input reflection coefficient

At the output port $a_2 = 0$ is selected since it is terminated in R_2. In other words, $V_2 = -R_2 I_2$ and therefore

$$a_2 = \frac{v_2}{\sqrt{R_2}} + \sqrt{R_2} I_2 = 0$$

On the other hand,

$$b_2 = \frac{v_2}{\sqrt{R_2}} - \sqrt{R_2} I_2 = 2\frac{v_2}{\sqrt{R_2}}$$

In terms of the actual incident and reflected waves A and B, $V_2 = A + B$. Then, using Equation (4.16),

$$b_2 = \frac{2}{\sqrt{R_2}}[A + B] = \frac{2A}{\sqrt{R_2}}[1 + S_L(Z_0)]$$

or

$$b_2 = \frac{V_G}{\sqrt{R_2}}[Z_0(R_2 + Z_0)][1 + S_L(Z_0)] = \frac{\sqrt{1 - \lambda^2}}{C_1 + D_1\lambda} \tag{4.22}$$

Example 4.2: Consider a one-port resistor of $Z_L = 100\,\Omega$ driven by an ideal voltage source of $V_G(t) = 10\cos(\omega t) = \mathrm{real}\{10e^{j\omega t}\}$ V.

(a) For the normalization number $R_0 = 50\,\Omega$, find the normalized impedance z_L, load reflection coefficient S_L, the port incident and reflected waves and the input reflection coefficient S_{in}.
(b) For the normalization number $R_0 = 100\,\Omega$, find the normalized impedance z_L, the port incident and reflected waves and the input reflection coefficient S_{in}.
(c) Comment on the above results.

Solution:

(a) For $R_0 = 50\,\Omega$:

- The normalized termination with respect to R_0 is defined as $z_L(R_0 = 50\,\Omega) = Z_L/R_0 = 100/50 = 20$ (dimensionless).
- Load reflection coefficient

$$S_L(R_0 = 50\,\Omega) = \frac{z_L - 1}{z_L + 1} = \frac{20 - 1}{20 + 1} = 0.9048$$

- The normalized port voltage which is considered as the excitation of the port is given by[7]

$$v_{in}(R_0 = 50\,\Omega) = \frac{v_{in}}{\sqrt{R_0}} = \frac{10}{\sqrt{50}} = 1.4142 \text{ V}/\Omega^{1/2}$$

- The normalized port current which is considered as the response to the excitation is given as

$$i_{in}(R_0 = 50\,\Omega) = \sqrt{R_0}I_{in} \quad I_{in} = \frac{v_{in}}{R_L} = \frac{10}{100} = 0.1 \text{ A}$$

Then, $i_{in} = 50 \times 0.1 = 0.707 \text{ A}\Omega^{1/2}$.

- Thus, the incident wave $a_{in} = a_1 = v_{in} + i_{in} = (1.4142) + (0.7071) = 2.1213 \text{ V}/\Omega^{1/2}$ and the reflected wave $b_{in} = b_1 = v_{in} - i_{in} = 0.7071 \text{ V}/\Omega^{1/2}$.

---

[7] Here, in this example, we use the phasor forms of the normalized voltage and current.

- Finally, the input reflection coefficient is given by

$$S_{in}(R_0 = 50\,\Omega) = \frac{b_1}{a_1} = \frac{0.7071}{2.1213} = 0.5000$$

(b) For $R_0 = 100\,\Omega$, which is equal to the actual termination impedance Z_L, we can proceed as above:

- The normalized termination resistance is $z_L(R_0 = 100\,\Omega) = 100/100 = 1$ (dimensionless).
- Load reflection coefficient $S_L(R_0 = 100\,\Omega) = (z_L - 1)/(z_L + 1) = 0$.
- The normalized port voltage is $v_{in} = v_{in}/R_0 = 10/100 = 1.0\,\mathrm{V}/\Omega^{1/2}$.
- The normalized port current is

$$i_{in} = \sqrt{R_0}I_{in} \quad I_{in} = \frac{v_{in}}{R_L} = \frac{10}{100} = 0.1\,\mathrm{A}$$

Then, $i_{in} = 100 \times 0.1 = 1.0\,\mathrm{A}\Omega^{1/2}$.
- The incident wave is $a_{in} = a_1 = v_{in} + i_{in} = 1 + 1 = 2\,\mathrm{V}/\Omega^{1/2}$ and the reflected wave $b_{in} = b_1 = v_{in} - i_{in} = 1 - 1 = 0\,\mathrm{V}/\Omega^{1/2}$.
- Thus, $S_{in} = 0$ since $b_{in} = b_1 = 0$.

(c) Comments:

(i) Since the driving voltage excitation is ideal, it has zero internal impedance. Therefore, $V_{in} = V_G$ is selected. This situation may be very dangerous since it may create unexpectedly high input mismatch which may burn out the driving generator due to the high reflective power.

(ii) For case (i), the normalization number R_0 is different than that of the termination impedance Z_L. Therefore, we observe a considerable amount of mismatch which in turn results in reflection at the load end. The amount of mismatch may be measured by means of the amplitude of S_L or, equivalently, by means of the load voltage standing wave ratio given by Equation (3.26) as

$$\mathrm{VSWR} = \frac{1 + |S_L|}{1 - |S_L|}$$

For this case, $|S_L| = 0.908$ or equivalently $\mathrm{VSWR} = 20$. It should be mentioned that the ideal value of S_L must be zero, which corresponds to the perfect matched case.[8]

(iii) For case (ii), we see that there is no reflection since $R_0 = Z_L$, which in turn results in $S_L = 0$ or equivalently a unity value of VSWR, which corresponds to a perfect match.

Example 4.3: Consider an ideal TEM line with a fixed physical length l and characteristic impedance $Z_0 = 10\,\Omega$. Let the port normalization numbers be $R_1 = 50\,\Omega$ and $R_2 = 50\,\Omega$.

(a) Find a_1, b_1 when $a_2 = 0$.
(b) Find b_2 when $a_2 = 0$.
(c) Find the scattering parameters of the TEM line.
(d) Show that $S_{11}(\lambda)S_{11}(-\lambda) + S_{21}(\lambda)S_{21}(-\lambda) = 1$.
(e) Find the input reflection coefficient $S_{in}(R_1)$ normalized with respect to R_1.
(f) Find the input reflection coefficient $S_{in}(Z_0)$ normalized with respect to the line characteristic impedance Z_0.
(g) Find the scattering parameters with Z_0 normalization.
(h) Comment on the above results.

---

[8] As opposed to the ideal case, the worst value of S_L is unity or, equivalently, the worst value of VSWR is infinity. Under worst case condition, there should be no power transfer to the load.

Solution:

(a) Let us first calculate the relevant parameters which are essential to generate a_1, b_1 and b_2:

$$Z_0^2 - R_1 R_2 = 10 \times 10 - 50 \times 50 = 100 - 2500 = -2400$$

$$Z_0^2 + R_1 R_2 = 10 \times 10 + 50 \times 50 = 100 + 2500 = +2600$$

$$Z_0(R_2 - R_1) = 0$$

$$Z_0(R_2 + R_1) = 10 \times (50 + 50) = 1000$$

$$Z_0(R_2 + Z_0) = 10 \times (50 + 10) = 600$$

$$S_L(Z_0) = \frac{R_2 - Z_0}{R_2 + Z_0} = \frac{50 - 10}{50 + 10} = \frac{40}{60} = \frac{2}{3} = 0.6667$$

$$1 + S_L(Z_0) = 1 + \frac{2}{3} = \frac{5}{3} = 1.66667$$

$$C_1 = Z_0(R_2 + R_1) = 1000$$

$$D_1 = Z_0^2 + R_1 R_2 = 2600$$

Now, we can determine the input incident and reflected waves as follows. The input reflection wave is given by Equation (4.21):

$$b_1 = \frac{V_G}{\sqrt{R_1}} \left[\frac{Z_0(R_2 - R_1) + (Z_0^2 - R_1 R_2)\lambda}{C_1 + D_1\lambda} \right] = -\frac{V_G}{\sqrt{R_1}} \frac{2400\lambda}{1000 + 2600\lambda}$$

$$= -\frac{V_G}{\sqrt{R_1}} \frac{2.4\lambda}{1 + 2.6\lambda}$$

The input incident wave is given by Equation (4.20) as

$$a_1 = \frac{V_G}{\sqrt{R_1}}$$

(b) The TEM line is terminated in its output port normalization number R_2 which makes $a_2 = 0$. Under this condition, at the output port, the reflected wave is given by Equation (4.22) as

$$b_2 = \frac{V_G}{\sqrt{R_2}} [Z_0(R_2 + Z_0)][1 + S_L(Z_0)] \frac{\sqrt{1 - \lambda^2}}{C_1 + D_1\lambda} = \frac{V_G}{\sqrt{R_2}} [600][5/3] \frac{\sqrt{1 - \lambda^2}}{1000 + 2600\lambda}$$

$$= \frac{V_G}{\sqrt{R_2}} \frac{\sqrt{1 - \lambda^2}}{1 + 2.6\lambda}$$

(c) Hence, the scattering parameters are found as

$$S_{11} = \frac{b_1}{a_1} \Big|_{a_2 = 0} = -\frac{2.4\lambda}{1 + 2.6\lambda}$$

and

$$S_{21} = \frac{b_2}{a_1}\bigg|_{a_2=0} = \left[\sqrt{\frac{R_1}{R_2}}\right]\left[\frac{\sqrt{1-\lambda^2}}{1+2.6\lambda}\right] = \frac{\sqrt{1-\lambda^2}}{1+2.6\lambda}$$

Since the TEM line is electrically reciprocal, $S_{12} = S_{21}$, and by geometric symmetry, $S_{22} = S_{11}$. Thus,

$$S = \begin{bmatrix} -\dfrac{2.4\lambda}{1+2.6\lambda} & \dfrac{\sqrt{1-\lambda^2}}{1+2.6\lambda} \\[3mm] \dfrac{\sqrt{1-\lambda^2}}{1+2.6\lambda} & -\dfrac{2.4\lambda}{1+2.6\lambda} \end{bmatrix}$$

(d)

$$
\begin{aligned}
S_{11}(\lambda)S_{11}(-\lambda) + S_{21}(\lambda)S_{21}(-\lambda) &= \left[\frac{-2.4\lambda}{1+2.6\lambda}\right]\left[\frac{2.4\lambda}{1-2.6\lambda}\right] + \left[\frac{1-\lambda^2}{(1+2.6\lambda)(1-2.6\lambda)}\right] \\[3mm]
&= \frac{-5.76\lambda^2}{1-6.76\lambda^2} + \frac{1-\lambda^2}{1-6.76\lambda^2} = \frac{1-6.76\lambda^2}{1-6.76\lambda^2} = 1
\end{aligned}
$$

This condition is known as the losslessness condition.

As will be seen later, for any lossless two-port, the scattering parameters must satisfy the losslessness condition such that

$$S(\lambda)S^{\dagger}(\lambda) = I$$

or

$$\begin{bmatrix} S_{11}(\lambda) & S_{12}(\lambda) \\ S_{21}(\lambda) & S_{22}(\lambda) \end{bmatrix}\begin{bmatrix} S_{11}(-\lambda) & S_{21}(-\lambda) \\ S_{12}(-\lambda) & S_{22}(-\lambda) \end{bmatrix} = \begin{bmatrix} 1 & 0 \\ 0 & 1 \end{bmatrix} \tag{4.23}$$

where $\dagger$ denotes the dagger operation which takes the transpose of a para-conjugated matrix.

(e) For this part of the problem let us first compute the actual input impedance of the line when the other end is terminated in R_2:

$$Z_{in} = Z_0\frac{R_2 + Z_0\lambda}{Z_0 + R_2\lambda} = 10\frac{50 + 10\lambda}{10 + 50\lambda} = \frac{50 + 10\lambda}{1 + 5\lambda}$$

Then,

$$S_{in}(R_1) = \frac{Z_{in} - R_1}{Z_{in} + R_1} = -\frac{240\lambda}{100 + 260\lambda} = -\frac{2.4\lambda}{1 + 2.6\lambda}$$

$$S_{in}(R_1) = S_{11}$$

(f) The Z_0-based input reflection coefficient is given by

$$S_{in}(Z_0) = \frac{Z_{in} - Z_0}{Z_{in} + Z_0} = 0.6667\frac{1 - \lambda}{1 + \lambda} = S_L(Z_0)\left[\frac{1 - \lambda}{1 + \lambda}\right]$$

as it should be.

(g) When the normalization numbers are equal to the line characteristic impedance $Z_0 = R_1 = R_2$ then

$$S_{11}(Z_0) = S_{22}(Z_0) = \frac{Z_{in} - Z_0}{Z_{in} + Z_0} = \frac{Z_0 - Z_0}{Z_0 + Z_0} = 0$$

and

$$S_{21}(Z_0) = \frac{b_2(Z_0)}{a_1(Z_0)} = \frac{\sqrt{1 - \lambda^2}}{1 + \lambda} = \sqrt{\frac{1 - \lambda}{1 + \lambda}} = S_{12}(Z_0)$$

Or the scattering parameters are given by

$$S(Z_0) = \begin{bmatrix} 0 & \sqrt{\dfrac{1 - \lambda}{1 + \lambda}} \\ \sqrt{\dfrac{1 - \lambda}{1 + \lambda}} & 0 \end{bmatrix} \tag{4.24}$$

These are the reflection-free or perfectly matched UE scattering parameters.

(h) Comments:

   (i) For the problem under consideration, the line characteristic impedance is different than that of the normalization numbers of the scattering parameters. Therefore, the results obtained in this problem are quite general. Important points are summarized as follows.
  (ii) Normalization numbers are crucial in generating reflection coefficients and scattering parameters. Obviously, for different normalization numbers, different reflection coefficients and scattering parameters are obtained as in the above cases of (e) and (f).
 (iii) For a single UE in cascade configuration, the input scattering parameter can be considered as a first-order rational function such that

$$S_{11} = \frac{h(\lambda)}{g(\lambda)}$$

where
$$h(\lambda) = h_0 + h_1\lambda \quad h_0 = 0, h_1 = -2.4$$
$$g(\lambda) = g_0 + g_1\lambda \quad g_0 = 1, g_1 = 2.6$$

 (iv) On the other hand, it is striking that the denominator of $S_{21}(\lambda)$ is the same as the one given for $S_{11}(\lambda)$, which is $g(\lambda) = g_0 + g_1\lambda = 1 + 2.6\lambda$.
  (v) The numerator $f(\lambda)$ of the forward scattering parameter $S_{21} = f(\lambda)/g(\lambda)$ is irrational and given by $f(\lambda) = (1 - \lambda^2)^{n/2}$ with $n = 1$, where n designates the number of cascaded UEs. However,

$$S_{21}(\lambda)S_{21}(-\lambda) = \frac{1 - \lambda^2}{g(\lambda)g(-\lambda)}$$

Here, it is interesting to observe that the numerator polynomial $f(\lambda)f(-\lambda) = 1 - \lambda^2$ is same as the numerator polynomial of $r_{in}(\lambda)$ in Equation (4.4).
 (vi) In general, it can be shown that for n cascaded UE sections,

$$f(\lambda) = (1 - \lambda^2)^{n/2}$$
$$h(\lambda) = h_0 + h_1\lambda + h_2\lambda^2 + \ldots + h_n\lambda^n$$
$$g(\lambda) = g_0 + g_1\lambda + g_2\lambda^2 + \ldots + g_n\lambda^n$$

In this regard, it can be shown that

$$S_{11}(\lambda) = \frac{h(\lambda)}{g(\lambda)} \quad \text{and} \quad S_{21}(\lambda) = \frac{(1 - \lambda^2)^{n/2}}{g(\lambda)}$$

and by reciprocity

$$S_{12}(\lambda) = S_{21}(\lambda)$$

Furthermore, by the losslessness condition, it can be shown that

$$S_{22}(\lambda) = -\frac{h(-\lambda)}{g(\lambda)} \qquad .$$

(vii) In case (d) it was found that

$$S_{11}(\lambda)S_{11}(-\lambda) + S_{21}(\lambda)S_{21}(-\lambda) = 1$$

or, equivalently,

$$g(\lambda)g(-\lambda) = h(\lambda)h(-\lambda) + f(\lambda)f(-\lambda)$$

Similarly,

$$S_{22}(\lambda)S_{22}(-\lambda) + S_{21}(\lambda)S_{21}(-\lambda) = 1$$

The above conditions are known as losslessness conditions.
In general, the losslessness condition is stated as in Equation (4.23):

$$S(\lambda)S^\dagger(\lambda) = I$$

(viii) It is important to emphasize that for a perfectly matched UE in cascade configuration with characteristic impedance Z_0, the Z_0-based scattering matrix is specified by Equation (4.23) as

$$S(Z_0) = \begin{bmatrix} 0 & \sqrt{\dfrac{1-\lambda}{1+\lambda}} \\ \sqrt{\dfrac{1-\lambda}{1+\lambda}} & 0 \end{bmatrix}$$

(ix) On the other hand, for arbitrary normalization numbers R_1 and R_2, the scattering matrix of a UE in cascade configuration is given by

$$S(R_1, R_2) = \begin{bmatrix} \dfrac{h_0 + h_1\lambda}{g_0 + g_1\lambda} & \dfrac{\sqrt{1-\lambda^2}}{g_2 + g_1\lambda} \\ \dfrac{\sqrt{1-\lambda^2}}{g_2 + g_1\lambda} & \dfrac{h_0 + h_1\lambda}{g_0 + g_1\lambda} \end{bmatrix} \tag{4.25}$$

such that the losslessness condition of Equation (4.23) is satisfied. Details of the lossless two-ports will be given in the following chapters.

4.7 Lossless Two-ports Constructed with Commensurate Transmission Lines

As described in Section 3.9, introducing the complex Richard variable $\lambda = \Sigma + j\Omega$, it is shown that:

- a commensurate line of characteristic impedance $Z_k = L$ can be utilized as an inductor with impedance $Z_L = \lambda L$ when it is shorted; or
- it may be employed as a capacitor with admittance $Y_C = \lambda C$ such that $C = Y_k = 1/Z_K$, when it is left open.

Thus, we can create microwave circuits with commensurate lines by properly connecting the distributed elements as shown in Figure 4.2.

Regarding the connections of the commensurate lines, it is appropriate to make the following definitions.

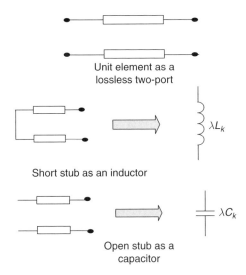

Unit element as a
lossless two-port

λL_k

Short stub as an inductor

λC_k

Open stub as a
capacitor

Figure 4.2   Distributed one-ports and a two-port consisting of a single UE in the λ domain

Definition 4 (Short stub): A shorted UE is called a short stub.

Definition 5 (Open stub): An open-ended UE is called an open stub.

Statement 1: The input impedance of a short stub of characteristic impedance $Z_k = L_k$ must exhibit an impedance of $Z_{in}(\lambda) = \lambda L_k$. In the frequency domain $\lambda = j\Omega = j \tan(\omega\tau)$, this impedance resembles a parallel resonant circuit. However, for short lines it acts like an inductor as shown in Section 3.9.

Statement 2: The input admittance of an open stub of characteristic admittance $Y_k = C_k$ must exhibit an input admittance $Y_{in}(\lambda) = \lambda C_k$. In the frequency domain $\lambda = j\Omega = j \tan(\omega\tau)$, this impedance resembles a series resonant circuit. However, for short lines it acts like a capacitor as shown in Section 3.9.

Statement 3: Distributed circuits can be constructed by properly growing the connections of UEs, open and short stubs in series and/or parallel configuration in a sequential manner as shown in Figure 4.3.

Assume that all the lines are TEM, lossless, have equal physical lengths with the same and constant propagation constant β and phase velocity ν_p (i.e. they are all commensurate) with different positive real characteristic impedancess Z_k. We will investigate the property of the input impedance $Z_{in}(\lambda)$ as we build it by connecting the distributed elements in a sequential manner as described below.

In step 1, assume that a UE with characteristic impedance Z_1 is terminated by a positive real resistance R_2. In this case, the actual input impedance is given by

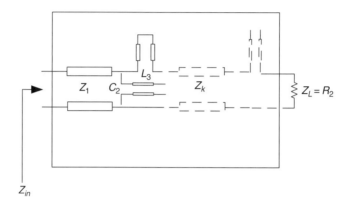

Figure 4.3 Generation of a lossless distributed two-port by connecting commensurate transmission lines

$$[Z_{in}(\lambda)]^{(1)} = Z_1 \frac{R_2 + Z_1 \lambda}{Z_1 + R_2 \lambda}$$

Equivalently, we may generate the normalized input impedance $z_{in}(Z_1)$ with respect to a characteristic impedance Z_1

$$z_{in}(Z_1) = \frac{a_0^{(1)} + a_1^{(1)} \lambda}{b_0^{(1)} + b_1^{(1)} \lambda}$$

where[9]

$$a_0^{(1)} = R_2 \geq 0 \quad a_1^{(1)} = Z_1 \geq 0$$

$$b_0^{(1)} = 1 > 0 \quad b_1^{(1)} = R_2 \geq 0$$

This impedance is positive real (PR) in the λ domain since it has a simple pole at $p = -b_0^{(1)}/b_1^{(1)}$ and a simple zero at $z = -a_0^{(1)}/a_1^{(1)}$.

Then, we can make the following statement.

---

[9] Note that it may be convenient to normalize all the above coefficients with respect to $a_0^{(1)}$.

Statement 4: The input impedance of a single UE terminated in a positive real resistor yields simple, rational, PR impedance in the λ *domain*.[10] In this context what we mean by simple is that the PR function $z_{in}(Z_1)$ has first-order numerator and denominator polynomials in the complex variable λ as above.

In step 2, we can connect a series or shunt stub to $[Z_{in}(\lambda)]^{(1)}$ or we may also connect another UE with characteristic impedance Z_2. In any case, the resulting input impedance $[Z_{in}(\lambda)]^{(2)}$ must be a PR function since the addition of two PR functions is also PR.

We can grow the circuit as above by connecting ideal UEs, open and short stubs in shunt or series configuration in a sequential manner, while the far end is terminated in the resistor R_2. In this case, the input impedance of the resulting circuit becomes a rational PR function in a complex variable $\lambda = \Sigma + j\Omega$. Hence, we can make the following general statement.

Statement 5: Referring to Figure 4.3, any lossless two-port constructed with ideal commensurate TEM lines by properly connecting UEs in cascade, series and shunt configurations, while terminating the load end in a resistor R_2, yields a PR, rational input impedance function $Z_{in}(\lambda) = Z(\lambda)/D(\lambda)$ such that

$$N(\lambda) = a_1 \lambda^m + a_2 \lambda^{m-1} + \ldots + a_m \lambda + a_{m+1}$$

$$D(\lambda) = b_1 \lambda^n + b_2 \lambda^{n-1} + \ldots + b_n \lambda + b_{n+1}$$

with

$$|n - m| = 1$$

and with the even part of $R_{in}(\lambda)$

$$R_{in}(\lambda) = \frac{1}{2} [Z_{in}(\lambda) + Z_{in}(-\lambda)]$$

or

$$R_{in}(\lambda) = F(\lambda)F(-\lambda) = \left[\frac{A(\lambda)}{D(\lambda)}\right]\left[\frac{A(-\lambda)}{D(-\lambda)}\right]$$

or in open form

$$R_{in}(\lambda) = C_{even}^2 \frac{\lambda^{2q}[1 - \lambda^2]^k}{D(\lambda)D(-\lambda)} \geq 0, \text{ for all real}\{\lambda\} \geq 0$$

where k is the total number of UEs in casacde configuration, q is the total number of series–open and shunt–short stubs,[11] and $n = q + k + r$ is total number of distributed elements such that r is the number of series–short (i.e. λL) and shunt-open (i. e. λC) stubs.[12]

---

[10] In general it can be shown that the input impedance of a UE terminated in a PR impedance is also PR.

[11] Note that these stubs act like a series capacitor and shunt inductor in the λ domain.

[12] These stubs act like a series inductor and shunt capacitor in the λ domain.

For specified q, k, n the integer r is given by $r = n - q - k$ and

$$F(\lambda) = \frac{A(\lambda)}{D(\lambda)}$$
$$A(\lambda) = C_{even}\lambda^q[1 - \lambda^2]^{k/2}$$

The proof of this statement is given by induction while $F(\lambda)$ and $F_{even}(\lambda)$ are developed in a sequential manner as shown by H. J. Carlin.[13]

4.8 Cascade Connection of Two UEs

In this section, we will investigate the input impedance property of the cascade connection of two UEs.

Referring to Figure 4.4, using a cascade connection of two UEs, we form a lossless two-port with standard terminations $R_0 = R_1 = R_2 = 50\,\Omega$. However, it is assumed that all the line characteristic impedances, and other impedances, are normalized with respect to R_0. Therefore, the normalized value of R_0 is set to 1. Thus, for the problem under consideration we work with all normalized impedance values. Eventually, they can be renormalized by R_0 via simple multiplication.[14]

Z_{in2} is considered as the load for UE 1 with characteristic impedance Z_1. In this case, the input impedance of the first line is given by

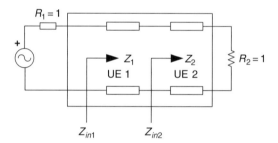

Figure 4.4   Cascade connection of two UEs

$$Z_{in1} = Z_1 \frac{Z_{in2} + Z_1\lambda}{Z_1 + Z_{in2}\lambda}$$

where

$$Z_{in2} = Z_2 \frac{1 + Z_2\lambda}{Z_2 + \lambda}$$

---

[13] H. J. Carlin, 'Distributed Circuit Design with Transmission Line', *Proceedings of the IEEE*, Vol. 59, No. 7, pp. 1059–1081, 1971.
[14] Renormalization of any impedance is given by $Z = R_0 z$ where z is the normalized impedance and Z is the actual impedance.

Using Z_{in2} in Z_{in1} and by simple algebraic manipulations, we obtain

$$Z_{in1} = \frac{N(\lambda)}{D(\lambda)} = \frac{a_2\lambda^2 + a_1\lambda + a_0}{b_2\lambda^2 + b_1\lambda + b_0}$$

such that

$$N(\lambda) = a_2\lambda^2 + a_1\lambda + a_0$$
$$D(\lambda) = b_2\lambda^2 + b_1\lambda + b_0$$

(4.26)

where

$$\begin{array}{ll} a_2 = Z_1^2 & b_2 = Z_2^2 \\ a_1 = Z_1Z_2(Z_1 + Z_2) & b_1 = (Z_1 + Z_2) \\ a_0 = Z_1Z_2 & b_0 = Z_1Z_2 \end{array}$$

(4.27)

and

$$\text{even}\{Z_{in1}\} = \frac{1}{2}\left[Z_{in1}(\lambda) + Z_{in1}(-\lambda)\right] = \frac{A_2\lambda^4 + A_1\lambda^2 + A_0}{B_2\lambda^4 + B_1\lambda^2 + B_0}$$

(4.28)

where

$$\begin{array}{ll} A_2 = a_2b_2 = [Z_1Z_2]^2 & B_2 = b_2^2 = Z_2^4 \\ A_1 = a_2b_0 - a_1b_1 + a_0b_2 & B_1 = 2b_2b_0 - b_1^2 \\ \text{or} & \text{or} \\ A_1 = -2[Z_1Z_2]^2 & B_1 = 2Z_2^2(Z_1Z_2) - (Z_1 + Z_2)^2 \\ A_0 = (Z_1Z_2)^2 & B_0 = (Z_1Z_2)^2 \end{array}$$

(4.29)

Hence, we see that

$$R_{in}(\lambda) = \text{even}\{Z_{in1}\} = \frac{(Z_1Z_2)^2[\lambda^2 - 1]^2}{B_2\lambda^4 + B_1\lambda^2 + B_0}$$

or

$$R_{in}(\lambda) = (Z_1Z_2)^2\frac{(1 - \lambda^2)^2}{B_2\lambda^4 + B_1\lambda^2 + B_0}$$

Considering Statement 5, we can write that

$$R_{in}(\lambda) = \text{even}\{Z_{in}\} = F(\lambda)F(-\lambda)$$

where

$$F(\lambda) = \frac{A(\lambda)}{D(\lambda)}$$

with
$$A(\lambda) = C_{even}[\lambda^2 - 1]^{2/2} = [1 - \lambda^2]$$
$$C_{even} = \sqrt{Z_1 Z_2}$$
$$D(\lambda) = b_2 \lambda^2 + b_1 \lambda + b_0$$

(4.30)

Equation (4.30) confirms Statement 5.

Employing Equations (4.26) and (4.27) we can easily derive the input reflection coefficient for the standard normalization $R_0 = R_1 = R_2 = 1$. In this case,

$$S_{in} = S_{11} = \frac{Z_{in} - 1}{Z_{in} + 1} = \frac{N - D}{N + D} = \frac{h(\lambda)}{g(\lambda)} = \frac{h_2 \lambda^2 + h_1 \lambda + h_0}{g_2 \lambda^2 + g_1 \lambda + g_0}$$

(4.31)

For the case under consideration, the coefficients are given as

$$
\begin{aligned}
h_0 &= a_0 - b_0 = 0 & g_0 &= a_0 + b_0 = 2Z_1 Z_2 \\
h_1 &= a_1 - b_1 & g_1 &= a_1 + b_1 \\
&\text{or} & &\text{or} \\
h_1 &= (Z_1 Z_2 - 1)(Z_1 + Z_2) & g_1 &= (Z_1 Z_2 + 1)(Z_1 + Z_2) \\
h_2 &= a_2 - b_2 = Z_1^2 - Z_2^2 & g_2 &= a_2 + b_2 = Z_1^2 + Z_2^2
\end{aligned}
$$

(4.32)

4.9 Major Properties of the Scattering Parameters for Passive Two-ports

Referring to Figure 4.5, assume that the input power is transferred to a load over a lossless two-port which is constructed by proper connections of commensurate TEM lines.

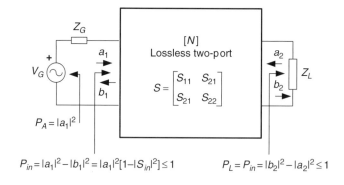

$$P_{in} = |a_1|^2 - |b_1|^2 = |a_1|^2[1 - |S_{in}|^2] \le 1 \qquad P_L = P_{in} = |b_2|^2 - |a_2|^2 \le 1$$

Figure 4.5   Power transfer P_{in} over a lossless two-port

Let us also assume that the two-port is terminated in Z_G and Z_L at the input and output ends respectively. Let the two-port be reciprocal, yielding $S_{21}(\lambda) = S_{12}(\lambda)$. In general, we can make the following statements which stem from the physical facts and previously made definitions and statements.

Physical Fact 1: The available power of the generator is defined as the maximum deliverable power and given by

$$[P_{in}]_{max} = P_A = |a_1|^2 = \frac{|V_G|^2}{4R_G}$$

where $R_1 = R_G = Z_G$ is taken as the normalization number of the input port, which yields a reflection-free transmission from input to output if the input impedance is $Z_{in} = R_1$. This condition maximizes the power delivery and also makes the input reflection wave zero.[15] That is, $b_1 = 0$.

Physical Fact 2 (Maximum energy transfer): Power delivered to the input port must be less than or equal to that of the available power of the generator. Mathematically speaking, this statement is expressed as

$$P_{in} = |a_1|^2 - |b_1|^2 = |a_1|^2[1 - |S_{in}|^2] \le P_A$$

which forces $|S_{in}(j\omega)| \le 1$. In other words, for a passive system, the power delivered to the input port must always be less than or equal to the available power of the generator yielding

$$\frac{P_{in}}{P_A} = 1 - |S_{in}(j\Omega)|^2 \le 1$$

Maximum energy is transferred from the generator to the input port if and only if $P_A = P_{in}$, which makes $S_{in} = 0$. This is called the perfect match, established between the generator and the input port. If $P_{in} = 0$ or equivalently if $S_{in} = 1$, then there will be no power transfer to the system. In this case, all the power supplied by the generator must be dissipated on itself which, practically, may burn out the generator. Therefore, one has to be very careful with the reflected power.

All the above conditions require that, over the entire frequency axis ω or $\Omega = \tan(\omega\tau)$, the amplitude of $S_{in}(j\Omega)$ must be bounded by one.

Hence, by analytic continuation, we can state that

$$\boxed{S_{in}(\lambda) \le 1, \text{for all real}\{\lambda\} \ge 0}$$ (4.33)

and, therefore, $S_{in}(\lambda)$ must be free of open right half plane (RHP) and $j\Omega$ poles.

Recall that for a lossless two-port constructed with commensurate transmission lines, the input impedance $Z_{in}(\lambda) = N(\lambda)/D(\lambda)$ must be a rational PR function in complex variable $\lambda = \Sigma + j\Omega$ if the termination $Z_L(\lambda)$ is positive real in λ. As described in Section 1.18 of Chapter 1, PR of $Z_{in}(\lambda)$ requires that polynomials $N(\lambda)$ and $D(\lambda)$ must be free of open RHP roots, and the roots on the $j\Omega$ axis must be simple.[16] In order to facilitate further explanations, we will make the following definitions.

---

[15] In general, $R_G = real\{Z_G\}$. The reflection-free transmission is obtained when $Z_G(j\omega) = R_G + jX_G = Z_{in}^*$ where '*' designates the complex conjugate of a complex quantity.

[16] That is, they must be of order 1 and not of multiple orders.

Definition 6 (Hurwitz polynomial): A polynomial $D(p)$, which is free of open RHP roots, but possesses only simple $j\Omega$ roots, is called a Hurwitz polynomial. In other words, a Hurwitz polynomial must be free of open RHP poles.

Definition 7 (Strictly Hurwitz polynomial): A polynomial $g(p)$, which is free of closed RHP roots, is called strictly Hurwitz. In other words, a strictly Hurwitz polynomial $g(p)$ must have all its roots only in the closed LHP and must be free of $j\Omega$ and *RHP* roots.

Statement 6: The addition of two Hurwitz polynomials, such as $N(\lambda)$, and $D(\lambda)$, results in strictly Hurwitz polynomials such as $g(\lambda) = N(\lambda) + D(\lambda)$ which is strictly free of RHP and $j\Omega$ roots. Proof of this statement can be found in many textbooks such as that written by Carlin and Civalleri.[17] Actually, by Physical Fact 2, one can deduce the same result.

Definition 8 (Bounded real (BR) function): A reflection coefficient

$$S_{in}(\lambda) = \frac{Z_{in} - 1}{Z_{in} + 1} = \frac{h(\lambda)}{g(\lambda)}$$

is called a bounded real (BR) rational function if and only if:

(a) $S_{in}(\lambda)$ is real when is real.
(b) $S_{in}(\lambda) \leq 1$ when real$\{\lambda\} \geq 0$.

The definition of bounded realness reveals that, if $Z_{in}(\lambda) = N(\lambda)/D(\lambda)$ is a PR function, then:

(a) $h(\lambda) = N(\lambda) - D(\lambda)$.
(b) $g(\lambda) = N(\lambda) + D(\lambda)$.
(c) By Statement 6, $g(\lambda)$ must be strictly Hurwitz.
(d) By Equation (4.33), $|S_{in}(j\Omega)| \leq 1$ for all Ω.

Based on the above results we can make the following statement.

Statement 7: A rational function $S_{in}(\lambda) = h(\lambda)/g(\lambda)$ is BR if and only if:

(a) $S_{in}(\lambda)$ is real when λ is real.
(b) $g(\lambda)$ is strictly Hurwitz.
(c) $|S_{in}(j\Omega)| \leq 1$ for all Ω.

Physical Fact 3 (Lossless two-ports): If the two-port under consideration is lossless, then the input power must be delivered to the load without any loss. Therefore, the power delivered to the load P_L must be equal to the input power P_{in}. As far as the load power is concerned, we can state that

$$\boxed{P_L = |b_2|^2 - |a_2|^2 = P_{in} = |a_1|^2[1 - |S_{in}|^2]} \qquad (4.34)$$

Statement 8: Let a lossless two-port, which is constructed with commensurate TEM lines, be terminated in its port normalization numbers $R_1 = Z_G$ and $R_2 = Z_L$. In this case, by definition of the input and output incident and reflected waves:

---

[17] H. J. Carlin and P. P. Civalleri, *Wideband Circuit Design*, CRC Press, 1998.

(a) The output incident wave must be zero, $a_2 = 0$, and the input scattering parameter is given by

$$S_{11} = S_{in} = \frac{b_1}{a_1}$$

(b) The forward transfer scattering parameter is given by

$$S_{21} = \frac{b_2}{a_1}$$

(c) In general, for the lossless two-port under consideration, TPG is defined as the ratio of the power delivered to load $[P_L(j\omega)]$ to the available power of the generator $[P_A(j\omega)]$. For the present case, since a_2 is equal to zero, then, by Equation (4.34), $P_L = |b_2(j\omega)|^2 = |a_1|^2[1 - |S_{11}|^2]$ or from the definition of TPG it is found that

$$0 \leq TPG = T(j\Omega) = \frac{|b_2(j\Omega)|^2}{|a_1(j\Omega)|^2} = |S_{21}(j\Omega)|^2 \leq 1, \text{ for } \forall\Omega \tag{4.35}$$

or equivalently

$$|S_{21}|^2 = 1 - |S_{11}|^2$$

which is the losslessness requirement of the two-port. By analytic continuation,

$$S_{11}(\lambda)S_{11}(-\lambda) + S_{21}(\lambda)S_{21}(-\lambda) = 1 \tag{4.36}$$

(d) If we excite the two-port from the load end with a generator, with an internal resistance of R_2, while terminating the input port in its port normalization number R_1, we obtain

$$T = \frac{P_L}{P_A} = |S_{12}|^2 = 1 - |S_{22}|^2 = 1 - |S_{11}|^2 = |S_{21}|^2 \leq 1$$

By analytic continuation the above equations yield the following results:

$$S_{12}(\lambda)S_{12}(-\lambda) = S_{21}(\lambda)S_{21}(-\lambda)$$

or

$$\frac{S_{21}(\lambda)}{S_{12}(-\lambda)} = \frac{S_{12}(\lambda)}{S_{21}(-\lambda)} \tag{4.37}$$

and

$$S_{22}(\lambda)S_{22}(-\lambda) + S_{12}(\lambda)S_{12}(-\lambda) = 1 \tag{4.38}$$

Physical Fact 4 (Property of the scattering parameters for passive two-ports): In general, the total power delivered to a passive two-port must be positive since the two-port consumes energy. However, if it is lossless, then there should be no power dissipation. Thus, mathematically speaking, this statement can be expressed as follows:

$$P_{total} = [|a_1|^2 - |b_1|^2] + [|a_2|^2 - |b_2|^2] \geq 0 \qquad (4.39)$$

By rearranging the above equation we can write that

$$\begin{bmatrix} a_1 & a_2 \end{bmatrix} \begin{bmatrix} a_1^* \\ a_2^* \end{bmatrix} - \begin{bmatrix} b_1 & b_2 \end{bmatrix} \begin{bmatrix} b_1^* \\ b_2^* \end{bmatrix} \geq 0 \qquad (4.40)$$

Let

$$a = \begin{bmatrix} a_1 \\ a_2 \end{bmatrix} \quad \text{and} \quad a^\dagger = [a_1^* a_2^*]$$

Similarly,

$$b = \begin{bmatrix} b_1 \\ b_2 \end{bmatrix} \quad \text{and} \quad b^\dagger = [b_1^* b_2^*]$$

Employing the above notation, we can also write that

$$a^\dagger a - b^\dagger b \geq 0 \qquad (4.41)$$

Remember that, by definition,

$$b = Sa = \begin{bmatrix} S_{11} & S_{12} \\ S_{21} & S_{22} \end{bmatrix}$$

and

$$b^* = S^* a^*$$

or

$$b^\dagger = a^\dagger S^\dagger$$

Then, using analytic continuation, and by replacing $-j\Omega$ with $-\lambda$ (i.e. $-j\Omega = -\lambda$), Equation (4.39) reveals that

$$a^\dagger \begin{bmatrix} 1 & 0 \\ 0 & 1 \end{bmatrix} a - a^\dagger S^\dagger S a \geq 0$$

or

$$a^\dagger [I - S^\dagger S] a \geq 0$$

Thus, the scattering matrix S must satisfy

$$S^\dagger S = S S^\dagger \leq I = \begin{bmatrix} 1 & 0 \\ 0 & 1 \end{bmatrix} \tag{4.42}$$

$$\begin{bmatrix} S_{11}(-\lambda) & S_{21}(-\lambda) \\ S_{12}(-\lambda) & S_{22}(-\lambda) \end{bmatrix} \begin{bmatrix} S_{11}(\lambda) & S_{12}(\lambda) \\ S_{21}(\lambda) & S_{22}(\lambda) \end{bmatrix}$$

$$\leq \begin{bmatrix} 1 & 0 \\ 0 & 1 \end{bmatrix}, \text{ for all real } \{\lambda\} \geq 0$$

Equation (4.40) is called the general form of the passivity condition. The open form of this condition is given by[18]

$$\begin{aligned} S_{11}(\lambda)S_{11}(-\lambda) + S_{21}(\lambda)S_{21}(-1) &\leq 1 \\ S_{22}(\lambda)S_{22}(-\lambda) + S_{12}(\lambda)S_{12}(-\lambda) &\leq 1 \\ S_{12}(-\lambda)S_{11}(\lambda) + S_{22}(-\lambda)S_{21}(\lambda) &\leq 0 \end{aligned} \tag{4.43}$$

For lossless two-ports, all of the above equations must be satisfied with the equality sign. Thus, we make the following definition.

Definition 9 (BR scattering matrix): The scattering matrix $S(\lambda)$, which satisfies Equation (4.43), is called the BR scattering matrix.

Statement 9 (Properties of BR scattering matrix): A BR scattering matrix must satisfy the following conditions:

(a) $S(\lambda)$ is real when λ is real.
(b) $S(\lambda)$ must be analytic in the closed RHP, which means that, if all the elements of S are specified in rational form, then they must all have regular denominator polynomials in the closed RHP.
(c) $S^\dagger(\lambda)S(\lambda) \leq I$.

---

[18] Note that Equation (4.40) yields only three independent equations. The fourth one can be obtained by taking the para-conjugate of the third one.

Statement 10 (Lticlessness condition): For any lossless two-port, scattering parameters must satisfy the equation set given by Equation (4.43) with the equality sign. That is,

$$\boxed{\begin{aligned} &S_{11}(\lambda)S_{11}(-\lambda) + S_{21}(\lambda)S_{21}(-1) = 1 \\ &S_{22}(\lambda)S_{22}(-\lambda) + S_{12}(\lambda)S_{12}(-\lambda) = 1 \\ &S_{22}(\lambda) = -\frac{S_{12}(\lambda)}{S_{21}(-\lambda)}S_{11}(\lambda) \end{aligned}}$$

(4.44)

4.10 Rational Form of the Scattering Matrix for a Resistively Terminated Lossless Two-port Constructed by Transmission Lines

Consider a lossless two-port, constructed by proper connections of ideal commensurate transmission lines. Assume that the two-port is terminated in unit resistance, i.e. in $R_0 = 1$, which is also taken as the common port normalization number. In other words, $R_1 = R_2 = R_0 = 1\,\Omega$.

In this case, the input impedance $Z_{in}(\lambda) = N(\lambda)/D(\lambda)$ must be a PR impedance in complex variable λ. Since the output is terminated in its port normalization number, then the R_0 normalized input scattering coefficient is given by

$$S_{11} = \frac{Z_{in} - 1}{Z_{in} + 1} = \frac{N(\lambda) - D(\lambda)}{N(\lambda) + D(\lambda)} = \frac{[N(\lambda) - D(\lambda)]}{[N(\lambda) + D(\lambda)]}$$

or

$$\boxed{\begin{aligned} &S_{11}(\lambda) = \frac{h(\lambda)}{g(\lambda)} \\ &\text{where} \\ &h(\lambda) = [N(\lambda) - D(\lambda)] \\ &g(\lambda) = [N(\lambda) + D(\lambda)] \end{aligned}}$$

(4.45)

Note that, in Equation (4.45), the degrees of the polynomials $h(\lambda)$ and $g(\lambda)$ are the same and equal to the degree of $N(\lambda)$ or $D(\lambda)$, whichever is higher.[19] When working with scattering parameters in the design and analysis of microwave circuits, we usually take the degree of $g(\lambda)$ as the reference degree and denote if by n. Then, the degrees of the rest of the polynomials are determined as they are generated.

Now, let us derive $S_{21}(\lambda)S_{21}(-\lambda)$ using Equation (4.44):

$$S_{21}(\lambda)S_{21}(-\lambda) = 1 - S_{in}(\lambda)S_{in}(-\lambda)$$

---

[19] Recall that for a positive real function $Z(\lambda) = N(\lambda)/D(\lambda)$, the degree of difference between the numerator and denominator must be 1. That is to say, $|m - n| = 1$ where m denotes the degree of $N(\lambda)$ and n denotes the degree of $D(\lambda)$.

In terms of the input impedance Z_{in} we have

$$S_{21}(\lambda)S_{21}(-\lambda) = \frac{2[Z_{in}(\lambda) + Z_{in}(-\lambda)]}{[Z_{in}(\lambda) + 1][Z_{in}(-\lambda) + 1]} = \frac{4 \text{ even}\{Z_{in}\}}{[Z_{in}(\lambda) + 1][Z_{in}(-\lambda) + 1]}$$

(4.46)

By Statement 5,

$$\text{even}\{Z_{in}\} = F(\lambda)F(-\lambda) = \frac{A(\lambda)}{D(\lambda)} \frac{A(-\lambda)}{D(-\lambda)}$$

Then,

$$S_{21}(\lambda)S_{21}(-\lambda) = \left[\frac{2A(\lambda)}{N(\lambda) + D(\lambda)} \right] \left[\frac{2A(-\lambda)}{N(-\lambda) + D(-\lambda)} \right]$$

(4.47)

By Statement 9, Equation (4.47) suggests that

$$S_{21}(\lambda) = \frac{f(\lambda)}{g(\lambda)}$$

where

$$f(\lambda) = 2A(\lambda) = 2C_{even}\lambda^q (1 - \lambda^2)^{k/2}$$

and

$$g(\lambda) = N(\lambda) + D(\lambda)$$

(4.48)

Since the lossless two-port is constructed on the ideal commensurate TEM lines, it must be electrically reciprocal. Therefore,

$$S_{12}(\lambda) = S_{21}(\lambda) = \frac{f(\lambda)}{g(\lambda)}$$

Finally, by Equation (4.44), $S_{22}(\lambda)$ is found as

$$S_{22}(\lambda) = -\frac{S_{12}(\lambda)}{S_{21}(-\lambda)} S_{11}(\lambda)$$

$$S_{22} = -\frac{f(\lambda)}{g(\lambda)} \frac{g(-\lambda)}{f(-\lambda)} \frac{h(-\lambda)}{g(-\lambda)}$$

$$= -\frac{(\lambda)^q (1 - \lambda^2)^{k/2}}{(-\lambda)^q (1 - \lambda^2)^{k/2}} \frac{h(-\lambda)}{g(\lambda)}$$

$$S_{22} = -(-1)^q \frac{h(-\lambda)}{g(\lambda)}$$

(4.49)

In summary, for a lossless two-port, constructed with ideal commensurate TEM lines, the reciprocal scattering matrix is given by

$$S = \begin{bmatrix} \dfrac{h(\lambda)}{g(\lambda)} & 2C_{even}\dfrac{\lambda^q(1-\lambda^2)^{k/2}}{g(\lambda)} \\ 2C_{even}\dfrac{\lambda^q(1-\lambda^2)^{k/2}}{g(\lambda)} & (-1)^{1+q}\dfrac{h(-\lambda)}{g(\lambda)} \end{bmatrix} \tag{4.50}$$

Now, let us look at several examples to verify some of the above results and to show some practical applications.

Example 4.4: Referring to Figure 4.6, let the line characteristic impedance be $Z_1 = 4\ \Omega$ Let the λ domain inductor be $L = 1$. Then, using the unit normalization port numbers (i.e. $R_0 = R_1 = R_2$):

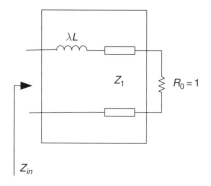

Figure 4.6 Circuit layout for Example 4.4

(a) Find the input impedance.
(b) Generate the even part of the input impedance.
(c) Compute the full scattering parameters of the lossless two-port.
(d) Comment on the results.

Solution:

(a) The Input impedance is

$$Z_{in} = \lambda L + Z_1\frac{R_0 + Z_1\lambda}{Z_1 + R_0\lambda}$$

$$Z_{in} = \frac{(LR_0)\lambda^2 + (LZ_1 + Z_1^2)\lambda + Z_1R_0}{Z_1 + R_0\lambda} = \frac{N(\lambda)}{D(\lambda)}$$

where

$$N(\lambda) = (LR_0)\lambda^2 + (LZ_1 + Z_1^2)\lambda + Z_1R_0$$
$$D(\lambda) = Z_1 + R_0\lambda$$

Or the open form of $Z_{in}(\lambda)$ is found as

$$Z_{in}(\lambda) = \frac{\lambda^2 + 20\lambda + 4}{4 + \lambda}$$

(b) The even part of the input impedance is

$$\text{even}\{Z_{in}(\lambda)\} = \frac{1}{2}[Z_{in}(\lambda) + Z_{in}(-\lambda)] = Z_1^2 R_0 \left[\frac{1 - \lambda^2}{Z_1^2 - R_0^2 \lambda^2}\right] \text{even}\{Z_{in}(\lambda)\} = C_{even}^2 \frac{1 - \lambda^2}{D(\lambda)D(-\lambda)}$$

where

$$C_{even} = Z_1 \sqrt{R_0} = 4$$

(c) By Equation (4.50), explicit forms of the scattering parameters are given in terms of the input impedance $Z_{in} = N/D$ related polynomials, namely $h(\lambda), g(\lambda)$ and $A = \lambda^q(1 - \lambda^2)^{k/2}$. For the problem under consideration, it is found that

$$h(\lambda) = N(\lambda) - D(\lambda) = (LR_0)\lambda^2 + [LZ_1 + Z_1^2 - R_0]\lambda + Z_1[R_0 - 1]$$

or

$$h(\lambda) = \lambda^2 + 19\lambda$$

Similarly,

$$g(\lambda) = N(\lambda) + D(\lambda) = (LR_0)\lambda^2 + [LZ_1 + Z_1^2 + R_0]\lambda + Z_1[R_0 + 1]$$

or

$$g(\lambda) = \lambda^2 + 21\lambda + 8$$

Thus, the explicit form of $S_{11}(\lambda)$ is given by

$$S_{11}(\lambda) = \frac{\lambda^2 + 19\lambda}{\lambda^2 + 21\lambda + 8}$$

The even part of the input impedance is computed as

$$\text{even}\{Z_{in}(\lambda)\} = C_{even}^2 \frac{1 - \lambda^2}{D(\lambda)D(-\lambda)}, \quad \text{with } C_{even} = Z_1 \sqrt{R_0} = 4$$

Then,

$$S_{21} = [2Z_1 \sqrt{R_0}] \frac{(1 - \lambda^2)^{1/2}}{g(\lambda)}$$

or the open form of $S_{21}(\lambda)$ is found as

$$S_{21}(\lambda) = \frac{8(1 - \lambda^2)^{1/2}}{\lambda^2 + 21\lambda + 8}$$

Finally,

$$S_{22}(\lambda) = -(-)^q \frac{h(-\lambda)}{g(\lambda)}$$

or, by setting $q = 0$, we have

$$S_{22} = -\frac{\lambda^2 - 19\lambda}{\lambda^2 + 21\lambda + 8}$$

Note that the same result can be obtained by generating the output impedance

$$Z_{out} = Z_1 \frac{Z_L + Z_1\lambda}{Z_1 + Z_L\lambda} \quad Z_L = L\lambda + 1$$

or, in the open form,

$$Z_{out} = \frac{(Z_1 L + Z_1^2)\lambda + Z_1}{L\lambda^2 + \lambda + Z_1} = \frac{20\lambda + 4}{\lambda^2 + \lambda + 4}$$

Then, the back end reflection coefficient is given by

$$S_{22}(\lambda) = \frac{Z_{out} - 1}{Z_{out} + 1} = -\frac{\lambda^2 - 19\lambda}{\lambda^2 + 21\lambda + 8}$$

which confirms the previous result.

(d) Comments:

(i) In this problem, it has been shown that the full scattering parameters of a resistively terminated lossless two-port can be generated from its positive real input impedance.

(ii) The numerator polynomial $f(\lambda) = C_{even}(1 - \lambda^2)$ of $S_{21}(\lambda)$ has the same form as that of a single UE since they both have the same even part of the input impedance.

Example 4.5: Referring to Figure 4.4, for a Cascade connection of two commensurate TEM lines, the characteristic impedances are given by $Z_1 = 1, Z_2 = 2$ as they are normalized with respect to the standard port normalization number $R_0 = 50\,\Omega$.

(a) Find the normalized input impedance.
(b) Find the even part of the input impedance.
(c) Find the complete scattering parameters of the lossless two-port.

Solution:

(a) For the problem under consideration, by Equations (4.27) and (4.30), explicit expressions are derived for the input impedance and also for its even part. Hence, we have

$$Z_{in} = \frac{a_2\lambda^2 + a_1\lambda + a_0}{b_2\lambda^2 + b_1\lambda + b_0}$$

with

$$
\begin{array}{ll}
a_2 = Z_1^2 = 1 & b_2 = Z_2^2 = 4 \\
a_1 = Z_1 Z_2 (Z_1 + Z_2) = 6 & b_1 = (Z_1 + Z_2) = 3 \\
a_0 = Z_1 Z_2 = 2 & b_0 = Z_1 Z_2 = 2
\end{array}
\tag{4.51}
$$

or

$$Z_{in} = \frac{N}{D} = \frac{\lambda^2 + 6\lambda + 2}{4\lambda^2 + 3\lambda + 2}$$

(b) The even part is

$$R_{in} = \frac{A_2\lambda^4 + A_1\lambda^2 + A_0}{B_2\lambda^4 + B_1\lambda^2 + B_0}$$

with

$$
\begin{array}{ll}
A_2 = [Z_1 Z_2]^2 = 4 & B_2 = b_2^2 = Z_2^4 = 16 \\
A_1 = -2(Z_1 Z_2)^2 = -8 & B_1 = 2Z_2^2(Z_1 Z_2)^2 - (Z_1 + Z_2)^2 = 7 \\
A_0 = (Z_1 Z_2)^2 = 4 & B_0 = (Z_1 Z_2)^2 = 4
\end{array}
\tag{4.52}
$$

or, in open form,

$$R_{in} = (C_{even})^2 \frac{\left[1 - \lambda^2\right]^2}{16\lambda^4 + 7\lambda^2 + 4} = 4 \frac{\left[1 - \lambda^2\right]^2}{16\lambda^4 + 7\lambda^2 + 4}$$

yielding $C_{even} = 2$.

(c) To generate the scattering parameters, first of all we should obtain the polynomials h, f and g:

$$h = N - D = -3\lambda^2 + 3\lambda$$

$$g = N + D = 5\lambda^2 + 9\lambda + 4$$

$$f = 2C_{even}(1 - \lambda^2) = 4(1 - \lambda^2)$$

Hence, by Equation (4.50), we have

$$S_{11}(\lambda) = \frac{-3\lambda^2 + 3\lambda}{5\lambda^2 + 9\lambda + 4}$$

$$S_{12} = S_{21}(\lambda) = \frac{4(1 - \lambda^2)}{5\lambda^2 + 9\lambda + 4}$$

$$S_{22}(\lambda) = -\frac{-3\lambda^2 - 3\lambda}{5\lambda^2 + 9\lambda + 4}$$

So far we have worked with the scattering parameters for circuits constructed with commensurate transmission lines. However, we may also generate the impedance matrix directly from the two-ports under consideration, as shown in the following examples.

Example 4.6 (Impedance matrix of a single UE in cascade configuration): Determine the impedance parameters of a UE with the characteristic impedance Z_0 in cascade configuration.

Solution: Here, the UE is considered as a lossless two-port and its open-circuit impedance parameters are defined as

$$\boxed{\begin{aligned} V_1 &= Z_{11}I_1 + Z_{12}I_2 \\ V_2 &= Z_{21}I_1 + Z_{22}I_2 \end{aligned}}$$

(4.53)

In open form, the impedance parameters are given by

$$Z_{11} = \frac{V_1}{I_1}\bigg|_{I2=0} = Z_{in} = Z_0\frac{Z_L + Z_0\lambda}{Z_0 + \lambda Z_L}, \quad \text{when } Z_L = \infty$$

or

$$Z_{11} = \frac{Z_0}{\lambda} = \frac{1}{C\lambda}$$

which is a Foster function with $C = 1/Z_0$.
By symmetry

$$\boxed{Z_{22} = Z_{11} = \frac{Z_0}{\lambda}}$$

(4.54)

It should be noted that, by Equation (4.17),

$$S_L = \frac{B}{A} = \frac{Z_L - Z_0}{Z_L + Z_0} = 1$$

Then, we see that $A = B$. By Equation (4.53),

$$Z_{21} = \frac{V_2}{I_1}\bigg|I_2 = 0$$

On the other hand,

$$V_2 = A + B = 2A$$

and

$$I_1 = \frac{1}{Z_0}\left[Ae^{\gamma l} - Ae^{-\gamma l}\right]$$

Then,

$$Z_{21} = \frac{V_2}{I_1} = 2Z_0\frac{1}{e^{\gamma l}[1 - e^{-2\gamma l}]}$$

such that $I_2 = 0$.

Noted that

$$\lambda = \tanh(\gamma l) = \frac{e^{\gamma l} - e^{-\gamma l}}{e^{\gamma l} - e^{-\gamma l}} = \frac{1 - e^{-2\gamma l}}{1 + e^{-2\gamma l}}$$

or

$$e^{-2\gamma l} = \frac{1 - \lambda}{1 + \lambda} \quad \text{and} \quad e^{\gamma l} = \sqrt{\frac{1 + \lambda}{1 - \lambda}}$$

Using the above identities in Z_{21}, we have

$$Z_{21}(\lambda) = 2Z_0 \sqrt{\frac{1 - \lambda}{1 + \lambda}} \frac{1}{1 - \frac{1 - \lambda}{1 + \lambda}} = Z_0 = \frac{\sqrt{1 - \lambda^2}}{\lambda}$$

and by reciprocity

$$Z_{12} = Z_{21} \tag{4.55}$$

Hence, in matrix form,

$$\begin{bmatrix} V_1 \\ V_2 \end{bmatrix} = [Z] \begin{bmatrix} I_1 \\ I_2 \end{bmatrix} = \begin{bmatrix} Z_{11} & Z_{12} \\ Z_{21} & Z_{22} \end{bmatrix} \begin{bmatrix} I_1 \\ I_2 \end{bmatrix}$$

where the impedance matrix $[Z]$ of a UE is given by

$$[Z] = \begin{bmatrix} \dfrac{Z_0}{\lambda} & Z_0 \dfrac{\sqrt{1 - \lambda^2}}{\lambda} \\ Z_0 \dfrac{\sqrt{1 - \lambda^2}}{\lambda} & \dfrac{Z_0}{\lambda} \end{bmatrix} \tag{4.56}$$

Example 4.7 (Impedance matrix of a single UE with capacitive loading): Referring to Figure 4.7(a), find the impedance matrix of the lossless two-port constructed with commensurate transmission lines such that a single UE of characteristic impedance Z_A is terminated in capacitive load $Z_L = 1/C\lambda = Z_{CAP}/\lambda$ where Z_{CAP} is the characteristic impedance of the open stub

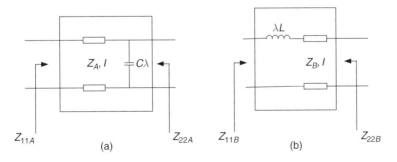

Figure 4.7   Kuroda identity I

Solution: For this problem Z_{11A} is given as the input impedance of the UE which is terminated in $Z_L = 1/c\lambda = Z_{CAP}/\lambda$ such that

$$Z_{11A} = Z_A \frac{Z_L + Z_A\lambda}{Z_A + Z_L\lambda} = \left(\frac{Z_A Z_{CAP}}{Z_A + \dfrac{1}{C}}\right)\left(\frac{1}{\lambda}\right) + \left(\frac{Z_A^2}{Z_A + \dfrac{1}{C}}\right)\lambda$$

or

$$Z_{11A}(\lambda) = \left(\frac{Z_A^2}{Z_A + Z_{CAP}}\right)\lambda + \left(\frac{Z_A Z_{CAP}}{Z_A + Z_{CAP}}\right)\left(\frac{1}{\lambda}\right) \tag{4.57}$$

On the other hand, Z_{22A} will be the parallel combination of the impedances Z_A/λ^{20} and Z_{CAp}/λ. Thus,

$$Z_{22A} = \left(\frac{Z_A Z_{CAP}}{Z_A + Z_{CAP}}\right)\left(\frac{1}{\lambda}\right) \tag{4.58}$$

As in the previous problem, when $I_2 = 0$

$$Z_{21} = \frac{V_2}{I_1} = \frac{A + B}{\dfrac{1}{Z_A}\left[Ae^{\gamma l} - Be^{-\gamma l}\right]} = Z_A \frac{1 + B/A}{e^{\gamma l}\left[1 - \dfrac{B}{A}e^{-2\gamma l}\right]}$$

such that by Equation (4.17), $B = S_L(Z_A)A$. Therefore, using the identities given for $e^{\gamma l}$ and $e^{-2\gamma l}$ in terms of λ, we have

$$Z_{21} = Z_A\sqrt{\frac{1-\lambda}{1+\lambda}}\,\frac{1 + S_L}{1 - \dfrac{1-\lambda}{1+\lambda}S_L} = Z_A\sqrt{1-\lambda^2}\,\frac{1 + S_L}{(1 - S_L) + \lambda(1 + S_L)}$$

Note that the load reflection coefficient is given by

$$S_L = \frac{Z_L - Z_A}{Z_L + Z_A} = \frac{\dfrac{Z_{CAP}}{\lambda} - Z_A}{\dfrac{Z_{CAP}}{\lambda} + Z_A} = \frac{Z_{CAP} - Z_A\lambda}{Z_{CAP} + Z_A\lambda\lambda}$$

---

[20] Remember that Z_A/λ is the output impedance of the UE when the input is left open.

and

$$\frac{1 - S_L}{1 + S_L} = \frac{1 - \dfrac{Z_{CAP} - Z_A \lambda}{Z_{CAP} + Z_A \lambda}}{1 + \dfrac{Z_{CAP} - Z_A \lambda}{Z_{CAP} + Z_A \lambda}} = \frac{Z_A \lambda}{Z_{CAP}}$$

Thus,

$$Z_{21} = Z_A \sqrt{1 - \lambda^2} \frac{1}{\dfrac{1 - S_L}{1 + S_L} + \lambda}$$

or

$$\boxed{Z_{21} = \left[\frac{Z_A Z_{CAP}}{Z_A + Z_{CAP}} \right] \left[\frac{\sqrt{1 - \lambda^2}}{\lambda} \right]}$$

and by reciprocity

$$\boxed{Z_{12} = Z_{21}}$$

(4.59)

Hence, the impedance matrix is given by

$$[Z] = \begin{bmatrix} \left(\dfrac{Z_A^2}{Z_A + Z_{CAP}} \right) \lambda + \left(\dfrac{Z_A Z_{CAP}}{Z_A + Z_{CAP}} \right) \left(\dfrac{1}{\lambda} \right) & \left[\dfrac{Z_A Z_{CAP}}{Z_A + Z_{CAP}} \right] \left[\dfrac{\sqrt{1 - \lambda^2}}{\lambda} \right] \\ \left[\dfrac{Z_A Z_{CAP}}{Z_A + Z_{CAP}} \right] \left[\dfrac{\sqrt{1 - \lambda^2}}{\lambda} \right] & \left(\dfrac{Z_A Z_{CAP}}{Z_A + Z_{CAP}} \right) \left(\dfrac{1}{\lambda} \right) \end{bmatrix}$$

(4.60)

Example 4.8 (Impedance matrix of a single UE loaded with a series short stub): Referring to Figure 4.7(b), find the impedance matrix of the lossless two-port constructed with commensurate transmission lines such that a series inductor $L\lambda$ with characteristic impedance Z_{IND} is connected to a UE of characteristic impedance Z_B.

Solution: For this problem, life is a little bit easier. It is straightforward to write Z_{11B} and Z_{22B} as

$$Z_{11B} = L\lambda + \frac{Z_B}{\lambda} = Z_{IND}\lambda + \frac{Z_B}{\lambda}$$

$$Z_{22B} = \frac{Z_B}{\lambda}$$

and, as in the single cascade UE configuration case,

$$Z_{21} = Z_{12} = Z_B \sqrt{1 - \lambda^2}$$

Hence, the [Z] matrix is given by

$$[Z] = \begin{bmatrix} L\lambda + \dfrac{Z_B}{\lambda} & \dfrac{Z_B\sqrt{1-\lambda^2}}{\lambda} \\[3mm] \dfrac{Z_B\sqrt{1-\lambda^2}}{\lambda} & \dfrac{Z_B}{\lambda} \end{bmatrix} \tag{4.61}$$

Note that a comparison of Equations (4.60) and (4.61) reveals that they are in the same generic form.

Example 4.9 Referring to Figure 4.8:

(a) Determine the [Z] matrix for Figure 4.8(a).
(b) Determine the [Z] matrix for Figure 4.8(b).
(c) Comment on the results.

Solution:

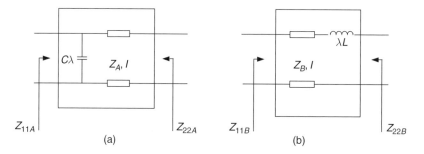

Figure 4.8   Kuroda identity II

(a) Employing the results of the previous examples, the $[Z_A]$ matrix for Figure 4.8(a) is given in a straightforward manner as

$$[Z_A] = \begin{bmatrix} \left(\dfrac{Z_A Z_{CAP}}{Z_A + Z_{CAP}}\right)\left(\dfrac{1}{\lambda}\right) & \left[\dfrac{Z_A Z_{CAP}}{Z_A + Z_{CAP}}\right]\left[\dfrac{\sqrt{1-\lambda^2}}{\lambda}\right] \\[4mm] \left[\dfrac{Z_A Z_{CAP}}{Z_A + Z_{CAP}}\right]\left[\dfrac{\sqrt{1-\lambda^2}}{\lambda}\right] & \left(\dfrac{Z_A^2}{Z_A + Z_{CAP}}\right)\lambda + \left(\dfrac{Z_A Z_{CAP}}{Z_A + Z_{CAP}}\right) \end{bmatrix} \tag{4.62}$$

(b) In a similar manner, the $[Z_B]$ matrix for Figure 4.8(b) is given by

$$[Z_B] = \begin{bmatrix} \dfrac{Z_B}{\lambda} & \dfrac{Z_B\sqrt{1-\lambda^2}}{\lambda} \\ \dfrac{Z_B\sqrt{1-\lambda^2}}{\lambda} & Z_{IND}\lambda + \dfrac{Z_B}{\lambda} \end{bmatrix} \tag{4.63}$$

(c) Close examinations of the impedance matrices of Equations (4.60), (4.61) and (4.62) reveal that they have the same generic form. Therefore, they may be replaced by each other leading to the Kuroda identities as described in the following section.

4.11 Kuroda Identities

Based on the results of Examples 4.8 and 4.9, we propose the following statements.

Statement 11 (Capacitive loading to inductive loading): Referring to Figure 4.7, a single UE of characteristic impedance Z_A loaded with an open stub of characteristic impedance Z_{CAP} in shunt configuration at the output can be replaced by a single UE of characteristic impedance Z_B loaded with a short stub of characteristic impedance Z_{IND} in series configuration at the input.

Statement 12 (Inductive loading to capacitive loading): Referring to Figure 4.8, a single UE of characteristic impedance Z_A loaded with an open stub of characteristic impedance Z_{CAP} in shunt configuration at the input can be replaced by a single UE of characteristic impedance Z_B loaded with a short stub of characteristic impedance Z_{IND} in series configuration at the output.

The above statements were proposed by Kuroda in 1952.[21] Verification of these statements directly follows from the solutions of Examples 4.8 and 4.9 respectively.

The equivalence between the configurations can be obtained by comparing the impedance matrices of Equations (4.60), (4.61) and (4.62), yielding

$$Z_B = \frac{Z_A Z_{CAP}}{Z_A + Z_{CAP}}$$

and

$$Z_{IND} = \frac{Z_A^2}{Z_A + Z_{CAP}} \tag{4.64}$$

By dividing both sides of the above equations it is found that

$$\frac{Z_B}{Z_{IND}} = \frac{Z_{CAP}}{Z_A}$$

The above equations simply say that 'a single UE with characteristic impedance Z_A loaded with an open stub of characteristic impedance Z_{CAP} at the input plane' can be replaced by that of 'a single UE with characteristic impedance $Z_B = Z_A Z_{CAP}/(Z_A + Z_{CAP})$ loaded with a series short stub of characteristic impedance $Z_{IND} = Z_A^2/(Z_A + Z_{CAP})$ at the output port'.

[21] K. Kuroda, 'A Method to Derive Distributed Constant Filters from Constant Filters', Joint Convention of Electrical Engineering Institutes, Japan, Kansai, October 1952, Ch. 9–10.

Similarly, by employing Equation (4.64) for given Z_B and Z_{IND}, one can determine the equivalent circuit characteristic impedances Z_A and Z_{CAP}:

$$\boxed{\begin{aligned} Z_A &= Z_B + Z_{IND} \\ Z_{CAP} &= Z_B \left[1 + \frac{Z_B}{Z_{IND}} \right] \end{aligned}} \qquad (4.65)$$

Equation (4.64) and Equation (4.65) are called the Kuroda identity pairs. They are very useful for manufacturing practical circuits with commensurate transmission lines. From the practical implementation point of view, it may be difficult to implement a short stub connected to UEs in series configuration. On the other hand, open stubs are natural. Therefore, it is always preferable to work with open stubs connected with UEs in shunt configurations. This type of circuit structure is excellent for designing and manufacturing low-pass filters and amplifiers at microwave frequencies using microstrip lines.

It should also be mentioned that, unfortunately, we are not able to derive Kuroda-type identities by connecting series capacitors or shunt inductors with UEs in cascade configuration. The reason is that a single UE in cascade configuration provides a low-pass-like circuit structure, but when it is connected to a series capacitor or to a shunt inductor in the λ domain, which is a high-pass form, we do not obtain proper λ cancellations to end up with Kuroda-like identities.

4.12 Normalization Change and Richard Extractions

Consider the cascaded connections of several commensurate transmission lines with different characteristic impedances $\{Z_1, Z_2,\ Z_k\}$ as shown in Figure 4.9.

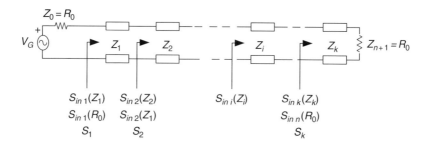

Figure 4.9   Cascaded connections of n commensurate transmission lines

In designing power transfer networks in cascade connections, we usually generate the R_0 normalized input reflection coefficient $S_{11}(\lambda) = h(\lambda)/g(\lambda)$ of a lossless two-port in the λ domain. Then, it is synthesized by extracting UE sections one by one until we end up with the final termination R_0. For example, the first UE with characteristic impedance Z_1 is extracted from $S_{11}(R_0) = S_{in\,1}(R_0)$. However, by Equation (4.14), it is straightforward to express $S_{in\,1}(Z_1)$ in terms of $S_{in\,2}(Z_1)$ as

$$S_{in\ 1}(Z_1) = \left[\frac{1-\lambda}{1+\lambda} \right] S_{in\ 2}(Z_1)$$

Therefore, it may be appropriate to change the normalization from R_0 to Z_1, which leads to characteristic impedance Z_1 in the extraction process, yielding $S_{in\,2}(Z_1)$. The same process goes on.

Now, let us see how we shift the normalization from one section of a UE to the next section.

Normalization Change

Let Z_{i-1} and Z_i be the normalization numbers or equivalently characteristic impedances of two adjacent UE sections indexed by $(i-1)$ and (i) respectively. The input reflectance $S_{in\,i}(Z_{i-1})$ of the UE section is given by

$$S_{in\,i}(Z_{i-1}) = \frac{Z_{in\,i} - Z_{i-1}}{Z_{in\,i} + Z_{i-1}}$$

Similarly,

$$S_{in\,i}(Z_i) = \frac{Z_{in\,i} - Z_i}{Z_{in\,i} - Z_i}$$

On the other hand,

$$Z_{in\,i} = Z_{i-1}\frac{1 + S_{in\,i}(Z_{i-1})}{1 - S_{in\,i}(Z_{i-1})}$$

Using the above driving point impedance expression in $S_{in\,i}(Z_i)$ we have,

$$S_{in\,i}(Z_i) = \frac{Z_{i-1}\dfrac{1 + S_{in\,i}(Z_{i-1})}{1 - S_{in\,i}(Z_{i-1})} - Z_i}{Z_{i-1}\dfrac{1 + S_{in\,i}(Z_{i-1})}{1 - S_{in\,i}(Z_{i-1})} + Z_i}$$

or

$$S_{in\,i}(Z_i) = \frac{(Z_{i-1} - Z_i) + (Z_{i-1} + Z_i)S_{in\,i}(Z_{i-1})}{(Z_{i-1} + Z_i) + (Z_{i-1} - Z_i)S_{in\,i}(Z_{i-1})}$$

Let

$$
\boxed{
\begin{aligned}
&K(i-1,i) \triangleq \frac{Z_{i-1} - Z_i}{Z_{i-1} + Z_i}\\
&\text{or, similarly,}\\
&K(i,i-1) \triangleq \frac{Z_i - Z_{i-1}}{Z_i + Z_{i-1}} = -K(i-1,i)\\
&\text{In general, the quantity}\\
&\qquad K(i,j) = \frac{Z_i - Z_j}{Z_i + Z_j}\\
&\text{may be referred to as the fictitious reflectance.}^{22}
\end{aligned}
}
\tag{4.66}
$$

---

[22] In Equation (4.66), the notation $(i-1,i)$ or (i,j) is used to designate the indexes of K. Here, in this context, we may also use the indexing notations as in $K_{i,j} = K(i,j) = (Z_i - Z_j)/(Z_i + Z_j)$.

Then,

$$S_{in\ i}(Z_i) = \frac{K(i-1,i) + S_{in\ i}(Z_{i-1})}{1 + K(i-1,i)S_{in\ i}(Z_{i-1})} \tag{4.67}$$

The above equations can be expressed in the sequential connection order of Figure 4.9 as a function of the Richard variable λ such that

$$S_{in\ i}(\lambda, Z_i) = \frac{K(i-1,i) + S_{in\ i}(\lambda, Z_{i-1})}{1 + K(i-1,i)S_{in\ i}(\lambda, Z_{i-1})}, i = 1, 2, \ldots, k \tag{4.68}$$

Line Extraction and Degree Reduction

Normalization shift and Equation (3.29) lead the line extractions in a sequential manner as follows. The input reflectance of the section i is expressed in terms of the input reflectance of the section $i + 1$ as

$$S_{in\ i}(\lambda, Z_i) = \left[\frac{1-\lambda}{1+\lambda}\right] S_{in\ (i+i)}(\lambda, Z_i), i = 1, 2, 3, \ldots, k$$

In this case, we can extract section i, leaving a lower degree input reflectance for section $i + 1$ as described below:

$$S_{in\ (i+i)}(\lambda, Z_i) = \left[\frac{1+\lambda}{1-\lambda}\right] S_{in\ i}(\lambda, Z_i), i = 1, 2, 3, \ldots, k \tag{4.69}$$

In Equation (4.68), $S_{in\ i}(\lambda, Z_{i-1})$ can be selected in such a way that we can always introduce a zero at $\lambda = 1$ in the numerator:

$$[K(i-1,i) + S_{in\ i}(\lambda, Z_{i-1})]$$

yielding a possible cancellation with the term $(1 - \lambda)$ in the denominator of Equation (4.69). In order to obtain this cancellation it is sufficient to set

$$K(i-1,i) + S_{in\ i}(\lambda, Z_{i-1})|_{\lambda=1} = 0, \qquad i = 1, 2, 3, \ldots, k$$

or at $\lambda = 1$

$$K(i-1,i) = -S_{in\ i}(1, Z_{i-1}), \qquad i = 1, 2, 3, \ldots, k \tag{4.70}$$

Furthermore, by employing Equation (4.70) in the denominator term of Equation (4.68), it is seen that at $\lambda = -1$

$$1 + K(i-1,i)S_{in\ i}(\lambda, Z_{i-1}) = 1 - S_{in\ i}(1, Z_{i-1})S_{in\ i}(-1, Z_{i-1})$$

Since the input reflectance $S_{in\ i}(\lambda, Z_{i-1})$ defines a lossless two-port which consists of commensurate transmission lines, then

$$S_{21i}(\lambda)S_{21i}(-\lambda) = \frac{(1-\lambda^2)^{n-1+1}}{gi(\lambda)} = 1 - S_{in\ i}(\lambda, Z_{i-1})S_{in\ i}(-\lambda, Z_{i-1})$$

This equation is valid for all values of λ including $\lambda = 1$. Therefore,

$$1 + K(i-1, i)S_{in\ i}(\lambda, Z_{i-1}) = 1 - S_{in\ i}(1, Z_{i-1})S_{in\ i}(-1, Z_{i-1}) = 0$$

Thus, we see that by selecting, $K(i-1, i) = -S_{in\ i}(1, Z_{i-1})$, we simultaneously introduce the terms $(1-\lambda)$ and $(1+\lambda)$ into the numerator and the denominator of $S_{in\ i}(\lambda, Z_i)$ in Equation (4.69), which leads to cancellations in

$$S_{in(i+i)}(\lambda, Z_i) = \left[\frac{1+\lambda}{1-\lambda}\right]S_{in\ i}(\lambda, Z_i)$$

of Equation (4.69), resulting in lower degree numerator and denominator polynomials in $S_{in\ i}(\lambda, Z_i)$. For example, when we set $i = 1$, Equation (4.70) reveals that

$$K(0, 1) = \frac{Z_0 - Z_1}{Z_0 + Z_1} = -S_{in\ 1}(1, Z_0)$$

or

$$Z_1 = Z_0 \frac{1 + S_{in\ 1}(1, Z_0)}{1 - S_{in\ 1}(1, Z_0)}$$

where $Z_0 = R_0$.

Richard Extractions

We can simplify the nomenclature of the above equations by setting $S_i = S_{in\ i}$ and then describe the extraction of a UE in cascade configuration as follows.

Let the $Z_0 = R_0$ normalized input reflectance of a lossless two-port consisting of commensurate transmission lines be

$$S_{11}(\lambda) = S_1(\lambda, Z_0) = \frac{h(\lambda)}{g(\lambda)} = \frac{h_1\lambda^n + h_2\lambda^{(n-1)} + \ldots + h_n\lambda + h_{(n+1)}}{g_1\lambda^n + g_2\lambda^{(n-1)} + \ldots + g_n\lambda + g_{(n+1)}}$$

Further assume that this lossless two-port includes a k section of UEs in cascade configurations. In other words, the lossless two-port designated by $S_{11}(\lambda) = h(\lambda)/g(\lambda)$ reveals a transfer scattering parameter $S_{21}(\lambda) = f(\lambda)/g(\lambda)$ such that the polynomials $h(\lambda)$ and $g(\lambda)$ result in

$$f(\lambda)f(-\lambda) = g(\lambda)g(-\lambda) - h(\lambda)h(-\lambda) = \lambda^{2q}(1-\lambda^2)^k$$

Then, the Richard extraction of the characteristic impedances of UEs is given by

$$Z_i = Z_{i-1} \frac{1 + S_i(1, Z_{i-1})}{1 - S_i(1, Z_{i-1})}, i = 1, 2, \ldots, k$$

where initially the input reflection coefficient is specified as

$$S_1(\lambda, Z_0) = \frac{h(\lambda)}{g(\lambda)}$$

$$= \frac{h_1 \lambda^n + h_2 \lambda^{(n-1)} + \ldots + h_n \lambda + h_{(n+1)}}{g_1 \lambda^n + g_2 \lambda^{(n-1)} + \ldots + g_n \lambda + h_{(n+1)}}$$

and follow-up reflection coefficients are given in sequential

order by   (4.71)

$$S_{(i+i)}(\lambda, Z_i) = \left[\frac{1 + \lambda}{1 - \lambda}\right] S_i(\lambda, Z_i), i = 1, 2, \ldots, k$$

where

$$S_i(\lambda, Z_i) = \frac{K(i-1, i) + S_i(\lambda, Z_{i-1})}{1 + K(i-1, i) S_i(\lambda, Z_{i-1})}, i = 1, 2, \ldots k$$

and

$$K(i-1, i) = \frac{Z_{i-1} - Z_i}{Z_{i-1} + Z_i}, i = 1, 2, \ldots, k$$

Equation (4.71) is called the Richard extraction set, which synthesizes the given input reflection coefficient $S_{11}(\lambda, R_0) = S_1(\lambda, R_0)$ in a sequential manner as cascaded connections of UEs with characteristic impedances Z_i.

The above computational steps may be summarized in the following algorithm.

Richard Algorithm: Extraction of UEs from the Given Input Reflectance $S_{11}(\lambda) = h(\lambda)/g(\lambda)$

Here, it is assumed that the input reflectance includes cascaded connections of k UE sections yielding a transfer scattering parameter $S_{21}(\lambda)$ such that

$$S_{21}(\lambda)S_{21}(-\lambda) = 1 - S_{11}(\lambda)S_{11}(-\lambda) = g(\lambda)g(-\lambda) - h(\lambda)h(-\lambda) = f(\lambda)f(-\lambda)$$
$$= \lambda^{2q}(1 - \lambda^2)^{2k}$$

Input: The input reflectance is

$$S_1(\lambda, Z_0) = \frac{h_1 \lambda^n + h_2 \lambda^{n-1} + \ldots + h_n \lambda + h_{n+1}}{g_1 \lambda^n + g_2 \lambda^{n-1} + \ldots + g_n \lambda + g_{n+1}}$$

Computational Steps:

Step 1: Compute $S_1(\lambda, Z_0)|_{\lambda=1} = S_{11}(1)$ which is specified with respect to the normalization impedance $R_0 = Z_0 = 1$. In this case,

$$S_1(1, Z_0) = \frac{h_1 + h_2 + \ldots + h_n + h_{n+1}}{g_1 + g_2 + \ldots + g_n + g_{n+1}}$$

Step 2: Compute Z_i for $i = 1, 2, \ldots, k$, as

$$Z_i = Z_{i-1} \frac{1 + S_{i+1}(1, Z_{i-1})}{1 - S_{i+1}(1, Z_{i-1})}$$

Step 3: Compute

$$K(i-1, i) = \frac{Z_{i-1} - Z_i}{Z_{i-1} + Z_i}, \text{ for } i = 1, 2, \ldots, k$$

Step 4: Generate

$$S_i(\lambda, Z_i) = \frac{K(i-1, i) + S_i(\lambda, Z_{i-1})}{1 + K(i-1, i)S_i(\lambda, Z_{i-1})}, \ i = 1, 2, \ldots, k$$

Remember that, in the λ domain, the reflectance $S_i(\lambda, Z_i)$ must include the term $(1 - \lambda)/(1 + \lambda)$ to reduce the degree.

Step 5: Determine the reflection coefficient of the next section as in

$$S_{i+1}(\lambda, Z_i) = \left[\frac{1 + \lambda}{1 - \lambda}\right] S_i(\lambda, Z_i), \ i = 1, 2, \ldots, k$$

Step 6: This is the final step of the algorithm. After extracting k cascaded UE sections, we will end up with a reflectance S_{k+1} which in turn yields the termination impedance Z_{k+1}. If $k = n$, this impedance must be a simple resistor. However, if $n > k$, then it must be a positive real function which may be synthesized as a lossless two-port in the form of open/short stub connections in series and parallel configurations. Obviously, it should be free of UE sections since they are extracted in the previous steps.

Let us examine the use of the Richard extraction algorithm by means of a simple example.

Example 4.10: Let

$$S_{11}(\lambda) = \frac{h(\lambda)}{g(\lambda)} = \frac{\lambda^2 + 19\lambda}{\lambda^2 + 21\lambda + 8}$$

be the unit-normalized (i.e. $Z_0 = 1$) input reflection coefficient of a lossless two-port constructed with ideal commensurate TEM lines. Synthesize $S_{11}(\lambda)$.

Solution: First by, let us investigate the topology of the lossless two-port by deriving the forward transfer scattering parameter $S_{21}(\lambda) = f(\lambda)/g(\lambda)$.

Equation (4.44) reveals that

$$f(\lambda)f(-\lambda) = g(\lambda)g(-\lambda) - h(\lambda)h(-\lambda)$$

or

$$f(\lambda)f(-\lambda) = [\lambda^2 + 21\lambda + 8][\lambda^2 - 21\lambda + 8] - [\lambda^2 + 19\lambda][\lambda^2 - 19\lambda]$$

$$f(\lambda)f(-\lambda) = \lambda^4 - 425\lambda^2 + 64 - [\lambda^4 - 361\lambda^2] = -64\lambda^2 + 64 = 64(1 - \lambda^2)$$

The above result yields $f(\lambda) = 8(1 - \lambda^2) = 8(1 - \lambda^2)^{k/2}$ with $k = 1$.

Hence, we see that $S_1(\lambda, Z_0) = S_{11}(\lambda)$ must include only one UE section. Now let us implement the algorithm.

Step 1: Compute $S_1(\lambda, Z_0)|_{\lambda = 1} = S_{11}(1)$ with $Z_0 = 1$:

$$S_{11}(1) = \frac{20}{1 + 21 + 8} = \frac{20}{30} = 2/3$$

Step 2: Compute Z_i for $i = 1, 2, .., k$ as

$$Z_i = Z_i \frac{1 + S_i(1, Z_0)}{1 - S_i(1, Z_0)}$$

$$Z_1 = \frac{1 + \dfrac{2}{3}}{1 - \dfrac{2}{3}}$$

$$Z_1 = 5\,\Omega$$

Step 3: Compute

$$K(i - 1, i) = \frac{Z_{i-1} - Z_i}{Z_{i-1} + Z_i}, \text{ for } i = 1, 2, .., k$$

Setting $i = 1$, we have

$$K(0, 1) = \frac{Z_0 - Z_1}{Z_0 + Z_1} = \frac{1 - 5}{1 + 5} = -4/6$$

Step 4: Generate

$$S_i(\lambda, Z_i) = \frac{K(i - 1, i) + S_i(\lambda, Z_{i-1})}{1 + K(i - 1, i)S_i(\lambda, Z_{i-1})}, i = 1, 2, .., k$$

Recalled that, in the λ domain, $S_i(\lambda, Z_i)$ must include the term $(1 - \lambda)/(1 + \lambda)$. Then, for $i = 0$,

$$S_1(\lambda, Z_1) = \frac{-(4/6) + S_1(\lambda, Z_0)}{1 - (4/6)S_1(\lambda, Z_0)} = \frac{-\dfrac{4}{6} + \dfrac{\lambda^2 + 19\lambda}{\lambda^2 + 21\lambda + 8}}{1 - \dfrac{4}{6}\dfrac{\lambda^2 + 19\lambda}{\lambda^2 + 21\lambda + 8}} = \frac{2\lambda^2 + 30\lambda - 32}{2\lambda^2 + 50\lambda + 48}$$

$$= -\frac{2(1 - \lambda)(16 + \lambda)}{2(1 + \lambda)(24 + \lambda)}$$

Hence, we get

$$S_1(\lambda, Z_1) = -\frac{(1-\lambda)(16+\lambda)}{(1+\lambda)(24+\lambda)}$$

Step 5: Determine the reflection coefficient of the next section as in

$$S_{i+1}(\lambda, Z_i)\left[\frac{1+\lambda}{1-\lambda}\right]S_i(\lambda, Z_i), \ i=1, 2, \ldots, k$$

Hence, for $i=1$ we have

$$S_2(\lambda, Z_1) = \left[\frac{1+\lambda}{1-\lambda}\right]S_1(\lambda, Z_1) = -\left[\frac{1+\lambda}{1-\lambda}\right]\frac{(1-\lambda)(16+\lambda)}{(1-\lambda)(24+\lambda)}$$

or

$$S_2(\lambda, Z_1) = \frac{h(\lambda)}{g(\lambda)} = -\frac{\lambda+16}{\lambda+24}$$

Step 6: At this stage we know that $(n=2) > (k=1)$. Therefore, we expect to obtain a PR termination impedance $Z_2(\lambda)$ which is given by

$$Z_{k+1} = Z_k\frac{1+S_{k+1}(\lambda, Z_k)}{1-S_{k+1}(\lambda, Z_k)}$$

or

$$Z_2 = 5\frac{1-\dfrac{\lambda+16}{\lambda+24}}{1+\dfrac{\lambda+16}{\lambda+24}} = \frac{1}{\left(\dfrac{1}{20}\right)\lambda+1}$$

Actually, the above impedance can simply be realized in the λ domain as the parallel combination of a capacitance $C = 1/20$ F and a resistance $R = 1\ \Omega$. Obviously these are all normalized values with respect to the given center frequency $\Omega_0 = \tan(\omega_0\tau)$ and standard normalization number R_0. Hence, the final synthesis is as shown in Figure 4.10.

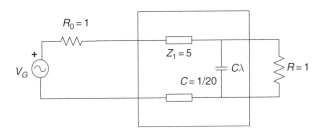

Figure 4.10 Synthesis of the input reflectance for Example 4.10

We should note that the above structure can easily be replaced by a shorted series stub connected at the input plane using the Kuroda identities. In this case, the UE's characteristic impedance Z_B is given by Equation (4.64) as

$$Z_B = \frac{Z_1 Z_{CAP}}{Z_1 + Z_{CAP}}$$

such that $Z_{CAP} = 1/C = 20$ and $Z_1 = 5$, or $Z_B = 4$. Similarly, the characteristic impedance of the shorted stub Z_{IND} is given by

$$Z_{IND} = \frac{Z_1^2}{Z_1 + Z_{CAP}} = 1$$

4.13 Transmission Zeros in the Richard Domain

In designing lossless two-ports, the structure of circuit topology can be controlled via the numerator polynomial of the transfer scattering parameter $S_{21}(\lambda) = f(\lambda)/g(\lambda)$.

With R_1 and R_2 being the port normalization numbers at the input and output ports respectively, $|S_{21}(j\Omega)|^2$ specifies TPG of a lossless two-port when the two-port is excited with a voltage generator of internal resistance R_1 while it is terminated in R_2, the output port.

Over the real frequency axis, TPG is specified by Equation (4.35) as

$$T(\Omega) = |S_{21}(j\Omega)|^2 = \frac{P_L}{P_A} = S_{21}(j\Omega)S_{21}(-j\Omega) = \frac{f(j\Omega)f(-j\Omega)}{g(j\Omega)g(-j\Omega)} \qquad (4.72)$$

Here, it is interesting to note that the power transfer capability of the two-port is determined by the shape and the zeros of $S_{21}(j\Omega)S_{21}(-j\Omega)$.

If $S_{21}(j\Omega)S_{21}(-j\Omega) = 0$, then there is no power transferred to the load. Therefore, we make the following broad definitions.

Definition 10 (Transmission zeros in the λ domain): For a lossless two-port the zeros of

$$T(\lambda) = S_{21}(\lambda)S_{21}(-\lambda) = \frac{f(\lambda)f(-\lambda)}{g(\lambda)g(-\lambda)} = \frac{F(\lambda^2)}{G(\lambda^2)}$$

are called the transmission zeros.

Definition 11 (Finite zeros of transmission): The zeros of $f(\lambda)f(-\lambda)$ are called the finite zeros of transmission of the lossless two-port.

Definition 12 (Transmission zeros at infinity): Let the degree of the numerator polynomial $F(\lambda^2) = f(\lambda)f(-\lambda)$ be $2m$ and the degree of the denominator polynomial $G(\lambda^2) = g(\lambda)g(-\lambda)$ be $2n$; then, the count for the transmission zeros at infinity is $2n_\infty = 2n - 2m$.

For a lossless two-port constructed with commensurate transmission lines, a typical practical form of $F(\lambda^2) = f(\lambda)f(-\lambda)$ is given by

$$F(\lambda^2) = (-1)^q \lambda^{2q} \prod_{l=1}^{n_z} [\lambda^2 + \Omega_l^2]^2 (1 - \lambda^2)^k$$

in association with

$$G(\lambda^2) = G_1 \lambda^{2n} + G_2 \lambda^{2(n-1)} + \ldots + G_n \lambda^2 + G_{n+1}$$

(4.73)

In Equation (4.73):

- k is the total count of cascaded UEs.
- $2q$ is the total count of transmission zeros at DC and they are realized as series short stubs or shunt open stubs.
- $2n_z$ is the count of the finite zeros of transmission on the $j\Omega$ axis.
- n is the total number of elements in the lossless two-port.
- Note that $n \geq q + 2n_z + k$.
- $2n_\infty = 2(n - q + 2n_z + k)$ is the total count of transmission zeros at ∞.

4.14 Rational Form of the Scattering Parameters and Generation of $g(\lambda)$ via the Losslessness Condition

As was shown in Section 4.10, for a reciprocal lossless two-port constructed with commensurate transmission lines, the scattering parameters are given in rational form as

$$S_{11}(\lambda) = \frac{h(\lambda)}{g(\lambda)} \quad S_{12}(\lambda) = \frac{f(\lambda)}{g(\lambda)}$$
$$S_{21}(\lambda) = \frac{f(\lambda)}{g(\lambda)} \quad S_{22}(\lambda) = -\left[\frac{f(\lambda)}{f(-\lambda)}\right]\left[\frac{h(\lambda)}{g(\lambda)}\right] \tag{4.74}$$

with

$$g(\lambda)g(-\lambda) = h(\lambda)h(-\lambda) + f(\lambda)f(-\lambda) \tag{4.75}$$

The above rational form of the scattering parameters is called the Belevitch form.[23]

Obviously, Equation (4.75) stems from the losslessness condition of $SS^\dagger = I$ of the BR scattering parameters. The equation can be viewed from a different perspective as expressed in the following statement.

Statement 13 (Fundamental relations between polynomials g, h, f): No matter what the polynomials $g(\lambda), h(\lambda)$ and $f(\lambda)$ are, if the triplet $\{g(\lambda), h(\lambda) \text{ and } f(\lambda)\}$ satisfy the losslessness condition of Equation (4.75), then the following inequalities must be satisfied:

$$g(j\Omega)g(-j\Omega) \geq h(j\Omega)h(-j\Omega) \geq 0, \forall\Omega$$
$$g(j\Omega)g(-j\Omega) \geq f(j\Omega)f(-j\Omega) \geq 0, \forall\Omega \tag{4.76}$$

Proof of this statement is directly from the definition of complex numbers and losslessness.

4.15 Generation of Lossless Two-ports with Desired Topology

Many practical microwave design problems, such as filters and matching network designs, demand the construction of special types of lossless two-ports consisting of commensurate transmission lines. Among these, the most practical ones are the cascade connection of UEs. The reason is that cascaded

---

[23] V. Belevitch, *Classical Network Theory*, Holden Day, 1968.

UEs are easily manufactured on selected substrates in the form of microstrip lines with different characteristic impedances. Therefore, it is essential to control the design topology of cascaded connection of UEs.

In order to design a specific topology, one has to fix $F(\lambda^2)$ in advance. In other words, transmission zeros at DC and at finite frequencies Ω_l are specified based on the passband and stopband requirements. By selecting integer k, a total number of k UEs are introduced in cascade connections. Foremost of these, no matter how the transmission zeros are selected, is that the common denominator polynomial $g(\lambda)$ must be strictly Hurwitz. Furthermore, the scattering parameters of the lossless two-port must be BR.

Considering all the above points, Equation (4.75) and Statement 14, we conclude that, in designing a lossless two-port for a specific problem, the only free parameter is that of the numerator polynomial $h(\lambda)$. In this case, we can propose the following statement.

Statement 14 (Construction of lossless two-ports for specified transmission zeros in the Richard domain): Let the lossless two-port which consists of commensurate transmission lines be described in terms of its real normalized scattering parameters as in Equation (4.74). Then the complete scattering parameters of the two-port can be generated by means of the numerator polynomial

$$h(\lambda) = h_1 \lambda^n + h_2 \lambda^{n-1} + \ldots + h_n \lambda + h_{n+1}$$

of the input reflectance $S_{11}(\lambda) = h(\lambda)/g(\lambda)$ if and only if the transmission zeros and the total number of cascaded connections of UEs are specified, satisfying the condition that $h(j\Omega)h(-j\Omega) + f(j\Omega)f(-j\Omega)$ is strictly positive for all Ω.[24]

Proof: Once the transmission zeros and total number of cascade connections of UEs are specified, it is straightforward to construct $f(\lambda)$ by means of Equation (4.73) as

$$f(\lambda) = \lambda^q \prod_{l=1}^{n_z} (\lambda^2 + \Omega_l^2)(1 - \lambda^2)^{k/2}$$

This form of $f(\lambda)$ may be irrational if the integer k is odd. However, the function $F(\lambda^2) = f(\lambda)f(-\lambda) = F(\lambda^2) = (-1)^q \lambda^{2q} \prod_{l=1}^{n_z} [\lambda^2 + \Omega_l^2]^2 (1 - \lambda^2)^k$ is always an even polynomial in the Richard variable λ.[25]

On the other hand, from the given condition of the statement, the even polynomial $G(-\Omega^2) = g(j\Omega)g(-j\Omega) = h(j\Omega)h(-j\Omega) + f(j\Omega)f(-j\Omega)$ is always positive. Therefore, it must be free of $j\Omega$ zeros.

Let $X = \Omega^2 = -\lambda^2$. Then the roots of the real polynomial $G(X) > 0$ must satisfy the following conditions:

(a) The real roots must be negative since $G(X) = (X + X_r)\tilde{G}(X)$ must be strictly positive for all X.
(b) A complex root must be paired with its conjugate since $G(X)$ is real.

Note that $\lambda = j\Omega$ requires that $\lambda^2 = -\Omega^2 = -X$. Therefore, in the λ domain, a root of $G(\lambda^2) = g(\lambda)g(-\lambda)$ must be determined as $\lambda_r = \pm\sqrt{-X_r}$. Thus, the roots are mirror symmetric with respect to the $j\Omega$ axis as shown in Figure 4.11.

For example, let $X_r = -\sigma_r$ with $\sigma_r > 0$ be a real root of $G(X)$; then, $\lambda_r = \pm\sqrt{\sigma_r}$ must be the mirror image symmetric roots of the even polynomial $G(\lambda^2)$. Similarly, if a complex conjugate paired root of

---

[24] Statement 14 can be equivalently expressed in terms of the backend reflection coefficient $S_{22}(\lambda)$.
[25] For almost all the practical power transfer network designs we skip finite transmission zeros on the $j\Omega$ axis. Therefore, we have the freedom to set $n_z = 0$.

$G(X)$ is given by $X_i = \gamma_i \pm j\delta_1$, then the roots of $G(\lambda^2)$ are determined as $\lambda_i = \pm\sqrt{\gamma_i \pm j\delta_1} = \pm[\alpha_i \pm j\beta_i]$ which are also mirror image symmetric as shown in Figure 4.11. In this case, one can always construct a Strictly Hurwitz polynomial on the LHP roots of $G(\lambda^2)$.

Thus, as shown above, once $f(\lambda)$ is selected based on the design specifications and the coefficients of $h(\lambda)$ are initialized, then $g(\lambda)$ is generated immediately as in the following algorithm.

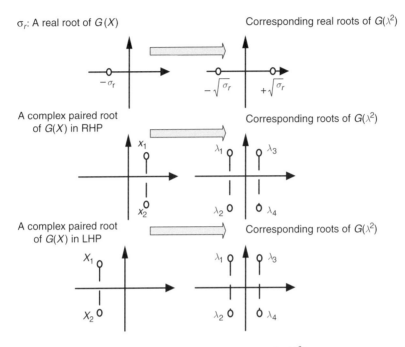

Figure 4.11 Mirror image roots of $G(\lambda^2)$

Algorithm: Generation of the Strictly Hurwitz Denominator Polynomial g(λ)

This algorithm generates the strictly Hurwitz polynomial $g(\lambda)$ from the initialized coefficients of $h(\lambda)$ and for the specified transmission zeros and number of cascade UE connections.

Inputs:

- A row vector $\mathbf{h} = [h_1 h_2 h_3 \ldots h_n h_{n+1}]$, the coefficients of $h(\lambda)$.
- Integer k, the total number of cascade connections of UEs.
- Integer q, the transmission zeros at DC.
- Integer n_z and $\Omega_i, i = 1, 2, \ldots, n_z$, finite transmission zeros of the $j\Omega$ axis.

Note that degree 'n' of the numerator polynomial $h(\lambda)$ must be equal to or larger than $q + 2n_z + k$. For many practical design problem we set $n_z = 0$.

Computational Steps:

Step 1: Generate the polynomial $H(\lambda^2) = h(\lambda)h(-\lambda)$ in full polynomial coefficient form as

$$H(\lambda^2) = H_1\lambda^{2n} + H_2\lambda^{2(n-1)} + \ldots + H_n\lambda^2 + H_{n+1}$$

In MATLAB®, this polynomial is generated as a row vector by taking the convolution of $h(\lambda)$ and $h(-\lambda)$.

Step 2: From the given integers q, n_z, k and finite transmission zeros, generate the full coefficient form for $F(\lambda^2)$ as in Equation (4.73)

$$F(\lambda^2) = F_1\lambda^{2n} + F_2\lambda^{2(n-1)} + \ldots + F_n\lambda^2 + F_{n+1}$$

In MATLAB, $F(\lambda^2)$ is represented by a row vector $\mathbf{F} = [F_1 F_2 \ldots F_n F_{n+1}]$.

Step 3: Generate $G(\lambda^2) = H(\lambda^2) + F(\lambda^2)$ in full coefficient form. In MATLAB, $G(\lambda^2)$ is represented by a row vector $\mathbf{G} = \mathbf{H} + \mathbf{F}$.

Step 4: In step 3, the polynomial $G(\lambda^2)$ is an even function of the Richard variable λ. Therefore, we set $X = -\lambda^2$ and find the roots of $G(X)$. In MATLAB, this operation is completed by flipping the signs of the vector $\mathbf{G}$ such that $\mathbf{GX} = [(-1)^n \ldots (G_{n-1}) - (G_n)(G_{n+1})]$.

Step 5: Find the roots of $G(X)$. In MATLAB, the roots of a polynomial are simply found by the function root. In this case, we set $X_r = \text{roots}(GX)$.

Step 6: Set $\lambda_r = \pm\sqrt{-X_r}$ and by selecting LHP roots generate $g(\lambda)$ as

$$g(\lambda) = \left[\sqrt{\text{abs}(G_1)}\right] \prod_{i=1}^{n} [\lambda + \alpha_i + j\beta_i] = g_1\lambda^n + g_2\lambda^{n-1} + \ldots + g_n\lambda + g_{n+1}$$

In MATLAB, the above operations can simply be completed by means of readily available polynomial functions.

Let us show the implementation of the above steps by means of an example.

Example 4.11: Find the complete scattering parameters of the lossless two-port described by $q = 0, n_z = 0, k = 2$, and $h(\lambda) = 3\lambda^2 + 2\lambda$.

Solution: Using the above algorithm, we can generate $f(\lambda)$ and $g(\lambda)$ step by step, as follows.

Step 1: Since $h(\lambda) = 3\lambda^2 + 2\lambda$, then the row vector $\text{h} = [3\ 2\ 0]$.

Now $h(-\lambda) = 3\lambda^2 - 2\lambda$ is the para-conjugate of $h(\lambda)$ and it may be specified by a row vector $\text{h}\_ = [3\ -2\ 0]$. Thus, in MATLAB the row vector H is computed as $\text{H} = \text{conv}(\text{h}, \text{h}\_)$ which yields $\text{H} = [9\ 0\ -4\ 0\ 0]$. This vector also includes odd terms of a polynomial convolution. Therefore, we should suppress these odd terms. Hence, $\text{H} = [9\ -4\ 0]$, meaning that $H(\lambda^2) = 9\lambda^4 - 4\lambda^2$.

At this point we should note that, in MATLAB, we developed two functions to perform the above computations. They are listed as follows.

Program Listing 4.1 Para-conjugation; generation of $h(-\lambda)$ from $h(\lambda)$

```
function h_=paraconj(h)
% This function generates the para-
% conjugate of a polynomial h=[]
na=length(h);
n=na-1;sign=-1;h_(na)=h(na);
for i=1:n
h_(n-i+1)=sign*h(n-
i+1);%Para_conjugate coefficients
sign=-sign;

end
```

Program Listing 4.2   Selection of even terms of a given polynomial H

```
function He=poly_eventerms (H)
% This function eliminates the zeros of a
% polynomial H=conv(h,h_)
% Here h_ is the para-conjugate of
% polynomial h.
% h_ is computed from "function
% h_=paraconj (h)"
Na=length(H);%Degree of H including zeros
n=(Na-1)/2;
na=n+1;
He(na)=H(Na);
for i=1:n
He(n-i+1)=H(Na-2*i);

end
```

Step 2: Since $q = 0$ and $n_z = 0$, then $F = f(\lambda)f(-\lambda) = [1 - \lambda^2]^2$. The open form of $F(\lambda)$ can be generated by a MATLAB function which is listed below.

Program Listing 4.3   Computation of binomial coefficients of
$F = (1 - \lambda^2)^k = [F(1)F(2)F(3)\ldots F(k)F(k+1)]$

```
function F=cascade(k)
% This function generates the F=f*f_
polynomial of
% Cascade connections of k UEs in Richard
% Domain.
% F=(1-lamda^2)^k
%
F=F(1)(Lambda)^2n+...F(k)(lambda)^2+F(k+1)
F=[1]; %Unity Polynomial;
UE=[-1 1];% Single Unit-Element; [1-
% (lambda)^2]
for i=1:k
F=conv(F,UE);

end
```

Hence, it is found that for $k = 2$, F= [1 -2 1]. In other words, $F(\lambda^2) = (1 - \lambda^2)^2 = \lambda^4 - 2\lambda^2 + 1$.

Step 3: Generate $G(\lambda^2) = H(\lambda^2) + F(\lambda^2)$ in full coefficient form. This is easy since H and F have already been generated in MATLAB. Thus, G=H+F. In MATLAB, we can add only vectors of the same size. However, we developed a function called vector_sum which adds two vectors starting from the right end. Function vector_sum is listed as follows.

Program Listing 4.4   Row vector summation with different lengths

```
function C=vector_sum(A,B)
% This function adds two vectors with
% different sizes
% Size of A is na; Size of B is nb
% A=[A(1) A(2)...A(na) A(na + 1)]
% B=[B(1) B(2)...B(nb) B(n + 1)]
% m=max(na,nb);
% C(m+1)=A(na+1)+B(nb+1),
% C(m)=A(na)+B(nb),
% C(m-1)=A(na-1)+B(nb-1)....
%
na=length(A);
nb=length(B);
m=max(na,nb);
% flip the vectors from left to right
for i=1:na;AA(i)=A(na-i+1);end
for i=1:nb;BB(i)=B(nb-i+1);end
%
if(na>nb);
for i=(nb+1):na;BB(i)=0; end
end

if (nb>na)
for i=(na+1):nb;AA(i)=0;end

end

for i=1:m;
CC(i)=AA(i)+BB(i)

end

for i=1:m

C(i)=CC(m-i+1)

end
```

Hence, G=[10 -6 1]. In other words, $G(\lambda^2) = 10\lambda^4 - 6\lambda^2 + 1$.

Step 4: In this step, by setting $X = -\lambda^2$ the polynomial $G(X)$ is formed. In MATLAB, this action can easily be performed automatically using the function `paraconj(G)`.
Hence, we have >>GX=[10 6 1], meaning that $G(X) = 10X^2 + 6X + 1$.

Step 5: Find the roots of $G(X)$. This is easy. In MATLAB, the function root directly computes the roots of a given polynomial. Hence, >>Xr=roots(GX)

```
X(1)= -0.3000 + 0.1000i; X(2)= -0.3000 - 0.1000i
```
As can be seen, we have two complex conjugate paired roots in the LHP plane of X which leads to four roots of $G(\lambda^2)$ as depicted in Figure 4.11.

Step 6: Set $\lambda_r = \pm\sqrt{-X_r}$ and by selecting the LHP roots generate $g(\lambda)$ as

$$g(\lambda) = \left[\sqrt{abs(G_1)}\right] \prod_{i=1}^{n} [\lambda + \alpha_i + j\beta_i] = g_1\lambda^n + g_2\lambda^{n-1} + \ldots + g_n\lambda + g_{n+1}$$

The roots λ_r (lambda$_r$) are computed directly in MATLAB as

$$\text{lambda}(1,3) = \pm(0.5551 - 0.0901i)$$

and

$$\text{lambda}(2,4) = \pm(0.5551 + 0.0901i)$$

Then, $g(\lambda)$ is constructed on the LHP roots of $G(\lambda^2)$ as

$$g(\lambda) = \sqrt{10}(\lambda + 0.5551 - 0.0901i)(0.5551 + 0.0901i)$$

or

$$g(\lambda) = 3.1623\lambda^2 + 3.5106\lambda + 1.0000$$

Actually, all the above steps are programmed within a function called function g=htog(q,k,h). In this function, a MATLAB code called function F=lambda2q_UE2k(k,q) generates the open form of

$$F(\lambda^2) = f(\lambda)f(-\lambda) = (-1)^q \lambda^{2q}(1 - \lambda^2)^k = F_1\lambda^{2(q+k)} + \ldots + F_{q+k} + F_{q+k+1}$$

Obviously, the integer n must be $n \geq q + k$.

Now, let us test the MATLAB functions to generate a selected topology.

Example 4.12: Let $h = 3\lambda^5 + 2\lambda^4 + \lambda^3 + 5\lambda^2 - 8\lambda + 1$ and $q = 2, k = 3$. Find the open forms of:

(a) $F(\lambda^2) = (1)^q \lambda^{2q}(1 - \lambda^2)^k = F_1\lambda^{2(q+k)} + \cdot + F_{q+k} + F_{q+k+1}$.
(b) $g(\lambda)$.

Solution:

(a) Using the MATLAB function function F=lambda2q_UE2k(k,q) we can immediately generate the open form of $F(\lambda^2) = f(\lambda)f(-\lambda)$ as F=[-1 3 -3 1 0 0], meaning that

$$F(\lambda^2) = -\lambda^{10} + 3\lambda^8 - 3\lambda^6 + \lambda^4$$

(b) Using the main program CH4COMTRL_Example12.m we can immediately generate $g(\lambda)$. Here is the result:

g=[3.1623 9.6098 14.4433 17.2807 9.4107 1.0000]

meaning that

$$g(\lambda) = 3.1623\lambda^5 + 9.6098\lambda^4 + 14.4433\lambda^3 + 17.2807\lambda^2 + 9.4107\lambda + 1$$

In the following:

- the main program CH4COMTRL_Example12.m,
- function F=lambda2q_UE2k(k,q) and
- function g=htog(q,k,h)

Program Listing 4.5   Main program for Example 4.12

```
% Main Program for Chapter 4
% CH4COMTRL_Example12.m


% CIRCUITS CONSTRUCTED WITH COMMENSURATE
% TRANSMISSION LINES
% Chapter 4
% Example (4.12)
clear
h=[3 2 1 5 -8 1]
q=2
k=3
g=htog(q,k,h)
```

Program Listing 4.6   Computation of Strictly Hurwitz polynomial $g(\lambda)$

```
function g=htog(q,k,h)
%This function is written in Richard Domain.
%Given h(lambda) as a polynomial
%Given f(lambda)=(-1)^q*(lambda)^q*(1-lambda^2)^k
%Computes g(lambda)
%
F=lambda2q_UE2k(k,q);% Generation of
F=f(lambda)*f(-lambda)
h_=paraconj(h);
He=conv(h,h_);%Generation of h(lambda)*h(-lambda)
H=poly_eventerms(He);%select the even terms
```

```
n1=length(H);
G=vector_sum(H,F);% Generate G=H+F with different
sizes
GX=paraconj(G);% Generate G(X)
Xr=roots(GX);% Compute the roots og G(X)
z = sqrt(-Xr);% Compute the roots in lambda
%**************************************************
% Generation of g(lambda) from the given LHP
% roots
% Compute the first step k=1
n=length(z);
g=[1 z(1)]
for i=2:n
g=conv(g,[1 z(i)]);
end
Cnorm=sqrt(abs(G(1)));
g=Cnorm*real(g);
```

Program Listing 4.7 Generation of the numerator polynomial of $S_{21}(\lambda)S_{21}(-\lambda) = F(\lambda^2)/G(\lambda^2)$ such that $F(\lambda^2) = (-1)^q \lambda^{2q} (1 - \lambda^2)^k$

```
function F=lambda2q_UE2k(k,q)
% This function computes F=(-
1)^q(lambda)^2q(1-lambda^2)^k
Fa=cascade(k);
nf=length(Fa);
nq=nf+q;
for i=1:nq
F(i)=0.0;
end
if q==0;
F=Fa;
end
if(q>0)
for i=1:nf
F(i)=((-1)^q)*Fa(i);
end
for i=1:q
F(nf+i)=0.0;
end
end
end
```

The next example in this chapter constructs a lossless two-port consisting of several cascaded connections of UEs.

Example 4.13: Referring to Statement 14, a lossless two-port is constructed by Cascade connections of three UEs. Its $50\,\Omega$ normalized Input reflection coefficient $S_{11}(\lambda) = h(\lambda)/g(\lambda)$ is driven from the numerator polynomial $h(\lambda) = 11.627\lambda^3 + 2.6835\lambda$.

(a) Determine $g(\lambda)$.
(b) Synthesize $S_{11}(\lambda) = h(\lambda)/g(\lambda)$ by means of Richard extractions.
(c) Plot the transducer power gain $T = |S_{21}|^2$ over a wide frequency band and comment on the results.

Solution:

(a) Since the lossless two-port consists of three UEs in Cascade connections, then $F = (-1)^q(1 - \lambda^2)^k$, with $q = 0, k = 3$. Therefore, we can generate the open form of $F(\lambda^2)$ using the MATLAB code function F=lambda2q_UE2k(k,q) as in the previous example.
Thus, setting $q = 0, k = 3$ we compute $g(\lambda)$, employing the MATLAB function g=htog(q,k,h). For this example, the main program above is revised and saved under the main program CH4COMTRL_Example13.m. Hence, this program yields
g = [11.6699 7.6442 5.0487 1.0000]
meaning that

$$g(\lambda) = 11.6699\lambda^3 + 7.6442\lambda^2 + 5.0487\lambda + 1$$

(b) Remember, that Equation (4.71) describes the synthesis of the cascaded connection of UEs, which is also detailed in the Richard extraction algorithm step by step. The Richard extraction algorithm is programmed as a MATLAB function [z]=UE_sentez(h,g).[26] This function is called by the main program CH4COMTRL_Example13.m. It yields the line characteristic impedances as follows. Given

$$h = [11.6270 \quad 0 \quad 2.6835 \quad 0]$$
$$g = [11.6699 \quad 7.6442 \quad 5.0487 \quad 1.0000]$$

the line characteristic impedances are

$$Z = [3.5896 \quad 0.5531 \quad 3.5896 \quad 1.0000]$$

These are the 50 Ω normalized impedances. The last term is the termination.

(c) Under the normalization terminations $(R_0 = 50\ \Omega)$, TPG of the lossless two-port is specified by $|S_{21}(j\Omega)|^2 = F(\lambda^2)/G(\lambda^2)|_{\lambda = j\Omega}$. For the example under consideration, the numerator polynomial is given by

$$F(\lambda^2)|_{\lambda = j\Omega} = [1 - \lambda^2]^3|_{j\Omega} = (1 + \Omega^2)^3$$

and the denominator polynomial is

$$G(\lambda^2)|_{\lambda = j\Omega} = G(-\Omega^2) = g(\lambda)g(-\lambda)|_{\lambda = j\Omega}$$

---

[26] This function has been written by Dr Metin Sengul of the Department of Electrical–Electronics Engineering, Kadir Has University, Istanbul, Turkey.

or

$$G(-\Omega^2) = B_1\Omega^{2n} + B_2\Omega^{2(n-1)} + \ldots + B_n\Omega^2 + B_{n+1}$$

Hence, in general, for cascaded connections of k UEs, TPG is given by

$$T(\Omega^2) = \frac{(1 + \Omega^2)^n}{B_1\Omega^{2n} + B_2\Omega^{2(n-1)} + \ldots + B_n\Omega^2 + B_{n+1}} \qquad (4.77)$$

For this example, the polynomial $G(-\Omega^2)$ can easily be generated on the MATLAB command window as follows.

In the λ Domain:

g=[11.6699 7.6442 5.0487 1.0000]

The para-conjugate of $g(\lambda)$ is g_=paraconj(g) or g_ =[-11.6699 7.6442 -5.0487 1.0000].

Let $Ge(\lambda^2)$=conv(g,g_) and Gfull=[136.1871 0 -59.4021 0 -10.2012 0 1.0000] Clear the zeros of Gfull: G=poly_eventerms(Gfull), or

G=[-136.1871 -59.4021 -10.2012 1.0000]

In the Ω Domain:

$$B(\Omega^2) = \text{paraconj}(G)$$
$$B = [136.1871 - 59.4021\ 10.2012\ 1.0000]$$

meaning that

$$G(\Omega^2) = 136.1871\Omega^6 - 59.4021\Omega^4 + 10.2012\Omega^2 + 1$$

Then, Equation (4.77) becomes

$$T(\Omega^2) = \frac{(1 + \Omega^2)^3}{136.1871\Omega^6 - 59.4021\Omega^4 + 10.2012\Omega^2 + 1}$$

where $\Omega = \tan(\omega\tau)$.

A plot of TPG is shown in Figure 4.12. This figure reveals that the given $h(\lambda)$ describes a stepped line filter with three UE sections.

The main Program CH4COMTRL also generates TPG as described above.

When working with the Richard variable, we set $\lambda = j\tan(\omega\tau)$ or the Richard frequency $\Omega = \tan\theta$ with $\theta = \omega\tau$. In this case, if θ varies from $-\pi/2$ to $+\pi/2$, then the Richard frequency Ω varies from $-\infty$ to $+\infty$ which covers the entire Ω axis. Furthermore, network functions, generated in terms of the Richard frequency Ω, must be periodic in the actual frequencies $\omega\tau$ with periodicity π since $\Omega = \tan(\theta) = \tan(\omega\tau)$ is periodic with period π. Therefore, in Figure 4.12, the full period of TPG is observed over the interval of $[-\pi/2, +\pi/2]$. In order to see the periodicity of the gain, the plot is drawn over the interval of $[-3\pi, +3\pi]$ as shown in Figure 4.12.

This example clearly shows that Cascade connections of n UEs can be used to design step line transformers or filters, or in general to design power transfer networks as described in the following sections.

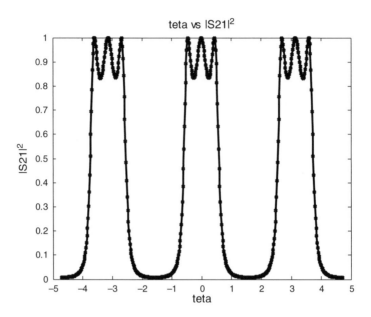

Figure 4.12   Periodic plot of TPG of Example 4.13

Below we give the listing of the main program and also the MATLAB code for the synthesis of cascade connections of UEs via Richard extractions.

Program Listing 4.8   Main program for Example 4.13

```
% Main Program for Chapter 4
% CH4COMTRL_Example13.m
% CIRCUITS CONSTRUCTED WITH COMMENSURATE
% TRANSMISSION LINES
% Chapter 4; Example (4.13)
clear
h=[11.627 0 2.6835 0]
q=0
k=3
g=htog(q,k,h)
[z]=UE_sentez(h,g)
j=sqrt(-1);
% Generate Transducer Power Gain:
% Step 1: Generate numerator polynomial F
F=lambda2q_UE2k(k,q);
teta_start=-3*pi/2;
teta_end=3*pi/2;
Nplot=251;
teta=teta_start;
delteta=(teta_end-teta_start)/(Nplot-1);
```

```
for i=1:Nplot
tetap(i)=teta;
% Computation of the Gain by closed forms:
lambda=j*tan(teta);
omega=tan(teta);
gp=polyval(g,lambda);
gp_=conj(gp);
G=real(gp*gp_);
Fp=real(polyval(F,lambda*lambda));
T(i)=Fp/G;
%
% Generation of gain by Equation (4.77)
% Alternative computation from the
% explicit equation
G2=136.1871*omega^6-
59.4021*omega^4+10.2012*omega^2+1
TPG(i)=(1+omega^2)^3/G2
teta=teta+delteta;
end
figure(1)
plot(tetap,T)
title('teta vs |S21| ^2')
xlabel('teta')
ylabel('|S21|^2')
%
figure(2)
plot(tetap,TPG)
title('teta vs |S21| ^2')
xlabel('teta')
ylabel('|S21| ^2')
```

Program Listing 4.9   Numerical Richard extractions

```
function [z]=UE_sentez(h,g)
% Richard Extraction written by Dr. Metin Sengul
% of Kadir Has University, Istanbul, Turkey.
% Inputs:
% h(lambda), g(lambda)as MATLAB Row vectors
% Output:
% [z] Characteristic impedances of UE lines
%─────────────────────────────────────────────
na=length(g);
for i=1:na
h0i(i)=h(na-i+1);
g0i(i)=g(na-i+1);
end
```

Program Listing 4.9   (Continued)

```
m=length(g0i);
for a=1:m
gt=0;
ht=0;
k=length(g0i);
for b=1:k
gt=gt+g0i(b);
ht=ht+h0i(b);

end
if a==1;
zp=1;
else

zp=zz(a-1);
end
zz(a)=((gt + ht)/(gt-ht))*zp;

x=0;
y=0;
for c=1:k-1
x(c)=h0i(c)*gt-g0i(c)*ht;
y(c)=g0i(c)*gt-h0i(c)*ht;
end

N=0;
D=0;
n=0;
for e=1:k-1
N(e)=n+x(e);
n=N(e);
d=0;
for f=1:e
ara=((-1)^(e + f))*y(f);
D(e)=ara+d;
d=D(e);

end

end

h0i=N;
g0i=D;

end
z=zz;
return
```

4.16 Stepped Line Butterworth Gain Approximation

We can approximate the desired gain form by employing cascade connection of n UE sections, which in turn yields the characteristic impedances of the Commensurate transmission lines in cascade configurations.

Gain approximation

As shown by Equation (4.77), for standard normalizations, the amplitude of the transfer scattering parameter describes TPG such that

$$
T(\Omega^2) = |S_{21}(j\Omega)|^2 = \frac{F(\lambda^2)}{G(\lambda^2)}\Big|_{\lambda=j\Omega}
$$

or

$$
T(\Omega^2) = \frac{(1+\Omega^2)^n}{B_1\Omega^{2n} + B_2\Omega^{2(n-1)} + \ldots + B_n\Omega^2 + B_{n+1}}
\tag{4.78}
$$

where

$$
\Omega = \tan(\theta) = \frac{\sin(\theta)}{\cos(\theta)}, \theta = \omega\tau
$$

Employing trigonometric identities on Equation (4.78),

$$
\Omega^2 = \tan^2(\theta) = \frac{\sin^2(\theta)}{\cos^2(\theta)} = \frac{1-\cos^2(\theta)}{\cos^2(\theta)} = \frac{1}{\cos^2(\theta)} - 1
$$

or

$$
1+\Omega^2 = 1 + \tan^2(\theta) = \frac{1}{\cos^2(\theta)}
$$

or

$$
(1+\Omega^2)^n = \frac{1}{\cos^{2n}(\theta)}
\tag{4.79}
$$

Let the denominator polynomial be designated by $B(\Omega^2)$:

$$
B = B_1\Omega^{2n} + B_2\Omega^{2(n-1)} + \ldots + B_n\Omega^2 + B_{n+1}
$$

Then,

$$
B(\Omega^2) = B_1\left[\frac{\sin^2(\theta)}{\cos^2(\theta)}\right]^n + B_2\left[\frac{\sin^2(\theta)}{\cos^2(\theta)}\right]^{n-1} + \ldots + B_n\left[\frac{\sin^2(\theta)}{\cos^2(\theta)}\right] + B_{n+1}
$$

Multiplying $B(\Omega^2)$ by $\cos^{2n}(\theta)$ we obtain

$$
B(\Omega^2) = B_1[\cos^2(\theta)]^0[\sin^2(\theta)]^n + B_2[\cos^2(\theta)]^1[\sin^2(\theta)]^{n-1} + \ldots + B_n[\cos^2(\theta)]^{n-1}[\sin^2(\theta)]
$$
$$
+ B_{n+1}
$$

Note that $\cos^2(\theta) = 1 - \sin^2(\theta)$. Then, using $\cos^2(\theta) = 1 - \sin^2(\theta), B(\Omega^2)$ can be expressed only in terms of the powers of $[\sin^2(\theta)]$. Thus, setting

$$x = \alpha[\sin(\theta)] \tag{4.80}$$

TPG can be expressed in terms of the new variable x as

$$T(x) = \frac{1}{P(x^2)} \tag{4.81}$$

where polynomial $P(x^2)$ must be strictly positive.

In Equation (4.80) the scalar α is determined by setting the bandwidth. For example, let the actual cut-off frequency of the filter be $\omega_c = 2\pi f\, c$ and be fixed at $\alpha[\sin(\pm\omega_c\tau)] = \pm 1$. Then, α is given by

$$\alpha = \frac{1}{\sin(\omega_c\tau)} \tag{4.82}$$

Once the filter design is made based on Equation (4.81), it can be directly transformed to the λ domain in the following manner.

x to λ Transformation

Here $x^2 = \alpha^2\sin^2(\theta)$ can be expressed in terms of $\tan^2(\theta)$ as

$$x^2 = \alpha^2\sin^2(\theta) = \alpha^2[1 - \cos^2(\theta)] = \alpha^2\left[1 - \frac{1}{1 + \tan^2(\theta)}\right] = \alpha^2\frac{\tan^2(\theta)}{1 + \tan^2(\theta)}$$

On the other hand, $\Omega^2 = \tan^2(\theta) = -\lambda^2$. Thus, x^2 is given by

$$x^2 = (-1)\alpha^2\frac{\lambda^2}{1 - \lambda^2} \tag{4.83}$$

Design of a Butterworth filter

In classical filter design theory, firstly the mathematical form of TPG is selected based on the design requirements.

For example, the Butterworth form is given by

$$T(x^2) = \frac{1}{1 + x^{2n}} = \frac{1}{P(x^2)}$$

This form can be transformed to the λ domain by means of Equation (4.83) as

$$T(\lambda^2) = \frac{1}{1 + (-1)^n (\alpha^2 \frac{\lambda^2}{1 - \lambda^2})^n}$$

$$\boxed{T(\lambda^2) = \frac{(1 - \lambda^2)^n}{(1 - \lambda^2)^n + (-1)^n \alpha^{2n} \lambda^{2n}} = \frac{(1 - \lambda^2)^n}{G(\lambda^2)}} \qquad (4.84)$$

which is the form of n cascaded UE sections. Now, let to consider an example to design a simple Butterworth filter.

Example 4.14: For standard normalization $R_0 = 50\,\Omega$, Let the length of the commensurate lines be fixed at the cut-off frequency $f_c = 1$ GHz yielding $\theta_c = \omega_c \tau = \pi/2$. Design a Butterworth filter with Cascade connections of two UEs printed on a dielectric with a speed of propagation of $\upsilon = 2 \times 10^8$ m/s.
Solution: Standard filter design can be completed step by step as follows.

Step 1 (Computation of commensurate lengths): Firstly, let us determine the fixed lengths of the commensurate lines at $f_c = 1$ GHz:

$$\omega_c \tau = 2\pi f_c \left(\frac{l}{\upsilon}\right) = \frac{\pi}{2}$$

or

$$l = \frac{1}{4}\left(\frac{\upsilon}{f_c}\right) = \frac{1}{4}\left(\frac{2 \times 10^8}{1 \times 10^9}\right) = 0.05\ \text{m} = 5\ \text{cm}$$

Step 2 (Determination of the scalar $\alpha = 1/\sin(\theta_c)$): Now we calculate the real constant α such that $x_c = \alpha . \sin(\theta_c) = 1$, or

$$\alpha = 1$$

Step 3 (Determination of the analytic form of TPG): Then we generate the explicit form of $T(\lambda^2)$ by employing Equation (4.84). For this example $n = 2$, so

$$T(\lambda^2) = \frac{(1 - \lambda^2)^2}{1 - 2\lambda^2 + \lambda^4} = \frac{(1 - \lambda^2)^2}{2\lambda^4 - 2\lambda^2 + 1} = \frac{(1 - \lambda^2)^2}{g(\lambda)g(-\lambda)}$$

Step 4 (Generation of strictly Hurwitz denominator polynomial g(λ)): Here, $g(\lambda)$ is constructed on the LHP roots of the polynomial

$$G(\lambda^2) = g(\lambda)g(-\lambda) = 2\lambda^4 - 2\lambda^2 + 1$$

The roots of $G(\lambda^2)$ are given as

$$\lambda_1 = -0.7769 + 0.3218i \qquad \lambda_2 = -0.7769 - 0.3218i$$

$$\lambda_3 = 0.7769 + 0.3218i \qquad \lambda_4 = 0.7769 - 0.3218i$$

Then, selecting LHP roots λ_1 and λ_2, we form $g(\lambda)$ as

$$g(\lambda) = \sqrt{2}(\lambda + 0.7769 + 0.3218i)(\lambda + 0.7769 - 0.3218i)$$

or

$$g = [1.4142\ 2.1974\ 1.0000]$$

meaning that

$$g(\lambda) = 1.412\lambda^2 + 2.1974\lambda + 1$$

Step 5 (Generation of the numerator polynomial $h(\lambda)$): By the Losslessness condition, $H(\lambda^2) = h(\lambda)h(-\lambda) = G(\lambda^2) - F(\lambda^2)$ is generated as

$$H(\lambda^2) = 2\lambda^4 - 2\lambda^2 + 1 - (1 - 2\lambda^2 + \lambda^4)$$

$$H(\lambda^2) = 2\lambda^4 - \lambda^2 + 1 - 1 + 2\lambda^2 - \lambda^4$$

$$H(\lambda^2) = h(\lambda)h(-\lambda) = \lambda^4$$

Hence, it is found that

$$h(\lambda) = \lambda^2$$

Step 6 (Synthesis of $S_{11}(\lambda) = \lambda^2/(1.4142\lambda^2 + 2.1974\lambda + 1)$ by means of Richard extractions): Using our synthesis function, normalized and actual characteristic impedances are found as in the following (the final design is depicted in Figure 4.13):

$$Z = [1.5538\ 0.6436\ 1.0000]$$

$$Z_{actual} = [77.6882\ 32.1797\ 50.0000]$$

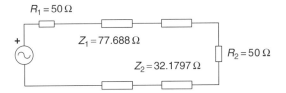

Figure 4.13   Synthesis of $S_{11}(\lambda) = \lambda^2/(1.4142\lambda^2 + 2.1974\lambda + 1)$ of Example 4.14

The gain plot of TPG is depicted in Figure 4.14.

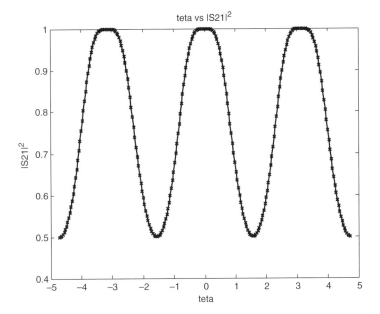

Figure 4.14   Gain plot for Example 4.14

This plot is completed using the MATLAB main program CH4COMTRL_Example14.m. This program generates the complete scattering parameters from the given numerator polynomial $h(\lambda) = \lambda^2$ as for two cascaded lines. The program is listed next.

Program Listing 4.10   Main program for Example 4.14

```
% Main Program for Chapter 4
% CH4COMTRL_Example14.m
% CIRCUITS CONSTRUCTED WITH COMMENSURATE
% TRANSMISSION LINES
% Chapter 4; Example (4.14)
clear
h=[1 0 0];
```

Program Listing 4.10   (Continued)

```
q=0
k=2
g=htog(q,k,h)
[z]=UE_sentez(h,g)
j=sqrt(-1);
% Generate Transducer Power Gain:
% Step 1: Generate numerator polynomial F
F=lambda2q_UE2k(k,q);
teta_start=-3*pi/2;
teta_end=3*pi/2;
Nplot=251;
teta=teta_start;
delteta=(teta_end-teta_start)/(Nplot-1);
for i=1:Nplot
tetap(i)=teta;
% Computation of the Gain by closed forms:
% lambda=j*tan(teta);
omega=tan(teta);
gp=polyval(g,lambda);
gp_=conj(gp);
G=real(gp*gp_);
Fp=real(polyval(F,lambda*lambda));
T(i)=Fp/G;
%
teta=teta+delteta;
end
figure(1)
plot(tetap,T)
title('teta vs |S21| ^2')
xlabel('teta')
ylabel('|S21|^2')
```

4.17 Design of Chebyshev Filters Employing Stepped Lines

Stepped line filters can be designed using Chebyshev polynomials in Gain approximation. First of all, let us introduce Chebyshev polynomials.

An n th-degree Chebyshev polynomial is defined by

$$
\begin{aligned}
&T_n(x) = \cos[n(\cos^{-1}(x))] \\
&\text{Setting } z = \cos^{-1}(x) \text{ we have} \\
&\quad -1 \leq x = \cos(z) \leq 1 \\
&\text{and} \\
&-1 \leq T_n(\cos(z)) = \cos(nz) \leq 1 \\
&\text{when } -\pi \leq z \leq \pi.
\end{aligned}
\tag{4.85}
$$

Equation (4.85) indicates that, $|T_n(x)|$ is bounded by 1 over the interval $-1 \le x = cos(z) \le 1$. Chebyshev polynomials for various values of n are listed in Table 4.1.

Table 4.1   Chebyshev polynomials for $n = 1, 2, 3, 4, 5$

$n=0$	$T_0 = cos(0)$	$T_0 = 1$
$n=1$	$T_1 = cos(Z)$	$T_1(x) = x$
$n=2$	$T_2 = cos(2z)$	$T_2(x) = 2cos^2(z) - 1 = 2x^2 - 1$
$n=3$	$T_3 = cos(3z)$	$T_3(x) = 4cos^3(z) - 3cos(z) = 4x^3 - 3x$
$n=4$	$T_4 = cos(4z)$	$T_4(x) = 8x^4 - 8x^2 + 1$
$n=5$	$T_5 = cos(5z)$	$T_5(x) = 16x^5 - 20x^3 + 5x$

Close examination of Table 4.1 reveals the following statements.

Statement 15 (Major properties of Chebyshev polynomials): An odd value of n results in an odd Chebyshev polynomial, while an even value of n yields an even Chebyshev polynomial. The Coefficient of the leading power term of a Chebyshev polynomial of degree n is 2^{n-1}.

Furthermore, these polynomials can be generated by means of recursive relations as detailed below.

Statement 16 (Recursive relation): Chebyshev polynomials can be generated using the recursive relation given by

$$
\boxed{
\begin{aligned}
T_{n+1}(x) &= 2xT_n(x) = T_{n-1}(x) \\
&\text{with initial polynomials} \\
&\quad T_0(x) = 1 \\
&\text{and} \\
&\quad T_1(x) = x
\end{aligned}
}
\tag{4.86}
$$

Proof: A Chebyshev polynomial of degree $n + 1$ can be expressed as

$$T_{n+1} = cos(n + 1)z = cos(nz + z) = cos(nz)\,cos(z) - sin(nz)sin(z)$$

Similarly, a Chebyshev polynomial of degree $n - 1$ is written as

$$T_{n-1} = cos(n - 1)z = cos(nz - z) = cos(nz)cos(z) + sin(nz)sin(z)$$

Then, summation of the above forms results in

$$T_{n+1} + T_{n-1} = cos(nz)cos(z) = 2xT_n(x)$$

or

$$T_{n+1}(x) = 2xT_n(x) - T_{n-1}(x)$$

A typical odd Chebyshev polynomial of degree $n = 9$ is depicted in Figure 4.15. This polynomial goes from $-\infty$ to $+\infty$. It is zero when $x = 0$. Over the interval $-1 \le x \le +1$, it exhibits an equal ripple property. Outside this interval, the polynomial booms sharply. Therefore, we can make the following statement.

Statement 17 (A property of odd Chebyshev polynomials): At x = 0, an odd Chebyshev polynomial of degree n is zero. That is,

$$\boxed{\text{when } n \text{ is odd } T_n(x)|_{x=0} = 0} \tag{4.87}$$

Proof: Let $n = (2k+1)$ be an odd integer. $x=0$ corresponds to $z=\pi/2$ since $z=\cos^{-1}(x)$ or $x=\cos(z)$. Therefore,

$$T_{2k+1}(x) = \cos\left[(2k+z)\frac{\pi}{2}\right] = 0, \text{ for all } k$$

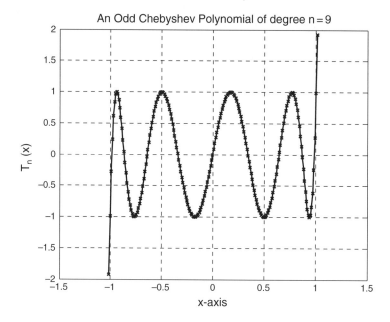

Figure 4.15 An odd-power Chebyshev polynomial of degree $n=9$

An even Chebyshev polynomial of degree $n=10$ is shown in Fig 4.16. As opposed to Figure 4.15, close examination of this figure reveals that $T_{10}(0) = -1$ while exhibiting an equal ripple property over the interval $-1 \leq x \leq +1$. Thus, we can make the following statement.

Statement 18 (A property of an even Chebyshev polynomial): Let $n=2k$. At $x=0$, an even Chebyshev polynomial $T_{2k}(x)$ reveals that

$$\boxed{\begin{aligned} T_{2k}(x)|_{x=0} &= -1 \quad \text{when } k \text{ is odd} \\ T_{2k}(x)|_{x=0} &= +1 \quad \text{when } k \text{ is even} \end{aligned}} \tag{4.88}$$

Proof: $x = 0$ corresponds to $z = \pi/2$. Therefore,

$$T_{2x}(0) = \cos(2k\frac{\pi}{2}) = \cos(k\pi)$$

which is -1 when k is odd and $+1$ when k is even.

Finally we can make the following general statement for Chebyshev polynomials.

Statement 19 (Fluctuation of Chebyshev polynomials over the interval $-1 \leq x \leq 1$ and the capture property)

(a) Ripples of the Chebyshev polynomials fluctuate between -1 and $+1$ over the interval of $-1 \leq x \leq 1$. This is easily seen from the definition of Chebyshev polynomials: $T_n(x) = \cos[n(\cos^{-1}(x))]$ or $T_n = \cos(nz)$ where $z = \cos^{-1}(x)$ or $x = \cos(z)$. When the variable x varies between -1 and $+1$, then the corresponding variable $z = \cos^{-1}(x)$ varies between $\pi(2k+1)$ and $2k\pi$. Therefore, $T_n(z) = \cos(nz)$ varies between -1 and $+1$ over the interval $(2k+1)\pi \leq z \leq 2k\pi$, which reveals the statement. This property of the Chebyshev polynomials is called the capture property.

(b) Outside the interval $-1 \leq x \leq 1$, a Chebyshev polynomial is dominated by the leading term $[2^{n-1}x^n]$.

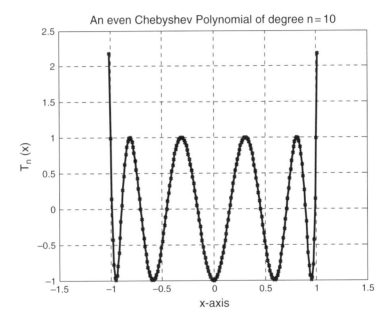

Figure 4.16   An even-power Chebyshev polynomial of degree $n = 10$

Definition 13 (Monotone roll-off Chebyshev gain function):

Fluctuations of the ripples in the Chebyshev Gain approximation can be controlled by means of a term ε^2 which is called the ripple factor. In this regard, a typical monotone roll-off Chebyshev gain function is given by

$$0 \leq T(x) = \frac{1}{1 + \varepsilon^2 T_n^2(x)} = \frac{1}{G(x^2)} \leq 1, \text{ for all } x \qquad (4.89)$$

No matter what the value of the integer n is, $\varepsilon^2 T_n^2(x)$ is always a non-negative even polynomial and bounded by ε^2.

If it is desired to force $T(0) = 1$, then n must be an odd integer to get $T_n(0) = 0$.

A typical Chebyshev gain function is shown in Figure 4.17 for $n = 5$ and $\varepsilon = 0.5$. In this figure, at $x = 0$, the gain is zero. This is a crucial property in designing microwave filters without an ideal transformer. Therefore, an odd integer n is a practical choice for designers.

Statement 20 (Minimum of the passband): No matter what the value of the integer n is, in the passband, the last minimum point of a monotone roll-off Chebyshev gain function occurs at $x = 1$ and is given by

$$T_n(x)|_{x=1} = \frac{1}{1 + \varepsilon^2} \tag{4.90}$$

Proof: By Statement 19, $T_n^2(x) = \cos^2(n\cos^{-1}(x))$ is always sinusoidal over the interval $0 \leq x \leq 1$ with zero minimums and unity maximums. Therefore, at $x = 1$ the gain function should take its minimum value as $T_n(x)|_{x=1} = 1/(1 + \varepsilon^2)$. Hence, the statement follows.

It should be noted that $x = 1$ can always be considered as the normalized cut-off frequency of the Chebyshev gain functions.

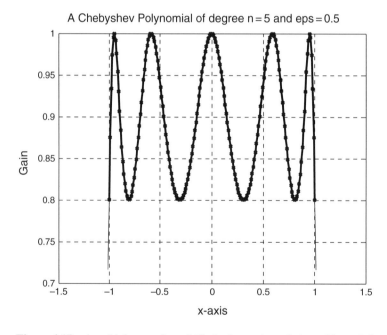

Figure 4.17   An odd degree of $n = 5$ Chebyshev gain variation with $\varepsilon = 0.5$

In order to analyze the properties of Chebyshev functions we developed a simple generic MATLAB code called "Chebyshev.com". This main program generates the Chebyshev polynomials and Chebyshev gain functions numerically for a specified degree n and ripple factor ε.

The user of the program enters the values of n and ε to obtain plots of the Chebyshev polynomial and the Chebyshev gain function. As a result, the Chebyshev polynomial and the gain variations are plotted. The listing of the program is as follows.

Program Listing 4.11   Numerical generation of Chebyshev polynomials

```
% Main Program Chebyshev
% This program Generates the Chebyshev
% polynomials
% and Monotone Roll-off Chebyshev gain
% forms
% numerically and plots these at the
% output.
% Inputs:
% n: Degree of the Chebyshev
% Polynomial
% eps: Ripple Factor
clear
%
N=201;
x1=-1.01;
x2=+1.01;
delx=(x2-x1)/(N-1);
x=x1;
n=input(' Enter degree of the Chebyshev
polynomial n=')
eps=input('Enter the ripple factor eps=')
n1=n+1;
for i=1:N
z(i)=x;
if n==1; T(1)=1;Tj=T(1); end
if n==2; T(2)=x;Tj=T(2); end
if n>2;
for j=3:n1
T(1)=1;%zero order Chebyshev
polynomial
T(2)=x;%First order Chebyshev
polynomial
T(j)=2*x*T(j-1)-T(j-2);%Recursive
Formula
Tj=T(j);
end
end
G(i)=Tj;

Gain(i)=1.0/(1+eps*eps*Tj*Tj);%Chebyshev
Gain
```

Program Listing 4.11   (Continued)

```
x=x+delx;

end
figure(1)
plot(z,G)
title('Chebyshev polynomials')
xlabel('x')
ylabel('T_n (x)')
%
figure(2)
plot(z,Gain)
title('Chebyshev polynomials')
xlabel('x')
ylabel('T_n (x)')
```

The reader can generate all the above plots using the MATLAB program `Chebyshev.m`.

As shown for a Butterworth gain function, a Chebyshev gain function of x can also be expressed in the Richard variable λ using the transformation specified by Equation (4.83):

$$x^2 = (-1)\alpha^2 \frac{\lambda^2}{1 - \lambda^2}$$

which in turn leads to an equal ripple, stepped line transformer gain form as

$$T(\lambda^2) = S_{21}(\lambda)S_{21}(-\lambda) = \frac{(1 - \lambda^2)^n}{G(\lambda^2)}$$

Definition 14 (Insertion loss): In classical filter theory, we usually deal with base 10 logarithms of gain functions. The negative of '10 times the base 10 logarithm of the gain function' is called the insertion loss (or in short IL) and is given by

$$\text{IL} = -10\log_{10}|S_{21}(j\omega)|^2 = 10\log_{10}\left|\frac{1}{S_{21}(j\omega)}\right|^2 \tag{4.91}$$

where S_{21} is the forward transfer scattering parameter of the filter. Usually, IL is expressed in units of decibels (dB).

Now, let us consider an example to show the design of a stepped line transformer with equal ripple analytic gain characteristics.

Example 4.15: Design a low-pass stepped line filter using a monotone roll-off Chebyshev gain function with the following specifications:

- The filter should include no ideal transformer.
- The filter cut-off frequency must be located at $f_c = 1\text{GHz}$.

- Up to the cut-off frequency, it is desired to have an ILR[27] less than or equal to 0.4 dB (which leads to determination of the ripple factor ε^2).
- The filter stopband frequency must be located at $f_e = 3\text{GHz}$ (which in turn yields the length of the Commensurate transmission lines such that $\omega_e \tau = \pi/2$, where $\omega_e = 2\pi f_e$).
- It is desired to have an IL greater than or equal to 40 dB in the stopband (i.e. for all frequencies $f \geq f_e$).

Due to periodicity of the gain functions in $\Omega = \tan(w\tau)$, it is known that this low-pass filter can be used as a bandpass filter. Determine the next passband frequency range $f_1 \leq f \leq f_2$ of the filter.

Solution: We can design the desired low-pass filter step by step as follows.

Step 1 (Choice of the transfer function): As specified, the choice of the gain function or analytic form of the transfer function is given by

$$T_n(x) = \frac{T_0}{1 + \varepsilon^2 T_n^2(x)}$$

with

$$x = \alpha.\sin(w\tau)$$

$$x^2 = (-1)\alpha^2 \frac{\lambda^2}{1 - \lambda^2} = \frac{\alpha^2 \lambda^2}{\lambda^2 - 1}$$

Step 2 (Determination of the unknown parameters of the transfer function based on the desired specifications):

(a) **Determine whether the degree n is odd or even**: Since we use no ideal transformer in the filter design, then the DC gain level must be unity, i.e. $T_0 = 1$. Therefore, degree n must be an odd integer.

(b) **Determine the ripple factor ε^2**: Since the IL must be less than or equal to 0.4 dB up to cut-off frequency f_c, which corresponds to $x = 1$, then at $x = 1$ it is desired to have $\text{ILR}(1) = 10\log_{10}(1 + \varepsilon^2) = 0.4 \text{ dB}$.

Thus, the ripple factor is given by

$$\boxed{\varepsilon^2 = 10^{\text{ILR}(1)/10} - 1} \tag{4.92}$$

or, for the problem under consideration,

$$\varepsilon^2 = (10)^{(0.04)} - 1 = 0.0965 \cong 0.1$$

(c) **Determine the commensurate delay length τ**: In the Richard domain, the gain characteristic must be periodic over the Richard frequencies Ω such that $-\pi/2 \leq \Omega \leq \pi/2$ since $\Omega = \tan(w\tau)$.

Therefore, the minimum gain must occur at the actual stopband frequency $f_e = 3 \text{ GHz}$. In this case, we must have $\omega_e \tau = \pi/2$ or the delay length of the commensurate transmission lines becomes

$$\boxed{\tau = \frac{\pi}{2\omega_e} = \frac{1}{4f_e}} \tag{4.93}$$

---

[27] $\text{ILR} = 10\log_{10}\left[1 + \varepsilon^2 T_n^2(x)\right]$ refers to insertion loss within the passband $0 \leq x \leq 1$ of the Chebyshev gain function. It basically covers the insertion loss of the ripples (or in short ILR) of the gain within the passband.

or

$$\tau = \frac{1}{4 \times 3 \times 10^9} = \frac{10^{-9}}{12} \, \text{s}$$

(d) **Computation of α in $x = \alpha \sin(\omega \tau)$ at the cut-off frequency f_c:** At the cut-off frequency f_c, the variable x must be equal to 1. Therefore,

$$\boxed{\begin{aligned}
\alpha &= \frac{1}{\sin(\omega_c \tau)} = \frac{1}{\sin(\theta_c)} \\
&= \frac{1}{\sin(\frac{\pi}{2} \frac{\omega_c}{\omega_e})} = \frac{1}{\sin(\frac{\pi}{2} \cdot \frac{f_c}{f_e})}
\end{aligned}} \tag{4.94}$$

Thus,

$$\alpha = \frac{1}{\sin(\frac{\pi}{6})} = \frac{1}{0.5} = 2$$

(e) **Computation of the degree n:** The minimum value of TPG in the Richard domain must take place at the stopband frequency f_e for which $x = \alpha \sin(w_e \tau) = \alpha$.
On the other hand, by definition, the insertion loss $\mathrm{IL}(x)$ is expressed as

$$\boxed{\begin{aligned}
\mathrm{IL} &= 10 \log_{10}\left[\frac{1}{T}\right] = 10 \log_{10}\left[1 + \varepsilon^2 T_n^2(x^2)|_{x^2 = \alpha^2}\right] \\
&= 10 \log_{10}\left[1 + \varepsilon^2 (2^{n-1}\alpha^n + \ldots)^2\right]
\end{aligned}} \tag{4.95}$$

In this case, for the problem under consideration, we must have $\mathrm{IL} \geq 40$ dB within the stopband, which is described by $f_e \geq 3\,\text{GHz}$. Employing Equation (4.95), the inequality $\mathrm{IL} \geq 40$ dB may be solved by trial and error.

For example, if we choose $n = 5$ then

$$T_5(x) = 16x^5 - 20x^3 + 5x$$

At $x = \alpha = 2$,

$$T_5(x)|_{x=2} = 16 \times 2^5 - 20 \times 2^3 + 5 \times 2 = 362$$
$$10 \log_{10}(1 + 0.1 \times 362 \times 362) = 41.1745 \, \text{dB}$$

which satisfies the inequality.

However, we can derive an approximate expression for degree n if we ignore the lower degree terms of Equation (4.95) beyond the term $2^{n-1}\alpha^n$. In this case, we can state that at $x = \alpha$

$$\mathrm{IL}(\alpha) \cong 10 \log_{10}\left[1 + \varepsilon^2 2^{2(n-1)}\alpha^{2n}\right] = 10 \log_{10}\left\{\left[1 + \left(\frac{\varepsilon}{\alpha}\right)^2 (2\alpha)^{2(n-1)}\right]\right\}$$

or

$$\left[\frac{\varepsilon}{\alpha}\right]^2 (2\alpha)^{2(n-1)} \cong 10^{\mathrm{IL}(\alpha)} - 1$$

or

$$n \cong \left[\frac{1}{2\log_{10}(2\alpha)}\right] \log_{10}\left[\frac{\alpha^2}{\varepsilon^2}\left(10^{\mathrm{IL}(\alpha)/10} - 1\right)\right] + 1 \qquad (4.96)$$

Equation (4.96) is programmed as a MATLAB function called `function n=Cheby_Degree(IL, ILR, fc, fe)`. The listing of this function is as follows.

Program Listing 4.12   Approximate computation of degree n

```
function n=Cheby_Degree(IL,ILR,fc,fe)
% This function determines the degree n
% of a Chebyshev polynomial for
% specified IL (in dB), ripple loss ILR,
% cut-off frequency fc and stop band
% frequency fe
eps_sq=10^(ILR/10)-1;
alfa=1/(sin((fc/fe)*(pi/2)));
y=IL/10;
N1=alfa*alfa*(10^y-1)/eps_sq;
N=log10(N1);
D=2*log10(2*alfa);
n=fix(N/D + 1);
```

Thus, for the inputs IL $= 40\,$dB, ILR $= 0.4\,$dB, $f_c = 1\,$GHz and $f_e = 3\,$GHz, it is found that $n = 5$.

It should be mentioned that in the above MATLAB function, the degree n is generated as a floating point number. Then it is converted to an integer using the MATLAB expression `fix (°)` So, in the computations, there is no restriction imposed on the integer n. However, if it is desired to work with odd numbers, then one should add 1 to or subtract 1 from n when it is computed as an even integer. Eventually, we must check the end result if the design specifications are satisfied using the exact polynomial expression of the IL specified by Equation (4.95).

Step 3 (Generation of the analytic form of TPG of the filter): So far, we have determined the unknown parameters of the gain function to obtain

$$T(x^2) = \frac{1}{1 + 0.1(16x^5 - 20x^3 + 5x)^2}$$

or

$$T(x^2) = \frac{1}{1 + (0.1)(x^2)[16(x^2)^2 - 20(x^2) + 5]^2}$$

Setting

$$x^2 = \frac{-\alpha^2 \lambda^2}{1 - \lambda^2} = \frac{(\alpha^2 \lambda^2)}{[\lambda^2 - 1]}$$

and replacing x^2 by $(\alpha^2 \lambda^2)/[\lambda^2 - 1]$, it is found that

$$T(\lambda^2) = S_{21}(\lambda) S_{21}(-\lambda) = \frac{F(\lambda^2)}{G(\lambda^2)}$$

$$= \frac{(1 - \lambda^2)^5}{(1 - \lambda^2)^5 - 0.1 \times 2^2 \times \lambda^2 [(16 \times 2^4)\lambda^4 + (20 \times 2^2)\lambda^2 (1 - \lambda^2)^2 + 5(1 - \lambda^2)^4]^2}$$

Straightforward algebraic manipulation yields

$$\boxed{G(\lambda^2) = -(13\,105\lambda^{10} + 10\,131\lambda^8 + 2694\lambda^6 + 270\lambda^4 + 15\lambda^2 - 1)}$$

and the open form of $F(\lambda^2)$ is given by

$$\boxed{F = -\lambda^{10} + 5\lambda^8 - 10\lambda^6 + 10\lambda^4 - 5\lambda^2 + 1}$$

At this point we should note that the analytic form of the gain function can be generically obtained from MATLAB. Details are provided in the following section.

Step 4 (Generation of the numerator polynomial $h(\lambda)$ and the denominator polynomial $g(\lambda)$):

(a) **Computation of $h(\lambda)$:** Using the Losslessness condition an even polynomial $H(\lambda^2) = h(\lambda)h(-\lambda)$ can be generated as

$$H = G(\lambda^2) - F(\lambda^2) = 13\,104\lambda^{10} + 1.0136\lambda^8 + 2684\lambda^6 + 280\lambda^4 + 10\lambda^2$$

Hence, by setting $z = \lambda^2$ as an explicit factorization, the roots of the above polynomial are generated in MATLAB as

$$z_1 = 0$$
$$z_2 = -0.2922 + 0.0000i$$
$$z_3 = -0.2922 - 0.0000i$$
$$z_4 = -0.0945 + 0.0000i$$
$$z_5 = -0.0945 - 0.0000i$$

Roots in the λ domain are found by taking the square root of the above roots. Thus, taking five of them as complex conjugate pairs,

$$\lambda_{1,2} = 0$$
$$\lambda_{3,4} = \pm 0.5406i$$
$$\lambda_{4,5} = \pm 0.3075i$$

we end up with the numerator polynomial

$$h(\lambda) = \sqrt{|H_1|}\lambda^2 (\lambda + \lambda_2)(\lambda + \lambda_3)(\lambda + \lambda_4)(\lambda + \lambda_5)$$

or

$$h(\lambda) = 114.4788\lambda^5 + 44.2736\lambda^3 + 3.1624\lambda$$

(b) **Generation of $g(\lambda)$**: As in (a), we can generate the strictly Hurwitz $g(\lambda)$ from the even polynomial

$$G(\lambda^2) = -[13\,150\lambda^{10} + 10\,131\lambda^8 + 2694\lambda^6 + 270\lambda^4 + 15\lambda^2 - 1]$$

In this case, we set $z = \lambda^2$ and compute the roots of G as

$$z_{G1} = -0.3348 + 0.1079i$$
$$z_{G2} = -0.3348 - 0.1079i$$
$$z_{G3} = 0.0693 + 0.1126i$$
$$z_{G4} = 0.0693 - 0.1126i$$
$$z_{G5} = 0.0353$$

By taking $\pm$ the square root of the above roots and selecting the ones on the LHP, we obtain

$$g(\lambda) = \sqrt{|G_1|}(\lambda - \lambda_1)(\lambda - \lambda_2)(\lambda - \lambda_3)(\lambda - \lambda_4)(\lambda - \lambda_5)$$

where

$$\lambda_1 = -0.0921 + 0.5859i$$
$$\lambda_2 = -0.0921 - 0.5859i$$
$$\lambda_3 = -0.1773 + 0.3174i$$

$$\lambda_4 = -0.1773 - 0.3174i$$

$$\lambda_5 = -0.1878$$

or

$$g(\lambda) = 114.4788\lambda^5 + 83.1831\lambda^4 + 74.4698\lambda^3 + 28.8806\lambda^2 + 8.5300\lambda + 1$$

Step 5 (Synthesis of the stepped line filter by means of Richard extractions): This step can be completed by employing the MATLAB function [z]=UE_sentez(h,g). Hence, we have the 50Ω normalized characteristic impedances given as z=[3.18 0.44 4.45 0.44 3.18 1] or the actual impedances given as
Z=[159 22 222 22 159 50]

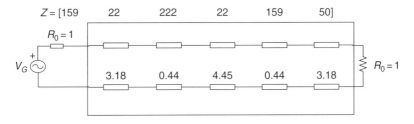

Figure 4.18 Monotone roll-off stepped line filter with five UEs

The gain performance of the stepped line filter (Figure 4.18) is depicted in Figure 4.19.

Close examination of Figure 4.19 reveals that all the design specifications are satisfied. Figure 4.19(a) indicates that ripples of the gain in the pass band are less than or equal to 0.4 dB, as imposed. Figure 4.19(b) clearly shows that the IL is about 40 dB at the stopband, as desired.

In Figure 4.19(c) three periods of the gain function are plotted:

- On the far left, we can see the gain characteristic over the negative frequencies (i.e. over $-9\,\text{GHz} \leq f \leq -3\,\text{GHz}$). Of course, the negative frequencies have no physical meaning.
- In the middle, we have the gain variation over $-3\,\text{GHz} \leq f \leq +3\,\text{GHz}$. Actually, this is the low-pass characteristic of the filter design. Its meaningful or physical bandwidth is described over $0 \leq f \leq f_c = 1\,\text{GHz}$, and its stopband starts at $f_e = 3\,\text{GHz}$.
- If f_b is the frequency over the baseband which is described within the frequency interval of $-f_e \leq f_b \leq +f_e$, in general the periodicity of the gain function is described by $\Omega = \tan(\omega_b\tau) = \tan[\omega_b\tau + (2k{\pm}1)\pi] = \tan(\omega_k\tau)$ for all $k = 0, 1, 2, 3, \ldots$ In this case, for any harmonic frequency ω_k we can write

$$\omega_b\tau + (2k{\pm}1)\pi = \omega_k\tau$$

or

$$\boxed{\omega_k = 2\pi f_k = \omega_b + (2k{\pm}1)\pi/\tau} \tag{4.97}$$

However, by Equation (4.93) $\tau = 1/4f_e$. Then,

$$\boxed{\begin{aligned} \omega_k &= 2\pi f_k = \omega_b + (2k{\pm}1)4\pi f_e \\ \omega_k &= 2\pi f_k = \omega_b + (2\omega_e)(2k{\pm}1) \\ &\text{For example, } k = 0 \text{ yields that} \\ \omega_k &= \omega_b \pm 2\omega_e \\ &\text{or} \\ f_k &= f_b \pm 2f_e \end{aligned}} \tag{4.98}$$

More specifically, let f_1 and f_2 be the lower and upper cut-off frequencies of the bandpass filter constructed on the low-pass prototype with cut-off frequency f_c. Then,

$$\boxed{\begin{aligned} f_1 &= 2f_e - f_c \\ f_2 &= 2f_e + f_c \end{aligned}} \tag{4.99}$$

Based on the above discussions, we can easily shift the baseband of $-3\,\text{GHz}$ to $+3\,\text{GHz}$ to the next period by adding $2f_e$ to it. Hence, we end up with

$$[-3\,\text{GHz} + 6\,\text{GHz}\ to + 3\,\text{GHz} + 6\,\text{GHz}] = [3\,\text{GHz}\ to\ 9\,\text{GHz}]$$

The passband of $-1\,\text{GHz} \leq f_b \leq +1\,\text{GHz}$ is then shifted to $5\,\text{GHz} \leq f_k \leq 7\,\text{GHz}$. Thus, we see that the low-pass filter designed can be used as a bandpass filter over the frequency range of $f_1 = 5\,\text{GHz}$ to $f_2 = 7\,\text{GHz}$.

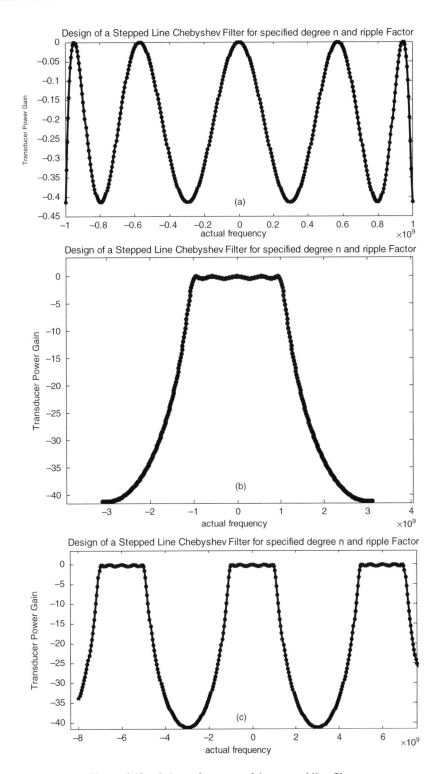

Figure 4.19 Gain performance of the stepped line filter

In Example 4.15, a systematic procedure is presented to design stepped line filters using Chebyshev approximations. In the following section we introduce MATLAB codes to design stepped line filters step by step.

4.18 MATLAB® Codes to Design Stepped Line Filters Using Chebyshev Polynomials

A filter is the simplest broadband network that can be constructed between a resistive generator and a resistive load termination. It only allows power transfer over a prescribed bandwidth. For many practical cases, it must have a sharp roll-off. With their frequency band selection property, filters are essential components of communication systems. Therefore, this section is devoted to developing a practical filter design code in MATLAB employing commensurate stepped lines with monotone roll-off Chebyshev polynomials.

Basic Filter Design Outline

Step 1 (Selection of the analytic form of the transfer function): In general, filter design starts with the selection of an analytic form of a transfer function T in the frequency domain or in a dummy variable x as

$$0 \leq T_n(x^2, \alpha, \beta, \gamma, \varepsilon \ldots \sigma) = \frac{F(x^2)}{G(x^2)} \leq 1 \qquad (4.100)$$

A realizable $T_n(x, \alpha, \beta, \gamma, \varepsilon \ldots \sigma)$ must be an even, rational, non-negative function in variable x and it must be bounded by 1. Here, the dummy variable x corresponds to a frequency variable and needs to be replaced by means of a proper complex variable transformation which makes the design possible. For example, if the design is made using the lumped circuit elements, then x^2 is replaces by $-p^2$, or if the design consists of Commensurate transmission lines then x^2 is replaced by

$$x^2 = \frac{[\alpha^2 \lambda^2]}{[\lambda^2 - 1]}$$

This transfer function must include unknown parameters, so they are determined to satisfy the design specifications. In Equation (4.100) the parameters $\alpha, \beta, \gamma, \varepsilon \ldots \sigma$ designate the unknowns of the transfer function.

Step 2 (Determination of the unknown parameters of the transfer function): In the second step of the design, the unknown parameters are determined to satisfy the design specifications. For example, in designing monotone roll-off Chebyshev filters with stepped commensurate lines, for specified stopband frequency f_e commensurate Delay lengths are given by Equation (4.93):

$$\tau = \frac{1}{4f_e}$$

or the fixed line lengths are given by

$$l = \frac{\nu_m}{4f_e} \qquad (4.101)$$

where ν_m is the speed of wave propagation in the medium of stepped lines.

For specified ripple loss ILR, the ripple factor ε^2 is given by Equation (4.92):

$$\varepsilon^2 = 10^{\text{ILR}/10} - 1$$

For specified cut-off frequency f_c, α is given by Equation (4.94):

$$\alpha = \frac{1}{\sin\left(\dfrac{\pi f_c}{2 f_e}\right)}$$

At the stopband frequency f_e, for a specified minimum value of the IL, the degree of the Chebyshev polynomial is given by Equation (4.96):

$$n \cong \left[\frac{1}{2\log_{10}(2\alpha)}\right] \log_{10}\left[\frac{\alpha^2}{\varepsilon^2}\left(10^{\text{IL}(\alpha)/10} - 1\right)\right] + 1$$

Step 3 (Construction of the scattering parameters in Belevitch form): Once the unknown parameters are determined, the explicit rational form of the transfer function is obtained in the complex frequency domain as

$$T = \frac{F(\gamma)}{G(\gamma)}$$

where γ denotes the complex frequency of the design domain.

For example, if the design is carried out in the complex Richard domain then $\gamma = \lambda$, and F and G are the full coefficient even polynomials in λ^2 satisfying the condition $0 \leq T(\lambda^2) \leq 1$, for all real $\{\lambda\} \geq 0$.

Then, by explicit factorization of the polynomials $G(\gamma)$ and $F(\gamma)$, strictly Hurwitz common denominator polynomials $g(\gamma)$ and $f(\gamma)$ are determined. Eventually, the numerator polynomial $h(\gamma)$ is constructed on the explicit factorization of the even polynomial $\mathbf{H} = \mathbf{G} - \mathbf{F}$ which refers to the Losslessness condition of the filter.

For example, in the Richard variable λ, polynomials $h(\lambda), g(\lambda)$ and $f(\lambda)$ are obtained as follows:

(a) $f(\lambda)$ is constructed on the explicit factorization of the even numerator polynomial $F(\lambda^2) = F_1\lambda^{2n} + F_2\lambda^{2(n-1)} + \ldots + F_n\lambda + F_{n+1}$.

(b) $g(\lambda)$ is constructed on the explicit factorization of the even denominator polynomial $G(\lambda^2) = G_1\lambda^{2n} + G_2\lambda^{2(n-1)} + \ldots + G_n\lambda + G_{n+1}$. It should be kept in mind that $g(\lambda)$ must be strictly Hurwitz.

(c) By the losslessness condition of $H(\lambda^2) = G(\lambda^2) - F(\lambda^2)$, $h(\lambda)$ is formed on the explicit factorization of $H(\lambda^2) = H_1\lambda^{2n} + H_2\lambda^{2(n-1)} + \ldots + H_n\lambda + H_{n+1}$ where $H_i = G_i - F_i$ for $i = 1, 2, 3, \ldots, (n+1)$.

(d) Then, $h(\lambda)$ is constructed on the selected roots of the even polynomial $H(\lambda^2)$ in an arbitrary fashion provided that it must be real, satisfying the Losslessness condition. In other words, $h(\lambda)$ can be constructed on the selected LHP and RHP roots of $H(\lambda^2)$ as an n-th-degree real polynomial which satisfies the losslessness condition.

Step 4 (Synthesis of the Filter): In this step, either the reflectance $S = h/g$ or the corresponding normalized driving point impedance $Z = (1 + S)/(1 - S)$ is synthesized.

Here, the reflectance S designates either the input S_{11} or the output S_{22} reflection coefficient of the filter. Thus, it is synthesized as a lossless two-port in unit termination, revealing the filter two-port.

In short, all the driving point synthesis procedures stem from the famous theorem of Darlington, which states that any bounded real reflectance or any positive real driving point impedance can be synthesized as a lossless two-port in resistive or unit termination. Thus, at the end of the synthesis process, the lossless filter is obtained.

MATLAB® Codes to Generate Stepped Line Filters Using Chebyshev Approximations

In this subsection, the analytic form of TPG is expressed in terms of the Chebyshev polynomials $T_n(x^2)$ as

$$T(x^2) = \frac{1}{1 + \varepsilon^2 T_n^2(x^2)} \tag{4.102}$$

Let us see how we generated the Chebyshev polynomials in MATLAB in coefficient form.

n = odd-integer case

In Equation (4.102), if n is odd then $T_n(x)$ is odd. In this case, let the even integer n_T be $n_T = n - 1$. Then, the odd-degree Chebyshev polynomial can be expressed in terms of the degree $n_k = n_T/2$ as

$$T_n^2(x^2) = x^2 [T_1 x^{2n_k} + T_2 x^{2(n_k - 1)} + \ldots + T_{n_k} x^2 + T_{n_k+1}]^2 \tag{4.103}$$

In order to facilitate MATLAB computations, we can redefine the transformation

$$x^2 = \frac{[\alpha^2 \lambda^2]}{[\lambda^2 - 1]}$$

as

$$x^2 = \frac{y}{z}$$

where

$$y = a^2 \lambda^2 \tag{4.104}$$

and

$$z = [\lambda^2 - 1]$$

Then, Equation (4.102) can be evaluated in MATLAB as

$$T_n^2(x^2) = \frac{y}{z^{2nk+1}} [T_1 y^{nk} + T_2 y^{nk-1} z + \ldots + T_{nk} y z^{nk-1} + T_{nk+1} z^{nk}]^2$$

Since $n = 2n_k + 1$, then

$$T_n^2(x^2) = \frac{y}{z^{2n}} [T_{nk}(z, y)]^2 \tag{4.105}$$

with

$$T_{nk}(z, y) = T_1 y^{nk} + T_2 y^{nk-1} z + \ldots + T_{nk} y z^{nk-1} + T_{nk+1} z^{nk}$$

Thus, the analytic form of the filter transfer function is given by

$$
\begin{array}{l}
T(z,y) = \dfrac{F(z)}{G(z,y)} = \dfrac{z^n}{z^n + \varepsilon^2 y T_n^2(z,y)} \\
\text{where} \\
T_n(z,y) = T_1 y^{nk} + T_2 y^{nk-1} z + \ldots + T_{nk} y z^{nk-1} + T_{nk+1} z^{nk}
\end{array}
\tag{4.106}
$$

Equation (4.106) is in a very handy form to be programmed in MATLAB.

The polynomial coefficients $\{T_i, i = 1, 2, \ldots, n_{k+1}\}$ are easily produced by means of a MATLAB function called `function Tn=poly_cheby (n)`. This function generates the coefficients of Chebyshev polynomials using the recursive formula of Equation (4.86) as

$$
T_{n+1} = 2T_n - T_{n-1}
$$

Then, Equation (4.106) is evaluated for given coefficients $\{T_i, i = 1, 2, \ldots, n_{k+1}\}$.

As far as programming in MATLAB is concerned, once the coefficients of the Chebyshev polynomial $[T_1 T_2 \ldots T_{nk} T_{nk+1}]$ are obtained, we can regard each term of Equation (4.106) as a separate polynomial as

$$
\left\{ [T_1 y^{nk}] [T_2 y^{nk-1}] \ldots [T_{nk} y] [T_{nk+1}] \right\}
$$

Then, we generate

$$
T_n(z,y) = T_1 y^{nk} + T_2 y^{nk-1} z + \ldots + T_{nk} y z^{nk-1} + T_{nk+1} z^{nk}
$$

by employing Horner's rule with convolution of polynomials in MATLAB.

n = even-integer case

In a similar manner to that for odd values of degree n, for even values the Chebyshev polynomials can be expressed as

$$
\begin{array}{l}
T_n^2(x^2) = \left[T_1 x^{2nk} + T_2 x^{2(nk-1)} + \ldots + T_{nk} x^2 + T_{nk+1} \right]^2 \\
\text{where} \\
n_k = n/2
\end{array}
\tag{4.107}
$$

In this case, the filter transfer function is given by

$$
\begin{array}{l}
T(z,y) = \dfrac{F(z)}{G(z,y)} = \dfrac{z^n}{z^n + \varepsilon^2 T_n^2(z,y)} \\
\text{where} \\
T_n(z,y) = T_1 y^{nk} + T_2 y^{nk-1} z + \ldots + T_{nk} y z^{nk-1} + T_{nk+1} z^{nk}
\end{array}
\tag{4.108}
$$

Thus, Equations (4.106) and (4.108) can easily be programmed using MATLAB.

In fact, we developed a function called `[F G g]=Denom_Cheby (n,eps_sq,alfa)` which generates the polynomial coefficients of $F(\lambda)$ and $G(\lambda)$ also generates the Strictly Hurwitz polynomial $g(\lambda)$.

In the Function `[F G g]=Denom_Cheby(n,eps_sq,alfa)` coefficients of the Chebyshev polynomial are generated using the recursive formula. This formula is programmed under the MATLAB function `Tn=poly_cheby(n)`. Eventually, we developed a MATLAB main program called `Cheby_Gain.m` which generates fixed Delay lengths and the characteristic impedances of a 'low-pass, stepped line, monotone roll-off Chebyshev filter' for specified degree n, ripple factor ε^2, cut-off frequency f_c and stopband frequency f_e.

Inputs to the main program `Cheby_Gain.m` are as follows:

Inputs:

- `eps_sq` or ε^2: ripple factor of the transfer function
- n: degree of the Chebyshev polynomial
- f_c: cut-off frequency of the filter
- f_e: stopband frequency of the filter (located at $\omega_e \tau = \pi/2$).

At this point, we must state that all the computations can be made generically as follows.

At the input, instead of specifying the actual cut-off frequency f_c we can generically specify the cut-off phase θ_c as a fraction of the stopband phase θ_e which is fixed by the nature of the problem at $\theta_e = \pi/2$.

Furthermore, the ripple factor ε^2 can be specified as the ripple loss ILR (in dB) at the cut-off frequency f_c or equivalently at θ_c[28] which in turn yields

$$\varepsilon^2 = 10^{\mathrm{ILR}/10} - 1$$

as in Equation (4.92).

Instead of specifying degree n we may specify the attenuation $10\log_{10}|S_{21}(j\omega_e)|$ at the stopband frequency or equivalently the insertion loss $\mathrm{IL(dB)} \cong 10\log_{10}(1 + \varepsilon^2 2^{n-1}\alpha^n)^2$ which approximately yields the degree n as in Equation (4.96):

$$n \cong \frac{1}{2\log_{10}(2\alpha)} \log_{10}\left[\frac{\alpha^2}{\varepsilon^2}\left(10^{\mathrm{IL}/10} - 1\right)\right] + 1$$

where α is given by

$$\alpha = \frac{1}{\sin\left[\dfrac{\pi}{2}\dfrac{\omega_c}{\omega_e}\right]} = \frac{1}{\sin[\theta_c]}, \text{ such that } 0 < \theta_c < \frac{\pi}{2}$$

In other words, all the computations can be made over the phase domain, and, eventually, the fixed lengths of the commensurate lines are determined as a function of the stopband frequency f_e and the propagation velocity of the medium

$$\nu_r = \frac{\nu_0}{\sqrt{\mu_r \varepsilon_r}} = \frac{3 \times 10^8}{\sqrt{\mu_r \varepsilon_r}}$$

---

[28] f_c corresponds to $x = 1$ yielding $T = 1/(1 + \varepsilon^2)$.

as

$$\tau = \frac{\pi}{2\omega_e}$$

$$l = \nu_r \tau$$

$$\nu_r = \frac{\nu_0}{\sqrt{\mu_r \varepsilon_r}} = \frac{3 \times 10^8}{\sqrt{\mu_r \varepsilon_r}}$$

or simply

$$l = \frac{\nu_r}{4f_e} = [\frac{1}{\sqrt{\mu\varepsilon}}][\frac{1}{4f_e}]$$

(4.109)

In return `Cheby_Gain.m` lists the following outputs.

Outputs:

- $\tau = 1/4f_e$: fixed delay length of the commensurate lines.
- $\alpha = 1/\sin(\theta_c)$: fixes cut off frequency at f_c.
- Even polynomials H, G, F in MATLAB polynomial vectors.
- Belevitch polynomials h, g as MATLAB row vectors.
- $[Z_{imp}]$: normalized characteristic impedances with respect to R_0 of the commensurate lines as they are connected in cascade UEs.
- Gain plot of the stepped line filter in absolute scale (Figure 1) and in dB (Figure 2).

Let us run a couple of examples to illustrate the utilization of the main program `Cheby_Gain`.

Example 4.16: Using the main program `Cheby_Gain`, design a stepped line bandpass filter for $f_1 = 5\,\text{GHz}$, $f_2 = 8\,\text{GHz}$, ILR $= 0.5\,\text{dB}$, IL $= 30\,\text{dB}$.

Solution: For this problem, firstly, we should find the stopband f_e and the cut-off f_c frequencies of the low-pass proto type filterly employing Equation (4.99).

Thus, for the stopband frequency f_e we have

$$f_e = \frac{f_1 + f_2}{4}$$

(4.110)

which reveals that

$$f_e = 3.25\,\text{GHz}$$

and for the cut-off frequency f_c it is found as

$$f_c = \frac{f_2 - f_1}{2}$$

(4.111)

yielding

$$f_c = 1.5\,\text{GHz}$$

The Delay length or equivalently the physical length of the stepped lines can be determined using Equation (4.109) as

$$\tau = \frac{\pi}{2\omega_e} = 7.6923 \times 10^{-11}\,\text{s}$$

Let us assume that stepped lines are printed on alumina of $\varepsilon_r = 9.8$ as microstrip lines. In this case, the speed of propagation becomes

$$\nu_r = \frac{3 \times 10^8}{\sqrt{\mu_r \varepsilon_r}} = \frac{3 \times 10^8}{\sqrt{9.8}} = 9.5831 \times 10^7$$

and

$$l = \nu_r \tau = 7.3716\,\text{mm}$$

On the other hand, Equation (4.92) yields the ripple factor ε^2

$$\varepsilon^2 = 10^{\text{ILR}/10} - 1 = 0.2589$$

Furthermore, by Equation (4.94) α is given by

$$\alpha = \frac{1}{\sin[\frac{\pi}{2}\frac{f_c}{f_e}]} = 1.5080$$

Then, the approximate degree n is determined by Equation (4.96) as

$$n \cong \frac{1}{2\log_{10}(2\alpha)}\log_{10}\left[\frac{\alpha^2}{\varepsilon^2}(10^{\text{IL}/10} - 1)\right] + 1 = 5$$

Hence, we can run the main program with the above values.
   The result is summarized as follows:

```
F=[1 -5 10 -10 5 -1]
G=1.0e+003*[1.0526 1.7608 1.0001 0.1989 0.0197 -0.0010]
g=[32.4445 27.7971 39.0437 18.1704 7.4873 1.0000]
h=[32.4445 0 27.2389 0 3.8384 0]
```

The open form of the TPG function is given by

$$T(\lambda) = \frac{[\lambda^2 - 1]^5}{10\,526\lambda^{10} + 17\,606\lambda^8 + 10\,001\lambda^6 + 1989\lambda^4 + 197\lambda^2 + 1}$$

and the input reflectance is given by

$$S_{11} = \frac{32.445\lambda^5 + 27.2389\lambda^3 + 3.833\,84\lambda}{32.344\lambda^5 + 27.7971\lambda^4 + 39.0437\lambda^3 + 18.1704\lambda^2 + 7.4873\lambda + 1}$$

Finally, Richard extractions reveal the normalized characteristic impedance of the stepped line filter
```
Z_imp=[3.0353 0.7330 3.7961 0.7330 3.0353 1.0000]
```

or the actual characteristic impedances are given by
```
Z=[152 37 190 37 152 50.0000]
```
TPG of the filters is depicted in Figure 4.20.

For this example, we revised the main program `Cheby_gain` under the program `CH4COMTRL_Example16.m`. Its program listing is given toward the end of this chapter.

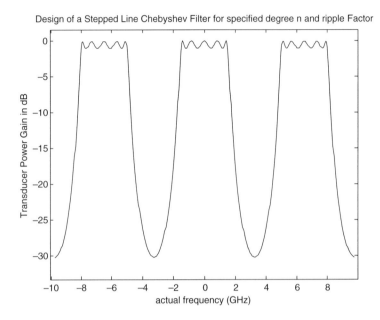

Figure 4.20   TPG plot of the stepped line filter of Example 4.16

Symmetrical Lossless Two-ports

A lossless two-port which satisfies the condition of $S_{11}(\lambda) = S_{22}(\lambda)$ is called symmetric. For cascaded stepped line structures, input and output reflectance are given in Belevitch form as

$$S_{11}(\lambda) = \frac{h(\lambda)}{g(\lambda)}$$

and

$$S_{22}(\lambda) = -\frac{h(-\lambda)}{g(\lambda)}$$

Then, for symmetrical structures, one must have

$$\boxed{h(\lambda) = -h(-\lambda)} \tag{4.112}$$

Equation (4.112) is satisfied if $h(\lambda)$ is odd. An odd polynomial can be constructed as

$$h(\lambda) = \lambda \tilde{h}(\lambda^2)$$

where $\tilde{h}(\lambda^2)$ is an even polynomial which is constructed on the purely imaginary roots such that

$$\tilde{h}(\lambda^2) = \prod_{k=1}^{n-1}(\lambda^2 + \gamma_k^2)$$

Note that $h(\lambda)$ is constructed on the roots of the even polynomial $H(\lambda^2) = h(\lambda)h(-\lambda) = G(\lambda^2) - F(\lambda^2)$ which has all its roots on the $j\Omega$ axis at $\lambda = 0$ and at $\lambda = \pm j\gamma_k$. In this case, from a numerical point of view, computation of the purely imaginary roots of $H(\lambda^2)$ is a little tricky and also construction of $h(\lambda) = \lambda \tilde{h}(\lambda^2)$ requires special care.

On the other hand, construction of $g(\lambda)$ from the roots of $G(\lambda^2)$ is straightforward numerically since it is strictly Hurwitz.

In this section we develop a MATLAB function called `function gtoh(F,G)` to generate the numerator polynomial $h(\lambda)$ from the given even polynomials $G(\lambda^2)$ and $F(\lambda^2)$, considering the possibility that $h(\lambda)$ is an odd polynomial. A listing of this function is also included later in the chapter.

Example 4.17: In this example, employing stepped line filters, the effect of the filter bandwidth on insertion loss is investigated. For this purpose, the main program `Cheby_Gain` is slightly modified in the program `CH4COMTRL_Example17.m` to run the following experiments.

For fixed degree $n = 7$ and for fixed ripple loss ILR $= 0.5$ dB:

(a) Select the band of operation as $f_1 = 5$ GHz and $f_2 = 7$ GHz, find the stopband frequency f_e and examine the insertion loss on it.
(b) Select the band of operation as $f_1 = 5$ GHz and $f_2 = 10$ GHz, find the stopband frequency f_e and examine the insertion loss on it.
(c) Select the band of operation as $f_1 = 5$ GHz *and* $f_2 = 13$ GHz, find the stopband frequency f_e and examine the insertion loss on it.

Solution:

(a) For $n = 7$, ILR $= 0.5$ dB and $f_1 = 5$ GHz and $f_2 = 7$ GHz, the main program `CH4COMTRL_Example17.m` results in the stepped line filter as summarized in Table 4.2. TPG of the filter is depicted in Figure 4.21(a).

Table 4.2 Design of a stepped line filter for $n = 7$, ILR $= 0.5$ dB and $f_1 = 5$ GHz and $f_2 = 7$ GHz, with $R_0 = 50\,\Omega$

f_{e1} (GHz) $= 3$ GHz	IL $(f_e) = 65$ dB	$\tau = 833\ ns$
f_{e2} (GHz) $= 9$ GHz		
$g = [1761.2 \quad 1178.9 \quad 1348.2 \quad 587.7 \quad 277.4 \quad 68 \quad 12.9 \quad 1]$		
$h = [1761.2 \quad 0 \quad 953.6 \quad 0 \quad 141.8 \quad 0 \quad 4.9 \quad 0]$		
Normalized characteristic impedances of the cascaded UEs:		
$Z_{imp} = [3.41 \quad 0.4395 \quad 4.854 \quad 0.4029 \quad 4.854 \quad 0.4395 \quad 3.41 \quad 1]$		

At the edge of the stopbands (i.e. at $f_e = 3$ GHz and $f_e = 9$ GHz) the insertion loss is about 65 dB.

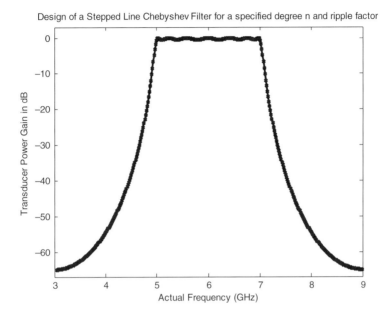

Figure 4.21a TPG of the stepped line filter designed for $n = 7$, ILR $= 0.5$ dB and $f_1 = 5$ GHz and $f_2 = 7$ GHz

(b) For $n = 7$, ILR $= 0.5$ dB and $f_1 = 5$ GHz and $f_2 = 10$ GHz the main program `CH4COMTRL_Example17.m` results in the stepped line filter as summarized in Table 4.3.

Table 4.3 Design of a stepped line filter for $n = 7$, ILR $= 0.5$ dB and $f_1 = 5$ GHz and $f_2 = 10$ GHz with $R_0 = 50\,\Omega$

$f_{e1} = 3.75$ GHz	IL $(f_e) = 18.3$ dB	$\pi = 666.67\ ns$
$g = [8.23\ 17.83\ 47.22\ 52.12\ 49.73\ 24.22\ 7.96\ 1.]$		
$h = [8.2325\ 0\ 28.7612\ 0\ 21.8079\ 0\ 2.8445\ 0]$		
Normalized characteristic impedances of the cascaded UEs:		
$Z_{imp} = [1.84\ 0.99\ 2.12\ 0.92\ 2.13\ 0.98\ 1.84\ 1]$		

At the edge of the stopbands (i.e. at $f_{e1} = 3.75$ GHz and $f_{e2} = 11.25$ GHz) the insertion loss is about 18.3 dB, which is much less than the 65 dB in Figure 4.21(a).

TPG of the filter is depicted in Figure 4.21(b).

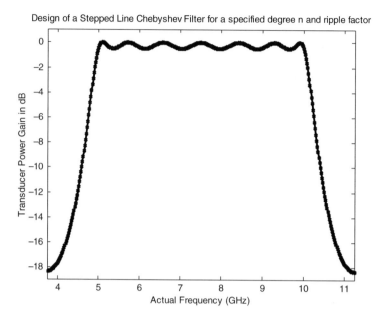

Figure 4.21b   TPG of the stepped line filter designed for $n = 7$, ILR $= 0.5$ dB and $f_1 = 5$ GHz and $f_2 = 10$ GHz

(c) For $n = 7$, ILR $= 0.5$ dB   and $f_1 = 5$ GHz   and $f_2 = 13$ GHz   the   main   program CH4COMTRL_Example17.m results in the stepped line filter as summarized in Table 4.4.

Table 4.4   Design of a stepped line filter for $n = 7$, ILR $= 0.5$ dB, and $f_1 = 5$ GHz, and $f_2 = 13$ GHz, with $R_0 = 50\,\Omega$

$f_{e1} = 4.5$ GHz	$IL\,(f_e) = 1.52$ dB	$\tau = 555.56\,ns$
$f_{e2} = 13.5$ GHz		
$g = [1.19\ 7.08\ 22.87\ 35.16\ 37.44\ 21.03\ 7.43\ 1]$		
$h = [1.19\ 0\ 16.048\ 0\ 23.97\ 0\ 4.56\ 0]$		
Normalized characteristic impedances of the cascaded UEs:		
$Z_{imp} = [2.05\ 1.78\ 2.07\ 1.77\ 2.11\ 1.751\ 2.00\ 1]$		

At the edge of the stopbands (i.e. at $f_{e1} = 4.5$ GHz and $f_{e2} = 13.5$ GHz) the insertion loss is about 1.521 dB, which is much less than the 18.3 dB in Figure 4.21(b).

TPG of the filter is depicted in Figure 4.21(c). This figure clearly shows that as the bandwidth of the filter increases, for the fixed number of elements, the sharpness of the roll-off of the filter is drastically penalized, as expected.

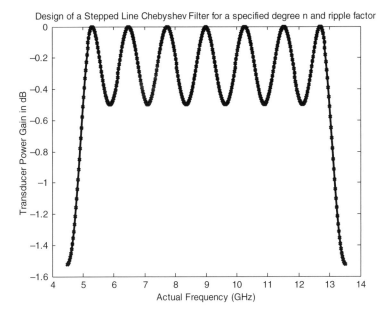

Design of a Stepped Line Chebyshev Filter for a specified degree n and ripple factor

Figure 4.21c TPG of the stepped line filter designed for $n = 7, \mathrm{ILR} = 0.5$ dB and $f_1 = 5$ GHz and $f_2 = 15$ GHz

4.19 Summary and Concluding Remarks on the Circuits Designed Using Commensurate Transmission Lines

In this chapter, classical network theory developed for the design of lumped element Lossless two-ports is expanded to construct Lossless two-ports with Commensurate transmission lines.

For this purpose, a complex frequency, called the Richard variable, $\lambda = \tanh(p\tau) = \Sigma + j\Omega$, is introduced. In this representation, $p = \sigma + j\omega$ designates the classical complex frequency or complex domain variable utilized in the design of lumped element passive networks. Furthermore, classical angular frequency ω is related to Ω by

$$\Omega = \tan(\omega\tau)$$

In the above representations, τ is called the Delay length of the Commensurate transmission lines and is related to the physical length l as

$$l = \nu\tau$$

where ν denotes the propagation velocity of the transmission medium.

It has been shown that network functions such as impedance, admittance and scattering parameters belonging to passive networks constructed with Commensurate transmission lines can be expressed in terms of the Richard variable λ, preserving all the outstanding properties developed in the complex variable p. This greatly facilitates the design of lossless two-ports employing commensurate transmission lines.

In practice, commensurate transmission lines can easily be implemented using microstrip lines to design broadband power transfer networks such as filters and matching networks at radio frequencies.

It was mentioned that lossless broadband power transfer networks can be constructed by utilizing scattering parameters. In this regard, for a lossless reciprocal two-port constructed with simple connections of Commensurate transmission lines, the scattering parameters are given in Belevitch form such that

$$S(\lambda) = \begin{bmatrix} S_{11} & S_{12} \\ S_{21} & S_{22} \end{bmatrix} = \begin{bmatrix} \dfrac{h(\lambda)}{g(\lambda)} & \dfrac{f(\lambda)}{g(\lambda)} \\ \dfrac{f(\lambda)}{g(\lambda)} & -\dfrac{f(\lambda)h(-\lambda)}{f(-\lambda)g(\lambda)} \end{bmatrix}$$

Many practical applications demand connections of cascaded UEs with open-ended or short-ended series and shunt stubs. In this case, the forms of polynomials are given by

$$f(\lambda) = \lambda^q (1 - \lambda^2)^{k/2}$$

$$h(\lambda) = h_1 \lambda^n + h_2 \lambda^{n-1} + \ldots + h^n \lambda + h_{n+1}$$

$$g(\lambda) = g_1 \lambda^n + g_2 \lambda^{n-1} + \ldots + g^n \lambda + g_{n+1}$$

In the above representation, n denotes the total number of commensurate lines in the lossless two-port, q is the total count of open-ended series stubs and short-ended shunt stubs and $r = n - q$ is the total count of short-ended series stubs and open-ended shunt stubs.

The losslessness condition of the two-port demands that

$$G(\lambda^2) = H(\lambda^2) - F(\lambda^2)$$

where

$$G(\lambda^2) = g(\lambda)g(-\lambda) \quad H(\lambda^2) = h(\lambda)h(\lambda-) \quad F(\lambda^2) = f(\lambda)f(-\lambda)$$

At this point it must be remember that open-ended stubs act like capacitors. Similarly, short-ended stubs may be regarded as inductors.

It is interesting to note that, for a lossless two-port, a measure of power transfer over the real frequencies $\Omega = \tan(\omega\tau)$ is given by the forward scattering parameter as

$$T(\Omega^2) = |S_{21}(j\Omega)|^2 = \frac{F(\Omega^2)}{G(\Omega^2)}$$

In this representation, $T(\Omega^2)$ or equivalenty $T(\lambda^2)$, which is obtained by replacing Ω^2 by $-\lambda^2$, is called the analytic form of the power transfer function. It must also be mentioned that over the actual angular frequency ω axis, $T(\Omega^2)$ is periodic due to the periodicity of $\Omega = \tan(\omega\tau)$ over $-\pi(2k+1) \leq \omega\tau \leq +\pi(2k+1), k = 0, 1, 2, \ldots$.

In designing filters, first of all, the analytic form of $T(x^2)$ is selected. Then, $T(\lambda^2)$ is obtained by setting $x^2 = \alpha^2 \lambda^2 / (\lambda^2 - 1)$. Here, the real variable α is determined by means of the cut-off frequency f_c of the filter such that

$$\alpha = \frac{1}{\sin(2\pi f_c \tau)}$$

The integer n for other unknown parameters of $T(x^2)$ is determined by means of design specifications. For example, degree n is related to the attenuation capability of the filter in the stopband region.

From a numerical point of view, the filter is constructed on the scattering parameters. Therefore, explicit factorization of the filter transfer function $T(\lambda^2)$ is inevitable, which in turn yields the scattering Hurwitz polynomials $g(\lambda)$ and $f(\lambda)$. Eventually, the numerator polynomial $h(\lambda)$ is generated on the explicit factorization of $H(\lambda^2) = G(\lambda^2) - F(\lambda^2)$. Here, it is very important to emphasize that $F(\lambda^2)$ is preselected by the design specification of the filter which describes the properties of the filter stopband region. In this case, in filter design, by selecting a proper transfer function one basically determines $g(\lambda)$ and then generates $h(\lambda)$. It is crucial to note that for a high degree of transfer functions $(n > 5)$ one may face severe numerical problems in generating the numerator polynomial $h(\lambda)$. Therefore, it would have been nice if we had a technique to generate filter transfer functions starting directly from the numerator polynomial $h(\lambda)$ and then determine the denominator polynomial $g(\lambda)$ from the previously fixed $F(\lambda^2)$. In fact, we can do it.

The approach, called the simplified real frequency technique (SRFT), can simply generate the filter transfer function directly from the numerator polynomial $h(\lambda)$ for selected $F(\lambda)$, which in turn yields $g(\lambda)$ as desired. SRFT for designing filters is presented in the next chapter.

Program Listing 4.13

```
% Main Program: CH4COMTRL_Example17.m
% This program generates the analytic form of the
%Chebyshev gain function for Stepped Line Band-Pass
%filters constructed with cascaded connections of
% UE.
%
% ─────────────────────────────────────────────
% Inputs:
% Fc1=Beginning of the pass band
% Fc2=End of the pass band
% ILR(dB)=Ripple Loss
% IL(dB)=Insertion Loss at the stop band
%
%─────────────────────────────────────────────
% Outputs
% n: Degree of the Chebyshev polynomial
% eps_sq: The ripple factor epsilon^2
% fe=Stop band frequency of the filter.
% Program generates T(lmbda^2)=S21(lmda)S21(-
%lmda)=F(lmda^2)/G(lmda^2)
% and plots over the frequencies.
% In the computations all the polynomials are given
% as a full polynomial in
% x=lmbda^2.
% That is,
```

Program Listing 4.13   (Continued)

```
% g(lmda)=g(1)lmbda^n+g(2)lmbda^(n-
%1)+...+g(n)lmbda+g(n+1)
% G(lmda)=G(1)lmda^2(n)+G(2)lmda^2(n-
%1)+...+G(n)lmda^2+G(n+1)
% F(lmda)=(lmda^2-
%1)^n=F(1)lmda^2(n)+...+F(n)lmda^2+F(n+1)
% It is noted that in its classical form F(lmda)=(-
%1)^n(1-lmda^2)^n
% Due to its periodic nature of tan(wtau), This
%program may as well
% be used to generate bandpass filters over the
%frequencies
%
% f1=2*fc to f2=4fc
%-----------------------------------------------
clear
%
F1=input('Beginning of the pass band frequency
F1(GHz)=')
F2=input(' End of the Pass Band Frequency F2
(GHz)=')
ILR=input(' Enter ripple loss in dB ILR=');
%IL=input('Enter insertion loss in dB IL=');
%-----------------------------------------------
fc=(F2-F1)/2
fe=(F2+F1)/4
eps_sq=10^(ILR/10)-1
wc=2*pi*fc;
we=2*pi*fe;
f1=fe;
f2=+3*fe;
w1=2*pi*f1;
w2=2*pi*f2;
%n=Cheby_Degree(IL,ILR,fc,fe)
n=input(' n=');
%At stop Band frequency we have we*tau=pi/2
% This is the periodicity of tan(w*tau) function.
tau=pi/2/we
% At cut-off frequency we must have
% x=1=alfa*sin(wc*tau)
alfa=1.0/(sin(pi*wc/we/2))
%
[F G g]=Denom_Cheby(n,eps_sq,alfa)
teta1=w1*tau;
teta2=w2*tau;
N=401;
del_teta=(teta2-teta1)/(N-1);
```

```
teta=teta1;
del_f=(f2-f1)/(N-1);
f=f1;
for i=1:N
z(i)=teta;
fr(i)=f;
omega=tan(teta);
omg2=omega*omega;
lmd2=-omg2;
lmd=j*omega;
Gval=polyval(G,lmd2);
Fval=polyval(F,lmd2);
Gain=Fval/Gval;
T(i)=Gain;
TdB(i)=10*log10(Gain);
teta=teta+del_teta;
f=f+del_f;
end
% Computation of h(lmda) polynomial
%————————————————————
% %
h =gtoh(F,G); %
% %
%————————————————————
Z_imp=UE_sentez(h,g);



figure(1)
plot(fr,T)
title('Design of a Stepped Line Chebyshev Filter
for specified degree n and ripple Factor')
xlabel('actual frequency')
ylabel('Transducer Power Gain')



figure(2)
plot(fr,TdB)
title('Design of a Stepped Line Chebyshev Filter
for a specified degree n and ripple factor')
xlabel('Actual Frequency (GHz)')
ylabel('Transducer Power Gain in dB')
```

Program Listing 4.14

```
% Main Program: CH4COMTRL_Example16.m
% This program generates the analytic form of the
% Chebyshev gain function
% for Stepped Line Band-Pass filters constructed
% with cascaded connections of UE.
% ─────────────────────────────────────────────
% Inputs:
% Fc1=Beginning of the pass band
% Fc2=End of the pass band
% ILR(dB)=Ripple Loss
% IL(dB)=Insertion Loss at the stop band
%
% ─────────────────────────────────────────────
% Outputs
% n: Degree of the Chebyshev polynomial
% eps_sq: The ripple factor epsilon^2
% fe=Stop band frequency of the filter.
% Program generates T(lmbda^2)=S21(lmda)S21(-
% lmda)=F(lmda^2)/G(lmda^2)
% and plots over the frequencies.
% In the computations all the polynomials are given
% as a full polynomial in
% x=lmbda^2.
% That is,
% g(lmda)=g(1)lmbda^n + g(2)lmbda^(n-
% 1)+...+g(n)lmbda+g(n+1)
% G(lmda)=G(1)lmda^2(n)+G(2)lmda^2(n-
% 1)+...+G(n)lmda^2+G(n+1)
% F(lmda)=(lmda^2-
% 1)^n=F(1)lmda^2(n)+...+F(n)lmda^2+F(n+1)
% It is noted that in its classical form F(lmda)=(-
1)^n(1-lmda^2)^n
% Due to its periodic nature of tan(wtau), This
% program may as well
% be used to generate bandpass filters over the
% frequencies
%
% f1=2*fc to f2=4fc
% ─────────────────────────────────────────────
clear
%
F1=input('Beginning of the pass band frequency F1=')
F2=input(' End of the Pass Band Frequency F2=')
ILR=input(' Enter ripple loss in dB ILR=');
IL=input('Enter insertion loss in dB IL=');
% ─────────────────────────────────────────────
```

```
fc=(F2-F1)/2
fe=(F2+F1)/4
eps_sq=10^(ILR/10)-1
wc=2*pi*fc;
we=2*pi*fe;
f1=-3*fe;
f2=+3*fe;
w1=2*pi*f1;
w2=2*pi*f2;
n=Cheby_Degree(IL,ILR,fc,fe)
n=input(' n=');
%At stop Band frequency we have we*tau=pi/2
% This is the periodicity of tan(w*tau) function.
tau=pi/2/we
% At cut-off frequency we must have
x=1=alfa*sin(wc*tau)
alfa=1.0/(sin(pi*wc/we/2))
%
[F G g]=Denom_Cheby(n,eps_sq,alfa)
teta1=w1*tau;
teta2=w2*tau;
N=401;
del_teta=(teta2-teta1)/(N-1);
teta=teta1;
del_f=(f2-f1)/(N-1);
f=f1;
for i=1:N
z(i)=teta;
fr(i)=f;
omega=tan(teta);
omg2=omega*omega;
lmd2=-omg2;
lmd=j*omega;
Gval=polyval(G,lmd2);
Fval=polyval(F,lmd2);
Gain=Fval/Gval;
T(i)=Gain;
TdB(i)=10*log10(Gain);
teta=teta+del_teta;
f=f+del_f;
end
% Computation of h(lmda) polynomial
%————————————————
% %
h =gtoh(F,G); %
% %
%————————————————
Z_imp=UE_sentez(h,g);
```

Program Listing 4.14   (Continued)

```
figure(1)
plot(fr,T)
title('Design of a Stepped Line Chebyshev Filter
for specified degree n and ripple Factor')
xlabel('actual frequency')
ylabel('Transducer Power Gain')


figure(2)
plot(fr,TdB)
title('Design of a Stepped Line Chebyshev Filter
for specified degree n and ripple Factor')
xlabel('actual frequency')
ylabel('Transducer Power Gain in dB')
```

Program Listing 4.15   The main program Cheby_Gain

```
% Main Program: Cheby_Gain
% This program generates the analytic form of the Chebyshev
gain function
% for Stepped Line filters constructed with cascaded
% connections of UE.
% ───────────────────────────────────────────

% Inputs:
% n: Degree of the Chebyshev polynomial
% eps_sq: The ripple factor epsilon^2
% fc=Cut=off frequency of the lowpass filter
% fe=Stop band frequency of the filter.
% It should be noted that due to the periodicity of the
% tangent function
% fc%─────────────────────────────────────────

% Outputs
%
% Program generates T(lmbda^2)=S21(lmda)S21(-
% lmda)=F(lmda^2)/G(lmda^2)
% and plots over the frequencies.
% In the computations all the polynomials are given as a full
% polynomial in
% x=lmbda^2.
% That is,
% g(lmda)=g(1)lmda^n+g(2)lmda^(n-1)+...+g(n)lmda+g(n+1)
```

```
% G(lmda)=G(1)lmda^2(n)+G(2)lmda^2(n-
% 1)+...+G(n)lmda^2+G(n+1)
% F(lmda)=(lmda^2-
% 1)^n=F(1)lmda^2(n)+...+F(n)lmda^2+F(n+1)
% It is noted that in its classical form F(lmda)=(-1)^n(1-
% lmda^2)^n
% Due to its periodic nature of tan(wtau), This program may as
% well
% be used to generate bandpass filters over the frequencies
%
% f1=2*fc to f2=4fc
%————————————————————————————————————————
————————
clear
n=input(' n=');
% eps_sq=input('eps_sq=');
eps_sq=input('ripple factor eps_sq=')
%
fc=input('Cut-off Frequency fc=')
fe=input(' Stop band frequency fe=')
wc=2*pi*fc;
we=2*pi*fe;
f1=-3*fe;
f2=+3*fe;
w1=2*pi*f1;
w2=2*pi*f2;
%At stop Band frequency we have we*tau=pi/2
% This is the periodicity of tan(w*tau) function.
tau=pi/2/we
% At cut-off frequency we must have x=1=alfa*sin(wc*tau)
alfa=1.0/(sin(pi*wc/we/2))
%
[F G g]=Denom_Cheby(n,eps_sq,alfa)
teta1=w1*tau;
teta2=w2*tau;
N=401;
del_teta=(teta2-teta1)/(N-1);
teta=teta1;
del_f=(f2-f1)/(N-1);
f=f1;
for i=1:N
z(i)=teta;
fr(i)=f;
omega=tan(teta);
omg2=omega*omega;
lmd2=-omg2;
lmd=j*omega;
Gval=polyval(G,lmd2);
Fval=polyval(F,lmd2);
```

Program Listing 4.15   (Continued)

```
Gain=Fval/Gval;
T(i)=Gain;
TdB(i)=10*log10(Gain);
teta=teta+del_teta;
f=f+del_f;
end
% Computation of h(lmda) polynomial
%——————————————————
%
%
h =gtoh(F,G);
%
%——————————————————
Z_imp=UE_sentez(h,g);


figure(1)
plot(fr,T)
title('Design of a Stepped Line Chebyshev Filter for specified
degree n and ripple Factor')
xlabel('actual frequency')
ylabel('Transducer Power Gain')
figure(2)
plot(fr,TdB)
title('Design of a Stepped Line Chebyshev Filter for specified
degree n and ripple Factor')
xlabel('actual frequency')
ylabel('Transducer Power Gain in dB')
```

Program Listing 4.16   Computation of $h(\lambda)$ from F and G

```
function h =gtoh(F,G)
% This function generates the h polynomial
% Inputs:
% F and G
% Output:
% h
% Note that generation of h is a little
% bit tricky
% H=G-F may have negative roots.
```

```
% which in turn yields pure imaginary
% roots on the jw axis
% Therefore, we must select the imaginary
roots.
H=G-F;
n1=length(H);
if abs(H(n1))n2=n1-1;
for i=1:n2;
A(i)=H(i);
end
end
if abs(H(n1))>=1e-5;
for i=1:n1;
A(i)=H(i);
end
end
r=roots(A);
n=length(A);
alfa=real(r);
beta=imag(r);
i=1;
for j=i:n-1
if abs(beta(i))zr(i)=sqrt(alfa(i));
zr(i+1)=-sqrt(alfa(i));
end
if abs(beta(i))>=1e-5;
zr(i)=sqrt(r(i));i=1+1;end
if i5;i=i+2;end;end


end

h1=abs(G(1));
h1=sqrt(h1);
h=h1*poly(-zr);
if abs(H(n1))p=[1 0];
h=conv(p,h);
end


return
```

Program Listing 4.17   Generation of the coefficients of Chebyshev functions

```
function Tn=poly_cheby (n)
% This program Generates the Coefficients
% of Chebyshev polynomials
%
% Inputs:
% n: Degree of the Chebyshev
Polynomial
% Output:
% Tn=[Coefficients' Vector]
n1=n+1;
if n==0; T0=[1]; Tn=T0; end
if n==1; T1=[1 0];Tn=T1; end
if n==2; T2=[2 0 -1];Tn=T2;end
if n>2;
TA=[1];%zero order Chebyshev
polynomial
TB=[1 0];%First order Chebyshev
% polynomial
for j=3:n1
%Recursive Formula
T=2*conv(TB,[1 0]);
TC=vector_sum(T,-TA);
TA=TB;
TB=TC;
end


Tn=TC;
end
return
```

Program Listing 4.18   Generation of the analytic form of the stepped line filter transfer function using Chebyshev polynomials

```
function [F G
g]=Denom_Cheby(n,eps_sq,alfa)
% Given n and alfa Compute G
% Here computations are carried out in
% several
% steps
% Step 1: find if n is odd or even.
```

```
% Step 2: if n is odd
% x=lambda^2
%—————————————————————
Tn=poly_cheby (n);
Y=[alfa*alfa 0];%Y=alfa^2*x; Row vector
z=[1 -1];%[x-1]; Row Vector
n1=length(Tn);
% check if n is odd or even
k=n1/2;k=fix(k);
% n is odd case
if n>1
if abs((k-n1/2))==0;%n is odd
nT=n1/2;T=poly_oddterms(Tn);
%Chebyshev poly with odd terms
% Y loop;
A=[T(1)];                        '
for i=1:nT-1;% Start Y loop
B=[T(i+1)];%polynomial
C=conv(A,Y);%C=AY
C=vector_sum(C,B);%C=AY+B
A=C;% Row vector
end;%for loop ends here
T=C;
% (x-1) loop
A=[T(nT)];
for i=1:nT-1
B=T(nT-i)*poly_xn(i);%B=T(i+1)*Y
A=conv(A,z);
C=vector_sum(A,B);
A=C;
end;% for loop ends here
G1=C;
G2=conv(G1,G1);
Z3=eps_sq*Y;
G3=conv(G2,Z3);
end;%if n is odd loop ends here
%——————————————————————
% if n is an even integer
if abs((k-n1/2))>0; % n is even
nk=n/2;
T=poly_eventerms(Tn);%Chebyshev poly
% with even terms
% Y loop;
A=[T(1)];
for i=1:nk;% Start Y loop
B=[T(i+1)];%polynomial
C=conv(A,Y);%C=AY
C=vector_sum(C,B);%C=AY+B
A=C;% Row vector
```

Program Listing 4.18   (Continued)

```
end;%for loop ends here
T=C;
% z loop
A=[T(nk+1)];
for i=1:nk
B=T(nk-i+1)*poly_xn(i);%B=T(nk-
i+1)*Y
A=conv(A,z);
C=vector_sum(A,B);
A=C;
end;% for loop ends here
G1=C;
G2=conv(G1,G1);
G3=conv(G2,[eps_sq]);
end
% end of n=even integer loop
end
%n=1 case
if n==1
G3=eps_sq*Y;
end
%
F=[1];
for i=1:n
F=conv(F,z);
end
G=vector_sum(F,G3);
%
% Construction of g(lambda)
r=roots(G);
zr=sqrt(r);
g1=abs(G(1));
g1=sqrt(g1);
g=g1*poly(-zr);
return
```

5

Insertion Loss Approximation for Arbitrary Gain Forms via the Simplified Real Frequency Technique: Filter Design via SRFT

5.1 Arbitrary Gain Approximation

In the previous chapter, we demonstrated the design process of classical microwave filters employing commensurate transmission lines. The first step in the design method was to choose a proper transfer function. Obviously the form of the transfer function must reflect the design requirements. For many practical cases, it may be sufficient to work with readily available low-pass kinds of prototype functions constructed by means of Butterworth or Chebyshev polynomials as shown in Chapter 4. However, for some applications, the desired shape of the transducer power gain (TPG) may be dictated by the problem under consideration in an arbitrary manner. For example, gain distortion of a communication channel may be equalized by means of a lossless filter which compensates the channel distortion.

A typical frequency-dependent channel attenuation is depicted in Figure 5.1. One can compensate the tapered gain of the channel by designing an equalizer which may have a TPG shape as shown in Figure 5.2.

Thus, the resulting gain shape of the new system which consists of the cascaded connection of the channel and the filter (or equalizer) is flat. There may be some other instances for which the gain and the phase of the system are equalized simultaneously. In this regard, the forward scattering parameter of the system is described over the actual angular frequencies as $S_{21}(j\omega) = |S_{21}(j\omega)|e^{j\varphi_{21}(\omega)}$. Then, TPG (or simply the 'gain') is specified as $T(\omega) = |S_{21}(j\omega)|^2$ and the phase of the system is given by $\emptyset(\omega) = \varphi_{21}(\omega)$. All the above-mentioned quantities may be measured as a data set by means of a network analyzer over the actual frequencies. In this case, $|S_{21}(j\omega)|$ and $\phi_{21}(\omega)$ are measured in a straightforward manner. Then, the approximation problem turns out to be a nonlinear curve-fitting problem of either only the amplitude $|S_{21}(j\omega)|$ or both the amplitude and phase of $S_{21}(j\omega)$.

Unfortunately, there is no well-established analytic procedure for solving the arbitrary gain and phase approximation problem. Therefore, we seek computer-aided solutions with well-behaved numerics.

Design of Ultra Wideband Power Transfer Networks   Binboga Siddik Yarman
© 2010 John Wiley & Sons, Ltd

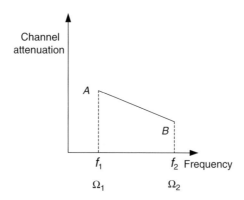

Figure 5.1   A typical channel attenuation as a function of frequency

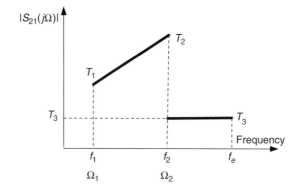

Figure 5.2   A typical arbitrary gain shape of a lossless equalizer which compensates the tapered gain of a communication channel

A straightforward design procedure may start with the selection of an equalizer topology with unknown element values designated by $\{x_1, x_2, \ldots, x_n\}$. Then, the forward scattering parameter $S_{21}(j\Omega)$ of the equalizer can be expressed in terms of these unknowns and the Richard frequency variable $\Omega = \tan(\omega\tau)$. Eventually, by means of a nonlinear optimization algorithm, one determines the unknown element values to reach the desired optimization targets.

At this point it should be mentioned that the amplitude function $|S_{21}(j\Omega| = M = M(\omega, x_1, x_2, \ldots, x_n)$ is highly nonlinear in terms of the unknown element values. Therefore, this approach may not result in successful optimization. On the other hand, the method known as the simplified real frequency technique (SRFT) always describes a convergent optimization scheme which in turn yields the desired equalizer scattering parameters. Eventually, the filter or the equalizer can be constructed on the numerically generated scattering parameters. SRFT is outlined in the following section.

5.2 Filter Design via SRFT for Arbitrary Gain and Phase Approximation

In SRFT, firstly the numerator polynomial $f(\lambda)$ of $S_{21}(\lambda) = f(\lambda)/g(\lambda)$ is selected based on the design specifications. Recall that the polynomial $f(\lambda)$ is constructed on the transmission zeros of the filter or the equalizer. Obviously, $f(\lambda)$ must be free of transmission zeros in the band of operation. Furthermore, from a practical implementation point of view, we do not prefer to utilize finite frequency transmission zeros outside the passband. Therefore, we usually prefer to select a simple form for $f(\lambda)$ as described in Chapter 4 as

$$f(\lambda) = \lambda^q (1 - \lambda^2)^{k/2} \tag{5.1}$$

Here, the integer q is the count of transmission zeros at $\Omega = 0$ (or equivalently $\omega = 0$)[1] and the integer k is the count of cascaded connection of unit elements (UEs) placed in the equalizer. As explained in Chapters 3 and 4, a DC transmission zero is realized either in the form of a series open stub (like a capacitance $C\lambda$) or in the form a parallel short stub (like an inductance $L\lambda$).

In the second step of the SRFT approach, coefficients $\{h_1, h_2, \ldots, h_n, h_{n+1}\}$ of the numerator polynomial $h(\lambda) = h_1 \lambda^n + h_2 \lambda^{n-1} + \ldots + h_n \lambda^n + h_{n+1}$ are selected as the unknowns of the approximation problem, are initialized in an ad hoc manner, and eventually are determined to optimize the predefined objective function to fulfill the design specifications. In the course of optimization, coefficients $\{g_1, g_2, \ldots, g_n, g_{n+1}\}$ of the denominator polynomial $g(\lambda)$ are determined using the losslessness condition given by

$$
\begin{aligned}
G(\lambda^2) &= g(\lambda)g(-\lambda) \\
&= G_1 \lambda^{2n} + G_2 \lambda^{2(n-1)} + \cdots + G_n \lambda^2 + G_{n+1} \\
&= H(\lambda^2) + F(\lambda^2)
\end{aligned} \tag{5.2}
$$

where

$$H(\lambda^2) = h(\lambda)h(-\lambda) = H_1 \lambda^{2n} + H_2 \lambda^{2(n-1)} + \cdots + H_n \lambda^2 + H_{n+1}$$

$$F(\lambda^2) = f(\lambda)f(-\lambda) = (-1)^q \lambda^{2q} (1 - \lambda^2)^k = F_1 \lambda^{2n} + F_2 \lambda^{2(n-1)} + \ldots + F_n \lambda^2 + F_{n+1}$$

$$G_i = H_i - F_i, \quad i = 1, 2, \ldots, n(n+1)$$

The open form of the denominator polynomial $g(\lambda) = g_1 \lambda^n + g_2 \lambda^{n-1} + \ldots + g_n \lambda^n + g_{n+1}$ is obtained on the left half plane (LHP) roots of $G(\lambda^2)$.

In the third step of the SRFT filter design method, once $g(\lambda)$ is determined, the input reflection coefficient $S_{11} = h(\lambda)/g(\lambda)$ is synthesized by means of known synthesis algorithms. In the course of synthesis, the crucial point is the extraction of transmission zeros from the given reflectance which in turn guarantees the form of S_{21} as $f(\lambda)/g(\lambda)$. In this regard, transmission zeros at DC can be extracted either as a series capacitance $C\lambda$ or as a shunt inductance $L\lambda$ directly from the normalized input impedance.

$$Z_{in} = \frac{1 + S_{11}}{1 - S_{11}} = \frac{g + h}{g - h} = \frac{N(\lambda)}{D(\lambda)}$$

---

[1] Note that, in practice, on the real frequency axis the $\omega = 0$ point is usually referred to as DC, which is the abbreviation for 'direct current'.

Extraction of UE can be completed by the Richard synthesis procedure as presented in Chapter 4. For this purpose, our MATLAB® function [Z]=UE_sentez(h,g) yields line characteristic impedances in the row vector [Z].

In MATLAB, all the above computations can be carried out with simple statements as explained in the following subsections.

Generation of $H(\lambda^2)$

Once the coefficients of $h(\lambda) = h_1\lambda^n + h_2\lambda^{n-1} + \ldots + h_n\lambda^n + h_{n+1}$ are initialized, $H(\lambda^2)=h(\lambda)h(-\lambda)= H_1\lambda^{2n} + H_2\lambda^{2(n-1)} + \ldots + H_n\lambda^2 + H_{n+1}$ is computed in two steps:

1. In MATLAB, the polynomial of $h(\lambda)$ is expressed as a row vector

$$h = [h(1)h(2)\ldots h(n) \ \ h(n+1)]$$

Then, $h(-\lambda)$ is computed using the MATLAB function paraconj as

$$h_- = parconj(h)$$

It should be noted that the function paraconj is given in Chapter 4.
2. Coefficients of $H(\lambda^2) = h(\lambda)h(-\lambda)$ can simply be generated in MATLAB by using function 'conv' as a row vector

$$H = [H(1) \ \ H(2)\ldots H(n) \ \ H(n+1)]$$

Generation of $F(\lambda^2)$

In Chapter 4, the open form of $F(\lambda^2) = f(\lambda)f(-\lambda) = (-1)^q\lambda^{2q}(1-\lambda^2)^k = F_1\lambda^{2n} + F_2\lambda^{2(n-1)} + \ldots + F_n\lambda^2 + F_{n+1}$ is generated as a row vector by means of the MATLAB function F=lambda2q_UE2k (k,q).

Recall that this function calls another MATLAB function, namely F=cascade(k), to generate cascaded connections of k UE.

Generation of $g(\lambda)$

Eventually, as shown in Chapter 4, the open form of $g(\lambda)$ is generated as a MATLAB row vector using function g = htog. This function combines the above functions to generate the realizable denominator polynomial $g(\lambda) = g_1\lambda^n + g_2\lambda^{n-1} + \ldots + g_2\lambda^n + g_{n+1}$ with a single MATLAB statement g=htog(q,k,h).

Optimization

MATLAB provides an excellent nonlinear optimization function called lsqnonlin. Function lsqnonlin minimizes the user-supplied objective function by employing the Levenberg–Marquardt (LEV-MQ) algorithm. This minimizes an objective function, which is described in the form of an error function at each target point i such that

$$\boxed{\varepsilon(i) = M(i,x) - \text{Target}(i), \ \ i = 1,2,3, \ldots, \text{Nsample}} \quad (5.3)$$

where the row vector $x = \{x_1, x_2 \ldots, x_m\}$ denotes the unknowns of the minimization problem.

For example, in arbitrary gain approximation problems, $M(i)$ may define the squared amplitude of the forward transfer scattering parameter at an actual frequency f_i such that[2]

$$M(i) = |S_{21}(j\Omega_i)|^2 = \frac{F(\lambda_i^2)}{H(\lambda_i^2) + F(\lambda_i^2)}\Big|_{\lambda_i^2 \ = \ -\Omega_i^2}$$

where

$$\Omega_i^2 = \tan^2(2\pi f_i \tau)$$

and

$$\tau = \frac{1}{4f_e}$$

is the delay length of the commensurate transmission lines in seconds. f_e is the beginning of the stopband frequency, which is specified by the designer.

Obviously, in the present case, coefficients $x = h = \{h_1, h_2, \ldots, h_n, h_{n+1}\}$ of the $h(\lambda)$ polynomial will be the unknowns of the optimization.

On the other hand, the quantity 'Target(i)' may designate the desired value of $|S_{21}(j\Omega_i)|^2$ at the sampling frequency f_i as shown in Figure 5.2. Thus, the unknowns of the equalizer design problem can be determined via minimization of Equation (5.3).

Now, let us consider some practical examples to illustrate the usage of the above in MATLAB functions.

Example 5.1 (Generation of $g(\lambda)$ from intialized $h(\lambda)$ and selected $f(\lambda)$): Let $q = 0$, $k = 5$ and, $h(\lambda) = 114.4788\ \lambda^5 + 44.2736\ \lambda^3 + 3.1624\ \lambda$, yielding a MATLAB row vector h = [114.4788 0 44.27336 0 3.1624 0]. Find $g(\lambda)$ as a strictly Hurwitz polynomial.

Solution: As you may recall, the above form of $f(\lambda) = \left(1 - \lambda^2\right)^{5/2}$ with h=[114.4788 44.27336 3.1624 0] describes the five-section stepped line Chebyshev filter of Example 4.15. Here, in this example, we will simply generate $g(\lambda)$ using the losslessness condition and employing the MATLAB function g = htog(q,k,h).

Hence, in the MATLAB command window we type
>> h=[114.4788 44.27336 3.1624 0]
h=114.4788 44.2734 3.1624 0
Then we run the function g=gtoh in the command window and obtain the following result:
>> g=htog(q,5,h);
>> g
g=114.4832 83.1849 74.4714 28.8809 8.5301 1.0000
This means that

$$g(\lambda) = 114.048\lambda^5 + 83.18\lambda^4 + 74.47\lambda^3 + 28.88\lambda^2 + 8.53\lambda + 1$$

which verifies our previous result.

It is necessary to point out that, in Example 4.15, we generated $h(\lambda)$ as an odd polynomial from the polynomials $f(\lambda)$ and $g(\lambda)$ which showed severe numerical problems. However, in SRFT, number crunching is very smooth since $G(\lambda^2)$ is always non-negative. Therefore, it is free of imaginary axis zeros, which in turn yields a full coefficient and strictly Hurwitz polynomial $g(\lambda)$. Therefore, we can confidently state that SRFT results in robust numerical stability in generating denominator polynomial

---

[2] M may also be selected as

$$M = |S_{21}(j\omega_i)| = \sqrt{\frac{F(\lambda_i^2)}{H(\lambda_i^2) + F(\lambda_i^2)}}\Big|_{\lambda_i^2 \ = \ -\Omega_i^2}$$

$g(\lambda)$ from $h(\lambda)$ and $f(\lambda)$ as long as $h(\lambda)h(-\lambda) + f(\lambda)f(-\lambda) \neq 0$ for all λ.

Example 5.2 (Arbitrary Gain Equalization via SRFT): Let the ideal amplitude variation $M = |S_{21}(j\omega)|$ of an equalizer be a straight line with positive slope as shown in Figure 5.2.

(a) Develop a general purpose MATLAB program to generate the analytic form of the scattering parameters of a lossless equalizer as a function of the Richard variable λ in Belevitch form to approximate the ideal amplitude form.

(b) Referring to Figure 5.2, choose $T_1 = 0$, $T_2 = 1$ and $T_3 = 0.01$ with $f_1 = 0$ GHz, $f_2 = 1$ GHz and $f_e = 3$ GHz generate the explicit form of the scattering parameters using your program developed in part (a).

(c) Compute the driving point input impedance

$$Z_{in}(\lambda) = \frac{N(\lambda)}{D(\lambda)} = \frac{g+h}{g-h}$$

Solution:

(a) Let us develop the main program step by step as follows.

Step 1: Since this is an arbitrary gain approximation problem with specified ideal amplitude variation $M = |S_{21}(j\omega)|$, we should first write a MATLAB function which generates the desired target values.

In Program Listing 5.6, the MATLAB code for function `Target` is given as

```
[fr_opt,modula,phase]=Target(Nopt,F1,F2,fe)
```

This function generates the measured or targeted forms of the amplitude and phase over the frequencies subject to optimization. Target values of the amplitude are written on the array called `modula` and the target values of the phase are registered on the array called 'phase'. Their corresponding frequencies are written on the array called `fr_opt`. The size of these arrays is given by the integer `Nopt` which is the first input of the `function Target`. End points T_1, T_2 and T_3 of the line shape amplitude are fixed inside the function `Target`. Therefore, the user of the code only needs to specify the beginning of the passband frequency `F1` and the end of the passband frequency `F2` as well as the beginning of the stop band frequency `fe`.

Step 2: Secondly, we need to develop a MATLAB code for the objective function subject to minimization. Program Listing 5.7, gives the user-supplied objective function to be employed for the Levenberg–Marquardt algorithm. This function is used in the following form: `Func = SRFT_objective(x,fe,ntr,q,k,fr_opt,modula, phase)`
The array `Func` specifies the error function of Equation (5.3).

For the problem under consideration, the error stored in array `moduale` is given by `error_modula(i)=1.5*abs(modula(i)-MS21(i))`, where `modula(i)` are the target values, `MS21(i)` are the actual values generated as a function of the unknowns and the frequency, and the constant 1.5 is a bias or weight on the error.

Hence, `Func(i)` is set to `error(i)`.

Step 3: We are now ready to develop the main program code to determine the unknown coefficients of the polynomial $h(\lambda)$ for specified $f(\lambda) = \lambda^q(1 - \lambda^2)^{k/2}$.
The inputs of the main program must be specified by the user as follows:

- `F1`: beginning of the passband.
- `F2`: end of the passband.
- `fe`: beginning of the stopband for the low-pass prototype equalizer to be designed.

- k: total number of the UE section in the equalizer.
- q: count for the multiple transmission zeros to be placed at DC. Note that low-pass equalizer designs require $q = 0$. On the other hand, the $q > 0$ case corresponds to bandpass designs.
- ntr: a control flag which is set according to the needs of the designer. If ntr is set to 1, the designer has extra freedom in the design by introducing an ideal transformer into the equalizer topology. If ntr is set to 0, then the designer must work with a low-pass prototype by setting $q = 0$. In this case, the equalizer topology includes no ideal transformer. Actually, this a desirable situation for many practical problems. Let us explain this case in detail.

In SRFT, when we set $q = 0$, we have a choice to control the DC gain level. If we do not wish to have an ideal transformer in the equalizer then we must set $h(n + 1) = 0$, yielding $g(n + 1) = 1$, which in turn results in a unity gain at DC. This fact can easily be seen in the following.

For, $q = 0$ at DC or equivalently at $\lambda = 0$, $f(\lambda) = \lambda^q(1 - \lambda^2) = 1$. If we desire to set the gain at unity, then $S_{11}(\lambda = 0) = h_{n+1}/g_{n+1}$ must be zero so that $S_{21}(\lambda)S_{21}(-\lambda) = 1 - S_{11}(\lambda)S_{11}(-\lambda) = 1$ at DC. Thus, for fixed unity gain at DC, one must set $h_{n+1} = 0$, which in turn yields $g_{n+1} = 1$.[3]

However, if we initialize h_{n+1} to a non-zero value, then, at DC, $S_{11}(0) = h_{n+1}/g_{n+1}$ which yields a non-unity gain. In this case, the termination resistance of the equalizer must be different than the standard normalization R_0. Therefore, the resulting termination resistance can be leveled to using an ideal transformer.

- $\mathrm{h} = [\mathrm{h}(1)\mathrm{h}(2) \cdot \mathrm{h}(n)\mathrm{h}(n + 1)]$: a row vector, unknown coefficients of the polynomial $h(\lambda)$. Here, it is reminded once again that, based on the above explanations, if we set $q = 0$, $\mathrm{ntr} = 0$, then $h(n + 1)$ must be zero. Otherwise, we can enjoy the freedom of having an extra ideal transformer in the equalizer topology to level up the terminating resistance at the far end to the standard normalization resistance R_0.

For the output of the main program, we can print the polynomials $h(\lambda)$ and $g(\lambda)$ and plot the fit between the target and actual values of the amplitude of S_{21}. Eventually, the designer can construct the scattering parameters as

$$S_{11}(\lambda) = \frac{h(\lambda)}{g(\lambda)} \quad S_{21}(\lambda) = \frac{\lambda^q(1 - \lambda^2)^{k/2}}{g(\lambda)} \quad \text{and} \quad S_{22}(\lambda) = (-1)^{q+1}\frac{h(-\lambda)}{g(\lambda)}$$

The MATLAB code for the main program $\mathrm{SRFT\_Approximation}$ is given in Program Listing 5.8. It should be kept in mind that this is not a professional program, yet it shows the reader how to implement SRFT for arbitrary gain approximation problems.

(b) By choosing, $T_1 = 0$, $T_2 = 1$ and $T_3 = 0.01$, let us run the main program $\mathrm{SRFT\_}$ $\mathrm{Approximation}$ with the following inputs.

```
Inputs:
F1=0Hz
F2=1GHz
fe=3 GHz
k=5
q=1
ntr=1
h=[1 1 1 1 1 1 1]
```

---

[3] Note that, at DC, the losslessness conditions is simplified as $g_{n+1}^2 = h_{n+1}^2 + 1$ and the gain is given by $|S_{21}(0)|^2 = 1/g_{n+1}^1$. Thus, at DC, the unity gain condition requires $g_{n+1} = 1$, which demands $h_{n+1} = 0$.

We should note that the first three inputs are dictated by the design specifications. The fourth one is our choice. By selecting k=5 we introduce five UEs in the equalizer topology. In the fifth input, by selecting q=1 we introduce a DC transmission zero since the design specification demand $T_1 = 0$ at DC.

In the sixth input, we must set ntr=1 since we have already selected q=1. In the seventh input, we initialized the row vector h with seven entries in an ad hoc manner. In this case, $n_h + 1 = 7$, which yields the degree of the $h(\lambda)$ polynomial as $n_h = 6$. On the other hand, degree n_g of the polynomial $g(\lambda)$ must be

$$n_g = \frac{\text{maximum of } \{2n_n + n_F\}}{6} = 6$$

where n_F is the degree of $F(\lambda^2) = f(\lambda)f(-\lambda)$. This means that in the final equalizer topology we will have five UEs and either a series open stub like a capacitance $C\lambda$ or a parallel short stub like an inductance $L\lambda$.

Thus, with the above inputs, the optimization of the main program converges resulting in

$$h = [473.6729 \quad 50.7737 \quad 197.5642 \quad 14.0791 \quad 16.2623 \quad -0.2274 \quad -0.2956]$$

$$g = [473.6740 \quad 222.7381 \quad 247.2068 \quad 80.9384 \quad 29.5170 \quad 5.3022 \quad 0.2956]$$

or, in explicit form, we have

$$h(\lambda) = 473.67\lambda^6 + 50.7737\lambda^5 + 197.5642\lambda^4 + 14.0791\lambda^3 + 16.2623\lambda^2 - 0.2274\lambda - 0.2956$$

$$g(\lambda) = 473.67\lambda^6 + 222.7381\lambda^5 + 247.2068\lambda^4 + 80.9384\lambda^3 + 29.5170\lambda^2 + 5.3022\lambda + 0.2956$$

Remember that at the input we had already provided the numerator polynomial of the transfer scattering parameter as

$$f(\lambda) = \lambda^1 (1 - \lambda^2)^{5/2}$$

Thus, the complete S parameters of the equalizer are expressed as

$$S_{11}(\lambda) = \frac{h(\lambda)}{g(\lambda)} \quad S_{21}(\lambda) = \frac{\lambda^q (1 - \lambda^2)^{k/2}}{g(\lambda)} \quad \text{and} \quad S_{22}(\lambda) = (-1)^{q+1} \frac{h(-\lambda)}{g(\lambda)}$$

The amplitude response of the equalizer is depicted in Figure 5.3.

(c) Since $S_{11}(\lambda)$ is generated, the driving point input impedance is given by

$$Z_{in} = \frac{N(\lambda)}{g(\lambda)} = \frac{g(\lambda) + h(\lambda)}{g(\lambda) - h(\lambda)}$$

or

$$N = [947.3469 \ 273.5117 \ 444.7710 \ 95.0175 \ 45.7793 \ 5.0749 \ 0.0000]$$
$$D = [0.0011 \ 171.9644 \ 49.6426 \ 66.8592 \ 13.2547 \ 5.5296 \ 0.5911]$$

$$Z_{in}(\lambda) = \frac{947.3469\lambda^6 + 273.5117\lambda^5 + 444.7710\lambda^4 + 95.0175\lambda^3 + 45.7793\lambda^2 + 5.0749\lambda}{0.0011\lambda^6 + 171.9644\lambda^5 + 49.6426\lambda^4 + 66.8592\lambda^3 + 13.2547\lambda^2 + 5.5296\lambda + 0.5911}$$

As can be seen from the above impedance, it has a zero at $\lambda = 0$ or equivalently the admittance function

$$Y_{in}(\lambda) = \frac{0.0011\lambda^6 + 171.9644\lambda^5 + 49.6426\lambda^4 + 66.8592\lambda^3 + 13.2547\lambda^2 + 5.5296\lambda + 0.5911}{947.3469\lambda^6 + 273.5117\lambda^5 + 444.7710\lambda^4 + 95.0175\lambda^3 + 45.7793\lambda^2 + 5.0749\lambda}$$

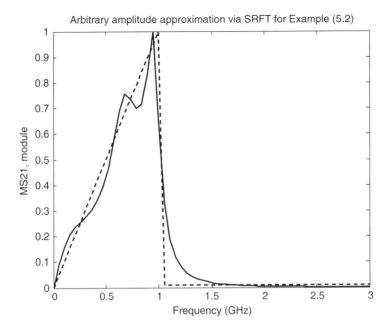

Figure 5.3   Approximation of an arbitrary amplitude form via SRFT

has a pole at $\lambda = 0$.

Example 5.3: The purpose of this example is to show the user how SRFT is employed to design standard low-pass stepped line filters employing the SRFT approach.

(a) Referring to Figure 5.2, choose `T1=1`, `T2=1`, `T3=0.01`. Then, using the main program `SRFT_Approximation`, design a stepped line filter for the following input parameters.

Inputs to main program, `SRFT_Approximation`:

- `F1 = 0`
- `F2 = 1`
- `fe = 3`
- `k = 5`
- `q = 0`
- `ntr = 0`
- `h = [1 -1 1 -1 1 0]`

(b) Synthesize the filter and plot its gain response

Solution:

(a) In the function `Target`, we set $T1 = 1, T2 = 1, T3 = 0.01$. Then, the result of the main program is given as follows.

Outputs:

- `tau=8.331e-11seconds`
- `h=[211.4043   −0.0263   68.5632   −0.0066   3.8793   0]`
- `g=[211.4067   123.6751   104.7262   34.8593   9.4746   1.0000]`

(b) Using Richard extraction, function `[z]=UE_sentez(h,g)` reveals that the Normalized characteristic impedances of the stepped line filter designed using SRFT are:

```
3.8194
0.4108
4.8916
0.4110
3.8210
1.0000
```

The gain performance of the stepped line filter is depicted in Figure 5.4. Further, the periodic behavior of the filters is shown in Figure 5.5.

Note that a comparison of the characteristic impedances and the gain performance of the SRFT filter to those of Example 4.15 exhibits similarities.

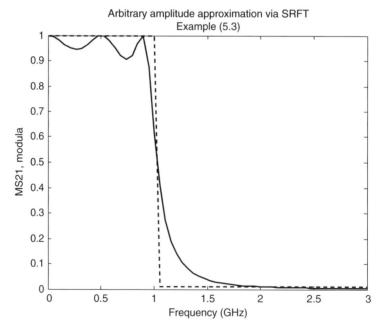

Figure 5.4   Gain performance of the stepped line filter designed using SRFT

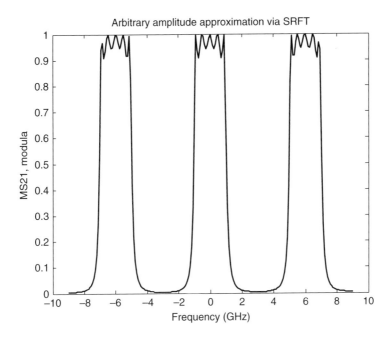

Figure 5.5 Periodic behavior of the stepped line filter designed via SRFT

Example 5.4:

(a) Referring to Figure. 5.2, design a stepped line filter employing SRFT for the following inputs:

F1 = 0	T1 = 1
F2 = 1	T2 = 0.1
Fe	T3 = 0.1
k = 5	
q = 0	
ntr = 0	
h = [-1 1 -1 1 -1 0]	

(b) Synthesize the filter and plot the gain response.

Solution:

(a) The function `Target` was changed with the above inputs and the main program
     `SRFT_Approximation` was run, yielding the following results.

Outputs:

- tau=8.331e−11 seconds
- h=−732.5951 − 26.6833 −152.3571 −27.0308 −8.12400]
- g=732.5958 421.2834 272.9984 74.2389 14.8148 1.0000]

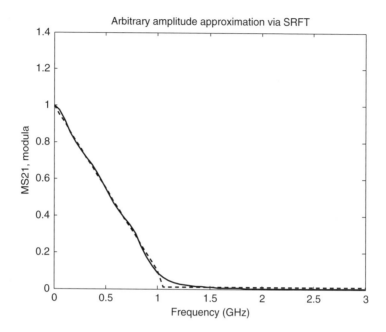

Figure 5.6   Gain performance of the stepped line filter designed with SRFT

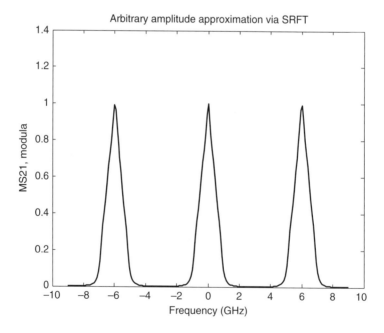

Figure 5.7   Periodicity of the amplitude performance

(b) Using Richard extraction, function `[z]=UE_sentez(h,g)` reveals that the Normalized characteristic impedances of the stepped line filter designed using SRFT are:

```
0.2314
1.4810
0.0702
4.6207
0.2876
1.0000
```

The performance of the stepped line filter is depicted in Figure 5.6 and in Figure 5.7, the periodic gain behavior of the stepped line filter is shown.

5.3 Conclusion

In this chapter, the SRFT design approach to construct microwave UE filters with arbitrary amplitude characteristics was introduced.

SRFT is a semi-analytic method, which requires in advance neither the selection of circuit topology nor an analytic form for the transfer function. The desired amplitude shape of the filter is described as a target function to the main program. In this approach, the filter is described in terms of the numerator polynomial $h(\lambda)$ of the input reflectance. Therefore, it is selected as the known quantity in the filter design problem. Then, it is determined via optimization of the amplitude performance to reach the preset target values.

In the course of the optimization, coefficients of $h(\lambda)$ are selected in an ad hoc manner since the objective function subject to optimization is quadratic in terms of the unknowns and, therefore, the optimization is convergent.

Several examples are presented to show the implementation of the SRFT design process.

Obviously, the same procedure can be used to design broadband lumped element filters, equalizers and matching networks.

Program Listing 5.1   MATLAB function `paraconj`

```
function h_=paraconj(h)
%This function generates the para-conjugate of a polynomial h=[]
na=length(h);
n=na-1;
sign=-1;
h_(na)=h(na);
for i=1:n
    h_(n-i+1)=sign*h(n-i+1);%Para_conjugate coefficients
    sign=-sign;
end
    return
```

Program Listing 5.2   MATLAB function `F = lambda2q_UE2k(k,q)`

```
function F=lambda2q_UE2k(k,q)
%This function computes F = (-1)^q(lambda)^2q(1-lambda^2)^k
Fa=cascade(k);
nf=length(Fa);
nq=nf+q;
    for i=1:nq
    F(i)=0.0;
    end
if q==0;
      F=Fa;
    end
if(q>0)
    for i=1:nf
      F(i)=((-1)^q)*Fa(i);
    end
for i=1:q
      F(nf+i)=0.0;
    end
end
return
```

Program Listing 5.3   MATLAB function F=cascade(K)

```
function F =cascade(k)
%This function generates the F=f*f_ polynomial of
%cascade connections of k UEs in Richard Domain.
%F = (1-1lambda^2)^k
%F = F(1)(Lambda)^2n+...F(k)(lambda)^2+F(k+1)
F =[1]; %Unity Polynomial;
UE=[-1 1];%Single Unit-Element; [1-(lambda)^2]
for i=1:k
    F =conv(F,UE);
end
return
```

Program Listing 5.4   Function g=htog(q,k,h)

```
function g=htog(q,k,h)
%This function is written in Richard Domain.
%Given h(lambda) as a polynomial
%Given f(lambda) = (-1)^q*(lambda)^q*(1-lambda^2)^k
%Computes g(lambda)
%
```

```
F=lambda2q_UE2k(k,q);%Generation of F=f(lambda)*f(-lambda)
h_=paraconj(h);
He=conv(h,h_);%Generation of h(lambda)*h(-lambda)
H=poly_eventerms(He);%select the even terms of He to set H
n1=length(H);
G=vector_sum(H,F);%Generate G=H+F with different sizes
GX=paraconj(G);%Generate G(X) by setting X=-lambda^2
Xr=roots(GX);%Compute the roots of G(X)
z=sqrt(-Xr);%Compute the roots in lambda
%*****************************************************************
%Generation of g(lambda) from the given LHP roots
%Compute the first step k=1
n=length(z);
g=[1 z(1)]
for i=2:n
    g=conv(g,[1 z(i)]);
end
Cnorm=sqrt(abs(G(1)));
g=Cnorm*real(g);
return
```

Program Listing 5.5

```
function [z]=UE_sentez(h,g)
%Richard Extraction written by Dr. Metin Sengul
%of Kadir Has University, Istanbul, Turkey.
%Inputs:
%h(lambda), g(lambda)as MATLAB Row vectors
%Output:
%[z] Characteristic impedances of UE lines
%------------------------------------
na=length(g);
for i=1:na
    h0i(i)=h(na-i+1);
    g0i(i)=g(na-i+1);
end
m=length(g0i);
for a=1:m
    gt=0;
    ht=0;
    k=length(g0i);
    for b=1:k
        gt=gt+g0i(b);
        ht=ht+h0i(b);
end
if a==1;
    zp=1;
```

Program Listing 5.5 (Continued)

```
else
    zp=zz(a-1);
end
zz(a)=((gt+ht)/(gt-ht))*zp;
x=0;
y=0;
for c=1:k-1
    x(c)=h0i(c)*gt-g0i(c)*ht;
    y(c)=g0i(c)*gt-h0i(c)*ht;
end
N=0;
D=0;
n=0;
for e=1:k-1
    N(e)=n+x(e);
    n=N(e);
    d=0;
    for f=1:e
        ara=((-1)^(e+f))*y(f);
        D(e)=ara+d;
        d=D(e);
    end
end
    h0i=N;
    g0i=D;
end
z=zz;
return
```

Program Listing 5.6 Function Target

```
function [fr_opt,modula,phase]=Target(Nopt,F1,F2,fe)
tau=1/4/fe;
teta1=2*pi*F1*tau;
tetae=2*pi*fe*tau;
del_teta=(tetae-teta1)/(Nopt-1);
teta=teta1;
%Tapered Gain formula
T1=0.0;T2=1.0;
a=(T1-T2)/(F1-F2);
b=(F1*T2-F2*T1)/(F1-F2);
for i=1:Nopt
    fr_opt(i)=teta/(2*pi*tau);
    if fr_opt(i)<=F2;
```

```
      modula(i) =a*fr_opt(i) +b;
      end
      if fr_opt(i)> F2;
         modula(i) =0.01;
      end
      phase(i) =sin(teta);
      teta =teta +del_teta;
end
return
```

Program Listing 5.7

```
function Func=SRFT_objective(x,fe,ntr,q,k,fr_opt,modula, phase)
%This function generates the objective function
%subject to optimization of a filter
%with specified TPG and Phase.
%Computations are carried out in Richard domain
%Inputs:
%    F1: Lower edge of the passband
%    F2: Upper edge of the passband
%    fe: Stop band frequency in Richard Domain
%    ntr:
%      If ntr=1> Yes, there is an ideal transformer in the filter
%      If ntr=0> There is no ideal transformer in the filter
%    q: In Richard Domain, count of transmission %%zeros at
%    lambda=0
%    k: Count of UEs in F(lmbda^2) = (-%1)^q*(lmbda^2q)*(1-
%    lambda^2)^k
%    x(i):Unknown coefficients of h(lambda)
%    modula(i): Targeted modula of S21 of the %filter
%    phase(i): Targeted phase of S21 in actual %frequency domain
%Output:
%    error subject to minimization
%- - - - - - - - - - - - - - - - - - - - - - - - - - -
if ntr= =1; %Yes transformer case
h=x;
end
if ntr= =0;%No transformer, h(n+1) =0 case
  n =length(x);
  h(n+1) =0;
    for i=1:n
    h(i) =x(i);
    end
end
[G,H,F,g] =SRFT_htoG(h,q,k);
j=sqrt(-1);
```

Program Listing 5.7   (Continued)

```
tau = 1/4/fe;
%
Nopt = length(modula);
for i = 1:Nopt
  teta = 2*pi*tau*fr_opt(i);
  lmbda = j*tan(teta);
  %
  fval = (-1)^q*(lmbda)^q*(1-lmbda^2)^(k/2);
  freal = real(fval);
  fimag = imag(fval);
  %
  gval = polyval(g,lmbda);
  greal = real(gval);
  gimag = imag(gval);
%
  S21 = fval/gval;
  MS21(i) = abs(S21);
%
phase_f = atan(fimag/freal);
phase_g = atan(gimag/greal);
phase_S21(i) = phase_f-phase_g;
%
error_phase(i) = 1.1*abs(phase(i)-phase_S21(i));
error_modula(i) = 1.5*abs(modula(i)-MS21(i));
%E(i) = error_phase(i) + error_modula(i);
E(i) = error_modula(i);
Func(i) = E(i);
end
return
```

Program Listing 5.8

```
%Main Program: SRFT_Approximation
%This main program provides the general gain and phase approxi-
%mation
%SRFT Insertion Loss Approximation for gain and phase
%with specified TPG and Phase.
%Computation are carried out in Richard domain
%Inputs:
%    F1: Lower edge of the passband
%    F2: Upper edge of the passband
%    ntr:If ntr = 1> Yes, there is an ideal %transformer in the
%    filter
%    If ntr = 0> There is no ideal transformer in the filter
%    q: In Richard Domain, count of transmission zeros at lambda = 0
```

```
%        k: Count of UEs in F(lmbda^2) = (-%1)^q*(lmbda^2q)*(1-
%lambda^2)^k
%     x(i):Unknown coefficients of h(lambda
%     modula(i): Targeted modula of S21 of the %filter
%     phase(i): Targeted phase of S21 in actual %frequency domain
%Output:
%     error subject to minimization
%----------------------------%
clear
F1 = input('Enter beginning of the passband F1 = ')
F2 = input('Enter end of passband F2 = ')
fe = input('Enter stop band frequency fe = ')
k = input('Enter total number of UEs, k = ')
q = input('Enter number of transmission zeros at DC q = ')
ntr = input('Enter transformer flag, Yes transformer> = ntr = 1; No
transformer> = ntr = 0; ntr = ')
Nopt = 51;
h = input('Enter initials for h = [114.4788 0 44.2736 0 3.1624 0]
h = [ ]');
%--------------------------------------------
[fr_opt,modula,phase] = Target(Nopt,F1,F2,fe);
%
%--------------------------------------------
if ntr = = 1; %Yes transformer
   x0 = h;
   % Call optimization
OPTIONS = OPTIMSET('MaxFunEvals',20000,'MaxIter',50000);
x = lsqnonlin('SRFT_objective',x0,[],[],OPTIONS,fe,ntr,q,k,fr_op-
t,modula, phase);
h = x;
end
%
if ntr = = 0;%No transformer, h(n+1) = 0 case
   n1 = length(h);
   n = n1-1;
   h(n+1) = 0;
   for i = 1:n
   x0(i) = h(i);
   end
%Call optimization with no transformer

OPTIONS = OPTIMSET('MaxFunEvals',20000,'MaxIter',50000);
x = lsqnonlin('SRFT_objective',x0,[],[],OPTIONS,fe,ntr,q,k,fr_op-
t,modula, phase);
   h(n+1) = 0;
   for i = 1:n
   h(i) = x(i);
   end
end
```

Program Listing 5.8   (Continued)

```
[G,H,F,g] = SRFT_htoG(h,q,k);
tau = 1/4/fe;
N = length(modula);
j = sqrt(-1);
for i = 1:N
  teta = 2*pi*tau*fr_opt(i);
  omega = tan(teta);
  lmbda = j*tan(teta);
  %
  fval = (-1)^q*(lmbda)^q*(1-lmbda^2)^(k/2);
  freal = real(fval);
  fimag = imag(fval);
  %
  gval = polyval(g,lmbda);
  greal = real(gval);
  gimag = imag(gval);
%
  S21 = fval/gval;
  MS21(i) = abs(S21);
%
phase_f = atan(fimag/freal);
phase_g = atan(gimag/greal);
%
phase_S21(i) = phase_f-phase_g;
%
ph(i) = teta;
fr(i) = fr_opt(i);
end
%------------------------------------
if q == 0
[Z_imp] = UE_sentez(h,g)
end
figure(1)
plot(fr,phase_S21, fr_opt,phase)
xlabel('frequency')
ylabel('phase-S21')
figure(2)
plot(fr,MS21, fr_opt,modula)
xlabel('frequency')
ylabel('MS21, modula')
f1 = -3*fe;f2 = +3*fe;
N = 201
teta1 = 2*pi*f1*tau;
teta2 = 2*pi*f2*tau;
del_teta = (teta2-teta1)/(N-1);
teta = teta1;
for i = 1:N
```

```
    fr(i) = teta/2/pi/tau;
    omega = tan(teta);
    lmbda = j*tan(teta);
    %
    fval = (-1)^q*(lmbda)^q*(1-lmbda^2)^(k/2);
    freal = real(fval);
    fimag = imag(fval);
    %
    gval = polyval(g,lmbda);
    greal = real(gval);
    gimag = imag(gval);
%
    S21 = fval/gval;
    MS21(i) = abs(S21);
%
phase_f = atan(fimag/freal);
phase_g = atan(gimag/greal);
%
phase_S21(i) = phase_f-phase_g;
%
ph(i) = teta;
teta = teta + del_teta;
end
%
figure(3)
plot(fr,MS21)
xlabel('frequency')
ylabel('MS21, modula')
```

6

Formal Description of Lossless Two-ports in Terms of Scattering Parameters: Scattering Parameters in the p Domain

6.1 Introduction

In Chapters 3 and 4, we worked with commensurate transmission lines to construct lossless two-ports and introduced the natural description of lossless two-ports in terms of wave parameters, which in turn leads to the definition of scattering parameters in the Richard domain. Then, filters were constructed with commensurate transmission lines using the analytic form of transfer function generated by means of Butterworth and Chebyshev polynomials. In this chapter, incident and reflected wave concepts are extended to describe lossless two-ports built with lumped elements, which in turn leads to the formal definition of scattering parameters in the classical complex variable $p = \sigma + j\omega$ as introduced in Chapter 1.

For many engineering applications, two-ports may be constructed to optimize the system performance under consideration. In this case, the descriptive parameters of two-ports must be chosen in such a way that they are easily generated on the computer to reach the preset targets or goals linked to system performance.

Impedance, admittance or hybrid parameters are somewhat idealized since they are measured under open- or short-circuit conditions. Furthermore, they may be unbounded, such as the impedance of an open circuit or admittance of a short circuit. On the other hand, scattering parameters are bounded and are a practical tool to describe linear active and passive systems since they are defined or measured under operational conditions. As far as computer-aided design or synthesis problems are concerned, they present excellent numerical stability in number-crunching processes.

In many engineering applications, there is a demand to construct lossless two-ports for various kinds of problems in such things as filters, power transfer networks or equalizers. Therefore, in the following sections we will first give a formal definition of scattering parameters to specifically describe linear active and passive two-ports. However, these definitions can easily be extended to describe n-ports. Then, some important properties of these parameters will be summarized in the classical complex variable $p = \sigma + j\omega$.

Design of Ultra Wideband Power Transfer Networks   Binboga Siddik Yarman
© 2010 John Wiley & Sons, Ltd

6.2 Formal Definition of Scattering Parameters

Linear two-ports may be described in terms of two wave quantities called incident and reflected waves. Literally speaking, these wave quantities are defined by means of port voltages and currents.

Referring to Figure 6.1, let us define the incident waves for port 1 and port 2 as functions of time t as follows:

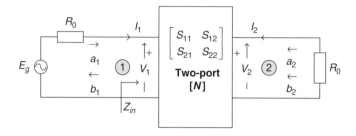

Figure 6.1   Definition of wave quantities for a linear two-port $[N]$

$$a_1(t) = \frac{1}{2}\left[\frac{\nu_1(t)}{\sqrt{R_0}} + \sqrt{R_0}i_1(t)\right]$$

$$a_1(t) = \frac{1}{2\sqrt{R_0}}[\nu_1(t) + R_0 i_1(t)]$$

$$a_2(t) = \frac{1}{2}\left[\frac{\nu_2(t)}{\sqrt{R_0}} + \sqrt{R_0}i_2(t)\right]$$ (6.1)

$$a_2(t) = \frac{1}{2\sqrt{R_0}}[\nu_2(t) + R_0 i_2(t)]$$

Where $\nu_1(t)$ and $\nu_2(t)$ and $i_1(t)$ and $i_2(t)$ designate the port voltages and currents with selected polarities and directions as shown in Figure (6.1). R_0 is the normalization resistance which is essential for defining the wave quantities. It may also be utilized to terminate the ports under consideration. It is also known as the port normalization number.

Note that the above relations are preserved under linear transformations such as the Fourier or Laplace transformation.

Let the Laplace transformation of $a_1(t)$ and $a_2(t)$ be $a_1(p)$ and $a_2(p)$ respectively. Then, these are given by

$$a_j(p) = \int_{-\infty}^{+\infty} a_j(t)e^{-pt}dt; j = 1, 2$$ (6.2)

where $p = \sigma + j\omega$ is the conventional complex frequency of the Laplace domain.

Denoting the time and the Laplace domain pairs as $a_j \leftrightarrow a_j(p)$, $\nu_j \leftrightarrow V_j(p)$ and $i_j \leftrightarrow I_j(p)$, the above equations can be directly written in the p domain:

$$
\begin{aligned}
a_1(p) &= \frac{1}{2}\left[\frac{V_1(p)}{\sqrt{R_0}} + \sqrt{R_0}I_1(p)\right] \\
a_2(p) &= \frac{1}{2}\left[\frac{V_2(p)}{\sqrt{R_0}} + \sqrt{R_0}I_2(p)\right]
\end{aligned}
\tag{6.3}
$$

Specifically, at port 1, the incident wave quantity can be given in terms of the excitation $e_g(t) = \nu_1(t) + R_0 i_1(t)$ or in its Laplace domain pair $E_g(g) = V_1(p) + R_0 I_1(p)$. Thus, it is found that

$$
a_1(t) = \frac{e_g(t)}{2\sqrt{R_0}}
$$

or in the p domain

$$
a_1(p) = \frac{E_g(p)}{2\sqrt{R_0}}
\tag{6.4}
$$

For many applications, the actual frequency response of a system is determined by setting $p = j\omega$. In this case, the complex port quantities, $a_j(j\omega)$, $V_j(j\omega)$ and $I_j(j\omega)$ are measured with properly designed equipment, namely network analyzers. Here, in this representation, f is the actual operating frequency measured in hertz; $\omega = 2\pi f$ represents the actual angular frequency in radians per second.

For passive reciprocal two-ports which consist of lumped circuit elements such as capacitors, inductors, resistances and ideal transformers, the voltage and current quantities are expressed by means of the ratio of polynomials in the complex variable p, known as rational functions. Obviously, for passive systems, these functions must be bounded for all bounded excitations.

Similarly, reflected wave quantities are defined as time and Laplace domain pairs $b_j \leftrightarrow b_j(p)$ as follows:

$$
b_1(t) = \frac{1}{2}\left[\frac{\nu_1(t)}{\sqrt{R_0}} - \sqrt{R_0}i_1(t)\right]
$$

$$
b_2(t) = \frac{1}{2}\left[\frac{\nu_2(t)}{\sqrt{R_0}} - \sqrt{R_0}i_2(t)\right]
$$

In the p domain

$$
\begin{aligned}
b_1(p) &= \frac{1}{2}\left[\frac{V_1(p)}{\sqrt{R_0}} - \sqrt{R_0}I_2(p)\right] \\
b_2(p) &= \frac{1}{2}\left[\frac{V_2(p)}{\sqrt{R_0}} - \sqrt{R_0}I_2(p)\right]
\end{aligned}
\tag{6.5}
$$

At this point, it is appropriate to mention that, based on the maximum power theorem, a complex exponential (or equivalently a sinusoidal) excitation $e_g(t) = E_m e^{j\omega t}$ (or equivalently $e_g(t) = E_m \cos(\omega t)$

which is the real part of $\{E_m e^{j\omega t}\}$), with internal resistance R_0, delivers its maximum power to the input port when it sees a driving point input impedance $Z_{in}(j\omega) = R_0$. In this case, the maximum average power P_A over a period $T = 1/f$ s is given by

$$P_A = \text{real}\left\{\frac{1}{T}\int_0^T v_1(t)i_1^*(t)dt\right\} \tag{6.6}$$

where

$$v_1(t) = \frac{e_g(t)}{2}$$

$$i_1(t) = \frac{e_g(t)}{2R_0}$$

and the * denotes the complex conjugate of a complex quantity. Thus, Equation (6.6) yields

$$P_A = \frac{E_m^2}{4R_0} \tag{6.7}$$

On the other hand, the squared amplitude of the incident wave is given by

$$a_1(t)a_1^*(t) = |a_1|^2 = \frac{E_m^2}{4R_0}$$

which also measures the maximum deliverable power of the input excitation. Therefore, we say that $P_A = |a_1|^2$ is the available power of the generator.

As a matter of fact, the incident wave is defined in such a way that, for $p = j\omega$, the incident or available power of the excitation is equal to the squared amplitude of the incident wave $a_1(j\omega)$, or simply $P_A = |a_1(j\omega)|^2$.

Based on the above introduction, for any linear two-port in the Laplace domain $p = \sigma + j\omega$, reflected waves $b_1(p)$ and $b_2(p)$ can be expressed in terms of the incident waves $a_1(p)$ and $a_2(p)$ such that

$$\begin{aligned} b_1 &= S_{11}a_1 + S_{12}a_2 \\ b_2 &= S_{21}a_1 + S_{22}a_2 \end{aligned} \tag{6.8}$$

or, in matrix form,

$$\begin{bmatrix} b_1(p) \\ b_2(p) \end{bmatrix} = \begin{bmatrix} S_{11}(p) & S_{12}(p) \\ S_{21}(p) & S_{22}(p) \end{bmatrix} \begin{bmatrix} a_1(p) \\ a_2(p) \end{bmatrix} \tag{6.9}$$

where $S_{ij}(p)$, $i,j = 1,2$, are called the scattering parameters and the matrix $S(p)$ is given by

$$S(p) = \begin{bmatrix} S_{11}(p) & S_{12}(p) \\ S_{21}(p) & S_{22}(p) \end{bmatrix} \tag{6.10}$$

and called the scattering matrix.

Specifically, $S_{11}(p)$ and $S_{22}(p)$ are called the input and output reflectance under R_0 terminations and $S_{21}(p)$ and $S_{12}(p)$ are called the forward and backward transfer scattering parameters from port 1 to port 2 and from port 2 to port 1, respectively.

It is important to note that for any linear active and passive two-ports, for a preselected normalization port number R_0, for any excitation applied to either port 1 or port 2, and under arbitrary port terminations, Equation (6.11) holds for all measured sets of $\{v_i(t) \leftrightarrow V_i(p)$ and $j_i(t) \leftrightarrow I_i(p), i = 1, 2\}$. In other words, the scattering parameters must be invariant under any excitation and terminations.

Therefore, we can say that real normalized (or normalized with respect to $R_0\Omega$) scattering parameters which are measured or derived over the entire frequency band $(-\infty < \omega = 2\pi f < +\infty)$ completely describe the linear two-ports. Hence, proper generation of the scattering parameters is essential.

In the following, we introduce a straightforward technique to generate or measure the scattering parameters for linear networks.

Definition 1 (Reflectance of a port): Referring to Figure 6.2, for a one-port, we can always define an incident $a(t) \leftrightarrow a(p)$ and reflected $b(t) \leftrightarrow b(p)$ wave in terms of the one-port's input voltage $v_{in}(t) \leftrightarrow V_{in}(p)$ and current $i_{in}(p) \leftrightarrow I_{in}(p)$ pair. In the Laplace domain 'p', the incident wave $a(p)$ is defined by

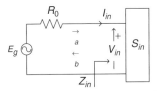

Figure 6.2   Definition of wave quantities for a one-port network

$$a(p) = \frac{1}{2}\left[\frac{V_{in}(p)}{\sqrt{R_0}} + \sqrt{R_0}I_{in}(p)\right]$$
$$= \frac{1}{2\sqrt{R_0}}\left[V_{in}(p) + R_0 I_{in}\right] \tag{6.11}$$

Similarly, the reflected wave $b(p)$ is given by

$$b(p) = \frac{1}{2}\left[\frac{V_{in}(p)}{\sqrt{R_0}} - \sqrt{R_0}I_{in}(p)\right] = \frac{1}{2\sqrt{R_0}}\left[V_{in}(p) - R_0 I_{in}\right]$$
$$= \frac{E_g(p)}{2\sqrt{R_0}} \tag{6.12}$$

Then, the input reflectance (or the input reflection coefficient) $S_{in}(p)$ of the one-port is defined by

$$S_{in}(p) \triangleq \frac{b(p)}{a(b)}$$

or

$$S_{in}(p) = \frac{V_{in} - R_0 I_{in}}{V_{in} + R_0 I_{in}} = \frac{Z_{in}(p) - R_0}{Z_{in}(p) + R_0} \tag{6.13}$$

where

$$Z_{in}(p) = \frac{V_{in}(p)}{I_{in}(p)} \tag{6.14}$$

is called the input impedance of the one-port under consideration.

Definition 2 (**Normalized impedance**): The ratio

$$z_{in} = \frac{Z_{in}}{R_0} \tag{6.15}$$

is called the normalized impedance with respect to normalization resistance R_0. Then, using Equation (6.16) in Equation (6.14), the input reflectance is given by

$$S_{in}(p) = \frac{z_{in} - 1}{z_{in} + 1} \tag{6.16}$$

Now, let us investigate the power relations for the 'one-port' under consideration.

As discussed above, the available power or incident power of the generator is given by $P_A = |a(j\omega)|^2$. Similarly, it may be appropriate to define the reflected power from the one-port as $P_R = |b(j\omega)|^2$. In this case, at a given angular frequency ω, the net power P_{in} delivered to the 'one-port' must satisfy the relation

$$P_{in} = P_A - P_R$$

or

$$P_{in} = P_A \left[1 - \frac{P_R}{P_A} \right] = |a|^2 \left[1 - \left| \frac{b}{a} \right|^2 \right] \tag{6.17}$$

or

$$P_{in} = |a|^2 [1 - |S_{in}(j\omega)|^2]$$

Let the power transfer ratio $T(\omega)$ be defined as the ratio of the input power to the available power. Then,

$$T = \frac{P_{in}}{P_A} = 1 - |S_{in}(j\omega)|^2 \tag{6.18}$$

Equation (6.18) can be expressed in terms of the normalized input impedance as

$$z_{in}(j\omega) = r_{in}(\omega) + jx_{in}(\omega)$$

Then,

$$T = 1 - \left| \frac{z_{in} - 1}{z_{in} + 1} \right|^2$$

or   (6.19)

$$T = 1 \frac{(r_{in} - 1)^2 + x_{in}^2}{(r_{in} + 1)^2 + x_{in}^2}$$

or

$$T = \frac{4r_{in}}{(r_{in} + 1)^2 + x_{in}^2}$$

Note that the above relation is valid if the definition given for the reflected power $P_R = |b(j\omega)|^2$ results in the actual power P_{ac} delivered to the one-port being equal to $P_{in} = P_A - P_R$. In other words, one has to show that the actual power delivered to the one-port is $P_{ac} = \text{real}\{V_{in}I_{in}^*\} = P_{in}$.

In fact, this is true, as shown by the following derivations.[1]

Definition 3 (**Normalized voltage**): The ratio

$$v_{in} = \frac{V_{in}}{\sqrt{R_0}}$$

(6.20)

is called the normalized voltage.

Definition 4 (**Normalized current**): The ratio

$$i_{in} = \frac{I_{in}}{\sqrt{R_0}}$$

(6.21)

is called the normalized current.

Based on the above definition, $e_g = E_g/\sqrt{R_0}$ defines the normalized voltage of the input excitation source.

Considering the definitions of the incident and reflected waves, the normalized voltage and current expressions can be given as

$$v_{in}(j\omega) = a(j\omega) + b(j\omega)$$

$$i_{in}(j\omega) = a(j\omega) - b(j\omega)$$

(6.22)

---

[1] Note that in terms of the normalized port voltage v_{in} and current i_{in}, the actual power delivered to port 1 is invariant, i.e. $P_{ac} = \text{real}\{V_{in}I_{in}^*\} = \text{real}\{v_{in}i_{in}^*\}$.

Thus, the actual power delivered to the one-port is given by

$$P_{ac} = real\{(a+b)(a+b)^*\} = real\{aa^* - bb^* - + ab^* + a^*b\}$$

Note that the quantity $a^*b - ab^*$ is purely imaginary.[2]

Therefore,

$$\boxed{P_{ac} = |a|^2 - |b|^2}$$

(6.23)

which is described as the input power P_{in}.

Thus, we conclude that the definition given for the reflected power $P_R = |b|^2$ is appropriate.

Now, let us investigate some important properties of the input reflection coefficient S_{in}.

Property 1 (Positive real function): If the one-port consists of passive lumped elements such as inductors, capacitors, resistors and transformers, then the input impedance $Z_{in}(p)$ is a positive real (PR) and rational function in the complex variable $p = \sigma + j\omega$ and is expressed as

$$\boxed{Z_{in}(p) = \frac{N(p)}{D(p)}}$$

(6.24)

where both the numerator $N(p)$ and denominator $D(p)$ polynomials are Hurwitz,[3] perhaps with simple roots on the imaginary axis $j\omega$ of the complex p plane. Thus, the input admittance

$$y_{in}(p) = \frac{1}{z_{in}(p)}$$

is also a PR rational function in p. In this case, the corresponding input reflectance

$$S_{in} = \frac{z_{in} - 1}{z_{in} + 1}$$

must be a real rational function in p such that

$$
\begin{aligned}
S_{in}(p) &= \frac{h(p)}{g(p)} \\
\text{where} \\
h(p) &= N(p) - D(p) \\
&= h_1 p^n + h_2 p^{n-1} + \cdots + h_n p + h_{n+1} \\
g(p) &= N(p) - D(p) \\
&= g_1 p^n + g_2 p^{n-1} + \cdots + g_n p + g_{n+1}
\end{aligned}
$$

(6.25)

---

[2] If we define $C = a^*b = A + jB$, then $C^* = ab^* = A - jB$. Thus, $a^*b - ab^* = 2jB$ is purely imaginary. Therefore, $P_{ac} = |a|^2 - |b|^2$.

[3] Hurwitz polynomial must have all its roots in the closed left half plane (LHP) in the complex domain $p = \sigma + j\omega$. Equivalently, a Hurwitz polynomial must be free of open right half plane (RHP) roots. These definitions imply that a root on $j\omega$ is allowable.

In Equation (6.25), the degree n of polynomials $h(p)$ and $g(p)$ refers to the total number of reactive elements (i.e. capacitor and inductors) in the one-port under consideration.

Let us examine the following example to find the input reflectance of a simple one-port consisting of a parallel combination of a capacitor $C = 1$ F and resistor $R = 1$ Ω as shown in Figure 6.3.

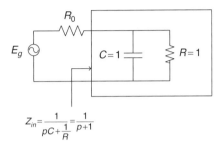

$$Z_{in} = \frac{1}{pC + \frac{1}{R}} = \frac{1}{p+1}$$

Figure 6.3   A one-port consisting of R//C

Example 6.1: Determine the input reflectance of the one-port shown in Figure 6.3.

Solution: The input impedance of the one-port is given as

$$Z_{in} = \frac{1}{p+1} = \frac{N(p)}{D(p)}$$

with

$$N(p) = 1 \quad D(p) = p + 1$$

Then

$$h(p) = N(p) - D(p) = 1 - p - 1 = -p$$

and

$$g(p) = N(p) + D(p) = 1 + p + 1 = 2 + p$$

Thus,

$$S_{in}(p) = -\frac{p}{p+2}$$

Property 2 (Subtraction of Hurwitz polynomials): Subtraction of Hurwitz polynomials results in an arbitrary polynomial with real coefficients. Therefore, the numerator polynomial $h(p)$ is an ordinary polynomial with real coefficients.

Property 3 (Addition of Hurwitz polynomials): Addition of two Hurwitz polynomials results in a strictly Hurwitz polynomial which has all its roots in the closed LHP.[4]

According to Property 3, the denominator polynomial $g(p)$ must be strictly Hurwitz, which makes the input reflection coefficient $S_{in}(p)$ regular on the $j\omega$ axis as well as in the open RHP of the complex

---

[4] By definition, in strictly Hurwitz polynomials, $p = j\omega$ roots are not permissible.

domain $p = \sigma + j\omega$. In other words, $S_{in}(p)$ is bounded over the entire frequency band as well as in the open RHP.

It is also important to note that, on the $j\omega$ axis, PR functions must always yield non-negative real parts. Therefore, if $z_{in}(j\omega) = r_{in}(\omega) + jx_{in}(\omega)$ is a PR function, then $r_{in}(\omega)$ must be non-negative over the entire frequency axis which bounds the squared amplitude of the input reflectance by 1 as detailed below:

$$
|S_{in}(j\omega)|^2 = \left| \frac{z_{in} - 1}{z_{in} + 1} \right| = \frac{(r_{in} - 1)^2 + x_{in}^2}{(r_{in} + 1)^2 + x_{in}^2}
$$

or

$$
|S_{in}(j\omega)|^2 = \left| \frac{h(j\omega)}{g(j\omega)} \right|^2 \leq 1, \text{ for all } \omega
$$

(6.26)

or equivalently

$$
|h(j\omega)| \leq |g(j\omega)|
$$

By analytic continuation,[5] in Equation (6.26), complex variable $p = j\omega$ can be replaced by its full version $p = \sigma + j\omega$ yielding $S_{in}(p) = h(p)/g(p) \leq 1$ for all $p = \sigma \leq 0$, since $g(p)$ is strictly Hurwitz (free of closed RHP zeros).

Now let us verify Equation 6.26 for the one-port shown in Figure 6.3.

Example 6.2: Show that for the one-port depicted in Figure 6.3, Equation (6.26) holds.

Solution: From Example 6.1, $h(j\omega)$ and $g(j\omega)$ are given as $h(j\omega) = -j\omega$ and $g(j\omega) = 2 + j\omega$ or $|h(j\omega)|^2 = \omega^2$ and $|g(j\omega)|^2 = 4 + \omega^2$. Obviously, $\omega^2 < 4 + \omega^2$ for all ω. Thus, Equation 6.26 holds.

Property 4 (Lossless one-ports): If there is no reflection from the one-port network ($P_R = |b(j\omega)|^2 = 0$), then the incident power is dissipated on the one-port yielding $T = P_{in}/P_A = 1 - |S_{in}|^2 = 1$ or equivalently $S_{in} = 0$. In this case, $z_{in} = 1$ or the actual input impedance z_{in} must be equal to the normalization resistance R_0. This situation describes a perfect match between the excitation with internal resistance R_0 and the one-port network.

On the other hand, if the one-port network is purely reactive ($r_{in} = 0$), then $|S_{in}|^2 = 1$. In this case, the one-port reflects the incident power back to the generator. Obviously, this power must be dissipated on the internal impedance of the generator. In practice, this may burn out the excitation source. Therefore, one has to take the necessary measures to avoid such an undesirable situation, perhaps by introducing attenuation over a circulator to dissipate the reflected power.

It should be noted that the above properties can be derived from Equation (6.24) in a straightforward manner as follows.

For a passive one-port network, the actual power delivered to its port must be non-negative. Therefore, $P_{ac} = |a|^2 - |b|^2 \geq 0$. If the one-port is lossless (i.e. purely reactive), yielding $z_{in}(j\omega) = jx_{in}(\omega)$, then there will be no power dissipated on it. In this case $P_{ac} = 0$ and the equality holds. By analytic continuation (i.e. replacing $j\omega$ by $p = \sigma + j\omega, \sigma \geq 0$), we can state that $a(p)a(-p) - b(p)b(-p) \geq 0$ for all $p = \sigma \geq 0$. Replacing $b(p)$ by $S_{in}a(p)$, we find that $a(p)[1 - S_{in}(p)S_{in}(-p)]a(-p) \geq 0$ for all $p = \sigma \geq 0$ or, equivalently, $1 - S_{in}(p)S_{in}(-p) \geq 0$ for all $p = \sigma \geq 0$. For a lossless one-port,

---

[5] That is to say, by expanding $p = j\omega$ to the RHP by setting $p = \sigma + j\omega$ with $\sigma \geq 0$.

$S_{in}(p)S_{in}(-p) = 1$. Or, in general, for a passive one-port, the inequality $S_{in}(p)S_{in}(-p) \leq 1$ must be satisfied for all $p = \sigma \geq 0$.

Because of all these properties, the rational form of

$$S_{in}(p) = \frac{h(p)}{g(p)}$$

is called the bounded real (BR) function.

Example 6.3: Referring to Figure 6.4, let us compute the reflectance for a one-port consisting of a single inductor.

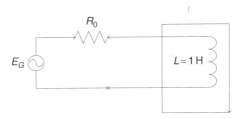

Figure 6.4   A one-port consisting of a single inductor (a purely imaginary input impedance)

In this case, $z_{in} = p$. On the $j\omega$ axis,

$$|S_{in}|^2 = \left|\frac{b(j\omega)}{a(j\omega)}\right|^2 = \left|\frac{j\omega - 1}{j\omega + 1}\right|^2 = \left|\frac{1 + \omega^2}{1 + \omega^2}\right| = 1$$

This result indicates that the reflected power is equal to the incident power. In other words, the one-port dissipates no power if it is purely reactive, as expected.

Property 5 (Bounded realness of the input reflectance): In Equation (6.26), the squared amplitude functions $|h(j\omega)|^2$ and $|g(j\omega)|^2$ of ω define non-negative even polynomials $H(\omega^2)$ and $G(\omega^2)$ such that

$$
\boxed{
\begin{aligned}
H(\omega^2) &= h(j\omega)h(j\omega) \\
&= H_1\omega^{2n} + H_2\omega^{2(n-1)} + \cdots + H_n\omega^2 + H_{n+1} \\
G(\omega^2) &= g(j\omega)g(j\omega) \\
&= G_1\omega^{2n} + G_2\omega^{2(n-1)} + \cdots + G_n\omega^2 + G_{n+1} \\
H(\omega^2) &\leq G(\omega^2)
\end{aligned}
}
\tag{6.27}
$$

At this point, we can comfortably state that one can always find a non-negative even polynomial

$$F(\omega^2) = f(j\omega)f(j\omega) = F_1\omega^{2n} + F_2\omega^{2(n-1)} + \cdots + F_n\omega^2 + F_{n+1}$$

such that the last inequality can be transformed to an equality as follows:

$$G(\omega^2) = H(\omega^2) + F(\omega^2) \qquad (6.28)$$

In fact, the physical existence of $F(\omega^2)$ can easily be verified by close examination of the power transfer ratio (PTR) given by Equation (6.19).

Obviously, the power transfer ratio $T(\omega^2)$, which is defined by Equation (6.19), describes an even, non-negative, bounded and real function over the entire frequency axis as follows:

$$T = 1 - |S_{in}|^2 = 1 - \frac{H(\omega^2)}{G(\omega^2)}$$

or

$$T = \frac{G(\omega^2) - H(\omega^2)}{G(\omega^2)} \qquad (6.29)$$

or

$$T(\omega^2) = \frac{F(\omega^2)}{G(\omega^2)} \le 1, \text{ for all } \omega$$

Thus, $F(\omega^2)$ appears as the numerator polynomial of $T(\omega^2)$. Clearly, the zeros of $F(\omega^2)$ are the zeros of the power transfer ratio.

Here, it is interesting to note that if the passive one-port consists of lumped elements and transformers, then one can immediately generate the power transfer ratio function $T(\omega^2)$ step by step in terms of network descriptive functions as follows.

In the first step, normalized input impedance $z_{in}(p)$ is generated as a PR function by $z_{in}(p) = N(p)/D(p)$ with $N(p) = a_1 p^{na} + a_2 p^{na-1} + \cdots + a_{na} + a_{na+1}$ and $D(p) = b_1 p^{nb} + b_2 p^{nb-1} + \cdots + b_{nb} + b_{nb+1}$ such that $0 \le |na - nb| \le 1$. Here, the real coefficients $\{a_i, b_j\}$ are explicitly computed by means of the lumped element values of the one-port under consideration.

In the second step, the input reflection coefficient is derived as $S_{in}(p) = h(p)/g(p)$ where $h(p) = N(p) - D(p)$ and $g(p) = N(p) + D(p)$. In this step, all the coefficients of $h(p)$ and $g(p)$ are computed in terms of the coefficients of $N(p)$ and $D(p)$. Obviously, degree 'n' is determined as max $\{na, nb\}$.

In the third step, even functions $H(p^2) = h(p)h(-p)$ and $G(p^2) = g(p)g(-p)$ are derived.

In the fourth step, $F(p^2)$ of Equation (6.30) is generated as $F(p^2) = G(p^2) - H(p^2)$.

Finally, in the fifth step, on replacing p^2 by $-\omega^2$, $T(\omega^2)$ is generated as in Equation (6.30).

Let us consider a simple exercise to generate the power ratio for a one-port.

Example 6.4: Derive the power transfer ratio function for the one-port depicted in Figure 6.3.

Solution: In Example 6.1, we obtained

$$S_{in}(p) = \frac{h(p)}{g(p)} = \frac{-p}{p+2}$$

yielding

$$H(p^2) = h(p)h(-p) = -p^2$$

and

$$G(p^2) = (p+2)(-p+2) = -p^2 + 4$$

Thus,

$$F(p^2) = G(p^2) - H(p^2) = 4$$

yielding

$$T(\omega^2) = \frac{4}{\omega^2 + 4}$$

Property 6 (Darlington's theorem): In his remarkable PhD dissertation, Sydney Darlington showed that any PR immittance function[6] can be realized as a lossless two-port in resistive termination. However, this resistive termination can always be leveled to the normalizing resistance R_0 by using a transformer which is also lossless. This property is known as Darlington's theorem, which constitutes the heart of classical filter theory.[7]

Now, let us use Darlington's theorem in the following example.

Example 6.5: Show that

$$z(p) = \frac{2p^2 + 8p + 1}{p + 4}$$

is a PR impedance function. Find the zeros of the even part of $z(p)$. Synthesize $z(p)$ as a lossless two-port in unit termination.

Solution:
In order for $z(p)$ to be a PR real function the following must be satisfied:

- When $p = \sigma$, $z(\sigma)$ must be real. In fact this is the case.
- When $\sigma \geq 0$, $z(\sigma)$ must be non-negative. As a matter of fact

$$z(\sigma) = \frac{2\sigma^2 + 8\sigma + 1}{\sigma + 1} \geq 0, \text{ for all } \sigma \geq 0$$

Now let us find the even part of $z(p)$ which is designated by $r(p^2)$:

$$r(p^2) = \frac{1}{2}[z(p) + z(-p)] = \frac{4}{-p^2 + 16}$$

Close examination of the above results reveals that that $r(p^2)$ is free of finite zeros and its zeros are all at infinity of order 2. Zeros at infinity can easily be extracted by continued fraction expansion of $z(p)$, which is also known as long division. Thus,

$$z(p) = 2p + \frac{1}{p + 4}$$

---

[6] Immittance function is a generic name referring to either an impedance or admittance function.
[7] S. Darlington, PhD dissertation, University of New Hampshire, 1939.

which yields a series inductance of $L = 2\,H$ and a parallel capacitance of $C = 1\,F$ and is finally terminated in a conductance of $4\,S$. Eventually, the resistive termination can be shifted to $1\,\Omega$ by means of a transformer as shown in Figure 6.5.

We can extend the above properties given for the one-port reflectance $S_{in}(p)$ to cover the complete scattering parameters of lossless two-ports. However, in the following section, first let us consider the generation procedures of the scattering parameters for two-ports, then study the properties of the lossless two-ports.

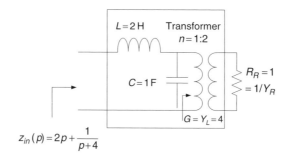

$$z_{in}(p) = 2p + \frac{1}{p+4}$$

Figure 6.5   Darlington synthesis of $z(p)$

6.3 Generation of Scattering Parameters for Linear Two-ports

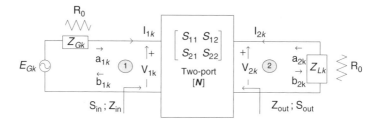

Figure 6.6   Two-port under arbitrary terminations

Referring to Figure 6.6, let $\{S_{i,j}(p),\ i,j = 1,2\}$ be the scattering parameters of a linear passive two-port. Let E_{Gk} be an arbitrary excitation with internal complex impedance $Z_{Gk}(p)$ applied to either port 1 or port 2. Let $Z_{Lk}(p)$ be any complex impedance which terminates the other end of the two-port. Since the scattering parameters are invariant under any terminations, the port reflected waves $b_{1k}(p)$ and $b_{2k}(p)$ are always linearly related to the port incident waves $a_{1k}(p)$ and $a_{2k}(p)$ by means of scattering parameters:

$$
\boxed{
\begin{aligned}
b_{1k}(p) &= S_{11}(p)a_{1k}(p) + S_{12}(p)a_{2k}(p) \\
b_{2k}(p) &= S_{21}(p)a_{1k}(p) + S_{22}(p)a_{2k}(p)
\end{aligned}
}
\tag{6.30}
$$

for any k (i.e. for any source and load terminations).

Obviously, under arbitrary port terminations $Z_{Gk}(p)$ and $Z_{Lk}(p)$, one can always measure port voltages and currents or related incident and reflected wave quantities for a selected common port normalization number R_0 satisfying the equation set given by Equation (6.31) .Then, we can always compute the input and output reflectance yielding

$$S_{in} = \frac{b_{1k}}{a_{1k}} = \frac{Z_{in} - R_0}{Z_{in} + R_0}$$

and

$$S_{out} = \frac{b_{2k}}{a_{2k}} = \frac{Z_{out} - R_0}{Z_{out} + R_0}$$

respectively.

In this representation, Z_{in} is the driving point input impedance when the two-port is terminated in Z_{Lk}. Similarly, Z_{out} is the output impedance when the input port is terminated in Z_{Gk}.

On the other hand, Equation (6.31) below implies that S_{11} and S_{22} can be directly measured for any excitation applied to port 1 (with an arbitrary internal impedance Z_{Gk}) when a_{2k} is zero.

This can be easily achieved by choosing $Z_{Lk} = R_0$, meaning that

$$V_{2k} = -Z_{Lk}I_{2k} = R_0 I_{Lk}$$

or

$$a_{2k} = \frac{1}{2\sqrt{R_0}} \left[V_2k + R_0 I_{2k} \right] = 0$$

In this case, the ratio b_{1k}/a_{1k} measures S_{11}, and S_{21} is measured as the ratio of b_{2k}/a_{1k}. Similarly, when port 1 is terminated in R_0 forcing $a_{1k} = 0$ and port 2 is derived by an arbitrary excitation, then S_{22} and S_{12} are measured as $S_{22} = b_{2k}/a_{2k}$ and $S_{12} = b_{1k}/a_{2k}$.

The above measurements can be summarized as follows:

$$\boxed{\begin{aligned} S_{11} &= \frac{b_{1k}}{a_{1k}} \quad \text{and} \quad S_{21} = \frac{b_{2k}}{a_{1k}} \quad \text{when} \quad a_{2k} \\ &= 0, \text{ which requires } Z_{Lk} = R_0 \\ S_{22} &= \frac{b_{2k}}{a_{2k}} \quad \text{and} \quad S_{12} = \frac{b_{1k}}{a_{2k}} \quad \text{when} \quad a_{1k} \\ &= 0, \text{ which requires } Z_{Gk} = R_0 \end{aligned}} \tag{6.31}$$

Once the scattering parameters are measured, the input (S_{in}) and output (S_{out}) reflectance of the two-port can be easily obtained in terms of these parameters.

Referring to Figure 6.7, let us first determine the output reflection coefficient $S_{out} = b_2/a_2$ when port 1 is terminated in arbitrary impedance Z_G.

Considering (6.31), by definition,

$$S_{out} = \frac{b_2}{a_2} = S_{21} \left(\frac{a_1}{a_2} \right) + S_{22}$$

On the other hand,

$$\frac{b_1}{a_1} = S_{11} + S_{12} \left(\frac{a_1}{a_1} \right)$$

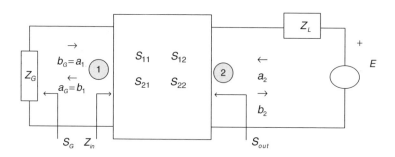

Figure 6.7 Derivation of output reflectance $S_{out} = b_2/a_2$.

Now, let us consider the input termination Z_G as a passive one-port with incident wave a_G and reflected wave b_G. By definition, the reflectance S_G is given by $S_G = b_G/a_G$ and in terms of the impedance Z_G, $S_G = (Z_G - R_0)/(Z_G + R_0)$, as shown in the previous section. However, close examination of Figure 6.7 reveals that, at port-1, $a_G = b_1$ and $b_G = a_1$. Hence,

$$\frac{b_1}{a_1} = \frac{1}{S_G}$$

Then, one obtains that

$$\frac{1}{S_G} = S_{11} + S_{12}\left(\frac{a_2}{a_1}\right) \quad or \quad \frac{a_1}{a_2} = \frac{S_{12}S_G}{1 - S_{11}S_G}$$

Employing this result in S_{out}, we have

$$\boxed{S_{out} = S_{22} + \frac{S_{12}S_{21}}{1 - S_{11}S_G}S_G} \tag{6.32}$$

Similarly,

$$\boxed{S_{in} = S_{11} + \frac{S_{12}S_{21}}{1 - S_{22}S_L}S_L} \tag{6.33}$$

6.4 Transducer Power Gain in Forward and Backward Directions

At this point it is interesting to look at the ratio specified by

$$|S_{21}(j\omega)|^2 = \left|\frac{b_2(j\omega)}{a_1(j\omega)}\right|^2$$

when $a_2 = 0$.

By definition,

$$b_2 = \frac{1}{2\sqrt{R_0}} = [V_2 - R_0 I_2]$$

and the voltage V_L across the termination R_0 is $V_L = -R_0 I_2$. Then, $|b_2|^2 = |V_L|^2/R_0$ which is the power P_L delivered to termination R_0 at port 2.[8] On the other hand, we know that $|a_1|^2$ is the available power P_A. Thus, $|S_{21}|^2$ measures the rate of power transfer from port 1 to port 2 (forward power transfer ratio), as the ratio of power delivered to the termination R_0 to the available power of the generator with internal resistance R_0.

In fact, this is the formal definition of transducer power gain $(\text{TPG})_F$ under resistive terminations R_0 at both ends, yielding

$$(\text{TPG})_F = \frac{P_L}{P_A} = |S_{21}|^2 \tag{6.34}$$

where subscript F refers to power transfer in a forward direction (or from port 1 to port 2).

Similarly, referring to Figure 6.7, transducer power gain $(\text{TPG})_B$ in a backward direction is given by

$$(\text{TPG})_B = \left|\frac{b_1}{a_2}\right|^2 = |S_{12}|^2 \tag{6.35}$$

when $a_1 = 0$.

6.5 Properties of the Scattering Parameters of Lossless Two-ports

A lumped element lossless two-port, which consists of ideal inductors, capacitors, coupled coils and transformers, can be constructed by cascading the sections of type A, B, C and D as shown in Figure 6.8.

Let $S_{i,j}(p), i, j = 1, 2$, designate the scattering parameters of the lumped lossless two-port under consideration. Let R_0 be the normalizing resistance for both input and output ports. Since the two-port is lossless, there will be no power dissipation on it. In other words, the total power delivered to its ports must be zero. Let us elaborate this statement by mathematical equations as follows.

Referring to Figure 6.9, let $P_1 = |a_1|^2 - |b_1|^2$ be the net power delivered to port 1. Similarly, let $P_2 = |a_2|^2 - |b_2|^2$ be the net power delivered to port 2. Thus, the total power delivered to $[N]$ given by

$$P_T = P_1 + P_2 = \begin{bmatrix} a_1^* & a_2^* \end{bmatrix} \begin{bmatrix} a_1 \\ a_2 \end{bmatrix} - \begin{bmatrix} b_1^* & b_2^* \end{bmatrix} \begin{bmatrix} b_1 \\ b_2 \end{bmatrix}$$

must be zero.

Recall that by definition.

$$b = \begin{bmatrix} b_1 \\ b_2 \end{bmatrix} = S \begin{bmatrix} a_1 \\ a_2 \end{bmatrix} = Sa$$

---

[8] It is interesting to note that $|b_2|^2$ is the reflected power of port 2 which is directly dissipated on the termination R_0 when $a_2 = 0$.

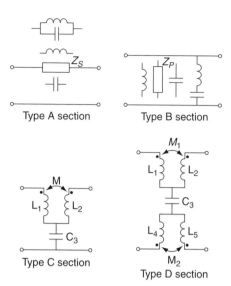

Figure 6.8   Simple lumped element building blocks which constitute lossless two-ports

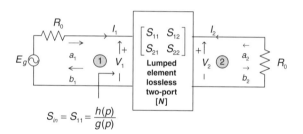

Figure 6.9   Scattering parameters for lossless two-ports

Then Equation (6.36) becomes

$$P_T = a^\dagger [I - S^\dagger S] a \qquad\qquad (6.36)$$

Since the two-port $[N]$ is lossless, P_T must be zero. Thus,

$$I - S^\dagger S = 0$$

$$\text{or}$$

$$S^\dagger S = I \qquad\qquad (6.37)$$

where the superscript † indicates the complex conjugate of the transposed matrix and

$$I = \begin{bmatrix} 1 & 0 \\ 0 & 1 \end{bmatrix}$$

is the identity matrix. It should be noted that the matrix $S^\dagger S$ must be symmetric since the identity matrix is symmetric. Hence,

$$\boxed{S^\dagger S = SS^\dagger = I} \tag{6.38}$$

The expression $SS^\dagger = I$ yields four equations, but the one with zero equality is redundant. Therefore, the open form of Equation (6.38) is written as

$$\boxed{\begin{aligned} S_{11}(p)S_{11}(-p) + S_{12}(p)S_{12}(-p) &= 1 \\ S_{22}(p)S_{22}(-p) + S_{21}(p)S_{21}(-p) &= 1 \\ S_{11}(-p)S_{21}(p) + S_{12}(-p)S_{22}(p) &= 0 \end{aligned}} \tag{6.39}$$

or simply,

$$\boxed{\begin{aligned} S_{11}(p)S_{11}(-p) &= 1 - S_{12}(p)S_{12}(-p) \\ S_{22}(p)S_{22}(-p) &= 1 - S_{21}(p)S_{21}(-p) \\ S_{22}(p) &= -\frac{S_{21}(p)}{S_{12}(-p)} S_{11}(-p) \end{aligned}} \tag{6.40}$$

Similarly, the open form of $S^\dagger S = I$ yields

$$\boxed{\begin{aligned} S_{11}(p)S_{11}(-p) &= 1 - S_{21}(p)S_{21}(-p) \\ S_{22}(p)S_{22}(-p) &= 1 - S_{12}(p)S_{12}(-p) \\ S_{22}(p) &= -\frac{S_{12}(p)}{S_{21}(-p)} S_{11}(-p) \end{aligned}} \tag{6.41}$$

By substituting $p = j\omega$, it is interesting to note that a comparison of Equations (6.41) and (6.42) reveals that

$$\boxed{\begin{aligned} |S_{11}(j\omega)| &= |S_{22}(j\omega)| \\ \text{and} \\ \left| \frac{S_{12}(j\omega)}{S_{21}(-j\omega)} \right| &= \left| \frac{S_{21}(j\omega)}{S_{12}(-j\omega)} \right| \end{aligned}} \tag{6.42}$$

For lumped element two-ports, reflection coefficients $S_{11}(p)$ and $S_{22}(p)$ can be regarded as one-port reflection coefficients $S_{in}(p) = S_{11}(p)$ when port 2 is terminated in R_0, and $S_{out}(p) = S_{22}(p)$ when port 1 is terminated in R_0. Therefore, they must be BR rational functions. In other words, they must both be regular in the closed RHP, i.e. $(\sigma \geq 0)$.

Now $S_{11}(p) = h(p)/g(p)$. If then Equation (6.41) reveals that

$$S_{12}(p)S_{12}(-p) = 1 - \frac{h(p)}{g(p)}\frac{h(-p)}{g(p)}$$

or

$$S_{12}(p)S_{12}(-p) = \frac{g(p)g(-p) - h(p)h(-p)}{g(p)g(-p)} = \frac{F(p^2)}{G(p^2)} \tag{6.43}$$

then Equation (6.44) suggests that

$$S_{12}(p) = \frac{f_{12}(p)}{g(p)} \tag{6.44}$$

where $f_{12}(p)$ is constructed on propery selected roots of

$$F(p^2) = g(p)g(-p) - h(p)h(-p)$$

Clearly, the solution is not unique.

Similarly, Equation (6.42) suggests that

$$S_{21}(p) = \frac{f_{21}(p)}{g(p)} \tag{6.45}$$

where $f_{21}(p)$ is also obtained by factorization of $F(p^2)$.

In this case,

$$S_{22}(p) = -\frac{f_{12}(p)}{f_{21}(p)}\frac{h(-p)}{g(p)} \tag{6.46}$$

Similarly,

$$S_{22}(p) = -\frac{f_{21}(p)}{f_{12}(-p)}\frac{h(-p)}{g(p)} \tag{6.47}$$

Hence, for lossless two-ports, one must have

$$\boxed{\frac{f_{12}(p)}{f_{21}(-p)} = \frac{f_{21}(p)}{f_{12}(-p)}} \tag{6.48}$$

Bounded realness of the scattering parameters requires that the ratios given by Equation (6.48) should be free of open RHP poles. This could be achieved with possible cancellations in

$$\eta_a = \frac{f_{12}(p)}{f_{21}(-p)} \quad \text{and} \quad \eta_b = \frac{f_{21}(p)}{f_{21}(-p)}$$

Furthermore, these ratios must have unity amplitude on the $j\omega$ axis, $|S_{11}(j\omega)| = |S_{22}(j\omega)|$. Henceforth, we can say that $f_{12}(p) = \mu f_{21}(-p)$.

In this case, let us set $f(p) = f_{12}(p)$; then, $f_{21}(p) = \mu f(-p)$ where μ is a unimodular constant (i.e. $\mu = \pm 1$).

Moreover, if the lossless two-port is reciprocal, then

$$\boxed{S_{12}(p) = S_{21}(p) \quad \text{or} \quad f_{12}(p) = f_{21}(p)} \tag{6.49}$$

which automatically satisfies Equation (6.49). In this case, $f(p)$ cannot be a full polynomial. Rather, it must be even or odd. If this is not true, the lossless two-port cannot be reciprocal.

If $f(p)$ is even, i.e. $f(p) = f(-p)$, then $f_{21}(p) = \mu f_{21}(p) = \mu f_{12}(-p) \triangleq \mu f(-p) = \mu f(p)$.

Thus, the reciprocity condition given by Equation (6.48) is satisfied with $\mu = +1$. Therefore, for reciprocal networks, if $f(p)$ is an even polynomial then the unimodular constant must be $\mu = +1$.

On the other hand, if $f(p)$ is an odd polynomial then reciprocity demands a negative unimodular constant such that

$$f_{12}(p) = f(p) \triangleq -f(-p)$$

Then, by Equation (6.50),

$$f_{21}(p) = \mu f_{12}(-p) \triangleq \mu f(-p) = -\mu f(p), \text{which requires } \mu = -1$$

Thus, the scattering matrix of a lossless reciprocal two-port is given by

$$\boxed{S = \frac{1}{g}\begin{bmatrix} h & f \\ \mu f_* & \dfrac{f}{\mu f_*} h_* \end{bmatrix} = \frac{1}{g}\begin{bmatrix} h & f \\ f & -\mu h_* \end{bmatrix}} \tag{6.50}$$

where the subscript * designates the para-conjugate of a function $x(p)$ such that $x(p) = x(-p)$.

In summary, for a reciprocal lossless two-port $f(p)$ is either an even or odd polynomial and the unimodular sign μ is given by

$$\mu = \frac{f(p)}{f(-p)} = \begin{cases} +1 \text{ when } f(p) \text{ is even} \\ -1 \text{ when } f(p) \text{ is odd} \end{cases}$$

The above results are summarized as follows.[9]

Referring to Figure 6.9, for a lumped element, reciprocal lossless two-port, if the BR input reflection coefficient has the following form

$$S_{11}(p) = \frac{h(p)}{g(p)} \qquad\qquad (6.51)$$

with real polynomials $h(p)$ and $g(p)$ of degree n, then the rest of the scattering parameters are given as

$$S_{12}(p) = S_{21}(p) = \frac{f(p)}{g(p)} \qquad\qquad (6.52)$$

$$S_{22}(p) = \frac{f(p)}{f(-p)} \frac{h(p)}{g(p)} \qquad\qquad (6.53)$$

provided that $f(p)$ is either an even or odd polynomial satisfying

$$F(p^2) = g(p)g(-p) - h(p)h(-p) \qquad\qquad (6.54)$$

In the above forms, all the scattering parameters must be BR rational functions in complex variable $p = \sigma + j\omega$. Using Equation (6.55), construction of $f(p)$ requires a little care. Therefore, let us investigate the possible situations in the following discussion.

For a given lossless two-port, one may wish to obtain the full scattering matrix, perhaps starting from the driving point input impedance $Z_{in}(p)$ which directly specifies the input reflection coefficient

$$S_{11}(p) = \frac{h(p)}{g(p)} = \frac{z_{in} - 1}{z_{in} + 1}$$

and then, by spectral factorization of

$$F(p^2) = f(p)f(-p) = g(p)g(-p) - h(p)h(-p)$$

$f(p)$ is generated on the selected roots of $F(p^2)$ which in turn yields

$$S_{21}(p) = \frac{f(p)}{g(p)}$$

Eventually, $S_{22}(p)$ is generated as

$$S_{22}(p) = -\frac{f(p)}{f(-p)} \frac{h(-p)}{g(p)} = -\mu \frac{h(-p)}{g(p)}$$

as in Equation (6.54). However, this process requires attention as discussed below.

---

[9] A comprehensive discussion of the subject can be found in 'Wideband Circuit Design' by H. J. Carlin and P. P. Civalleri, CRC Press, 1997, pp. 231–235.

In the literature, the above forms of the scattering parameters are known as the Belevitch[10] canonical form.

As we have shown in the previous section, under the standard terminations of Figure (6.9), $|S_{21}(j\omega)|^2 = |b_2|^2/|a_1|^2$ measures TPG in the forward direction (from port 1 to port 2). In this equation, $|a_1|^2$ is the measure of the available power P_A of port 1, and $|b_2|^2$ specifies the power P_L delivered to load R_0.[11]

Let us now investigate some properties of the lossless two-ports.

Property 7: As was proved by Darlington, any PR function Z_{in} which is specified as a rational function $Z_{in}(p) = N(p)/N(p)$ or the corresponding input reflection coefficient

$$S_{in}(p) = S_{11}(p) = \frac{Z_{in} - 1}{Z_{in} + 1} = \frac{h(p)}{g(p)}$$

can be realized as a reciprocal lossless two-port in unit termination.

Proof of Darlington's theorem is straightforward: since $h(p)$ and $g(p)$ are specified, then the complete scattering parameters of the lossless two-port can be determined by factorization of the even polynomial $F(p^2) = f(p)f(-p) = g(p)g(-p) - h(p)h(-p)$. Then, $f(p)$ is constructed on the selected roots of $F(p^2)$. At this point, it is very important to note that if the factorization process does not yield an odd or even polynomial for $f(p)$, we can always augment $S_{21}(p) = f(p)/g(p)$ with unity products in the form of $(\gamma_r + p)/(\gamma_r + p)$ with γ_r real and positive such that

$$
\begin{aligned}
S_{21}(p) &= \hat{f}(p) \\
\hline
\hat{g}(p) &= \frac{f(p)(\gamma_r + p)}{g(p)(\gamma_r + p)}
\end{aligned}
$$

yielding odd or even $\hat{f}(p)$. The same augmentation is also applied to $S_{11}(p)$ and $S_{22}(p)$ to give the common denominator polynomial

$$\hat{g}(p) = g(p)(\gamma_r + p)$$

in all the scattering parameters.

Now, let us consider an example to clarify the above situation.

Example 6.6: Let

$$S_{11}(p)\frac{1 - p}{\sqrt{2} + \sqrt{2}p}$$

Generate the scattering parameters for the corresponding lossless reciprocal lumped element two-port.

Solution: Here,

$$h(p) = 1 - p \quad \text{and} \quad g(p) = \sqrt{2} + \sqrt{2}p$$

Let

$$G(p^2) = g(p)g(-p) = 2 - 2p^2$$

---

[10] V. Belevitch, *Classical Network Theory*, Holden Day, 1968, p. 277.
[11] In the literature, it is common to designate transducer power gain as TPG $= P_L/P_A$.

and

$$H(p^2) = h(p)\, h(-p) = 1 - p^2$$

In this case,

$$F(p^2) = f(p)f(-p) = G - H = 1 - p^2$$

Since $\hat{g}(p) = f(-p)g(p)$ must be strictly Hurwitz, then $f(p) = 1 - p$.

At this point we have to be careful since $f(p)$ is neither even nor odd in complex variable p. Rather, it is a full polynomial of degree 1. Therefore, it cannot belong to a reciprocal lossless network.

However, by augmenting with the product $(1 + p)/(1 + p)$, $S_{21}(p)$ can be written as

$$S_{21}(p) = \frac{1-p}{\sqrt{2} + \sqrt{2}} \frac{1+p}{1+p} = \frac{1 - p^2}{\sqrt{2} + 2\sqrt{2}p + \sqrt{2}p^2} = \frac{\hat{f}(p)}{\hat{g}(p)}$$

In this case, $\hat{f}(p) = 1 - p^2$, which is an even polynomial, and $\hat{g}(p) = \sqrt{2} + 2\sqrt{2}p + \sqrt{2}p^2$, which is a strictly Hurwitz common denominator polynomial. Eventually, augmented $S_{11}(p)$ and $S_{22}(p)$ are give as follows:

$$S_{11}(p) = \frac{\hat{h}(p)}{\hat{g}(p)} = \frac{1 - p^2}{\sqrt{2} + 2\sqrt{2}p + \sqrt{2}p^2}$$

$$S_{11}(p) = -\mu \frac{\hat{h}(-p)}{g}(p) = -\frac{1 - p^2}{\sqrt{2} + 2\sqrt{2}p + \sqrt{2}p^2}$$

6.6 Blashke Products or All-Pass Functions

As can be understood from the above discussions, the complex function $\eta(p) = f(p)/f(-p)$ must define a regular function in the closed RHP (i.e. $p = \sigma \geq 0$) since the denominator polynomial $\hat{g}(p) = f(-p)g(p)$ must be strictly Hurwitz to make $S_{22}(p)$ regular in the closed RHP.[12] Furthermore, on the $j\omega$ axis, the amplitude function $|\eta(j\omega)|^2 = 1$. Thus, a rational regular (or analytic) function with unity amplitude is called a Blashke product or an all-pass function

Definition 5 (Proper polynomial $f(p)$):[13] A real polynomial in complex variable 'p' is called proper if it yields a regular Blashke product $\eta(p) = f(p)/f(-p)$.

At this point it is crucial to note that for a reciprocal lossless two-port, the numerator polynomial $f(p)$ of $S_{21} = S_{12} = f(p)/g(p)$ must be proper to end up with BR scattering parameters.

---

[12] A complex variable function $\eta(p)$ is called analytic in a complex region R if it has no singularity in R. In other words it has to be differentiable in R. If $\eta(p)$ is rational and analytic in R, then it should be free of poles in R because poles lying in R introduce singularities for which $\eta(p)$ becomes infinite.

[13] To our knowledge, this is the first description of the proper polynomial definition of $f(p)$.

6.7 Possible Zeros of a Proper Polynomial *f(p)*

In general, the zeros of a proper polynomial $f(p)$ can be classified as in the following cases.

Case A: If $f(p)$ includes a real positive zero which is denoted by $\alpha > 0$ such that $f(p) = (\alpha - p)\hat{f}(p)$, then the term $(\alpha - p)$ in $f(p)$ appears as a factor $(\alpha + p)$ in $f(-p)$, yielding $\hat{g}(p) = (\alpha + p)\hat{f}(-p)g(p)$ which preserves the scattering Hurwitz property of $\hat{g}(p)$. Therefore, a real RHP zero of $f(p)$ is called a proper zero. In this case, the Blashke product $\eta(p)$ must include the term $(\alpha - p)/(\alpha + p)$ or in general

$$\eta(p) = \frac{\alpha - p}{\alpha + p} \frac{\hat{f}(p)}{\hat{f}(-p)}$$

Obviously, in this representation, it is assumed that the term $\hat{f}(p)/\hat{f}(-p)$ is regular in closed RHP to make $S_{22}(p)$ regular (or BR) in closed RHP. This can only be achieved with possible cancellations in $\hat{\hat{f}}(p)/\hat{\hat{f}}(-p)$ as discussed in Case C.

Case B: If $f(p)$ includes a complex zero (such as $p = \alpha + j\beta$ with $\alpha > 0$ in the closed RHP), it must be accompanied by its conjugate in order to make $S_{21}(p)$ a real function. In this case, $f(p)$ takes the following form:

$$f(p) = [(\alpha + j\beta) - p][(\alpha - j\beta) - p]\hat{f}(p)$$

which is a real polynomial in p. Again, assuming the regular term $\hat{f}(p)/\hat{f}(-p)$ with possible cancellations, the Blashke product $\eta(p)$ is written as

$$\eta(p) = \frac{(\alpha - p)^2 + \beta^2}{(\alpha + p)^2 + \beta^2} \frac{\hat{f}(p)}{\hat{f}(-p)}$$

Therefore, a complex closed RHP zero of $f(p)$ must be matched with its conjugate pair in order to make $f(p)$ real and proper.

Case C: Assume that $f(p)$ includes a purely real closed LHP zero such that $p_{z1} = \delta < 0$. Obviously, this zero is not proper since it introduces a closed RHP zero in $f(-p)$ yielding unregular $\eta(p)$.[14]

However, if $p_{z1} = \delta$ is matched with its mirror image $p_{z2} = -\delta$ in the closed RHP, then $\eta(p)$ must include the term

$$\left[\frac{\delta - p}{\delta + p}\right]\left[\frac{-\delta - p}{-\delta + p}\right]$$

which is equal to 1 by cancellation. Thus, we say that a proper $f(p)$ may include a purely real closed LHP zero if it is paired with its mirror image to make $\eta(p)$ regular with cancellations.

Case D: Assume that $f(p)$ includes a complex closed LHP zero $p_{z1} = \delta + j\beta$ with $\delta < 0$. As explained in the previous case, this zero cannot be proper unless it is paired with its mirror image, which results in cancellations in $\eta(p)$. Furthermore, in order to assure the reality of the polynomial $f(p)$, complex LHP zeros must be accompanied by their conjugate pairs yielding quadruple symmetry as shown in Figure 6.10. Thus, we say that complex LHP zeros of a proper $f(p)$ must have quadruple symmetry. In this case, $f(p)$ should take the following form:

$$f(p) = [(\delta - p)^2 + \beta^2][(\delta + p)^2 + \beta^2]\hat{f}(p)$$

where $\delta < 0$ and $\hat{f}(p)$ describes a regular all-pass function $\hat{f}(p)/\hat{f}(-p)$.

Case E: Multiple DC zeros of $f(p)$, which is specified by p^k in $f(p) = p^k\hat{f}(p)$, are always proper since they cancel in $\eta(p)$.

---

[14] In this context, the term 'unregular' is used to indicate a singularity in $\eta(p)$ at a point $p = -\delta > 0$ in the closed RHP.

Case F: Finite non-zero $j\omega$ zeros of $f(p)$ are always proper as long as they are paired with their complex conjugates. In this case, $f(p)$ is expressed as $f(p) = (p^2 + \omega_k^2)\hat{f}(p)$ where ω_k denote non-zero but finite jw zeros of $f(p)$.

All the above zeros of $f(p)$ are summarized in Figure 6.10.

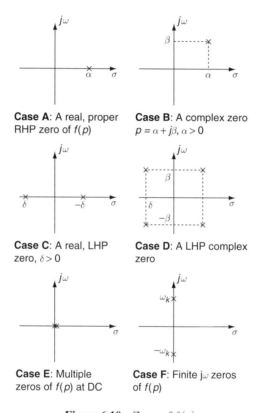

Figure 6.10 Zeros of $f(p)$

6.8 Transmission Zeros[15]

The zeros of $f(p)$ coincide with the finite zeros of $S_{21}(p)$. Moreover, they also coincide with the zeros of the transducer power gain TPG $= |S_{21}(j\omega)|^2$ over the real frequency axis $j\omega$.

Obviously, signal transmission stops at the zeros of TPG $= |S_{21}(j\omega)|^2$. Therefore, the zeros of $S_{21}(p)S_{21}(-p)$ are called the transmission zeros of the lossless two-port. Such zeros may be introduced at finite values of p or at infinity.

---

[15] The definition of the transmission zeros introduced in this book is somewhat different from those introduced in the existing literature by Fano, Youla, Carlin and Yarman.

The circuit structure (or the circuit topology) of the lossless two-port imposes the transmission zeros. Therefore, in the following properties some practical circuit topologies are introduced in connection with transmission zeros.

Property 8: In the complex p plane, the general form of $f(p)$ may be written as follows:

Let $f_{DC}(p) = p^k$ be the term with multiple DC zeros

Let $f_\omega(p) = \prod_{r=1}^{n_\omega} (p^2 + \omega_r^2)$ be the term with finite $j\omega$ zeros.

Let $f_\sigma(p) = \prod_{r=1}^{n_\sigma} (\sigma_r - p)$ be the term with pur real RHP zeros[16].

Let $f_{MLHP}(p) = (-1)^{n_\delta} \prod_{r=1}^{n_\delta} (\delta_r^2 - p^2)$ be the term with pure real LHP zeros
which are matched with mirror image pairs.

Let $f_{CRHP}(P) = \prod_{r=1}^{n_R} [(\sigma_r - p)^2 + \beta_r^2]$ be the term with complex RHP zeros[17].

Let $f_{MCLHP}(P) = \prod_{r=1}^{n_L} [(\delta_r - p)^2 + \beta_r^2][(\delta_r + p)^2 + \beta_r^2]$ be the term with complex
LHPzeros which are matched with their mirror image pairs.

Then, the general form of $f(p)$ may be specified as

$$f(p) = f_{DC}(p) f_\omega(p) f_\sigma(p) f_{MLHP}(p) f_{CRHP}(p) f_{MCLHP}(p)$$

This form of $f(p)$ is not proper due to terms with un matched RHP zeros. However, by augmentation one can add LHP mirror image zeros to be paired with RHP zeros, which makes the terms $f_\sigma(p)$ and $f_{CRHP}(p)$ even. In this case, the proper generic form of $f(p)$ is given by

$$f(p) = p^k \prod_{r=1}^{n_\omega} (p^2 + \omega_r^2) \prod_{r=1}^{n_\delta} (\delta_r^2 - p^2) \prod_{r=1}^{n_L} [(\delta_r - p)^2 + \beta_r^2][(\delta_r + p)^2 + \beta_r^2] \qquad (6.55)$$

The full form of Equation (6.55) corresponds to a highly complicated network structure which may be impractical to realize. In practice, however, we generally deal with lossless ladder forms as summarized in Properties 9–10.

---

[16] This form of $f_\sigma(p)$ is not proper. However, it may be augmented by its mirror image.
[17] This form of $f_{CRHP}(p)$ is not proper. However, it may be augmented by its mirror image.

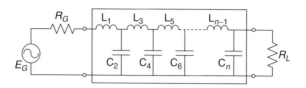

Figure 6.11 $f(p)$ yields an LC low-pass ladder structure

Property 9 (Low-pass structures): The simplest proper form of $f(p)$ includes no zero. In this case, $f(p)$ is a constant and may be normalized at unity such that $f(p) = 1$. This form corresponds to a low-pass L C ladder as shown in Figure 6.11.

In this case, transmission zeros of the system will be at infinity of degree $2n$, where 'n' is the degree of the denominator polynomial $g(p) = g_1 p^n + g_2 p^{n-1} + \cdots + g_n p + g_{n+1}$ which also specifies the total number of reactive elements[18] in the lossless two-port under consideration.

Property 10 (Bandpass structures): A more complicated form of $f(p)$ may have only multiple zeros at DC such that $f(p) = p^k$. This form corresponds to a band pass structure as shown in Figure 6.12.

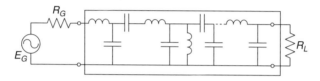

Figure 6.12 Lossless ladder with $f(p) = p^k$ which includes zeros of transmissions at DC of order $2k$ and at infinity of order $2(n - k)$

In this case, integer 'k' is a count of the total number of series capacitors and shunt inductors which introduce transmission zeros at DC of degree '$2k$'. Furthermore, the difference $(n - k)$ is the total number of series inductors and shunt capacitors which introduce transmission zeros at infinity of degree $2(n - k)$.

Property 11: Finite real frequency or $j\omega$ zeros of $f(p) = \prod_{r=1}^{n_\omega}(p^2 + \omega_r^2)$ may be realized either as a parallel resonance circuit in series configuration, or as a series resonance circuit in shunt configuration– or even as a Darlington C section with coupled coils as depicted in Figure 6.13.

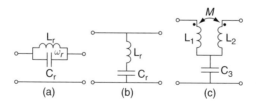

Figure 6.13 For the $f(p) = p^2 + \omega_r^2$ case, realization of $j\omega_r$ zeros: (a) as a parallel resonance circuit is series configuration; (b) as a series resonance circuit in parallel configuration; (c) as a Darlington C section

---

[18] In this case, reactive elements will be the elements of the low-pass ladder (LPL) structure which is constructed as cascade connections of series inductors and shunt capacitors as depicted in Figure 6.11.

Example 6.7: For a lossless two-port let the unit normalized BR input reflection coefficient be given by

$$S_{11}(p) = \frac{h(p)}{g(p)} = \frac{-p^4 + p^3 + 2p}{p^4 + 3p^3 + 4p + 3p + 1}$$

(a) Derive $f(p)$ and check if it is proper.
(b) Find the transmission zeros of the lossless two-port.
(c) Comment on the possible network topology of the two-port.
(d) Construct the full scattering parameters from $S_{11}(p)$.
(e) Synthesize $S_{11}(p)$ as a Darlington lossless two-port in resistive termination which yields the transmission zeros of $T(p^2) = S_{21}(p)S_{21}(-p)$.

Solution:

(a) The transmission zeros of the two-port are determined from

$$S_{21}(p)S_{21}(-p) = 1 - S_{11}(p)S_{11}(-p) = \frac{F(p^2)}{G(p^2)} = \frac{g(p)g(-p) - h(p)h(-p)}{g(p)g(-p)}$$

where

$$h(p) = -p^4 + 3p^2 + 2p + 0$$

and

$$g(p) = p^4 + 3p^3 + 4p + 3p + 1$$

Then,

$$G(p^2) = p^8 - p^6 + 0p^4 - p^2 + 1$$

$$H(p^2) = p^8 - p^6 - p^4 + p^2 + 0$$

or

$$F(p^2) = G(p^2) - H(p^2) = p^4 - 2p^2 + 1 = (p^2 + 1)^2$$

Thus,

$$f(p) = f(-p) = p^2 + 1$$

and it is proper with one finite zero of transmissions located at $\omega_{1,2} = \pm j1$.

Furthermore, one can readily see that $\eta(p) = f(p)/f(-p) = 1$.

(b) Since the degree of the denominator $g(p)$ is $n = 4$ and the degree of the proper polynomial $f(p)$ is $n_f = 2$, then the lossless two-port must have transmission zeros of order $2(n - n_f) = 2(4 - 2) = 4$ at infinity.

(c) The real frequency zero can be realized as either a parallel resonance circuit in series configuration or a series resonance circuit in shunt configuration if and only if

$$z_{in}(p) = \frac{1 + s_{11}}{1 - s_{11}} \quad \text{or} \quad y_{in}(p) = \frac{1 - s_{11}}{1 + s_{11}}$$

satisfies Fujisawa's theorem.[19] If this is the case the term $(p^2 + 1)$ appears as a multiplying factor in denominator polynomials of the impedance or admittance functions as we go along with long division of the driving point impedance function. This fact can be seen during the synthesis process. Otherwise, this transmission zero can be extracted using a zero shifting technique yielding a Darlington C section.

(d) Since $f(p) = p^2 + 1$ and $\eta(p) = 1$, then

$$S_{21}(p) = \frac{p^2 + 1}{p^4 + 3p^3 4p + 3p + 1}$$

and

$$S_{21}(p) = \frac{-p^4 + p^3 + 2p^2}{p^4 + 3p^3 4p + 3p + 1}$$

(e) Let us first generate the driving point input impedance of the lossless two-port when the output port is terminated in 1 Ω:

$$z_{in}(p) = \frac{1 + S_{11}}{1 - S_{11}} = \frac{g(p) - h(p)}{g(p) + h(p)} = \frac{2p^3 + 2p^2 + 2p + 1}{p^4 + p^3 + p^2 + p + 1}$$

which satisfies the condition of Fujisawa's theorem to be synthesized as a low-pass ladder (LPL) since $z_{in}(0) = 1 > 0$.

Furthermore, it is found that

$$Z_{in}(p) = \frac{2p^3 + 2p^2 + 2p + 1}{p^4 + p^3 + p^2 + p + 1} = \frac{2p^3 + 2p^2 + 2p + 1}{(p^2 + 1)(p^2 + p + 1)}$$

which includes the term $(p^2 + 1)$ in the denominator. This term can be immediately extracted from the input impedance, yielding

$$z_{in}(p) = \frac{2p}{(p^2 + 1)} + \frac{p + 1}{p^2 + p + 1}$$

---

[19] Fujisawa's theorem indicates that for a given PR impedance function $z_{in}(p)$, if $f(p) = \prod_{r=1}^{n_\omega}(p^2 + \omega_r^2)$, and if $z_{in}(0) = c > 0$, or equivalently if $f_{in}(0) = c > 0$, and if $z_{in}(p)$ has either a pole or a zero at infinity, then the zero of transmission ω_r can be realized as either a parallel resonance circuit in mid-series configuration or a series resonance circuit in mid-shunt configuration by extracting the term $2L_r p/(L_r C_r p^2 + 1)$ during the long division (or, equivalently, continued fraction expansion)process of the driving point impedance.

Then, we can continue with the long division process:

$$z_{in}(p) = \frac{2p}{(p^2+1)} + \cfrac{1}{p + \cfrac{1}{p+1}}$$

Hence, the above form of the driving point impedance indicates that the first term $2p/(p^2+1)$ is a parallel resonance circuit with $L_1 = 2\,H$ and $C_1 = \frac{1}{2}\,F$ in series configuration. It is followed with a shunt capacitor of $C_1 = 1\,F$ and then a series inductor of $L_3 = 1\,H$ follows. Eventually, the lossless ladder is terminated in 1Ω resistance as shown in Figure 6.14.

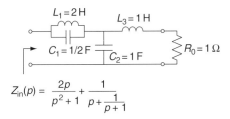

$$Z_{in}(p) = \frac{2p}{p^2+1} + \cfrac{1}{p + \cfrac{1}{p+1}}$$

Figure 6.14   Synthesis of the input impedance $Z_{in}(p)$ of Example 6.7

6.9 Lossless Ladders

If a lumped element reciprocal lossless two-port is free of coupled coils, it exhibits a special kind of circuit topology called a lossless ladder. In this configuration, simple sections of Figure 6.10 are in tandem connection as shown in Figure 6.15. In this figure Z_i and Y_i designate the series arm impedances and the shunt arm admittances respectively. For lossless ladders, it is straightforward to see that the proper $f(p)$ is either as even or odd polynomial in the complex variable $p = \sigma + j\omega$ having all its zeros on the $j\omega$ axis, yielding $\eta(p) = f(p)/f(-p) = \mu = \pm 1$. This can easily be seen by erasing the real and

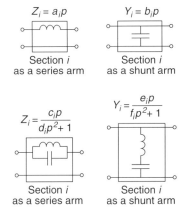

Figure 6.15   Realization of lossless ladders

complex zeros of $f(p)$ from Equation (6.56). Thus, for a lossless ladder

$$f(p) = p^k \prod_{r=1}^{n_k} (p^2 + \omega_r^2)$$ (6.56)

Obviously, in Equation 6.56, $f(p)$ is either an even or odd polynomial in p depending on the value of the integer k which is associated with the multiple zeros at DC. Furthermore, all the zeros of $f(p)$ must appear as poles of Z_i and Y_i.

Let the input impedance of the lossless ladder be $Z_{in}(p)$. Then, by continued fraction expansion,

$$Z_{in}(p) = Z_1 + \cfrac{1}{Y_2 + \cfrac{1}{Z_3 + \cfrac{1}{Y_4 + \cfrac{1}{\ldots + \cfrac{1}{Z_n(or\ Y_n) + \xi}}}}}$$ (6.57)

where ξ is the constant resistive termination (resistance or conductance) of the lossless two-port which can be removed with a transformer closed in unity resistance and

$$\{Z_i \text{ or } Y_i\} = \begin{cases} a_i p \\ 1 \\ \dfrac{b_i p}{c_i p} \\ \dfrac{}{d_i p_2 + 1} \end{cases}$$ (6.58)

At this point it is appropriate to elaborate the synthesis of driving point functions in a different chapter.

6.10 Further Properties of the Scattering Parameters of Lossless Two-ports

For many practical manipulations, it is interesting to evaluate the determinant of a para-unitary scattering matrix ΔS. This derivation is introduced below.

Derivation of ΔS
Let

$$S = \begin{bmatrix} S_{11} & S_{12} \\ S_{21} & S_{22} \end{bmatrix}$$

be a para-unitary scattering matrix. That is,

$$SS^\dagger = \begin{bmatrix} 1 & 0 \\ 0 & 1 \end{bmatrix}$$

Determinant S is given by

$$\Delta S = S_{11}S_{22} - S_{12}S_{21} \qquad (6.59)$$

The para-unitary condition yields that

$$
\begin{aligned}
S_{11}S_{11*} + S_{12}S_{12*} &= 1 \\
S_{22} &= -\frac{S_{21}}{S_{12*}}S_{11*}
\end{aligned}
\qquad (6.60)
$$

Inserting Equation (6.60) in to Equation (6.59) we have

$$\Delta S = -S_{11}S_{11*}\frac{S_{21}}{S_{12*}} - S_{12}S_{21}$$

$$= -\frac{S_{21}}{S_{12*}}\left(S_{11}S_{11*} + S_{12}S_{12*}\right)$$

or

$$\Delta S = -\frac{S_{21}}{S_{12*}}$$

(6.61)

For reciprocal lossless two-ports it is well known that $S_{21} = S_{12}$. Therefore,

$$\Delta S = -\frac{S_{21}}{S_{21*}} = -\frac{S_{12}}{S_{12*}}$$

This result can be summarized in the following property.

Property 12 (**Determinant of the scattering matrix**): For a reciprocal lossless two-port, determinant $\Delta S(p)$ is an all-pass function expressed as[20]

$$\eta_\Delta = \Delta S = -\frac{f(p)}{f(-p)}\frac{g(-p)}{g(p)} = -\eta_f(p)\eta_G(p)$$

where $\eta_f(p) = f(p)/f(-p)$ is constructed on the RHP zeros of $S_{21}(p)$, and $\eta_g(p) = g(-p)/g(p)$.

Obviously, on the real frequency axis $\Delta S(j\omega)$ has unitary amplitude. That is,

$$|\Delta S(j\omega)| = 1$$

---

[20] In this expression, Belevitch notation is used. That is, $S_{11} = h(p)/g(p)$ and $S_{12} = S_{21} = f(p)/g(p)$.

6.11 Transfer Scattering Parameters

Most communication systems are constructed by tandem connection of subsystems. Each of these may be described by means of scattering parameters. In order to assess the overall performance such as TPG, we should be able to generate the scattering parameters of the complete structure. Thus, the transfer scattering parameters are introduced to facilitate computation of the overall performance of communication systems as follows.

Referring to Figure 6.16, the transfer scattering parameters are defined in such a way that, at the input, reflected b_1 and incident a_1 waves are related to the output port's incident a_2 and reflected b_2 waves such that

$$\begin{bmatrix} b_1 \\ a_1 \end{bmatrix} = \begin{bmatrix} \tau_{11} & \tau_{12} \\ \tau_{21} & \tau_{22} \end{bmatrix} \begin{bmatrix} a_2 \\ b_2 \end{bmatrix}$$

or, in short,   (6.62)

$$\begin{bmatrix} b_1 \\ a_1 \end{bmatrix} = T \begin{bmatrix} a_2 \\ b_2 \end{bmatrix}$$

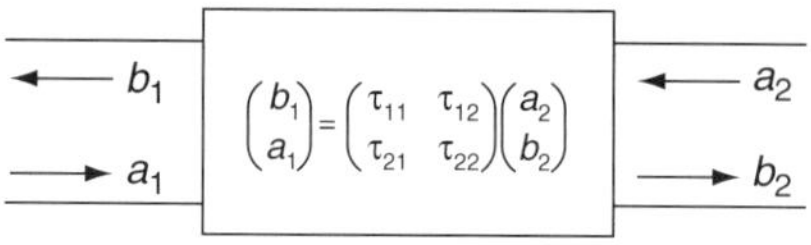

Figure 6.16   Definition of transfer scattering parameters

From Equation (6.62), the relation between scattering parameters and transfer scattering parameters can easily be derived.

To recap, scattering parameters were defined as

$$b_1 = S_{11}a_1 + S_{12}a_2$$

which is equal to   (6.63)

$$b_1 = \tau_{11}a_2 + \tau_{12}b_2$$

A comparison of these equations reveals that one needs to express a_1 in terms of a_2. On the other hand,

$$b_2 = S_{21}a_1 + S_{22}a_2$$

or

$$a_1 = \frac{b_2 - S_{22}a_2}{S_{12}} = -\frac{S_{22}}{S_{21}}a_2 + \frac{1}{S_{21}}b_2$$

or

$$b_1 = \left(S_{12} - \frac{S_{11}S_{22}}{S_{21}} \right)a_2 + \left(\frac{S_{11}}{S_{21}} \right)b_2$$

Thus   (6.64)

$$\tau_{11} = S_{12} - \frac{S_{11}S_{22}}{S_{21}} \qquad \tau_{21} = \frac{S_{11}}{S_{21}}$$

and

$$\tau_{21} = -\frac{S_{22}}{S_{21}} \qquad \tau_{22} = \frac{1}{S_{21}}$$

It is interesting to note that, in Equation (6.64), τ_{22} is directly related to S_{21} which is the measure of TPG. In other words, for an arbitrary two-port, TPG is given by

$$T(\omega) = \left| \frac{1}{\tau_{22}(j\omega)} \right|^2 \tag{6.65}$$

6.12 Cascaded (or Tandem) Connections of Two-ports

Referring to Figure 6.17, let us consider the cascaded connection of two-ports $[N_1]$ and $[N_2]$.

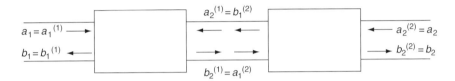

Figure 6.17   Tandem connection of two-ports

Let $T^{(1)}$ and $T^{(2)}$ be the transfer scattering parameters of $[N_1]$ and $[N_2]$ respectively. Then,

$$\begin{bmatrix} b_1^{(1)} \\ a_1^{(1)} \end{bmatrix} = \begin{bmatrix} \tau_{11}^{(1)} & \tau_{12}^{(1)} \\ \tau_{21}^{(1)} & \tau_{22}^{(1)} \end{bmatrix} \begin{bmatrix} a_2^{(1)} \\ b_2^{(1)} \end{bmatrix} = T^{(1)} \begin{bmatrix} a_2^{(1)} \\ b_2^{(1)} \end{bmatrix} \tag{6.66}$$

and

$$\begin{bmatrix} b_1^{(2)} \\ a_1^{(2)} \end{bmatrix} = \begin{bmatrix} \tau_{11}^{(1)} & \tau_{12}^{(2)} \\ \tau_{21}^{(2)} & \tau_{22}^{(2)} \end{bmatrix} \begin{bmatrix} a_2^{(2)} \\ b_2^{(2)} \end{bmatrix} = T^{(2)} \begin{bmatrix} a_2^{(2)} \\ b_2^{(2)} \end{bmatrix}$$

From Figure 6.17, it can be observed that the cascade connection of $[N_1]$ and $[N_2]$ reveals that

$$\begin{bmatrix} a_2^{(1)} \\ b_2^{(1)} \end{bmatrix} = \begin{bmatrix} b_1^{(2)} \\ a_1^{(2)} \end{bmatrix} T^{(2)} \begin{bmatrix} a_2^{(2)} \\ b_2^{(2)} \end{bmatrix} \tag{6.67}$$

Thus, inserting Equation (6.67) into Equation (6.66), we obtain

$$\begin{bmatrix} b_1^{(1)} \\ a_1^{(1)} \end{bmatrix} = T^{(1)} T^{(2)} \begin{bmatrix} a_2^{(2)} \\ b_2^{(2)} \end{bmatrix} = T \begin{bmatrix} a_2^{(2)} \\ b_2^{(2)} \end{bmatrix} \tag{6.68}$$

Therefore, we conclude that overall the transfer scattering matrix T of the network $[N]$ which is formed on the cascaded connection of $[N_1]$ and $[N_2]$ is given by

$$T = \begin{bmatrix} \tau_{11} & \tau_{12} \\ \tau_{21} & \tau_{22} \end{bmatrix} = T^{(1)}T^{(2)}$$

or

$$\tau_{11} = \tau_{11}^{(1)}\tau_{11}^{(2)} + \tau_{12}^{(1)}\tau_{21}^{(2)} \quad \tau_{12} = \tau_{11}^{(1)}\tau_{11}^{(2)} + \tau_{12}^{(1)}\tau_{21}^{(2)}$$

$$\tau_{21} = \tau_{21}^{(1)}\tau_{11}^{(2)} + \tau_{22}^{(1)}\tau_{21}^{(2)} \quad \tau_{22} = \tau_{21}^{(1)}\tau_{12}^{(2)} + \tau_{22}^{(1)}\tau_{22}^{(2)}$$

(6.69)

At this point, it is interesting to evaluate $\tau_{22} = 1/S_{21}$ of the composite structure. Since

$$\tau_{21}^{(1)} = -\frac{S_{22}^{(2)}}{S_{21}^{(1)}} \quad \tau_{12}^{(2)} = -\frac{S_{11}^{(2)}}{S_{21}^{(2)}} \quad and \quad \tau_{22}^{(1)} = \frac{1}{S_{21}^{(1)}} \quad \tau_{22}^{(2)} = \frac{1}{S_{21}^{(2)}}$$

Then,

$$\tau_{22} = \frac{1 - S_{22}^{(1)}S_{11}^{(2)}}{S_{21}^{(1)}S_{21}^{(2)}}$$

or, equivalently,

$$S_{21} = \frac{S_{21}^{(1)}S_{21}^{(2)}}{1 - S_{22}^{(1)}S_{11}^{(2)}}$$

(6.70)

We should stress that Equation (6.70) is quite an important equation since it directly reveals the TPG $T_c(\omega)$ of the composite structure:

$$T_c(\omega) = \left| \frac{S_{21}^{(1)}S_{21}^{(2)}}{1 - S_{22}^{(1)}S_{11}^{(2)}} \right|^2$$

(6.71)

Assume that $[N_1]$ and $[N_2]$ are lossless reciprocal two-ports and that they possess the scattering matrices in Belevitch canonical form such that

$$S_{11}^{(1)} = \frac{h_1}{g_1} \quad S_{12}^{(1)} = S_{21}^{(1)} = \frac{f_1}{g_1} \quad S_{22}^{(1)} = -\frac{f_{1*}}{f_1}\frac{h_{1*}}{g_1} = -\mu_1\frac{h_{1*}}{g_1}$$

and

$$S_{11}^{(2)} = \frac{h_2}{g_2} \quad S_{12}^{(2)} = S_{21}^{(2)} = \frac{f_2}{g_2} \quad S_{22}^{(2)} = -\frac{f_{2*}}{f_2}\frac{h_{2*}}{g_2} = -\mu_2\frac{h_{2*}}{g_2}$$

(6.72)

Then,

$$TC(p) = S_{21}(p)S_{21}(-p)$$

$$= \frac{f_1 f_{1*} f_2 f_{2*}}{(g_1 g_2 + \mu_1 h_{1*} h_2)(g_{1*} g_{2*} + \mu_1 h_1 h_{2*})}$$

$$= \frac{f f_*}{g g_*} \tag{6.73}$$

where

$$\mu_1 = +1 \text{ if } f_1(p) \text{ is even}$$

$$\mu_1 = -1 \text{ if } f_1(p) \text{ is odd}$$

Equation (6.71) and Equation (6.73) are quite important for the design of matching networks. Therefore, it is appropriate to make the following comments.

6.13 Comments

The composite structure $[N]$ must include all the transmission zeros of $[N_1]$ and $[N_2]$.

If $[N_1]$ and $[N_2]$ are reciprocal lossless two-ports having scattering matrices in the Belevitch canonical form, then $[N]$ must also have its scattering matrix in Belevitch form:

$$S_{11} = \frac{h}{g} \quad S_{12} = S_{21} = \frac{f}{g} \quad S_{22} = -\mu \frac{h_*}{g}$$

Equation (6.73) demands that $f(p) = f_1(p)f_2(p)$.

It should be emphasized that $f_1(p)$ and $f_2(p)$ are proper. Therefore, $f(p) = f_1(p)f_2(p)$ is proper.

The denominator polynomial $g(p)$ can be constructed on the closed LHP roots of the polynomial $G(p)$ which is specified as

$$G(p) = (g_1 g_2 + \mu_1 h_{1*} h_2)(g_{1*} g_{2*} + \mu_1 h_1 h_{2*})$$

Similarly, the numerator polynomial $h(p)$ is constructed on the explicit factorization of

$$H(p) = G(p) - F(p) = (g_1 g_2 + \mu_1 h_{1*} h_2)(g_{1*} g_{2*} + \mu_1 h_1 h_{2*}) - f_1 f_{1*} f_2 f_{2*}$$

where

$$F(p) = f_1 f_{1*} f_2 f_{2*}$$

TPG of the 'Tandem Connection of m Sections'

Introduction of the transfer scattering matrix (TSM) greatly facilitates computation of TPG of the tandem connection of m sections.

Referring to Figure 6.18, we can generate TPG of the m sections step by step in a sequential manner.

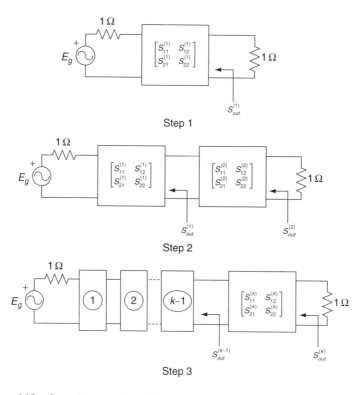

Figure 6.18 Cascade connection of m sections in a sequential manner (step by step)

At step 1, the procedure is initialized. Here, we take section 1 and set

$$T_1 = \left| S_{21}^{(1)} \right|^2$$

and   (6.74)

$$S_{out}^{(1)} = S_{22}^{(1)}$$

At step 2, the second section is connected to section 1. Then, we set

$$T_2 = T_1 \frac{|S_{21}^{(2)}|^2}{|1 - S_{out}^{(1)} S_{11}^{(2)}|^2}$$

(6.75)

$$S_{out}^{(2)} = S_{out}^{(1)} + \frac{S_{12}^{(2)} S_{21}^{(2)}}{1 - S_{out}^{(1)} S_{11}^{(2)}} S_{11}^{(2)}$$

At step k,

$$
\boxed{
\begin{aligned}
T_k &= T_{(k-1)} \frac{|S_{21}^{(k)}|^2}{|1 - S_{out}^{(k-1)} S_{11}^{(k)}|^2} \\[2em]
S_{out}^{(k)} &= S_{out}^{(k-1)} + \frac{S_{12}^{(k)} S_{21}^{(k)}}{1 - S_{out}^{(k-1)} S_{11}^{(k)}} S_{11}^{(k)}
\end{aligned}
}
\qquad (6.76)
$$

Hence, the above steps continues for $k = 3, 4, \ldots$ up to $k = m$.
The last step will be $k = m$. At this step, the mth section is connected to the $(m-1)^{\text{th}}$ section. Thus,

$$
T_m = T_{(m-1)} \frac{|S_{21}^{(m)}|^2}{|1 - S_{out}^{(m-1)} S_{11}^{(m)}|^2}
$$

which halfs the sequential procedure.

6.14 Generation of Scattering Parameters from Transfer Scattering Parameters

It is straightforward to show that the scattering parameters of a two-port can be generated from the transfer scattering parameters as follows:

$$
\boxed{
\begin{aligned}
S_{11} &= \frac{\tau_{12}}{\tau_{22}} \qquad S_{12} = \frac{\tau_{11}\tau_{22} - \tau_{12}\tau_{21}}{\tau_{22}} \\[1.5em]
S_{21} &= \frac{1}{\tau_{22}} \qquad S_{22} = -\frac{\tau_{21}}{\tau_{22}}
\end{aligned}
}
\qquad (6.77)
$$

The above form will be useful for generating idealized scattering parameters for equalizers in various matching problems as will be shown in Chapter 10.

7

Numerical Generation of Minimum Functions via the Parametric Approach

7.1 Introduction

One of the major problems of circuit design is the generation of proper positive real functions to optimize the predefined system performance. In this regard, we can utilize minimum functions. In this case, generation of a positive real function becomes straightforward. A minimum function must have all its poles in the open left half plane. Corresponding residues must possess positive real parts. In other words, the function $F(p)$

$$F(p) = F_0 + \sum_{j=1}^{n} \frac{k_j}{p - p_j} \tag{7.1}$$

is positive real if

- the poles $p = p_j$ are all in the left half plane (LHP);
- the real constant F_0 is non-negative; and
- Even part of $F(p)$ is non-negative.

This form of a positive real function is called the 'Bode form[1]' or the 'parametric representation' of the minimum function $F(p)$. Obviously, one can always generate a minimum reactance function $Z(p) = R(p^2) + \text{odd}(Z)$ or a minimum susceptance function $Y(p) = G(p^2) + \text{odd}(Y)$ by means of a parametric representation of Equation (7.1) if the even parts $R(p^2)$ and $G(p^2)$ are known. Equation (7.1) is very practical for generating positive real functions on a computer.

---

[1] H.W. Bode, '*Network Analysis and Feedback Amplifier Design*,' Van Nostrand, 1945.

Design of Ultra Wideband Power Transfer Networks   Binboga Siddik Yarman
© 2010 John Wiley & Sons, Ltd

7.2 Generation of Positive Real Functions via the Parametric Approach using MATLAB®

In general, the real part of a minimum immittance function is specified as an even rational function in the frequency domain either in 'all-pole' or in open form as

$$R(\omega^2) = \frac{A(\omega^2)}{B_1(\omega^2 + p_1^2)(\omega^2 + p_2^2)\ldots(\omega^2 + p_n^2)} = \frac{A(\omega^2)}{B(\omega^2)} \geq 0 \qquad (7.2)$$

where $\{p_1, p_2, \ldots, p_n\}$ are the strictly LHP (or open LHP) poles of $F(p)$,[2] and

$$A(\omega^2) = A_1\omega^{2n} + A_2\omega^{2(n-1)} + \ldots + A_n\omega^2 + A_{n+1} \geq 0$$
$$B(\omega^2) = B_1\omega^{2n} + B_2\omega^{2(n-1)} + \ldots + B_n\omega^2 + B_{n+1} > 0 \qquad (7.3)$$

are even polynomials in ω with coefficients $\{B_1 > 0 \text{ and } B_{n+1)} > 0\}$ and $\{A_1 \geq 0 \text{ and } A_{n+1} \geq 0\}$. Obviously, $R(\omega^2)$ must be finite and non-negative.

The above form can be expressed in the p domain by replacing ω^2 with $-p^2$ as follows:

$$R(-p^2) = \frac{A(-p^2)}{(-1)^n B_1(p^2 - p_1^2)(p^2 - p_2^2)\ldots(p^2 - p_n^2)}$$
$$= \frac{A(-p^2)}{B(-p^2)} \qquad (7.4)$$

On the other hand,

$$R(-p^2) = \frac{1}{2}[F(p) + F(-p)] = F_0 + \sum_{i=1}^{n} \frac{k_i p}{p^2 - p_i^2}$$

where

$$F_0 = \frac{A_1}{B_1} \geq 0$$
$$k_i = (-1)^n \frac{p^2 - p_i^2}{p} R(-p^2) \mid_{p = p_i}$$

or

$$k_i = (-1)^n \frac{A(p_i^2)}{p_i B_1 \prod_{\substack{i=1 \\ i \neq j}}^{n}(p_i^2 - p_j^2)}, \quad i = 1, 2, \ldots, n \qquad (7.5)$$

Once the residues k_i are determined, the minimum function $F(p)$ can be generated in a straightforward manner using MATLAB as described in the following example.

---

[2] Open LHP does not include the $j\omega$ axis.

Example 7.1: Let the real part of an admittance function be

$$G(\omega^2) = \frac{1}{\omega^4 + \omega^2 + 1}$$

which is always non-negative for all values of the angular frequency ω.

(a) Using MATLAB, generate the analytic form of a minimum susceptance admittance function $Y(p)$, from $G(\omega^2)$.
(b) Synthesize the admittance function $Y(p)$ as a lossless two-port in resistive termination.

Solution:

(a) Let us develop an algorithmic approach to generate $Y(p)$ step by step, employing MATLAB.

Step 1 (Generation of the even part of function $G(p^2) = A(p^2)/B(p^2)$ in the complex p domain):
Parametric representation of the admittance function $Y(p)$ is specified by Equation (7.1) as

$$F(p) = Y(p) = Y_0 + \sum_{j=1}^{n} \frac{k_j}{p - p_j}$$

For this problem, the real part of the admittance function is $G(\omega^2) = 1/(\omega^4 + \omega^2 + 1)$ which specifies the integer $n = 2$.

By substituting $\omega^2 = -p^2$, we find the numerator polynomial $A(p^2) = 1$ with $\{A_1 = 0, A_2 = 0, A_3 = 1\}$ and $B(p^2) = B_1 p^4 + B_2 p^2 + 1$ with $\{B_1 = 1, B_2 = -1, B_3 = 1\}$.

Hence, we will be able to generate residues

$$k_j = (-1)^n \frac{A(p_j^2)}{p_j B_1 \prod_{\substack{i=1 \\ i \neq j}}^{n} (p_j^2 - p_i^2)} \quad \text{and} \quad Y_0 = \frac{A_1}{B_1} = 0$$

Step 2 (Computation of LHP poles of $Y(p)$: Once $G(p^2) = A(p^2)/B(p^2)$ is generated, we can find the LHP poles p_i of $B(p^2)$.
Here we set $x = p^2$ and $B = [1 \ -1 \ 1]$ to compute the roots of $B(x)$ using MATLAB as $>>$
$x = roots(B)$ Thus,

```
X(1)=0.5000 + 0.8660i
X(2)=0.5000 - 0.8660i
```
Then, by setting $p = - sqrt(x)$ we find the LHP poles as

$$p_1 = -0.8660 - 0.5000i$$
$$p_2 = -0.8660 + 0.5000i$$

It should be noted that, in MATLAB, the roots of $B(p^2)$ can be directly computed from the augmented polynomial vector B. In this case, the full polynomial form of B is utilized by inserting zeros for the missing odd terms. In other words, let BA be the augmented B vector. Then BA $= [1\ 0\ -1\ 0\ 1]$ and the full roots are obtained as
```
roots(BA)
ans =
 0.8660 + 0.5000i
 0.8660 - 0.5000i
-0.8660 + 0.5000i
-0.8660 - 0.5000i
```

Step 3 (Computation of residues): Using Equation (7.5), we compute

$$k_1 = (-1)^n \frac{A(p_1^2)}{p_1 B_1 \prod_{\substack{i=1 \\ i \neq j}}^{n} (p_1^2 - p_i^2)} = \frac{1}{p_1 . 1 . (p_1^2 - p_2^2)} = 0.2887 + 0.5000i$$

$$k_2 = (-1)^n \frac{A(p_1^2)}{p_1 B_1 \prod_{\substack{i=1 \\ i \neq j}}^{n} (p_1^2 - p_i^2)} = \frac{1}{p_2 . 1 . (p_2^2 - p_1^2)} = 0.2887 - 0.5000i$$

Note that $real\{k_1\} = real\{k_2\} = 0.2887$ and $k_1 = k_2^*$ as expected.

Step 4 (Computation of analytic form of $Y(p) = Y_0 + \Sigma_{j=1}^{n} k_j/(p - p_j) = a(p)/b(p)$):

$$Y(p) = \frac{k_1}{p - p_1} + \frac{k_2}{p - p_2} = \frac{p(k_1 + k_2) - (k_1 p_1 + k_2 p_2)}{p^2 - (p_1 + p_2) - p_1 p_2} = \frac{a_1 p + a_2}{b_1 p^2 + b_2 p + b_3}$$

Here, $a_1(k_1 + k_2) = 0.5774$, which is purely real as expected, and $a_2 = -(k_1 p_2 + k_2 p_1) = 1$; also $b_1 = 1, b_2 = -(p_1 + p_2) = 1.7321, b_3 = p_1 p_2 = 1$. Thus,

$$Y(p) = \frac{0.5774p + 1}{p^2 + 1.7321p + 1}$$

(b) To synthesize $Y(p)$, firstly, it would be appropriate to write the impedance $Z(p) = 1/Y(p)$. Then, using simple expansion and long division on $Z(p)$, we have

$$Z(p) = \frac{p^2 + 1.7321p + 1}{0.5774p + 1} = \left[\frac{1}{0.5774}\right] p + \frac{1}{0.5774p + 1} = 1.7321p + \frac{1}{0.5774p + 1}$$

Hence, the impedance $Z(p)$ starts with a series inductor of $L_1 = 1.7321 = 1/0.5774$ and ends with a shunt capacitor of $C_2 = 0.5774$ in parallel with a 1 S conductance as shown in Figure 7.1.

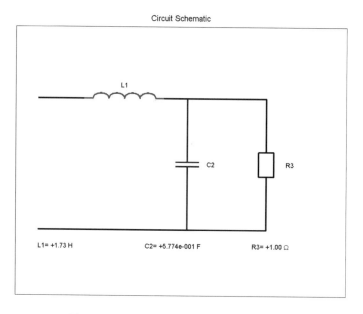

Circuit Schematic

Figure 7.1   Synthesis of $Y(p)$ for Example 7.1

In order to formalize the above computational steps, we will propose a MATLAB algorithm to generate minimum functions via the parametric approach. However, firstly, it would be useful to review some major polynomial operations in MATLAB.

7.3 Major Polynomial Operations in MATLAB®

The Computation of residues and generation of minimum functions in rational form involve product evaluations and polynomial multiplications in MATLAB. These operations must be performed with special care. Let us review these operations by means of examples run on MATLAB.

Example 7.2: Let $A(\omega^2) = 3\omega^6 + 4\omega^4 + 5\omega^2 + 1$ and $B(\omega^2) = \omega^6 + 2\omega^4 + \omega^2 + 1$ be two polynomials.

(a) Express the above polynomials in MATLAB.
(b) Generate the polynomial $C(\omega^2) = A(\omega^2) + B(\omega^2)$ on MATLAB.
(c) Generate the polynomial $C(\omega^2) = A = (\omega^2)B(\omega^2)$ on MATLAB.

Solution:

(a) In MATLAB, polynomials are represented by vectors with ascending coefficient orders. Thus, setting $x = \omega^2$, polynomials A and B are specified as MATLAB row vectors such that A = [3 4 5 1] and B = [1 2 1 1], meaning that $A(x) = 3x^3 + 4x^2 + 5x + 1$ and $B(x) = x^3 + 2x^2 + x + 1$.
(b) In MATLAB, two row vectors can be added in a straightforward manner if they have the same dimensions. Thus, C = A + B = [4 6 6 2], which means that $C(x) = 4x^3 + 6x^2 + 6x + 2$. Here, we should note that MATLAB does not recognize an even polynomial. Rather, it treats full coefficient polynomials. Therefore, in this example, we simplified the computations by setting $x = \omega^2$.
(c) Considering the general forms of polynomials $A(x) = \sum_{k=1}^{m+1} A_j x^{(m+1-j)}$ of degree m and $B(x) = \sum_{j=1}^{n+1} B_j x^{(n+1-j)}$ of degree n, coefficients C_i of $C(x) = A(x)B(x) = \sum_{k=1}^{m+n+1} C_k x^{(m+n+1-i)}$ are given by

$$C_k = \sum_{j=1}^{k} A_j B_{k+1-j}, k = 1, 2 \ldots, (m+n+1)$$

This equation is called the convolution operation and it is performed in MATLAB by a built-in function called C = conv(A, B).

Thus, MATLAB output of the above operation is given as
>> C = conv(A, B)
C = [3 10 16 18 11 6 1]
meaning that

$$C(x) = 3x^6 + 10x^5 + 16x^4 + 18x^3 + 11x^2 + 6x + 1$$

We should keep in mind that $x = \omega^2$. Therefore,

$$C(\omega) = 3\omega^{12} + 10\omega^{10} + 16\omega^8 + 18\omega^6 + 11\omega^4 + 6\omega^2 + 1$$

The MATLAB standard convolution function does not handle matrix row and column vectors with indices; rather it operates on single row vectors. Therefore, using the convolution operation, nested loops must be handled with special care.

For example, when multiplying simple polynomials `[p-p(1)][p-p(2)]...[p-p(n)]`, firstly we define a unity polynomial as `pr = [1]`. Then, a MATLAB loop is initiated employing the MATLAB loop expression `'for...end'` where we define a simple polynomial as `simple = [1 -p(j)]`.

Within this loop, continuous multiplication of polynomials is performed using the convolution function in a nested manner as `pr = conv(pr, simple)`

Thus, we implement the following code to perform polynomial multiplications in MATLAB:

Program Listing 7.1   MATLAB polynomial multiplication

```
pr=[1]; %Initialize the polynomial multiplication with unitary
polynomial pr=1.
for j=1:n
  simple=[1 -p(j)];%this is the simple polynomial [p-p(j)]
% Generate multiplication of polynomials starting with pr=1.
  pr=conv(pr,simple);
end
```

In this MATLAB code, roots `p(j)` can be real or complex. Obviously, a complex root `p(j)` must be accompanied by its conjugate to yield a real coefficient polynomial as it is specified by the row vector `pr`. In general, results of the convolution operation include complex entries in the row vector `pr`. However, in our case, the imaginary parts of all the complex entries must be zero since we always deal with real polynomials and complex conjugate root pairs along with real roots.

Let us verify the above statement by a simple example.

Example 7.3: Let $p(1) = -1.3247$, $p(2) = -0.6624 - 0.5623i$, $p(3) = -0.6624 + 0.5623i$

Find the full coefficient form of `[p-p(1)][p-p(2)][p-p(3)]`.

Solution: For this purpose we wrote a simple test program to implement the above code as listed below:

Program Listing 7.2   Program `Test_product.m`

```
% Test_product
clear
p=[-1.3247 -0.6624 - 0.5623i -0.6624 + 0.5623i]
%
pr=[1]; %Define unitary polynomial pr=1.
n=3
for j=1:n
```

```
  simple=[1 -p(j)];%this is the simple polynomial [p-p(j)]
% Generate multiplication of polynomials starting with pr=1.
  pr=conv(pr,simple)
end
```

Execution of the above program yields
`Pr=[ 1.0000  2.6495  2.5099 + 0.0000i  1.0001 - 0.0000i]`
meaning that

$$pr(x) \approx x^3 + 2.6495p^2 + 2.5099p + 1$$

As we can see from this example, the imaginary parts of the entries of `pr` are all equal to zero or negligibly small as expected.

7.4 Algorithm: Computation of Residues in Bode Form on MATLAB®

Equation (7.1) specifies the Bode form or so-called 'open parametric form' of a minimum function $F(p)$ in terms of residues and pole pairs $\{k_i, p_i; i = 1, 2, \ldots n\}$.

In order to carry out numerical derivations with conventional nomenclature without loss of generality, let

$$F(p) = Z(p) = Z_o + \sum_{j=1}^{n} \frac{k_j}{p - p_j}$$

be a minimum reactance impedance function. On the $j\omega$ axis it is specified in terms of its real and imaginary parts as

$$Z(j\omega) = R(\omega^2) + jX(\omega)$$

Assume that real part $R(\omega^2) = A(\omega^2)/B(\omega^2)$ is specified by means of MATLAB row vectors

$$A = [A(1) A(2) A(3) \ldots A(n) A(n+1)]$$

and

$$B = [B(1) B(2) B(3) \ldots B(n) B(n + 1)]; B(1) > 0$$

Here, we wish to generate the rational form of the impedance function

$$Z(p) = \frac{a(p)}{b(p)} = \frac{a_1 p^n + a_2 p^{(n-1)} + \ldots + a_n + a_{n+1}}{b_1 p^n + b_2 p^{(n-1)} + \ldots + b_n + b_{n+1}}$$

from its given real part $R(\omega^2) = A(\omega^2)/B(\omega^2)$ by developing a MATLAB algorithm. Actually, in Example 7.1, computations are described step by step. In this section, we will generalize the above computational steps within a MATLAB program consisting of proper MATLAB functions.

Step 1 (p domain representation of $R(\omega^2) = A(\omega^2)/B(\omega^2)$): In this step we replace ω^2 by $-p^2$ to generate $R(-p^2) = A(-p^2)/B(-p^2)$.
This is accomplished by a MATLAB function called `polarity`, developed by us.

Example 7.4: Let $A = \begin{bmatrix} 0 & 00 & 1 \end{bmatrix}$ and $B = \begin{bmatrix} 1 & 21 & 1 \end{bmatrix}$. Find the p domain version of A and B.

Solution: In MATLAB, we generate a test program called `Test_polarity`:

Program Listing 7.3   Program `Test_polarity.m`

```
% Test_Polarity
A=[0 0 0 1]; B=[1 2 1 1];
A=polarity(A)
B=polarity(B)
```

Then, we run this program to get p domain versions of vectors A and B as

$$A = \begin{bmatrix} 0 & 0 & 0 & 1 \end{bmatrix} \text{ and } B = \begin{bmatrix} -1 & 2 & -1 & 1 \end{bmatrix}$$

Then, we set $Z_0 = (A(n+1))/(B(n+1)) = 0$. It should be noted that $n1 = (n+1)$ is found using the MATLAB built-in function called `length`. Thus,

$$n1=length(B)=4$$

which corresponds to $n = 3$.
In the following we provide the listing of function `polarity`.

Program Listing 7.4   Function `polarity.m`

```
function B=polarity(A)
% This function changes the sign of the coefficients for a given
polynomial
% Polynomial A is given as an even polynomial in w domain.
% By setting p=-w^2 we compute the coefficients of the new even
polynomial in p domain.
% Polynomial A is an even polynomial in w domain
%P(w^2)=A(1)[w^(2n)] + A(2)[w^2(n-1)] + A(3)[w^2(n-2) + ... +
A(n) [w] + A(n + 1)
% Inputs
%--------------- MATLAB Vector A=[A(1) A(2) A(3)...A(n) A(n + 1)] :
%               Coefficients of the even polynomial in w domain
% Outputs
%--------------- MATLAB Vector B=[B(1) B(2) B(3) ...B(n) B(n + 1)
sigma=-1;
NN=length(A);
n=NN-1;
for i=1:NN
  j=NN-i + 1;
  B(j)=-sigma*A(j);
  sigma=-sigma;
end
```

Step 2 (Computation of residues): Now we have the p domain version of $R(-p^2)$, the following actions must be performed to determine the LHP poles of $R(-p^2)$, and then computations of the residues at these poles.

For this purpose, we programmed Equation (7.5) under the function `residue_Z0`. In order to test function `residue_Z0`, we developed a simple program called `Test_Residue`. The function `residue_Z0(A,B)` takes the frequency domain vectors of A and B.

Example 7.5: Let us run `Test_Residue` for $A = [0\ 0\ 0\ 1]$ and $B = [1\ 2\ 1\ 1]$ which are specified in the frequency domain.

The listing of `Test_Residue` is given as follows.

Program Listing 7.5

```
% Test_Residue
A=[0 0 0 1];%input specified in w-domain.
B=[1 2 1 1];%input specified in w-domain.
[p,k]=residue_Z0(A,B)
```

The results of the above program is given as

$$p = -1.3247 \quad -0.6624 - 0.5623i \quad -0.6624 + 0.5623i$$

$$k = 0.2345 \quad 0.1172 + 0.4143i \quad 0.1172 - 0.4143i$$

We now provide the listing of the function `residue_Z0(A,B)`.

Program Listing 7.6   function `residue_Z0.m`

```
function [p,k]=residue_Z0(A,B)
% This function computes the residues k of a given real part
R(w^2)
% R(w^2)=A(w^2)/B(w^2)
% It should be noted the R(w^2) is given in w domain as an even
function of
% w rather than complex variable p=sigma + jw.
% Inputs:
% ------- A(w^2); numerator polynomial
% ------- B(w^2); denominator polynomial
%Step 1: Find p domain versions of A and B
% That means we set w^2=-p^2
AA=polarity(A);
BB=polarity(B);
% Step 2: Find the LHP poles of R(p^2)
x=roots(BB);
p=-sqrt(x);
```

Program Listing 7.6   (Continued)

```
% Step 3: Compute the product terms
prd=product(p);
n=length(p);
% Step4: Generate residues
for j=1:n
  y=p(j)*p(j);
  Aval=polyval(AA,y);
  k(j)=(-1)^n*Aval/p(j)/B(1)/prd(j);
end
```

Step 3 (Computation of the rational form of $Z(p)$: Once the residues are computed, the rational form of $Z(p)$ can be generated using Equation (7.1). This is a straightforward evaluation of both numerator $a(p)$ and denominator polynomials. In this case,

$$b(p) = \prod_{i=1}^{n} (p - p_1) = p^n + b_2 p^{(n-1)} + \ldots + b_n p + b_{n+1} \tag{7.6}$$

Numerator polynomial $a(p)$ can also be evaluated as

$$a(p) = Z_0 b(p) + \sum_{i=1}^{n} \prod_{\substack{j=1 \\ i \neq j}}^{n} k_i (p - p_i)$$
$$= a_1 p^n + a_2 p^{(n-1)} + \ldots + a_n p + a_{n+1} \tag{7.7}$$

Equation (7.6) is implemented within MATLAB functions denom and it is tested by means of MATLAB program Test_denom.

Example 7.6: Let the real part $R(\omega^2) = A(\omega^2)/B(\omega^2)$ be specified as A = [0 0 0 1] and B = [1 2 1 1]. Find the denominator polynomial of $Z(p) = a(p)/b(p)$.

Solution: The following test program is implemented to compute the denominator polynomial b(p) = denom.

Program Listing 7.7   Test_denom.m

```
% Test_denom
A=[0 0 0 1]
B=[1 2 1 1]
[p,k]=residue_Z0(A,B)
[denom,errord]=denominator(p)
```

In this program all the computations are carried using complex algebra. Therefore, coefficients of the denominator are found as complex numbers. However, their imaginary parts must be zero. In function denom, the imaginary parts of the coefficients are collected in the row vector errord. Vector **b** designates the real part of the computed polynomial coefficients of Equation (7.6). Thus, execution of Test_denom yields the following results:

$$\text{denom} = [1.000 \quad 2.6494 \quad 2.5098 \quad 1.000]$$
$$\text{errord} = 1.0e\text{-}015 * [0 \quad 0 \quad -0.2220 \quad -0.0555]$$

It is observed that the imaginary parts of the coefficients are in the order of 10^{-15} as specified by the vector errod. Thus, the coefficients of the denominator polynomial are collected under the vector denom meaning that

$$b(p) = p^3 + 2.694p^2 + 2.5098p + 1$$

Now, let us list function denominator.

Program Listing 7.8:   Function denominator.m

```
function [denom,errord]=denominator(p)
% This function computes the denominator polynomial of an
% immitance function from the given poles.
% It should be noted that this form of the denominator is nor-
% malized with
% the leading coefficient b(1)=1: denom=(p-p(1))(p-p(2))....(p-
% p(n))
% Inputs:
%------ poles p(i) as a MATLAB row vector p
%------ Residues k(i) of poles p(i)
% Output:
%------ denom; MATLAB Row-Vector
% which includes coefficients of the denominator polynomial of
% an immitance function.
% denom=product[p-p(i)]
%------ Step 1: Determine n
n=length(p);
%------ Step 2: Form the product term.
pr=[1]; %Define a simple "unity" polynomial of pr=1.
for j=1:n
simple=[1 -p(j)];%this is the simple polynomial [p-p(j)]
% Generate multiplication of polynomials starting with 1.
pr=conv(pr,simple);
end
denom=real(pr);
errord=imag(pr);
```

Similarly, Equation (7.7) is implemented under the MATLAB function num_Z0 and tested by means of program Test_num which is listed below.

Example 7.7: Let the real part $R(\omega^2) = A(\omega^2)/B(\omega^2)$ be specified as $A = [0\ 0\ 0\ 1]$ and $B = [1\ 2\ 1\ 1]$. Find the numerator polynomial of $Z(p) = a(p)/b(p)$.

Solution: For this purpose we developed the following MATLAB code.

Program Listing 7.9 Program `Test_num.m`

```
% Test_num
A=[0 0 0 1]
B=[1 2 1 1]
[p,k]=residue_Z0(A,B)
[denom,errord]=denominator(p)
b=sqrt(B(1))*denom
Z0=A(1)/B(1)
[num,errorn]=num_Z0(p,Z0,k)
```

Execution of `Test_num` yields the following results:

$$Z0 = 0$$
$$num = [0\quad 0.4690\quad 1.2425\quad 1.0000]$$
$$errorn = [0\quad 0\quad 0\quad 0]$$

meaning that

$$a(p) = 0.p^4 + 0.4690p^3 + 1.2425p^2 + 1$$

The row vector `errorn` represents the imaginary part of the numerator coefficients which is supposed to be zero. In fact it is a 'zero vector'.

In the following we list the function `num_Z0`.

Program Listing 7.10 Function `num_Z0.m`

```
    function [num,errorn]=num_Z0(p,Z0,k)
% This function computes the numerator polynomial of an
% immitance function: Z(p)=Z0 + sum{k(1)/[p-p(i)]}
% where we assume that Z0=A(1)/B(1)which is provided as input.
% Input:
%------ poles p(i) of the immitance function Z(p)
%as a MATLAB row vector p
%------ Residues k(i) of poles p(i)
% Output:
%------ num; MATLAB Row-Vector
% which includes coefficients of numerator polynomial of an
% immitance function.
% num=Sum{k(j)*product[p-p(i)]} which skips the term when j=i
%----- Step 1: Determine total number of poles n
n=length(p);
nn=n-1;
%
```

```
%----- Step 2: Generation of numerator polynomials:
% numerator polynomial=sum of
% sum of
% {Z0*[p-p(1)].[p-p(2)]...(p-p(n)]; n the degree-full product
% + k(1)*[p=p(2)].[p-p(3)]..[p-p(n)];degree of (n-1); the term
% with p(1)is skipped.
% + k(2)*[p-p(1)].[p-p(3)]..[p-p(j-1)].[p-p(j + 1)]..[p-p(n)];
%                                degree of(n-1)-the term
% with p(2)is % skipped
% ...............................................................
% + k(j)*[p-p(1)].[p-p(2)]..[p-p(j-1)].[p-p(j + 1)]..[p-p(n)];
%                                degree of (n-1)-the term
% with p(j)is skipped.
%...............................................................
% + k(n)[p-p(1)].[p-p(2)]...[p-p(n-1)];
%                                degree of (n-1)-the term
% with p(n) is skipped.
% Note that we generate the numerator polynomial within 4 steps.
% In Step 2a, product polynomial pra of k(1)is evaluated.
% In Step 2b, product polynomial prb of k(j)is evaluated by
% skipping the term when i=j.
% In Step 2c, product polynomial prc of k(n)is evaluated.
% In Step 2d, denominator of Z0 is generated.
% -------------------------------------------------
% Step 2a: Generate the polynomial for the residue k(1)
pra=[1];
for i=2:n
  simpA=[1 -p(i)];
% pra is a polynomial vector of degree n-1; total number of
entries are n.
pra=conv(pra,simpA);% This is an (n-1)th degree polynomial.
end
na=length(pra);
% store first polynomial onto first row of A i.e. A(1,:)
for r=1:na
  A(1,r)=pra(r);
end
% compute the product for 2<j<(n-1)
for j=2:nn
  prb1=[1];
    for i=1:j-1
    simpB=[1 -p(i)];
    prb1=conv(prb1,simpB);
    end
% Skip j th term
prb2=[1];
for i=(j + 1):n
  simpB1=[1 -p(i)];
  prb2=conv(prb2,simpB1);
```

Program Listing 7.10   (Continued)

```
end
prb=conv(prb1,prb2);
nb=length(prb);
% Store j polynomials on to j th row of A; i.e. A(j,:)
for r=1:nb
A(j,r)=prb(r);
end
    end
% Compute the product term for j=n
prc=[1];
for i=1:nn
  simpC=[1 -p(i)];
  prc=conv(prc,simpC);
end
nc=length(prc);
% store n the polynomial onto n the row of A(n,:)
for r=1:nc
A(n,r)=prc(r);
end
%-------------------------------------------------------------
for i=1:n
  for j=1:n
    C(i,j)=k(i)*A(i,j);
end
end
%------ Step 4: Generate the numerator as a MATLAB row vector.
for i=1:n
D(i)=0;%Perform the sum operation to compute numerator polyno-
mial
end
for j=1:n
  for r=1:n
  D(j)=D(j) + C(r,j);
  end;% Here is the numerator polynomial of length n.
end
[denom,errord]=denominator(p);
prd_n=Z0*denom; % this is n th degree polynomial vector with
length n + 1
a(1)=prd_n(1);
for i=2:(n + 1)
  a(i)=D(i-1) + prd_n(i);
end
%
num=real(a);
errorn=imag(a);
```

Note that in the function `denominator`, the leading term coefficient is 1. In other words, $b(1) = 1$. This is due to the end result of Equation (7.6). On the other hand, when we specify the real part $R(\omega^2)$ in the ω domain as

$$R(\omega^2) = \frac{A(\omega^2)}{B(\omega^2)} = \frac{A_1\omega^{2n} + A_2\omega^{2(n-1)} + \ldots + A_n\omega^2 + A_{(n+1)}}{B_1\omega^{2n} + A_2\omega^{2(n-1)} + \ldots + B_n\omega^2 + B_{(n+1)}}$$

It yields the even part of

$$Z(p) = \frac{a(p)}{b(p)} = \frac{\text{num}(p)}{\text{denom}(p)}$$

in the p domain as

$$R(p^2) = \frac{A(-p^2)}{b(p)b(-p)}$$

where

$$b(p) = \sqrt{B_1}[\text{denom}(p)]$$

and

$$a(p) = \sqrt{B_1}[\text{num}(p)]$$

Actually, the above representation of $a(p)$ and $b(p)$ does not change the result; rather, it is just a gimmick to obtain the even part $R(-p^2) = \frac{1}{2}[Z(p) + Z(-p)]$ as specified in the original statement.

Let us see this with an example.

Example 7.8: In this example, the rational form of $R(\omega^2) = A(\omega^2)/B(\omega^2)$ is specified with MATLAB vectors `A=[0 0 0 0 1]` and `B=[4 1 2 1 9]`. Find the minimum reactance impedance function $Z(p) = a(p)/b(p)$ from its given real part.

Solution: For this example, we implemented a simple test program called `Test_Immitance` which directly generates $a(p) = \sqrt{B_1}[\text{num}(p)]$ and $b(p) = B_1[\text{denom}(p)]$ and then, checks the end result by forming the even part

$$R(p^2) = \frac{AA}{BB} = \frac{1}{2}[Z(p) + Z(-p)]$$

$$= \frac{1}{2}\left[\frac{a(p)}{b(p)} + \frac{a(-p)}{b(-p)}\right] = \frac{0.5[a(p)b(-p) + a(-p)b(p)]}{b(p)b(-p)}$$

where $AA = 0.5[a(p)b(-p) + a(-p)b(p)]$ and $BB = b(p)b(-p)$.

Here is the listing of the Program

Program Listing 7.11 `Test_immitance.m`

```
% Test_Immitance
A=[0 0 0 0 1] ;% Input
B=[4 1 2 1 9 ];%Input
[p,k]=residue_Z0(A,B)
[denom,errord]=denominator(p)
b=sqrt(B(1))*denom; % OUTPUT-1
Z0=A(1)/B(1)
[num,errorn]=num_Z0(p,Z0,k)
a=sqrt(B(1))*num;% OUTPUT-2
%
% Check the result
a_p=polarity(a)
b_p=polarity(b)
AA=0.5*(conv(a,b_p) + conv(a_p,b)); % in p domain
BB=conv(b,b_p);                 % in p domain
% Convert AA and BB to w domain using function polarity:
AA=polarity(AA)
BB=polarity(BB)
```

Execution of this program yields the following results:

```
b=[ 2.0000  6.0511  8.9040  7.3773  3.0000]
             a=[0  0.1459  0.4414  0.5818  0.3333]
```

Finally, as a check step, from `a(p)` and `b(p)` we obtained AA and BB in the p domain as

$$AA=[0 \quad 0.0000 \quad 0 \quad 0.0000 \quad 0 \quad 0.0000 \quad 0 \quad 1.0000]$$
$$BB=[4.0000 \quad 0 \; {-1.0000} \quad 0 \quad 2.0000 \quad 0 \; {-1.0000} \quad 0 \quad 9.0000]$$

Here it is interesting to observe that, at the final step, we take the convolutions of the full polynomials $a(p)b(-p)$, $a(-p)b(p)$ and $b(p)b(-p)$ to compute $AA = 0.5[a(p)b(-p) + a(-p)b(p)]$ $= 0.5*[\mathtt{conv(a,b\_p)} + \mathtt{conv(a\_p,b)}]$ and $BB = \mathtt{conv(b,b\_p)}$. Of course, the resulting polynomials must also be in full coefficient form. Therefore, in the above representation, vectors AA and BB have zeros for the odd-power terms of p as expected. However, the odd-power terms can easily be cleaned by means of a simple code.

Example 7.9: Develop a MATLAB function which clears the odd-power terms of a given polynomial and test your results for the above example.

Solution: The MATLAB program `Test_clearoddpower` performs the desired operation together with the MATLAB function `clear_oddpower`.

Here is the main program:

Program Listing 7.12 `Test_clearoddpower.m`

```
% Test_clearoddpower
% Let AA and BB be two-polynomials specified in p-domain with
% full
% coefficients.
% This program clears the odd power terms of given polynomials
% AA and BB
% Then converts the results to w-domain.
clear
AA=[ 0  0  0.0000  0  0.0000  0  0.0000  0  1.0000]
BB=[ 4.0000  0 -1.0000  0  2.0000  0 -1.0000  0  9.0000]
AT=clear_oddpower(AA)
BT=clear_oddpower(BB)
ACheck=polarity(AT)
BCheck=polarity(BT)
```

Execution of the above program yields the following results:

$$AA=[0\ 0\ 0\ 0\ 0\ 0\ 0\ 0\ 1]$$
$$BB=[4\ 0\ -\ 1\ 0\ 2\ 0\ -\ 1\ 0\ 9]$$
$$AT=[0\ 0\ 0\ 0\ 1]$$
$$BT=[4\ -\ 1\ 2\ -\ 1\ 9]$$
$$Acheck=[0\ 0\ 0\ 0\ 1]$$
$$Bcheck=[4\ 1\ 2\ 1\ 9]$$

Thus, as can be seen from the above results, generated $Z(p)$ verifies the original input version of the real part which was specified by vectors $A = [0\ 0\ 0\ 0\ 1]$ and $B = [4\ 1\ 2\ 1\ 9]$ in the w domain.

Let us run a final test example to generate a positive real function from its real part.

Example 7.10: This is a tough example with real part polynomials of degree 19:

$$A = [20\ 19\ 18\ 17\ 16\ 15\ 14\ 13\ 12\ 11\ 10\ 9\ 8\ 7\ 6\ 5\ 4\ 3\ 2\ 1]$$
$$B = [1\ 2\ 3\ 4\ 5\ 6\ 7\ 8\ 9\ 10\ 11\ 12\ 13\ 14\ 15\ 16\ 17\ 18\ 19\ 20]$$

Find the minimum immitance function $F = a(p)/b(p)$ using the parametric approach.

Solution: For this purpose, we run the MATLAB main program `Test_Immitance` with the above inputs. The results are listed in Table 7.1 to describe the Bode or parametric representation of the immittance function as specified by Equation (7.1).

In order to check the correctness of the results we reconstruct the even part function from the given positive real form of $Z(p)$. Hence we find the frequency domain version of $R(w^2) = ACheck(w^2) / BCheck(w^2) =$ as in Table 7.2.

In the last row of Table 7.2, we included the norms of the error (or difference) vectors

$EA = [A] - [ACheck]$ and $EB = [B] - [BCheck]$. As can be seen, both norms are less than 10^{-6}, which gives some idea about the numerical accuracy of the computations. Based on our experience, we can safely state that this accuracy is tolerable in designing practical power transfer networks via the parametric approach.

Table 7.1   Results of Example 7.10

Outputs		$Z0 = a(p)/b(p)$	
Poles	Residues k_i	Numerator $a(p)$ $1.0e+005*$	Denominator $b(p)$ $1.0e+004*$
-1.0950	-2.0082 + 0.0000i	0.0002	0.0001
-1.0818 - 0.1676i	-1.9845 + 0.0277i	0.0026	0.0014
-1.0818 + 0.1676i	-1.9845 - 0.0277i	0.0161	0.0098
-1.0424 - 0.3309i	-1.9139 + 0.0543i	0.0669	0.0447
-1.0424 + 0.3309i	-1.9139 - 0.0543i	0.2045	0.1506
-0.9780 - 0.4859i	-1.7981 + 0.0785i	0.4877	0.3968
-0.9780 + 0.4859i	-1.7981 - 0.0785i	0.9408	0.8476
-0.8901 - 0.6286i	-1.6397 + 0.0992i	1.4999	1.5015
-0.8901 + 0.6286i	-1.6397 - 0.0992i	2.0030	2.2386
-0.7808 - 0.7552i	-1.4425 + 0.1152i	2.2571	2.8335
-0.7808 + 0.7552i	-1.4425 - 0.1152i	2.1512	3.0574
-0.6530 - 0.8625i	-1.2109 + 0.1255i	1.7302	2.8130
-0.6530 + 0.8625i	-1.2109 - 0.1255i	1.1660	2.1984
-0.5097 - 0.9471i	-0.9503 +0.1284i	0.6497	1.4472
-0.3545 - 1.0060i	-0.9503 - 0.1284i	0.2930	0.7912
-0.1906 - 1.0348i	-0.6661 - 0.1218i	0.1033	0.3512
-0.5097 + 0.9471i	-0.1906 + 1.0348i	0.0269	0.1222
-0.3627 + 0.1000i	-0.3627 - 0.1000i	0.0046	0.0314
		0.0004	0.0053
		0.0000	0.0004

Table 7.2   Check of the results of Example 7.10

A(w^2) as given at the input	B(w^2) as given at the input	ACheck: computed	BCheck: computed
20	1	20.0000	1.0000
19	2	19.0000	2.0000
18	3	18.0000	3.0000
17	4	17.0000	4.0000
16	5	16.0000	5.0000
15	6	15.0000	6.0000
14	7	14.0000	7.0000
13	8	13.0000	8.0000
12	9	12.0000	9.0000
11	10	11.0000	10.0000
10	11	10.0000	11.0000
9	12	9.0000	12.0000
8	13	8.0000	13.0000

Table 7.2 (*continued*)

A(w^2) as given at the input	B(w^2) as given at the input	ACheck: computed	BCheck: computed
7	14	7.0000	14.0000
6	15	6.0000	15.0000
5	16	5.0000	16.0000
4	17	4.0000	17.0000
3	18	3.0000	18.0000
2	19	2.0000	19.0000
1	20	1.0000	20.0000
Size of the numerical errors in computations		norm(A-ACheck)= 3.5214e-006	norm(B-BCheck)= 2.6679e-007

7.5 Generation of Minimum Functions from the Given All-Zero, All-Pole Form of the Real Part

In many practical problems, we wish to specify the even part of a minimum immittance function from its given finite real frequency axis zeros and open LHP poles in the complex frequency domain p. In this case, the all-zero, all-pole form of $R(-p^2)$ is given by

$$R(-p^2) = \frac{(-1)^{ndc} p^{2*ndc} \prod_{i=1}^{nz}(p^2 + \omega_i^2)^2}{(-1)^n B_1 \prod_{i=1}^{n}(p^2 - p_i^2)} \geq 0, \forall p = j\omega \qquad (7.8)$$

where

- ndc is referred to as the total number of zeros at $p = 0$. In the classical theory of power transfer networks, these zeros are known as the transmission zeros at DC.
- nz is referred as the total zeros at finite real frequencies located at $p = \pm j\omega_1$ beyond zero.
- ω_i is called the finite transmission zero on the real frequency axis $j\omega$. In the classical theory of power transfer networks, these zeros are referred to as the finite transmission zeros.
- n is the total number of LHP poles of the minimum immittance $F(p)$.
- B_1 is a positive real number which refers to the leading coefficient of the denominator polynomial of the real part

$$R(\omega^2) = \frac{A(\omega^2)}{B(\omega^2)} = \frac{A_1 \omega^{2n} + \cdots + A_{n+1}}{B_1 \omega^{2n} + \cdots + B_{n+1}}$$

  in the ω domain.
- $p_i = -\alpha_i + j\beta_i$ is the open LHP poles of $F(p)$ with $\alpha_i < 0$.

Using Equation (7.8) we can generate the above equation for $R(\omega^2)$ from the given zeros and poles of $R(-p^2)$. For this purpose we developed a MATLAB function called function

$[A, B] = R\_allpolezero\ (ndc, W, B1, p)$, which generates the coefficients of the equation for $R(\omega^2)$. To generate the real part $R(\omega^2)$ from its zeros and poles, let us use $ndc = 1$, $W = [0]$, $B_1 = 2$, $p(1) = -1$, $p(2) = -2$, $p(3) = -1.5 + i$, $p(4) = -1.5 - i$, in the following example.

Example 7.11: Write a MATLAB code to generate MATLAB polynomials $[A]$ and $[B]$ for the real part $R(\omega^2)$ as specified above.

Solution: The name of the MATLAB main program is `Test_Rpolezero` and it is listed as follows.

Program Listing 7.13 `Test_Rpolezero.m`

```
% Test_Rpolezero
%This Program Computes the Rational form of Even-Part
%        from the given zeros and poles of R(-p^2)
% Even-Part: R(-p^2)=RA(-p^2)/RB(-p^2)
% All zero form of the numerator:
%        RA(-p^2)=(-1)^(ndc)*(p*p)^(ndc)*{[(p^2 +
W(1)^2]^2}...{[p^2 + W(n)^2]^2}
% All pole form of the denominator:
%        RB(-p^2)=B1*(-1)^n*p^n*[(p^2-p(1)^2]...[p-p(n)^2n]
% Inputs for RA(-p^2):
%        ndc=transmission zeros at DC
%        W (i):Finite Frequency Transmission zeros beyond zero.
%        Note that total number of finite transmission zeros nz is
%        specified by nz=length(W). Therefore, one does not need to
enter it.
%        W(i)=Transmission zeros at W(i)
%        Note: If there is no finite transmission zero then enter
W=[0]
% Inputs for RB(-p^2):
%        p(i)=poles of immitance function Z. They have to be open
LHP
%        B1: Leading coefficient of RB(-p^2) in p-domain
%
% Outputs:
%        Apoly: Even polynomial coefficients of the numerator
polynomial
%        R(-p^2)
%        Bpoly: Even polynomial coefficients of the denominator
polynomial
%        of R(-p^2)
%
% Computation Steps
%
clear
% Inputs for RA(-p^2)
ndc=1
W=[0];%Real Frequency Transmission zeros
nz=length(W);
```

```
if(W== [0])nz=0;end
nz
%
B1=2; p(1)=-1;p(2)=-2; p(3)=-1.5 + i;p(4)=-1.5-i;
%
[A,B]=R_allpolezero(ndc,W,B1,p)
```

Execution of the program yields

$$A = [0 \ 1 \ 0]$$
$$B = [2.000 \ 15.000 \ 54.1250 \ 125.6250 \ 84.5000]$$

In this regard, we list below the function `[A,B]=R_allpolezero(ndc,W,B1,p)`.

Program Listing 7.14 Function `R_allpolezero.m`

```
function [A,B] = R_allpolezero(ndc,W,B1,p)
%This function computes the Rational form of Even-Part in w
% domain.
%      from the given transmission zeros and poles of R(-p^2)
% Even-Part: R(-p^2) = RA(-p^2)/RB(-p^2)
% All zero form of the numerator:
%      RA(-p^2) = (-1)^(ndc)*(p*p)^(ndc)*{[(p^2 +
% W(1)^2]^2}...{[p^2 + W(n)^2]^2}
% All pole form of the denominator:
%      RB(-p^2) = B1*(-1)^n*p^n*[(p^2-p(1)^2]...[p-p(n)^2n]
% Inputs for RA(-p^2):
%      ndc = transmission zeros at DC
%      W (i):Finite Frequency Transmission zeros
%      nz = Transmission zeros at W(i)
%      W(i) = Transmission zeros at W(i)
%
% Inputs for RB(-p^2):
%      p(i) = poles of immitance function Z. They have to be LHP
%      B1: Leading coefficient of RB(-p^2)
%
% Outputs:
% Even numerator polynomial in w-domain: A(w^2) = A(1)w^2n +
% A(2)w^2(n-1) + ... + A(n)w^n + A(n + 1)
%
% Even denominator polynomial in w-domain:B(w^2) = B(1)w^2n +
% B(2)w^2(n-1) + ... + B(n)w^n + B(n + 1)
%
% Computation Steps
%
```

Program Listing 7.14   (Continued)

```
nz = length(W);%if there is no finite transmission zero set nz =
% 0
if (W = = [0]) nz = 0;end
% Generate Numerator polynomial of RA(-p^2) in p-domain
Apoly = R_allzero(ndc,nz,W);
% Generate Denominator polynomial of RB(-p^2) in p-domain
Bpoly = R_allpole(B1,p);
%
% Clear odd indexed terms from Apoly
AP = clear_oddpower(Apoly);
BP = clear_oddpower(Bpoly);
%
A = polarity(AP);
B = polarity(BP);
```

Finally, we present an example to generate a minimum function from its even part which is specified in terms of its finite real frequency axis zeros and open LHP poles.

Example 7.12: Let ndc=1, W=[2], meaning that $A(-p^2) = -p(p^2 + 4)^2$.

Let B1=2 and the six open LHP poles of F(p) be

```
p=
-1.0000
-2.0000
-1.5000 - 1.0000i
-1.5000 + 1.0000i
-3.0000 - 1.5000i
-3.0000 + 1.5000i
```

meaning that $B(-p^2) = (-1)^6 B_1 \prod_{i=1}^{6}(p^2 - p_i^2)$.

Find the full form of F(p).

Solution: For this example, we have developed a function called function [a,b] = parametric(ndc,W,B1,p) which directly generates $F(p) = a(p)/b(p)$ from the given zeros and poles of the even part. This function is executed by means of the test program called Test_parametric. The result is summarized as in Table 7.3.

Or, in rational form,

$$F(p) = \frac{0.0003p^5 + 0.0034p^4 + 0.0160p^3 + 0.06296p}{1.4142 + 16.97p^5 + 86.97p^4 + 238.64p^3 + 369.55p^2 + 305.73p + 103.41}$$

Table 7.3 Results of Example 7.12

	a(p)		b(p)
a(1)	0	b(1)	1.4142
a(2)	0.0003	b(2)	16.9706
a(3)	0.0034	b(3)	86.9741
a(4)	0.0160	b(4)	238.6485
a(5)	0.0313	b(5)	369.5517
a(6)	0.0629	b(6)	305.7353
a(7)	0	b(7)	103.4144

A quick test on MATLAB yields the even part as

$$R(-p^2) = \frac{1}{2}[Z(p) + Z(-p)] = \frac{\text{ACheck}(-p^2)}{\text{BCheck}(-p^2)}$$

where

```
ACheck =
    -1.0000
    0
    -8.0000
    0
    -16.0000
    0
```

Similarly,

```
BCheck =
.0e + 005 *

    0.00002
    0.0005
    0.0053
    0.0363
    0.1671
    0.5492
    1.3191
    2.3308
    3.0048
    2.7533
    1.6991
    0.6323
    0.1069
```

One can readily find the roots of ACheck and BCheck by using MATLAB built-in functions by setting ACheck = as -1 0 -8 0 -16 0 0 and writing >>Zeros = roots(ACheck) we find that the zeros of the even part are

```
0
0
-0.0000 + 2.0000i
-0.0000 - 2.0000i
 0.0000 + 2.0000i
 0.0000 - 2.0000i
```

Similarly, the poles of F(p) are determined from Poles = roots(BCheck) to give the poles of the even part as

```
-3.0000 + 1.5000i
-3.0000 - 1.5000i
-3.0000 + 1.5000i
-3.0000 - 1.5000i
-1.5000 + 1.0000i
-1.5000 - 1.0000i
-1.5000 + 1.0000i
-1.5000 - 1.0000i
-2.0000 + 0.0000i
-2.0000 - 0.0000i
-1.0000 + 0.0000i
-1.0000 - 0.0000i
```

Thus, the above confirms our initial start.

Below, we list our test program Test-Parametric and the function parametric.

Program Listing 7.15   Main Program Test_Parametric.m

```
% Test_Parametric
%This Program Computes the Parametric representation of a mini-
% mum PR
%immitance function from the given all zero-all pole form of the
% even part.
%
% Even-Part: R(-p^2)=RA(-p^2)/RB(-p^2)
% All zero form of the numerator:
%      RA(-p^2)=(-1)^(ndc)*(p*p)^(ndc)*{[(p^2 +
% W(1)^2]^2}...{[p^2 + W(n)^2]^2}
% All pole form of the denominator:
%      RB(-p^2)=B1*(-1)^n*p^n*[(p^2-p(1)^2]...[p-p(n)^2n]
% Inputs for RA(-p^2):
%      ndc=transmission zeros at DC
%      W (i):Finite Frequency Transmission zeros
%      Note that total number of finite transmission zeros nz is
```

```
%        specified by nz=length(W). Therefore, does not need to
%  enter it.
%        W(i)=Transmission zeros at W(i)
%
%  Inputs for RB(-p^2):
%        p(i)=poles of immitance function Z. They have to be LHP
%        B1: Leading coefficient of RB(-p^2)
%
%  Outputs:
%        Apoly: Even polynomial coefficients of the numerator
%  polynomial
%        R(-p^2)
%        Bpoly: Even polynomial coefficients of the denominator
%  polynomial
%        of R(-p^2)
%
%  Computation Steps
%
clear
%  Inputs for RA(-p^2)
ndc=1
W=[2];%Real Frequency Transmission zeros
nz=length(W);
if(W== [0])nz=0;end
nz
%
B1=2
p(1)=-0.1;
p(2)=-0.22;
p(3)=-0.15 + 0.1i;
p(4)=-0.15-0.1i;
p(5)=-0.33 + 0.15i;
p(6)=-0.3-0.15i;
%
%
[a,b,A,B,k]=parametric(ndc,W,B1,p);
a=sqrt(B1)*a
b=sqrt(B1)*b
% Check the real part if it is OK.
ap=polarity(a);
bp=polarity(b);
A1=conv(a,bp);
A2=conv(ap,b);
AT=A1 + A2;
AT=clear_oddpower(AT);
AN=0.5*AT;
nT=length(AN);
for i=1:nT
    ACheck(i)=AN(i)
```

Program Listing 7.15   (Continued)

```
      if(abs(AN(i))<1e-7) ACheck(i)=0;end
end
BCheck=conv(b,b)
xz=roots(ACheck);
pz1=sqrt(xz);
pz2=-sqrt(xz);
%
```

Program Listing 7.16   Function `parametric.m`

```
function [a,b,A,B,k]=parametric(ndc,W,B1,p)
% This function computes the Bode-or-Parametric form of an
% minimum immitance
%      function from the given all-zero-all pole form of the eve
% part
% Even-Part: R(-p^2)=RA(-p^2)/RB(-p^2)
% All zero form of the numerator:
%      RA(-p^2)=(-)^(ndc)*(p*p)^(ndc)*{[(p^2 +
% W(1)^2]^2}...{[p^2 + W(n)^2]^2}
% All pole form of the denominator:
%      RB(-p^2)=B1*(-1)^n*p^n*[(p^2-p(1)^2]...[p-p(n)^2n]
% Inputs for RA(-p^2):
%      ndc=transmission zeros at DC
%      W (i):Finite Frequency Transmission zeros
%      Note that total number of finite transmission zeros nz is
%      specified by nz=length(W). Therefore, does not need to
% enter it.
%      W(i)=Transmission zeros at W(i)
%
% Inputs for RB(-p^2):
%      p(i)=poles of immitance function Z. They have to be LHP
%      B1: Leading coefficient of RB(-p^2)
%
% Outputs:
%      Apoly: Even polynomial coefficients of the numerator
% polynomial
%      R(-p^2)
%      Bpoly: Even polynomial coefficients of the denominator
% polynomial
%      of R(-p^2)
%
% Computation Steps
% Inputs for RA(-p^2)
[A,B]=R_allpolezero(ndc,W,B1,p);
```

```
% Note that here A and B are given in w-domain.
m1=length(A);
n1=length(B);
Z0=0;
if(n1==m1)Z0=A(1)/B(1);end
% Generate residues
AA=polarity(A);% Convert A to p-domain vector.
BB=polarity(B);% Convert B to p-domain vector.
prd=product(p);
n=length(p);
for j=1:n
  y=p(j)*p(j);
  Aval=polyval(AA,y);
  k(j)=(-1)^n*Aval/p(j)/B(1)/prd(j);
end
[a,b]=Bode(Z0,k,p)
```

Let us test the above program with the following example.

Example 7.13: Let the LHP poles of $F(p)$ be

$$
p =
$$
$$
-1.0000
$$
$$
-2.0000
$$
$$
-3.0000 - 1.0000i
$$
$$
-3.0000 + 1.0000i
$$

Let B1=1, let R(-p2) posses a finite real frequency zero located at W(1)=1.5, and let ndc=0 (no DC zero). Find the rational form of F(p).

Solution: Using our program Test_Parametric, we immidiatetly obtain the solution. The MATLAB outputs are listed as follows:

```
a'

ans =
   0
   0.0074
   0.0670
   0.1056
   0.2531
   >> b'
ans =
   1
   9
   30
```

```
      42
      20
meaning that
```

$$F(p) = \frac{0.0074p^3 + 0.067p^2 + 0.150\ 56p + 0.2531}{p^4 + 9p^3 + 30p^2 + 42p + 20}$$

Backward checks verify the above result.

Generation of a Minimum Function F(p) for Practical Problems

From the reality condition of positive real functions, complex poles must be paired with their conjugates. In this case, in generating minimum functions employing the parametric approach, one only needs to know the unpaired form of complex poles such that $p_1 = \alpha_i + j\beta_i$. Then, they can be automatically paired with their complex conjugates within the algorithm. Therefore, for practical applications, we develop a simple function called function p=poles (real, alfa, beta) to generate complex conjugate paired poles of a minimum function from the given real and unpaired LHP poles.

In function p = poles (real, alfa, beta):

- input 'real' designates the pure real LHP poles;
- input 'alfa' designates the real part of an LHP pole $p_i = \alpha_i j\beta_i$;
- input 'beta' designates the imaginary part of an LHP pole $p_i = \alpha_i j\beta_i$;
- output p includes all real and complex conjugate paired poles.

The listing of function p=poles (real, alfa, beta) is given as follows.

Program Listing 7.17

```
function p=poles(real, alfa, beta)
% Generation of the poles of real Part R(p)=RA(p)/RB(p) as com-
% plex conjugate
% pairs from given alfa(i) and beta(i).
% complex roots are given as
% p(i)=-|alfa(i)| + jbeta(i); p=-|alfa(i)|-jbeta(i)
% This function generates the complex conjugate poles for the
% parametric
% approach.
% inputs:
%    real: real root of RB(p) if it exist.
% if real=0, then, that there is no real root.
% alfa(i)=real part of the pole which is positive.
%    imaginary part of the pole
% outputs:
% complex conjugate roots, the real roots is being %the last
% one.
%———————————————————————————————————————————
cmplx=sqrt(-1);
n=length(alfa);
```

```
   for i=1:n
    r1(i)=-abs(alfa(i)) + cmplx*beta(i);
 end
 for i=1:n
    r2(i)=-abs(alfa(i))-cmplx*beta(i);
 end
    p=[r1 r2];
 if norm(real)>0
 p=[r1 r2 -abs(real)];
 end
```

Eventually, the analytic form of a positive real minimum function $F(p) = a(p)/b(p)$ can be directly generated from its transmission zeros and real and unpaired complex LHP poles using the MATLAB function [a b] = F_parametric (ndc, W, B1, real, alfa, beta) where

- ndc is the total number of DC transmission zeros.
- $W = [W(1)W(2)...W(nz)]$ represent finite transmisison zeros beyond zero.
- It should be noted that if there is no finite transmission zero, then we set $W = 0$.
- $\sqrt{(B1)}$ is the leading coefficient of the denominator polynomial $b(p)$.
- real is an array which includes the real roots of $b(p)$.
- alfa is an array which includes the real part of the complex roots of $b(p)$.
- beta is an array which includes the imaginary parts of the complex roots of b(p).

The listing of function F_parametric is given below.

Program Listing 7.18

```
function [a b]=F_parametric(ndc,W,B1,real,alfa,beta)
% This function generates an immitance function F(p) using
% Parametric Approach
%          from its strictly Hurwitz poles and transmission zeros.
% Here, the poles are given in a simplified manner such that
%                p(i)=-alfa(i) + jbeta(i)
%                p(i + 1)=-alfa(i)-jbeta(i)
%                p(r)=-[real]
% Zeros are given as transmission zeros of F(p)as follows.
% Even {F(p)}=R(p^2)=RA(p^2)/RB(p^2) given in p-domain
% RA(-p^2)=(-1)^(ndc)*(p*p)^(ndc)*{[(p^2 + W(1)^2]^2}...{[p^2 +
% W(n)^2]^2}
% RB(-p^2)=B1*(-1)^n*p^n*[(p^2-p(1)^2]...[p-p(n)^2n] See
% Equation(7.8)
% Inputs:
%          ndc: zeros at DC
%          W: Finite transmission zeros beyond zero
%          if there is no finite tr-zero then, set [W]=0.
%          B1: leading Coefficient of the denominator RB(p^2)
% RB(p^2)
```

Program Listing 7.18   (Continued)

```
%           [real]: real LHP roots of F(p); this is an array
%           [alfa(i)]: real part of the poles; this is an array
%           [beta(i)]: imaginary part of the poles; this is an array
% Outputs:
%           a(p)=a(1)p^n + a(2)p^(n-1) + ... + a(n)p + a(n + 1)
%           b(p)=b(1)p^n + b(2)p^(n-1) + ... + b(n)p + b(n + 1)
%
p=poles(real, alfa, beta)
[a,b]=parametric(ndc,W,B1,p)
```

Now let us run an example using the above functions.

Example 7.14: Let

$$ndc=0,W=0,B1=1$$
$$real(1)=0.1, real(2)=0.2$$
$$alfa(1)=0.1, alfa(2)=0.3$$
$$beta(1)=0.5, beta(2)=0.7$$

(a) Generate the minimum function $F(p) = a(p)/b(p)$.
(b) Synthesize the immittance function as an impedance.

Solution:

(a) For this example, we developed a simple MATLAB code called 'Example7_14. In this program, firstly, complex conjugate paired and real poles are stored on a vector 'p' by means of function p=poles(real, alfa, beta). Then, function [a,b] =parametric(ndc,W,B1,p) is called to generate the immittance function $F(p) = a(p)/b(p)$.
Thus, the solution is obtained as follows.
Complete poles of the denominator polynomial:

```
Poles p(i)=sigma(i) or p=alfa(i)+/-jbeta(i)


-0.1 - 0.5i
-0.3 - 0.7i
-0.1 + 0.5i
-0.3 + 0.7i
-0.1
-0.2
```

Numerator polynomial $a(p)$:

$$a(p) = a(1)p^5 + a(2)p^4 + a(3)p^2 + a(4)p + a(5)$$

```
0
9150.10423124014
10065.1146543642
10810.7405267649
4882.83473831141
1936.62944423594
331.564986737401
```

Denominator polynomial $b(p)$:

$$b(p) = b(1)p^5 + b(2)p^4 + b(3)p^2 + b(4)p + b(5)$$

```
1
1.1
1.22
0.576
0.2516
0.05068
0.003016
```

The listing of MATLAB code `Example7_14.m` is as follows.

Program Listing 7.19

```
% Main Program Example7_14
% Generation of a positive real immittance function
% Via parametric approach
% This program provides the practical means for the computation.
% The user of the program only deals with real and single com-
%plex poles.
% Then, Complex Conjugate pairs are generated by the
% function p=poles(real,alfa, beta)
% and they are fed into function
% [a,b,A,B,k]=parametric(ndc,W,B1,p)
% The above operation is completed within the function
% [a b]=F_parametric(ndc,W,B1,real,alfa,beta)
% Inputs:
% zeros of the real part R(p)=A(p)/B(p) of F(p)=a(p)/b(p)
% ndc: Total number of zeros at DC
% W: Finite transmission zeros on the jw axis
%      B1: Leading coefficient of B(p)
%      real(i): Array. Pure real valued poles of b(p)
```

Program Listing 7.19   (Continued)

```
%       alfa: Array. Real part of a complex pole
%       beta: Array. Imaginary part of a complex pole
% Output:
%       p: poles which are Complex Conjugate paired
%       a(p): Array. Numerator polynomial of F(p)
%       b(p): Array. Denominator polynomial of F(p)
clc
clear
ndc=input('Enter transmission zeros at DC ndc=')
W=input('Enter finite transmisison zeros at [W]=[....]=')
B1=input('Enter B1=')
[real]=input('Enter [real]=[....]=')
[alfa]=input('Enter [alfa]=[....]=')
[beta]=input('Enter [beta]=[....]=')
p=poles(real, alfa, beta);
[a b]=F_parametric(ndc,W,B1,real,alfa,beta)
```

(b) Using long division, a synthesis of the impedance function is obtained as in Figure 7.2.

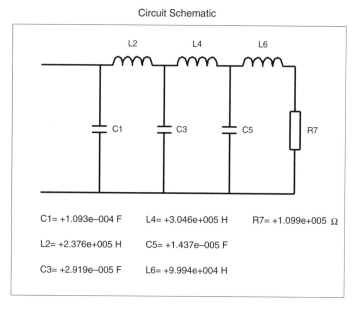

Circuit Schematic

C1= +1.093e-004 F   L4= +3.046e+005 H   R7= +1.099e+005 Ω

L2= +2.376e+005 H   C5= +1.437e-005 F

C3= +2.919e-005 F   L6= +9.994e+004 H

Figure 7.2   Synthesis of F(p) as an impedance function using long division

7.6 Immittance Modeling via the Parametric Approach

In designing ultra wideband power transfer networks, the designer may generate numerical data to construct the lossless power transfer networks. Most probably, data can be obtained as the real part of the driving point immittance function which is either minimum susceptance or minimum reactance–or, in short, the minimum function. In this case, positive real function modeling becomes an important issue. Therefore, in this section we present an algorithm to construct a minimum function $F(p)$ from its even part $R(p)$ by employing the parametric approach.

Algorithm: Generation of a Minimum Function from its Real Part Data via the Parametric Approach

The purpose of this algorithm is to find the best fit to measured real part data which belongs to a minimum immittance function. In general, measured immittance data may not necessarily be a minimum function. However, data can always be represented as a minimum function with a Foster part such that

$$\boxed{F(j\omega) = R(\omega) + jX(\omega) = F_{min}(j\omega) + F_{Foster}(j\omega)} \tag{7.9}$$

where

$$F_{min}(j\omega) = R(\omega) + jX_{min}(\omega) \tag{7.10a}$$

and

$$F_{Foster}(j\omega) = jX_F(\omega) \tag{7.10b}$$

where the p domain representation of $F_{Foster}(p) = F_{Foster}(j\omega)|_{p\,=\,j\omega}$

is given by Equation (1.63) as follows :

$$F_{Foster}(p) = K_\infty p + \frac{k_o}{p} + \sum_{j\,=\,1}^{N_\omega} \frac{k_j p}{p^2 + \omega_j^2} \tag{7.10c}$$

Thus, in general, we can state that a complete immittance modeling problem involves two major steps. In the first one, the real part data is modeled as in Equation (7.8) such that

$$R(-p^2) = \frac{(-1)^{ndc} p^{2*ndc} \prod_{i\,=\,1}^{nz}(p^2 + \omega_i^2)^2}{(-1)^n B_1 \prod_{i\,=\,1}^{n}(p^2 - p_i^2)} \geq 0, \quad \forall p = j\omega$$

which in turn yields the minimum immittance function via the parametric approach as

$$F_{min}(p) = \frac{a(p)}{b(p)}$$

or equivalently

$$F_{min}(j\omega) = R(\omega) + jX_{min}(\omega)$$

Then, in the second step, the resulting Foster data is generated as

$$\boxed{X_F(\omega) = X(\omega) - X_{min}}$$ (7.11)

and it is modeled properly.

It is crucial to point out that, for Foster reactance or susceptance data, the derivative of $X_F(\omega)$ must always be non-negative and it should fit the mathematical form specified by Equation (7.10c). Otherwise, it is not realizable and we can say that the data does not belong to a physically realizable passive device.[3]

Let us now return to our specific case.

In the MATLAB code, which was developed for real part modeling purposes, the model builder is provided with the real part data called `REA(w)` measured over the normalized angular frequencies.[4] For practical purposes, it is assumed that the real part has only real frequency zeros as described by Equation (7.8). From the given data, one can always locate theses zeros precisely. Therefore, the real frequency zeros of the measured data `REA(w)` are taken as the input to the algorithm. Thus, the modeler only determines the poles of the given immittance data via a nonlinear curve-fitting program. Fortunately, MATLAB provides an excellent function called '`lsqnonlin`'. This function employs the well-known Levenberg–Marquardt method to solve the nonlinear least-squares (nonlinear data-fitting) problems. Details of the optimization are skipped here. Interested readers are referred to MATLAB and related references [1–6] at the end of the chapter.

The MATLAB code consists of the following steps.

Step 1 (Load data): This step is devoted to reading the input data for the real part to be modeled. Within the main program, the normalized angular frequencies and the real part data are stored in text files called `WA.txt` and `REA.txt` respectively. Furthermore, if the imaginary part of the immittance data is specified, then we also read it and eventually use it to test the natural fit between the given immittance data and the analytic form generated from the immittance function.

Step 2 (Determine zeros of the real part data): In this step, the zeros of the real part data are determined and entered in to the main program as input. Furthermore, poles p_i of $R(p)$ are initialized as the unknown in the fitting problem. In the main program, initialized real poles are denoted by `p(i) = sigma(i)` which is an array `sigma = [...]`.

Initialized complex poles are denoted by `p(i) = alfa(i) + jbeta(i)`. Both `alfa(i)` and `beta(i)` are arrays. The user of the program specifically enters `alfa` and `beta` arrays separately.

Step 3 (Generate inverse of the specified immittance data): The real part data fitting can be carried out either directly on the measured immittance data or on its inverse. Therefore, in this step, we take the inverse of the measured data and store it in the arrays called `GEA` and `BEA`, where `GEA` and `BEA` are the real and imaginary parts of the inverse immittance data. Obviously, if the given function `F(jw) = REA(w) + jXEA(w)` is an impedance then, `Y(jw) = 1/F(jw) = GEA(w) + jBEA(w)` is an admittance.

Step 4 (Call optimization for nonlinear data fitting): In this step, we call the optimization function `lsqnonlin` to determine the unknown poles. The optimized poles are placed in vector `[x] = [sigma alfa beta]`.

Step 5 (Generate the analytic forms of the even part $R(\omega^2) = A(\omega^2)/B(\omega^2)$ **and** $F_{min}(p) = a(p)/b(p)$**):** Once the poles are determined, we can immediately generate the analytic form of

---

[3] Fundamental properties of Foster functions are presented in Chapter 9.

[4] In this section, the nomenclature matches that utilized in the main program which was developed to model the real part data.

$R(\omega^2) = A(\omega^2)/B(\omega^2)$ as in Section 7.5 by employing the MATLAB function `[A,B]=`
`R_allpolezero (ndc,W,B1,poles)`.

In a similar manner, the analytic form of $F_{min}(p) = a(p)/b(p)$ is obtained by the MATLAB function `[a b]=F_parametric(ndc,W,B1,sigma,alfa,beta)`. In this step, over the normalized angular frequencies w, the real part `RE(w)` and imaginary part `XE(w)` of $Fmin(jw)$ are generated and the error is computed.

Step 6: (Plot the results for comparative purposes): Eventually, in the last step, plots are provided for both the real parts `RE/REA` and the imaginary parts `XE/XEA` vs. normalized angular frequencies for comparison purposes.

Remarks:

• Optimization works on an objective function which includes the error.
  The error is produced as the difference between the given data and the analytic form of the real part as specified by Equation (7.8). Optimization routine `lsqnonlin` calls the objective function which in turn returns the computed value of the unknown vector `[x]`. Obviously, `[x]` includes new values of poles which must be distributed into arrays `'sigma, alfa` and `beta'` to generate the rational form of $F_{min}(p) = a(p)/b(p)$.
• In the course of optimization we directly determine the poles, which in turn imposes the constant term $b(n+1)$ of $b(p)$ as

$$b(n+1) = \sqrt{(B_1)}p(1)p(2)\ldots p(n)$$

On the other hand, $a(n+1)$ of $a(p)$ is pre-fixed by the zeros of $R(p)$. Thus, the DC value of $R(p)$ is obtained as the result of optimization as

$$R(0) = \frac{a(n+1)}{b(n+1)} = \frac{a(n+1)}{\sqrt{(B_1)}p(1)p(2)\cdot p(n)}$$

So, we have no control over it. This situation introduces a transformer in the resulting network when $F(p)$ is synthesized as the driving point immittance of a lossless two-port in unit or standard termination.

However, we can always fix the value of $R(0)$ by selecting the last pole as

$$p(n) = \frac{a(n+1)}{R(0) \times \sqrt{(B_1)} \times p(1) \times p(2) \times \ldots \times p(n-1)} = \texttt{alfa(n)} + \texttt{jbeta(n)}$$

In this case, we only use $n-1$ variables in the optimization; the last pole is determined as above.

Let us run an example to implement the above algorithm.

Example 7.15: Let the measured immittance data be given as in Table 7.4.

(a) Investigate the real part data `REA` if it has a zero over the real frequency axis.
(b) Find the analytic form $R(p)$ for the real part data `REA`.
(c) Generate the analytic form of the corresponding minimum immittance function $F_{min}(p) = a(p)/b(p)$.
(d) Comment on the quality of the real part data fitting.
(e) Synthesize $F_{min}(p)$ as an impedance function.

Table 7.4 Data for a measured immittance

WA (frequency)	REA (real part)	XEA (imaginary part)
0	1	0
0.2	1.1866	-0.26399
0.4	1.7123	-0.67099
0.6	2.0941	-0.62335
0.8	0.27862	0.1536
1	0.0065277	0.59638
1.2	0.00028068	0.52581
1.4	2.0535e-005	0.43698
1.6	2.1843e-006	0.36913
1.8	3.0707e-007	0.31865
2	5.3628e-008	0.28025

Solution:

(a) The plot of the given real part data is depicted in Figure 7.3.

Close examination of this plot reveals that REA(0) = 1. Then, it monotonically goes down to zero as the frequency approaches infinity. Therefore, for the numerator polynomial of $R(p)$, we must choose

ndc = 0, w = 0.

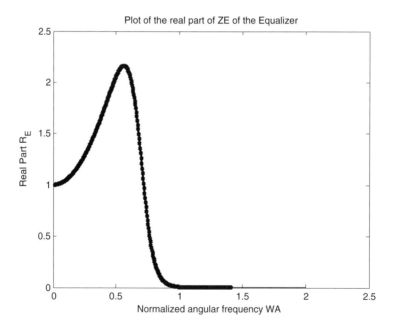

Figure 7.3 Plot of the given real part data

(b) In order to find the analytic form $R(p)$ as in Equation (7.8), we developed the MATLAB code
`Example7_15.m`. As outlined in the above algorithm, this code minimizes the error function
which in turn determines the unknown poles $p(i)$.

Inputs to `Example7_15.m` are given as follows: `ndc=0,W=0`, meaning that we have no real
frequency zero in $R(p)$.

Let us choose `B1=1` and initialize the poles as `Sigma(1)=[1]`, `alfa(1)=[1]`,
`beta(1)=[1]`; and the corresponding poles are given by $p(1) = 1 + j1, p(2) = 1 - j1, p(3) = 1$.

The above choice of poles places three reactive elements in the synthesis of $F_{min}(p)$.
Then, we should expect to end up with a CLC type of low-pass ladder structure terminated in a resistor.

Thus, optimization returns with vector `[x]` such that `x=[3.1981 0.17967 0.55105]`,
yielding `sigma(1)=3.1981, alfa(1)=0.17967, beta(1)=0.55105`.

Or, equivalently, optimized poles are given by

$$p(1) = -0.17967 - 0.55105i$$
$$p(2) = -0.17967 + 0.55105i$$
$$p(3) = -3.1981$$

Then, the analytic form of the even part $R(-p^2) = A(p)/B(p)$ is generated by means of the function
`[A,B]=R_allpolezero(ndc,W,B1,pole)`, which is introduced in Section 7.5.

Hence, we find that

$$A = \begin{bmatrix} 0 & 0 & 0 & 1 \end{bmatrix}$$

$$B = \begin{bmatrix} 1 & 9.6852 & -5.4384 & 1.1543 \end{bmatrix}$$

meaning that

$$R(\omega^2) = \frac{1}{\omega^2 + 9.6852\omega^2 + 1.1543}$$

(c) Assume that $F_{min}(p)$ is a minimum reactance impedance function. Then, the driving point input
impedance is generated by means of the MATLAB function `[a b]=F_parametric(ndc, W,
B1, sigma, alf a, beta)`.

Hence, we have

$$a = \begin{bmatrix} 0 & 0.78668 & 2.7986 & 0.93077 \end{bmatrix}$$

$$b = \begin{bmatrix} 1 & 3.5575 & 1.4852 & 1.0744 \end{bmatrix}$$

yielding the analytic form of $F_{min}(p) = a(p)/b(p)$ as

$$F(p) = \frac{0.786\ 68p^2 + 2.7986p + 0.930\ 77}{p^3 + 3.5575p^2 + 1.4852p + 1.0744}$$

(d) In Figure 7.4, the original data and the data obtained from the analytic model of the real part
are depicted. The fit is acceptable for degree $n = 3$. However, it can be improved by increasing
the number of unknown poles up to 5, 6 or 7. For example, with five poles we have the following
results:

```
inputs :  ndc = 0,  W = 0,  B1 = 1
```

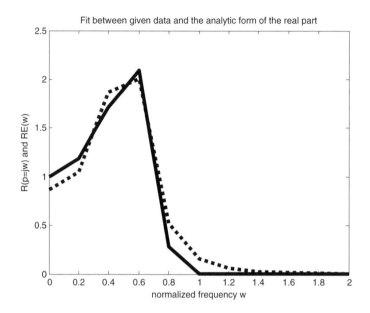

Figure 7.4 Plots of the real parts: solid line, original data; dashed line, data obtained from the analytic model

$$\texttt{Initial values for poles:}$$

$$\texttt{Sigma} = [1]$$

$$\texttt{Alfa} = [0.2\ 0.9];\ \texttt{beta} \ = [0.5\ 1]$$

The result of the optimization is

$$\texttt{x} = [8.9656 \quad 0.13929 \quad 0.35175 \quad 0.63338 \quad -0.37657]$$

The optimized poles are

$$p(1) = -0.13929 - 0.63338i$$

$$p(2) = -0.35175 + 0.37657i$$

$$p(3) = -0.13929 + 0.63338i$$

$$p(4) = -0.35175 - 0.37657i$$

$$p(5) = -8.9656$$

for the real part

$$R(\omega^2) = \frac{A(\omega^2)}{B(\omega^2)}$$

such that

$$A = [0\ 0\ 0\ 0\ 0\ 1]$$

$$B = [1\ 79.582\ -64.006\ 22.044\ -4.8289\ 1.0025]$$

Similarly, the corresponding minimum reactance function $F_{min}(p) = a(p)/b(p)$ is given by

$a(p)$	$b(p)$
0	1
0.72667	9.9477
7.2287	9.687
6.8791	8.2783
4.4217	3.4276
0.99877	1.0012

The resulting real part plots are depicted in Figure 7.5. Close examination of this figure reveals that the fit between the given real part data and the model is very close.

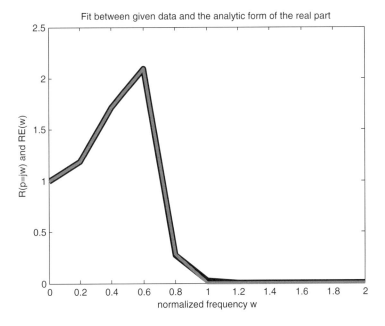

Figure 7.5   Solution to Example 7.15 with five poles

(e) Here

$$F(p) = Z_{in}(p) = \frac{0.78668p^2 + 2.7986p + 0.93077}{p^3 + 3.5575p^2 + 1.4852p + 1.0744}$$

is synthesized as a CLC ladder as depicted in Figure 7.6.

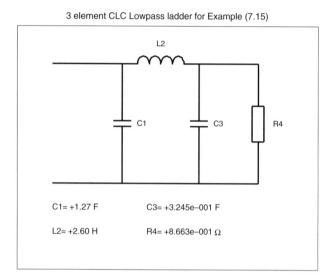

3 element CLC Lowpass ladder for Example (7.15)

C1= +1.27 F   C3= +3.245e−001 F

L2= +2.60 H   R4= +8.663e−001 Ω

Figure 7.6   Synthesis of $F_{min}(p)$

We list below the main Program `Example7_15.m` and related MATLAB functions.

Program Listing 7.20   Main program `Example7_15.m`

```
% Main Program Example7_15
clc
clear
% Nonlinear Immitance Modeling via Parametric Approach
% Immintance F(p)=a(p)/b(p)
% Inputs:
%      ndc, W(i), B1: Transmission zeros of the model immitance
%      sigma0: real pole of F(p); Initial value
%      p=alfa(i) + jbeta(i); Complex poles of F(p)
%      alfa0(i); Initial value for the optimization.
%      beta0(i); Initial value for the optimization.
%
% Step 1: Load Data to be modeled
load WA.txt; % Sampling Frequency Array
```

```
load REA.txt; % Real Part of the Immitance data
load XEA.txt; % Imaginary part of the immitance data
% Step 2: Enter zeros of the real part data
ndc=input('Enter transmission zeros at DC ndc=')
W=input('Enter finite transmission zeros at [W]=[....]=')
B1=input('Enter B1=')
%
[sigma0]=input('Enter initials for [sigma0]=[....]=')%sigma(i):
real roots of b(p)
[alfa0]=input('Enter initials for [alfa0]=[....]=')
[beta0]=input('Enter initials for [beta0]=[....]=')
%
Nsigma=length(sigma0);Nalfa=length(alfa0);Nbeta=length(beta0);
%
% Step 3: Generate inverse of the immitance data
NA=length(WA);
cmplx=sqrt(-1);
for i=1:NA
    Z(i)=REA(i) + cmplx*XEA(i);
    Y(i)=1/Z(i);
    GEA(i)=real(Y(i));
    BEA(i)=imag(Y(i));
end
% Step 4: Call nonlinear data fitting function
%%%%%%%%%% Preparation for the optimization
%%%%%%%%%%%%%%%%%%%%%%%%%%
OPTIONS=OPTIMSET('MaxFunEvals',20000,'MaxIter',50000);
%
%%%%%%%%%%%%%%%%%%%%%%%%%%%%%%%%%%%%%%%%%%%%%%%%%%%%
x0=[sigma0,alfa0,beta0];
% Call optimization function lsqnonlin:
x=lsqnonlin('objective',x0,[],[],OPTIONS,ndc,W,B1,Nsigma,Nalfa,-
Nbeta,WA,REA,XEA);
% (x0,ndc,W,B1,Nsigma,Nalfa,Nbeta,WA,REA,XEA)
% After optimization separate x into sigma, alfa and beta
for i=1:Nsigma; sigma(i)=x(i);end;for i=1:Nalfa; alfa(i)=x(i +
Nsigma);end
for i=1:Nbeta; beta(i)=x(i + Nsigma + Nalfa);end
%
% Step 5: Generate analytic form of R(p)=A(p)/B(p) and
Fmin(p)=a(p)/b(p)
% Generate poles to compute analytic form of R(p)
    pole=poles(sigma, alfa, beta);
% Generate analytic form of R(p)
    [A,B]=R_allpolezero(ndc,W,B1,pole);
% Generate analytic form of Fmin(p)=a(p)/b(p)
    [a b]=F_parametric(ndc,W,B1,sigma,alfa,beta);
    NA=length(WA);
    cmplx=sqrt(-1);
```

Program Listing 7.20   (Continued)

```
% Step 6: Print and Plot results
for j=1:NA
    w=WA(j);
    p=cmplx*w;
    aval=polyval(a,p);
    bval=polyval(b,p);
    F=aval/bval;
    %
    RE(j)=real(F);
    XE(j)=imag(F);
    XF(j)=XEA(j)-XE(j);%Foster part of the given impedance
    %
    error1(j)=abs(REA(j)-RE(j));
    error2(j)=abs(XEA(j)-XE(j));
    fun(j)=error1(j) + error2(j);
end
figure(1)
plot(WA,RE,WA,REA)
title('Fit between given data and the analytic form of the real
part')
xlabel('normalized frequency w')
ylabel('R(p=jw) and RE(w)')
%
figure(2)
plot(WA,XE,WA,XEA)
title('Comparison of XEA and XE')
xlabel('normalized frequency w')
ylabel('XEA(w) and XE(w)')
%
figure(3)
plot(WA,XF)
title('Foster part of the given immitance')
xlabel('normalized angular frequency w')
ylabel('X_F(w)=XEA-XE Foster Part of the immitance')
CVal=general_synthesis(a,b)
```

Program Listing 7.21   Objective function for optimization

```
function
fun=objective(x0,ndc,W,B1,Nsigma,Nalfa,Nbeta,WA,REA,XEA)
    % Modeling via Parametric approach
    for i=1:Nsigma; sigma(i)=x0(i);end;
    for i=1:Nalfa; alfa(i)=x0(i+Nsigma);end
for i=1:Nbeta; beta(i)=x0(i+Nsigma+Nalfa);end
```

```
%
     [a b] = F_parametric(ndc,W,B1,sigma,alfa,beta);
     NA = length(WA);
     cmplx = sqrt(-1);
%
for j = 1:NA
     w = WA(j);
     p = cmplx*w;
     aval = polyval(a,p);
     bval = polyval(b,p);
     Z = aval/bval;
     RE = real(Z);
     XE = imag(Z);
     error1 = abs(REA(j)-RE);
     %  error2 = abs(XEA(j)-XE);
     fun(j) = error1;
end
```

7.7 Direct Approach for Minimum Immittance Modeling via the Parametric Approach

In the previous section, the real part of the given minimum immittance data is modeled by means of its poles. This approach has its own merits since LHP poles directly yields the minimum immittance function without any restriction. On the other hand, it is hard to initialize. From the designer's point of view, there is not much indication about how the poles are placed in the LHP. Furthermore, the degree of nonlinearity $n_{non-lin}$ is equal to the total number of poles since the denominator polynomial is directly expressed as multiplication of the poles. However, these drawbacks of the parametric approach can be overcome by means of what we call the 'direct approach'.

In the 'direct modeling' approach, the real part of the immittance is directly expressed in its coefficient form such that

$$R(\omega^2) = \frac{A_0\omega^{2ndc}\prod_{i=1}^{nz}(\omega_i^2 - \omega^2)^2}{B_1\omega^{2n} + B_2\omega^{2(n-1)} + \dots + B_n\omega^2 + 1} = \frac{A(\omega^2)}{B(\omega^2)} \geq 0, \forall\omega \qquad (7.12)$$

In Equation (7.12), we assume that the even part $R(-p^2)$ possesses only real frequency zeros. In other words, it is free of RHP zeros. The real frequency zeros of $A(-p^2) = (-1)^{ndc}p^{2n}\prod_{i=1}^{nz}(p^2 + \omega_i^2)^2$ can be directly read from the given data and are fixed in advance by the modeler. In this case, the modeling problem becomes a simple linear regression problem of the specified data $\text{Data}(\omega)$ expressed as

$$\begin{aligned}\text{Data}(\omega^2) &= \frac{A(\omega^2)}{R(\omega^2)} \\ &= \frac{1}{A_0}(B_1\omega^{2n} + B_2\omega^{2(n-1)} + \dots + B_n\omega^2 + 1) > 0\end{aligned} \qquad (7.13)$$

or

$$Data(\omega^2) = \widehat{B}_1\omega^{2n} + \widehat{B}_2\omega^{2(n-1)} + \ldots + \widehat{B}_n\omega^2 + \widehat{B}_{n+1} > 0, \forall \omega$$

where

(7.14)

$$\widehat{B}_{n+1} = \frac{1}{A_0} \quad \widehat{B}_k = \frac{B_k}{A_0} \quad k = 1, 2, \ldots, n$$

Here, we should emphasize that the even polynomial $Data(\omega^2)$ must be strictly positive; however, a straightforward linear regression or a simple linear curve-fitting algorithm cannot satisfy this condition. Nevertheless, this problem can easily be solved in the following manner.

Let $c(\omega)$ be an arbitrary 'auxiliary polynomial' with real coefficients such that

$$c(\omega) = c_1\omega^n + c_2\omega^{n-1} + \ldots + c_n\omega + c_{n+1}$$

(7.15)

Then, we can always define a strictly positive even polynomial $B(\omega^2)$ such that

$$B(\omega^2) = \frac{1}{2}\left[c^2(\omega) + c(-\omega)\right] + \epsilon^2$$

(7.16)

In this representation, ϵ^2 is a negligibly small positive number, extracted from the specified data if required. Otherwise, it is set to $\epsilon^2 = 0$, which is the usual case.

Thus, one can express the real part $R(\omega^2)$ as a function of new variables $\{c_1, c_2, \ldots, c_n\}$ such that

$$R(\omega^2, c_1, c_2, \ldots, c_n) = \frac{A_0\omega^{2ndc} \prod_{i=1}^{nz}(\omega_i^2 - \omega^2)^2}{\frac{1}{2}[c^2(\omega) + c^2(-\omega)]}$$

perhaps with $c_{n+1} = 1$.

(7.17)

This equation can easily be programmed in MATLAB.

Let us illustrate implementation of the above equations by means of an example.

Example 7.16 Solve Example 7.15 using the direct parametric approach.

Solution: For this problem, we developed the MATLAB main program called `Example7_16.m`. In this program, the even part of the minimum immittance function $F(p)$ is represented in direct form as

$$R(-p^2) = \frac{A(p^2)}{B(p^2)} = \frac{(-1)^{ndc}p^{2*ndc}\prod_{i=1}^{nz}(p^2 + \omega_i^2)^2}{B(p^2)}$$

The coefficient of $B(p^2)$ is evaluated as in Equation (7.16), replacing ω^2 by $-p^2$.

The strictly positive polynomial $B(\omega^2)$ of Equation (7.16) is generated by means of function $B = \text{Poly\_Positive}(c)$. Thus, the first set of inputs to the main program are provided as follows.

Input set 1:

- WA: sampling frequencies (normalized angular frequency).
- REA: real part of the immittance to be modeled.
- XEA: imaginary part of the immittance to be modeled.

Note that the above data set is loaded into the main program by text files as in the previous example:

- ndc: total number of zeros at DC.
- W $= [\text{W}(1)\text{W}(2)...\text{W}(\text{nz})]$: finite zeros of $R(-p^2)$.

The non-negative constant A0 is expressed as

$$\text{A0} = a_0^2 \geq 0$$

Depending on the model requirements, a_0 may be fixed in advance or included among the unknowns of the nonlinear curve-fitting problem.

We should mention that, at the end of the optimization, when the analytic form of the minimum immittance F(p) is synthesized as a lossless two-port in resistive termination, one may end up with a transformer depending on the value of the termination resistance. If 'no transformer' is desired in the final synthesis, then a_0 must be fixed at a level which yields the desired terminating resistance. This can only be achieved for low-pass design problems. In the MATLAB code, this situation is controlled by an input flag which is an integer denoted by ntr.

If a_0 is part of the unknowns, then we set ntr $= 1$. Otherwise it is zero.

The second set of inputs is listed as follows.

Input set 2:

- a0(or a_0) initialized real constant associated with $A(-p^2)$ as in Equation (7.17) such that $A_0 = a_0^2 < 0$. It is determined as the result of nonlinear curve fitting.
- c0(or c_i): initialized real coefficients of the polynomial $B(\omega^2) = \frac{1}{2} [c^2(\omega) + c^2(-\omega)]$ as described by Equation (7.16). These coefficients are determined as the result of nonlinear curve fitting.
- ntr: control flag. If ntr=1 then a0 is among the optimization variables. If ntr=0 then a0 is pre-fixed by the modeler.

Optimization is carried out as in Example 7.15, employing the MATLAB function lsqnonlin.

For this example, optimization calls the objective function named direct. In function fun = direct(x0,ntr,ndc,W,a0,WA,REA,XEA) the real part is generated in direct form and then the minimum function $F(p) = a(p)/b(p)$ is obtained via the parametric approach by means of the following MATLAB functions.

Generation of strictly positive $B(p^2)$:

- BB = Poly_Positive(C) (Note that $BB(\omega^2)$ is a strictly positive polynomial in the ω domain.)
- B = polarity(BB) (Note that $B(p^2)$ is in the p domain.)

Generation of the numerator polynomial $A(p^2)$:

- nB = length(B)
- A = (a0*a0)*R_Num(ndc,W)
- nA = length(A)
- A = fullvector(nB,A)

It should be mentioned that MATLAB function `fullvector` matches the size of vectors A and B for mathematical operations. Here, it is assumed that the size of B is greater than the size of A.

Generation of the minimum function $F(p) = a(p)/b(p)$ via the parametric approach:

- `[a,b] = RtoZ(A,B)` (Note that in the above function polynomials $A(p)$ and $B(p)$ are the same size.)

Once the optimization is completed on the unknown coefficients of the fabricated polynomial $C(p)$ and $a0$, then $R(-p^2) = A(p^2)/B(p^2)$ and $F(p)$ are generated as above. Eventually, the results of optimization are plotted and `F(p)` is synthesized by means of the MATLAB function[5] `CVal=general_synthesis(a,b)`. The function `'general_synthesis'` constructs lossless ladders in resistive termination from the given driving point immittance function $F(p) = a(p)/b(p)$ by means of long division.

Now, let us run the main program `Example7_16.m` with the following initial values:

- Enter transmission zeros at DC, `ndc=0`.
- Enter finite transmission zeros at $[W]=[...]=[0]$.
- Enter `a0=1`.
- Enter initial values for $[c0] = [....] = [1\ 1\ 1]$.
- `ntr=1`, yes, model with transformer; `ntr=0`, no transformer `ntr=0`.

Notice that before optimization we set `ntr=0`, which means that $a0$ is fixed as above ($a0=1$), and it is not included among the unknowns of the problem.

Thus, the program is run with the above initialvalues. Optimization ends successfully and returns the following results.

$$A = [0\ 0\ 0\ 1]$$
$$C = [5.2137\ -1.5163\ -1.0078\ 1]$$
$$B = [-27.182\ -8.2097\ 2.0169\ 1]$$
$$a = [0\ 0.7346\ 0.60755\ 0.1918]$$
$$b = [1\ 0.82704\ 0.49301\ 0.1918]$$

or, in open analytic form, we have

$$R(-p^2) = \frac{1}{-27p^6 - 8.2097p^4 + 2.0169p^2 + 1}$$

and

$$F(p) = Z(p) = \frac{0.7346p^2 + 0.60755p + 0.1918}{p^3 + 0.82704p^2 + 0.49301p + 0.1918}$$

The fit between the given data and optimized real part is shown in Figure 7.7.
The synthesis of `F(p)` is shown in Figure 7.8.

Remarks: Immittance modeling via the direct form of the even part presents better numerical stability in optimization over the pole approach since the real part is quadratic in terms of the real coefficients $c(i)$ no matter what the degree of the polynomial $b(p)$ is, whereas, in the pole approach, $b(p)$ is expressed as the

---

[5] Function `'general_synthesis'` was developed by Dr Ali Kilinc of Okan University.

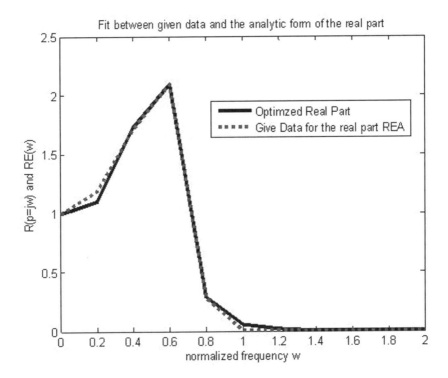

Figure 7.7 Fit between the given data and optimized real part for Example 7.16

product of pole terms $(p - p_j)$, which introduces an n th degree of dependence on poles p_i in the even part. Therefore, ad hoc initialization of coefficients usually results in successful optimization. In this regard, initial values $c_0(i)$ are always set to $c_0(i) = \pm 1$.

In order to facilitate the programming, we developed a function called $[\mathtt{A,B}]=\mathtt{Evenpart\_Direct(ndc, W, a0, c)}$ which generates a non-negative even part in the p domain from the given transmission zeros p^{ndc} and W, $a0$ coefficients $c(i)$ of the auxiliary polynomial $C(p)$, which in turn yields the even part $R(p^2) = A(p^2)/B(p^2)$.

For example, for the inputs $\mathtt{ndc} = 1, \mathtt{W} = [4], \mathtt{a0} = 1, \mathtt{c} = [1\ 1\ 1\ 1\ 1]$ then, $[\mathtt{A,B}] = \mathtt{Evenpart\_Direct(ndc, W, a0, c)}$ yields that

$$A = [0\ 0\ \text{-}1\ \text{-}32\ \text{-}256\ 0] \quad B = [\text{-}1\ 3\ \text{-}5\ 5\ \text{-}3\ 1]$$

meaning that

$$R(p^2) = \frac{-p^6 - 32p^4 - 256p^2}{-p^{10} + 3p^8 - 5p^6 + 5p^4 - 3p^2 + 1}$$

Similarly, the MATLAB function $[a, b] = \mathtt{Immitance\_Direct(ndc, W, a0, c)}$ generates the realizable minimum function $F(p) = a(p)/b(p)$ directly from the given transmission zeros, $a0$ and $c(i)$ via the parametric approach.

Circuit Schematic

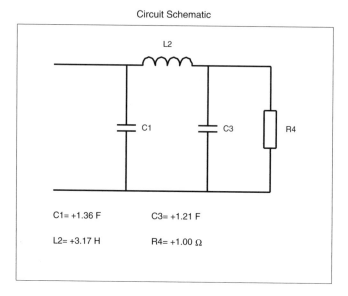

C1= +1.36 F   C3= +1.21 F

L2= +3.17 H   R4= +1.00 Ω

Figure 7.8 Synthesis of

$$\mathbf{F(P)} = \mathbf{Z(P)} = \frac{0.7346p^2 + 0.60755p + 0.19818}{p^3 + 0.82704p^2 + 0.49301p + 0.1918}$$

For this example, function `immitance_Direct` results in a minimum function $F(p) = a(p)/b(p)$ as

$$a = [0 \ 13.82 \ 61.695 \ 108.86 \ 81.731 \ 0]$$

$$b = [1 \ 4.4641 \ 8.4641 \ 8.4641 \ 4.4641 \ 1]$$

meaning that

$$F(p) = \frac{13.82p^4 + 61.695p^3 + 108.86p^2 + 81.731p}{p^5 + 4.4641p^4 + 8.4641p^3 + 8.4641p^2 + 4.4641p + 1}$$

Now let us end this chapter by listing the relevant MATLAB programs employed for this example.

Program Listing 7.22 Main program for Example 7.16

```
% Main Program Example7_16
clc
clear
% Nonlinear Immitance Modeling via Positive Polynomial Approach:
%
%        Direct Modeling Method via Parametric Approach
```

```
%
% Immitance F(p)=a(p)/b(p)
% Real Part of F(p)is expressed as in direct computational
% techniques R(w^2)=AA(w^2)/BB(w^2)
% Where AA(w^2) if fixed by means of transmission zeros
% BB(w^2)=(1/2)[C^2(w) + C^2(-w)] where C(w)=C(1)w^n +
% C(2)w^(n-1) + ... + C(n)w + 1
% Note that AA(w^2) is computed by means of function R_num which
% is
% specified in p-domain as
% A(-p^2)=(a0^2)*R_Num=(ndc,W)=A(1)p^2m + A(2)p^2(m-1) + ... +
% A(m)p^2 + A(m + 1)
% Inputs:
%         ndc, W(i), B1: Transmission zeros of the model immitance
%         a0: Constant multiplier of the denominator as a0^2
%         c0; Initial coefficients of BB(w^2)
%         ntr: Model with or without transformer/
%         ntr=1> Yes Transformer. ntr=0>No transformer
% Ouput:
%         AA(w^2)=polarity(A)
%         c: optimized coefficients.
% Step 1: Load Data to be modeled
load WA.txt; % Sampling Frequency Array
load REA.txt; % Real Part of the Immitance data
load XEA.txt; % Imaginary part of the immitance data

% Step 2: Enter zeros of the real part data
ndc=input('Enter transmission zeros at DC ndc=')
W=input('Enter finite transmission zeros at [W]=[....]=')
%
a0=input('Enter a0=')
[c0]=input('Enter initials for [c0]=[....]=')%sigma(i): real
roots of b(p)
ntr=input('ntr=1, Yes, model with Transformer, ntr=0 No
Transformer ntr=')
%
Nc=length(c0);
%

% Step 3: Generate inverse of the immitance data
NA=length(WA);
cmplx=sqrt(-1);
for i=1:NA
      Z(i)=REA(i) + cmplx*XEA(i);
      Y(i)=1/Z(i);
      GEA(i)=real(Y(i));
      BEA(i)=imag(Y(i));
end
```

Program Listing 7.22   (Continued)

```
% Step 4: Call nonlinear data fitting function
%%%%%%%%%% Preparation for the optimization
%%%%%%%%%%%%%%%%%%%%%%%%%%%%
OPTIONS=OPTIMSET('MaxFunEvals',20000,'MaxIter',50000;
%%%%%%%%%%%%%%%%%%%%%%%%%%%%%%%%%%%%%%%%%%%%%%%%%%%%%
if ntr==1; x0=[c0 a0];Nx=length(x0);end;%Yes transformer case
if ntr==0; x0=c0;Nx=length(x0);end;%No transformer case
%
% Call optimization function lsqnonlin:
x=lsqnonlin('direct',x0,[],[],OPTIONS,ntr,ndc,W,a0,WA,REA,XEA);
%
% After optimization separate x into sigma, alfa and beta
if ntr== 1; %Yes transformer case
   for i=1:Nx-1;
   c(i)=x(i);
   end;
   a0=x(Nx);
end
if ntr== 0;% No Transformer case
   for i=1:Nx;
     c(i)=x(i);
   end;
end
%

% Step 5: Generate analytic form of Fmin(p)=a(p)/b(p)
% Generate B(-p^2)
      C=[c 1];
      BB=Poly_Positive(C);% This positive polynomial is in w-
domain
      B=polarity(BB);% Now, it is transferred to p-domain
% Generate A(-p^2) of R(-p^2)=A(-p^2)/B(-p^2)
nB=length(B);
      A=(a0*a0)*R_Num(ndc,W);% A is specified in p-domain
nA=length(A);
if (abs(nB-nA)>0)
  A=fullvector(nB,A);% work with equal length vectors
end
      [a,b]=RtoZ(A,B);% Here A and B are specified in p-domain
% Generate analytic form of Fmin(p)=a(p)/b(p)
      NA=length(WA);
      cmplx=sqrt(-1);
% Step 6: Print and Plot results
   for j=1:NA
      w=WA(j);
      p=cmplx*w;
```

```
        bval=polyval(b,p);
        F=aval/bval;
        %
        RE(j)=real(F);
        XE(j)=imag(F);
        XF(j)=XEA(j)-XE(j);%Foster part of the given
        impedance
        %
        error1(j)=abs(REA(j)-RE(j));
        error2(j)=abs(XEA(j)-XE(j));
        fun(j)=error1(j) + error2(j);
end
        figure(1)
        plot(WA,RE,WA,REA)
        title('Fit between given data and the analytic form of the
real part')
        xlabel('normalized frequency w')
        ylabel('R(p=jw) and RE(w)')
        %
        figure(2)
        plot(WA,XE,WA,XEA)
        title('Comparison of XEA and XE')
        xlabel('normalized frequency w')
        ylabel('XEA(w) and XE(w)')
        %
        figure(3)
        plot(WA,XF)
        title('Foster part of the given immitance')
        xlabel('normalized angular frequency w')
        ylabel('X_F(w)=XEA-XE Foster Part of the immitance')
CVal=general_synthesis(a,b)
```

Program Listing 7.23

```
        function
fun=direct(x0,ntr,ndc,W,a0,WA,REA,XEA)
% Modeling via Direct - Parametric approach
if ntr==1; %Yes transformer case
  Nx=length(x0);
    for i=1:Nx-1;
      c(i)=x0(i);
    end;
    a0=x0(Nx);
end
if ntr==0;% No Transformer case
  Nx=length(x0);
```

Program Listing 7.23   (Continued)

```
      for i=1:Nx;
        c(i)=x0(i);
      end;
  end
  %
  % Step 5: Generate analytic form of Fmin(p)=a(p)/b(p)
  % Generate B(-p^2)
        C=[c 1];
        BB=Poly_Positive(C);% This positive polynomial is in w-
  % domain
        B=polarity(BB);% Now, it is transferred to p-domain
  % Generate A(-p^2) of R(-p^2)=A(-p^2)/B(-p^2)
  nB=length(B);
        A=(a0*a0)*R_Num(ndc,W);% A is specified in p-domain
  nA=length(A);
  if (abs(nB-nA)>0)
        A=fullvector(nB,A);
  % Note that function RtoZ requires same length vectors A and B
  end
        [a,b]=RtoZ(A,B);% Here A and B are specified in p-domain
  %
        NA=length(WA);
        cmplx=sqrt(-1);
  %
        for j=1:NA
          w=WA(j);
          p=cmplx*w;
          aval=polyval(a,p);
          bval=polyval(b,p);
          Z=aval/bval;
          RE=real(Z);
          XE=imag(Z);
          error1=abs(REA(j)-RE);
          % error2=abs(XEA(j)-XE);
          fun(j)=error1;
        end
```

Program Listing 7.24

```
function [a,b]=RtoZ(A,B)
% This MATLAB function generates a minimum function Z(p)=a(p)/
% b(p)
%       from its even part specified as R(p^2)=A(p^2)/B(p^2)
```

```
%        via Bode (or Parametric)approach
% Inputs: In p-domain, enter A(p) and B(p)
% A=[A(1) A(2) A(3)...A(n + 1)]; for many practical cases we set
% A(1)=0.
% B=[B(1) B(2) B(3)...B(n + 1)]
% Output:
%       Z(p)=a(p)/b(p) such that
%       a(p)=a(1)p^n + a(2)p^(n-1) + ... + a(n)p + a(n + 1)
%       b(p)=b(1)p^n + b(2)p^(n-1) + ... + b(n)p + a(n + 1)
% Generation of an immitance Function Z by means of Parametric
% Approach
% In parametric approach Z(p)=Z0 + k(1)/[p-p(1)] + ... + k(n)/
% [p-p(n)]
% R(p^2)=Even{Z(p)}=A(-p^2)/B(-p^2) where Z0=A(n + 1)/B(n + 1).
%
% Given A(-p^2)>0
% Given B(-p^2)>0
%
BP=polarity(B);%BP is in w-domain
AP=polarity(A);%AP is in w-domain
% Computational Steps
% Given A and B vectors. A(p) and B(p) vectors are in p-domain
% Compute poles p(1),p(2),...,p(n)and the residues k(i) at poles
p(1),p(2),...,p(n)
[p,k]=residue_Z0(AP,BP);
%
% Compute numerator and denominator polynomials
Z0=abs(A(1)/B(1));
[num,errorn]=num_Z0(p,Z0,k);
[denom,errord]=denominator(p);
%
a=num;
b=denom;
```

Program Listing 7.25   function B = Poly_Positive(b)

```
function B=Poly_Positive(b)
% This function computes a positive polynomial in w domain
% B(w^2)=(1/2){b^2(w) + b^2(-w)}
% Input:
%       b=[b(1) b(2)...b(n) b(n + 1)]
% Output:
%       B=[B(1)*w^2n + B(2)*w^2(n-1)...B(n)w^2 B(n + 1)]
b2=conv(b,b);% Take the square of b(w)
b2_=polarity(b2);
```

Program Listing 7.25   (Continued)

```
C1=b2 + b2_;
C2=C1/2;
B=clear_oddpower(C2);
```

Program Listing 7.26

```
function B=fullvector(n,A)
na=length(A);
for i=1:na
    B(n-i + 1)=A(na-i + 1);
end
for i=1:(n-na-1)
    B(i)=0;
End
```

Program Listing 7.27

```
    function [A,B]=Evenpart_Direct(ndc,W,a0,c)
% Construction of an immitance function F(p)=a(p)/b(p)
% employing the direct form of even part
% R(-p^2)=a0^2*(-1)^ndc*(-p^2 + w(i)^2)B(-p^2)
% B(w^2)=(1/2)*[c^2(w) + c^2(-w)]>0
% Inputs:
% ndc: zeros at DC
% W=[...] finite zeros of A(-p^2)
% a0: leading coefficient of A(-p^2)
% c(p)=[c(1) c(2) ...c(n)]; Coefficients of c(p)
% ─────────────────────────────────────────────
C=[c 1];
BB=Poly_Positive(C);% This positive polynomial is in w-domain
B=polarity(BB);% Now, it is transferred to p-domain
% Generate A(-p^2) of R(-p^2)=A(-p^2)/B(-p^2)
nB=length(B);
A=(a0*a0)*R_Num(ndc,W);% A is specified in p-domain
nA=length(A);
if (abs(nB-nA)>0)
A=fullvector(nB,A);
% Note that function RtoZ requires same length vectors A and B
end
```

Program Listing 7.28

```
function [a,b]=Immitance_Direct(ndc,W,a0,c)
% Construction of an immitance function F(p)=a(p)/b(p)
% employing the direct form of even part
% R(-p^2)=a0^2*(-1)^ndc*(-p^2 + w(i)^2)B(-p^2)
% B(w^2)=(1/2)*[c^2(w) + c^2(-w)]>0
% Inputs:
% ndc: zeros at DC
% W=[...] finite zeros of A(-p^2)
% a0: leading coefficient of A(-p^2)
% c(p)=[c(1) c(2) ...c(n)]; Coefficients of c(p)
% _____
C=[c 1];
BB=Poly_Positive(C);% This positive polynomial is in w-domain
B=polarity(BB);% Now, it is transferred to p-domain
% Generate A(-p^2) of R(-p^2)=A(-p^2)/B(-p^2)
nB=length(B);
A=(a0*a0)*R_Num(ndc,W);% A is specified in p-domain
nA=length(A);
if (abs(nB-nA)>0)
A=fullvector(nB,A);
% Note that function RtoZ requires same length vectors A and B
end
[a,b]=RtoZ(A,B);% Here A and B are specified in p-domain
```

References

[1] T. F. Coleman, and Y. Li, "An Interior, Trust Region Approach for Nonlinear Minimization Subject to Bounds", *SIAM Journal on Optimization*, Vol. **6**, pp. 418–445, 1996.

[2] T. F. Coleman, and Y. Li, "On the Convergence of Reflective Newton Methods for Large-Scale Nonlinear Minimization Subject to Bounds", *Mathematical Programming*, Vol. **67**, No. 2, pp. 189–224, 1994.

[3] J. E. Dennis, Jr, "Nonlinear Least-Squares", in *State of the Art in Numerical Analysis*, ed. D. Jacobs, Academic Press, pp. 269–312, 1977.

[4] K. Levenberg, "A Method for the Solution of Certain Problems in Least-Squares", *Quarterly of Applied Mathematics*, Vol. **2**, pp. 164–168, 1944.

[5] D. Marquardt, "An Algorithm for Least-Squares Estimation of Nonlinear Parameters", *SIAM Journal of Applied Mathematics*, Vol. **11**, pp. 431–441, 1963.

[6] J. J. Moré, "The Levenberg-Marquardt Algorithm: Implementation and Theory", in *Numerical Analysis*, ed. G. A. Watson, Lecture Notes in Mathematics 630, Springer-Verlag, pp. 105–116, 1977.

8

Gewertz Procedure to Generate a Minimum Function from its Even Part: Generation of Minimum Function in Rational Form

8.1 Introduction

In the previous chapter, a minimum function is generated in Bode form (or so-called parametric form) from its given even part. In dealing with Bode forms, it is appropriate to express the even part $R(-p^2)$ in terms of its zeros and poles. This can easily ensure the condition of $R(-p^2) \geq 0$, for all $p = \sigma \geq 0$, for the realizibility. Then, the minimum function is generated by means of a Bode procedure which yields a realizable positive real immittance function. Eventually, this function is synthesized as a lossless two-port in resistive termination by employing the Darlington method, which in turn results in the desired power transfer function.

However, one may also express the even part $R(-p^2)$ in rational form and then generate the minimum immittance function directly from the given coefficients of the rational even part. This method of construction of minimum functions is called the Gewertz procedure. This procedure has its own merits in numerical computation. In short, what we can say is that, in practice, we constructed many power transfer networks using the Gewertz procedure to generate minimum functions. This chapter is devoted to the generation of minimum functions via the Gewertz method.

Design of Ultra Wideband Power Transfer Networks   Binboga Siddik Yarman
© 2010 John Wiley & Sons, Ltd

8.2 Gewertz Procedure

In this method, the minimum function $F(p)$ is given as[1]

$$F(p) = \frac{x(p)}{y(p)} = \frac{x_0 + x_1 p + x_2 p^2 + \ldots + x_n p^n}{y_0 + y_1 p + y_2 p^2 + \ldots + y_n p^n}$$
$$= \frac{m_1(p) + n_1(p)}{m_2(p) + n_2(p)}$$

(8.1)

where the unknown polynomials $x(p)$ and $y(p)$ will be determined from the given even part function $R(-p^2)$ which is specified in coefficient form as

$$R(-p^2) = \frac{1}{2}[Z(p) + Z(-p)] \geq 0, \forall p = \sigma \geq 0$$

(8.2)

or

$$R(-p^2) = \frac{N(-p^2)}{D(-p^2)} = \frac{\alpha_0 + \alpha_1 p^2 + \alpha_2 p^4 + \ldots + \alpha_n p^{2n}}{Y_0 + Y_1 p^2 + Y_2 p^4 + \ldots + Y_n p^{2n}}$$

(8.3)

In Equation (8.1), m_1 and n_1 represent the even and odd parts of the numerator polynomial $x(p)$ respectively and are given as

$$m_1 = x_0 + x_2 p^2 + x_4 p^4 + \ldots + x_r p^r$$

(8.4)

and

$$n_1 = x_1 p + x_3 p^3 + x_5 p^5 + \ldots + x_r p^r$$

(8.5)

Note that the integer is chosen properly depending if $n =$ even or odd. For example in Equation (8.4) $r = n$ if n is even. Or in Equation (8.5), $r = n - 1$ if $n =$ even.

Similarly, m_2 and n_2 are the even and odd parts of the denominator polynomial:

$$m_2 = y_0 + y_2 p^2 + y_4 p^4 + \ldots + y_r p^{2r}$$

(8.6)

and

$$n_2 = y_1 p + y_3 p^3 + y_5 p^5 + \ldots + y p^r$$

(8.7)

---

[1] Noted that in MATLAB® standard form

$$F(p) = \frac{a(p)}{b(p)} = \frac{a_1 p + a_2 p^2 + \ldots + a_n p + a_{n+1}}{b_1 p + b_2 p^2 + \ldots + b_n p + b_{n+1}}$$

However, in presenting the Gewertz method we prefer the classical representation of polynomials such as $x(p) = x_0 + x_1 p + \ldots + x_n p^n$.

From the polynomial properties of even and odd parts, one can readily see that

$$m_i(p) = m_i(-p) \quad \text{and} \quad n_i(p) = -n_i(-p), i = 1, 2 \tag{8.8}$$

Considering Equation (8.8) and employing Equations (8.4–8.5) and (8.6–8.7), Equation (8.3) can be written as

$$R(-p^2) = \frac{m_1 m_2 - n_1 n_2}{m_2^2 - n_2^2} \tag{8.9}$$

From Equation (8.9) it can immediately be seen that

$$Y(p) = m_2^2 - n_2^2 = (m_2 + n_2)(m_2 - n_2) = y(p)y(-p) > 0, \forall p = \sigma \geq 0 \tag{8.10}$$

Equation (8.10) must have all its roots in quadruple symmetry since $y(p)$ has all its roots in the open LHP, which makes $y(-p)$ have all its roots in the open RHP.

Thus, one can readily construct $y(p)$ on the open LHP roots of $Y(p)$ as

$$y(p) = \sqrt{|Y_n|} \prod_{i=1}^{n} (p - p_i) = y_0 + y_1 p + \ldots + y_n p^n \tag{8.11}$$

where the p_i are the open LHP roots of $Y(p^2)$.

Once $\{y_0, y_1, \ldots, y_n\}$ are computed, we can find the unknown coefficients of $\{x_0, x_1, \ldots, x_n\}$ by Equation (8.3) and Equation (8.9):

$$\alpha_0 + \alpha_1 p^2 + \ldots + \alpha_n p^{2n} = m_1 m_2 - n_1 n_2 = \left(x_0 + x_2 p^2 + x_4 p^4 + \ldots\right)\left(y_0 p + y_2 p^2 + y_4 p^4 + \ldots\right)$$
$$- \left(x_1 p + x_3 p^3 + x_5 p^5 + \ldots\right)\left(y_1 p + y_3 p^3 + y_5 p^5 + \ldots\right)$$

Thus, the above open form yields

$$\alpha_0 = x_0 y_0 \quad \text{or} \quad x_0 = \frac{\alpha_0}{y_0} \tag{8.12}$$

and

$$\alpha_i = y_{2i} x_0 + \sum_{j=1}^{n} (-1)^j y_{2i-j} x_j \tag{8.13}$$
$$\text{such that } y_{2i-j} = 0 \text{ for } 2i - j > n, \text{ and where indices } (i, j) = 1, 2, \ldots, n.$$

Since the term $y_{2i} x_0$ is known, then $\beta_i = \alpha_i - y_{2i} x_0$ is also known with $x_0 = \alpha_0 / y_0$.

Then, we can set up the following linear equations:

$$
\begin{aligned}
\beta_1 &= -y_1 x_1 + y_0 x_2 - 0.\ x_3 + 0.\ x_4 - 0\ \dots \\
\beta_2 &= -y_3 x_1 + y_2 x_2 - y_1\ x_3 + y_0\ x_4 - 0\ \dots \\
\beta_3 &= -y_5 x_1 + y_4 x_2 - y_3\ x_3 + y_2\ x_4 - 0\ \dots \\
\beta_4 &= -y_7 x_1 + y_6 x_2 - y_5\ x_3 + y_4\ x_4 - 0\ \dots \\
\beta_5 &= -y_9 x_1 + y_8 x_2 - y_7\ x_3 + y_6\ x_4 - 0\ \dots
\end{aligned}
$$

In matrix form

$$
\begin{bmatrix} \beta_1 \\ \beta_2 \\ \beta_3 \\ \beta_4 \\ \beta_5 \\ \vdots \end{bmatrix}
=
\begin{bmatrix}
-y_1 & y_0 & 0 & 0 & 0 & 0.. \\
-y_3 & y_2 & -y_1 & y_0 & 0 & 0.. \\
-y_5 & y_4 & -y_3 & y_2 & -y_1 & y_0.. \\
-y_7 & y_6 & -y_5 & -y_4 & -y_3 & y_2.. \\
-y_9 & y_8 & -y_7 & -y_6 & -y_5 & y_4.. \\
\vdots & \vdots & \vdots & \vdots & \vdots & \vdots
\end{bmatrix}
\begin{bmatrix} x_1 \\ x_2 \\ x_3 \\ x_4 \\ x_5 \\ \vdots \end{bmatrix}
$$

where the left hand side is known, as specified by

$$
\beta_i = \alpha_i - y_{2i} x_0 \quad \text{and} \quad x_0 = \alpha_0 / y_0
$$

Let

$$
\beta_i = \sum_{j=1}^{n} (-1)^j y_{2i-j} x_j, \quad \text{with } y_{2j-i} = 0 \text{ for } 2i - j > n \tag{8.14}
$$

and let the entries of the matrix C be given by

$$
C(i,j) = (-1)^j y_{2i-j}, \quad \text{with } y_{2j-i} = 0 \text{ for } 2i - j > n \tag{8.15}
$$

Then,

$$
\beta = CX
$$

or

$$
X = C^{-1} \beta \tag{8.16}
$$

is the solution for the unknown vector $X^T = [x_1\ x_2\ x_3 \dots x_n]$.

This way of determining denominator polynomial $y(p)$ and numerator polynomial $x(p)$ was first introduced by Gewertz in 1933.[2]

Equations (8.12–8.16) can easily be programmed on MATLAB using the algorithm presented in the next section.

---

[2] C. Gewertz, 'Synthesis of a Finite Four Terminal Network from its Prescribed Driving Point Functions and Transfer Function', *Journal of Mathematics and Physics*, Vol. 12, pp. 1–257, 1932–1933.

8.3 Gewertz Algorithm

In this section, we will summarize the above computational steps to generate a positive real minimum function $F(p) = x(p)/y(p) = a(p)/b(p)$ from its even part function

$$R(-p^2) = \frac{A(p^2)}{B(p^2)} = \frac{A_1 p^{2n} + A_2 p^{2(n-1)} + \ldots + A_n p^2 + A_{n+1}}{B_1 p^{2n} + B_2 p^{2(n-1)} + \ldots + B_n p^2 + B_{n+1}}$$

Inputs:

- $A = [A(1)\ A(2)\ldots A(n)\ A(n+1)]$, the polynomial row vector of the numerator.
- $B = [B(1)\ B(2)\ldots B(n)\ B(n+1)]$, the polynomial row vector of the denominator.

Outputs:

- $x0$, $x = [x(1)\ x(2)\ldots x(n)]$
- $y0$, $y = [y(1)\ y(2)\ldots y(n)]$
- $a = [a(1)\ a(2)\ldots a(n)\ a(n+1)]$
- $b = [b(1)\ b(2)\ldots b(n)\ b(n+1)]$

where

$$F(p) = \frac{a(p)}{b(p)} = \frac{a_1 p^n + a_2 p^{n-1} + \ldots + a_n p + a_{n+1}}{b_1 p^n + b_2 p^{n-1} + \ldots + b_n p + b_{n+1}} = \frac{x(p)}{y(p)} = \frac{x_0 + x_1 p + \ldots + x_n p^n}{y_0 + y_1 p + \ldots + y_n p^n}$$

Computational Steps:

Step 1 (Find the roots $p(i)$ of the denominator $B(p^2) = b(p)b(-p)$):

(a) Shuffle vector A to obtain vector α: $\alpha_0 = A(n+1), \alpha_1 = A(n), \alpha_2 = A(n-1), \ldots, \alpha_n = A(1)$.
(b) Set $x = p^2$.
(c) Find the roots xp of $B(x)$.
(d) Set $p_i = -\sqrt{[(xp_i)]}$.

Step 2 (Compute the coefficients of $y(p)$ from the given poles):

$$y(p) = \sqrt{|B_1|} \prod_{i=1}^{n} (p - p_i) = y_0 + y_1 p + \ldots + y_n p^n$$

Step 3 (Compute the entries of β and matrix C):

(a) Set $x_0 = A_{n+1}/y_0$.
(b) Set $\beta_i = \alpha_i - y_{2i} x_0$ with $y_{(2j-i)} = 0$ for $2i - j > n$.
(c) Set $C(i,j) = (-1)^j y_{2i-j}$ with $y_{2j-i} = 0$ for $2i - j > n$.

Step 4 (Solve $\beta = CX$ to obtain $X = C^{-1}\beta$): Here

$$X = \begin{bmatrix} x_1 \\ x_2 \\ \vdots \\ x_n \end{bmatrix}$$

Step 5 (Generate `a(p)` **and** `b(b)` **):**

(a) Shuffle vector `X` upside down to obtain `a(p)`.
(b) Shuffle vector `y` to obtain `b(p)`.

Note that
```
a(1)=x(n), a(2)=x(n-1)...a(n+1)=x0
b(1)=y(n), b(2)=y(n-1)...b(n+1)=y0
```
We have implemented the above steps under MATLAB function `Gewertz`. The function `Gewertz` calls the following related functions:

(a) The function `[V0,V]=swap_down(Q)` shuffles vector `Q` into `V0`, `V(1),V(2),...,V(n)`. In this representation, polynomial `V(p)` is expressed in its classical coefficient form as

$$V(p) = V0 + V(1)p + V(2)p^2 + \ldots + V(n)p^n$$

(b) The function `Q=swap_up(V0,V)` shuffles `V0`, `V` into MATLAB standard form.
(c) The function `[denom,errord]=denominator(p)` generates denominator polynomial `b(p)` in standard MATLAB form with leading coefficient `b(1)=1`.
(d) The function `[beta]=betta(alfa0,alfa,y0,y)` computes the entries of vector β from given $\alpha_0 = A(n+1), \alpha_1 = A(n), \alpha_2 = A(n-1), \ldots, \alpha_n = A(1)$ and $y0$ and y.
(e) The function `C=Mtrx_Gwrt(y0,y)` computes the matrix C as in Equation (8.15)

$$C(i,j) = (-1)^j y_{2i-j} \text{ with } y_{2j-i} = 0 \text{ for } 2i - j < n$$

A listing of function `Gewertz` and its related subfunctions follows.

8.4 MATLAB® Codes for the Gewertz Algorithm

Program Listing 8.1   Function `Gewertz.m`

```
function [a,b,x0,x,y0,y]=Gewertz(A,B)
% Inputs:
% Given Real Part R(-p^2)=A(p^2)/B(p^2)
% where
%    A(p^2)=A(1)p^2n+A(2)p^2(n-1)+...A(n)p^2+A(n+1)
%    B(p^2)=B(1)p^2n+B(2)p^2(n-1)+...B(n)p^2+B(n+1)
```

```
% Outputs: Find Z(p)=X(p)/Y(p)using Gewertz Procedure
% where
%    X(p)=x0+x(1)p+x(2)p^2+...+x(n)p^n
%    Y(p)=y0+y(1)p+y(2)p^2+...+y(n)p^n
% or
% Find Z(p)=a(p)/b(p)
% where
%    a(p)=a(1)p^n+a(2)p^(n-1)+...+a(n)p+a(n+1)
%    b(p)=b(1)p^n+b(2)p^(n-1)+...+b(n)p+b(n+1)
%
[alfa0,alfa]=swap_down(A);
r=roots(B);
poles=-sqrt(r);
[denom,errord]=denominator(poles);
Yn=sqrt(abs(B(1)));
Q=Yn*denom;
[y0,y]=swap_down(Q);
[beta]=betta(alfa0,alfa,y0,y);
C=Mtrx_Gwrt(y0,y);
D=inv(C);
x0=alfa0/y0;
x=D*beta';
a=swap_up(x0,x);
b=swap_up(y0,y);
```

Program Listing 8.2 Function `swap_down.m`

```
function [V0,V]=swap_down(Q)
% This function swaps down a vector
% Input:
%    Vector Q=[Q(1) Q(2) Q(3)...Q(n) Q(n+1)]
% Output
%  V0, Vector V=[V(1)=Q(n) V(2)=Q(n-1)...V(n)=Q(1)]
% Note that if Q(p)=Q(1)p^n+Q(2)p^(n-1)+..+Q(n)p+Q(n+1)
%            then, V(p)=V0+V(1)p+V(2)p^2+...+V(n)p^n
% This operation is called swap-down of a vector.
n=length(Q);
nn=n-1;
V0=Q(n);
for i=1:nn
    V(i)=Q(nn-i+1);
end
```

Program Listing 8.3   Function `swap_up.m`

```
function Q=swap_up(V0,V)
% This function swaps up down a vector.
% Input:
%   V0, Vector V=[V(1)=Q(n) V(2)=Q(n-1)...V(n)=Q(1)]
% Output:
%   Vector Q=[Q(1) Q(2) Q(3)...Q(n) Q(n+1)]
% Note that if V(p)=V0+V(1)p+V(2)p^2+...+V(n)p^n
%          then,Q(p)=Q(1)p^n+Q(2)p^(n-1)+..+Q(n)p+Q(n+1)
% This operation is called swap-up of a vector.
n=length(V);
n1=n+1;
Q(n+1)=V0;
for i=1:n
    Q(i)=V(n-i+1);
end
```

Program Listing 8.4   Function `denominator.m`

```
function [denom,errord]=denominator(p)
% This function computes the denominator polynomial of an
% immitance function from the given poles.
% It should be noted that this form of the denominator is nor-
malized with
% the leading coefficient b(1)=1: denom=(p-p(1))(p-p(2))....(p-
% p(n))
% Input:
%-------- poles p(i) as a MATLAB row vector p
%-------- Residues k(i) of poles p(i)
% Output:
%-------- denom; MATLAB Row-Vector
% ---which includes coefficients of the denominator polynomial
% of an immitance function.
% denom=product[p-p(i)]
%
% --------- Step 1: Determine n
n=length(p);
% -------- Step 2: Form the product term.
pr=[1]; %Define a simple polynomial pr=1.
for j=1:n
   simple=[1 -p(j)];%this is the simple polynomial [p-p(j)]
% Generate multiplication of polynomials starting with 1.
   pr=conv(pr,simple);
```

```
end
denom=real(pr);
errord=imag(pr);
```

Program Listing 8.5   Function betta.m

```
function [beta]=betta(alfa0,alfa,y0,y)
% Z(p)=X(p)/Y(p)
% Given Y(p)=y0+y(1)p+y(2)p^2+y(3)p^3+...+y(n)p^n
% find beta
% Note that
%       alfa(i) are the numerator coefficients of Even-Part
%       R(-p^2)=ALFA(-p^2)/y(p)y(-p)
%       ALFA(P^2)=alfa0+alfa(1)p^2+alfa(2)p^4+...+alfa(n)p^2n
%       beta(i)=alfa(i)-y(2*i)*x0
%    where x0=alfa0/y0
x0=alfa0/y0;
n=length(y);
for i=1:n
  j=2*i;
  if(j<=n)ya=y(j);
  end
  if(j>n)ya=0.0;
  end
 beta(i)=alfa(i)-ya*x0;
end
```

Program Listing 8.6   Function Mtrx_Gwrt.m

```
function C=Mtrx_Gwrt(y0,y)
% This function computes Gewertz Matrix C
% Z(p)=X(p)/Y(p)
% R(-p2)=alfa(-p^2)/[y(p)y(-p)]; Even Part in p domain.
% alfa(p)=alfa0+alfa(1)p+alfa(2)p^2+...+alfa(n)p^n
% Inputs:
%
% y0,y(j) swapped coefficients of the denominator polynomial
%
% Output C Matrix for the Gewertz procedure
%
n=length(y);
```

Program Listing 8.6   (Continued)

```
% Initialization of the first row
%
C(1,1)=-y(1);
C(1,2)=y0;
%
for j=3:n
  C(1,j)=0.0;
end
%
for j=1:n
  k=even_odd(j);
  for i=2:n
     if(k==1)
     ya=0.0;
     r=2*i-j;
     if(r<=n)
      if(r>0)ya=-y(r);end;
      C(i,j)=ya;
      end
    end
    %
     if(k==0)
        ya=0.0;
        r=2*i-j;
        if(r==0)ya=y0;end
        if(r<=n)
        if(r>0)ya=y(r);end;
        C(i,j)=ya;
      end
    end
   end
end
```

Now, let us consider some examples to generate a minimum function from its real part, step by step.

Example 8.1: Let $B(p^2) = -1\,p^{10} + 4p^8 - 2p^6 + 3p^4 - 1\,p^2 + 1$ be the denominator polynomial of the even part of a minimum reactance function $Z(p)$.

Using MATLAB:

(a) Find the roots of $B(p^2)$.
(b) Construct $y(p)$ using Equation (8.11).

(c) Convert the coefficients of y(p) to yield standard MATLAB polynomialvector b(p) which describes Z(p) in MATLAB standard form as

$$Z(p) = \frac{a_1 p + a_2 p^{n-1} + \ldots + a_n p + a_{n+1}}{b_1 p^n + b_2 p^{n-1} + \ldots + b_n p + b_{n+1}}$$

Solution:

(a) In MATLAB the full coefficient form of $B(p)$ is given as a polynomial row vector:

$$B(p^2) = -1\, p^{10} + 4p^8 - 2p^6 + 3p^4 - 1\, p^2 + 1$$

The complete roots of B are generated by writing polynomial coefficients in full vector form by inserting zeros between even terms. Thus, BFull = [-1 0 4 0 -2 0 3 0 -1 0 1] describes a full polynomial of degree $n = 10$ with missing coefficients set to zero. Hence, the roots of Y are found as >> p = roots(Bfull)

$$
\begin{aligned}
&-1.9138 \\
&0.7320 + 0.4164i \\
&0.7320 - 0.4164i \\
&0.5210 + 0.6821i \\
&0.5210 - 0.6821i \\
&-0.5210 + 0.6821i \\
&-0.5210 - 0.6821i \\
&-0.7320 + 0.4164i \\
&-0.7320 - 0.4164i
\end{aligned}
$$

In this case, the open LHP roots p(i) are selected as

$$
\begin{aligned}
p = \\
&-1.9138 \\
&-0.5210 - 0.6821i \\
&-0.5210 + 0.6821i \\
&-0.7320 - 0.4164i \\
&-0.7320 + 0.4164i
\end{aligned}
$$

(b) Then, simple polynomial multiplications [p-p(i)] of Equation (8.11) can be performed utilizing 'function [Q,errord] = denominator(p)'.
    Q(p) is a standard MATLAB polynomial specified as

$$Q(p) = Q_1 p^n + Q_2 p^{n-1} + \ldots + Q_n p + Q_{n+1}$$

In this case, we should shuffle the coefficients of vector Q. This could easily be done using MATLAB function [V0,V] = swap_down(Q). Finally, the coefficients of y = [y0,y(1), y(2)...y(n)] are determined by multiplying [V0 V] by $|B_1| = 1$.
    Hence, we have the following results:
    Q = [1.0000 4.4199 7.7677 7.5047 4.0012 1.0000]

Here, we should note that in function `denominator` the output vector **errord** designates the numerical error in performing sequential polynomial multiplication. For this problem it is given by
```
errord=1.0e-015*[ 0 0 0.2220 0 0 0.0555]
```
So, it is negligible.
```
>> [V0,V]=swap_down (denom)
```
$$V0 = 1.0000$$
$$V = [4.0012 \quad 7.5047 \quad 7.7677 \quad 4.4199 \quad 1.0000]$$
```
>> y=sqrt(abs(Y(1)))*[V0 V]
```
Thus, we find

$$y0 = 1.0000$$
$$y = [4.0012 \quad 7.5047 \quad 7.7677 \quad 4.4199 \quad 1.0000]$$

(c) In order to put the $Z(p)$ in MATLAB standard form we have to topple the coefficients of $x(p)$ and $y(p)$. For this purpose we wrote a MATLAB function called `swap_up`. Hence, using function `swap_up`, we have
```
>> b=swap_up(y0,y)
```

$$b = [1.0000 \quad 4.4199 \quad 7.7677 \quad 7.5047 \quad 4.0012 \quad 1.0000]$$

Example 8.2: Let

$$A(p^2) = 1$$

and

$$B(p^2) = -1\, p^{10} + 4p^8 - 2p^6 + 3p^4 - 1\, p^2 + 1$$

Find vector β as described by Equation (8.14).
Solution: Five terms of A are missing. Therefore, $A = [0 \, 0 \, 0 \, 0 \, 0 \, 1]$.
Firstly, we shuffle the entries of A upside down to generate vector α (or `alfa`). Hence, `alfa0 = 1`, `alfa = [0 0 0 0 0]`. Then, we must generate $y(p)$ as in Example 8.1. Thus, `y0=1`, `y = [ 4.0012 7.5047 7.7677 4.4199 1.0000]`.
At this point we need to compute $x0 = $ `alfa0/y0 = 1`.
Now, we are ready to generate

$$\beta_i = \alpha_i - y_{2i}x_0$$

Note that `y(6) =y(7) =y(8) =y(9) =y(10) = 0.`
Thus,

```
beta(1)=alfa(1)-y(2)*x0=-7.5047
beta(2)=alfa(2)-y(4)*x0=-4.4199
beta(3)=alfa(3)-y(6)*x0=0
beta(4)=alfa(4)-y(8)*x0=0
beta(5)=alfa(5)-y(10)*x0=0
```

For this example we wrote a MATLAB program called `Test_beta` which yields vector `beta`, as follows.

Program Listing 8.7 `Test_beta.m`

```
% Test betta
clear
y0=1.0
y=[ 4.0012 7.5047 7.7677 4.4199 1.0000]
A=[0 0 0 0 0 1]
[alfa0,alfa]=swap_down(A)
[beta]=betta(alfa0,alfa,y0,y)
```

Execution of `Test_beta` yields that
alfa0 = 1, alfa = [0 0 0 0 0]
and
beta = [−7.5047 −4.4199 0 0 0]
which verifies the above result.

Example 8.3: Let a minimum function $F(p)$ be specified with its real part $R(-p^2) = A(p^2)/B(p^2)$ such that $A(p^2) = 1$, $B(p^2) = -1\,p^{10} + 4p^8 - 2p^6 + 3p^4 - 1\,p^2 + 1$.

Find the matrix C as specified by Equation (8.15).
Solution: For this example we have written a MATLAB program called `Test_MatrixC`.

Program Listing 8.8

```
% Test_Matrix C
clear
A=[0 0 0 0 0 1];
B=[-1 4 -2 3 -1 1];
%
[alfa0,alfa]=swap_down(A);
r=roots(B);
poles=-sqrt(r);
[denom,errord]=denominator(poles);
Yn=sqrt(abs(B(1)));
Q=Yn*denom;
[y0,y]=swap_down(Q);
[alfa0,alfa]=swap_down(A);
[beta]=betta(alfa0,alfa,y0,y)
C=Mtrx_Gwrt(y0,y);
```

In this program the entries of matrix C are computed in function `Mtrx_Gwrt` as

$$C(i,j) = (-1)^j y_{2i-j} \text{ with } y_{2j-i} = 0 \text{ for } 2i - j > n$$

Thus, execution of `Test_MatrixC` yields

$C =$

-4.0012 1.0000 0 0 0

-7.7677 7.5047 - 4.0012 1.0000 0

-1.0000 4.4199 - 7.7677 7.5047 - 4.0012

0 0 -1.0000 4.4199 -7.7677

0 0 0 0 -1.0000

Example 8.4: Let `A = [0 0 0 0 0 1]` and `B = [-1 4 -2 3 -1 1]` be the numerator and denominator polynomials of the even part of a minimum reactance function `Z(p)`. Find the analytic form of `Z(p) = a(p)/b(p)`.

Solution: For this example we have already computed `beta` and matrix `C`. Thus, we can immediately generate `X = [C⁻¹][beta]`. Thus, we find

X =
2.6316

3.0246

1.7689

0.4002

0

with `x0 = A(n+1)/sqrt(B(n+1)) = 1`.

For this example, we wrote a MATLAB program called `Test_Gewertz`.

```
% Test_Gewertz
clear
A = [0 0 0 0 0 1];
B = [-1 4 -2 3 -1 1];
%
[a,b,x0,x,y0,y] = Gewertz(A,B)
```

which yields

```
a = [0 0.4002 1.7689 3.0246 2.6316 1.0000]
b = [1.0000 4.4199 7.7677 7.5047 4.0012 1.0000]
```

8.5 Comparison of the Bode Method to the Gewertz Procedure

In Chaper 7, the Bode method is employed to generate a minimum function $F(p) = a(p)/b(p)$ from its even part by means of the residues evaluated at the LHP poles. Then, by straightforward multiplications and additions, the rational form of the immittance function is determined. In the Gewertz procedure, denominator polynomial $b(p)$ is generated as in the Bode method. On the other hand, construction of the numerator polynomial $a(p)$ requires the solution of a linear equation set. From the numerical point of view, solution of this linear equation set may demand somewhat more elaboration in the computation of residues. Obviously, each method has its own pros and cons.

In the residue evaluation, one is required to deal with complex algebra which in turn results in numerical errors. In the Gewertz procedure, however, one only deals with real numbers, which simplifies the computations, but matrix inversion is inevitable and requires special attention when dealing with near singular matrices.

Nevertheless, for many practical problems we deal with polynomials of degree less than 20, which does not cause any serious problems in matrix inversion. The effect of computational differences may be observed in some numerical tests.

Example 8.5: The minimum reactance driving point function of Figure 8.1 is given by

$$Z(p) = \frac{30p^5 + 30p^4 + 23p^3 + 17p^2 + 4p + 1}{24p^6 + 24p^5 + 68p^4 + 44p^3 + 23p^2 + 3p + 1}$$

which yields the even part

$$R(p^2) = \frac{A(p^2)}{B(p^2)} = \frac{576p^8 + 672p^6 + 244p^4 + 28p^2 + 1}{576p^{12} + 2688p^{10} + 3616p^8 + 1096p^6 + 401p^4 + 37p^2 + 1}$$

(a) Find $Z(p)$ from $R(p^2)$ via the Bode method.
(b) Find $Z(p)$ from $R(p)$ via the Gewertz procedure.
(c) Compare the results.

Solution: For this example, a MATLAB program called `Test_Ch8Example5` was written to produce the minimum reactance driving point impedance function from the given numerator `[A]` and denominator `[B]` polynomials of the even part using both the Gewertz and Bode methods. Function `RtoZ` employs the Bode method to generate minimum functions. Similarly, function Gewertz employs the `Gewertz` method.

The results are given as follows:

(a) `Z(p)` via the Bode procedure yields

```
aBode = [0  30.0000  30.0000  23.0000  17.0000  4.0000  1.0000]
bBode = [24.0000 24.0000 68.0000 44.0000 23.0000 3.0000 1.0000]
```
as expected.

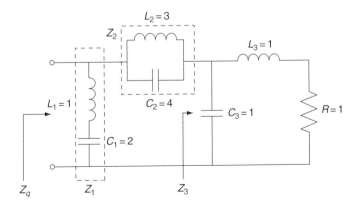

Figure 8.1   Circuit described in Example 8.5

(b) Using the Gewertz procedure, we have

```
aGew = [0  30.0000  30.0000  23.0000  17.0000  4.0000  1.0000]
bGew = [24.0000  24.0000  68.0000  44.0000  23.0000  3.0000  1.0000]
```

(c) In order to get some idea about the size of the numerical error occurring throughout the computations, we look at the norm of the error vectors formed on the difference vectors of both the numerator
and denominator polynomials.

Error_num is defined as the norm of the difference vector of [a-aBode] where [a] is the original
numerator polynomial of Z(p) = a(p)/b(p).

In other words, Error_num = norm(a-aBode). Similarly, Error_denom = norm(b-bBode).
Thus, ErrorT = Error_num + Error_denom gives an idea of the total error.

Hence, Error_T for the Bode method is found as
Error_Bode = 1.6365e-013
Similarly, Error_T for the Gewertz procedure is found as
Error_Gew = 1.4980e-013

As we can see from the above computations, the Gewertz procedure results in a slightly better error
than that of the Bode method. However, the difference is negligible.

Overall, the norm of error is on the order of 10^{-13} for both cases, which is acceptable for most practical
problems.

Here is the list of the MATLAB code Test_Ch4Example5 together with its functions.

Program Listing 8.9 Test_Ch8Example5

```
% Test_Ch8Example5
% Comparison of Bode and Gewertz Methods
% Inputs: Even Part of Z(p)=a(p)/b(p)
   A=[0 0 576 672 244 28 1];% given in p-domain
   B=[576 2688 3616 1096 401 37 1];% given in p-domain
% Original minimum reactance driving point impedance
   a=[0 30 30 23 17 4 1];% Provided for comparison purpose
   b=[24 24 68 44 23 3 1];%provided for comparison purpose
% Generate minimum reactance impedance via Gewertz Procedure:
% ZGew(p)=aGew(p)/bGew(p)
[aGew,bGew,x0,x,y0,y]=Gewertz(A,B);%Gewertz Method
%Given R(p^2)=A(p^2)/B(p^2) find Z via Bode Method.
[aBode,bBode]=RtoZ(A,B);%A and B are specified in p-domain.
b1=sqrt(abs(B(1)));
% Note: function RtoZ yields a(p) and b(p) via Bode Method with
leading
% coefficient b(1). Therefore, we have to multiply aBode and
bBode by b1.
aBode=b1*aBode;
bBode=b1*bBode;
% Error in Gewertz:
Error_Gew=norm(a-aGew)+norm(b-bGew)
Error_Bode=norm(a-aBode)+norm(b-bBode)
```

The related MATLAB functions will be presented at the end of this chapter.

Let us now, consider a more complicated problem to investigate the order of error in computations.

Example 8.6: Let the real part $A(\omega^2)/B(\omega^2)$ be specified in the ω domain with standard MATLAB vectors

$$A = [1\ 2\ 3\ 4\ 5\ 6\ 7\ 8\ 9\ 10\ 11\ 12\ 13\ 14\ 15\ 16\ 17\ 18\ 19\ 20]$$

$$B = [20\ 19\ 18\ 17\ 16\ 15\ 14\ 13\ 12\ 11\ 10\ 9\ 8\ 7\ 6\ 5\ 4\ 3\ 2\ 1]$$

(a) Find the corresponding minimum function $ZBode(p) = aBode(p)/bBode(p)$ using the Bode method.

(b) Find the corresponding minimum function $ZGew(p) = aGew(p)/bGew(p)$ using the Gewertz Procedure.

(c) Obtain the even part

$$RBode(p^2) = \frac{1}{2}[ZBode(p) + ZBode(-p)] = \frac{ABode(p^2)}{BBode(p^2)}$$

using the Bode method.

Table 8.1 Results of Example 8.6 minimum functions $ZBode(p) = aBode(p)/bBode(p)$ and $ZGew(p) = aGew(p)/bGew(p)$

Computed numerator vector aBode(p)	Computed denominator vector bBode(p)	Computed numerator vector aGew(p)	Computed numerator vector bGew(p)
1.0e + 005*	1.0e + 004 *	1.0e + 005*	1.0e + 004*
0.0000	0.0004	0.0000	0.0004
0.0004	0.0053	0.0004	0.0053
0.0046	0.0314	0.0046	0.0314
0.0269	0.1222	0.0269	0.1222
0.1033	0.3512	0.1033	0.3512
0.2930	0.7912	0.2930	0.7912
0.6497	1.4472	0.6497	1.4472
1.1660	2.1984	1.1660	2.1984
1.7302	2.8130	1.7302	2.8130
2.1512	3.0574	2.1512	3.0574
2.2571	2.8335	2.2571	2.8335
2.0030	2.2386	2.0030	2.2386
1.4999	1.5015	1.4999	1.5015
0.9408	0.8476	0.9408	0.8476
0.4877	0.3968	0.4877	0.3968
0.2045	0.1506	0.2045	0.1506
0.0669	0.0447	0.0669	0.0447
0.0161	0.0098	0.0161	0.0098
0.0026	0.0014	0.0026	0.0014
0.0002	0.0001	0.0002	0.0001

(d) Obtain the even part

$$RGew(p^2) = \frac{1}{2}[ZGew(p) + ZGew(-p)] = \frac{AGew(p^2)}{BGew(p^2)}$$

employing the Gewertz procedures.

(e) Compute the error vectors

```
Error_ABode = norm{A-Abode} + norm{B-BBode}
Error_Gew = norm{A-AGew} + norm{B-BGew}
```
and comment on the differences.

Solution: For this example we wrote a MATLAB program called `Test_Ch8Example6`. In this program, firstly the w domain vectors A and B are converted to p domain vectors as AP and BP using MATLAB function `polarity`. Then, {`aBode(p)`, `bBode(p)`} and {`aGew,bGew`} pairs were generated by means of MATLAB functions [`aBode,bBode`] = `RtoZ(AP,BP)` and [`aGew,bGew`] = `Gewertz(AP,BP)` respectively.

Eventually, the error terms are given as follows:

```
Error_ABode = norm{AP-Abode} + norm{BP-BBode}
```
and
```
Error_Gew = norm{AP-AGew} + norm{BP-BGew}
```

Table 8.2   Computation of error vectors `Error_Bode`

Vector $AP(-p^2)$	Vector $BP(-p^2)$	Computed numerator vector A_Bode(p)	Computed denominator vector B_Bode(p)	ErrorA_Bode 1.0e-005*	ErrorB_Bode 1.0e-006*
-1	-20	-1.0000	-20.0000	0.0000	0.0000
2	19	2.0000	19.0000	0.0000	0.0000
-3	-18	-3.0000	-18.0000	-0.0000	0.0000
4	17	4.0000	17.0000	0.0000	0
-5	-16	-5.0000	-16.0000	-0.0010	-0.0001
6	15	6.0000	15.0000	-0.0086	0.0023
-7	-14	-7.0000	-14.0000	-0.0401	0.0054
8	13	8.0000	13.0000	-0.0784	-0.0784
-9	-12	-9.0000	-12.0000	-0.2130	-0.1426
10	11	10.0000	11.0000	-0.0709	-0.3663
-11	-10	-11.0000	-10.0000	-0.0489	-0.5731
12	9	12.0000	9.0000	0.0341	-0.2957
-13	-8	-13.0000	-8.0000	-0.0536	-0.1554
14	7	14.0000	7.0000	-0.0060	-0.0284
-15	-6	-15.0000	-6.0000	-0.0044	-0.0042
16	5	16.0000	5.0000	-0.0022	-0.0011
-17	-4	-17.0000	-4.0000	-0.0002	-0.0001
18	3	18.0000	3.0000	-0.0000	-0.0000
-19	-2	-19.0000	-2.0000	0	0
20	1	20.0000	1.0000	0.0000	0.0000

```
Error_Bode = norm(AP-A_Bode) + norm(BP-B_Bode)
Error_Bode = 3.3197e-006
```

Table 8.3 Computation of error vectors `Error_Gewertz`

Vector $AP(-p^2)$	Vector $BP(-p^2)$	Computed numerator vector `A_Gew(p)`	Computed denominator vector `B_Gew(p)`	ErrorA_Gew 1.0e-005*	ErrorB_Gew 1.0e-006*
-1	-20	-1.0000	-20.0000	0	0.0000
2	19	2.0000	19.0000	-0.0000	0.0000
-3	-18	-3.0000	-18.0000	0.0004	0.0000
4	17	4.0000	17.0000	0.0048	0
-5	-16	-5.0000	-16.0000	-0.0579	-0.0001
6	15	6.0000	15.0000	-0.9942	0.0023
-7	-14	-6.9999	-14.0000	-5.7320	0.0054
8	13	8.0002	13.0000	-22.4905	-0.0784
-9	-12	-8.9993	-12.0000	-70.5285	-0.1426
10	11	10.0016	11.0000	-158.2742	-0.3663
-11	-10	-10.9976	-10.0000	-235.1600	-0.5731
12	9	12.0023	9.0000	-229.7737	-0.2957
-13	-8	-12.9985	-8.0000	-146.3642	-0.1554
14	7	14.0006	7.0000	-57.4546	-0.0284
-15	-6	-14.9999	-6.0000	-11.5073	-0.0042
16	5	16.0000	5.0000	-0.2979	-0.0011
-17	-4	-17.0000	-4.0000	0.2705	-0.0001
18	3	18.0000	3.0000	0.0406	-0.0000
-19	-2	-19.0000	-2.0000	0.0022	0
20	1	20.0000	1.0000	0	0.0000

Error_Gew = norm(AP-A_Gew) + norm(BP-B_Gew)
Error_Gew = 4.0445*e-003

Here, the results are summarized as in the following tables (Tables 8.1–8.3).

Clearly, in generating the minimum function from its even part, the Bode method yields much better numerical stability than that of the Gewertz procedure, as expected.

The reason for the improved numerical stability in the Bode method is that the Gewertz procedure requires the solution of a set of linear equations whereas the Bode method simply reaches a the solution via the computation of complex residues. Obviously, in equation solvers, one requires more laborious numerical computations than in straightforward residue computations. For Example 8.6, it is observed that the size of the error in the Gewertz procedure is approximately 1000 times more than that of the error for the Bode method.

In other words, we are comparing Error_Gew = 4.0445*e-003 to that of Error_Bode = 3.3197e-006 or Error_Ratio = Error_Gew/Error_Bode = Error_Ratio = 1.2183e+003.

Finally, we should make the following important comment: In generating minimum functions, if one is dealing with a high degree of rational functions of degree greater that $n = 7$, we strongly recommend employing the Bode method over the Gewertz procedure.

Before we close this chapter, let us model the measured minimum immittance data by means of the Gewertz procedure as we did in Chapter 7 employing the Bode method.

8.6 Immittance Modeling via the Gewertz Procedure

At the end of Chapter 7, when employing the 'parametric method', immittance data modeling via the direct approach was introduced. In this section, we use the same generic approach to model the measured minimum immittance data from its real part; then, the analytic form of the immittance function is obtained using the Gewertz procedure.

For this purpose, we developed a MATLAB program which, from an algorithmic point of view, includes two major parts.

In the first part of the algorithm, the real part data is modeled as a non-negative, even, rational function in the angular frequency domain using the direct approach, exactly or implemented in Chapter 7. In other words, using the least mean square optimization routine lsqnonlin, $R(-p^2) = A(p^2)/B(p^2)$ is generated by means of the MATLAB function [A,B] = Evenpart_Direct(ndc,W,a0,c).

In the second part of the algorithm, the analytic form of the minimum immittance function is derived via the Gewertz method by employing the MATLAB function [a,b,x0,x,y0,y] = Gewertx(A,B).

Now, let us consider an example to show immittance data modeling via the Gewertz procedure.

Example 8.7: Repeat Example 7.16 using the Gewertz procedure.

Solution: For this example, we developed a MATLAB program called Example8_7. The code is almost the same as Example7_16 except that the function [a,b]=RtoZ(A,B) is replaced with the function [a,b,x0,x,y0,y] = Gewertz(A,B) as listed in Program Listing 8.15.

Optimization is run with the following inputs:

- Enter transmission zeros at DC ndc = 0.
- Enter finite transmission zeros at [W] = [...] = 0.
- Enter a0 = 1.
- Enter initial values for [c0] = [...] = [1 1 1].
- ntr = 1, yes, model with transformer; ntr = 0, no transformer: ntr = 0. In this example we selected the no transformer case. Therefore, we set ntr = 0.

Thus, the result of optimization reveals the following real part $R = A(p^2)/B(p^2)$:

$$A = [0 \quad 0 \quad 0 \quad 1]$$
$$B = [-27.183 \quad -8.2099 \quad 2.0169 \quad 1]$$

and the minimum reactance function $Z = a(p)/b(p)$ is obtained as

$$a = [0 \quad 3.83 \quad 3.1676 \quad 1]$$
$$b = [5.2137 \quad 4.3119 \quad 2.5704 \quad 1]$$

Or, in open form,

$$Z(p) = \frac{3.83p^2 + 3.1676p + 1}{5.2137p^3 + 4.3119p^2 + 2.5704p + 1}$$

In order to compare the above result to that of Example 7.16, the leading coefficient of the denominator polynomial $b(p)$ must be set to 1.

Hence, dividing the numerator and denominator polynomials of $Z(p)$ by 5.213 we obtain

$$Z(p) = \frac{0.7346p^3 + 0.607\ 54p^2 + 0.1918}{p^4 + 0.82704p^3 + 0.49301p^2 + 0.1918}$$

which is almost the same result as obtained in Example 7.16.

From a numerical stability point of view, we must emphasize once again that the Bode approach results in more reliable solutions than the Gewertz method.

We will conclude this chapter by providing the MATLAB programs for Examples 8.6 and 8.7.

Program Listing 8.10

```
% Test_Ch8Example6
clear
% Numerical Comparison of Bode and Gewertz Methods
% Inputs: R=A(p^2)/B(p^2)=Even Part of {Z(p)=a(p)/b(p)}
        A=[1 2 3 4 5 6 7 8 9 10 11 12 13 14 15 16 17 18 19 20];%
% given in w-domain
        B=[20 19 18 17 16 15 14 13 12 11 10 9 8 7 6 5 4 3 2 1];%
% given in w-domain
%Step 1: Convert w-domain vectors to p-domain
        AP=polarity(A);
        BP=polarity(B);
%
% Step 2:
%Generate minimum reactance impedance via Gewertz Procedure:
% ZGew(p)=aGew(p)/bGew(p)
[aGew,bGew,x0,x,y0,y]=Gewertz(AP,BP);%Gewertz Method
%
% Step 3:
%Given R(p^2)=A(p^2)/B(p^2) find Z via Bode Method.
[aBode,bBode]=RtoZ(AP,BP);%A and B are specified in p-domain.
b1=sqrt(abs(B(1)));
% Step 4:
% Function RtoZ yields a(p) and b(p) via Bode Method with leading
% coefficient b(1). Therefore, we have to multiply aBode and
bBode by b1.
aBode=b1*aBode;
bBode=b1*bBode;
%
% Step 5:Error in Bode Method
% Step 5a: Compute even part
aBode_p=polarity(aBode);
bBode_p=polarity(bBode);
%
ABode=0.5*conv(aBode,bBode_p)+0.5*conv(aBode_p,bBode);
```

Program Listing 8.10   (Continued)

```
A_Bode=clear_oddpower(ABode);
BBode=conv(bBode,bBode_p);
B_Bode=clear_oddpower(BBode);
% Step 5b: Compute Eroor_Bode
ErrorA_Bode=(AP-A_Bode);
ErrorB_Bode=(BP-B_Bode);
Error_Bode=norm(ErrorA_Bode)+norm(ErrorB_Bode);
%
% Step 6: Error in Gewertz Method
% Step 6a: Compute even part
aGew_p=polarity(aGew);
bGew_p=polarity(bGew);
AGew=0.5*conv(aGew,bGew_p)+0.5*conv(aGew_p,bGew);
A_Gew=clear_oddpower(AGew);
BGew=conv(bGew,bGew_p);
B_Gew=clear_oddpower(BGew);
% Step 6b: Compute Eroor_Gewertz
ErrorA_Gew=(AP-A_Gew);
ErrorB_Gew=(BP-B_Gew);
Error_Gew=norm(ErrorA_Gew)+norm(ErrorB_Gew);
```

Program Listing 8.11

```
function [a,b]=RtoZ(A,B)
% This MATLAB function generates a minimum function Z(p)=a(p)/
% b(p)
%    from its even part specified as R(p^2)=A(p^2)/B(p^2)
%    via Bode (or Parametric)approach
% Inputs:
% A=[A(1) A(2) A(3)...A(n+1)]; for many practical cases we set
A(1)=0.
% B=[B(1) B(2) B(3)...B(n+1)]
% Output:
%     Z(p)=a(p)/b(p) such that
%     a(p)=a(1)p^n+a(2)p^(n-1)+...+a(n)p+a(n+1)
%     b(p)=b(1)p^n+b(2)p^(n-1)+...+b(n)p+a(n+1)
% Generation of an immittance Function Z by means of Parametric
% Approach
% In parametric approach Z(p)=Z0+k(1)/[p-p(1)]+...+k(n)/[p-
p(n)]
%         R(p^2)=Even{Z(p)}=A(-p^2)/B(-p^2) where Z0=A(n+1)/
B(n+1).
%
% Given A(-p^2)>0
```

```
% Given B(-p^2)>0
%
BP=polarity(B);
AP=polarity(A);
% Computational Steps
% Given A and B vectors in p-domain
% Compute poles p(1),p(2),...,p(n)and the residues k(i) at poles
% p(1),p(2),...,p(n)
[p,k]=residue(AP,BP); % Note function residue takes w domain A
% and B
%
% Compute numerator and denominator polynomials
Z0=abs(A(1)/B(1));
[num,errorn]=num_Z0(p,Z0,k)
[denom,errord]=denominator(p);
%
a=num;
b=denom;
```

Program Listing 8.12

```
function [p,k]=residue(A,B)
% This function computes the residues k of a given real part
R(w^2)
% R(w^2)=A(w^2)/B(w^2)
% It should be noted the R(w^2) is given in w domain as an even
% function of
% w rather than complex variable p=sigma+jw.
%
% Inputs:
% ------- A(w^2); numerator polynomial
% ------- B(w^2); denominator polynomial
%
%Step 1: Find p domain versions of A and B
% That means we set w^2=-p^2
AA=polarity(A);
BB=polarity(B);
% Step 2: Find the LHP poles of R(p^2)
x=roots(BB);
p=-sqrt(x);
% Step 3: Compute the product terms
prd=product(p);
n=length(p);
```

Program Listing 8.12   (Continued)

```
% Step4: Generate residues
for j=1:n
    y=p(j)*p(j);
    Aval=polyval(AA,y);
    k(j)=(-1)^n*Aval/p(j)/B(1)/prd(j);
end
```

Program Listing 8.13

```
function [denom,error]=denominator(p)
% This function computes the denominator polynomial of an
% immitance function from the given poles.
% It should be noted that this form of the denominator is nor-
% malized with
% the leading coefficient b(1)=1: denom=(p-p(1))(p-p(2))....(p-
% p(n))
% Input:
% -------- poles p(i) as a MATLAB row vector p
% -------- Residues k(i) of poles p(i)
% Output:
% -------- denom; MATLAB Row-Vector
% ---which includes coefficients of the denominator polynomial
% of an immitance function.
% denom=product[p-p(i)]
%
% --------- Step 1: Determine n
n=length(p);
% --------- Step 2: Form the product term.
pr=[1]; %Define a simple polynomial pr=1.
for j=1:n
    simple=[1 -p(j)];%this is the simple polynomial [p-p(j)]
% Generate multiplication of polynomials starting with 1.
    pr=conv(pr,simple);
end
denom=real(pr);
errord=imag(pr);
```

Program Listing 8.14

```
function [num,errorn]=num_Z0(p,Z0,k)
% This function computes the numerator polynomial of an
% immitance function: Z(p)=Z0+sum{k(1)/[p-p(i)]
% where we assume that Z0=A(n+1)/B(n+1)which is provided as
% input.
%
% Input:
% -------- poles p(i) of the immitance function Z(p)
% as a MATLAB row vector p
% -------- Residues k(i) of poles p(i)
% Output:
% -------- num; MATLAB Row-Vector
% which includes coefficients of numerator polynomial of an
% immitance function.
% num=Sum{k(j)*product[p-p(i)]} which skips the term when j=i
%
% ----- Step 1: Determine total number of poles n
%
n=length(p);
nn=n-1;
%
% ----- Step 2: Generation of numerator polynomials:
% numerator polynomial=sum of
% sum of
% {Z0*[p-p(1)].[p-p(2)]...(p-p(n)]; n the degree-full product
% +k(1)*[p=p(2)].[p-p(3)]..[p-p(n)];degree of (n-1); the term
% with p(1)is skipped.
% +k(2)*[p-p(1)].[p-p(3)]..[p-p(j-1)].[p-p(j+1)]..[p-p(n)];de-
% gree of(n-1)-the term with p(2)is skipped
% +....................................
% +k(j)*[p-p(1)].[p-p(2)]..[p-p(j-1)].[p-p(j+1)]..[p-p(n)];de-
% gree of (n-1)-the term with p(j)is skipped.
% +..............................
% +k(n)[p-p(1)].[p-p(2)]...[p-p(n-1)];degree of (n-1)-the term
% with p(n)is
% skipped.
%
% Note that we generate the numerator polynomial within 4 steps.
% In Step 2a, product polynomial pra of k(1) is evaluated.
% In Step 2b, product polynomial prb of k(j) is evaluated by
% skipping the term when i=j.
% In Step 2c, product polynomial prc of k(n) is evaluated.
% In Step 2d, denominator of Z0 is generated.
```

Program Listing 8.14   (Continued)

```
%
% Step 2a: Generate the polynomial for the residue k(1)
pra=[1];
for i=2:n
  simpA=[1 -p(i)];
% pra is a polynomial vector of degree n-1; total number of
entries are n.
  pra=conv(pra,simpA);% This is an (n-1)th degree polynomial.
end
na=length(pra);
% store first polynomial onto firs row of A i.e. A(1,:)
for r=1:na
  A(1,r)=pra(r);
end
% Step 2a: Compute the product for 2<j<(n-1)
for j=2:nn
  prb1=[1];
    for i=1:j-1
    simpB=[1 -p(i)];
    prb1=conv(prb1,simpB);
    end
  % Skip j th term
  prb2=[1];
    for i=(j+1):n
    simpB1=[1 -p(i)];
    prb2=conv(prb2,simpB1);
  end
  prb=conv(prb1,prb2);
  nb=length(prb);
%
% Store j polynomials on to j th row of A; i.e. A(j,:)
for r=1:nb
A(j,r)=prb(r);
end
%
end
% Step 2c: Compute the product term for j=n
prc=[1];
for i=1:nn
  simpC=[1 -p(i)];
  prc=conv(prc,simpC);
end
nc=length(prc);
% store n the polynomial onto n the row of A(n,:)
for r=1:nc
A(n,r)=prc(r);
```

```
end
%
%
for i=1:n
   for j=1:n
      C(i,j)=k(i)*A(i,j);
   end
end
%
% -------- Step 4: Generate the numerator as a MATLAB row vec-
tor.
for i=1:n
D(i)=0;%Perform the sum operation to compute numerator polyno-
mial
end
for j=1:n
   for r=1:n
   D(j)=D(j)+C(r,j);
   end;% Here is the numerator polynomial of length n.
end
[denom,errord]=denominator(p);
prd_n=Z0*denom; % this is n th degree polynomial vector with
length n+1
a(1)=prd_n(1);
for i=2:(n+1)
   a(i)=D(i-1)+prd_n(i);
end
%
num=real(a);
errorn=imag(a);
```

Program Listing 8.15

```
% Main Program Example8_7
clc
clear
% Nonlinear Immitance Modeling via Gewertz Approach:
%
%         Direct Modeling Method via Gewertz Approach
%
% Immitance F(p)=a(p)/b(p)
% Real Part of F(p)is expressed as in direct computational
% techniques R(w^2)=AA(w^2)/BB(w^2)
% AA(w^2) it is fixed by means of transmission zeros
% BB(w^2)=(1/2)[C^2(w)+C^2(-w)] where C(w)=C(1)w^n+C(2)w^(n-
% 1)+...+C(n)w+
```

Program Listing 8.15   (Continued)

```
% Note that AA(w^2) is computed by means of function R_num which
% is
% specified in p-domain as
% A(-p^2)=(a0^2)*R_Num=(ndc,W)=A(1)p^2m+A(2)p^2(m-
% 1)+...+A(m)p^2+A(m+1)
% Inputs:
% Input Set #1:
% WA.txt; % Sampling Frequency Array to be loaded
% REA.txt;% Real Part of the Immitance data to be loaded
% XEA.txt;% Imaginary part of the immitance data to be loaded
%        ndc, W(i), B1: Transmission zeros of the model immitance
%        W=[...] Finite transmission zeros of even part
% Input Set #2:
% a0: Constant multiplier of the denominator as a0^2
%        c0; Initial coefficients of BB(w^2)
%        ntr: Model with or without transformer/
%        ntr=1> Yes Transformer. ntr=0>No transformer
% Output:
% Optimized values of a0, and coefficients of c(p)
% Direct form of Even Part R(p^2)=A(-p^2)/B(-p^2)
% Polynomials a(p) and b(p) of Minimum function F(p)
% Step 1: Load Data to be modeled
load WA.txt; % Sampling Frequency Array
load REA.txt;% Real Part of the Immitance data
load XEA.txt;% Imaginary part of the immitance data
% Step 2: Enter zeros of the real part data
ndc=input('Enter transmission zeros at DC ndc=')
W=input('Enter finite transmission zeros at [W]=[....]=')
%
a0=input('Enter a0=')
[c0]=input('Enter initials for [c0]=[....]=')%sigma(i): real
roots of b(p)
ntr=input('ntr=1, Yes, model with Transformer, ntr=0 No
Transformer ntr=')
%
Nc=length(c0);
%
% Step 3: Generate inverse of the immitance data
NA=length(WA);
cmplx=sqrt(-1);
for i=1:NA
    Z(i)=REA(i)+cmplx*XEA(i);
    Y(i)=1/Z(i);
    GEA(i)=real(Y(i));
    BEA(i)=imag(Y(i));
end
% Step 4: Call nonlinear data fitting function
```

```
%%%%%%%%%%% Preparation for the optimization
OPTIONS=OPTIMSET('MaxFunEvals',20000,'MaxIter',50000;
%%%%%%%%%%%%%%%%%%%%%%%%%%%%%%%%%%%%%%%%%%%%%%%%%%%%%%%%%%%
if ntr==1; x0=[c0 a0];Nx=length(x0);end;
% Yes transformer case
if ntr==0; x0=c0;Nx=length(x0);end;
% No transformer case
%
% Call optimization function lsqnonlin:
x=lsqnonlin('direct_Gewertz',x0,[],[],OPTIONS,ntr,ndc,W,a0,WA,R-
EA,XEA);
%
% Separate x into sigma, alfa and beta
if ntr==1; %Yes transformer case
    for i=1:Nx-1;
    c(i)=x(i);
    end;
    a0=x(Nx);
end
if ntr==0;% No Transformer case
    for i=1:Nx;
        c(i)=x(i);
    end;
end
%
% Step 5: Generate analytic form of Fmin(p)=a(p)/b(p)
% Generate B(-p^2)
        C=[c 1];
        BB=Poly_Positive(C);
% This positive polynomial is in w-domain
        B=polarity(BB);
% Now, it is transferred to p-domain
% Generate A(-p^2) of R(-p^2)=A(-p^2)/B(-p^2)
nB=length(B);
        A=(a0*a0)*R_Num(ndc,W);
% A is specified in p-domain
nA=length(A);
if (abs(nB-nA)>0)
  A=fullvector(nB,A);% work with equal length vectors
% Given nB>nA; nA=length(A), fill zeros on A
end
% Generation of minimum immitance function using
% Gewertz Procedure
% Here A and B are specified in p-domain
% Generate analytic form of Fmin(p)=a(p)/b(p)
[a,b,x0,x,y0,y]=Gewertz(A,B);
% Generate a(p) and b(p) using Gewertz Method
        NA=length(WA);
        cmplx=sqrt(-1);
```

Program Listing 8.15 (Continued)

```
% Step 6: Print and Plot results
        for j=1:NA
            w=WA(j);
            p=cmplx*w;
            aval=polyval(a,p);
            bval=polyval(b,p);
            F=aval/bval;
            %
            RE(j)=real(F);
            XE(j)=imag(F);
            XF(j)=XEA(j)-XE(j);
% Foster part of the given impedance
%
            error1(j)=abs(REA(j)-RE(j));
            error2(j)=abs(XEA(j)-XE(j));
            fun(j)=error1(j)+error2(j);
        end
        figure(1)
        plot(WA,RE,WA,REA)
        title('Fit between given data and the analytic form of the
real part')
        xlabel('normalized frequency w')
        ylabel('R(p=jw) and RE(w)')
        %
        figure(2)
        plot(WA,XE,WA,XEA)
        title('Comparison of XEA and XE')
        xlabel('normalized frequency w')
        ylabel('XEA(w) and XE(w)')
%
        figure(3)
        plot(WA,XF)
        title('Foster part of the given immitance')
        xlabel('normalized angular frequency w')
        ylabel('X_F(w)=XEA-XE Foster Part of the immittance')
    CVal=general_synthesis(a,b)
```

Program Listing 8.16 Objective function for Example 8.7

```
function
fun=direct_Gewertz(x0,ntr,ndc,W,a0,WA,REA,XEA)
% Modeling via Direct - Parametric approach
```

```
if ntr==1; %Yes transformer case
    Nx=length(x0);
        for i=1:Nx-1;
            c(i)=x0(i);
        end;
        a0=x0(Nx);
end
if ntr==0;
% No Transformer case
    Nx=length(x0);
    for i=1:Nx;
      c(i)=x0(i);
    end;
end
%
% Step 5: Generate analytic form of Fmin(p)=a(p)/b(p)
% Generate B(-p^2)
C=[c 1];
BB=Poly_Positive(C);
% This positive polynomial is in w-domain
B=polarity(BB);% Now, it is transferred to p-domain
% Generate A(-p^2) of R(-p^2)=A(-p^2)/B(-p^2)
nB=length(B);
      A=(a0*a0)*R_Num(ndc,W);
% A is specified in p-domain
nA=length(A);
if (abs(nB-nA)>0)
    A=fullvector(nB,A);
end
% Minimum immitance is generated using Gewertz Method
[a,b,x0,x,y0,y]=Gewertz(A,B);
      NA=length(WA);
      cmplx=sqrt(-1);
%
    for j=1:NA
      w=WA(j);
      p=cmplx*w;
      aval=polyval(a,p);
      bval=polyval(b,p);
      Z=aval/bval;
      RE=real(Z);
      XE=imag(Z);
      error1=abs(REA(j)-RE);
      % error2=abs(XEA(j)-XE);
      fun(j)=error1;
    end
```

9

Description of Power Transfer Networks via Driving Point Input Immittance: Darlington's Theorem

9.1 Introduction

A power transfer network is a lossless two-port. It couples an independent source, which is described by means of its complex internal immittance, to the output port, terminated in a complex passive load.

As described earlier the design of a lossless power transfer network is completed to achieve 'optimum power transfer' from input to output. The optimum power transfer requirements are specified by the designer when considering the practical problem under consideration.

For many applications, however, the optimum power transfer is defined as the optimization of the transducer power gain over a specified frequency band which is as flat and as high as possible.

In this chapter, the mathematical or numerical optimum power transfer problem is introduced via a description of lossless two-ports.

9.2 Power Dissipation P_L over a Load Impedance Z_L

Referring to Figure 9.1, let $E_G(j\omega)$ be a voltage source with internal impedance $Z_G(j\omega) = R_G + jX_G(\omega)$, which is specified in the frequency domain ω. Let this voltage source drive the load impedance $Z_L = R_L(\omega) + jX_L(\omega)$. Let $V_L(j\omega)$ and $I_L(j\omega)$ be the voltage and current of the load Z_L, and, let P_L be the average power dissipation on the load Z_L.

Then,

$$\boxed{P_L = \text{real}\left\{V_L I_L^*\right\}} \tag{9.1}$$

where superscript * designates the complex conjugate of a complex quantity.

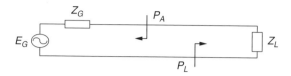

Figure 9.1   Power transfer from a complex generator to a complex load

For the circuit shown in Figure 9.1, the load current I_L and the load voltage V_L are given by

$$I_L = \frac{E_G}{Z_G + Z_L}$$
$$V_L = Z_L I_L$$

(9.2)

Thus,

$$P_L = \text{real} \left\{ E_G \frac{Z_L}{Z_G + Z_L} \frac{E_G^*}{(Z_G + Z_L)*} \right\}$$

or

$$P_L = |E_G|^2 \frac{R_L}{|Z_G + Z_L|^2} = |E_G|^2 \frac{R_L}{(R_G + R_L)^2 + (X_G + X_L)^2}$$

(9.3)

In the classical power transfer network literature, P_L is referred to as the power delivered to load Z_L.

9.3 Power Transfer

Again referring to Figure 9.1 and keeping the voltage source fixed, one can easily derive the maximum power transfer conditions as a function of the load impedance Z_L by solving the following equations:

$$\frac{\partial P_L}{\partial R_L} = |E_G|^2 \frac{|Z_G + Z_L|^2 - 2(R_G + R_L)}{|Z_G + Z_L|^4}$$
$$= |E_G|^2 \frac{(R_G + R_L)^2 + (X_G + X_L)^2 - 2(R_L + R_G)R_L}{|Z_G + Z_L|^4} = 0$$

(9.4)

and

$$\frac{\partial P_L}{\partial X_L} = |E_G|^2 \frac{-2(X_G + X_L)}{|Z_G + Z_L|^4} = 0$$

(9.5)

Thus, Equation (9.5) yields that

$$X_L = -X_G$$

(9.6)

Substituting Equation (9.6) in Equation (9.4), one obtains

$$R_L = R_G$$

(9.7)

Clearly, under maximum power transfer conditions, Equation (9.6) and Equation (9.7) can be written in compact form as

$$Z_L(j\omega) = R_G(\omega) - jX_G(\omega) = Z_G^*$$

(9.8)

Thus, we can state the maximum power transfer theorem as follows.

9.4 Maximum Power Transfer Theorem

A voltage source E_G with a passive internal impedance Z_G delivers its maximum power to a load impedance Z_L if and only if, on the real frequency axis, the load impedance Z_L is equal to the complex conjugate of the internal impedance Z_G.

Proof of this theorem follows from Equations (9.4–9.7).

Available Power

The maximum deliverable power of an independent generator is called the Available power and designated by P_A. Thus, for any voltage source with internal impedance $Z_G = R_G + jX_G$ such that $R_G \neq 0$, the available power is given by

$$P_A = P_L \text{ when } R_L = R_G \text{ and } X_L = -X_G$$
or
$$P_A = \frac{|E_G|^2}{4R_G}$$

(9.9)

Transducer Power Gain

The transducer power gain (TPG) is defined as the ratio of power delivered to a load (P_L) to the available power (P_A) of the generator. Thus, TPG is

$$\text{TPG} \triangleq \frac{P_L}{P_A}$$

The $\triangleq$ sign designates equality by definition. Obviously, TPG is non-negative and bounded by 1. Thus one can write
$$0 \leq \text{TPG} \leq 1$$

(9.10)

Referring to Figure 9.1, Equation (9.3) yields

$$\text{TPG} = \frac{4R_G R_L}{(R_G + R_L)^2 + (X_G + X_L)^2}$$

or in compact form

$$\text{TPG} = \frac{4R_G R_L}{|Z_G + Z_L|^2}$$

(9.11)

Analytic continuation of the above reveals that

$$\text{TPG} = T(p^2) = \frac{4R_G(p^2)R_L(p^2)}{[Z_G(p) + Z_L(p)][Z_G(-p) + Z_L(-p)]}$$

This last equation indicates that TPG is an even function of complex variable $p = \sigma + j\omega$

9.5 Transducer Power Gain for Matching Problems

Referring to Figure 9.2, let $[E]$ be a lossless two-port which transfers power from a complex generator $\{E_G, Z_G = R_G + jX_G; R_G \neq 0\}$ to a complex load $\{Z_L = R_L + jX_L; R_L \neq 0\}$.

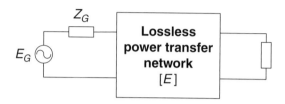

Figure 9.2   Power transfer over a lossless coupling network $[E]$

Since the power transfer network $[E]$ is lossless, the input power P_{in} must be transferred to the output port without any loss. In this case, let us represent the output port by its Thévenin equivalent circuit with a generator E_Q and its internal impedance $Z_Q = R_Q + JX_Q$ as shown in Figure 9.3.

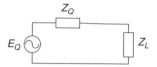

Figure 9.3   Thévenin equivalent of the output port

Then, TPG is obtained as in Equation (9.11) by replacing Z_G with Z_Q. Thus,

$$\text{TPG} = \frac{4R_Q R_L}{(R_Q + R_L)^2 + (X_Q + X_L)^2} \tag{9.12}$$

In Equation (9.12), Z_Q is the Thévenin impedance, which is the impedance seen from the output port of the lossless two-port while the input port is terminated in Z_G.

With this understanding, the lossless matching network $[E]$ transforms the positive real impedance Z_G to Z_Q. If the matching network is ideal, then it should perform such a kind of transformation that yields the complex conjugate of the load impedance Z_L within the band of operation.

9.6 Formal Definition of a Broadband Matching Problem

Literally, the broadband matching problem is defined as one of constructing a lossless two-port $[E]$ between strictly positive real impedances[1] Z_G and Z_L so that TPG of Equation (9.12) is optimized as high and as flat as possible over the frequency band of operation.

---

[1] What we mean by strictly positive real impedance is that it is a non-Foster impedance which dissipates power on it. In this case, it's even part is strictly positive in the closed RHP of the complex p domain.

As for a procedural understanding of the broadband matching problem, we structure the following description.

Conceptual Understanding of Broadband Matching Problems

Consider the following:

(a) Given frequency band of operation $B = \omega_{low} - \omega_{high}$.
(b) Given positive real impedance Z_G and Z_L.
(c) Find the best possible flat gain level T_0 which maximizes the power transfer from complex generator Z_G to a complex load Z_L.[2]
(d) Having obtained the ideal flat gain level T_0, construct a practical matching network $[E]$ which approximates T_0, meeting preset design specifications.

In the above, (a) and (b) are considered as the inputs of the broadband matching or power transfer problem. For simple problems the answers for (c) and (d) may be generated as a result of the implementation of the analytic theory of broadband matching.[3] However, for complicated problems, the analytic theory may not be accessible. In this case, one must carefully employ known CAD techniques to design power transfer networks between a complex generator and a complex load.

In the literature it has been shown that the real frequency techniques provide optimum solutions to gain–bandwidth problems.[4] Therefore, part of this book is solely devoted to designing power transfer networks via real frequency techniques (RFT).

In all the real frequency techniques, a lossless two-port is described in terms of either its positive real (PR) driving point input impedance $Z_Q(p)$ or equivalently corresponding bounded real (BR) input reflectance $S_Q(p) = (Z_Q - 1)/(Z_Q + 1)$.[5]

Eventually, the lossless matching network is obtained by synthesizing the driving point input immittance using the Darlington procedure. Therefore, in the following sections Darlington's description of lossless two-ports is presented with relevant background.

---

[2] In the theory of broadband matching, it is well established that one can always consider an ideal lossless two-port $[E]$ with infinitely many elements, which yields an upper limit for the ideal flat transducer power gain level T_0. This is the best one can do for given generator Z_G and load Z_L impedances to be matched over the specified frequency band B. This fact is known as the gain–bandwidth limit imposed by the impedances Z_G and Z_L. In this case, total energy dissipated on the load is given by $\int_{-\infty}^{+\infty} P_L(\omega)d\omega = 2T_0 \int_{\omega_1}^{\omega} P_A(\omega)d\omega$. Moreover, if $Z_G = R_G = $ constant, then $W = 2T_0 \left[|E_G|^2 / 4R_G \right] B = $ constant. Details of this theory can be found in the references given below: R. M. Fano, 'Theoretical Limitations on the Broadband Matching of Arbitrary Impedances', *Journal of the Franklin Institute*, Vol. 249, pp. 57–83, 1950. D. C. Youla, 'A New Theory of Broadband Matching', *IEEE Transactions on Circuit Theory*, Vol. 11, pp. 30–50, March, 1964.

[3] A 'simple matching problem' is defined as the one which consists of generator and load impedances with only one single reactive element. For example, a generator impedance Z_G which includes a pure resistance R_G in series with an inductance L_G and a load impedance Z_L which consists of a parallel connection of a pure resistance R_L with and a capacitor C_L constitutes a simple matching problem.

[4] H. J. Carlin and P. Amstutz, 'On Optimum Broadband Matching', *IEEE Transactions on Circuits and Systems*, Vol. 28, pp. 401–405, May, 1981.

[5] Or, equivalently, a two-port may be defined in terms of its driving point input admittance $Y_Q(p)$ or corresponding reflectance $S_Q = (1 + Y_Q)/(1 + Y_Q)$

9.7 Darlington's Description of Lossless Two-ports

In his famous work,[6] Sydney Darlington describes a lossless two-port by means of its driving point input immittance $F_Q(p) = a(p)/b(p)$ while the other end of the two-port is terminated in unit resistance. Obviously, this immitance designates either the impedance $Z_Q(p)$ or the admittance $Y_Q = 1/Z_Q$ without loss of generality.

Before we introduce Darlington's theorem, let us first review some basic properties of reactance functions.

Foster Functions

Let

$$F(p) = \frac{n(p)}{d(p)} = \frac{n_e + n_o}{d_e + d_o}$$

be a rational positive real immittance function, where n_e and d_e are the even parts and n_o and d_o are the odd parts of the numerator and denominator polynomials, respectively.

Let

$$R(p) = \frac{1}{2}[F(p) + F(-p)] = \frac{n_e d_e - n_o d_o}{d_e^2 = d_o^2} = \frac{A(p^2)}{D(p^2)}$$

be the even part of $F(p)$.

Similarly, let

$$O(p) = \frac{1}{2}[F(p) - F(-p)] = \frac{n_e d_o - n_o d_e}{d_e^2 - d_o^2} = \frac{B(p^2)}{D(p^2)}$$

be the odd part of $F(p)$.

Clearly, polynomials $A(p^2) = n_e d_e - n_o d_o$ and $D(p^2) = d_e^2 = d_o^2$ are even and $B(p^2) = n_e d_o - n_o d_e$ is odd in complex variable p.

Definition 1 (Foster function): A positive real function $F(p)$ is said to be Foster if its even part is zero (i.e. $R(p) = 0$). In this case,

$$F(p) = O(p) = \frac{B(p)}{D(p)} = \frac{\text{odd ploynomial}}{\text{even polynomial}}$$

Property 1: A Foster immittance cannot dissipate power by itself. Rather it stores energy and delivers it back. Therefore it is also called a lossless one-port.

To verify this, let the Foster function be an impedance function. Since its even part is zero for all $p = \sigma + j\omega$, then the power delivered to the impedance function is given by $P_p = \text{real}\{F(j\omega) \cdot I(-j\omega)\} = R(\omega)|I(j\omega)|^2 = 0$. Thus, a Foster immittance is said to be loss free or it is lossless.

Property 2 (Inverse of a Foster function): The inverse of a Foster function is also a Foster function. Since a Foster function is positive real, then its inverse must also be a positive real function. For example, if a Foster function is an impedance function such as $Z_F = $ even polynomial/odd ploynomial, then its inverse defines an admittance function such that $Y_F = 1/Z_F = $ odd ploynomial/even polynomial, which must also be odd and a positive real function. In other words, Y_F is free of an even part. Thus, it is Foster.

[6] S. Darlington, 'Synthesis of Reactance 4-poles', *Journal of Mathematics and Physics*, Vol. 18, pp. 257–353, 1939.

Property 3 (Reactance theorem): Let

$$F(p)\frac{\text{even polynomial}}{\text{odd ploynomial}} \text{ or } F(p) = \frac{\text{odd ploynomial}}{\text{even polynomial}}$$

be a Foster function of complex frequency variable $p = \sigma + j\omega$. On the real frequency axis let $F(j\omega) = jX(\omega)$. Then, it is proven that

$$\boxed{\frac{dX(\omega)}{d\omega} \geq 0, \forall \omega} \tag{9.13}$$

Proof of this theorem can found in the classical paper by R. M. Foster.[7]

In the classical network theory literature, $X(\omega)$ is known as the Foster reactance function. In fact Foster's reactance theorem is stated as follows.

Foster's reactance theorem: A rational function $F(p)|_{p\,=\,j\omega} = jX(\omega)$ is a Foster function if it satisfies Equation (9.13).

Property 4: All poles and zeros of a Foster function must be simple and they should be placed on the real frequency axis.

As verification, this property follows from the definition of a Foster function.

Since a Foster function is also a positive real function, then it must possess simple poles and zeros on the $j\omega$ axis. On the other hand, it cannot have any closed LHP poles and zeros since its even part is zero.

Property 5: Referring to Figure 9.4, the poles and zeros of a Foster reactance function must interlace. In other words, each pole must be followed by a zero and each zero must be followed by a pole etc.

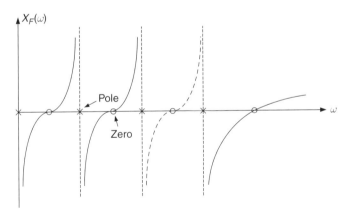

Figure 9.4 Foster function

Verification actually follows from Property 3. Assume that $X(\omega)$ starts from zero and increases continuously. In this case, it should go to infinity rather than zero since $dX(\omega)/d\omega \geq 0$, $\forall \omega$. Therefore a zero of the reactance function cannot be followed by a neighboring zero.

---

[7] R.M. Foster, 'A Reactance Theorem', *Bell System Technical Journal*, Vol. 3, pp. 259–267, 1924.

On the other hand, if the reactance function starts from minus infinity, then it should keep increasing toward plus infinity to satisfy the condition $dX(\omega)/d\omega \geq 0, \forall \omega$. Thus as it increases from minus infinity toward plus infinity it must cross the frequency axis. Hence a pole must be followed by a neighboring zero.

Property 6 (Parametric representation of Foster Functions): Foster functions can be written in terms of their poles as

$$
\begin{aligned}
F(p) &= K_\infty p + \frac{k_0}{p} + \sum_{i=1}^{N} \frac{k_i}{p = j\omega_i} \\
&\quad k_\infty p + \frac{k_0}{p} + \sum_{i=1}^{N/2} \frac{2k_{iR}p}{p^2 + \omega_i^2}
\end{aligned}
\tag{9.14}
$$

where
$$
k_\infty = lim_{p \to \infty} \frac{1}{p} F(p) \geq 0 \quad k_0 = lim_{p \to 0} pF(p) \geq 0
$$
$$
k_i = k_{iR} + jk_{iX} = lim_{p \to j\omega_i}(p - j\omega_i)F(p), \ k_{iR} \geq 0
$$

In the above equations, k_∞ is the residue of the pole at infinity (i.e. $p = \infty$), k_0 is the residue of the pole at zero (i.e. $p = 0$) and k_{iR} is the real part of the residues at the complex conjugate poles located at $p = \pm j\omega_i$. They are real and must be non-negative, i.e. $\{k_\infty, k_0, k_{iR} \geq 0, i = 1, 2 \ldots, n\}$.

Property 6 is the result of Equation (9.13) or Foster's reactance theorem, which also leads to the following theorem.

Foster's residue theorem: A rational function which is described by Equation (10.14) is Foster if and only if all the residues k_∞, k_0 and k_i are real and non-negative.

    The proof of this theorem directly follows from Equation (10.13).

Property 7 (Synthesis of a Foster impedance function): If Equation (9.14) describes an impedance function, then it can be realized as series connections of:

- an inductance $L = k_\infty$;
- a capacitance $C = 1/k_0$; and
- parallel tank circuits with resonance frequencies $\omega_i 1/(L_i C_i)$ with $2k_i = 1/C_i$ as shown in Figure 9.5.

Property 8 (Synthesis of a Foster admittance function): In a similar manner to that of Property

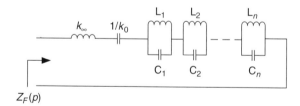

$Z_F(p)$

Figure 9.5   Foster I realization of a Foster impedance function

7, Equation (10.14) can be synthesized as an admittance function as a parallel combination of:

- a capacitance $C = k_\infty$;
- an inductance $L = 1/k_0$; and
- series resonance circuits with resonance frequencies $\omega_i = 1/(L_i C_i)$ with $2k_i = 1/L_i$ as shown in Figure 9.5.

Property 9: A Foster immittance function can be realized as a low-pass ladder with series inductors and shunt capacitors connected in a sequential manner as shown in Figure 9.6.

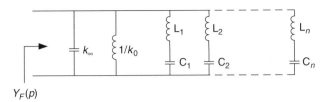

Figure 9.6   Foster II realization of a Foster admittance function

In this case, $F(p)$ is expressed in continued fraction form:

$$F(p) = r_1 p + \cfrac{1}{r_2 p + \cfrac{1}{r_3 p + \cfrac{1}{r_4 p + \cfrac{1}{...+\cfrac{1}{r_n p}}}}} \qquad (9.15)$$

In Equation (9.15), r_i designates the non-negative values of series inductors or shunt capacitors. This form of Foster immittance realization is called Cauer I and is shown in Figure 9.7.

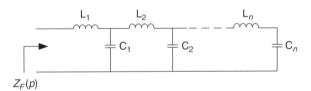

Figure 9.7   Cauer I realization of a Foster impedance function

Property 10: In a similar manner to that of Property 9, a Foster immittance function can be realized as a high-pass ladder with series capacitors and shunt inductors connected in a sequential manner as shown in Figure 9.8. In this case, $F(p)$ is expressed as

$$F(p) = \left[\frac{1}{r_1 p} \right] + \cfrac{1}{\left[\dfrac{1}{r_2 p} \right] + \cfrac{1}{\left[\dfrac{1}{r_3 p} \right] + \cfrac{1}{...+\cfrac{1}{...+\cfrac{\vdots}{\left[\dfrac{1}{r_n p} \right]}}}}} \qquad (9.16)$$

where r_i designates the values of series inductors and shunt capacitors connected in ladder configuration. This form of realization is called Cauer II.

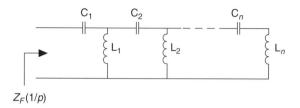

Figure 9.8   Cauer II realization of a Foster impedance function

Now, let us verify the above properties in an example.

Example 9.1: Let

$$Z(p) = \frac{5p^5 + 71p^3 + 70p}{2p^6 + 35p^4 117p^2 + 36}$$

be a positive real impedance.

(a) Show that $Z(p)$ does not dissipates power.
(b) Find the zeros and poles of $Z(p)$.
(c) Is $Z(p)$ Foster?
(d) Verify your result of (c) by computing the poles and residues of the admittance function

$$Y(p) = \frac{2p^6 + 35p^4 + 117p^2 + 36}{5p^5 71p^3 + 70p}$$

(e) Synthesize $Z(p)$ using Foster I.
(f) Synthesize $Z(p)$ using Foster II.

Solution:

(a) It is straightforward to see that $Z(p) = $ odd ploynomial/even polynomial is an odd rational function. Therefore its even part is zero. Thus, it is a lossless one-port.
(b) Let us look at the zeros and poles of $Z(p)$. Using MATLAB's function `roots` we can find the zeros and the poles of `Z(p)` as in Table 9.1.
(c) In order to understand if `Z(p)` is Foster, the easiest way is to find the residues at given poles. Thus, employing Equation (9.14), the residues are found as in Table 9.2.
    Since all poles are simple on the real frequency axis with positive residues, then $Z(p)$ must be a Foster function as stated by Foster's residue theorem.
(d) Obviously, the admittance function
$$Y(p) = \frac{2p^6 35p^4 117p^2 + 36}{5p^5 + 71p^3 + 70p}$$

and $k_\infty = 2/5 = 0.4$. Furthermore, it has four finite simple poles on the $j\omega$ axis beyond zero These poles and corresponding residues are listed in Table 9.3.

Table 9.1 Impedance function for Example 9.1

$$Z(p) = \frac{5p^5 + 71p^3 + 70p}{2p^6 + 35p^4 + 117p^2 + 36}$$

Zeros of Z(p)	Poles of Z(p)
0 (zero at DC)	0 + 3.62739614723677i
0 + 3.64409570054249i	0-3.627396147236677i
0-3.64409570054249i	0 + 2i
0 + 1.03243889087043i	0-2i
0-1.03243889087043i	0 + 5848052590049413i
	0-5848052590049413i

Table 9.2 Residues of the poles of the impedance function specified by Example 9.1

$$Z(p) = \frac{5p^5 + 71p^370p}{2p^6 + 35p^4117p^2 + 36}$$

$k_0 = 0$ (no pole at DC)	$k_\infty = 0$ (no pole at infinity)
Finite poles of Z(p)	Residue at the pole
0 + 3.62739614723677i	0.00308213833736778
0-3.62739614723677i	0.00308213833736778
0 + 2i	1
0-2i	1
0 + 0.584805259049413i	0.246917861662636
0-0.584805259049413i	0.246917861662636

Table 9.3 Admittance function $Y(p)$ as specified by Example 9.1

$$Y(p) = \frac{2p^6 + 35p^4 + 117p^2 + 36}{5p^5 + 71p^3 + 70p}$$

$k_0 = 36/70$	$k_\infty = 2/5 = 0.4$
= 0.514 285 71 285 714	

Finite poles	Residues
0	0.514285714285714
0 + 3.64409570054249i	0.00352850237300706
0-3.64409570054249i	0.00352850237300706
0 + 1.03243889087043i	0.399328640484137
0-1.03243889087043i	0.399328640484137

Thus, $Y(p)$ is a Foster function with simple real frequency axis poles and positive residues. Hence the above result of part (c) is verified.

(e) In this part of the problem we will synthesize a Foster impedance function as in the Foster I type. Since there is no pole at DC and infinity, the synthesis must consist of only three parallel tank circuits in series connection with resonance frequencies $\omega_1 = 3.627\ 396\ 147\ 236\ 77$, $\omega_2 = 2$, $\omega_3 = 0.584\ 805\ 259\ 049\ 413$ as given in Table 9.4. Thus, we end up with the synthesis form of Foster I as depicted in Figure 9.9.

Table 9.4   Foster i Design of $Z(p)$ for Example 9.1

$k_0 = 0$	Series capacitance $C = 1/k_0 = \infty$ (short circuit)		$k_\infty = 0$	Series inductance $L = k_\infty = 0$ (short circuit)
Resonance Frequency ω_i	Residue k_i		$C_i = \frac{1}{2k_i}$	$L_i = \dfrac{1}{C_i \omega_i^2}$
3.627 4	0.003 082 1		162.23	0.000 468 48
2	1		0.5	0.5
0.584 81	0.246 92		2.025	1.444

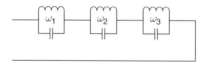

Figure 9.9   Synthesis of the Foster impedance as specified by Table 9.2

(f) By taking the inverse of the impedance, the Foster admittance function

$$Y(p) = \frac{1}{Z} = \frac{zp^6 + 35p^4 + 117p^2 + 36}{5p^5 + 71p^3 + 70p}$$

can be synthesized as in Foster II.

In this case, we immediately see that $Y(p)$ has poles at both zero and infinity as well as at finite frequencies of $\omega_1 = 3.6241$ and $\omega_2 = 1.0324$.

The result of synthesis yields the element values given in Table 9.5 and the final circuit shown in Figure 9.10.

All the above computations were completed using the MATLAB® program called Ch9Example9_1. The listing of this program is given below with related functions. Noted that, in this program, firstly we generate the Foster impedance

$$Z(p) = \frac{5p^5 + 71p^3 + 70p}{2p^6 + 35p^4 + 117p^2 + 36}$$

from the circuit shown in Figure 9.11.

Table 9.5   Foster II design of $Y(p)$ for Example 9.1

	Shunt Inductance		Shunt capacitance
$k_0 = 0.514\ 29$	$L = 1/k_0 = 1.9444$	$k_\infty = 0.4$	$C = k_\infty = 0.4$
Resonance frequency ω_i	Residue k_i	$L_i = \frac{1}{2k_i}$	$C_i = \frac{1}{C_i \omega_i^2}$
3.6241	0.003 528 5	141.7	0.000 537 31
1.0324	0.399 33	1.2521	0.749 26

Figure 9.10   Foster II synthesis of $Y(p)$ as specified in Table 9.5

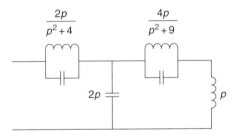

Figure 9.11   Generation of a Foster impedance from the given element values

In this case, the impedance is written as

$$Z(p) = \left[\frac{2p}{p^2+4}\right] + \cfrac{1}{2[p] + \cfrac{1}{\left[\dfrac{4p}{p^2+9}\right] + \cfrac{1}{\left[\dfrac{1}{p}\right]}}}$$

$$= \frac{5p^5 + 71p^3 + 70p}{2p^6 + 35p^4 + 117p^2 + 36}$$

Obviously, this impedance function can also be realized as in Figure 9.9 and Figure 9.10. In the main program of `Ch9Example9_1`, a simple component block is described as

$$F_i(p) = \begin{cases} \dfrac{N_{i2}p}{p^2 + D_{i2}} & \text{with } N_{i1} = N_{i3} = D_{i2} = 0 \text{ and } D_{i1} = 1 \\[2ex] \dfrac{N_{i2}p}{1} & \text{with } N_{i1} = N_{i3} = D_{i2} = D_{i1} = 0 \text{ and } D_{i3} = 1 \\[2ex] \dfrac{1}{D_{i2}p} & \text{with } N_{i1} = N_{i2} = D_{i1} = D_{i3} = p \text{ and } N_{i3} = 1 \end{cases}$$

and the Foster impedance is computed by the function `component`.

In the second part of the program the Foster I form is generated and finally, in the third part, the Foster II form.

Program Listing 9.1   Generation of a Foster function in rational form

```
% Program Ch9Example9_1:
% Part I:
%Generation of Foster functions from the component values
% Computational process starts from bottom-up
% At each step, simple rational forms N(i)/D(i) is put in com-
% putation.
% The last component is assumed to be N(n)/D(n)+R with R=0.
% In this case, immitance at the bottom will take the following
% form:
%      N(n-1)/D(n-1)+1/[N(n)/D(n) + alfa]
% Here, we designate the immitance function as F(p)=P(p)/R(p)
%At first we initialize PXB=N(n)+alfa*D(n)
%                 RXB=D(n)
% then, using sequential formula
%                   PX(n-i)=N(n-j)*PXB(n-i+1)+D(n-j)*RXB(n-i+1)
%                   RX(n-i))=D(n-i)*PXB(n-i+1)
% It should be noted that the above operations designate the
% polynomial
% multiplications. In MATLAB, polynomial multiplications are
% carried out by
% convolution operation.
% Simple form of N=[0 0 1] meaning that N=1; or N=[0 1 0] means
% that N=p.
% N=1 corresponds to N=[0 0 1]
% Similarly D(p)=p^2+wk^2 corresponds to [1 0 w^2]
% In this program, N is a matrix where all numerators;
% N(1,:)=[N(1,1)
% N(1,2), N(1,3)]
clear
% Inputs: Components as simple rational forms:
N(1,1)=0;N(1,2)=2;N(1,3)=0;
D(1,1)=1;D(1,2)=0;D(1,3)=4;
%
N(2,1)=0;N(2,2)=2;N(2,3)=0;
D(2,1)=0;D(2,2)=0;D(2,3)=1;
%
N(3,1)=0;N(3,2)=4;N(3,3)=0;
D(3,1)=1;D(3,2)=0;D(3,3)=9;
%
%
N(4,1)=0;N(4,2)=0;N(4,3)=1;
D(4,1)=0;D(4,2)=1;D(4,3)=0;
% end of inputs
%
n=length(N);
nn=n-1;
```

```
R=0;% Generation of Foster function
[a,b]=component(N,D,R);
[A,B]=ZtoR(a,b);
xz=roots(A);
pz_positive=sqrt(xz);
pz_negative=-sqrt(xz);
poles_b=roots(b);
zeros_a=roots(a);
aodd=clear_evenpower(a);
beven=clear_oddpower(b);
%
% Part II
% Foster I Solution for F(p)=Impedance
% Compute residues of F(p)at finite frequencies
[p_F1,k_F1,CZ]=Foster_residue(a,b);
n_F1=length(p_F1);
%%Compute residue at zero
pzero_F1=1e-15;%p=0 compute residue
avalzero_F1=polyval(a,pzero_F1);
bvalzero_F1=polyval(b,pzero_F1);
kzero_F1=pzero_F1*avalzero_F1/bvalzero_F1;
if(kzero_F1<=1e-10)kzero_F1=0;end
%Compute Residue at infinity
pinf_F1=1e+15;%p=infinity compute residue
avalinf_F1=polyval(a,pinf_F1);
bvalinf_F1=polyval(b,pinf_F1);
kinf_F1=(1/pinf_F1)*avalinf_F1/bvalinf_F1;
if(kinf_F1<=1e-9)kinf_F1=0;end
% Computation of Element values for Foster I
r=0;
for i=1:n_F1
   wi=imag(p_F1(i));
   if(wi>1e-10)%Select positive frequencies
     r=r+1;
     wr_F1(r)=wi;%Resonance frequencies
     kr_F1(r)=k_F1(i);%Corresponding Residues
     C_F1(r)=1/2/kr_F1(r);%Capacitance of the tank circuit
     L_F1(r)=1/wr_F1(r)/wr_F1(r)/C_F1(r);%Inductance of the tank
circuit
   end
end
% Part III
% Foster II Solution
%Compute residue at zero
pzero_F2=1e-15;%p=0 compute residue
avalzero_F2=polyval(a,pzero_F2);
bvalzero_F2=polyval(b,pzero_F2);
kzero_F2=pzero_F2*bvalzero_F2/avalzero_F2;
if(kzero_F2<=1e-10)kzero_F2=0;end
[p_F2,k_F2,CY]=Foster_residue(b,a);
```

Program Listing 9.1   (Continued)

```
%Compute Residue at infinity
pinf_F2=1e+15;%p=infinity compute residue
avalinf_F2=polyval(a,pinf_F2);
bvalinf_F2=polyval(b,pinf_F2);
kinf_F2=(1/pinf_F2)*bvalinf_F2/avalinf_F2;
if(kinf_F2<=1e-9)kinf_F2=0;end
% Element Values of Foster II
n_F2=length(p_F2);
r=0;
for i=1:n_F2
wi=imag(p_F2(i));
if(wi>1e-10);%Select positive frequencies
   r=r+1;
   wr_F2(r)=wi;%Resonance frequencies
   kr_F2(r)=k_F2(i);%Corresponding Residues
   L_F2(r)=1/2/kr_F2(r);%Capacitance of the tank circuit
C_F2(r)=1/wr_F2(r)/wr_F2(r)/L_F2(r);%Inductance of the tank
circuit
   end
end
% Print Results
% Foster I
KZERO_F1=kzero_F1
CSERIES=1/KZERO_F1%SERIES CAPACITOR
KINF_F1=kinf_F1
LSERIES=KINF_F1%SERIES INDUCTOR
WR_F1=wr_F1'
Residue_F1=kr_F1',
Cap_F1=C_F1',
Induc_F1=L_F1'
%Foster-II
KZERO_F2=kzero_F2
LSHUNT=1/KZERO_F2%SHUNT INDUCTOR
KINF_F2=kinf_F2
CSHUNT=KINF_F2%SHUNT CAPACITOR
WR_F2=wr_F2'
Residue_F2=kr_F2',
Cap_F2=C_F2',
Induc_F2=L_F2'
```

Program Listing 9.2

```
function [p,k,C]=Foster_residue(a,b)
% This function computes the residues k of a given Foster function F(p)
```

```
% F(p)=a(p)/b(p)
% It should be noted the F(p) is given in p domain as an odd
% function of p
%
% Inputs:
% ------- a(p); numerator polynomial
% ------ b(p); denominator polynomial
%
% Step 1: Find jw poles of F(p)
p=roots(b);
%
% Step 2: Compute the product terms
prd=Foster_product(p);
n=length(p);
N=length(b);
% Step3 : Generate residues
% Find lowest non zero coefficient of denominator b(p)
r=0;
for i=1:N
    if(abs(b(i))>=1e-10)
    r=r+1;
    C(r)=b(i);
    end
end
for j=1:n
    y=p(j);
    aval=polyval(a,y);
    k(j)=aval/prd(j)/C(1);
end
```

Program Listing 9.3

```
function prd=Foster_product(p)
% Input:
%-------- poles p(i) as a MATLAB row vector p
% Output:
%-------- product prd as a MATLAB row vector
%      prd(j)=p(j)=(p(j)-p1)(p-p2). . .(p(j)-p(j-1)).(p(j)-
% p(j+1)). . .(p(j)-p(n)
% which multiplies the terms [p(j)-p(i)] skipping the term when
% j=i.
%
% This function computes the product of
%      prd(j)=p(j)=(p(j)-p1)(p-p2). . .(p(j)-p(j-1)).(p(j)-
% p(j+1)). . .(p(j)-p(n)
```

Program Listing 9.3   (Continued)

```
% The above product form is used to compute residues for para-
% metric
% approach
n=length(p);
nn=n-1;
%
   for i=1:n
     ps(i)=p(i);
   end
% Step 1: Compute the first product term for j=1
% pra=(p(1)-p(2)).(p(1)-p(3). . .(p(1)-p(n))
%
pra=1;
for i=2:n
   pra=pra*(ps(1)-ps(i));
end
prd(1)=pra;
%
% Step 2: Compute the product for 2<j<(n-1)skipping the j th
% term.
%        prb=(p(j)^2-p(1)^2). . .((p(j)^2-p(j-1)^2).(p(j)^-
% p(j+1)^2)..(p(j)^2-p(n)^2)
%
for j=2:nn
   prb1=1;
     for i=1:j-1
     prb1=prb1*(ps(j)-ps(i));
     end
     prb2=1;
     for i=(j+1):n
   prb2=prb2*(ps(j)-ps(i));
     end
   prb=prb1*prb2;
   prd(j)=prb;
end
%
% Step 3: Compute the product term for j=n
%       prc=(p(n)^2-p(1)^2).(p(n)^2-p(2)^2. . .(p(n)^2-p(n-1)^2)
%
prc=1;
for i=1:nn
  prc=prc*(ps(n)-ps(i));
end
prd(n)=prc;
```

9.8 Description of Lossless Two-ports via Z Parameters

Referring to Figure 9.12, let us consider a lossless two-port terminated in a resistance $R_L = R$ and driven by a resistive generator $R_G = R$. Let it be described by means of its impedance matrix

$$Z = \begin{bmatrix} Z_{11} & Z_{12} \\ Z_{21} & Z_{22} \end{bmatrix}$$

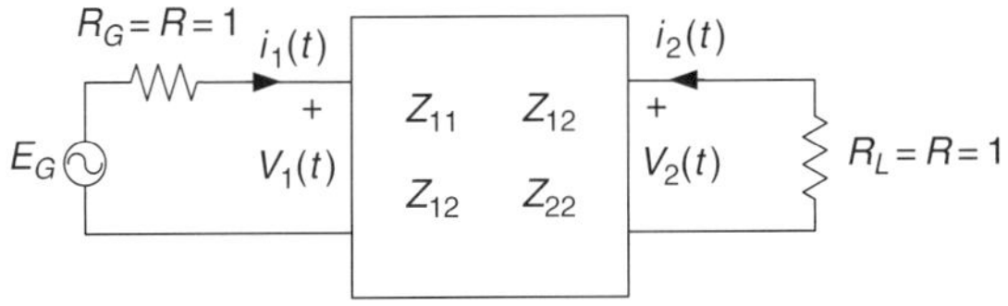

Figure 9.12   Description of lossless two-port in Z parameters

In this case,

$$V_1 = Z_{11}I_1 + Z_{12}I_2$$

$$V_2 = Z_{21}I_1 + Z_{22}I_2$$

or in compact form

$$\begin{bmatrix} V_1 \\ V_2 \end{bmatrix} = \begin{bmatrix} Z_{11} & Z_{12} \\ Z_{21} & Z_{22} \end{bmatrix} \begin{bmatrix} I_1 \\ I_2 \end{bmatrix} \tag{9.17}$$

or in matrix notation

$$[V] = [Z][I]$$

For reciprocal lossless two-ports it is well known that $Z_{12} = Z_{21}$

Total power delivered to the lossless two-port which is

$$P_T = \text{real}\{V_1 I_1^* + V_2 I_2^*\} = [V_1 \quad V_2] \begin{bmatrix} I_1 \\ I_2 \end{bmatrix}^*$$

$$= \text{real}\{[V]^T [I]^*\} \tag{9.18}$$

or

$$P_T = \text{real}\left\{ Z_{11}|I_1|^2 + Z_{12}\left(I_1^* I_2 I_1 I_2^*\right) + Z_{22}|I_2|^2 \right\}$$

must be zero.

Note that $\left(I_1^* I_2 + I_1 I_2^*\right) = 2\,\text{real}\{I_1 I_2^*\} = 2\,\text{real}\{I_1^* I_2\} = $ a pure real number.

Thus, Equation (9.18) requires that real parts of all the Z parameters $\{Z_{ij}, i, j = 1, 2\}$ must be zero. Therefore, Equation (9.18) can be written in compact form as

$$P_T\left\{\left[I(j\omega)^T\right]\right\}^* \left\{[Z(j\omega)] + [Z(j\omega)]^*\right\}[I(j\omega)] = 0 \tag{9.19}$$

Thus, for a lossless two-port, the impedance matrix must satisfy

$$[Z(j\omega)] + [Z(j\omega)]^* = 0$$

By analytic continuation, the general form of Equation (9.19) can be written as   (9.20)

$$[Z(p)] + [Z(p)]^\dagger = 0$$

where the dagger † indicates the transpose of the para-conjugate of a matrix $[Z(p)]$ specified in complex variable p. That is, $[Z(p)]^\dagger = [Z(-p)]^T$.

The impedance matrix $[Z(p)] = -[Z(p)]^\dagger$ is called a 'skew Hermitian matrix (SHM)'. Hence, for a lossless two-port the impedance matrix is SHM.

For a lossy two-port the total power delivered

$$P_T(\omega) = \text{real}\{[V(j\omega)]^T [I]^*\} = \text{real}\{[I]^T [Z][I]^*\}$$

or in compact form $[I]$ (9.21)

$$P_T(\omega) = [I]^\dagger \{[Z(j\omega)] + [Z(j\omega)]^\dagger\}$$

must be positive. In general we can say that two-ports consisting of passive components must satisfy the condition of $P_T(\omega) \geq 0$. By analytic continuation, this condition requires that

$$[Z(p)] + [Z(p)]^\dagger \geq \text{ for all } p = \sigma + j\omega \text{ with } \sigma \geq 0$$ (9.22)

The impedance matrix which satisfies Equation (9.22) is called positive definite.

All the above derivations can he summarized in the following properties.

Property 11: A passive two-port must possess a positive definite impedance matrix. Similarly its admittance matrix $[Y(p)]$ must also be positive definite.[8]

Property 12: The impedance parameters of a two-port can be expressed as

$$Z_{11} = \left.\frac{V_1}{I_1}\right|_{I_2=0} \qquad Z_{21} = \left.\frac{V_2}{I_1}\right|_{I_2=0}$$

$$Z_{22} = \left.\frac{V_2}{I_2}\right|_{I_1=0} \qquad Z_{12} = \left.\frac{V_1}{I_2}\right|_{I_1=0}$$ (9.23)

The Z parameters of a two-port are also called open-circuit parameters since they are measured under open-circuit conditions. More specifically, by setting $I_2 = 0$ we measure Z_{11} and Z_{21}. Similarly, Z_{22} and Z_{12} are measured when the input port is open (i.e. $I_1 = 0$).

Property 13: $Z_{11}(p)$ is called the open-circuit input impedance of the two-port.

Clearly, for a lossless two-port, $Z_{11}(p)$ must be a Foster function. Similarly, $Z_{22}(p)$ is called the open-circuit output impedance and it must be Foster function if the two-port is lossless.

---

[8] Note that all the above derivations can be carried out in terms of admittance parameters. In this case,

$$\begin{bmatrix} I_2 \\ I_2 \end{bmatrix} = [I] = \begin{bmatrix} Y_{11} & Y_{12} \\ Y_{21} & Y_{22} \end{bmatrix} \begin{bmatrix} V_1 \\ V_2 \end{bmatrix} = [Y][V] \quad \text{and} \quad P_T = [V]^\dagger \{Y + Y^\dagger\}[V] \geq 0$$

For lossless two-ports consisting of lumped elements, open-circuit input and output impedances must be rational functions in Foster form as

$$
Z_{11}(p) \ or \ Z_{22}(p)
$$
$$
= \left\{ \frac{\text{even polynomial}}{\text{odd ploynomial}} \ or \ \frac{\text{odd ploynomial}}{\text{even polynomial}} \right\} \tag{9.24}
$$

with positive residues.

Furthermore, for a reciprocal lossless two-port consisting of passive lumped circuit elements, transfer impedance parameters $Z_{12} = Z_{21}$ must be an odd rational function in complex variable p, as in Equation (9.24), to satisfy the SHM condition of Equation 9.(20). However, real-valued residues of Z_{12} are not restricted to being non-negative.

Property 14: For a lossless two-port, the impedance parameters can be expressed as

$$
Z_{11}(p) = k_{11\infty}p + \frac{k_{110}}{p} + \sum_{i=1}^{N_{z11}} \frac{2k_{11i}p}{p^2 + \omega_{11i}^2}
$$
$$
Z_{22}(p) = k_{22\infty}p + \frac{k_{220}}{p} + \sum_{i=1}^{N_{z22}} \frac{2j_{22i}p}{p^2 + \omega_{22i}^2} \tag{9.25}
$$
$$
Z_{12}(p) = k_{12\infty}p + \frac{k_{120}}{p} + \sum_{i=1}^{N_{z12}} \frac{2k_{12i}p}{p^2 + \omega_{12i}^2}
$$

As will be shown later, the poles of $Z_{12}(p)$ must overlap with the poles of Z_{11} and Z_{22}. However, Z_{11} and Z_{22} may have non-overlapping poles, which are called the private poles of Z_{11} and Z_{22}.

Description of Two-ports via Admittance Parameters
Referring to Figure 9.13, a two-port can be described in terms of its admittance parameters.

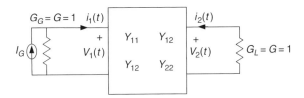

Figure 9.13   Description of two-port via admittance parameters

Admittance parameters are also known as Y parameters. In this case, the input and output currents, namely currents I_1 and I_2 of a linear two-port can be expressed in terms of the input and output voltages V_1 and V_2 as

$$
\begin{bmatrix} I_1 \\ I_2 \end{bmatrix} = [I] = \begin{bmatrix} Y_{11} & Y_{12} \\ Y_{21} & Y_{22} \end{bmatrix} \begin{bmatrix} V_1 \\ V_2 \end{bmatrix} = [Y][V] \tag{9.26}
$$

Here, we see that the impedance and admittance parameters can easily be related to each other since $[V] = [Z][I]$. Thus, we immediately see that

$$
[Y] = [Z]^{-1}
$$

$$
\begin{bmatrix} Y_{11} & Y_{12} \\ Y_{21} & Y_{22} \end{bmatrix} = \begin{bmatrix} Z_{11} & Z_{12} \\ Z_{21} & Z_{22} \end{bmatrix}^{-1}
$$

(9.27)

or, in open form,

$$
Y_{11} = \frac{Z_{22}}{\Delta Z} \quad Y_{12} = \frac{Z_{21}}{\Delta Z}
$$

$$
Y_{21} = \frac{Z_{12}}{\Delta Z} \quad Y_{22} = \frac{Z_{11}}{\Delta Z}
$$

(9.28)

where $\Delta Z = Z_{11}Z_{22} - Z_{12}Z_{21}$ is the determinant of the Z matrix. For reciprocal two-ports, $Y_{12} = Y_{21}$. From Equation (9.28), it is interesting to observe that

$$
\frac{Y_{11}}{Z_{22}} = \frac{1}{\Delta Z} = \frac{Y_{22}}{Z_{11}}
$$

(9.29)

Furthermore, all the derivations carried out in the previous section can be completed in Y parameters. For example, total power delivered to the two-port under consideration can be given as

$$
P_T(\omega) = \text{real}\left\{ [V]^{T*}[I] \right\} = \text{real}\left\{ [V]^{T*}[Y][V] \right\} \geq 0
$$

or

$$
P_T = [V]^{\dagger}\left\{ [Y] + [Y]^{\dagger} \right\}[V] \geq 0
$$

(9.30)

Equation (9.30) indicates that the admittance matrix of a passive two-port must be positive semi-definite, i.e. $[Y] \geq 0$.

In the meantime, from Equation (9.30) we see that for a lossless two-port

$$
P_T = [V]^{\dagger}\left\{ [Y] + [Y]^{\dagger} \right\}[V] = 0
$$

Hence, for lossless two-ports $[Y]$ must be skew Hermitian, which reveals that Y_{11} and Y_{22} must be Foster functions. Furthermore, Y parameters should be expressed as

$$
Y_{ij}\left\{ \frac{\text{even polynomial}}{\text{odd ploynomial}} \quad \text{or} \quad \frac{\text{odd ploynomial}}{\text{even polynomial}} \right\}, i,j = 1, 2
$$

(9.31)

9.9 Driving Point Input Impedance of a Lossless Two-port

Referring to Figure 9.14, the input impedance of the resistively terminated lossless two-port may be expressed in terms of Z parameters.

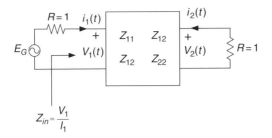

Figure 9.14   Driving point impedance of a lossless two-port

In this case, using Equation (9.17) one can write

$$Z_{in} = \frac{V_1}{I1} = Z_{11} + Z_{12} \left[\frac{I_2}{I_1}\right]$$ (9.32)

Equation (9.17) reveals that $V_2 = -RI_2 = Z_{21}I_1 + Z_{22}I_2$ or $(R + Z_{22})I_2 = -Z_{21}I_1$ or

$$\left[\frac{I_2}{I_1}\right] = -\frac{Z_{21}}{Z_{22} + R}$$

Using this result in Equation (9.32) one can obtain

$$Z_{in} = Z_{11} - \frac{Z_{12}Z_{21}}{Z_{22} + R} = \frac{Z_{11}R + Z_{11}Z_{22} - Z_{12}Z_{21}}{Z_{22} + R}$$ (9.33)

Let all the impedance components of the lossless two-port be normalized with respect to terminating resistance R. Then,

$$Z_{in} = [Z_{11}]\frac{1 + \left[\dfrac{\Delta Z}{Z_{11}}\right]}{[Z_{22}] + 1}$$ (9.34)

where $\Delta Z = Z_{11}Z_{22} - Z_{12}Z_{21}$ is the determinant of the impedance matrix Z. Z_{11} and Z_{22} are the open-circuit input and output Foster impedances.

Substituting Equation (9.28) in Equation (9.34), one can readily obtain

$$Z_{in} = [Z_{11}]\frac{1 + \left[\dfrac{1}{Y_{22}}\right]}{1 + [Z_{22}]}$$ (9.35)

For a reciprocal lossless two-port, consisting of lumped circuit elements, the driving point input impedance can be written as

$$Z_{in}(p) = \frac{a(p)}{b(p)} = \frac{m_1(p) + n_1(p)}{m_2(p) + n_2(p)}$$ (9.36)

where $m_i(p)(i = 1, 2)$ and $n_i(p)(i = 1, 2)$ are the even and odd parts of the numerator and denominator polynomials respectively.

Equation (9.36) can be written as

$$Z_{in}(p) \left[\frac{m_1}{n_2} \right] \frac{1 + \left[\frac{n_1}{m_1} \right]}{1 + \left[\frac{m_2}{n_2} \right]}$$

or

$$Z_{in}(p) = \left[\frac{n_1}{m_2} \right] \frac{1 + \left[\frac{m_1}{n_1} \right]}{1 + \left[\frac{n_2}{m_2} \right]}$$

(9.37)

By comparing Equation (9.35) and Equation (9.37) in his classical work, Darlington proposed that a lossless two-port can be constructed from its driving point impedance by setting[9]

$$\text{Case A}: Z_{11} = \left[\frac{m_1}{n_2} \right] \quad Z_{22} = \left[\frac{m_2}{n_2} \right] \quad Y_{22} = \frac{m_1}{n_1} \quad Z_{12} = \frac{\sqrt{m_1 m_2 - n_1 n_2}}{n_2}$$

$$\text{Case B}: Z_{11} = \left[\frac{n_1}{m_2} \right] \quad Z_{22} = \left[\frac{n_2}{m_2} \right] \quad Y_{22} = \frac{n_1}{m_1} \quad Z_{12} = \frac{\sqrt{-(m_1 m_2 - n_1 n_2)}}{m_2}$$

(9.38)

Note that in the above cases

$$Z_{12} = \sqrt{Z_{11} Z_{22} - \Delta Z} = \sqrt{Z_{11} Z_{22} - \frac{Z_{22}}{Y_{11}}}$$

As in Equation (8.9) of Chapter 8, the even part $R(p^2)$ of $Z_{in}(p)$ is given by

$$R(p^2) = \frac{m_1 m_2 - n_1 n_2}{m_2^2 - n_2^2} = \frac{N(p^2)}{D(p^2)} = \frac{\text{even polynomial}}{\text{even polynomial}}$$

(9.39)

In the meantime Equation (9.38) reveals that

$$\text{Case A}: \quad Z_{12}(p) Z_{12}(-p) = \frac{N(p^2)}{n_2(p) n_2(-p)}$$

$$\text{Case B}: \quad Z_{12}(p) Z_{12}(-p) = \frac{N(p^2)}{m_2(p) m_2(-p)}$$

(9.40)

A comparison of Equation (9.39) to Equation (9.40) indicates that the zeros of $Z_{12}(p) Z_{12}(-p)$ co-incide with the zeros of $R(p^2)$.

On the other hand, by Equation (9. 11), we have shown that analytic continuation of TPG yields

$$T(p^2) = \frac{4 R_G(p^2) R(p^2)}{[Z_G(p) + Z_{in}(p)][Z_G(-p) + Z_{in}(-p)]}$$

or, by setting $Z_G = 1$,

$$T(p^2) = \frac{4 R(p^2)}{[Z_{in}(p) + 1][Z_{in}(-p) + 1]}$$

(9.41)

---

[9] S. Darlington, 'Synthesis of Reactance 4-poles', *Journal al Mathematics and Physics*, Vol. 18, pp. 257–353, 1939.

Definition 2 (Transmission zeros of lossless two ports): The zeros of TPG described in complex variable $p = \sigma + j\omega$ as in Equation (9.41) are called the zeros or transmission zeros of the lossless two-port under consideration.

Note that, due to the nature of Equation (9.41), count for the zeros of $T(p^2)$ must be even integers and exhibit mirror image symmetry with doubled zeros on the $j\omega$ axis. In this case, it is convenient to consider only zeros of transmissions in the RHP as well as the ones on the $j\omega$ axis with a single count. Therefore, we can simply refer to half of the zeros of $T(p^2)$ as the transmission zeros of the lossless two-port without loss of generality.

From Equations (9.39–9.41), it is interesting to observe that the zeros of $T(p^2)$, $R(p^2)$ and $Z_{12}(p)Z_{12}(-p)$ coincide and are basically governed by the driving point impedance $Z_{in}(p)$. Clearly, finite transmission zeros are specified by the zeros of the even part $R(p^2)$. Furthermore, the poles of $Z_{in}(p)$ are also among the zeros of transmission.[10]

Now, there are several issues that need to be clarified.

We know that for a realizable lossless two-port, $Z_{11}, Z_{22}, Y_{11}, Y_{22}$ must be Foster functions. Is this true for the functions specified by Equation (9.38)?

It is well established that Z and Y parameters must be rational functions in complex variable p. In this case, one should find out if the even polynomial $N(p^2) = \pm(m_1 m_2 - n_1 n_2)$ is a perfect square yielding a rational odd function for $Z_{12}(p)$.

The answers to these questions are affirmative and summarized in the following properties, skipping the detailed proofs.

Property 15: Equation (9.38) reveals that for both Case A and Case B, finite poles of $Z_{12}(p)$ must overlap with those finite poles of Z_{11}, Z_{12} and Z_{22}. For Case A, these poles are specified by the zeros of odd polynomial $n_2(p)$. Similarly, for Case B, finite poles are given as the zeros of even polynomial $m_2(p)$. Note that for Case A, one of the finite poles must be located at DC since $n_2(p)$ is an odd polynomial.

Property 16: Let $p = j\omega_r$ be a simple pole of Z parameters such that, in the neighborhood of this pole, we can approximate Z parameters as

$$
\boxed{
\begin{aligned}
Z_{11}(p) &\approx \frac{2k_{11}p}{p^2 + \omega_r^2} \\
Z_{22}(p) &\approx \frac{2k_{22}p}{p^2 + \omega_r^2} \\
Z_{12}(p) &\approx \frac{2k_{12}p}{p^2 + \omega_r^2}
\end{aligned}
}
\qquad (9.42)
$$

Since the impedance matrix $[Z]$ is positive semi-definite, then determinants formed with cofactors must be non-negative for all $\{p = \sigma + j\omega; \sigma \geq 0\}$ yielding

$$
\boxed{
\begin{aligned}
k_{11} &\geq 0 \\
k_{22} &\geq 0 \\
k_{11}k_{22} - k_{12}^2 &\geq 0
\end{aligned}
}
\qquad (9.43)
$$

---

[10] Note that, equivalently, transmission zeros can also be specified in terms of the driving point input admittance $Y_{in}(p) = G(p^2) + \text{odd}\{Y_{in}(p)\}$. In this case, TPG is given by

$$
T(p^2) = \frac{4G(p^2)}{[Y_{in}(p) + 1][Y_{in}(-p) + 1]}
$$

Thus, the zeros of the even part $G(p^2)$ and the poles of $Y_{in}(p)$ are the transmission zeros of the lossless two-port.

Property 17: It can be shown that any irreducible ratio[11]

$$\left\{ f(p) = \frac{m_i}{n_i} = \frac{\text{even polynomial}}{\text{odd polynomial}} \right\} \text{ or } \left\{ f(p) = \frac{n_i}{m_i} = \frac{\text{odd ploynomial}}{\text{even polynomial}} \right\}; \ i = 1, 2 \qquad (9.44)$$

generated from a positive real immitance function

$$F_{in}(p)\frac{a(p)}{b(p)} = \frac{m_1(p) + n_1(p)}{m_2(p) + n_2(p)} \qquad (9.45)$$

yields a Foster function.

Let us look at look at the term 'irreducible'. In order for Equation (9.44) to yield a Foster function there should be no cancellation between numerator and denominator polynomials, which in turn requires no missing terms in these polynomials.

Verification of this property can be found in Balabanian's book.[12]

Property 18: In Property 17, if the driving point immittance is defined as an admittance function, then all the results obtained in Z parameters can be generated for Y parameters.

Hence, for

$$Y_{in}(p) = \frac{a(p)}{b(p)} = \frac{m_1(p) + n_1(p)}{m_2(p) + n_2(p)}$$

admittance parameters of the lossless two-port can be obtained as in Equation (9.38):

$$\begin{array}{llll}
\textbf{Case A}: Y_{11} = \left[\dfrac{m_1}{n_2}\right] & Y_{22} = \left[\dfrac{m_2}{n_2}\right] & Z_{22} = \dfrac{m_1}{n_1} & Y_{12} = \dfrac{\sqrt{m_1 m_2 - n_1 n_2}}{n2} \\[4mm]
\textbf{Case B}: Y_{11} = \left[\dfrac{n_1}{m_2}\right] & Y_{22} = \left[\dfrac{n_2}{m_2}\right] & Z_{22} = \dfrac{n_1}{m_1} & Y_{12} = \dfrac{\sqrt{-(m_1 m_2 - n_1 n_2)}}{m_2}
\end{array} \qquad (9.46)$$

9.10 Proper Selection of Cases to Construct Lossless Two-ports from the Driving Point Immittance Function

From the section above, we can see that there is the opportunity to construct a lossless two-port from either its driving point impedance or admittance function.

Once the impedance or admittance decision is made, one may proceed in two ways: either with Case A or Case B. Actually, either case is okay as long as the ratios constructed to form the triplets $\{Z_{11}, Z_{22} \text{ and } Y_{22}\}$ or $\{Y_{11}, Y_{22} \text{ and } Z_{22}\}$ are irreducible as dictated by Property 17.

For example, assume that we start with Case A by forming the triplet $\{Z_{11}, Z_{22} \text{ and } Y_{22}\}$. In this case, we should check if there is a cancellation in the ratios of $Z_{11} = [m_1/n_2], Z_{22} = [m_2/n_2], Y_{22} = m_1/n_1$. If not, we can proceed with this case as described above. On the other hand, if there is a cancellation, then we should switch to Case B. We should still check if the triplet $Z_{11} = [n_1/m_2], Z_{22} = [n_2/m_2], Y_{22}$ is irreducible. If not, we can proceed with this case. Similarly, all the above checks can be carried out on the driving point admittance function as we wish.

---

[11] Irreducible means that no cancellation occurs between the numerator and denominator polynomial.
[12] Norman Balabanian, *Network Synthesis*, Prentice Hall, 1958.

Example 9.2: Let

$$Z(p) = \frac{p^2 + 2p}{p + 1}$$

be a driving point impedance.

(a) Find the transmission zeros of the lossless two-port dictated by $Z(p)$.
(b) Find the proper case to construct a lossless two-port with the Darlington procedure.
(c) Synthesize $Z(p)$ as a lossless two-port in unit termination.

Solution:

(a) For the given driving point impedance

$$Z(p) = \frac{p^2 + 2p}{p + 1}$$

numerator and denominator polynomials yield that $m_1 = p^2, n_1 = 2p, m_2 = 1, n_2 = p$. Thus, the even part $R(p^2)$ of $Z(p)$ is given by

$$R(p^2) = \frac{m_1 m_2 - n_1 n_2}{m_2^2 - n_2^2} = \frac{p^2 - 2p^2}{1 - p^2} = \frac{-p^2}{1 - p^2}$$

By definition, the transmission zeros of the lossless two-port are the zeros of

$$T(p^2) = \frac{4R(p^2)}{[Z_{in}(p) + 1][Z_{in}(-p) + 1]}$$

Therefore, they are located at $p = 0$ and infinity (i.e. $p = \infty$).

(b) Firstly, let us consider Case A:

$$Z_{11} = \left[\frac{m_1}{n_2}\right] = \frac{p^2}{p} = p \quad Z_{22} = \left[\frac{m_2}{n_2}\right] = \frac{1}{p} \quad Y_{22} = \frac{m_1}{n_1} = \frac{p^2}{p} = p$$

$$Z_{12} = \frac{\sqrt{m_1 m_2 - n_1 n_2}}{n_2} = \frac{\sqrt{p^2 - 2p^2}}{p} = \frac{\sqrt{-p^2}}{p}$$

In this case we have cancellations in both Z_{11} and Y_{22}. Furthermore, Z_{12} is irrational. Therefore, we should try Case B.

For Case B we have

$$Z_{11} = \left[\frac{n_1}{m_2}\right] = 2p \quad Z_{22} = \left[\frac{n_2}{m_2}\right] = p \quad Y_{22} = \frac{n_1}{m_1} = 2p$$

$$Z_{12} = \frac{\sqrt{-(m_1 m_2 - n_1 n_2)}}{m_2} = p$$

As we can see, Case B reveals no cancellation. Furthermore, $Z_{12}(p)$ is proper. Therefore, a skew Hermitian Z matrix is realizable.

(c) For the synthesis, let us consider the common poles of $Z_{11}(p)$, $Z_{12}(p)$ and $Z_{22}(p)$. They are all at infinity. Furthermore, residues $k_{22\infty}$ and $k_{12\infty}$ at infinity are equal to 1 (i.e. $k_{22\infty} = 1$ and $k_{12\infty} = 1$). In this case, it may be proper to express $Z_{11}(p)$ as $Z_{11}(p) = p + k_{11\infty}p$ such that $k_{11\infty} = 1$.

In this case, a shunt inductance of $L_2 = 1$ H can realize the Z Matrix specified by

$$Z(p)\begin{bmatrix} k_{11\infty}p & k_{12}p \\ k_{12}p & k_{22\infty}p \end{bmatrix} \text{ such that } k_{11\infty}k_{22\infty} - k_{12}^2 = 0 \qquad (9.47)$$

as shown in Figure 9.15.

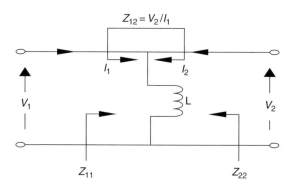

Figure 9.15   Synthesis of Z matrix specified by Equation (9.47) where $L = K_{11\infty} = k_{22\infty} = k_{12\infty} = 1$

Then, it is straightforward to complete the lossless two-port by adding the separate term of $Z_{11}(p)$ as a private pole as depicted in Figure 9.16. A quick check of the impedance parameters reveals that

$$Z_{11}(p) = \frac{V_1(p)}{1_1(p)}\bigg|_{I_2 = 0} = L_1p + L_2p = p + p = 2p$$

$$Z_{22}(p) = \frac{V_2(p)}{I_2(p)}\bigg|_{I_1 = 0} = L_2p = p$$

$$Z_{12}(p) = \frac{V_1(p)}{I_2(p)}\bigg|_{I_1 = 0} = L_2p = p$$

Example 9.3: Let

$$Z(p) = \frac{a(p)}{b(p)} = \frac{a_1p^2 + a_2p + a_3}{b_1p^3 + b_2p^2 + b_3p + b_4} = \frac{p^2 + 2p + 1}{p^3 + 2p^2 + 2p + 1}$$

(a) Verify that Z(p) is positive real.
(b) Construct the Z parameters of a lossless two-port using the Darlington method for Case A.
(c) Check if the above derived form (i.e. form of part (b)) is realizable as it is, or does it require augmentation? If the latter, carry out the augmentation process to end up with realizable Z parameters.
(d) Check if the final Z matrix of the lossless two-port is positive semi-definite.

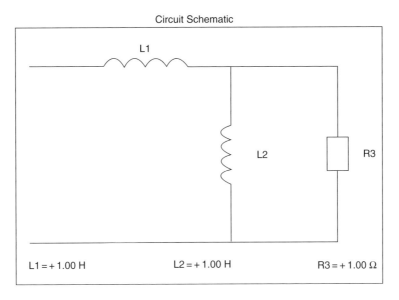

Circuit Schematic

L1

L2

R3

$L1 = +1.00$ H $L2 = +1.00$ H $R3 = +1.00$ Ω

Figure 9.16   Synthesis of Z matrix for Case B of Example 9.2

Solution:

(a) Let us find out if all the zeros and poles of $Z(p)$ if are in the LHP.

Two roots of $a(p)$ are located at $p = -1$. Note that $a(p) = (p+1)^2$. There is no real frequency zero of $a(p)$, which is okay. The zeros of $a(p)$ are the zeros of the impedance function $Z(p)$.

Similarly, the zeros of $b(p)$ are located at

$$p_1 = -1$$
$$p_2 = -0.5 + j0.866\ 025$$
$$p_3 = -0.5 - j0.866\ 025$$

They are also placed in the LHP.

Therefore, $Z(p)$ must be positive real since its all poles and zeros are located in the open LHP.

(b) By Darlington's expression,

$$Z(p) = \frac{m_1(p) + n_1(p)}{m_2(p) + n_2(p)}$$

where

$$m_1(p) = p^2 + 1 \quad n_1(p) = 2p \quad m_2(p) = 2p^2 1 \quad n_2(p) = p^3 + 2p$$

Furthermore,

$$N(p) = m_1 m_2 - n_1 n_2 - (p^2+1)(2p^2+1) - (2p)(p^3+2p) = -p^2 + 1$$

or $m_1 m_2 - n_1 n_2 = (1-p)(1+p)$.

Case A is described by Equation (9.38) and the Z parameters are given as

$$Z_{11}(p)\frac{m_1(p)}{n_2(p)} = \frac{p^2+1}{p^3+2p}$$

$$Z_{22}(p) = \frac{m_2(p)}{n_2(p)} = \frac{2p^2+1}{p^3+2p}$$

$$Z_{12}(p) = \frac{\sqrt{m_1 m_2 - n_1 n_2}}{n_2(p)} = \frac{\sqrt{(1-p)(1+p)}}{p^3+2p}$$

Partial fraction expansion of the Z parameters yields

$$Z_{11} = \frac{k_{110}}{p} + \frac{k_{11}p}{p^2+2} = \frac{0.5}{p} + \frac{0.5p}{p^2+2}$$

and

$$Z_{22} = \frac{k_{220}}{p} + \frac{k_{22}p}{p^2+2} = \frac{0.5}{p} + \frac{1.5p}{p^2+2}$$

Since the residues of the above forms are all positive, then $Z_{11}(p)$ and $Z_{22}(p)$ must be Foster function with simple poles at $p=0$ and $p=\pm j\sqrt{(2)}$.

Unfortunately, the numerator of $Z_{12}(p)$ is not rational. Therefore, $Z(p)$ must be augmented by LHP zeros of $N(p) = m_1 m_2 - n_1 n_2$.

(c) In this case, the augmented version of $Z(p)$ is given by

$$Z(p) = \left[\frac{p^2+2p+1}{p^3+2p^2+2p+1}\right]\left[\begin{matrix}p+1\\p+1\end{matrix}\right]$$

or

$$Z(p) = \frac{m_1(p)+n_1(p)}{m_2(p)+n_2(p)} = \frac{p^3+3p^2+3p+1}{p^4+3p^3+4p^2+3p+1}$$

where

$$m_1(p) = 3p^2+1 \quad n_1(p) = p^3+3p$$
$$m_2(p) = p^4+4p^2 1 \quad n_2 = 3p^3+3p$$

Now, let us consider $N(p) = m_1 m_2 - n_1 n_2$:

$$N(p) = m_1 m_2 - n_1 n_2 = 4p^4 = 2p^2 - 1 = (p^2-1)^2$$

Hence, the augmented Z parameters are given by

$$Z_{11}(p) = \frac{3p^2+1}{3p^3+3p} = \frac{k_{110}}{p} + \frac{k_{11}p}{p^2+1} = \frac{\left(\dfrac{1}{3}\right)}{p} + \frac{\left(\dfrac{2}{3}\right)p}{p^2+1}$$

$$Z_{22}(p) = \frac{p^4+4p^2+1}{3p^3 3p} = k_{22\infty}p + \frac{k_{220}}{p} + \frac{k_{22}p}{p^2+1}$$

$$= \left(\frac{1}{3}\right)p + \frac{\left(\dfrac{1}{3}\right)}{p} + \frac{\left(\dfrac{2}{3}\right)p}{p^2+1}$$

and

$$Z_{12} = \frac{p^2 - 1}{3p^3 + 3p} = \frac{k_{120}}{p} + \frac{k_{12}p}{p^2 + 1} = \frac{\left(-\frac{1}{3}\right)}{p} + \frac{\left(\frac{2}{3}\right)p}{p^2 + 1}$$

(d) From part (c) we see that $Z_{22}(p)$ has a private pole at infinity with a positive residue, as it should. In other words, $k_{22\infty} = \left(\frac{1}{3}\right) > 0$; thus, the residue condition is satisfied at this pole. Furthermore, at common poles $p = 0$ and $p = \pm j$ the residues are all positive, which makes $Z_{11}(p)$ and $Z_{22}(p)$ Foster functions.

Here, what is left for the residue is to check the conditions derived from the positive semi-definite property of the impedance matrix. Let us first evaluate the residue condition at the pole $p = 0$ if $k_{110}k_{220} - k_{120}^2 \geq 0$ is true.

Well, we have $\frac{1}{3} \times \frac{1}{3} - \left[-\frac{1}{3}\right]^2 = 0$. Yes indeed, it is satisfied with equality.

Similarly, we should look at the residue condition given at the common poles of $p = \pm j$. In this case, $k_{11}k_{22} - k_{12}^2 = \frac{2}{3} \times \frac{2}{3} - \left[\frac{2}{3}\right]^2 = 0$. Hence, common poles at $p = \pm j$ are also compact. This is nice and allows us to make the following definition.

Definition 3 (Compact pole): A common pole which satisfies the residue condition under the equality sign is called compact. As we have shown above, the poles located at $p = 0$ and $p = \pm j$ are compact poles. The importance of compactness is found during the synthesis process of lossless two-ports.

9.11 Synthesis of a Compact Pole

It is straightforward to show that a finite compact pole p_r of a lossless two-port can be synthesized as shown in Figure 9.17. At the Compact Pole the residue condition is satisfied by the equality sign. In this case, $k_{11}k_{22} - k_{12}^2 = 0$ yielding $k_{22} = k_{12}^2/k_{11}$ or

$$Z_{22} = \left(\frac{k_{12}}{k_{11}}\right)^2 \frac{k_{11}p}{p^2 + \omega_r^2} = n^2 Z_{11} = \frac{k_{22}p}{p^2 + \omega_r^2} \text{ with } n = \frac{k_{12}}{k_{11}} \tag{9.48}$$

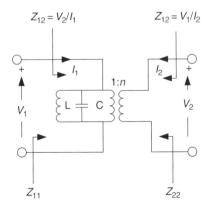

Figure 9.17   Synthesis of a finite Compact Pole of the impedance matrix

Equivalently, synthesis of the Compact Pole can be realized from the output port as depicted in Figure 9.18. In this case, the transformer ratio will be $n = k_{12}/k_{22}$ and at the input port the impedance

$$Z_{11} = \left(\frac{k_{12}}{k_{22}}\right)^2 \frac{k_{22}p}{p^2 + \omega_r^2} = n^2 Z_{22} = \frac{k_{11}p}{p^2 + \omega_r^2} \text{ with } n = \frac{k_{12}}{k_{22}}$$

The configuration shown in Figure 9.18 also yields the desired form of transfer impedance

$$Z_{12}(p) = nZ_{22} \text{ or } Z_{12} = \frac{k_{12}}{k_{22}} \cdot \frac{k_{22}p}{p^2 + \omega_r^2} = \frac{k_{12}p}{p^2 + \omega_r^2}$$

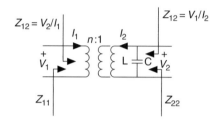

Figure 9.18 Compact Pole realization from the output port

9.12 Cauer Realization of Lossless Two-ports

For a given non-minimum immittance function $F(p)$, it is always possible to separate simple real frequency axis poles as Foster function $F_f(p)$ from $F(p)$. Then, the resulting minimum function $F_M(p)$ can be realized as described in this section. We should note that for the sake of easy understanding we use impedance notation for the synthesis. However, all the derivations and explanation can be carried out equivalently for admittance matrix synthesis.

Assume that we start the synthesis process from the input side. In this case, we can always separate private poles and compact poles from the given impedance matrix Z such that

$$\begin{aligned} Z_{11}(p) &= Z_{11P}(p) + Z_{11C}(p) \\ Z_{22}(p) &= Z_{22P}(p) + Z_{22C}(p) \\ Z_{12}(p) &= Z_{12C}(p) \end{aligned} \tag{9.49}$$

where the subscripts P and C designate private and compact portions of the Z parameters, respectively. Note that the Foster input and output impedances, namely $Z_{11}(p)$ and $Z_{22}(p)$, may have compact poles separately. At each compact pole one must satisfy the residue condition such that $k_{11}k_{22} - k_{12}^2 \geq 0$. However, we can always find positive integers k_{11P}, k_{11C} and k_{22P}, k_{22C} to satisfy the residue inequality as an equality such that

$$\begin{aligned} k_{11} &= k_{11P} + k_{11C} \\ k_{22} &= k_{22P} + k_{22C} \\ k_{11C}k_{22C} - k_{12}^2 &= 0 \end{aligned}$$

or, similarly,

$$\begin{aligned} Z_{11}(p) &= Z_{11P} + Z_{11C} \quad Z_{11P} = k_{11P}Z, Z_{11C} = k_{11C}Z \\ Z_{22}(p) &= Z_{22P} + Z_{22C} \quad Z_{22P} = k_{22P}Z, Z_{22C} = k_{22C}Z \end{aligned} \tag{9.50}$$

In Equation (9.47), the generic form of impedance Z can be expressed as

$$Z = \left\{ p \text{ or } \frac{1}{p} \text{ or } \frac{p}{p^2 + \omega_r^2} \right\} \tag{9.51}$$

Then, a compact pole with its private portions can be realized as in Figure 9.19.

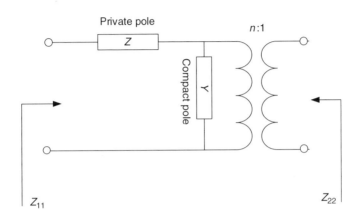

Figure 9.19   Cauer realization of a pole

On the other hand, as extracted in Equation (9.47), it may be possible to collect private poles in series with either $Z_{11}(p)$ or $Z_{22}(p)$ as shown in Figure 9.20.

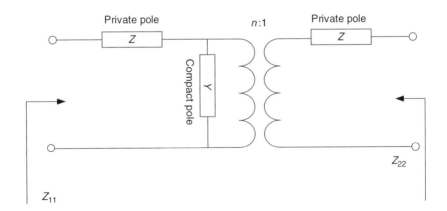

Figure 9.20   Cauer realization of a pole as a compact pole and private poles of Z_{11} and Z_{22}

In conclusion, we have shown that any positive real function can be realized as a lossless two-port in unit termination. This conclusion is known as Darlington's theorem, which constitutes the essence of power transfer networks.

The crux of the idea is that TPG of a power transfer problem can be expressed in terms of a positive real driving point immittance function, say $Z_Q(p) = a(p)/b(p)$. This PR function describes Darlington's two-port which is essentially the lossless power transfer network in unit termination.

Thus, a power transfer network design problem becomes a proper description of a positive real function which optimizes TPG. However, there are a lot of detailed practical issues that need to be clarified. We will tackle these issues as we design power transfer networks in the following chapters.

10

Design of Power Transfer Networks: A Glimpse of the Analytic Theory via a Unified Approach

10.1 Introduction

The problem of broadband matching is a major concern in high-frequency communication systems. All broadcasting networks such as radio and TV, wireless communication networks such as cellular telephones and satellite networks are well-known examples of systems where this problem is frequently encountered.

A typical high-frequency wireless communication system comprises two major components: a transmitter and a receiver (Figure 10.1).

On the transmitter side, the generated signal must be properly transferred to the antenna over a preferably non-dissipative (lossless) network so that maximum power of the generated signal is conveyed to the antenna. Similarly, on the receiver side, the received signal of the antenna is transferred over a lossless matching network and dissipated at the user's end. This may be a radio or a TV set or headphones etc. In this case, again the role of the matching network is to provide maximum power transfer of the received signal to the user's end. In the literature, several terms are associated with the non-dissipative power transfer network, such as 'impedance matching network', 'equalizer', 'lossless two-port', 'lossless network' or 'interstage equalizer'. All these terms are inter changeable. Classical broadband matching theory deals with the proper design of lossless matching networks between predetermined terminations.

It is common for the signal generation section of the transmitter to be simply modeled as an ideal signal generator in series with an internal impedance Z_G. The transmitter antenna will behave as a typical passive load termination Z_L to the lossless power transfer two-port $[E]$ (see Figure 10.2).

Likewise, the receiver antenna can be considered as an ideal signal source with an internal impedance Z_G. The user's end of the receiver site can thus be considered as a dissipative load Z_L to the lossless two-port $[E]$.

Design of Ultra Wideband Power Transfer Networks Binboga Siddik Yarman
© 2010 John Wiley & Sons, Ltd

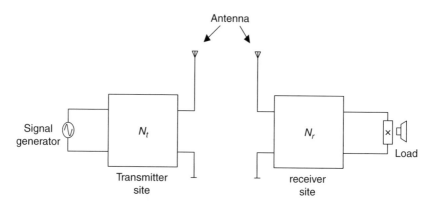

Figure 10.1   Description of a typical power transfer problem

In the discussion above it is evident that both the transmitter and the receiver components present a similar model as far as signal flow is concerned. In both cases, the crucial issue is the maximum power, transferred from the generator Z_G to the load Z_L. Therefore, once the signal generator and the load are given, the system performance can be optimized with the proper design of the non-dissipative or lossless two-port [E].

As a matter of fact, in all cascaded high-frequency systems, one is faced with the problem of power transfer between cascaded sections or the so-called 'inter stages'. As a principle using Thévenin's theorem, the left side of the interstage can be modeled as an ideal signal generator E_G in series with an internal impedance Z_G. Similarly, the right side is simply regarded as a passive load Z_L as shown in Figure 10.2.

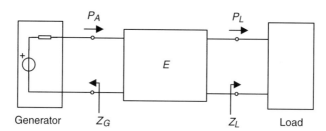

Figure 10.2   Thévenin equivalent of the transmitter which is terminated in a complex load: antenna over an equalizer [E]

Hence, the classical broadband matching problem is defined as one of 'constructing a lossless reciprocal two-port or equalizer so that the power transfer from source (or generator) to load is maximized over a prescribed frequency band'.

The power transfer capability of the lossless equalizer or the so-called 'matching' network is best measured with the transducer power gain T, which is defined as the ratio of the power delivered to the load P_L to the available power P_A of the generator, over a wide frequency band. That is,

$$T = \frac{P_L}{P_A}$$

(10.1)

Ideally, the designer requires the transfer of the available power of the generator to the load; this in turn requires a flat transducer power gain characteristic in the band of operation at a unitary gain level with a sharp rectangular roll-off, as illustrated in Figure 10.3.

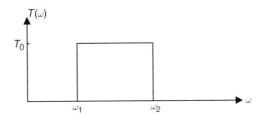

Figure 10.3   Ideal form of the transducer power gain over the specified angular frequency band $[\omega_1, \omega_2]$

However, the physics of the problem unfortunately permits the ideal power transfer at only a single frequency. In this case, the equalizer input impedance Z_{in} is matched in conjugation with the generator impedance Z_G. Therefore, the design of a matching equalizer over a wide frequency band with 'high' and 'flat' gain characteristics presents a very complicated theoretical problem. It is well known that the terminating impedances Z_G and Z_L impose the possible highset flat gain level over frequency band B, i.e. the theoretical 'gain–bandwidth limitation' of the matched system.

Before we introduce the design methodologies, let us classify the broadband matching problem:

- Single Matching: This is a matching problem where one of the passive terminations of the equalizer is resistive, whereas the other is complex or frequency dependent (see Figure 10.4(a)).

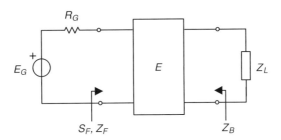

Figure 10.4(a)   Definition of single matching problem

- Double Matching: This is a matching problem where both passive terminations of the equalizer are complex (see Figure 10.4(b)).
- Active Matching: This is a matching problem of active devices. A typical example of active matching is the design of a microwave amplifier (see Figure 10.4(c)).

It should be mentioned that the 'filter' or the 'insertion loss' problem might also be considered as a very special form of the broadband matching problem that deals with a resistive generator and a resistive

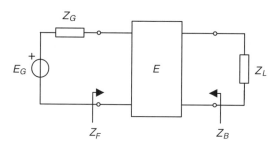

Figure 10.4(b)   Definition of double matching problem

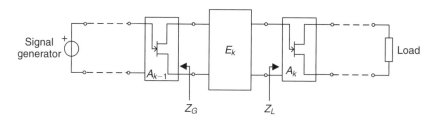

Figure 10.4(c)   Definition of active matching problem

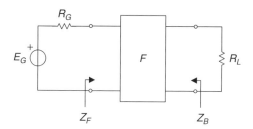

Figure 10.4(d)   Definition of filter or insertion loss problem

load (see Figure 10.4(d)). In this respect, well-established filter design techniques may be employed for broadband matching problems where appropriate.

There are two main approaches to the solution of broadband matching problems, namely analytical and computer-aided solutions. The classical procedure is through analytic gain–bandwidth theory [1]. The second type of solution is accomplished by numerical optimization and is referred to as real frequency techniques after Carlin [2]. Both cases seek optimal achievement of the maximum level of minimum gain within the passband.

Analytic gain–bandwidth theory is essential to understand the nature of the matching problem, but in general it is not accessible beyond simple problems. The real frequency computer-aided solutions, however, are rather practical and easy to carry out for more complicated problems.

In this chapter, we briefly discuss the analytic matching theory and then, in the following chapter, several 'real frequency' approaches to the broadband matching problem are introduced.

Generally, the lossless matching network to be designed can be described in terms of two-port parameters (such as impedance, admittance, chain, real or complex, normalized, regularized scattering or transmission parameters) or by means of the driving point; that is, the so-called Darlington immittance or bounded real (real normalized) reflection coefficient.

At this point, it is appropriate to state the modified version of Darlington's famous theorem [3].

Darlington's Theorem states that any positive real impedance (Z) or admittance (Y) function or corresponding bounded real reflection coefficient $S = (Z - 1)/(Z + 1)$ or $S = (1 - Y)/(1 + Y)$ can be represented as a lossless two-port terminated in unit resistance. The resulting lossless two-port is called the Darlington equivalent (see Figure 10.5).

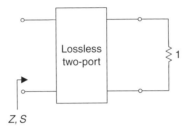

Figure 10.5   Darlington equivalent of a driving point impedance Z or reflectance $S = (Z - 1)/(Z + 1)$

Based on the fundamental gain–bandwidth theory introduced by Bode [1], the analytic approach to single matching problems was first developed by Fano [4] using the concept of the Darlington equivalent of the passive load impedance Z_L. In Fano's approach, the problem is handled as a 'pseudo-filter' or 'pseudo-insertion loss' problem, since the tandem connection of the lossless equalizer $[E]$ and the Darlington load equivalent $[L]$ is considered as a whole lossless filter $[F]$ (see Figure 10.6).

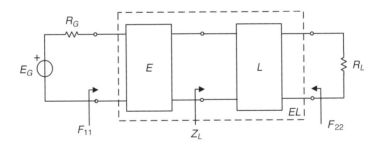

Figure 10.6   Fano's description of a single matching problem

Later, Youla [5] proposed a rigorous solution to the problem using the concept of complex normalization. In order to solve the double matching problem, Youla described the lossless matching network in terms of complex normalized scattering parameters with respect to frequency-dependent impedances of generator and load terminations. Youla's theory provided an excellent solution to handle the single matching problems, but was not practical to solve the double matching problems since the realizablilty conditions based on the complex normalized scattering parameters of the matching equalizer were complicated to implement.

The complete analytic solution to the double matching problem has been accomplished by the main theorem of Yarman and Carlin [6,7], [12] which relates to the 'real' and the 'complex normalized–regularized' generator and load reflection coefficients of the doubly matched system. This theorem enables the designer to fully describe the doubly matched system in terms of the 'realizable'–real normalized (or unit normalized) scattering parameters after replacing the generator and load with their Darlington equivalents, as in filter design theory.

Instructional accounts of the gain–bandwidth theory for both single and double matching problems have been elaborated by Chen [8].

In the following sections, the essence of Fano's and Youla's theories is reviewed. Subsequently, an attempt is made to introduce analytic solutions for single and double matching problems under the 'unified approach'. Then, modern computer-aided design or the 'real frequency' techniques which are employed to construct wideband matching networks are presented in the following chapter [21]. Finally, practical design techniques to construct matching networks with mixed lumped and distributed elements will be covered briefly in the last chapter of this book.

In order to understand the analytic theory of broadband matching, it may be appropriate firstly to review the filter or insertion loss problem, which constitutes the core of the unified approach and clarifies the basic properties of lossless two-ports.

10.2 Filter or Insertion Loss Problem from the Viewpoint of Broadband Matching

A typical filter or insertion loss problem is depicted in Figure 10.4(d). In view of broadband matching, the problem is stated as follows: 'Given the resistive generator R_1 and the resistive load R_2, construct the reciprocal lossless filter two-port [F] to transfer the maximum power of the generator to the load R_2 only over the passband ω_1 to ω_2.'

In this problem, it is convenient to describe the reciprocal lossless filter two-port [F] in terms of its real (or equivalently unit) normalized scattering matrix F with respect to port normalization numbers R_1 and R_2. For unit normalisation

$$F = \begin{bmatrix} F_{11} & F_{12} \\ F_{21} & F_{22} \end{bmatrix}$$

such that $F_{12} = F_{21}$ due to reciprocity.

(10.2)

The system performance of the filter two-port [F] is measured by transducer power gain $T(\omega)$ which is given by

$$T(\omega) = |F_{21}(j\omega)|^2$$

(10.3)

If the filter consists of one kind of element (i.e. either lumped or distributed elements), the real normalized bounded real (BR) scattering parameters are given in the following, so-called 'Belevitch' canonic form [9]:

$$F_{11} = \frac{h}{g} \quad F_{21} = \frac{f}{g} \quad F_{22} = -\frac{f\,h_*}{f_*\,g}$$

(10.4)

where h, f, g are real polynomials in complex variable $p = \sigma + j\omega$ for lumped element design or in $\lambda = \Sigma + j\Omega$ if $[F]$ is constructed with equal length or commensurate transmission lines. Here, λ designates the Richards variable and is given by $\lambda = \tanh(p\tau)$ as in Chapter 3 and Chapter 4.

As shown in previous chapters, a lossless two-port must possess a bounded real para-unitary scattering matrix. Therefore,

$$\boxed{FF^{\dagger} = I}$$

(10.5)

or

$$\boxed{F_{11}F_{11*} + F_{21}F_{21*} = 1, \text{ or on } j\omega \text{ axis } |F_{21}|^2 = 1 - |F_{11}|^2}$$

(10.6a)

$$\boxed{F_{22}F_{22*} + F_{21}F_{21*} = 1, \text{ or on } j\omega \text{ axis } |F_{21}|^2 = 1 - |F_{22}|^2}$$

(10.6b)

$$\boxed{F_{22} = -\frac{F_{21}}{F_{21*}}F_{11*} \quad F_{11} = -\frac{F_{21}}{F_{21*}}F_{22*}}$$

(10.6c)

where I designates a 2×2 unitary matrix, superscript $\dagger$ indicates the para conjugate and transpose of a matrix, and $*$ denotes either para-conjugate (as a subscript) or complex conjugate (as a superscript).

Having used the complex frequency variable, namely p, which is associated with the lumped filter design, the equation set (10.6a–10.6c) can be written in terms of the canonical polynomials h, f and g:

$$\boxed{\begin{array}{c} hh_* = gg_* - ff_* \\ \text{or} \\ h(p)h(-p) = g(p)g(-p) - f(p)f(-p) \end{array}}$$

(10.7)

In terms of the canonic polynomials f and g, the transducer power gain is given by

$$\boxed{T(\omega^2) = \frac{f(j\omega)f(-j\omega)}{g(j\omega)g(-j\omega)}}$$

(10.8a)

$$\boxed{\begin{array}{c} \text{or in complex variable } p = j\omega \\ T(-p^2) = \frac{f(p)f(-p)}{g(p)g(-p)} \end{array}}$$

(10.8b)

In essence, Equation (10.8) dictate the performance measures of a lossless reciprocal filter. When the transducer power gain $T(-p^2)$ is different than zero, the lossless system allows signal transmission.

However, there are complex frequencies p_k such that $T(-p_k^2)$ is zero. In general, Equation (10.8b) specifies the status of signal transmission. Therefore, the function $T(-p^2)$ determines the forward and backward signal transmission of a lossless reciprocal filter $[F]$. The following definition of transmission zeros is given in the Carlin sense [12].

Definition 1: In a broad sense, the transmission zeros of a lossless two-port are the closed right half plane (RHP) zeros of the complex function $F(p)$ given by[1]

$$
\begin{aligned}
&\widetilde{F}(p) = \eta(p)\frac{f(p)f(-p)}{g^2(p)} \\
&\text{where} \\
&\eta(p) = \frac{f}{f_*} = \frac{f(p)}{f(-p)}
\end{aligned}
\tag{10.9}
$$

Note that, in Equation(10.9), all possible common factors between the numerator and the denominator are cancelled and the zeros on the $j\omega$ axis are counted for their half multiplicity.

For reciprocal structures, the complex function $\widetilde{F}(p)$ becomes $\widetilde{F}(p) = f^2(p)/g^2(p)$ since $F_{12}(p) = F_{21}(p)$. In this case, the transmission zeros of the lossless reciprocal two-port will simply be the closed RHP zeros of transmittance parameter $F_{21}(p) = f(p)/g(p)$ with even multiplicity, which obviously overlaps with the zeros of the transducer gain function $T(-p^2)$ of Equation (10.8b). Transmission zeros at infinity are considered as the real frequency zeros on the $j\omega$ axis and determined as the degree difference between the polynomials $g(p)$ and $f(p)$.

10.3 Construction of Doubly Terminated Lossless Reciprocal Filters

Based on the above theoretical overview and the details presented in Chapters 4, 5 and 6, the design steps of doubly terminated lossless reciprocal filters are straightforward:

Step 1: Choose an appropriate transducer power gain form $T(\omega^2)$ which includes all the desired transmission zeros of the doubly terminated system. Any readily available form such as the Butterworth, Chebyshev, elliptic or Bessel type of function may be suitable depending on the application.

Step 2: Using the Belevitch notation, spectral factorization of the numerator and denominator of the selected gain function $T(-p^2)$ is carried out to obtain the polynomials $f(p)$ and $g(p)$. At this stage, it should be pointed that the numerator $f(p)f(-p)$ must be of even multiplicity so that $F_{21}(p) = F_{12}(p)$. The polynomial $g(p)$ is uniquely determined by the spectral factorization of the denominator of $T(-p^2)$ since it must be strictly Hurwitz. Hence, $F_{21}(p) = F_{12}(p) = f(p)/g(p)$ is determined.

Step 3: The polynomial $h(p)$ is formed via spectral factorization of hh_* as given by Equation (10.7a). However, the zeros of hh_* are freely divided between the polynomials $h(p)$ and $h_* = h(-p)$. The only requirement is assurance of the reality of $h(p)$. Therefore, each complex root must be accompanied with its complex conjugate. Thus, $F_{11} = h/g$ and $F_{22} = -\eta h_*/g$ are determined within an analytic all-pass η which also includes RHP zeros of $f(p)$ and the general solution to the factorization problem is

$$F_{11} = \eta F_{11m}$$

where F_{11m} is the minimum phase solution.

---

[1] Recall that closed RHP zeros also cover the ones on the $j\omega$ axis.

Step 4: Finally, the filter is constructed by means of Darlington's synthesis procedure of driving point impedance $Z = (1 + F_{11})/(1 - F_{11})$ as a lossless two-port in unit termination [12].

10.4 Analytic Solutions to Broadband Matching Problems

In this section, basic guidelines of Fano's and Youla's approaches are given and linked by means of the main theorem of Yarman and Carlin, which in turn leads to the unified approach to designing broadband matching networks.

Analytic Approach to Single Matching Problems

The single matching problem deals with the construction of a broadband lossless equalizer [E], which is placed between a resistive generator and a complex load as shown in Figure 10.4(a).

In Fano's theory, the frequency-dependent 'non-Foster' load $Z_L(p)$ is replaced by its Darlington equivalent [L] as in Figure 10.6. Let $L = \{L_{ij}\}$ be its unit normalized scattering parameters. Let the unit normalized scattering parameters of the lossless equalizer [E] be $E = \{E_{ij}\}$. Then, cascaded connections of [E] and [L] constitute a lossless filter [F] with the desired gain performance. Thus, Fano determines the necessary and sufficient conditions to be able to extract the load network from the filter [F] or, equivalently, he constructs the filter [F] with desired gain characteristics in such a way that it includes the specified load for the single matching problem under consideration.

In the following, the scattering parameters for the load network [L] are derived in the Yarman–Carlin sense.

Scattering Description of the Darlington Equivalent of the Load Network[2]

Employing the para-unitary property of the scattering matrix of the lossless load network [L] as in Equation (10.6), its unit normalized scattering parameters are given in terms of the impedance $Z_L(p) = N_L(p)/D_L(p)$ as

$$L_{11} = \frac{Z_L - 1}{Z_L + 1} = \frac{N_L - D_L}{N_L + D_L} \tag{10.10a}$$

$$L_{21} = \frac{2W_L}{Z_L + 1} \tag{10.10b}$$

$$L_{22} = -\eta_L b_L \frac{Z_{L*} - 1}{Z_L + 1} \tag{10.10c}$$

where
$$b_L(p) = \frac{D_{L*}}{D_L} = \frac{D_L(-p)}{D_L(p)} \tag{10.10d}$$

$$\eta_L(p) = \frac{n_{L*}}{n_L} = \frac{n_L(-p)}{n_L(p)} \tag{10.10e}$$

---

[2] Details of this section also appear on page 158 of Chapter 8 in [56].

$$W_L = \frac{n_{L*}}{D_L} \tag{10.10f}$$

In Equation (10.10(e)), the numerator polynomial n_{L*} is formed on the closed RHP zeros of $R_L(p^2)$ such that $R_L(p^2) = W_L(p)W_L(-p) =$ the even part of $\{Z_L(p)\}$.

Actually, the above result can be obtained as follows.

By losslessness of $[L]$,

$$L_{21}(p)L_{21}(-p) = 1 - L_{11}(p)L_{11}(-p)$$

or

$$L_{21}(p)L_{21}(-p) = 1 - \frac{Z_L - 1}{Z_L + 1} \cdot \frac{Z_{L*} - 1}{Z_{L*} + 1} = \frac{2[Z_L + Z_{L*}]}{[Z_L + 1][Z_{L*} + 1]}$$

On the other hand,

$$R_L(p^2) = \frac{1}{2}[Z_L + Z_{L*}]$$

By setting

$$R_L(p^2) = W_L \cdot W_{L*} = \left(\frac{n_{L*}}{D_L}\right)\left(\frac{n_L}{D_{L*}}\right) \text{ with } W_L = \frac{n_{L*}}{D_L}$$

one can suggest that

$$L_{21} = \frac{2W_L}{Z_L + 1}$$

which is the expression given for L_{21} above.

Once again, by employing the para-unitary condition,

$$L_{22} = -\frac{L_{21}}{L_{21*}}L_{11*} = -\left[\left(\frac{W_L}{W_{L*}}\right)\left(\frac{Z_{L*} + 1}{Z_L + 1}\right)\right]\left[\frac{Z_{L*} - 1}{Z_{L*} + 1}\right] = -\left(\frac{n_{L*}}{n_L}\right)\left(\frac{D_{L*}}{D_L}\right)\left(\frac{Z_{L*} - 1}{Z_L + 1}\right)$$

or, by setting $\eta_L(p) = n_{L*}/n_L$ and $b_L(p) = D_{L*}/DL$,

$$L_{22} = -\eta_L b_L \frac{Z_{L*} - 1}{Z_L + 1}$$

The above equation set can be expressed in Belevitch form as

$$L_{11} = \frac{h_L}{g_L} \tag{10.11a}$$

$$L_{21} = \frac{f_L}{g_L} \tag{10.11b}$$

$$L_{22} = -\eta_L \frac{h_{L*}}{g_L}$$ (10.11c)

where
$$h_L(p) = N_L(p) - D_L(p)$$ (10.11d)

$$f_L(p) = 2n_L(-p)$$ (10.11e)

$$g_L(p) = N_L(p) + D_L(p)$$ (10.11f)

Now, let us define the transmission zeros of the load network.

Transmission Zeros of the Load Network

As in Equation (10.9), the transmission zeros of the load network are defined in the Carlin sense as the zeros of the following function:[3]

$$\tilde{F}_L(p) = L_{21}^2 = b_L(p) \frac{4R_L(p^2)}{[Z_L(p) + 1]^2}$$ (10.12a)

or
$$\tilde{F}_L(p) = \frac{4n_{L*}^2}{g_L^2}$$ (10.12b)

In Fano's work, the power performance of the singly matched system is measured in terms of the transfer scattering parameter F_{21} of $[F]$. In fact, the transducer power gain of the system is measured as in Equation (10.3):

$$T(\omega^2) = |F_{21}(j\omega)|^2 = 1 - |F_{22}(j\omega)|^2$$ (10.13)

---

[3] Carlin's definition of transmission zeros is slightly different than that of Youla's definition.

In terms of the scattering parameters of $[E]$ and $[L]$,

$$T\left(\omega^2\right) = \frac{|E_{21}|^2 |L_{21}|^2}{|1 - E_{22}L_{11}|^2} \tag{10.14}$$

Hence, one may analytically treat the single matching problem as a pseudo-filter problem as in Fano's theory, which can be stated as: 'Construct the lossless two-port $[F]$ with preferred gain characteristic as an insertion loss problem subject to load constraints so that the load two-port $[L]$ is extracted from $[F]$, yielding the desired matching network $[E]$.'

In Youla's theory, however, matching network $[E]$ and the load Z_L are treated as separate entities and the transducer power gain of the singly matched system is defined in terms of the complex normalized reflectance S_{CL} at the load end. That is,

$$T\left(\omega^2\right) = 1 - |S_{CL}(j\omega)|^2 \tag{10.15}$$

where S_{CL} is given by

$$S_{CL} = \frac{Z_B - Z_{L*}}{Z_B + Z_L} \tag{10.16}$$

where Z_B designates the driving point Darlington impedance of $[E]$ at the backend [5]. In this representation, the load impedance Z_L is regarded as the complex normalization number at the output port of $[E]$. Clearly, in the complex p domain, S_{CL} is not analytic due to the RHP poles of the term $Z_{L*} = Z_L(-p)$. In order to make S_{CL} analytic, an all-pass factor $b_L(p)$ is introduced into S_{CL} to cancel the RHP poles of Equation (10.16). Thus, the complex normalized–regularized reflectance $S_{YL}(p)$ is defined in the Youla sense as follows:

$$S_{YL}(p) = b_L(p)\frac{Z_B(p) - Z_L(-p)}{Z_B(p) + Z_L(p)} \tag{10.17}$$

In Youla's theory, instead of load extraction, complex normalized–regularized reflectance is directly constructed from the analytic form of the transducer power gain, satisfying the gain–bandwidth restriction. Then, the driving point impedance Z_B is obtained as a realizable positive real function as[4]

---

[4] In order to obtain Equation (10.17), first the difference

$$b_L(p) - S_{YL}(p) = b_L(p) - b_L(p)\frac{Z_B(p) - Z_L(-p)}{Z_B(p) + Z_L(p)}$$

is formed. Then, by setting $R_L = \frac{1}{2}(Z_L + Z_{L*})$, Equation(10.17) follows by straightforward algebraic manipulation.

$$Z_B(p) = \frac{2b_L R_L}{b_L - S_{YL}} - Z_L \qquad (10.18)$$

Finally, by employing the Darlington procedure, Z_B is synthesized, yielding the desired lossless matching network [E] in resistive termination.

Unfortunately, based on the definition of the complex normalized scattering parameters of the two-ports, extension of Youla's theory to double matching problems was not straightforward [5]. Therefore, in the following section, what we call the 'unified approach' to broadband matching problems is presented. The unified approach combines Youla's and Fano's works under a unique format by means of the main theorem of [6, 7] and makes the analytic theory accessible for many practical problems.

Before we deal with the main theorem, let us look at the definition of the bounded real (BR) analytic complex normalized reflectance S_{YCL} in the Yarman and Carlin sense [7].

Definition 2 (BR analytic complex normalized reflectance): The reflectance S_{YCL} defined by the expression

$$S_{YCL} = \frac{W_L}{W_{L*}} \frac{Z_B - Z_{L*}}{Z_B + Z_L} \qquad (10.19)$$

is called the BR analytic complex normalized reflectance in the Yarman and Carlin sense. Clearly, Equation (10.19) can be related to S_{YL} which is described in the Youla sense by Equation (10.17):

$$S_{YCL}(p) = \eta_L(p) S_{YL}(p) \qquad (10.20)$$

Furthermore, if the load is 'simple' and consists of a few reactive elements having all the transmission zeros at finite frequencies or at infinity, then all-pass product $\eta_L = 1$ and $S_{YCL}(p) = S_{YL}(p)$, which is the case in many engineering applications. Nevertheless, with proper augmentation of the load Z_L, one can make $n_L n_{L*}$ a perfect square, yielding $n_L/n_{L*} = 1$. Thus, by invoking superfluous factors, we can always obtain $S_{YCL}(p) = S_{YL}(p)$ as described in [12]. Therefore, in the following sections, $S_{YCL}(p)$ and $S_{YL}(p)$ will be used interchangeably.

Now we are ready to state the main theorem [6, 7], [12].

The Main Theorem of Yarman and Carlin

Referring to Figure 10.6, let F_{22} be the unit normalized backend reflectance of the system [F] constructed by the cascade connection of the lossless two-port [E] and the lossless two-port [L]. Then, F_{22} is equal to the analytic complex normalized reflectance, which is defined in the Yarman and Carlin sense at the input port of [L]. That is,

$$F_{22}(p) = S_{YCL}(p) = \eta_L(p) b_L(p) \frac{Z_B - Z_{L*}}{Z_B + Z_L} = \eta_L(p) S_{YL}(p)$$

Equivalently, Youla's complex normalized–regularized reflectance can be expressed in terms of $F_{22}(p)$ as

$$S_{YL}(p) = \eta_{L*}(p)S_{YCL}(p) = \eta_{L*}(p)F_{22}(p) = b_L(p)\frac{Z_B - Z_{L*}}{Z_B + Z_L}$$

The significant consequences of the main theorem may be summarized as follows.

The load network [L] is directly extracted from [F] by applying the well-established gain–bandwidth restrictions of Youla on the analytic complex normalized reflectance $S_{YCL}(p)$ defined in the Yarman and Carlin sense, in a straightforward manner.

In the course of the extraction process, the entire structure [F] is described in terms of its realizable, unit normalized, bounded real scattering parameters without hesitation, as in Fano's theory, since $F_{22}(p) = S_{YCL}(p)$ by the main theorem. Then, Youla's gain–bandwidth restriction is applied to the reflectance described by $S_{YL}(p) = \eta_{L*}(p)S_{YCL}(p) = \eta_{L*}(p)F_{22}(p)$. These unique results constitute the basis of the unified approach.

Let us now classify the transmission zeros of the load and then introduce Youla's theorem on gain–bandwidth restrictions, which will enable us to extract the load network [L] from the combined structure.

Classification of Transmission Zeros [12]

Let $p_0 = \sigma_0 + j\omega_0$ denote any transmission zero of the load Z_L. Then, p_0 belongs to one of the following mutually exclusive classes:

- Class A: Real$\{p_0\} = \sigma_0 < 0$, the transmission zero lies in the open RHP.
- Class B: Real$\{p_0\} = 0$ and $Z_L(j\omega_0) \neq \infty$, the transmission zero lies on the imaginary axis at a point where the load impedance Z_L is finite (it may also be zero, which is obviously finite).
- Class C: Real$\{p_0\} = \sigma_0 = 0$ and $Z_L(j\omega_0) = \infty$, the transmission zero lies on the imaginary axis at a point where the load impedance has a pole.

The basic gain–bandwidth theorem for single matching problems relates the properties of the system reflectance to the load zeros of transmission so that extraction of the load becomes possible.

The Basic Theorem of Gain–Bandwidth for Single Matching Problems [12]

Let
$$b_L(p) - S_{YL}(p) = \frac{2R_L b_L}{Z_B + Z_L}$$

be specified with the following Taylor series expansions :

$$b_L(p) = \sum_{r=1}^{\infty} b_{Lr}(p - p_0)^r$$

$$S_{YL} = \sum_{r=1}^{\infty} S_{Lr}(p - p_0)^r$$

$$F(p) = 2R_L b_L = \sum_{r=1}^{\infty} F_r(p - p_0)^r$$

(10.21)

Then, an even non-negative function $T(\omega^2)$ of $0 \leq T(\omega^2) \leq 1, \forall \omega$, can be realized as the transducer power gain of a finite lossless equalizer [E] inserted between a resistive generator and a frequency-dependent positive real load, if and only if, at each transmission zero $p_0 = \sigma_0 + j\omega_0$ of order k, the following constraints are satisfied:

Class of transmission zeros	Gain–bandwidth restrictions	
For Class A	$b_{Lr} = S_{Lr}, r = 0, 1, 2, \ldots, k-1$	(10.22a)
For Class B	$b_{Lr} = S_{Lr}, r = 0, 1, 2, \ldots, 2k-1$	(10.22b)
	and $$\frac{b_{L(2k-1)} - S_{L(2k-1)}}{F_{2k}} \geq 0$$	(10.22c)
For Class C	$b_{Lr} = S_{Lr}, r = 0, 1, 2, \ldots, 2k-2$	(10.22d)
	$$\frac{b_{L(2k-1)} - S_{L(2k-1)}}{F_{2k-2}} \leq \frac{1}{c_{-1}} \geq 0$$	(10.22e)

In Equation (10.22e), c_{-1} is the residue of $Z_L(p)$ at the pole $p_0 = j\omega_0$, otherwise the subscripts identify a Taylor coefficient.

Note that similar sets of constraints are given in logarithmic and integral forms. For details, interested readers are referred to [4,5].

10.5 Analytic Approach to Double Matching Problems

In order to handle the double matching problem analytically, the Darlington two-ports [G] and [L] replace the generator impedance Z_G and the load impedance Z_L respectively. Hence, the system which will be doubly matched is represented as the cascaded trio of the lossless two-ports [G]–[E]–[L] as shown in Figure 10.7.

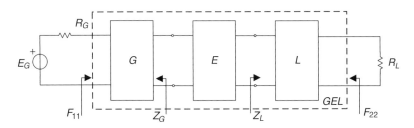

Figure 10.7 Double matching problem as represented by cascaded trio of $[F] = \{[G]-[E]-[L]\}$

The entire structure is combined under the lossless two-port $[F]$ and described by means of the unit normalised bounded real (BR) scattering parameters $F = \{F_{ij}, i, j = 1, 2\}$ as in the filter theory. The load network $[L]$ is described by equation sets (10.10) and (10.11). Similarly, the description of the generator network $[G]$ and its transmission zeros are given as follows.

Scattering Description of the Darlington Equivalent of the Generator Network

Unit normalized scattering parameters of the lossless generator network $[G]$ are given in terms of the impedance $Z_G(p) = N_G(p)/D_G(p)$ as follows:

$$G_{22} = \frac{Z_G - 1}{Z_G + 1} = \frac{N_G - D_G}{N_G + D_G} \tag{10.23a}$$

$$G_{21} = \frac{2W_G}{Z_G + 1} \tag{10.23b}$$

$$G_{11} = -\eta_G b_G \frac{Z_{G*} - 1}{Z_G + 1} \tag{10.23c}$$

where

$$b_G(p) = \frac{D_{G*}}{G} = \frac{D_G(-p)}{D_G(p)} \tag{10.23d}$$

$$\eta_G(p) = \frac{n_{G*}}{n_G} = \frac{n_G(-p)}{n_G(p)} \tag{10.23e}$$

$$W_G = \frac{n_{G*}}{D_G} \tag{10.23f}$$

with

$$R_G = \text{even}\{Z_G\} = W_G W_{G*}$$

In Belevitch form,

$$G_{22} = \frac{h_G}{g_G} \tag{10.24a}$$

$$G_{21} = \frac{f_G}{g_G} \qquad (10.24\text{b})$$

$$G_{11} = -\eta_G \frac{h_{G*}}{g_G} \qquad (10.24\text{c})$$

where

$$h_G(p) = N_G(p) - D_G(p) \qquad (10.24\text{d})$$

$$f_G(p) = 2n_G(-p) \qquad (10.24\text{e})$$

$$g_G(p) = N_G(p) + D_G(p) \qquad (10.24\text{f})$$

Transmission Zeros of the Generator Network

As in Equation (10.9), the transmission zeros of the generator network are defined in the Carlin sense as the zeros of the function

$$\widetilde{F}_G(p) = G_{21}^2 = b_G(p) \frac{4R_G(p^2)}{[Z_G(p) + 1]^2} \qquad (10.25\text{a})$$

or

$$\widetilde{F}_G(p) = \frac{4n_{G*}^2}{g_G^2} \qquad (10.25\text{b})$$

Before we introduce the unified procedure to construct the broadband matching equalizer [E], let us introduce the theorem for double matching problems.

Theorem for the double matching problem [12]: Referring to Figure 10.7, for a given double matching problem, let Z_G and Z_L be the given positive real (PR) dissipative driving point impedances of the generator [G] and load [L] networks respectively. Let $T(\omega^2)$ be a preset, realizable, rational transducer power gain function with the desired frequency characteristics including all the transmission zeros of [G] and [L]. Then, assuming no double degeneracy, the necessary and sufficient conditions imposed on $T(\omega^2)$ will be physically realizable as a single lossless two-port which consists of the cascaded connections of the trio [G]–[E]–[L], and mean that the single matching gain–bandwidth restrictions of Youla will be simultaneously satisfied at the generator and load ends of the equalizer.

Based on the above presentations, it is clear that for both single and double matching problems, the entire matched system can be described in terms of its unit normalized BR scattering parameters $F = \{F_{ij}\}$ as in filter theory, when the source (or generator) and the load are replaced with their Darlington equivalents. In order to construct the desired equalizer, Youla's gain–bandwidth restrictions are simultaneously applied at the source and load ends, and then extractions of $[G]$ and $[L]$ are accomplished.

At this point, it is very important to outline the basic properties of F_{ij} related to source and load networks so that an appropriate form of the transfer function $T(\omega^2) = |F_{21}(j\omega)|^2$ can be selected to give a realizable equalizer.

Essential Properties of the Scattering Parameters [F]

In principle, $F_{21}(p)$ must contain all the transmission zeros of $[G]$ *and* $[L]$ with at least the same multiplicity as well as the transmission zeros of the equalizer $[E]$. The following properties are dictated by the main theorem of Yarman and Carlin and the para-unitary condition imposed on the scattering matrix $[F]$.

Property 1 (Zeros of F_{21}(p)*:* Let

$$F_{21}(p) = \frac{f(p)}{g(p)}$$

be the transfer scattering parameter of an existing trio of $\{[G] - [E] - [L]\}$ which is considered as an all-pass free BR function. Then,

$$T(\omega^2) = |F_{21}(j\omega)|^2 = F_{21}(j\omega)F_{21}(-j\omega)$$

measures the transducer power gain of the matched system over the real frequencies. In the complex p domain, it is specified by

$$T(-p^2) = F_{21}(p)F_{21}(-p) = \frac{f(p)f(-p)}{g(p)g(-p)}$$

For the existing trio of $\{[G]-[E]-[L]\}$, $F_{21}(p)F_{21}(-p)$ must contain all the transmission zeros of $[G]$ and $[L]$ in the numerator polynomial $f(p)f(-p)$ while $F_{21}(p) = f(p)/g(p)$ is free of all-pass functions.

For example, let σ_G and σ_L be the RHP transmission zeros and let ω_G and ω_L be the real frequency transmission zeros of the generator $[G]$ and load $[L]$ networks respectively. Let $\eta_G(p) = (\sigma_G - p)/(\sigma_G + p)$ and $\eta_L(p) = (\sigma_L - p)/(\sigma_L + p)$ be the generator and load all-pass functions constructed on the RHP zeros of σ_G and σ_L. Then, the numerator polynomial $f(p)$ of degree n_f must take the following form:

$$f(p) = (\sigma_G - p)(\sigma_L - p)(p^2 + \omega_G^2)(p^2 + \omega_L^2)\hat{f}(p)$$

(a) In the above expression, the term $(\sigma_G - p)(\sigma_L - p)(p^2 + \omega_G^2)(p^2 + \omega_L^2)$ is due to the transmission zeros of $[G]$ and $[L]$ and the polynomial $\hat{f}(p)$ includes the finite transmission zeros of the equalizer $[E]$.

(b) The denominator polynomial $g(p)$ must be strictly Hurwitz and must possess a degree n_g such that $n_g \geq n_f$.

(c) Obviously, $T(\omega^2) = f(j\omega)f(-j\omega)/g(j\omega)g(-j\omega)$ must result in the desired gain characteristics, for which $F_{21}(p) = f(p)/g(p)$ is free of Blaschke products of $\eta_G(p) = (\sigma_G - p)/(\sigma_G + p)$ and $\eta_L(p) = (\sigma_L - p)/(\sigma_L + p)$.

(d) For the example under consideration, the main theorem reveals that

$$F_{11}(p) = \frac{h(p)}{g(p)} = \eta_G(p)S_{YG}(p)$$

while $F_{21}(p) = f(p)/g(p)$ is free of Blaschke products of $\eta_G(p) = (\sigma_G - p)/(\sigma_G + p)$ and $\eta_L(p) = (\sigma_L - p)/(\sigma_L + p)$ but $f(p)$ includes the term $(\sigma_G - p)(\sigma_L - p)(p^2 + \omega_G^2)(p^2 + \omega_L^2)$ as above. In this case, let

$$S_{YG}(p) = \frac{D_G(-p)}{D_G(p)} \frac{Z_F - Z_{G*}}{Z_F + Z_G} = \frac{h_Y(p)}{g_Y(p)}$$

Then, $h(p)$ and $g(p)$ are given by

$$h(p) = (\sigma_G - p)h_Y(p)$$
$$g(p) = (p + \sigma_G)g_Y(p)$$

and $F_{22}(p)$ is given by the losslessness condition as follows:

$$F_{22}(p) = -\frac{f(p)}{f(-p)} \frac{h(-p)}{g(p)}$$

or

$$F_{22}(p) = -\eta_G \eta_L \frac{\hat{f}}{f_*} \frac{h_{Y*}}{g_Y} = \eta_L S_{YL}$$

where

$$S_{YL}(p) = -\eta_G \frac{\hat{f}}{f_*} \frac{h_{Y*}}{g_Y}$$

Property 2 (Form of $F_{21}(p)$ *for preassigned transducer power gain):* Let $[E]$ be the lossless equalizer to be designed between the generator $[G]$ and load $[L]$ networks with preassigned transducer power gain form.

Let $T(\omega^2) = F_m(\omega^2)/G_m(\omega^2)$ be the preassigned transducer power gain function yielding the minimum form for $F_{21m}(p) = f_m(p)/g_m(p)$ as a result of the spectral factorization of $F_m(-p^2)$ as $f_m(p)f_m(-p)$ and $G_m(-p^2)$ as $g_m(p)g_m(-p)$.

In this case, the RHP transmission zeros of $[G]$ and $[L]$ must appear in $F_{21}(p)$ as all-pass functions preserving the shape of $|F_{21}(j\omega)|^2$. Thus, the transfer scattering parameter $F_{21}(p)$ must be expressed as

$$F_{21}(p) = \eta_G(p)\eta_L(p)\mu(p)F_{21m}(p) = \eta_G(p)\eta_L(p)\mu(p)\frac{f_m(p)}{g_m(p)} = \frac{f(p)}{g(p)}$$

where $\mu(p)$ is an arbitrary Blaschke product such as $\mu(p) = (\sigma - p)/(\sigma + p)$.

Considering the example presented in Property 1, the numerator polynomial $f(p) = (\sigma_G - p)(\sigma_L - p)(\sigma - p)f_m(p)$ must include all the RHP as well as the $j\omega$ zeros of transmission for both the generator and the load networks. However, minimum polynomial $f_m(p) = (p^2 + \omega_G^2)(p^2 + \omega_L^2)\hat{f}(p)$ includes only finite $j\omega$ zeros.[5]

Similarly, for the example under consideration, strictly Hurwitz polynomial $g(p)$ of degree n_g should be expressed as

$$g(p) = (p + \sigma_G)(p + \sigma_L)(p + \sigma)g_m(p)$$

such that $n_g \geq n_f$.

Property 3 (Blaschke products of $F_{11}(p)$ and $F_{22}(p)$ due to $F_{21}(p)$): In order to satisfy the losslessness condition imposed on the scattering parameters $\{F_{ij}\}$, a Blaschke product of order k of $F_{21}(p)$ should appear either in F_{11} or in F_{22} of order $2k$, or it appears in both F_{11} and F_{22} with respective orders k_{11} and k_{22} satisfying the condition $2k = k_{11} + k_{22}$.

For example, if $F_{21}(p) = \mu(p)f_m(p)/g_m(p)$, then one of the following equation must hold:

$$F_{11}(p) = \frac{h_m(p)}{g_m(p)} \quad F_{22}(p) = -\mu^2(p)\frac{h_m(-p)}{g_m(p)}$$

or

$$F_{11}(p) = \mu(p)\frac{h_m(p)}{g_m(p)} \quad F_{22}(p) = -\mu(p)\frac{h_m(-p)}{g_m(p)}$$

or

$$F_{11}(p) = \mu^2(p)\frac{h_m(p)}{g_m(p)} \quad F_{22}(p) = -\frac{h_m(-p)}{g_m(p)}$$

Property 4 (Analytic forms of $F_{11}(p)$ and $F_{22}(p)$ dictated by the main theorem): Let the transfer scattering parameter $F_{21}(p)$ be

$$F_{21}(p) = \frac{f(p)}{g(p)}$$

It may be free of Blaschke products or not. In any case, application of the main theorem to the generator and load ends reveals that

$$
\boxed{
\begin{aligned}
F_{11}(p) &= S_{YCG}(p) = \eta_G(p)S_{YG}(p) = \frac{h(p)}{g(p)} \\
\text{where} & \\
S_{YG}(p) &= b_G(p)\frac{Z_F - Z_{G*}}{Z_F + Z_G}
\end{aligned}
}
\tag{10.26}
$$

---

[5] Remember that, if polynomial $f(p)$ includes a LHP zero, it must be accompanied by its mirror symmetric image in the RHP. On the other hand, widely employed TPG functions such as the Chebyshev and elliptic forms include only real frequency zeros.

and

$$F_{22}(p) = S_{YCL(p)} = \eta_L(p)S_{YL}(p) = -\frac{f(p)}{f(-p)}\frac{h(-p)}{g(p)}$$

where

$$S_{YL}(p) = b_L(p)\frac{Z_B - Z_{L*}}{Z_B + Z_L}$$

(10.27)

In Equations (10.26) and (10.27):

- Z_F is the frontend or the input impedance of the equalizer [E] while the output or the backend is terminated in the load impedance Z_L.
- Z_B is the backend or the output impedance of the equalizer [E] while the input is terminated in the generator impedance Z_G.

Property 5 (Common all-pass functions of $F_{11}(p)$ and $F_{22}(p)$): Regarding Properties 1–4, all-pass function η_G of $F_{21}(p)$ must also appear in $F_{22}(p)$ since it already exists in $F_{11}(p)$ by the main theorem. Similarly, all-pass function η_L of $F_{21}(p)$ must appear in $F_{11}(p)$ due to the above properties. Therefore, for a preassigned minimum form of $F_{21m}(p) = f_m(p)/g_m(p)$, to obtain a realizable $\{[G] - [E] - [L]\}$ structure, the para-unitary condition imposed on the scattering parameters $\{F_{ij}\}$ demands the following forms for F_{11} and F_{22}:

$$F_{11}(p) = S_{YCG}(p), \text{ with } F_{11}(p) = \varepsilon\eta_G(p)\eta_L(p)[\mu(p)]^{k_{11}}\frac{h_m(p)}{g_m(p)}$$

(10.28a)

$$F_{22}(p) = S_{YCL}(p), \text{ with } F_{22}(p) = -\varepsilon\eta_G(p)\eta_L(p)[\mu(p)]^{k_{22}}\frac{h_m(-p)}{g_m(p)}$$

where ε is a unimodal constant, i.e. $\varepsilon = \mp1$. The polynomial $h_m(p)$ is free of RHP zeros obtained by the spectral factorization of the even polynomial

$$H_m(p^2) = h_m(p)h_m(-p) = G_m(-p^2) - F_m(-p^2)$$

(10.28b)

As far as extraction of the generator and load networks is concerned, Youla's reflectances become crucial. Therefore, in the following property, constructions of Youla's reflectances are given starting from the preassigned transducer power gain function $T(\omega^2) = F_m(\omega^2)/G_m(\omega^2)$.

Property 6 (Construction of Youla's reflectance functions): The main theorem of Yarman and Carlin relates the unit normalized reflectances $F_{11}(p)$ and $F_{22}(p)$ to Youla's reflection coefficient as

$$S_{YG}(p) = \eta_{G*}F_{11}(p)$$

and

$$S_{YL}(p) = \eta_{L*}(p)F_{22}(p)$$

Then, for a preassigned gain function $T(-p^2) = F_m(-p^2)/G_m(-p^2)$, the realizable analytic forms of Youla's reflection coefficients are obtained from Equation (10.28) as follows:

At the generator end,

> At the load end,
>
> $$S_{YG}(p) = b_G(p)\frac{Z_F(p) - Z_G(-p)}{Z_F(p) + Z_G(p)}$$
>
> and
>
> $$b_G(p) - S_{YG}(p) = \frac{2R_G(p^2)b_G(p)}{Z_F + Z_G}$$
>
> where
>
> $$S_{YG}(p) = \eta_{G*}F_{11}(p) = \varepsilon\eta_L(p)[\mu(p)]^{k_{11}}\frac{h_m(p)}{g_m(p)}$$

(10.29)

> At the load end,
>
> $$S_{YL}(p) = b_L(p)\frac{Z_B(p) - Z_L(-p)}{Z_B(p) + Z_L(p)}$$
>
> and
>
> $$b_L(p) - S_{YL}(p) = \frac{2b_L(p)R_L(p^2)}{Z_B + Z_L}$$
>
> where
>
> $$S_{YL}(p) = \eta_{L*}F_{22}(p) = -\varepsilon\eta_G(p)[\mu(p)]^{k_{22}}\frac{h_m(-p)}{g_m(p)}$$

(10.30)

Here, it is interesting to observe the following facts.

Fact 1 (Mandatory all-pass function of Youla reflectance at the generator end): For double matching problems, for a preassigned TPG function, the realizable complex normalized–regularized Youla's reflection coefficient $S_{YG}(p)$ at the generator end must include the all-pass function η_L of the load network [L].

Fact 2 (Mandatory all-pass function of Youla reflectance at the load end): Similarly, the realizable Youla reflectance $S_{YL}(p)$ at the load end must include the all-pass function $\eta_G(p)$ of the generator network [G].

Fact 3 (Gain–bandwidth restriction at the generator end): Let $p = p_{G0}$ be a transmission zero of order k at the generator end such that

$$Z_F(p_{G0}) + Z_G(p_{G0}) \neq 0$$

Then, Equation (10.29) imposes the gain–bandwidth restriction at the generator end by the following expression:

$$b_G(p) - S_{YG}(p) = \frac{2b_G(p)R_G(p^2)}{Z_F + Z_G} = K(p)(p - p_{G0})^k$$

If $k = 1$, then

$$b_G(p_{G0}) - S_{YG}(p_{G0}) = \frac{2b_G(p_{G0})R_G\left(p_{G0}^2\right)}{Z_F(p_{G0}) + Z_G(p_{G0})} = 0 \qquad (10.31)$$

which requires

$$b_G(p_{G0}) = S_{YG}(p_{G0}) \qquad (10.32)$$

This is the simplest form of the gain–bandwidth restriction imposed on the generator end. Moreover, if $2b_G(p)R_G(p^2)/(Z_F + Z_G)$ includes multiple orders of transmission zeros, then a series approach to derive the gain–bandwidth restrictions is appropriate as in Youla's theory. In this case, complete gain–bandwidth restriction is specified as follows.

Let
$$b_G(p) - S_{YG}(p) = \frac{2R_G b_G}{Z_F + Z_G}$$

be specified with the following Taylor series expansions :

$$b_G(p) = \sum_{r=1}^{\infty} b_{Gr}(p - p_0)^r$$

$$S_{YL} = \sum_{r=1}^{\infty} S_{Gr}(p - p_0)^r \qquad (10.33)$$

$$F_G(p) = 2R_G b_G = \sum_{r=1}^{\infty} F_{Gr}(p - p_0)^r$$

Then, Youla's gain–bandwidth equations at the generator end are given as in Equation (10.22):

Class of transmission zeros	Gain–bandwidth restrictions for the generator end:	
	Transmission zero at $p = p_G = \sigma_G + j\omega_G$ of $Z_G(p)$: (i) of order k if $\sigma_G < 0$ and $\omega_G = 0$; (ii) of order $2k$ if $\sigma_G = 0$ and $\omega_G \neq 0$.	
For Class A $p = \sigma_G > 0$	$b_{Gr} = S_{Gr}, r = 0, 1, 2, \ldots, k-1$	(10.34a)

For Class B $p = \text{Real}\{p_G\} = 0$ and $Z_L(j\omega_0) \neq \infty$	$b_{Gr} = S_{Gr}, r = 0, 1, 2, \ldots, 2k - 1$	(10.34b)
	and	
	$$\frac{b_{G(2k-1)} - S_{G(2k-1)}}{F_{G2k}} \geq 0$$	(10.34c)
For Class C $p = p_G = \text{real}\{p_G\} = \sigma_G = 0$ and $Z_G(j\omega_G) = \infty$	$b_{Gr} = S_{Gr}, r = 0, 1, 2, \ldots, 2k - 2$	(10.34d)
	$$\frac{b_{G(2k-1)} - S_{G(2k-1)}}{F_{G(2k-2)}} \leq \frac{1}{c_{G(-1)}} \geq 0$$	(10.34e)

Fact 4 (Gain–bandwidth restriction at the load end): Let $p = p_{L0}$ be a simple zero of transmission of the load network such that

$$Z_B(p_{G0}) + Z_L(p_{G0}) \neq 0$$

Then,

$$b_L(p_{L0}) - S_{YL}(p_{L0}) = \frac{2b_B(p_{L0})R_L(p_{L0}^2)}{Z_B(p_{L0}) + Z_L(p_{L0})} = 0$$

which requires

$$\boxed{b_L(p_{L0}) = S_{YL}(p_{L0})} \qquad (10.35)$$

However, if the order of a transmission zero is greater than 1, then gain–bandwidth restrictions, which are given by Equation (10.22), apply.

Fact 5 (Violation of gain–bandwidth constraints): If the generator [G] and load [L] lossless two-ports include the same RHP transmission zero as such $p = \sigma_0 > 0$, then Youla's reflectances are

$$S_{YG}(\sigma_0) = \varepsilon \eta_L(\sigma_0) [\mu(\sigma_0)]^{k_{11}} \frac{h_m(\sigma_0)}{g_m(\sigma_0)} = 0$$

and

$$S_{YL}(\sigma_0) = -\varepsilon \eta_G [\mu(\sigma_0)]^{k_{22}} \frac{h_m(-\sigma_0)}{g_m(\sigma_0)} = 0$$

since $\eta_G(\sigma_0) = \eta_L(\sigma_0) = 0$.

Therefore, the above equations clearly violate the gain–bandwidth restrictions derived for the Class A generator and load networks. For this case, the analytic solution does not exist for preassigned TPG functions. This fact is accepted as one of the major drawbacks of the analytic approach to broadband matching problems.

Let us now introduce the unified approach as a 'step-by-step design procedure' to construct matching equalizers for both single and double matching problems.

10.6 Unified Analytic Approach to Design Broadband Matching Networks

In practice, devices to be matched are specified by means of measured data. Therefore, the first step of broadband matching must start with device modeling.

In this section, considering the above analytic approach, the design steps of a lossless equalizer for broadband matching problems are given in an algorithmic way.

Step 1: Obtain analytic models for the impedances to be matched and find the Darlington lossless two-ports [G] and [L] for the generator and load impedances respectively. Then determine the transmission zeros of [G] and [L] as in Equation (10.12) and Equation (10.25). At this point, make sure that [G] and [L] do not have the common RHP zeros of transmission. Otherwise, the analytic theory is not accessible.

Step 2: Choose a desired form for the TPG function as in the filter design $T(\omega^2) = |F_{21}|^2 = 1 - |F_{11}|^2 = 1 - |F_{22}|^2$ with unknown parameters. Here, it is crucial that F_{21} includes all transmission zeros of [G] and [L] networks with at least the same multiplicity. In this step, the transfer scattering parameter F_{21} must be expressed as

$$F_{21}(p) = \eta_G(p)\eta_L(p)\mu(p)F_{21m}(p)$$

where

$$F_{21m}(p) = \frac{f_m(p)}{g_m(p)}$$

is a minimum phase function which includes all the transmission zeros of the generator and the load network on the $j\omega$ axis.

Step 3: Using the spectral factorization of $|F_{11}|^2 = |F_{22}|^2 = 1 - |F_{21}|^2$ as described in the filter design process, construct the unit normalized reflectances F_{11} and F_{22} in the Yarman and Carlin sense as follows:

$$F_{11}(p) = S_{YCG}(p), \quad \text{with } F_{11}(p) = \varepsilon\eta_G(p)\eta_L(p)\mu(p)\frac{h_m(p)}{g_m(p)} \tag{10.36a}$$

$$F_{22}(p) = S_{YCL}(p), \quad \text{with } F_{22}(p) = -\varepsilon\eta_G(p)\eta_L(p)\mu(p)\frac{h_m(-p)}{g_m(p)} \tag{10.36b}$$
$$\text{where } \varepsilon \text{ is a unimodal constant, i.e. } \varepsilon = \mp 1.$$

In the above equation set, polynomials $h_m(p)$ and $g_m(p)$ are obtained via spectral factorization of $F_{11}(p)F_{11}(-p) = 1 - T(-p^2)$ (or, equivalently, $F_{22}(p)F_{22}(-p) = 1 - T(-p^2)$). Certainly, $g_m(p)$ is strictly Hurwitz; $\mu(p)$ is an arbitrary Blaschke product and it might be necessary to satisfy the gain–bandwidth restrictions (GBRs), otherwise it is set to 1; and η_G and η_L are all-pass functions, constructed on the closed RHP zeros of even parts of the generator and the load impedances respectively.

Note that, in the course of spectral factorization, unknown parameters of T are carried into $F_{11}(p)$ and $F_{22}(p)$.

Step 4: Apply Youla's GBRs as detailed by Facts 3 and 4 at the generator and load ends to determine simultaneously the unknown parameters of the selected TPG function $T(\omega^2)$ as well as the unknowns of the Blaschke product $\mu(p)$.

Step 5: Finally, synthesize the equalizer using $Z_B(p)$ as introduced by Equation (10.18).

Remarks:

- Implementation of the above procedure is by no means unique. First of all, one may wish to start with expanded forms of the generator and load impedances to force the resulting η_G and η_G to unity, as described earlier. Spectral factorization of $T(-p^2)$ can be carried out in a variety of fashions. For the sake of brevity, details are omitted here. Interested readers are referred to [5–8] and [11, 12].[6]
- Use of an all-pass factor $\mu(p)$ in $T(-p^2)$ always penalizes the minimum gain level in the passband. Therefore, one should avoid employing any extra Blaschke product in step 3 if possible.
- Degeneracy: At the load end, if the equalizer driving point impedance $Z_B(p)$ has common transmission zeros with that of the load impedance, we say that matching is degenerate. It is well established that degeneracy reduces the gain level in the passband.
- Similarly, we can also talk about degeneracy at the generator end which also penalizes the gain performance of the matched system.
- Details of degenerate situations are elaborated by Youla, Carlin and Yarman [56].
- It is interesting to observe that whenever Z_G and Z_L possess the same transmission zeros of Class A, it is not possible to satisfy the gain–bandwidth equations simultaneously if they are inserted into F_{11} and F_{22} as all-pass functions. In this case, a proper form of F_{21} must be selected which naturally includes these RHP zeros of the load and generator, but not as all-pass products.
- It should be emphasized that it is not easy to utilize the analytic gain–bandwidth theory. It is generally the second step that imposes the most severe limitation on the practical applicability. Nevertheless, for simple terminations, it may be useful to determine the ideal highest flat gain level within the passband. For example, by invoking the analytic gain–bandwidth theory, for a typical R∥C load, the possible highest flat gain level G_{max} is found as

$$\boxed{G_{max} = 1 - e^{\frac{2\pi}{RCB}}} \tag{10.37}$$

where B designates the normalized angular frequency bandwidth [1, 4].

For double matching problems, however, the ideal flat gain over a finite passband cannot be obtained [19, 20].

Now, let us present examples for the application of the unified theory of broadband matching in conjunction with the lumped element filter design process.

10.7 Design of Lumped Element Filters Employing Chebyshev Functions

In Sections 4.16–4.18 of Chapter 4, the design of stepped line filters with Butterworth and Chebyshev functions was introduced. There, we used commensurate transmission lines to construct filters. In this section, lumped elements are utilized. Furthermore, considering the generator and load network models, which are lumped and simple enough to be imbedded into the filter structure, we investigate the appropriateness of filter theory to design an equalizer for single and double matching problems.

---

[6] A detailed discussion of the analytic theory of gain–bandwidth is presented in [56], Chapter 8.

Recall that the monotone roll-off Chebyshev gain function is specified by Equation (4.89) as

$$T(\omega) = \frac{1}{1 + \varepsilon^2 T_n^2(\omega)} \tag{10.38}$$

where $\omega = 2\pi f$ is the normalized angular frequency, $T_n(\omega)$ is the nth degree Chebyshev polynomial, and ε is a weight which specifies fluctuations or ripples of the gain function in the passband.

The Chebyshev polynomial $T_n(\omega)$ of degree n is given by

$$T_n(\omega) = \cos(nz)$$

where

$$\omega = \cos(z)$$

Let $p = \sigma + j\omega$ be the classical complex variable. For $p = j\omega$, one can set

$$p = j\omega = j\cos(z)$$

Then, replacing ω by p/j, we can state that

$$T_n\left(\frac{p}{j}\right) = \cos(nz)$$

and

$$T(p/j) = F_{21}(p)F_{21}(-p) = \frac{1}{1 + \varepsilon^2 T_n^2(p/j)}$$

$$= \frac{1}{g(p)g(-p)}$$

$\tag{10.39}$

Since[7] the filter is lossless,

$$F_{11}(p)F_{11}(-p) = 1 - T(p/j)$$

Therefore,

$$F_{11}(p)F_{11}(-p) = \frac{\varepsilon^2 T_n^2(p/j)}{1 + \varepsilon^2 T_n^2(p/j)} = \frac{\varepsilon^2 \cos^2(nz)}{1 + \varepsilon^2 \cos^2(nz)}$$

or

$$F_{11}(p)F_{11}(-p) = \frac{H_m(p^2)}{G_m(p^2)} = \frac{h_m(p)h_m(-p)}{g_m(p)g_m(-p)}$$

where

$$H_m(p) = \varepsilon^2 \cos^2(nz)$$

$$G_m(p) = 1 + \varepsilon^2 \cos^2(nz) \quad p = j\cos(z)$$

$\tag{10.40}$

---

[7] From Chapter 4, recall that, depending on whether the value of n is even or odd, $T_n(\omega)$ is either even or odd, then $T(P/j)$ must be an even real polynomial in p.

Roots of $G_m(p)$

By Equation (10.39), $p = j\omega = j\cos(z)$. Thus, $Z = \cos^{-1}(p/j)$ must describe a complex variable, which in turn makes z or nz a complex variable.

In this case, let

$$nz \triangleq \alpha + j\beta$$

Then, the roots of $G_m(p^2) = 1 + \varepsilon^2\cos^2(nz)$ with $Z = \cos^{-1}(p/j)$ can be found by setting $G_m(p^2) = 0$ as follows.

$G_m(p^2) = 0$ requires that at points (nz_k).

$$1 + \varepsilon^2\cos^2(nz_k) = 0$$

or

$$\varepsilon^2\cos^2(nz_k) = -1$$

or

$$\cos(nz_k) = \cos(\alpha_k + j\beta_k) = \mp j\frac{1}{\varepsilon}$$

or

$$\cos(\alpha_k + j\beta_k) = \cos(\alpha_k)\cos(j\beta_k) - \sin(\alpha_k)\sin(j\beta_k) = \mp j\frac{1}{\varepsilon}$$

Note that

$$\cos(j\beta_k) = \frac{1}{2}\left[e^{j(j\beta_k)} + e^{-j(j\beta_k)}\right] = \frac{1}{2}\left[e^{\beta_k} + e^{-\beta_k}\right] = \cosh(\beta_k)$$

Similarly,

$$\sin(j\beta_k) = \frac{1}{2j}\left[e^{j(j\beta_k)} - e^{-j(j\beta_k)}\right] = \frac{1}{j2}\left[e^{-\beta_k} - e^{\beta_k}\right] = j\sinh(\beta_k)$$

Then,

$$\cos(\alpha_k + j\beta_k) = \cos(\alpha_k)\cosh(\beta_k) - j\sin(\alpha_k)\sinh(\beta_k) = \mp j\frac{1}{\varepsilon}$$

Equating the real and imaginary parts of the above equation we end up with

$$\boxed{\begin{array}{l} \cos(\alpha_k) = 0 \\ \text{or} \\ \alpha_k = \left(k + \frac{1}{2}\right)\pi \text{ which requires } \sin(\alpha_k) = \mp 1 \end{array}} \tag{10.41}$$

$$\boxed{\beta_k = \mp\sinh^{-1}\left(\frac{1}{\varepsilon}\right)} \tag{10.42}$$

Thus, we obtain

$$z_k = \frac{\alpha_k + j\beta_k}{n}$$

Finally, we can set the roots of $G_m(p)$ at $p_k = j\cos(z_k)$ such that

$$p_k = \sin\left[\left(\frac{2k+1}{2n}\right)\pi\right]\sinh\left(\frac{\beta}{n}\right) + j\cos\left[\left(\frac{2k+1}{2n}\right)\pi\right]\cosh\left(\frac{\beta}{n}\right) \tag{10.43}$$

Here, it is interesting to observe that the roots $p_k = \sigma_k + j\omega_k$ of the above equation satisfy the following equation:[8]

$$\frac{\sigma_k^2}{\sinh^2\left(\frac{\beta}{n}\right)} + \frac{\omega_k^2}{\cosh^2\left(\frac{\beta}{n}\right)} = 1 \tag{10.44}$$

where

$$\sigma_k = \sin\left[\left(\frac{2k+1}{2n}\right)\pi\right]\sinh\left(\frac{\beta}{n}\right) \tag{10.45}$$

$$\omega_k = \cos\left[\left(\frac{2k+1}{2n}\right)\pi\right]\cosh\left(\frac{\beta}{n}\right)$$

Thus, the zeros of $g_m(p)g_m(-p)$ lie on an ellipse, centered at the origin, whose axes are $a = \sinh(\beta/n)$ and $a = \cosh(\beta/n)$.

In Equation (10.45) the integer k runs from zero to $2n - 1$, yielding a total number of $2n$ roots of $G_m(p)$. The ones on the LHP are selected to construct the denominator polynomial $g_m(p)$. Thence, we can state that for a lumped element low-pass filter the transfer scattering parameter is given by

$$F_{21m}(p) = \frac{1}{g_m(p)}$$

where

$$g_m(p) = \left(2^{n-1}\varepsilon\right)\prod_{r=1}^{n}(p - p_r) \tag{10.46}$$

or

$$g_m(p) = g_1 p^n + g_2 p^{n-1} + \cdots + g_n p + g_{n+1}$$

and p_r are the LHP roots of Equation (10.45).

---

[8] Recall that on the xy plane, a central elipse with axes a and b is described by

$$\frac{x^2}{a^2} + \frac{y^2}{b^2} = 1$$

Note that, in the Chebyshev gain approximation, the leading coefficient of $1 + \varepsilon^2 \cos^2(nz)$ must be equal to the leading coefficient of the polynomial $G_m(p) = 1 + \varepsilon^2 T_n^2(-p^2)$. As shown in Chapter 4, this coefficient must be equal to $(2^{n-1}\varepsilon)^2$. Therefore, in Equation (10.46) we use the multiplier $(2^{n-1}\varepsilon)$ to obtain $g_1 = (2^{n-1}\varepsilon)$, which will force $F_m(p) = 1$ (or equivalently $f_m(p) = 1$).

Example 10.1:

(a) Let $\varepsilon^2 = 0.1$. Using Equation (10.42) and Equation (10.45), generate the roots of $G_m(p) = g_m(p)g_m(-p)$ for $n = 3$ and 21 plot them.
(b) For $n = 3$, construct $g_m(p)$ as a strictly Hurwitz polynomial in coefficient form.

Solution:

(a) For this example, we developed the simple MATLAB® program `Example10_1.m`. The code contains the above equations.
Inputs to the program are $\varepsilon^2 = 0.1$ and $n = 3$.
Then, the $2n = 6$ roots of $G_m(p)$ are as listed in Table 10.1.

Table 10.1 Roots of even Chebyshev polynomial $G_m(p) = 1 + \varepsilon^2 T_n^2$ for $2n = 6$

$p_k = \sigma_k + j\omega_k$	
σ_k	ω_k
0.3320	1.0395
0.6639	0.0000
0.3320	-1.0395
-0.3320	-1.0395
-0.6639	-0.0000
-0.3320	1.0395

The locations of the roots are shown in Figure 10.8.
For $n = 21$, the locations of the roots are as depicted in Figure 10.9 on the p plane.
Clearly, these roots are on the ellipse as described by Equation (10.44).
(b) For $n = 3$, $g_m(p)$ is constructed on the LHP roots of Table (10.1). Thus, gm $= [1.2649$
1.6796   2.0638   1] is obtained, meaning that $g_m(p) = p^3 + 1.3278p^2 + 1.6316p + 0.7906$.
The MATLAB code for the example is as follows.

Program Listing 10.1

```
% Main Program for Example 10.1
% For monotone roll-off Chebyshev filters
% Generation of gm(p) explicitly.
% Note that for filter F11m=hm(p)/gm(p)
```

```
% Given eps^2=0.1
% Given n=3 or n=21
% Generate roots of 1+eps^2*Tn explicitly.
clear
eps_sq=0.1;
eps=sqrt(eps_sq);
beta=asinh(1/eps);
n=3;
n2=2*n;
i=1;
j=sqrt(-1);
for k=0:(n2-1)
    x(k+1)=k;
sigma(k+1)=sin(pi*(2*k+1)/2/n)*sinh(beta/n);
omega(k+1)=cos(pi*(2*k+1)/2/n)*cosh(beta/n);
    if(sigma(k+1)<0)
        z(i)=sigma(k+1)+j*omega(k+1);
        i=i+1;
    end
    end
gm=(2^(n-1)*eps)*real(poly(z))
    plot(sigma,omega, "x")
```

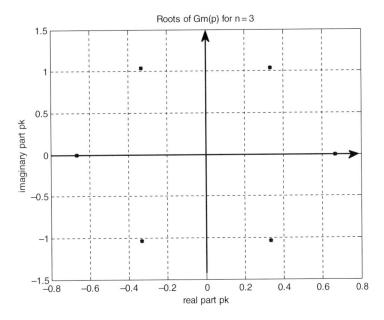

Figure 10.8 Roots of $G_m(p) = 1 + \varepsilon^2 T_n^2$ for $n = 3$

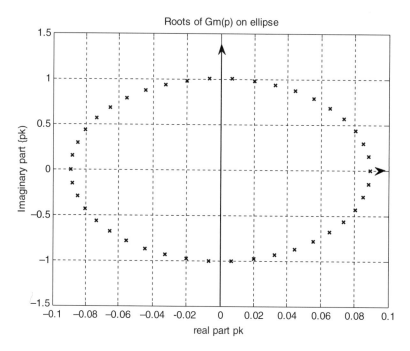

Figure 10.9   Roots of $G_m(p) = 1 + \varepsilon^2 T_n^2$ for $n = 21$

Roots of $H_m(p)$

The roots of $H_m(p)$ are given by the following equation:

$$H_m(p) = \varepsilon^2 \cos^2(nz) = 0$$

which requires that

$$\cos(\alpha_k + j\beta_k) = \cos(\alpha_k)\cos(j\beta_k) - \sin(\alpha_k)\sin(j\beta_k) = 0$$

or

$$\alpha_k = (2k+1)\pi/2$$

and

$$\beta_k = 0$$

In this case,

$$nz_k = \alpha_k = (2k+1)\frac{\pi}{2}$$

or

$$\boxed{z_k = \frac{\alpha_k}{n} = \frac{(2k+1)\pi}{2n}, k = 0, 1, 2, \ldots, 2n-1}$$

(10.47)

Thus, the roots are located at

$$p_k = j\cos(z_k)$$

$$p_k = j\cos\left[\frac{(2k+1)\pi}{2n}\right] = \sigma_k + j\omega_k; k$$

$$= 0, 1, 2, 3, \ldots, 2n-1$$

where

$$\sigma_k = 0$$

$$\omega_k = \cos\left[\frac{(2k+1)\pi}{2n}\right]$$

(10.48)

Thus, for a low-pass Chebyshev filter $F_{11}(p)$ is given by

$$F_{11m(p)} = \mp\frac{(2^{n-1}\varepsilon)\prod_{r=1}^{n}(p-p_r)}{g_m(p)} = \frac{h_m(p)}{g_m(p)}$$

where

$$h_m = \mp(2^{n-1}\varepsilon)\prod_{r=1}^{n}(p-p_r)$$

(10.49)

Construction of $h_m(p)$

As stated in Chapter 4, $h_m(p)$ is an even polynomial when n is even, and an odd polynomial when n is odd.

In this case, $H_m(p)$ must possess purely imaginary double roots on the envelope of the cosine function with respect to index k such that

$$p_k = j\omega_k = j\cos\left[\frac{(2k+1)\pi}{2n}\right]$$

Then:

(a) When n is even, $h_m(p)$ includes only non-zero anti-symmetric roots of $H_m(p)$ such that

$$h_m(p) = \mp(2^{n-1}\varepsilon)\prod_{r=1}^{n/2}(p^2+\omega_r^2), \quad \omega_r \neq 0$$

(10.50)

(b) When n is odd, one of the double zeros at $p = 0$ must be included in $h_m(p)$. Then,

$$h_m(p) = (2^{n-1}\varepsilon)p\prod_{r=1}^{(n-1)/2}(p^2+\omega_r^2), \quad \omega_r \neq 0$$

(10.51)

Let us run an example to clearly exhibit the usage of Equation, (10.50) and (10.51).

Example 10.2:

(a) Employing Equation (10.49), generate the roots of $H_m(p)$ for $n = 3$.
(b) Construct the polynomial $h_m(p)$ for $\varepsilon^2 = 0.1$.
(c) Show that the coefficients of $h_m(p)$ match those obtained using the recursive formula to generate the coefficients of Chebyshev polynomials as in Equation (4.86), utilizing the MATLAB function Tn `poly_cheby(n)` of Chapter 4.

Solution:

(a) For this example, we developed the simple MATLAB code `Example10_2` which generates the roots of $H_m(p)$ as

$$p_k = j\omega_k = j\cos\left[\frac{(2k+1)\pi}{2n}\right]$$

Hence, for $n = 3$ and $k = 0, 1, 2, \ldots, 2n - 1$, the above zeros are as listed in Table 10.2.

Table 10.2   Roots of $H_m(p)$

For $n = 3$

$$p_k = j\cos\left[\frac{(2k+1)\pi}{2n}\right] = j\omega_k$$

σ_k	ω_k
0	0.8660
0	0.0000
0	-0.8660
0	-0.8660
0	-0.0000
0	0.8660

(b) As can be seen from Table 10.2, we have a total number of $2n = 6$ roots which are double and purely complex conjugate on the envelope of the function

$$\omega_k = \cos\left(\frac{2k+1}{2n}\pi\right)$$

as shown in Figure 10.10. Obviously, two of these roots are placed at $p_k = 0$. In this case, $h_m(p)$ must be an odd polynomial in p.

From Table 10.2 and Equation (10.51), $(n-1)/2 = 1$ and the roots are located at $p_1 = j\omega_1 = \mp j0.866$. Thus, $h_m(p)$ may be written as,

$$\boxed{\begin{aligned} h_m(p) &= \mp 4\varepsilon p\left[p^2 + (0.886)^2\right] = \mp 4\varepsilon[p^3 + 0.75p] \\ &= \mp 4\varepsilon(p^3 + 0.75p) \end{aligned}}$$

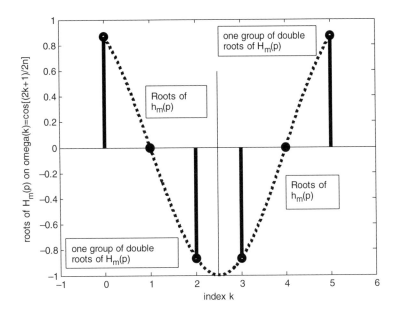

Figure 10.10   Double roots of $H_m(p)$

(c) For $n = 3$, function Tn = poly_Cheby(n) of Chapter 4 reveals that

$$\boxed{\begin{aligned} jT_n(p/j) &= -4(4p^3 + 3p) \\ &= -4(p^3 + 0.75p) \end{aligned}}$$

which verifies the above result.

Program Listing 10.2

```
% Main Program for Example 10.2
% Generation of roots of Hm(p)using Equation(10.48)
% Given eps^2=0.1
% Given n=21
% Generate roots of eps^2*Tn explicitly.
clear
j=sqrt(-1);
eps_sq=0.1;
eps=sqrt(eps_sq);
n=3;
n2=2*n;
fi=pi/n2;
t=0;
N=500;
```

Program Listing 10.2   (Continued)

```
dt=(n2-1)/(N-1);
for  r=0:N
     fi=(2*t+1)*pi/n2;
for  k=0:(n2-1)
x(k+1)=k;
     omega(k+1)=cos(pi*(2*k+1)/2/n);
     sigma(k+1)=0;
end
     for  k=1:n
          z(k)=j*omega(k);
     end
hm=poly(z);
Tn=poly_cheby (n);
SA(r+1)=cos(fi);
     t=t+dt;
     TA(r+1)=t;
          end
               plot(TA,SA,x,omega)
```

10.8 Synthesis of Lumped Element Low-Pass Chebyshev Filter Prototype

n = odd case

For the case $n =$ odd, Chebyshev polynomials $T_n(\omega)$ must be odd in ω and therefore, at $\omega =0$, $T_n(0) =0$, which in turn yields $T(0) = T_0 = 1/\left(1 + \varepsilon^2 T_n^2\right) =1$. This will make the constant termination of the filter unity (i.e. $x_0 = 1$).

Once the form for

$$F_{11}(p) = \mp \frac{j\varepsilon T_n\left(\dfrac{p}{j}\right)}{g_m} = \frac{h_m(p)}{g_m(p)}$$

is obtained, it is straightforward to synthesize driving point input impedance $Z_{in}(p) =(1 + F_{11})/(1 - F_{11})$ or admittance $Y_{in}(p) =1/Z_{in}(p)$ of the filter using continued fraction expansion.

In $F_{11}(p)$, if the '+' sign is chosen, the filter starts with an inductor of value x_1 since $lim_{p \to \infty}F_{11}(p) = +1$. Otherwise, the synthesis starts with a capacitor x_1. Corresponding low-pass ladder prototypes are shown in Figure 10.11(a) and Figure 10.11(b) respectively.

In general, the driving point immittance of Figure 10.11 is given by

$$F_{in} = \frac{N_{in}(p)}{D_{in}(p)}$$

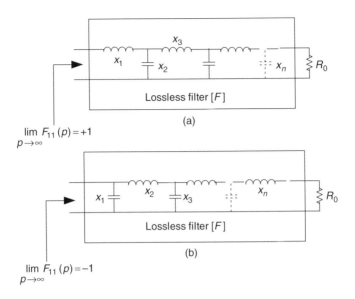

Figure 10.11 Low-pass ladder prototypes

as follows:

$$F_{in}(p) = \frac{N_{in}(p)}{D_{in}(p)} = x_1 p + \cfrac{1}{x_2 p + \cfrac{1}{x_3 p + \cfrac{1}{\ldots + \cfrac{1}{x_n p + x_0}}}} \qquad (10.52)$$

Note that Equation (10.52) ends with the term $x_n p + x_0$. However, the constant termination x_0 must be equal to $1 (x_0 = 1)$ since $T(0) = 1$.

n = even case

This situation is a little tricky since the even-degree Chebyshev polynomials yield non-unity gain at $\omega = 0$. Specifically, at DC, the gain of the filter must yield

$$T_0 = T(0) = \frac{1}{1 + \varepsilon^2}$$

Therefore, the constant term x_0 of Equation (10.52) must satisfy the gain equation at DC such that

$$T_0 = \frac{4x_0}{(x_0 + 1)^2} \qquad (10.53)$$

Just to be precise, for this case:

(a) If the filter starts with an inductor x_1, it must end with a capacitor x_n in parallel with a conductor $x_0 = G_0$.

(b) If the filter starts with a capacitor x_1, it must end with an inductor x_n in series with a resistance $x_0 = R_0$.

Thus, whether x_0 is a conductor or a resistor, its value can be calculated from Equation (10.53) as follows:

$$T_0 = T(0) = \frac{1}{1 + \varepsilon^2} = \frac{4x_0}{(x_0 + 1)^2} \quad (n = even \ \ case)$$

or

$$x_0^2 - 2\left[\frac{2 - T_0}{T_0}\right] x_0 + 1 = 0$$

or

$$x_0 = \left[\frac{2 - T_0}{T_0}\right] \mp \sqrt{\left[\frac{2 - T_0}{T_0}\right]^2 - 1}$$

Note that

$$\frac{2 - T_0}{T_0} = 1 + 2\varepsilon^2$$

and

$$\sqrt{\left[\frac{2 - T_0}{T_0}\right]^2 - 1} = 2\varepsilon\sqrt{2}$$

or

$$x_0 = 2\varepsilon^2 \mp 2\varepsilon\sqrt{\varepsilon^2 + 1} + 1$$

Thus, by taking the '−' sign, for the even values of integer n, constant termination is found as

$$\boxed{x_0 = 2\varepsilon^2 - 2\varepsilon\sqrt{\varepsilon^2 + 1} + 1} \qquad (10.54)$$

Explicit Element Values of the Chebyshev Filter

It can be shown that, no matter what the value of the integer n is, in the synthesis process of the driving point immittance, the first normalized reactive element x_1 is given by

$$\boxed{x_1 = \frac{2\sin\left(\dfrac{\pi}{2n}\right)}{\sinh\left[\left(\dfrac{1}{n}\right)\sinh^{-1}\left(\dfrac{1}{\varepsilon}\right)\right]}} \qquad (10.55)$$

Furthermore, the follow-up elements are given by the sequential formulas of Takahashi, made available to the western world by Levy,[9] as follows:

$$x_r x_{r+1} = \frac{4\sin\left[\dfrac{(2r-1)\pi}{2n}\right]\sin\left[\left(2r + \dfrac{1\pi}{2n}\right)\right]}{\sinh^2\left[\left(\dfrac{1}{n}\right)\sinh^{-1}\left(\dfrac{1}{\varepsilon}\right)\right] + \sin^2\left(\dfrac{r\pi}{n}\right)} \text{ where the index runs over } r = 1, 2, 3, \ldots, n-1$$

(10.56)

Obviously, if n is odd, the constant termination x_0 must be unity as detailed above. On the other hand, if n is even, then the constant termination must be specified by Equation (10.54).

All the above filter construction steps can be summarized in the following algorithm.

10.9 Algorithm to Construct Monotone Roll-Off Chebyshev Filters

A monotone roll-off filter can be constructed by employing two methods.

In the first one, for specified sign of $h_m(p)$, and for values of ε and n, we can directly determine the explicit element values using Equations (10.54), (10.55) and (10.56). The corresponding gain expression can be generated directly from

$$T(\omega) = \frac{1}{1 + \varepsilon^2 T_n^2(\omega)}$$

This is called the direct method.

On the other hand, the above results can be verified by constructing the polynomials $g_m(p)$ and $h_m(p)$ as in Equation (10.46) and (10.49) respectively, and synthesis of $F_{11}(p) = h_m(p)/g_m(p)$ reveals the explicit element values of the filter. Eventually, the gain function can be generated directly from the element values of the filter. This is called the indirect method. In this method, we set

$$h_m(p) = (\text{sign}) * \left(2^{n-1}\varepsilon\right)p \prod_{r=1}^{n}(p - p_r)$$

where the input quantity 'sign' controls the filter structure.

More specifically, the inputs of the algorithm are ε, n and sign such that:

1. If the designer wishes to have a low-pass Chebyshev filter which starts with an inductor x_1 and ends with an inductor x_n, then n must be chosen as an odd integer. Furthermore, the input quantity 'sign' must be set to sign $= +1$.
2. If the designer wishes to have a low-pass Chebyshev filter which starts with a capacitor x_1 and ends with a capacitor x_n, then n must be odd and sign $= -1$.
3. On the other hand, if the designer wishes to have a low-pass Chebyshev filter which starts with an inductor x_1 and ends with a capacitor x_n, then n must be an even integer and sign $= +1$.
4. If the designer wishes to have a low-pass Chebyshev filter which starts with a capacitor and ends with an inductor, then n must be even and sign $= -1$.

---

[9] R. Levy, 'Explicit Formulas for Chebyshev Impedance Matching Networks, Filters and Interstages,' *Proceedings of the IEE*, Vol. III, No. 6, pp. 1099–1106, 1964.

Thus, for instructive reasons, we developed a MATLAB code which generates the same Chebyshev filter by employing direct and indirect methods. The direct method generates the explicit element values from Equations (10.54–10.56).

Let us present some examples of how to design Chebyshev filters with monotone roll-off employing direct and indirect methods.

Example 10.3 (Construction of Chebyshev filter using direct method):

(a) Employing Equation (10.53) and Equation (10.54), design a monotone roll-off Chebyshev filter for $\varepsilon^2 = 0.1$ and $n = 3$.
(b) Plot the gain performance of the filter.
(c) Draw the possible filter circuit configurations.

Solution:

(a) Explicit element values of the filter are programmed in MATLAB within the main program `Example10_3.m`. The results are listed in Table 10.3.

Table 10.3 Design of monotone roll-off Chebyshev filter for $\varepsilon^2 = 0.1$ and $n = 3$

X(1) = 1.5062
X(2) = 1.1151
X(3) = 1.5062

(b) Let the filter's driving point input immittance be denoted by $F_{in}(p)$. Then, it is expressed as

$$F_{in}(p) = x_1 p + \cfrac{1}{x_2 p + \cfrac{1}{x_3 p + 1}}$$

over the real frequencies

$$F_{in}(j\omega) = R(\omega) + jX(\omega)$$

Then, by Equation (9.12), TPG of the filter can be directly expressed in terms of $F_{in}(j\omega) = R(\omega) + jX(\omega)$ such that

$$\boxed{\text{TPG}(\omega) = |F_{21}(j\omega)|^2 = \frac{4R}{(R+1)^2 + X^2}} \tag{10.57}$$

So, within the main program `Example10_3.m`, we also evaluated TPG and it is plotted in Figure 10.12. It should be mentioned that the continued fraction expansion form of $F_{in}(j\omega)$ was evaluated under the function `Lowpass_Ladder(W,x,,A0)`. In this representation W denotes the normalized angular frequencies, $x(r)$ is an array vector which includes all the element values of the filter, A0 is the conventional termination which is set to unity for odd values of n. In other words, $A0=R0=1/G0=1$.

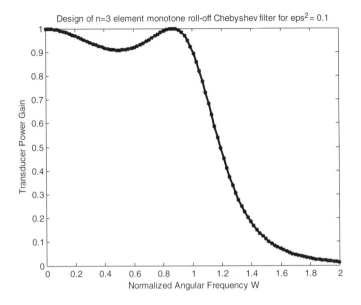

Figure 10.12 Gain plot of the filter design for Example 10.3

(c) As mentioned earlier, depending on the sign of the input scattering parameter,

$$F_{11}(p) = \mp \frac{\varepsilon T_n(p/j)}{g_m} = \frac{h_m(p)}{g_m(p)}$$

we obtain two configurations as shown in Figure 10.13.

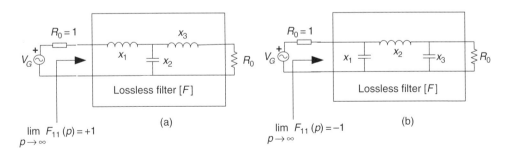

Figure 10.13 Two possible filter designs for Example 10.3.

The following program listings refer to Example 10.3.

Program Listing 10.3   Main program for Example 10.3

```
% Main Program Example 10_3.m
% Computation of explicit element values of the Chebyshev filter
% Using Levy formulas as in Carlin"s book:
% These formulas are good for only odd values of n. That is ,
% n=1,3,5,7,9,...
% R. Levy, "Explicit Formulas for Che-byshev
% impedance matching networks, filters and interstages,
% Proc. IEE, Vol.III, No.6, pp.1099-1106, June 1964.
clear
eps_sq=input("Enter eps_sq=")
eps_sq=0.1
n=input(" Enter degree of the Chebyshev polynomial n=")
eps=sqrt(eps_sq);
N1=2*sin(pi/2/n);
D1=sinh((1/n)*asinh(1/eps));
x(1)=N1/D1;
%
for r=1:(n-1)
Num=4*sin((2*r-1)*pi/2/n)*sin((2*r+1)*pi/2/n);
D2=sinh((1/n)*asinh(1/eps))*sinh((1/n)*asinh(1/eps));
D3=sin(r*pi/n)*sin(r*pi/n);
Denom=D2+D3;
x(r+1)=Num/Denom/x(r);
end
W1=0;
W2=2;
N=100;
DW=(W2-W1)/(N-1);
W=0;
A0=1;

for j=1:N
    WA(j)=W;
    [ZQ Z]=Lowpass_Ladder(W,x,A0);
    R=real(ZQ);
    X=imag(ZQ);
    T(j)=4*R/((R+1)*(R+1)+X*X);
    W=W+DW;
    end
plot(WA,T)
```

Program Listing 10.4   Computation of input immittance for low-pass ladders

```
function [immitance Z]=Lowpass_Ladder(W,x,A0)
% Computation of the input immitance of a lowpass ladder at a
% given
% normalized frequency W
% Inputs
%         W: Normalized angular frequency
%         x(r) elements of the lowpass ladder
%         A0: Termination resistance or conductance
% Output
%         value of the immitance at the chosen angular frequency
%
j=sqrt(-1);
p=j*W;
n=length(x);
% Initiate the computational steps:
Z(n)=x(n)*p+A0;
Z(n+1)=1;
for r=1:(n-1)
        Z(n-r)=Z(n-r+1)*x(n-r)*p+Z(n-r+2);
end
immitance=Z(1)/Z(2);
```

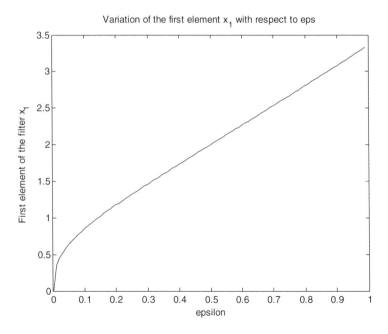

Figure 10.14   For $n = 3$, variation of the first element x_1 with respect to ε

Example 10.4 (Variation of the first element with respect to ripple factor ε): For Chebyshev filters, investigate the variation of the first element with respect to ε and degree $n = 3$ and $n = 21$.

Solution: For a fixed degree n, the simple MATLAB program `Example10_4.m` was developed. This program sweeps ε to generate the first element x_1 of the filter. Thus, for $n = 3$, we obtain Figure 10.14. Details of Figure 10.14 are given by Table 10.4. For example, $\varepsilon = 1$, $x_1 = 3.5353$.

Table 10.4 For $n = 3$, first-element variation with ε

ε	First element x_1
0.0	0.00000
0.1	0.85158
0.2	1.1772
0.3	**1.4617**
0.4	1.7328
0.5	2.0
0.6	2.267
0.7	2.5355
0.8	2.8063
0.9	3.0796
1.0	3.3553

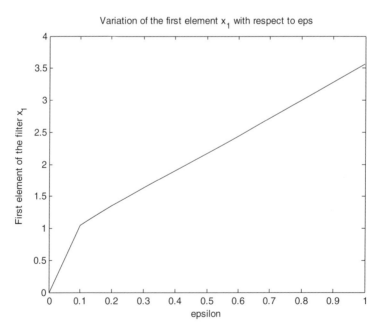

Figure 10.15 For $n = 21$, variation of the first component of the filter with ε

Similarly, for $n = 21$, variation of the first element is depicted in Figure 10.15 and detailed in Table 10.5.

Table 10.5   For $n = 21$, first-element variation with ε

ε	First element x_1
0	0.0000
0.1	1.0433
0.2	1.3546
0.3	**1.6334**
0.4	1.9035
0.5	2.1724
0.6	2.4433
0.7	2.7173
0.8	2.9948
0.9	3.2758
1.0	**3.5601**

It is interesting to point out that as the ripple factor ε increases, which in turn decreases the minimum of the passband gain of $T_{min} = T(\omega)|_{\omega = 1} = 1/(1 + \varepsilon^2)$, the value of the first component increases.

For example, for $n = 3$, x_1 varies from 0 to 3.35.

On the other hand, for fixed values of ε, the value of the first component x_1 slightly increases as degree n increases.

For example, when $\boldsymbol{\varepsilon = 0.3}$, $\boldsymbol{x_1(n = 3) = 1.4617}$. However, for the same ripple factor ε, the first component becomes $x_1 = 1.6334$ when $n - 21$.

The main program `Example10_4.m` is given in Program Listing 10.5.

Program Listing 10.5

```
% Main Program Example 10_4.m
% Investigation of eps vs. x1 for Chebyshev filter
clear
n=input(" Enter degree of the Chebyshev polynomial n=")
N1=2*sin(pi/2/n);
eps=0;
del_eps=0.1
for i=1:11
ep(i)=eps;
D1=sinh((1/n)*asinh(1/eps));
x1(i)=N1/D1;
    eps=eps+del_eps;
end
figure(1)
plot(ep,x1)
title("Variation of the first element x_1 with respect to eps")
ylabel("First element of the filter x_1")
xlabel("epsilon")
```

Example 10.5: By employing the indirect filter design method, for $\varepsilon^2 = 0.1$, design monotone roll-off Chebyshev filters as described below.

(a) The filter starts with a capacitor and ends with a capacitor for $8 < n < 10$.
(b) The filter starts with an inductor and ends with an inductor for $6 < n < 8$.
(c) The filter starts with an inductor but ends with a capacitor such that $9 < n < 11$.
(d) The filter starts with a capacitor and ends with an inductor such that $5 < n < 7$.

Solution: For this example we developed the MATLAB program `Example10_5.m` which designs the filters using the indirect method. The MATLAB code of the main program is given in Program Listing 10.6.

In this program, firstly the explicit roots of $G_m(p)$ are generated. Then, selecting those in the LHP, $g_m(p)$ is performed. In the second step, double and purely imaginary roots of $H_m(p)$ are generated and then $h_m(p)$ is constructed on the one set of purely imaginary and conjugate roots of $H_m(p)$. In the third step, $F_{11}(p)$ is constructed. In the fourth step, driving point impedance $Z_{in}(p)$ is generated. Finally, $Z_{in}(p)$ is synthesized using a long division algorithm. In the last step, gain is plotted.

Now, let us run this program for all the parts of the problem.

(a) Since the filter starts with a capacitor and ends with a capacitor, then the sign of $h_m(p)$ must be negative and degree n must be an odd integer. Therefore, we set sign $= -1$ and $n = 9$, which satisfies the inequality $8 < n < 10$. Thus, the filter is as given in Figure 10.16 and the gain plot is depicted in Figure 10.17.
(b) For this part of the problem, we must choose $n = 9$ and sign $= +1$ since the filter starts and ends with an inductor. Actually, the filter element values will be the same as the ones given in part (a) provided

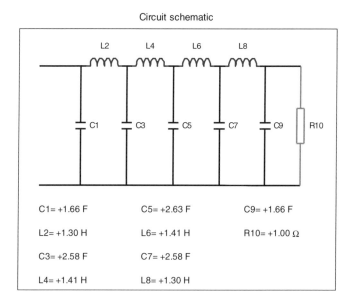

Circuit schematic

C1= +1.66 F   C5= +2.63 F   C9= +1.66 F

L2= +1.30 H   L6= +1.41 H   R10= +1.00 Ω

C3= +2.58 F   C7= +2.58 F

L4= +1.41 H   L8= +1.30 H

Figure 10.16   Design of a monotone roll-off filter for part (a) with $\varepsilon^2 = 0.1, n = 9$ and sign $= -1$

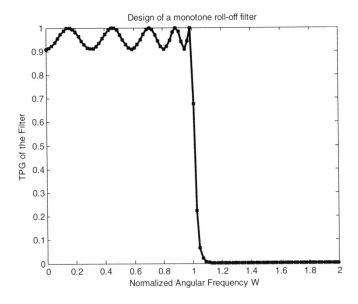

Figure 10.17 Gain performance of the filter shown in Figure 10.16

that shunt capacitors are replaced by series inductors and series inductors are replaced by shunt capacitors as shown in Figure 10.18.

Eventually, the gain performance of the filter must be the same as the one given in part (a).

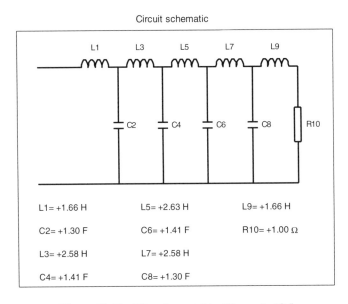

Figure 10.18 Filter for part (b) of Example 10.5

(c) In this part, the filter starts with an inductor and ends with a capacitor. Therefore, n must be an even integer such that $n = 10$, which satisfies the inequality of $p < n < 11$ and sign $= +1$.

Hence, when we run the main program `Example10_5.m`, we obtain the filter structure as shown in Figure 10.19. The gain performance of this filter is depicted in Figure 10.20.

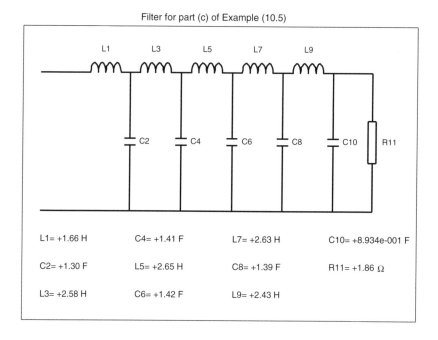

Filter for part (c) of Example (10.5)

L1= +1.66 H	C4= +1.41 F	L7= +2.63 H	C10= +8.934e-001 F
C2= +1.30 F	L5= +2.65 H	C8= +1.39 F	R11= +1.86 Ω
L3= +2.58 H	C6= +1.42 F	L9= +2.43 H	

Figure 10.19   Filter design for part (c) of Example 5.10

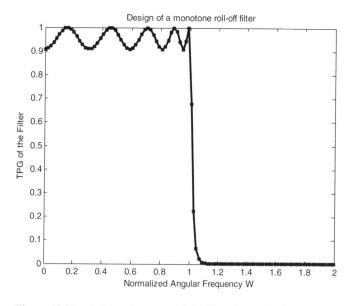

Figure 10.20   Gain performance of the filter shown in Figure 10.19

(d) In this part of the example, the first component of the filter is a capacitor and the last component is an inductor. Therefore, as the input of the program `Example10_5` we must choose `Eps_sq = 0.1`, `n = even = 6`, `sign = -;1`.

As a result we obtain the filter which is shown in Figure 10.21. Its gain is depicted in Figure 10.22.

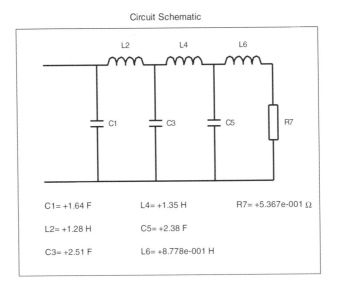

Circuit Schematic

C1= +1.64 F	L4= +1.35 H	R7= +5.367e-001 Ω
L2= +1.28 H	C5= +2.38 F	
C3= +2.51 F	L6= +8.778e-001 H	

Figure 10.21   Six-element Chebyshev filter with $\varepsilon^2 = 0.1, n = 6, \text{sign} = -1$

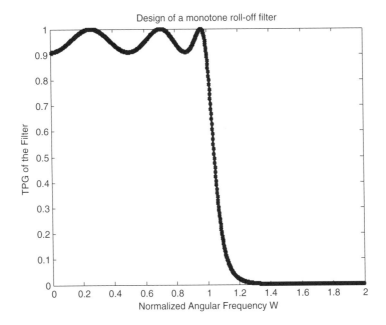

Figure 10.22   Performance of the filter shown in Figure 10.21

The MATLAB code for the main program Example10_5.m is as follows.

Program Listing 10.6   MATLAB code for `Example10_5.m`

```
% Main Program for Example 10.5
% Program Example10_5.m
% Construction of Monotone Roll-Off Chebyshev filters
% employing direct and indirect methods.
% Direct method uses explicit formulas for the element values.
% In direct method generates gm(p) and hm(p) then, synthesizes
% of F11(p)
% by means of continues fraction expansion of the input immit-
% tance is given
% F11(p)=hm(p)/gm(p)
% Inputs:
%              eps^2,n, sigma
%
% Outputs:
% For direct method:x(r); explicit element values
%            gm(p), hm(p), Hm, Gm, Fm,
%            CV: Components Values for the filter by CFE
% Note that roots of hm and gm are found explicitly
clear
% Enter inputs
eps_sq=input('Enter eps_sq=')
eps=sqrt(eps_sq);
beta=asinh(1/eps);
n=input("Enter n=")
sign=input("Enter sign of hm sign=")
% Step 1: Generates the Explicit roots of Gm(p)).
n2=2*n;
i=1;
j=sqrt(-1);
for k=0:(n2-1)
          x(k+1)=k;
sig(k+1)=sin(pi*(2*k+1)/2/n)*sinh(beta/n);%Real{pk}
omega(k+1)=cos(pi*(2*k+1)/2/n)*cosh(beta/n);%imag{pk}
     if(sig(k+1)<0)
          p(i)=sig(k+1)+j*omega(k+1);%roots of g(p)
          i=i+1;
     end
end
% Step 2: Construct gm(p)
gm=real(poly(p));%polynomial g(p)
gm=gm;%g(p) with eps*Tn(1)
%
%Step 3: Computation of hm(p)
for k=0:(n2-1)
x(k+1)=k;
          omega(k+1)=j*cos(pi*(2*k+1)/2/n);
```

```
    end
      for k=1:n
            ph(k)=omega(k);% Roots of hm(p)
    end
    hm=real(poly(-ph));%polynomial h(p)
            hm=sign*hm;% h(p) with eps*Tn(1)
    %
    % Step 4: Generation of F21m(p)
    hm_=paraconj(hm);
    gm_=paraconj(gm);
    Hm=conv(hm,hm_);
    Gm=conv(gm,gm_);
    Fm=Gm-Hm;
    % Generation of Zin(p)
    Nin=gm+hm;
    Din=gm-hm;
    %
    % Synthesis of Zin(p)
            CV=general_synthesis(hm,gm)
    % Step 7: Plot the gain performance of the filter
            W1=0;
W2=2;
N=1000;
DW=(W2-W1)/(N-1);
W=0;
    A0=1;
% Computation of n=even case
x0=2*eps_sq-2*eps*sqrt(eps_sq+1)+1;
if (n/2-fix(n/2)==0);%n=even case
    A0=x0;
end;%End of n=even case
N1=length (CV);
for i=1:N1-1
    component(i)=CV(i);
end

for j=1:N
    WA(j)=W;
    [ZQ Z]=Lowpass_Ladder(W,component,A0);
    R=real(ZQ);
    X=imag(ZQ);
    T(j)=4*R/((R+1)*(R+1)+X*X);
            W=W+DW;
end
figure(2)
plot(WA,T)
xlabel("Normalized Angular Frequency W")
ylabel("TPG of the filter")
title("Design of a monotone roll-off filter")
```

10.10 Denormalization of the Element Values for Monotone Roll-off Chebyshev Filters

All the above Chebyshev filters have been designed for a fixed degree n, a ripple constant ε and a cut-off frequency $\omega_c = 1$. Furthermore, the normalization number for the scattering parameters has been chosen as $R_0 = 1$. It can be shown that, for an actual cut-off frequency f_{ac}, and for an actual normalization R_{a0}, the element values are obtained by employing the following formulas:

$$\boxed{L_{ai} = \frac{L_{Ni}R_{a0}}{2\pi f_{ac}} \qquad C_{ai} = \frac{C_{Nj}}{2\pi f_{ac}R_{a0}}} \tag{10.58}$$

where $\{L_{ai}$ and $C_{ai}\}$ and $\{L_{Ni}$ and $C_{Ni}\}$ are the actual and normalized inductors and capacitors of the filter under consideration.

Example 10.6: Referring to Table 10.3 and Figure 10.13(a), let the actual cut-off frequency of the Chebyshev filter be $f_{ac} = 1$ GHz. Assume that actual normalization resistance is $R_{a0} = 50\ \Omega$.

(a) Find the actual element values of the filter by employing Equation (10.58).
(b) Plot the gain performance of the filter generated directly from the actual values given.

Solution:

(a) From Table 10.3 and Figure 10.13, the first element of the filter is an inductor $L_{N1} = 1.5062$ and the actual value of this inductor is given by

$$L_{a1} = \frac{L_{N1}R_{a0}}{2\pi f_{ac}} = 1.19861749168297 \times 10^{-8} \cong 11.99\ \text{nH}$$

The second component of the filter is a capacitor $C_{N2} = 1.115087728$ and its actual value is given by

$$C_{a2} = \frac{C_{N2}}{2\pi f_{ac}R_{a0}} = 3.54943448092532 \times 10^{-12} \cong 3.55\ \text{pF}$$

The filter is symmetric since the degree n is odd. Therefore, the last component must be equal to the first one. Hence,

$$L_{a3} = \frac{L_{N3}R_{a0}}{2\pi f_{ac}} = 1.19861749168297 \times 10^{-8} \cong 11.99\ \text{nH}$$

The termination resistance must be $R_{a0} = 50\ \Omega$.

(b) Specifically for this example, we developed the simple MATLAB code `denormalization.m` to generate TPG on the actual element values. The actual gain is depicted in Figure 10.23 and the code is given in Program Listing 10.7.

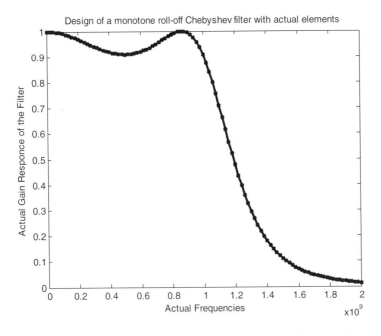

Figure 10.23 Computation of the gain response of the filter with actual elements

Program Listing 10.7

```
% Main Program for Example 10_6.m
% Denormalization of the element values
clear
eps_sq=0.1;
eps=sqrt(eps_sq);
x=           [1.50622716253418
              1.11508772896733
              1.50622716253418
                          1];

fac=1e9;%Actual cut-off frequency
Ra0=50;% Standard termination
E=(eps)^(1/3);
y(1)=x(1)*Ra0/2/pi/fac;%Inductance
y(2)=x(2)/Ra0/2/pi/fac;%Capacitance
y(3)=x(3)*Ra0/2/pi/fac;%Inductance
y(4)=x(4)*Ra0;%Resistance
%
j=sqrt(-1);
F1=0;
```

Program Listing 10.7   (Continued)

```
F2=2*fac;
N=101;
DF=(F2-F1)/(N-1);
F=F1;
for r=1:N
        p=j*(2*pi*F);
        FA(r)=F;
        Z3=y(3)*p+y(4);
        Y2=y(2)*p*Z3+1;
        ZQ=(y(1)*p*Y2+Z3)/Y2;
        R=real(ZQ);
        X=imag(ZQ);
        RA(r)=R;
        XA(r)=X;
        T=4*R*Ra0/((R+Ra0)*(R+Ra0)+X*X);
        TPG(r)=T;
        F=F+DF;
end
figure(1)
plot(FA,TPG)
xlabel("Actual Frequencies")
ylabel("Actual Gain Response of the filter")
title("Design of a monotone roll-off Chebyshev filter with
actual elements")
```

10.11 Transformation from Low-Pass LC Ladder Filters to Bandpass Ladder Filters

As demonstrated in Chapter 4, a typical low-pass filter characteristic can be shifted to a bandpass filer characteristic by frequency shifting.

Let ω_1 and ω_2 be the edge frequencies of a bandpass prototype filter such that cut-off frequency ω_c of the low-pass prototype is given by [10]

$$\omega_c = \sqrt{\omega_1 \omega_2}$$

and the passband is specified by

$$\Delta\omega = \omega_1 - \omega_2$$

(10.59)

---

[10] Here, all the frequencies are considered as normalized angular frequencies. Therefore, filters are prototypes.

Considering unit normalization $R_0 = 1$ Ω for both low-pass and bandpass prototypes, the series inductors L_{lpi} are replaced by parallel resonance circuits with element values specified by

L_{lpi} = series inductors of the low-pass protytpe

and the equivalent parallel resonance circuit elements are

$$L_{bpi} = \frac{L_{lpi}}{\Delta\omega}$$ (10.60)

$$C_{bpi} = \frac{\Delta\omega}{L_{lpi}\omega_c^2}$$

Similarly, the shunt capacitors C_{lpj} of the low-pass prototype are replaced by series resonance circuits with the following element values:

C_{lpi} = shunt capacitors of the low-pass protytpe

and the equivalent series resonance circuit elements are

$$C_{bpi} = \frac{C_{lpj}}{\Delta\omega}$$ (10.61)

$$L_{bpj} = \frac{\Delta\omega}{C_{lpj}\omega_c^2}$$

Low-pass to bandpass element transformations are depicted in Figure 10.24.

Here, it is important to note that element values of the bandpass filter double without changing the filter characteristic. This fact is considered as a serious drawback of analytic filter theory in designing bandpass filters.

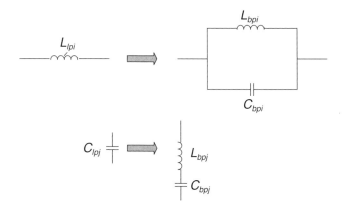

Figure 10.24 Low-pass to bandpass transformations

10.12 Simple Single Matching Problems

Until now, we have worked on several filter problems, but in this section we investigate the solution of simple single matching problems by employing the filter theory. First of all, let us define the simple single matching (SSM) problem.

Definition 3 (SSM problem): The SSM problem is where the load includes only one reactive element.

SSM problems can be solved using the filter theory.

Let us consider the design of a monotone roll-off Chebyshev filter. The first element of this filer is specified by Equation (10.55) as

$$x_1 \frac{2\sin\left(\frac{\pi}{2n}\right)}{\sinh\left[\left(\frac{1}{n}\right)\sinh^{-1}\left(\frac{1}{\varepsilon}\right)\right]}$$

In this case, for specified values of the simple reactive element x_1 and degree n, the ripple coefficient ε can be determined as

$$\sinh\left[\left(\frac{1}{n}\right)\sinh^{-1}\left(\frac{1}{\varepsilon}\right)\right] = \frac{2\sin\left(\frac{\pi}{2n}\right)}{x_1} \qquad (10.62)$$

or

$$\sinh^{-1}\left(\frac{1}{\varepsilon}\right) = n \times \sinh^{-1}\left[\frac{2\sin\left(\frac{\pi}{2n}\right)}{x_1}\right]$$

or

$$\left(\frac{1}{\varepsilon}\right) = \sinh\left\{n \times \sinh^{-1}\left[\frac{2\sin\left(\frac{\pi}{2n}\right)}{x_1}\right]\right\}$$

or

$$\varepsilon = \frac{1}{\sinh\left\{n \times \sinh^{-1}\left\{\frac{2\sin\left(\frac{\pi}{2n}\right)}{x_1}\right\}\right\}}$$

Thus, SSM problems can be solved using filter theory, perhaps for example by employing monotone roll-off Chebyshev approximations without any complication, provided that it yields a fixed ripple coefficient ε as dictated by Equation (10.62) for a specified number of filter elements n.

Let us consider an example.

Example 10.7 (SSM problem):

(a) For a given complex load R∥C with R $= 1\Omega$ and C $= 3$ F (all normalized element values), design an equalizer with three elements over the normalized angular frequency band of $0 \leq \omega \leq (\omega_c = 1)$ by employing the monotone roll-off Chebyshev approximation and plot the performance
(b) Using the above load and approximation method, design equalizers with n 5,7,9,11,13,15,17,19,21 and investigate the variation of the second element, which is an inductor, with respect to degree n.

Solution:

(a) Equation (10.62) is programmed in MATLAB, under `Example10_7.m`. This program solves the SSM problem using the monotone roll-off Chebyshev gain approximation. In the main program, for a given first reactive element x_1, the ripple coefficient ε is generated for a specified degree n. Then, the complete equalizer elements are computed by employing the recursive formula given by Equation (10.56). Eventually, TPG is plotted on the computed element values of the equalizer.

Thus, for this example, the first component of the filter is fixed as a capacitor $x_1 = 3$. Then, for $n = 3$, the ripple coefficient is computed as $\varepsilon \cong 0.871$. All the reactive elements of the filter are stored in vector x such that

$$x = [3 \quad 0.7742 \quad 3]$$

meaning that the first reactive element is $x(1) = 3$ as it should be, and the second one is $x(2) = 0.7742$, which must be an inductor.

With, the second element as an inductor, the third element is a capacitor. As shown earlier, the filter structure is symmetric since n is an odd integer. The third element is a capacitor $x(3) = 3$ which is equal to the first element since n is odd.

The plot of the resulting gain performance is depicted in Figure 10.25.

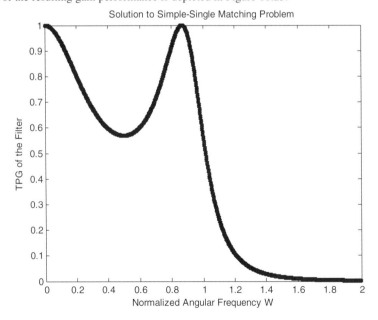

Figure 10.25 Gain performance of the filter in Example 10.7

(b) Using the main program `Example10_7.m`, and setting the first element value at $x_1 = 3$, equalizers were constructed for the desired values of degree n. The results are summarized in Table 10.6.

Table 10.6 Design of monotone roll-off Chebyshev equalizers for an R‖C load with R $= 1$ Ω, C $= x_1 = 3$ for various values of n

x_1 (first reactive element)	n	ε	x_2 (second reactive element)
3	3	0.8709	0.774 20
3	5	0.8258	0.859 20
3	7	0.8134	0.879 78
3	9	0.8082	0.887 91
3	11	0.8056	0.891 95
3	13	0.8041	0.894 25
3	100	0.8004	0.899 99

Table 10.6 indicates that, for a fixed value of the first element x_1, as the total number of equalizer elements increases, the ripple coefficient ε decreases and the second element x_2 increases. It is also observed that they both approach their asymptotic values. For this example, when ε falls to 0.8, the value of x_2 reaches 0.9 as shown in Figure 10.26.

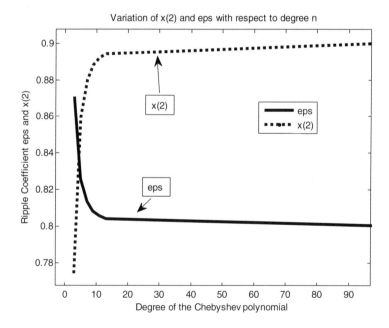

Figure 10.26 Variation of the second element x_2 and ε with respect to n

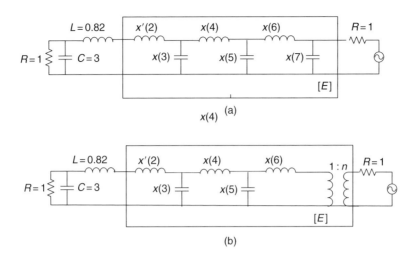

$x(4)$ (a)

(b)

Figure 10.27 Two possible equalizer design configurations for Example 10.8

Example 10.8:

(a) Using the monotone roll-off Chebyshev filter design method for a ripple coefficient ε less than 0.82, design a low-pass prototype matching network for the complex load, which is specified as $R//C + L$ with $R = 1, C = 3, L = 0.6$.

(b) Comment on your design if $L \leq 0.9$.

Solution:

(a) This load is not a simple one since it includes two reactive elements. However, we may still use the filter approach for this single matching problem by fixing the first element such that $C = x_1 = 3$, and investigate the solutions for various values of n for the Chebyshev gain approximation. In fact, we have done it already in Example 10.7. Close examination of Table 10.6 reveals that, for $x_1 = 3$ and $n = 7$, we have eps $= \varepsilon = 0.8134$ and $x_2 = 0.87978$ which includes the load inductor of $L = 0.82$ satisfying the design specifications with degeneracy.[11] Thus, extracting L from x_2, we can construct a seven-element equalizer using the main program `Example10_7.m`. Hence, we obtain the solution shown in Figure 10.27(a). Here, in this solution, the element values of the equalizer are given as follows:

$C = x(1) = 3 \ \ x(2) = 0.879788$, which includes $L = 0.82$ H with reminder $x'(2) \cong 0.05978$

and

$$x(3) = 4.033 \quad x(4) = 0.918\,86 \quad x(5) = x(3) \quad x(6) = x(4) \quad x(7) = x(1)$$

Obviously, the filter structure is symmetric since n is odd.

---

[11] Here, the load has double the transmission zeros at infinity due to shunt capacitor C and series inductor L. Since the load inductor is extracted from x_2, then the equalizer driving point impedance Z_B starts with an inductor $x'_2 = x_2 - L$ which in turn introduces a degeneracy. As mentioned in the remarks of Section 10.6, degeneracy penalizes the gain performance of the matched system.

On the other hand, when we run the same MATLAB code for $n = 6$, the design requirements are still satisfied with degeneracy as shown in Figure 10.27(b). Thus, the details are as follows:

$$x'(2) = 0.05219 \quad x(3) = 4.017 \quad x(4) = 0.90219 \quad x(5) = 3.88347 \quad x(6) = 0.67377$$

and the transformer ratio is specified by

$$1/n^2 = x_0 = 0.22459 = \frac{1}{1 + \varepsilon^2}$$

where $\varepsilon = 0.818$, which satisfies the design specifications.

Clearly, the above design employs a transformer with a turns ratio of $(1/n) = x_0 = 0.4739$ due to the mismatch introduced at DC since degree n of the Chebyshev polynomial is 6 which is an even integer. Design performances of the equalizers are depicted in Figure 10.28.

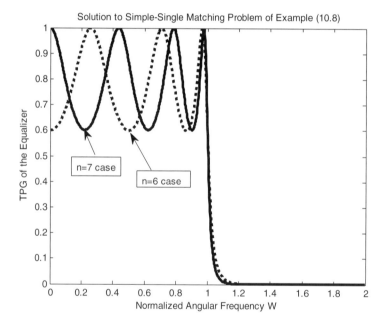

Figure 10.28   Gain performances of the matching problem of Example 10.8

(b) As pointed out in Example 10.7, for a normalized R//C load with R = 1 and C = 3, the second element can only reach up $x(2) = 0.9$. Therefore, for $L \geq 0.9$, there is no straightforward solution for the single matching problem via the filter design approach. Rather, GBRs must be satisfied with an all-pass function introduced in the input scattering parameter of the filter such that[12]

$$F_{11}(p)\eta_L(p)\eta(p)\frac{h_m(p)}{g_m(p)} = \left[\frac{p - \sigma}{p + \sigma}\right]^k \left[\frac{h_m(p)}{g_m(p)}\right] = \frac{h(p)}{g(p)}$$

---

[12] This load has all its transmission zeros at infinity. Therefore, it is Class C, which in turn yields the load all-pass function $\eta_L(p) = 1$.

By Property 3 of Section 10.5, k is chosen either as $k = 1$ or $k = 2$. For example, if we select a complex load with $R = 1$, $C = 3$, $L = 1.43$ and choose $k = 2$, the GBRs yield that $\varepsilon^2 = 3.44$ and $\sigma = 0.08$ which force the minimum passband gain to be $T_{min} = 1/(1 + \varepsilon^2) = 0.225$.

Actually, the above result can be verified by synthesis of $F_{11}(p)$. In other words, for a fixed value of degree n, we may also satisfy the GBSs by trial and error when we synthesize $F_{11}(p) = h(p)/g(p)$ for various values of ε^2 and σ.

It should be mentioned that for the complex load of $R//C + L$, explicit analytic solutions are as given by Chen [46, 47], but by the nature of analytic theory, the results are impractical and must include coupled coils as in Darlington C sections. As a live example, just consider the above solution, which yields the minimum passband gain of $T_{min} = 0.225$. This solution has no practical meaning at all. Furthermore, if the load is more complicated than that of a $R//C + L$ structure, who knows how to select a transfer function for the single or double matching problems under consideration?

10.13 Simple Double Matching Problems

Let us first define the simple double matching (SDM) problem.

Definition 4 (SDM problem): SDM problems include only single reactive elements in both generator and load networks, like a single series inductor at the generator end and a single shunt capacitor in the load network.

For SDM problems, we may be able to employ a filter approach with degeneracy if and only if the resulting filter includes generator and load reactive elements with termination resistors.

Let us consider Figure 10.27 in Example (10.8).

Figure 10.27(a) may be used to construct a double matching equalizer for the complex generator and load where the generator and load networks are restricted by the following element values:

$$\text{Generator end}: R_G = 1 \times N \text{ and } x(7) = C_G \leq 3 \times N$$
$$\text{Load end}: R_L = 1, C_L = 3 \text{ and } x(2) = L_L \leq 0.9$$

We choose $N = 1$ if there is no transformer in the equalizer, otherwise it is adjusted accordingly to produce the desired generator specifications.

Similarly, the structure shown in Figure 10.27(b) can also be employed for restricted generator and load networks which are specified by the following component values:

$$\text{Generator end}: R_G = N \times x0 \text{ and } x(6) = L_G \leq N \times x(6) = 0.67377$$
$$\text{Load end}: R_L = 1, C_L = 3 \text{ and } x(2) = L_L \leq 0.9$$

where N is a real number which specifies the transformer ratio to be employed at the generator end. Obviously, for both cases the ripple coefficient must be $\varepsilon^2 = 0.818$.

For more complicated double matching problems, bandwidth restrictions must be satisfied as detailed by Equation (10.21) and (10.34) simultaneously for both load and generator ends. However, this is not worth doing due to complicated equalizer structures and penalized gain responses.

Therefore, in the following section we propose a new numerical approach for the solution of double matching problems. The technique provides an excellent initialization for the complicated computer-aided design methods which involves nonlinear optimizations of the gain functions.

10.14 A Semi-analytic Approach for Double Matching Problems

In this approach, the double matching problem is represented by cascaded connections of generator [G], equalizer [E] and load [L] networks as shown in Figure 10.7. The scattering parameters of the cascaded trio are designated by [F] as in filter theory.

Over the real frequencies, we can always target or think of a reasonable realizable scattering matrix for [F] as a data set. For example, for selected ripple coefficient ε and degree n, we can always generate a realizable scattering matrix [F] for a monotone roll-off Chebyshev filter over the real frequencies, which may be considered as our target.

In this context, let us introduce transfer scattering matrices T_G, T_E, T_L and T_F for the generator [G], equalizer [E], load [L] and idealized filter [F] networks.

Then, we can write

$$T_F(j\omega) = T_G(j\omega)T_E(j\omega)T_L(j\omega)$$

or   (10.63)

$$T_E(j\omega) = T_G^{-1}(j\omega)T_F(j\omega)T_L^{-1}(j\omega)$$

$T_E(j\omega)$ simply describes an idealized equalizer data set in the form of a transfer scattering matrix which in turn yields the target data for the scattering matrix [E], or for the driving point input impedance $Z_E(j\omega) = R_E(\omega) + jX_E(\omega)$ of the equalizer such that

$$T_E = \begin{bmatrix} \tau_{11E} & \tau_{12E} \\ \tau_{21E} & \tau_{22E} \end{bmatrix}$$

and, by Equation (6.77),   (10.64)

$$E_{11}(j\omega) = \frac{\tau_{12E}}{\tau_{22E}} = \rho_E(\omega)e^{j\varphi E(\omega)} = E_{11R}(\omega) + jE_X(\omega)$$

or

$$Z_E(j\omega) = \frac{1 + E_{11}(j\omega)}{1 - E_{11}(j\omega)} = R_E(\omega) + jX_E(\omega),$$
$$R_E(\omega) \geq 0, \forall \omega$$

(10.65)

The only condition imposed on [E] is that $R_E(\omega) \geq 0$ for all ω. It can be shown that this is true if [F], G and [L] are BR scattering matrices

Once the ideal data set is produced for $Z_E(j\omega)$, then it can be modeled as a PR function which in turn yields the desired equalizer topology.

Similarly, the input reflection coefficient of the equalizer $E_{11}(j\omega) = E_{11R}(\omega) = E_{11R}(\omega) + jE_X(\omega)$ can be modeled as a realizable BR reflection function as described in [50–54].

Now, let us summarize the above computational steps in the following algorithm.

10.15 Algorithm to Design Idealized Equalizer Data for Double Matching Problems

This algorithm describes the computational steps to construct an idealized equalizer data set for double matching problems.

It has been experienced that the monotone roll-off Chebyshev approximation reasonably describes the electrical performance of the matched system, which in turn generates the target scattering matrix of the equalizer [E] over the real frequencies. Therefore, degree n and the ripple coefficient ε^2 of the Chebyshev approximation are taken as the inputs of the algorithm. Obviously, generator and load impedances are also included among the inputs of the algorithm. In this regard, for the sake of the experiments, generator and load networks are assumed to be two-reactive element $L + C\|R$ ladder impedances. In this context, at a given angular frequency ω, the MATLAB function $[S] = ZtoS(w,L,C,R)$ generates the scattering matrix $[S]$ for specified values of series inductor L, shunt capacitor C and termination resistor R.

Here is an outline of the algorithm.

Inputs:

- n: degree of the Chebyshev polynomial
- ε^2: ripple coefficient of the gain function

$$T(\omega) = \frac{1}{1 + \varepsilon^2 T_n^2(\omega)}$$

- sign $= +1$ or -1 for the input reflection coefficient of the filter

$$F_{11} = (\text{sign}) \times \frac{h_m(p)}{g_m(p)}$$

- analytic form of the generator impedance: $\{L_G, C_G, R_G\}$
- analytic form of the load impedance: $\{L_L, C_L, R_L\}$.

Computational Steps:

Step 1: Generate the target scattering parameters of the cascaded trio [F] over the real frequencies.
In this step,

$$F = \begin{bmatrix} F_{11}(j\omega) & F_{12}(j\omega) \\ F_{21}(j\omega) & F_{22}(j\omega) \end{bmatrix}$$

is generated over the normalized angular frequencies ω. For low-pass designs, the bandwidth is assumed to be $[\omega = 0 \ to \ \omega = 1]$.

Step 2: Generate the scattering matrix of the generator network by employing the function $[G] = ZtoS(w, L_G, C_G, R_G)$, and the generator transfer scattering matrix $[T_G]$ by employing the function $[TG] = Transfer\_Spar(G)$.
In this step we should note that from the given scattering matrix [S], transfer scattering matrix T is generated by Equation (6.64) as follows:

$$\tau_{11} = S_{12} - \frac{S_{11}S_{22}}{S_{21}}; \ \tau_{12} = \frac{S_{11}}{S_{21}}$$

$$\tau_{21} = -\frac{S_{22}}{S_{21}}; \ \tau_{22} = \frac{1}{S_{21}}$$

Step 3: In a similar manner to that of step 2, generate the scattering matrix of the load network by employing the function $[L] = ZtoS(w, L_L, C_L, R_L)$ and the load transfer scattering matrix by means of $[TL] = Transfer\_Spar(L)$.

Step 4: Generate the transfer scattering parameter [*TE*] as in Equation (10.63):

$$T_E(j\omega) = T_G^{-1}(j\omega)T_F(j\omega)T_L^{-1}(j\omega)$$

and generate the idealized scattering matrix for the equalizer by means of Equation (6.65) as in function `[E] = TtoS(TE)`. In function `TtoS` scattering parameters are given as

$$S_{11} = \frac{\tau_{12}}{\tau_{22}}; \; S_{12} = \frac{\tau_{11}\tau_{22} - \tau_{12}\tau_{21}}{\tau_{22}}$$
$$S_{21} = \frac{1}{\tau_{22}}; \; S_{22} = -\frac{\tau_{21}}{\tau_{22}}$$

Step 5: Generate the driving point input impedance $Z_E(j\omega) = (1 + E_{11})/(1 - E_{11}) = R_E(\omega) + jX_E(\omega)$ which in turn yields the equalizer when it is modeled and synthesized as a lossless two-port in resistive termination. Eventually, the resistive termination is removed and the equalizer is connected to its original generator and load impedances to be matched.

Let us consider an example to clarify the above steps.

Example 10.9: Let the complex generator and load networks be specified by an impedance structure which includes a series inductor L connected to a shunt C||R type of impedance. Generate idealized equalizer data by employing the above algorithm to match the complex generator of $\{[R_G = 1] \, || \, [C_G = 2] + [L_G = 1]\}$ to a complex load given by $\{[R_L = 1] \, || \, [C_L = 3] + [L_L = 1]\}$.

Solution: For this purpose, we developed the MATLAB program `Example10_9.m` which includes the above steps of the algorithm.

We have two reactive elements in both the generator and load networks, therefore the fictitious two-port [*F*] must include at least five elements. Thus, $n \le 5$. Let us choose $n = 5$. In the meantime, [*F*] must include both generator and load networks. Therefore, it is appropriate to select sign $= -1$.

Even though there is no filter-type solution for the double matching problem under consideration, just to end up with a good virtual design, let us choose $\varepsilon^2 = 0.15$.

At this point we should mention that for an R//C load with R = 1, C = 3, the ideal flat gain level must satisfy the inequality of $T \le 1 - e^{-2\pi/RC} = 0.876$. Therefore, with the above choice of ε^2, the minimum passband gain yields approximately $T_{min} = 1/(1 + \varepsilon^2) = 1/(1 + 0.15) \cong 0.87$, which is quite difficult to obtain for this double matching problem.

Hence, when we run MATLAB code `Example10_9.m`, the impedance and the input reflection coefficient variations are depicted as in Figure 10.29, Figure 10.30, Figure 10.31 and Figure 10.32 respectively.

From Figure 10.30, it is interesting to observe that, up to $\omega = 0.8$, the driving point impedance Z_E is capacitive and then it becomes inductive.

Since the data for the equalizer is fictitious, we are free to model it as long as it traces the generated data in a reasonable manner to yield the desired gain performance approximately. Obviously, the end result will deviate from the ideal Chebyshev gain approximation, as expected.

Impedance or equivalently BR reflection coefficient modeling can be carried out by means of different methods. There are linear and nonlinear interpolation techniques. We will introduce these techniques in the following chapters [50–55].

The gain plot of the fictitiously matched system is shown in Figure 10.33.

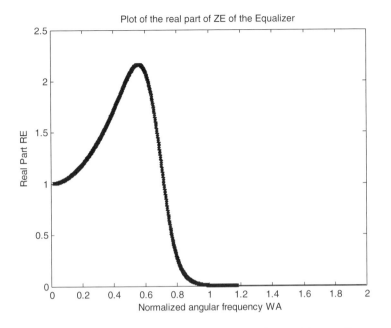

Figure 10.29   Real part of the driving point impedance of the equalizer

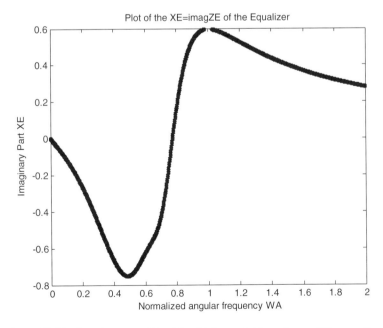

Figure 10.30   Imaginary part X_E of the driving point impedance of the equalizer

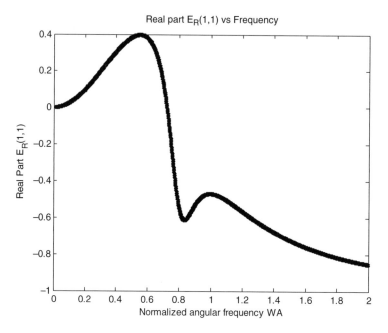

Figure 10.31   Real part of the input reflectance E_{11} of the equalizer

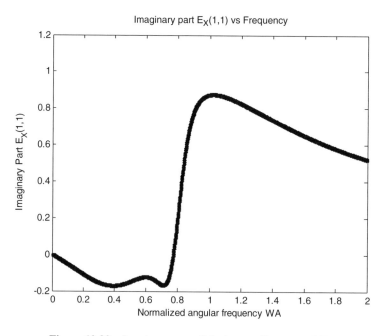

Figure 10.32   Imaginary part of the input reflectance of E_{11}

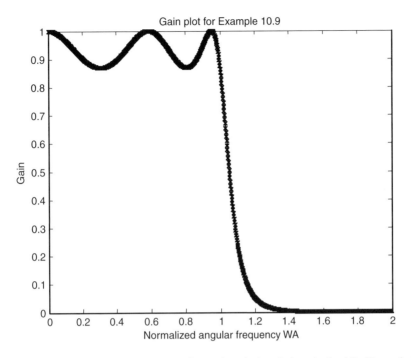

Figure 10.33 Idealized gain performance: the semi-analytic solution obtained for Example 10.9

Example 10.10: For the double matching problem introduced in Example 10.9, data for the virtual equalizer driving point impedance is summarized in Table 10.7.

Table 10.7 Data for a measured immittance

WA (frequency)	REA (real part)	XEA (imaginary part)
0	1	0
0.2	1.1866	−0.26399
0.4	1.7123	−0.67099
0.6	2.0941	−0.62335
0.8	0.27862	0.1536
1	0.0065277	0.59638
1.2	0.00028068	0.52581
1.4	2.0535e-005	0.43698
1.6	2.1843e-006	0.36913
1.8	3.0707e-007	0.31865
2	5.3628e-008	0.28025

(a) Model the real part data as a non-negative rational function using the direct approach presented in Section 7.8.

(b) Construct the minimum reactance function from the real part function obtained in part (a), and synthesize it as a lossless two-port in unit termination. That is, approximate the solution to the double matching problem presented in Example 10.9.

(c) Plot TPG of the matched structure as obtained in part (b).

Solution:

(a) The real part data specified in Table 10.7 is exactly the data given by Table 7.1. It was modeled using the MATLAB program developed for Example 7.16. There we obtained the following result:

$$R(-p^2) = \frac{1}{-27p^6 - 8.2097p^4 + 2.0169p^2 + 1}$$

(b) The corresponding minimum reactance driving point input impedance of the equalizer is found as

$$F(p) = Z(p) = \frac{0.7346p^2 + 0.607\,55p + 0.198\,18}{p^3 + 0.827\,04p^2 + 0.493\,01p + 0.1918}$$

Eventually, it is synthesized as a low-pass ladder as shown in Figure 10.34.

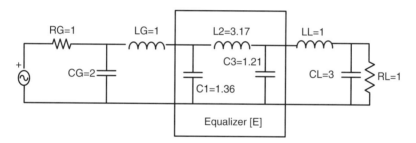

Figure 10.34 Approximate solution for the double matching problem of Example 10.9 using the semi-analytic method

Below, we present the MATLAB codes for Example10_9.m and related functions.

Program Listing 10.8

```
% Main Program Example10_9.m
% This program generates a realizable driving point input impe-
% dance data
% of an equalizer from the given Chebyshev filter
%
% Here the generator and the load networks are defined as R//C
% +L type
% Inputs:
```

```
%         eps_sq,n,sign
clear
eps_sq=input(" Enter eps_sq=")
n=input(" Enter degree of Chebyshev Approximation n=")
sign=input(" for hm=sign*hm(n) Enter sign=")
w1=0;w2=2;N=1000;
dw=(w2-w1)/(N-1);
w=w1;
% Generator impedance LG=1; CG=2, RG=1;
LG=1; CG=2; RG=1;
% Load Impedance: LL=1.2, C=3; RL=1;
LL=1.2, CL=3; RL=1;
for i=1:N
      WA(i)=w;
% Generation of idealized filter S and T-Parameters
% Step 1:
[F]=Cheby_Spar(w,eps_sq,n,sign);[TF]=Transfer_Spar(F);
Gain(i)=abs(F(2,1))^2;
% Generation of Generator S and T parameters
% Step 2:
      [G]=ZtoS(w,LG,CG,RG);[TG]=Transfer_Spar(G);
% Generation of Load S and T parameters
% Step 3:
      [L]=ZtoS(w,LL,CL,RL);[TL]=Transfer_Spar(L);
% Generation of equalizer T and S parameters
% Step 4:
      [TE]=inv(TG)*TF*inv(TL);[E]=TtoS(TE);
% Step 5:
      ZE=(1+E(1,1))/(1-E(1,1));
      RE=real(ZE);REA(i)=RE;
      XE=imag(ZE);XEA(i)=XE;
      ER=real(E(1,1)); ERA(i)=ER;
      EX=imag(E(1,1)); EXA(i)=EX;
      w=w+dw;
end
figure (1)
plot(WA,REA)
title("Plot of the real part of ZE of the Equalizer")
xlabel(" Normalized angular frequency WA")
ylabel(" Real Part R_E")
%
figure (2)
plot(WA,XEA)
title("Plot of the X_E=imag{ZE} of the Equalizer")
xlabel(" Normalized angular frequency WA")
ylabel(" Imaginary Part X_E")
%
figure (3)
plot(WA,ERA)
```

Program Listing 10.8   (Continued)

```
title("Real part E_R(1,1) vs Frequency")
xlabel(" Normalized angular frequency WA")
ylabel(" Real Part E_R(1,1)")
%
figure (4)
plot(WA,EXA)
title("Imaginary part E_X(1,1) vs Frequency")
xlabel(" Normalized angular frequency WA")
ylabel(" Imaginary Part E_X(1,1)")
%
figure (5)
plot(WA,Gain)
title("Gain plot for Example 10.9")
xlabel(" Normalized angular frequency WA")
ylabel("Gain")
```

Program Listing 10.9

```
function [T]=Transfer_Spar(F)
% Computation of Transfer Scattering Parameters
% Input:
%         [F]: Scattering Parameters of the filter
% Output:
%         [T] 2x2 Matrix
T(1,1)=F(1,2)-F(1,1)*F(2,2)/F(2,1);
T(1,2)=F(1,1)/F(2,1);
T(2,1)=-F(2,2)/F(2,1);
T(2,2)=1/F(2,1);
```

Program Listing 10.10

```
function [S]=ZtoS(w,L,C,R)
% This function generates the scattering parameters of an impe-
% dance
% specified as L+R//C
% where R=1 ohm.
% C
% Inputs:
```

```
%          a(p): Numerator of the impedance
%          b(p): Denominator of the impedance
% Outputs:
%          S parameters in Belevitch Form
%          h(p), f(p), g(p)
a=[L*C L 1/R]; b=[C 1/R];
a_=paraconj(a);b_=paraconj(b);
AB=conv(a,b_);BA=conv(b,a_);
AA=0.5*vector_sum(AB,BA);%pz=roots(AA);f=2*poly(pz);
h=vector_sum(a,-b)/2;g=vector_sum(a,b)/2;
h_=paraconj(h);g_=paraconj(g);
H=conv(h,h_);G=conv(g,g_);
F=G-H;f=1;
cmplx=sqrt(-1);
p=cmplx*w;
hval=polyval(h,p);gval=polyval(g,p);
S(1,1)=hval/gval; S(2,1)=f/gval;
S(1,2)=S(2,1);S(2,2)=-conj(hval)/gval;
```

Program Listing 10.11

```
function S=TtoS(T)
% Generation of Scattering Parameters from Transfer S-Parameters
% Inputs:
%          T: Transfer Scattering Parameters
%          T(1,1) T(1,2) T(2,1) T(2,2)
% Outputs:
%          S: Scattering Parameters
%          S(1,1) S(1,2) S(2,1) S(2,2)
%
S(1,1)=T(1,2)/T(2,2); S(1,2)=(T(1,1)*T(2,2)-T(1,2)*T(2,1))/
T(2,2);
S(2,1)=1/T(2,2);S(2,2)=-T(2,1)/T(2,2);
```

Program Listing 10.12

```
function [F]=Cheby_Spar(w,eps_sq,n,sign)
% Generation of the scattering parameters for a Chebyshev filter
% Inputs:
%          w: Normalized angular frequency
%          eps_sq: Ripple coefficient
%          n: Degree of the Chebyshev polynomial
%          sign: sign of hm(p); +/-1
```

Program Listing 10.12   (Continued)

```
% Output: 2x2 Matrix F
%          [F]=[F(1,1),  F(1,2)=F(2,1),  F(2,2)]
eps=sqrt(eps_sq);
beta=asinh(1/eps);
% Step 1: Generates the Explicit roots of Gm(p)).
n2=2*n;
i=1;
j=sqrt(-1);
for k=0:(n2-1)
    x(k+1)=k;
sig(k+1)=sin(pi*(2*k+1)/2/n)*sinh(beta/n);%Real{pk}
omega(k+1)=cos(pi*(2*k+1)/2/n)*cosh(beta/n);%imag{pk}
    if(sig(k+1)<0)
        p(i)=sig(k+1)+j*omega(k+1);%roots of g(p)
        i=i+1;
    end
end
% Step 2: Construct gm(p)
gm=real(poly(p));%polynomial g(p)
gm=2^(n-1)*eps*gm;%g(p) with eps*Tn(1)
%
%Step 3: Computation of hm(p)
    for k=0:(n2-1)
    x(k+1)=k;
        omega(k+1)=j*cos(pi*(2*k+1)/2/n);
    end
    for k=1:n
        ph(k)=omega(k);% Roots of hm(p)
    end
    hm=real(poly(-ph));%polynomial h(p)
    hm=sign*2^(n-1)*eps*hm;% h(p) with eps*Tn(1)
    fm=1;
    cmplx=sqrt(-1);
    p=cmplx*w;
    hval=polyval(hm,p);
    gval=polyval(gm,p);
    F11=hval/gval;
    F21=fm/gval;
    F22=-conj(hval)/gval;

F(1,1)=F11;
F(2,1)=F21;
F(1,2)=F21;
F(2,2)=F22;
```

10.16 General Form of Monotone Roll-Off Chebyshev Transfer Functions

So far we have dealt with monotone roll-off Chebyshev functions to build filters' single and double matched equalizers by employing the transfer function

$$T(\omega) = \frac{T_{max}}{1 + \varepsilon^2 T_n^2(\omega)}, \text{ with } T_{max} = 1$$

However, T_{max} can have any non-negative value satisfying the inequality

$$0 \leq T_{max} \leq 1$$

In this case, it is customary to express the monotone roll-off Chebyshev transfer function as

$$T(\omega) = \frac{T_{max}}{1 + \varepsilon^2 T_n^2(\omega)} = \frac{1}{1 + K^2 + \varepsilon^2 T_n^2(\omega)}$$
where
$$T_{max} = \frac{1}{1 + K^2}$$

(10.66)

Obviously, the minimum value of the transfer function T_{min} occurs at the frequencies ω_i which make $T_n^2(\omega_i) = 0$ and it is given by

$$T_{min} = \frac{T_{max}}{1 + \varepsilon^2} = \frac{1}{1 + K^2 + \varepsilon^2}$$

(10.67)

In this case,

$$F_{11}(p)F_{11}(-p) = 1 - T(p) = \frac{H_m(p)}{G_m(p)} = \frac{K^2 + \varepsilon^2 T_n^2}{1 + K^2 + \varepsilon^2 T_n^2}$$

where

$$H_m(p) = (K^2) + \varepsilon^2 T_n^2$$

$$G_m(p) = (1 + K^2) + \varepsilon^2 T_n^2$$

$$K^2 = \frac{1}{T_{max}} - 1$$

(10.68)

Recall that the nth degree Chebyshev polynomial $T_n(\omega)$ is defined as $T_n(\omega) = \cos(nz)$, where $\omega = \cos(z)$. In this case, the roots of $H_m(p)$ can be found as in Section 10.7 in a similar manner to that for the roots of $G_m(p)$ such that firstly we set

$$p = j\omega = j\cos(z)$$

or

$$z = \cos^{-1}\left(\frac{p}{j}\right)$$

and let

$$nz = \alpha + j\beta$$

Then, by Equation (10.68), the roots of $H_m(p)$ are found at points nz_{Hk} such that

$$K^2 + \varepsilon^2 T_n^2(nz_{Hk}) = 0$$

or

$$\cos(nz_{Hk}) - \cos(\alpha_{Hk} + j\beta_{Hk}) = \mp j\frac{K}{\varepsilon}$$

yielding

$$
\boxed{
\begin{aligned}
&\alpha_{Hk} = \left(k + \frac{1}{2}\right)\pi \\
&\text{and} \\
&\beta_{Hk} = \mp\sinh^{-1}\left(\frac{K}{\varepsilon}\right)
\end{aligned}
}
\tag{10.69}
$$

Hence, the roots of $H_m(p)$ are located at

$$p_k = j\cos(z_k) = j\cos\left(\frac{\alpha_k + j\beta_k}{n}\right)$$

and are given as in Equation (10.43):

$$
\boxed{
\begin{aligned}
p_{Hk} = {} & \sin\left[\left(\frac{2k+1}{2n}\right)\pi\right]\sinh\left(\frac{\beta_{Hk}}{n}\right) \\
& + j\cos\left[\left(\frac{2k+1}{2n}\right)\pi\right]\cosh\left(\frac{\beta_{Hk}}{n}\right)
\end{aligned}
}
\tag{10.70}
$$

In this equation, the integer k runs from 0 to $2n - 1$ yielding mirror image symmetric roots in the complex p plane. In other words, the roots p_{Hk} must be placed in the LHP and RHP or on the $j\omega$ axis with mirror image symmetry.

Eventually, numerator polynomial $h_m(p)$ of $F_{11m}(p) = h_m(p)/g_m(p)$ is generated on the LHP and the $j\omega$ roots of $H_m(p)$ as in Equation (10.46) such that

$$
\boxed{
\begin{aligned}
h_m(p) &= \mp\left(2^{n-1}\varepsilon\right)\prod_{r=1}^{n}(p - p_r) \\
&= h_1 p^n + h_2 p^{n-1} + \cdots + h_n p + h_{n+1}
\end{aligned}
}
\tag{10.71}
$$

Similarly, the roots p_{Gk} of $G_m(p)$ can be given by setting

$$
\begin{aligned}
\alpha_{Gk} &= \left(k + \frac{1}{2}\right)\pi \\[2mm]
&\text{and} \\[2mm]
\beta_{Gk} &= \mp \sinh^{-1}\left(\frac{\sqrt{1 + K^2}}{\varepsilon}\right)
\end{aligned}
\tag{10.72}
$$

and

$$
\begin{aligned}
p_{Gk} &= \sin\left[\left(\frac{2k + 1}{2n}\right)\pi\right]\sinh\left(\frac{\beta_{Gk}}{n}\right) \\[2mm]
&\quad + j\cos\left[\left(\frac{2k + 1}{2n}\right)\pi\right]\cosh\left(\frac{\beta_{Gk}}{n}\right)
\end{aligned}
\tag{10.73}
$$

Selecting the LHP roots, one can readily generate the open polynomial form of $g_m(p)$ as

$$
\begin{aligned}
g_m(p) &= \left(2^{n-1}\varepsilon\right)\prod_{r=1}^{n}(p - p_r) \\[2mm]
&= g_1 p^2 + g_2 p^{n-1} + \cdots + g_n p + g_{n+1}
\end{aligned}
\tag{10.74}
$$

Note that, for $T_{max} = 1$, or equivalently for $K = 0$, all the roots of $H_m(p)$ must be on the $j\omega$ axis as specified by Equation (10.44) and Equation (10.46).

In MATLAB, considering the general case with, $T_{max} \leq 1$, $h_m(p)$ and g_m are generated by employing the functions

```
gm = gm_ChebywithK(n, eps_sq, Tmax)
hm = hm_ChebywithK(n, eps_sq, Tmax)
     hm = hm_Cheby(n, eps_sq),
        for which Tmax = 1
```

Function $\text{gm} = \text{gm\_ChebywithK}(\text{n}, \text{eps\_sq}, \text{Tmax})$ yields the coefficient form of $g_m(p)$ as in Equation (10.74).

Function $\text{hm} = \text{hm\_Cheby}(\text{n}, \text{eps\_sq})$ yields the coefficient form of $h_m(p)$ as expressed in Equation (10.44–10.45).

Function $\text{hm} = \text{hm\_ChebywithK}(\text{n}, \text{eps\_sq}, \text{Tmax})$ yields the coefficient form of $h_m(p)$ as specified by Equation (10.71).

In view of the above presentation, real normalized (or unit normalized) input reflectance $F_{11m}(p)$ and transfer scattering parameter $F_{21m}(p)$ of the monotone roll-off Chebyshev filter are given as

$$F_{11m}(p) = \mp \frac{h_m(p)}{g_m(p)}$$

and   (10.75)

$$F_{21m}(p) = \frac{1}{g_m(p)}$$

In Equation (10.75) the '$-$' sign corresponds to a filter which starts with a capacitor $x_1 = C_1$. On the other hand, '$+$' corresponds to a filter starting with an inductor $x_1 = L_1$.

As detailed by Levy[13] for specified $T_{max} = 1/(1 + K^2)$ and ε^2, it is useful to give explicit element values of the general monotone roll-off Chebyshev filter as follows:

$$\left(\frac{1}{T_{max}}\right)\left(\frac{1}{\varepsilon^2}\right) = \frac{1 + K^2}{\varepsilon^2} = \sinh^2(na)$$

Or equivalently   (10.76)

$$a = \left(\frac{1}{n}\right)\sinh^{-1}\left(\sqrt{\frac{T_{max}}{\varepsilon^2}}\right)$$

$$\left(\frac{1 - T_{max}}{T_{max}}\right)\left(\frac{1}{\varepsilon^2}\right) = \frac{K^2}{\varepsilon^2} = \sinh^2(nb)$$

Or equivalently   (10.77)

$$b = \left(\frac{1}{n}\right)\sinh^{-1}\left(\sqrt{\frac{1 - T_{max}}{T_{max}\varepsilon^2}}\right)$$

$$x = \sinh(a)$$
$$y = \sinh(b)$$ (10.78)

The first reactive element of the filter is

$$x_1 = \frac{2\sin\left(\dfrac{\pi}{2n}\right)}{x - y}$$ (10.79)

---

[13] R. Levy, 'Explicit Formulas for Chebyshev Impedance Matching Networks, Filters and Interstages', *Proceedings of the IEE*, Vol. III, No. 6, pp. 1099–1106, 1964.

and the rest of the elements of the filter are given by

$$
x_r x_{r+1} = \frac{4\sin\left(\dfrac{2r-1}{2n}\pi\right)\sin\left(\dfrac{2r+1}{2n}\pi\right)}{x^2 + y^2 + \sin^2\left(\dfrac{r\pi}{n}\right) - 2xy\cos\left(\dfrac{r\pi}{n}\right)}
$$
$$
r = 1, 2, 3, \ldots, n-1
$$

(10.80)

In this design, it assumed that the generator side of the filter is driven by a normalized internal resistance $R_G = 1$. However, termination resistance R_L is no longer unity, even for the case $n = $ odd, since $T_{max} \le 1$.

Computation of the Resistive Termination $A0 = \{R_L$ *or* $G_L\}$

Resistive termination of the filter could be either a resistor R_L or a conductor G_L depending on the value of degree n and the sign of $F_{11m}(p) = \text{sign}\{h_m(p)/g_m(p)\}$.

In any case, we set the termination as $A0 = \{R_L$ *or* $G_L\}$ which in turn results in a DC gain $T(0) = 4A0/(1 + A0)^2$, whatever the value of degree n is.

For example, if n is an odd integer, and $\text{sign} = -1$, then the filter starts with a *capacitor (say x1 or x_1)* and ends with a *capacitor (say xn or x_n)*. In this case, resistive termination must be a conductor $A0 = G_L$. If the sign is chosen as $sign = +1$, then the filter starts with an *inductor (x1)* and ends with an *inductor (xn)*. In this case, resistive termination must be a resistor $A0 = R_L$. We should also mention that the DC value of TPG varies with *degree n* depending on whether it is even or odd. Therefore, we must classify the following cases.

$n = $ **odd integer case**

In this case, at DC the filter gain is given as

$$
T(0) = T_{max} = \frac{1}{1 + K^2}
$$
$$
= \frac{4A0}{(1 + A0)^2}
$$

or the termination $A0$ is given by

$$
A0 = \left[\frac{2 - T_{max}}{T_{max}}\right] + \sqrt{\left[\frac{2 - T_{max}}{T_{max}}\right]^2 - 1}
$$

(10.81)

$n = $ **even integer case**

In this case, $T_n^2(0) = 1$ and the DC gain level is given by

$$
T_0 = \frac{1}{1 + K^2 + \varepsilon^2}
$$

which is less than T_{max}.

Then, the termination is given by

$$
A0 = \left[\frac{2 - T_0}{T_0} \right] - \sqrt{ \left[\frac{2 - T_0}{T_0} \right]^2 - 1 }
\tag{10.82}
$$

As an exercise, all the above MATLAB functions and formulas are gathered together under the MATLAB code function [hm, gm, CV] = LevyGainwithK(n, sign, Tmax, eps_sq) This function constructs low-pass LC ladder filters for specified *degree* n, T_{max} andε^2, and

$$
\text{sign of } F_{11m}(p) = \text{sign} \frac{h_m(p)}{g_m(p)}
$$

such that if sign = −1 the filter starts with a capacitor, but if sign = +1 it starts with an inductor.

The output of the function is h_m, g_m and the component values of the filter which are obtained as a result of synthesis of the input reflectance $F_{11m}(p) - \text{sign}\{h_m(p)/g_m(p)\}$. The complete filter is displayed in Figure 10.35 and gain performance is shown in Figure 10.36. In MATLAB work space, this function can be directly run by typing the inputs.

Now let us run an exercise with $n = 5$, sign = +1(start with an inductor), $T_{max} = 0.8$ and $\varepsilon^2 = 0.4$. The results are given as they appear in the MATLAB work space.

Type the following:

```
n=5,sign=+1,Tmax=0.8,eps_sq=0.4

[hm, gm, CV]= levyGainwithK (n, sign, Tmax, eps_sq)
```

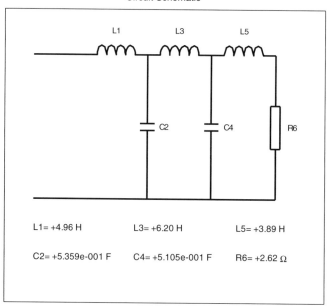

Circuit Schematic

L1 = +4.96 H   L3= +6.20 H   L5= +3.89 H

C2= +5.359e-001 F   C4= +5.105e-001 F   R6= +2.62 Ω

Figure 10.35 Monotone roll-off Chebyshev filter synthesis with $n = 5$, sign $= +1$, $T_{max} = 0.8$, eps_sq $= 0.4$

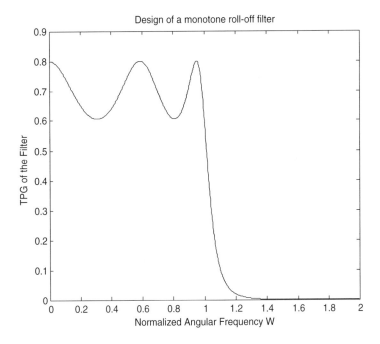

Figure 10.36 Gain performance of the filter shown in Figure 10.35

and the result is

```
hm = [10.119   4.7668    13.772   4.4751   3.8046   0.5]
gm = [4.9621   8.8454    16.515   9.0451   5.4978   1.118]
CV = [4.9621   0.53594   6.197    0.51048  3.8925   2.618]
```

Remark: The function [hm,gm,CV] = Levy_Gain(n,sign,Tmax,eps_sq) can be utilized by the reader to provide deep insight into matching problems for various inputs (i.e. n, sign, Tmax and eps_sq).

In fact, the following matching section is built on the above formulation of the filter problem.

10.17 LC Ladder Solutions to Matching Problems Using the General Form Chebyshev Transfer Function[14]

LC Ladder Solutions to Single Matching Problems with Two Reactive Elements in the Load

The beauty of using the general form of the monotone roll-off Chebyshev transfer function is that one can construct an equalizer without degeneracy for a single matching problem for which the load includes two reactive elements.

In this case, for specified reactive elements x_1 and x_2 the variables x and y are determined, which in turn yields T_{max} and ε^2. Thus, the rest of the element values are determined as in Equation (10.80).

The above statement can be quantified by the following equations.

---

[14] This section is based on the paper: R. Levy, 'Explicit Formulas for Chebyshev Impedance Matching Networks, Filters and Interstages', *Proceedings of the IEE*, Vol. III, No 6, pp. 1099–1106, 1964.

Given x_1, x_2 and degree n, we solve for x and y satisfying the following equations:

$$x - y = \frac{2\sin\left(\dfrac{\pi}{2n}\right)}{x_1} = A \tag{10.83}$$

$$x^2 + y^2 - 2xy\cos\left(\frac{\pi}{n}\right)$$
$$= \frac{4\sin\left(\dfrac{1}{2n}\pi\right)\sin\left(\dfrac{3}{2n}\pi\right)}{x_1 x_2} - \sin^2\left(\frac{\pi}{n}\right) = B \tag{10.84}$$

The above equations result in

$$a_1 y^2 + 2b_1 y + c_1 = 0$$

yielding

$$y = \frac{-b_1 \mp \sqrt{b_1^2 - a_1 c_1}}{a_1} \tag{10.85}$$

$$x = y + A$$

where

$$\begin{aligned} a_1 &= 2(1 - C) \\ b_1 &= A(1 - C) \\ c_1 &= A^2 - B \end{aligned} \tag{10.86}$$

and

$$A = \frac{2\sin\left(\dfrac{\pi}{2n}\right)}{x_1}$$
$$B = \frac{4\sin\left(\dfrac{1}{2n}\pi\right)\sin\left(\dfrac{3}{2n}\pi\right)}{x_1 x_2} - \sin^2\left(\frac{\pi}{n}\right) \tag{10.87}$$
$$C = \cos\left(\frac{\pi}{n}\right)$$

At this point we need to determine K^2 or equaivalently T_{max} and ε^2 to generate the termination R_L so that the complete equalizer and its performance are obtained.

By Equation (10.73) and Equation (10.74) we have

$$\frac{1 + K^2}{K^2} = \frac{\sinh^2(na)}{\sinh^2(nb)}$$

or

$$K^2 = \frac{\sinh^2(nb)}{\sinh^2(na) - \sinh^2(nb)}$$

and

$$T_{max} = \frac{1}{1 + K^2} \qquad (10.88)$$

where

$$a = \sinh^{-1}(x)$$
$$b = \sinh^{-1}(y)$$

Equation (10.74) reveals that

$$\varepsilon^2 = \frac{K^2}{\sinh^2(nb)} \qquad (10.89)$$

An Essential Remark

It should be emphasized that all the above equations are valid only for real values of x and y which are determined for the given values of x_1 and x_2.

By Equation (10.85), the real value of requires that

$$\Delta = b_1^2 - a_1 c_1 \geq 0 \qquad (10.90)$$

or, equivalently,

$$2B \geq (C + 1)A^2$$

or

$$2\left[\frac{4\sin\left(\frac{1}{2n}\pi\right)\sin\left(\frac{3}{2n}\pi\right)}{x_1 x_2} - \sin^2\left(\frac{\pi}{n}\right) \right]$$
$$\geq \left[\cos\left(\frac{\pi}{n}\right) + 1 \right]\left[\frac{2\sin\left(\frac{\pi}{2n}\right)}{x_1} \right] \qquad (10.91)$$

If the above inequality is not satisfied, then one needs to employ an all-pass function on the minimum transfer scattering parameter $F_{21m}(p)$ to invoke the GBRs. Details are skipped here.

Let us consider an example employing the above formulas.

Example 10.11:

(a) Repeat Example 10.8 using the general form of monotone roll-off Chebyshev transfer function.
(b) Comment on your result.

Solution:

(a) For this problem, all the above equations are programmed in MATLAB under the function `[x,A0,Tmax,eps_sq,Del]levy_Single(n,x1,x2)`.

Function `[x,A0,Tmax,eps_sq,Del]=levy_xy(n,x1,x2)` generates all the explicit element values of the singly matched equalizer for specified degree n and reactive components x_1 and x_2 with $R_G = 1$. In return, it gives complete element values $x(r)$ of the equalizer, with resistive termination $A0 = \{R_L\ or\ G_L\}$, maximum passband gain T_{max}, ripple coefficient ε^2 and the bound $\Delta=\text{Delta}$ for the ladder realization. Recall from Equation (10.90) that, if $\Delta \geq 0$, then the equalizer can be realized without an all-pass function.

Thus, for the example under consideration, for the impedance to be matched we select $x_1 = 3$, which is a shunt capacitor, $x_2 = 0.6$, which is a series inductor, and $n = 7$, as in Example 10.8. Then, we run the MATLAB main program `Example10_11.m` which includes the function `[x,A0,Tmax, eps_sq,Del]=levy_Single(n,x1,x2)`.

This program returns the following outputs:

- Element values of the equalizer with described generator network:

$$R_G = 1 \quad x(1) = 3(Capacitor\ C_G = 3) \quad x(2) = 0.6(inductor\ L_G = 0.6)$$
$$C_3 = x(3) = 3.72 \quad L_4 = x(4) = 0.555 \quad C_5 = x(5) = 2.97$$
$$L_6 = x(6) = 0.337 \quad C_7 = x(7) = 0.7858$$

- Termination conductor $G_L = A0 = 2.4984$ or $R_L = 1/A0 = 0.40026$
- Maximum passband gain $T_{max} = 0.81\,655$
- Ripple coefficient $\varepsilon^2 = 0.00\,210\,31$
- Bound for the ladder realization $\Delta = 0.019\,635$.

The resulting gain performance of the matched system is depicted in Figure 10.37.
(b) Comments are as follows:

- In the first place, this equalizer is not degenerate and results in a very small ripple coefficient $\varepsilon^2 = 0.000\,210\,31$, whereas in Example 10.8 the equalizer was degenerate and the ripple coefficient was $\varepsilon^2 = 0.669\,12$ (or $\varepsilon = 0.81$), which drastically reduces the minimum of the gain level. Hence, from the electrical performance, the currently designed equalizer is superior.
- In Example 10.8, the maximum value of the allowable inductor $L = x(2)$ was 0.879. In the present case, it is slightly higher, yielding $x(2) = 0.882$ which is about the maximum value of the second reactive element.
- In designing equalizers by employing the general form of monotone roll-off Chebyshev functions, it is inevitable that an ideal transformer is used in the matching network since a mandatory mismatch is introduced at DC by fixing T_{max}. This may be considered at the expense of having reduced ripples in the band of operation.

Hence, the final design is as depicted in Figure 10.38.

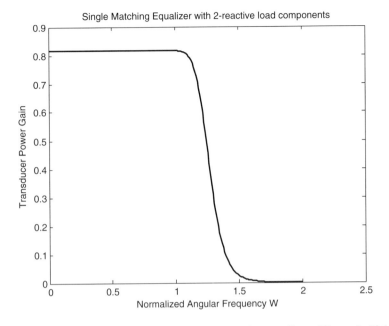

Figure 10.37 Gain performance of the single matching equalizer of Example 10.11

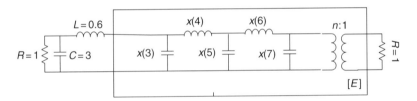

Figure 10.38 Equalizer designed with an ideal transformer $1/n^2 = A0 \approx 2.5$ for the single matching problem of Example 10.11

LC Ladder Solutions to Double Matching Problems with Single Reactive Elements in both Generator and Load Networks

By Equation (10.80), it is straightforward to show that the first and last elements of the filter constructed with the transfer function of Equation (10.66) are

$$
\begin{aligned}
x_1 &= \frac{2\sin\left(\dfrac{\pi}{2n}\right)}{x-y} \\[2ex]
x_n &= \frac{2\sin\left(\dfrac{\pi}{2n}\right)}{x+y}
\end{aligned}
\tag{10.92}
$$

Or equivalently

$$
\boxed{
\begin{aligned}
x &= \left(\frac{1}{x_n} + \frac{1}{x_1}\right)\sin\left(\frac{\pi}{2n}\right) \\
y &= \left(\frac{1}{x_n} - \frac{1}{x_1}\right)\sin\left(\frac{\pi}{2n}\right)
\end{aligned}
}
\tag{10.93}
$$

Insertion of the above equations into Equation (10.80) results in the element values of the desired double matched equalizer. However, when we insert x and y into Equation (10.80), the resulting last element $x(n)$ is terminated in the mismatch $A0$. Therefore, in order to obtain the original load component x_n, $x(n)$ must be normalized by $A0$. In other words, the final equalizer must include an ideal transformer at the far end with a turns ratio of $n^2 = A0$ as shown in Figure 10.39.

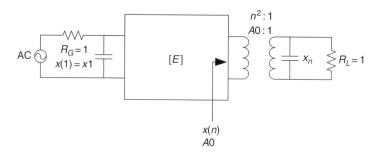

Figure 10.39 Simple double matching problem via monotone roll-off Chebyshev transfer function x(n) = [A0][xn]

Thus,

$$
\boxed{
\{xn \ or \ x_n\} = \frac{x(n)}{A0} \quad \text{terminated in} \quad \{R_L \ or \ G_L\} = 1
}
\tag{10.94}
$$

Now let us look at an example.

Example 10.12:

(a) Design a double matched equalizer with $n = 7$ (odd case), $x_1 = 3$ and $x_n = 2$. Here, assume that $x_1 = C_G$ is a capacitor which in turn yields the load reactive element as a capacitor since the degree n is odd.

(b) Solve the same double matching problem for $n = 6$ (even case). In this part it is also assumed that $x_1 = C_G = 3$, which in turn results in an inductor $xn = L = 2$ at the load end since the degree n is even.

Solution:

(a) For the example under consideration, we developed a MATLAB function called $[x, A0, Tmax, epssq] = LevyDouble(n, x1, xn)$ which generates x and y as in Equation (10. 90) and then produces the element values of the equalizer as specified by Equation (10.80). Finally, resistive termination $A0$ is determined as before, which in turn results in the turns ratio of an ideal transformer. Function `Levy_Double` is run under the main program `Example10_12.m`. The results are given as follows.

The components of the equalizer are

$x1 = x(1) = 3$, reactive element of the generator as specified at the input

$$x(2) = 0.874\,13$$
$$x(3) = 4.0263$$
$$x(4) = 0.9104$$
$$x(5) = 4.0013$$
$$x(6) = 0.856\,56$$
$$x(7) = 2.7406$$

Normalization number $A0$ or equivalently the turns ratio of the ideal transformer $n^2 = A0$ is found as

$$A0 = 1.3703$$

The single reactive component of the load is then given by Equation (10.94):

$$xn = \frac{x(n)}{A0} = \frac{2.7406}{1.3703} = 2$$

as it should be.
The maximum of the passband gain is found as

$$T_{max} = 0.9756$$

and the ripple coefficient is found as

$$\varepsilon^2 = 0.36311$$

The gain performance of the doubly matched equalizer is depicted in Figure 10.40.

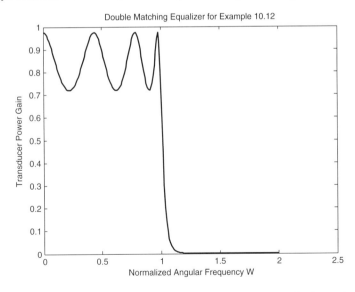

Figure 10.40   Gain performance of the doubly matched equalizer for Example 10.12

The result of the above design is checked via synthesis of the driving point impedance

$$Z_{in} = \frac{g_m(p) + h_m(p)}{g_m(p) - h_m(p)}$$

by employing the function $[hm, gm, CV] = LevyGainwithK(n, sign, Tmax, epssq)$.

The resulting circuit is shown in Figure 10.41, which confirms the above element values. In order to obtain the doubly matched equalizer, in Figure 10.41, capacitor $C7$ and termination resistor $R8$ must be removed and an ideal transformer must be inserted with a turns ratio $n^2 = A0 = 1/R_8$. Note that $xn = C_7 \times R_8 = 2.74 \times 0.7298 = 2$, as it should be.

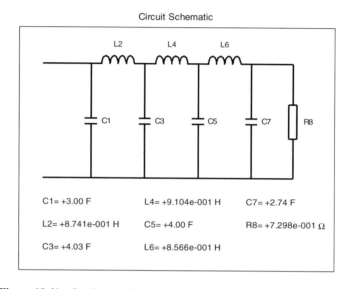

Figure 10.41   Synthesis of the driving point impedance for Example 10.12

(b) We run the same MATLAB program for $n = 6$ and obtain the following results:

$$x1 = x(1) = 3, \text{ as it should be}$$
$$x(2) = 0.8645$$
$$x(3) = 4.005$$
$$x(4) = 0.888\,84$$
$$x(5) = 3.8056$$
$$x(6) = 0.611\,65$$
$$A0 = 0.305\,82$$
$$xn = \frac{x(6)}{A0} = \frac{0.611\,65}{0.305\,82} = 2, \text{ as it should be}$$
$$T_{max} = 0.975\,88 \qquad \varepsilon^2 = 0.368\,68$$

The performance of the doubly matched equalizer is depicted in Figure 10.42.
The driving point impedance synthesis of this part is shown in Figure 10.43.

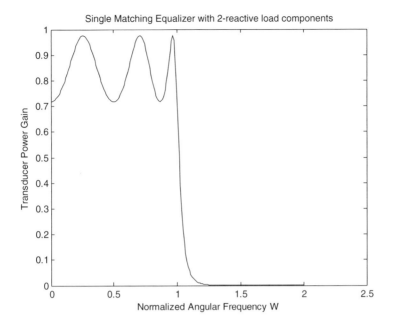

Figure 10.42   Gain performance of part (b) of Example 10.12

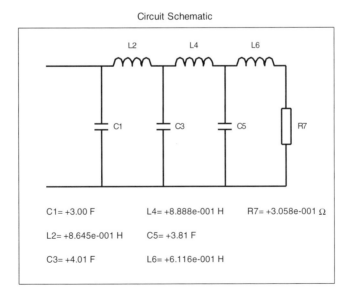

Figure 10.43   Synthesis of the driving point impedance of part (b) of Example 10.12

In order to obtain the doubly matched equalizer, we must remove L6 and R7 and insert an ideal transformer with a trans ratio $n^2 = A0 = R_7$ at the load end.

Clearly, the above circuit elements confirm the desired design.

10.18 Conclusion

In this chapter, a unified approach to analytic theory of broadband matching was presented. In this regard, the lumped element filter design technique was utilized to solve simple single and double matching problems.

It has been emphasized that, beyond simple problems, analytic theory is not accessible. Nevertheless, the semi-analytic approach, which is introduced in Section 10.14, may be employed to generate an initial numerical solution to complicated problems. It is well established that computer-aided design methods, especially the real frequency techniques, provide practical solutions to matching problems. Therefore, in the following chapters these techniques are presented with many examples.

Finally, we provide below the MATLAB programs and related function employed to solve the single and double matching problems.

Program Listing 10.13   Main program Example10_11

```
% Example10_11
% Design of a single matching equalizer with
% Inputs:
n=input("Enter n=")
x1=input("Enter the reactive component of the generator x1=")
x2=input("Enter the reactive Component of the load x2=")
sign=input("Enter sign=");% Start with a capacitor
% Outputs:
%           x: Component values of the equalizer
%           A0: Resistive termination of the equalizer
%           Tmax: Maximum gain level in the passband
%           eps_sq: Ripple coefficient/
%
[x,A0,Tmax,eps_sq,Delta]=Levy_Single(n,x1,x2)
% Check the results and plot the circuit:
[hm,gm]=Levy_GainwithK(n,sign,Tmax,eps_sq)
```

Program Listing 10.14   Main program Example10_12.m

```
% Main Program Example10_12.m
% This program solves the simple double matching problem
% Inputs:
%       x1=single reactive component of the generator.
%       xn=single reactive component of the load
% It should be noted here we assumed that RG=RL=1
```

```
%
% Outputs:
%         x(r): elements of the lowpass ladder
%         A0=n^2:Turn ration of the far end transformer;
%         and normalization number for the last component x(n)
clc
clear
n=input("Enter n=")
x1=input("Enter the reactive component of the generator x1=")
xn=input("Enter the reactive Component of the load xn=")
[x,A0,Tmax,eps_sq]=Levy_Double(n,x1,xn)
%Check the results and plot the circuit
sign=-1;
[hm,gm,CV]=Levy_GainwithK(n,sign,Tmax,eps_sq);
```

Program Listing 10.15   Solution to single matching problems using Levy's formulas. In this function, load includes only two reactive components

```
function [x,A0,Tmax,eps_sq,Delta]=Levy_Single(n,x1,x2)
% Computation of x and y for Levy formulas
% Solution for single matching problems with
% two-reactive elements.
% See section 10.17 of Wiley book by Yarman
%A=2sin(pi/2n)/x_1
A=2*sin(pi/2/n)/x1;
%B=4sin(pi/2n)sin(3pi/2n)/(x_1 x_2)-sin^2 (pi/n)
B=4*sin(pi/2/n)*sin(3*pi/2/n)/x1/x2-sin(pi/n)*sin(pi/n);
C=cos(pi/n);
a1=2*(1-C);
b1=2*(A-A*C);b1=b1/2;
c1=A*A-B;
% Check the realizability
Delta=b1*b1-a1*c1;
Del=2*B-A*A*(1+C);
%yb=[-b"(+/-)sqrt(b"^2-ac)]/a
%yb=(-(A-C)+sqrt((A-C)*(A-C)-2*(1-C)*A*A))/2/(1-C);
yb=(-b1+sqrt(b1*b1-a1*c1))/a1;
xa=A+yb;
%
%x(1)=2sin(pi/2n)/(xa-yb) (Normalized first element
% value)
x(1)=2*sin(pi/2/n)/(xa-yb);
%
% x(r)*x(r+1)=4sin((2r-1)/2n pi)sin(pi(2r+1)/2n)
```

Program Listing 10.15   (Continued)

```
%/(x^2+y^2 sin^2
%(rpi/n)-2xycos(pir/n) )
for r=1:(n-1)
Num=4*sin((2*r-1)*pi/2/n)*sin((2*r+1)*pi/2/n);
Denom=xa^2+yb^2+sin(r*pi/n)*sin(r*pi/n)-2*xa*yb*cos(r*pi/n);
x(r+1)=Num/Denom/x(r);
end
%a=sinh^(-1) (x)
a=asinh(xa);
%b=sinh^(-1) (y)
b=asinh(yb);
%K^2=(sinh^2 (nb))/(sinh^2 (na)-sinh^2 (nb))
K_sq=sinh(n*b)*sinh(n*b)/(sinh(n*a)*sinh(n*a)-sinh(n*b)*-
sinh(n*b));
eps_sq=K_sq/sinh(n*b)/sinh(n*b);
Tmax=1/(K_sq+1);
%
% Computation of the termination:
% n=odd case:
% A0=[(2-T_max)/T_max ]??([(2-T_max)/T_max ]^2-1)
A0=(2-Tmax)/Tmax+sqrt(((2-Tmax)/Tmax)^2-1);
% Computation of n=even case
if (n/2-fix(n/2)==0);%n=even case
T0=1/(1+K_sq+eps_sq);
A0=(2-T0)/T0-sqrt(((2-T0)/T0)^2-1);
end;%End of n=even case
W1=0;
W2=2;
N=200;
DW=(W2-W1)/(N-1);
W=0;
for j=1:N
        WA(j)=W;
        [ZQ Z]=Lowpass_Ladder(W,x,A0);
        R=real(ZQ);
        X=imag(ZQ);
        T(j)=4*R/((R+1)*(R+1)+X*X);
            W=W+DW;
end
figure( )
plot(WA,T)
xlabel("Normalized Angular Frequency W")
ylabel("TPG for Single Matching")
title("Single Matching Equalizer with 2-reactive load compo-
nents")
```

Program Listing 10.16 Solution to double matching problems using Levy's formulas. In this function, generator and load networks include only single reactive components

```
function
[x,A0,Tmax,eps_sq]=Levy_Double(n,x1,xn)

% Computation of x and y for double matching problem
% for specified x1 and xn.
% See section 10.17 of Wiley book by Yarman
%xa=(1/x_n +1/x_1 )sin(?/2n)
%yb=(1/x_n -1/x_1 )sin(?/2n)
%
xa=(1/xn+1/x1)*sin(pi/2/n);
yb=(1/xn-1/x1)*sin(pi/2/n);

x(1)=2*sin(pi/2/n)/(xa-yb);
% x(r)*x(r+1)=4sin(pi(2r-1)/2n)sin(pi(2r+1)/2n )/(x^2+y^2 sin^2
%pi(r/n)-2xycos(r?/n) )
for r=1:(n-1)
Num=4*sin((2*r-1)*pi/2/n)*sin((2*r+1)*pi/2/n);
Denom=xa^2+yb^2+sin(r*pi/n)*sin(r*pi/n)-2*xa*yb*cos(r*pi/n);
x(r+1)=Num/Denom/x(r);
end
%a=sinh^(-1) (x)
a=asinh(xa);
%b=sinh^(-1) (y)
b=asinh(yb);
%K^2=(sinh^2 (nb))/(sinh^2 (na)-sinh^2 (nb))
K_sq=sinh(n*b)*sinh(n*b)/(sinh(n*a)*sinh(n*a)-sinh(n*b)*-
sinh(n*b));
eps_sq=K_sq/sinh(n*b)/sinh(n*b);
Tmax=1/(K_sq+1);
%
% Computation of the termination:
% n=odd case:
% A0=[(2-T_max)/T_max ]+sqrt([(2-T_max)/T_max ]^2-1)
A0=(2-Tmax)/Tmax+sqrt(((2-Tmax)/Tmax)^2-1);
% Computation of n=even case
if (n/2-fix(n/2)==0);%n=even case
T0=1/(1+K_sq+eps_sq);
A0=(2-T0)/T0-sqrt(((2-T0)/T0)^2-1);
end;%End of n=even case
W1=0;
W2=2;
N=200;
DW=(W2-W1)/(N-1);
W=0;
for j=1:N
```

Program Listing 10.16   (Continued)

```
WA(j)=W;
[ZQ Z]=Lowpass_Ladder(W,x,A0);
R=real(ZQ);
X=imag(ZQ);
T(j)=4*R/((R+1)*(R+1)+X*X);
W=W+DW;
end
figure( )
plot(WA,T)
xlabel("Normalized Angular Frequency W")
ylabel("TPG for Double Matching")
title("Single Matching Equalizer with 2-reactive load
components")
```

Program Listing 10.17   Generation of denominator polynomial
gm = gm_Cheby(n,eps_sq) with Tmax = 1

```
function gm=gm_Cheby(n,eps_sq)
% For monotone roll-off Chebyshev filters
% Generation of gm(p) explicitly.
% Inputs:
%        n: Degree of the Chebyshev polynomial
%        eps_sq: Ripple coefficients of the Chebyshev gain
% Note that for filter F11m=hm(p)/gm(p)

eps=sqrt(eps_sq);
beta=asinh(1/eps);
n2=2*n;
i=1;
j=sqrt(-1);
for k=0:(n2-1)
        x(k+1)=k;
sigma(k+1)=sin(pi*(2*k+1)/2/n)*sinh(beta/n);
omega(k+1)=cos(pi*(2*k+1)/2/n)*cosh(beta/n);
        if(sigma(k+1)<0)
            z(i)=sigma(k+1)+j*omega(k+1);
            i=i+1;
end
end
gm=2^(n-1)*eps*real(poly(z));
```

Program Listing 10.18   Generation of denominator polynomial `gm(p)` with `Tmax<1`.
Note that $K^2 = \frac{1}{T_{max}} - 1$

```
function gm=gm_ChebywithK(n,eps_sq,Tmax)
% Computation of hm(p)when Tmax is specified
% to design Chebyshev Filters
% Design of Monotone roll-off Chebyshev filters
% Note that for filter F11m=hm(p)/gm(p)
% Inputs:
%        n: Degree of the Chebyshev polynomial
%        eps^2: Ripple Coefficient
%        T0: Constant Gain level
     K_sq=1/Tmax-1;
     K=sqrt(K_sq);
eps=sqrt(eps_sq);
beta=asinh(sqrt(1+K*K)/eps);
n2=2*n;
i=1;
j=sqrt(-1);
for k=0:(n2-1)
     x(k+1)=k;
sigma(k+1)=sin(pi*(2*k+1)/2/n)*sinh(beta/n);
omega(k+1)=cos(pi*(2*k+1)/2/n)*cosh(beta/n);
     if(sigma(k+1)<0)
         z(i)=sigma(k+1)+j*omega(k+1);
         i=i+1;
     end
end
gm=(2^(n-1)*eps)*real(poly(z));
```

Program Listing 10.19   Generation of numerator polynomial `hm(p)` with `Tmax<1`. Note
that $K^2 = \frac{1}{T_{max}} - 1$

```
function hm=hm_ChebywithK(n,eps_sq,Tmax)
% Computation of hm(p)when Tmax is specified
% to design Chebyshev Filters
% Design of Monotone roll-off Chebyshev filters
% Note that for filter F11m=hm(p)/gm(p)
% Inputs:
%        n: Degree of the Chebyshev
% polynomial
% eps^2: Ripple Coefficient
% T0: Constant Gain level
```

Program Listing 10.19   (Continued)

```
if Tmax<1
    K_sq=1/Tmax-1;
    K=sqrt(K_sq);
eps=sqrt(eps_sq);
beta=asinh(K/eps);
n2=2*n;
i=1;
j=sqrt(-1);
for k=0:(n2-1)
    x(k+1)=k;
sigma(k+1)=sin(pi*(2*k+1)/2/n)*sinh(beta/n);
omega(k+1)=cos(pi*(2*k+1)/2/n)*cosh(beta/n);
    if(sigma(k+1)<0)
        z(i)=sigma(k+1)+j*omega(k+1);
        i=i+1;
    end
end
hm=(2^(n-1)*eps)*real(poly(z));
end
if Tmax==1
    hm=hm_Cheby(n,eps_sq);
end
```

Program Listing 10.20   Generation of hm with Tmax = 1

```
function hm=hm_Cheby(n,eps_sq)
% Generation of Chebyshev gain for T0=1 Case
% Inputs:
%       n: Degree of Chebyshev polynomial
%       eps_sq: Ripple Coefficient
% Generation of roots of Hm(p)using Equation(10.48)
%
% Generate roots of eps^2*Tn explicitly.
j=sqrt(-1);
eps=sqrt(eps_sq);
n2=2*n;
%
        for k=0:(n2-1)
          x(k+1)=k;
              omega(k+1)=cos(pi*(2*k+1)/2/n);
              sigma(k+1)=0;
end
```

```
        for k=1:n
            z(k)=j*omega(k);
        end
   hm=2^(n-1)*eps*real(poly(z));
```

Program Listing 10.21   The function

$[hm,gm] = Levy\_GainwithK(n,sign,Tmax,eps\_sq)$. This function computes TPG of the filter with Tmax$<$1 directly from the given filter parameters n, sign, Tmax, eps_sq. Then it generates hm and gm and synthesizes the driving point impedance $Z_{in}(p) = \frac{g_m(p)+h_m(p)}{g_m(p)-h_m(p)}$ which in turn yields the element values of the filter. This program is used to check the results of single and double matching problems obtained using Levy's formulas

```
function
[hm,gm,CV]=Levy_GainwithK(n,sign,Tmax,eps_sq)
% Function Levy_Gain.m
% Construction of Monotone Roll-Off Chebyshev filters
% employing the form
% T(w)=Tmax/[1+eps_sq*Tn^2(w)].
% Using the above forms we obtain
% gm(p) and hm(p) then, synthesize of F11m(p)
% F11(p)=hm(p)/gm(p)
% Inputs:
%                Tmax, eps^2,n, sign of hm(p)
%     Note: if sign=-1 filter starts with C
%                if sign=+1 filter starts with an inductor
% Outputs:
%     For direct method:x(r); explicit element values
%                gm(p), hm(p), Hm, Gm, Fm,
%                CV: Components Values for the filter by CFE
% Note that roots of hm and gm are found explicitly
eps=sqrt(eps_sq);
% Step 1: Generation of gm(p)).
%gm=gm_Cheby(n,eps_sq);
gm=gm_ChebywithK(n,eps_sq,Tmax);
%
%Step 2: Generation of hm(p)
%hm=hm_ChebywithTmax(n,eps_sq,Tmax);
hm=hm_ChebywithK(n,eps_sq,Tmax);
hm=sign*hm;% h(p) with eps*Tn(1)
%
% Step 4: Generation of F21m(p)
hm_=paraconj(hm);
gm_=paraconj(gm);
Hm=conv(hm,hm_);
```

Program Listing 10.21   (Continued)

```
Gm=conv(gm,gm_);
Fm=Gm-Hm;
% Generation of Zin(p)
Nin=gm+hm;
Din=gm-hm;
%
% Synthesis of Zin(p)
CV=general_synthesis(hm,gm);
% Step 3: Plot the gain performance of the filter
W1=0;
W2=2;
N=1000;
DW=(W2-W1)/(N-1);
W=0;

for j=1:N
    p=sqrt(-1)*W;
    Fmval=polyval(Fm,p);
    Gmval=polyval(Gm,p);
    WA(j)=W;
  T(j)=real(Fmval/Gmval);
  %T(j)=10*log10(T(j));
    W=W+DW;
end
figure ( )
plot(WA,T)
xlabel("Normalized Angular Frequency W")
ylabel("TPG of the filter")
title("Design of a monotone roll-off filter")
```

Program Listing 10.22   Explicit element values for Chebyshev filter

```
function x=Elements_Levy(Tmax,n,eps_sq)
% Direct Method:
% Computation of explicit element values of the Chebyshev filter
% Using Levy formulas
% R. Levy, "Explicit Formulas for Chebyshev
% impedance matching networks, filters and
%interstages,
% Proc. IEE, Vol.III, No.6, pp.1099-1106, June 1964.
%eps_sq=input("Enter eps_sq=")
%
% See Book by Yarman for Wiley Section 10.16
```

```
%(1/T_max )(1/eps^2 )=(1+K^2)/eps^2 =sinh^2 (na)
%((1-T_max)/T_max )(1/eps^2 )=K^2/eps^2 =sinh^2 (nb)
%xa=sinh(a)
%yb=sinh(b)
%————————————————————————————————————
a=(1/n)*asinh(sqrt((1/Tmax)/eps_sq));
b=(1/n)*asinh(sqrt((1-Tmax)/Tmax/eps_sq));
%
xa=sinh(a);
yb=sinh(b);
%x(1)=2sin(pi/2n)/(xa-yb) first element)
x(1)=2*sin(pi/2/n)/(xa-yb);
%
% x(r)*x(r+1)=4sin(pi(2r-1)/2n)sin(pi(2r+1)/2n) /
%(x^2+y^2+ sin^2
%(rpi/n)-2xycos(rpi/n) )
for r=1:(n-1)
Num=4*sin((2*r-1)*pi/2/n)*sin((2*r+1)*pi/2/n);
Denom=xa^2+yb^2+sin(r*pi/n)*sin(r*pi/n)-2*xa*yb*cos(r*pi/n);
x(r+1)=Num/Denom/x(r);
end
% Computation of the termination:
% n=odd case:
% A0=R_L=[(2-T_max)/T_max ]sqrt[([(2-T_max)/T_max
%]^2-1)]
A0=(2-Tmax)/Tmax+sqrt(((2-Tmax)/Tmax)^2-1);
% Computation of n=even case
if (n/2-fix(n/2)==0);%n=even case
    K_sq=1/Tmax-1;
    T0=Tmax/(1+K_sq+eps_sq);
  A0=(2-T0)/T0-sqrt(((2-T0)/T0)^2-1);
end;%End of n=even case
W1=0;
W2=2;
N=200;
DW=(W2-W1)/(N-1);
W=0;
for j=1:N
    WA(j)=W;
    [ZQ Z]=Lowpass_Ladder(W,x,A0);
    R=real(ZQ);
    X=imag(ZQ);
    T(j)=4*R/((R+1)*(R+1)+X*X);
        W=W+DW;
end
figure(1)
plot(WA,T)
```

Bibliography

[1] H. W. Bode, *Network Analysis and Feedback Amplifier Design*, Van Nostrand, 1945.

[2] H. J. Carlin, 'A New Approach to Gain-Bandwidth Problems', *IEEE Transactions on Circuits and Systems*, Vol. **23**, pp. 170–175, April, 1977.

[3] S. Darlington, 'Synthesis of Reactance 4-poles', *Journal of Mathematics and Physics*, Vol.**18**, pp. 257–353, 1939.

[4] R. M. Fano, 'Theoretical Limitations on the Broadband Matching of Arbitrary Impedances', *Journal of the Franklin Institute*, Vol. **249**, pp. 57–83, 1950.

[5] D. C. Youla, 'A New Theory of Broadband Matching', *IEEE Transactions on Circuit Theory*, Vol. **11**, pp. 30–50, March, 1964.

[6] B. S. Yarman, 'Broadband Matching a Complex Generator to a Complex Load', PhD thesis, Cornell University, 1982.

[7] D. C. Youla, H. J. Carlin, and B. S. Yarman, 'Double Broadband matching and the Problem of Reciprocal Reactance 2n-port Cascade Decomposition', *International Journal of Circuit Theory and Applications*, Vol.**12**, pp. 269–281, February, 1984.

[8] W. K. Chen, *Broadband Matching: Theory and Implementations*, 2nd edition, World Scientific, 1988.

[9] V. Belevitch, *Classical Network Theory*, Holden Day, 1968.

[10] P. I. Richards, 'Resistor-transmission-line Circuits', *Proceedings of the IRE*, Vol. **36**, pp. 217–220, February, 1948.

[11] H. J. Carlin and A. B. Giordano, *Network Theory: An Introduction to Reciprocal and Nonreciprocal Circuits*, Prentice Hall, 1964.

[12] H. J. Carlin and P. P. Civalleri, 'Electronic Engineering Systems Series', in *Wideband Circuit Design* ed. J. K. Fidler, CRC Press, 1998.

[13] B. S. Yarman, 'A Simplified Real Frequency Technique for Broadband Matching Complex Generator to Complex Loads', *RCA Review*, Vol. **43**, pp. 529–541, September 1982.

[14] H. J. Carlin and B. S. Yarman, 'The Double Matching Problem: Analytic and Real Frequency Solutions', *IEEE Transactions on Circuits and Systems*, Vol. **30**, pp. 15–28, January, 1983.

[15] A. Fettweis, 'Parametric Representation of Brune Functions', *International Journal of Circuit Theory and Applications*, Vol. **7**, pp. 113–119, 1979.

[16] J. Pandel and A. Fettweis, 'Broadband Matching Using Parametric Representations', *IEEE International Symponium on Circuits and Systems*, Vol. **41**, pp. 143–149, 1985.

[17] B. S. Yarman and A. Fettweis, 'Computer-aided Double Matching via Parametric Representation of Brune Functions', *IEEE Transactions on Circuits and Systems*, Vol. **37**, pp. 212–222, February, 1990.

[18] B. S. Yarman and A. K. Sharma, 'Extension of the Simplified Real Frequency Technique and a Dynamic Design Procedure for Designing Microwave Amplifiers', *IEEE International Symponium on Circuits and Systems*, Vol. **3**, pp. 1227–1230, 1984.

[19] H. J. Carlin and P. Amstutz, 'On Optimum Broadband Matching', *IEEE Transactions on Circuits and Systems*, Vol. **28**, pp. 401–405, May, 1981.

[20] H. J. Carlin and P. P. Civalleri, 'On Flat Gain with Frequency-dependent Terminations', *IEEE Transactions on Circuits and Systems*, Vol. **32**, pp. 827–839, August, 1985.

[21] B. S. Yarman, 'Modern Approaches to Broadband Matching Problems', *Proceedings of the IEE*, Vol. **132**, pp. 87–92, April, 1985.

[22] A. Aksen, 'Design of Lossless Two-ports with Lumped and Distributed Elements for Broadband Matching', PhD thesis, Ruhr University at Bochum, 1994.

[23] M. Gewertz, 'Synthesis of a Finite, Four-terminal Network from its Prescribed Driving Point Functions and Transfer Functions', *Journal of Mathematics and Physics*, Vol. **12**, pp. 1–257, 1933.

[24] J. Pandel and A. Fettweis, 'Numerical Solution to Broadband Matching Based on Parametric Representations', *Archiv elektronische Übertragung*, Vol. **41**, pp. 202–209, 1987.

[25] E. S. Kuh and J. D. Patterson, 'Design Theory of Optimum Negative-resistance Amplifiers', *Proceedings of the IRE*, Vol. **49**, No. 6, pp. 1043–1050, 1961.

[26] W. H. Ku, M. E. Mokari-Bolhassan, W. C. Petersen, A. F. Podell and B. R. Kendall, 'Microwave Octave-band GaAs-FET Amplifiers', *Proceedings of the IEEE MTT-S International Microwave Symposium, Paolo Alto*, pp. 69–72, 1975.

[27] B. S. Yarman, 'Real Frequency Broadband Matching Using Linear Programming', *RCA Review*, Vol. **43**, pp. 626–654, December, 1982.

[28] H. J. Carlin and J. J. Komiak, 'A New Method of Broadband Equalization Applied to Microwave Amplifiers', *IEEE Transactions on Microwave Theory and Techniques*, Vol. **27**, pp. 93–99, February, 1979.

[29] B. S. Yarman and H. J. Carlin, 'A Simplified Real Frequency Technique Applied to Broadband Multi-stage Microwave Amplifiers', *IEEE Transactions on Microwave Theory and Techniques*, Vol. **30**, pp. 2216–2222, December, 1982.

[30] L. Zhu, B. Wu and C. Cheng, 'Real Frequency Technique Applied to Synthesis of Broadband Matching Networks with Arbitrary Nonuniform Losses for MMIC's', *IEEE Transactions on Microwave Theory and Techniques*, Vol. **36**, pp. 1614–1620, December, 1988.

[31] P. Jarry and A. Perennec, 'Optimization of Gain and VSWR in Multistage Microwave Amplifier Using Real Frequency Method', *European Conference on Circuit Theory and Design*, Vol. **23**, pp. 203–208, September, 1987.

[32] R. Perennec, P. Soares, P. Jarry, Legaud and M. Goloubkof, 'Computer-aided Design of Hybrid and Monolithic Broadband Amplifiers for Optoelectronic Receivers', *IEEE Transactions on Microwave Theory and Techniques*, Vol. **37**, pp. 1475–1478, September, 1989.

[33] L. Zhu, C. Sheng and B. Wu, 'Lumped Lossy Circuit Synthesis and its Application in Broadband FET Amplifier Design in MMIC's', *IEEE Transactions on Microwave Theory and Techniques*, Vol. **37**, pp. 1488–1491, September, 1989.

[34] B. S. Yarman, A. Aksen and A. Fettweis, 'An Integrated Design Tool to Construct Lossless Matching Networks with Lumped and Distributed Elements', *European Conference on Circuit Theory and Design*, Vol. **3**, pp. 1280–1290, September, 1991.

[35] C. Beccari, 'Broadband Matching Using the Real Frequency Technique', *IEEE International Symposium on Circuits and Systems*, Vol. **3**, pp. 1231–1234, May, 1984.

[36] G. Pesch, 'Breitbandanpassung mit Leitungstransformatoren', PhD thesis, Technische Hochschule, Aachen, 1978.

[37] R. Pauli, 'Breitbandanpassung Reeller und Komplexer Impedanzen mit Leitungsschaltungen', Doctoral thesis, Technische Universität München, 1983.

[38] B. S. Yarman and A. Aksen, 'An Integrated Design Tool to Construct Lossless Matching Networks with Mixed Lumped and Distributed Elements', *IEEE Transactions on Circuits and Systems*, Vol. **39**, pp. 713–723, September, 1992.

[39] B. S. Yarman, A. Aksen and A. Fettweis, 'An Integrated Design Tool to Construct Lossless Two-ports with Mixed Lumped and Distributed Elements for Matching Problems', *International Symposium on Recent Advances in Microwave Techniques*, Vol. **2**, pp. 570–573, August, 1991.

[40] A. Aksen and B. S. Yarman, 'Construction of Low-pass Ladder Networks with Mixed Lumped and Distributed Elements', *European Conference on Circuit Theory and Design*, Vol.**1**, pp. 1389–1393, September, 1993.

[41] A. Aksen and B. S. Yarman, 'A Semi-analytical Procedure to Describe Lossless Two-ports with Mixed Lumped and Distributed Elements', *ISCAS'94*, Vol. **5–6**, pp. 205–208, May, 1994.

[42] A. Aksen and B. S. Yarman, 'Cascade Synthesis of Two-variable Lossless Two-port Networks of Mixed Lumped Elements and Transmission Lines: A Semi-analytic Procedure', *NDS-98, The First International Workshop on Multidimensional Systems*, Poland, July 12–14, 1998.

[43] B. S. Yarman, 'A Dynamic CAD Technique for Designing Broadband Microwave Amplifiers', *RCA Review*, Vol. **44**, pp. 551–565, December, 1983.

[44] M. R. Wohlers, 'Complex Normalization of Scattering Matrices and the Problem of Compatible Impedances', *IEEE Transactions on Circuit Theory*, Vol. **12**, pp. 528–535, December, 1965.

[45] T. M. Chien, 'A Theory of Broadband Matching of a Frequency Dependent Generator and Load', *Journal of the Franklin Institute*, Vol. **298**, pp. 181–221, September, 1974.

[46] W. K. Chen and T. Chaisrakeo, 'Explicit Formulas for the Synthesis of Optimum Bandpass Butterworth and Chebyshev Impedance-matching Networks', *IEEE Transactions on Circuits and Systems*, Vol. **CAS-27**, No. 10, pp. 928–942, October, 1980.

[47] W. K. Chen and K. G. Kourounis, 'Explicit Formulas for the Synthesis of Optimum Broadband Impedance Matching Networks II', *IEEE Transactions on Circuits and Systems*, Vol. **25**, pp. 609–620, August, 1978.

[48] Y. S. Zhu, and W. K. Chen, 'Unified Theory of Compatibility Impedances', *IEEE Transactions on Circuits and Systems*, Vol. **35**, No. 6, pp. 667–674, June, 1988.

[49] C. Satyanaryana and W. K. Chen, 'Theory of Broadband Matching and the Problem of Compatible Impedances', *Journal of the Franklin Institute*, Vol. **309**, pp. 267–280, 1980.

[50] B. S. Yarman, A. Aksen and A. Kilinç, 'An Immitance Based Tool for Modelling Passive One-port Devices by Means of Darlington Equivalents', *International Journal of Electronic Communications (AEÜ)*, Vol. **55**, No. 6, pp. 443–451, 2001.

[51] B. S. Yarman, A. Kilinç and A. Aksen, 'Immitance Data Modeling via Linear Interpolation Techniques', *ISCAS 2002, IEEE International Symposium on Circuits and Systems*, Scottsdale, Arizona, May 2002.

[52] A. Kilinç, H. Pinarbaşi, M. Şengül, B. S. Yarman, 'A Broadband Microwave Amplifier Design by Means of Immitance Based Data Modelling Tool', *IEEE Africon 02, 6th Africon Conference in Africa*, George, South Africa, 2–4 October 2002, Vol. **2**, pp. 535–540.

[53] B. S. Yarman, A. Kilinç and A. Aksen, 'Immitance Data Modelling via Linear Interpolation Techniques: A Classical Circuit Theory Approach', *International Journal of Circuit Theory and Applications*, Vol. **32**, pp. 537–563, 2004.

[54] B. S Yarman, M. Sengul and A. Kilinc, 'Design of Practical Matching Networks with Lumped Elements Via Modeling', *IEEE Transactions on Circuits and Systems*, Vol. **54**, No. 8, pp. 1829–1837, August, 2007.

[55] M. Sengul and B. S. Yarman, 'Broadband Equalizer Design with Commensurate Transmission Lines via Reflectance Modelling', *IEICE Transactions: Fundamentals*, Vol. **E91-A**, No. 12, pp. 3763–3771, December, 2008.

[56] B. S. Yarman, Design of Ultra Wideband Antenna Matching Networks Via Simplified Real Frequency Techniques, Springer-Verlag, 2008.

11

Modern Approaches to Broadband Matching Problems: Real Frequency Solutions

11.1 Introduction

In the previous chapter, the analytic theory of broadband matching was presented. The analytic theory is essential to understand the gain–bandwidth limitations of the given impedances to be matched. However, its applicability beyond simple problems is limited. By 'simple' we mean those problems of single or double matching in which the generator and load networks include at most one reactive element, either a capacitor or an inductor. For simple impedance terminations, low-pass equal ripple or flat gain prototype networks, which are obtained by employing the analytic theory, may have practical usage. On the other hand, if the numbers of elements increase in the impedance models to be matched, the theory becomes inaccessible. If it is capable of handling the problem, the resulting gain performances turn out to be suboptimal. Equalizer structures become unnecessarily complicated, and it may not even be feasible to manufacture them. Therefore, in practice, computer-aided design (CAD) techniques are preferred; commercially available programs are employed to solve the matching problems. Readily available tools are very good in analyzing and optimizing the given structures, but they do not include network synthesis procedures in the literal sense. In other words, in designing a matching network or a microwave amplifier, a topology for the matching network with good initial element values should be supplied to a commercially available package. In this respect, many CAD packages work as fine trimming tools on the element values when the circuit is practically synthesized. Usually, a simple two-element, capacitor–inductor ladder network is a practical solution for narrow-bandwidth matching problems. However, if the optimum topology of the equalizer is unknown, or if substantial bandwidth is required, the design task becomes difficult. In this case, modern CAD techniques which are known as the 'real frequency techniques' are strongly suggested for designing matching networks.

In all single and double matching real frequency algorithms, the goal is to optimize the transducer power gain (TPG) as high and flat as possible in the band of operation. In the real frequency techniques, the matching network $[E]$, generator $[G]$ and load $[L]$ networks are considered as separate entities. TPG is expressed in terms of these entities. The lossless equalizer $[E]$ is described either in terms of its driving point 'back end impedance' Z_B (or, equivalently, admittance $Y_B = 1/Z_B$) or in terms of its unit normalized scattering parameters $[E_{ij}]$. Descriptive parameters of $[E]$ are chosen as the unknowns of the problem and

Design of Ultra Wideband Power Transfer Networks   Binboga Siddik Yarman
© 2010 John Wiley & Sons, Ltd

are determined as the result of an optimization process. In this way, analytic extraction of generator and load networks is simply omitted.

In the following, firstly, the real frequency line segment technique is introduced to handle single matching problems. Then, 'direct methods' via Bode or Gewertz procedures are discussed to solve double matching problems. Finally, the simplified real frequency technique (SRFT), which is suitable for designing single and multistage microwave amplifiers, is detailed.

11.2 Real Frequency Line Segment Technique

In 1977, a numerical approach known as the 'real frequency line segment technique (RFLT)' was introduced by H. J. Carlin for the solution of single matching problems.[1] RFLT utilizes measured load data, bypassing the analytic theory. Neither the equalizer topology nor the analytic form of a transfer function are assumed. They are the result of the design method. Measured data obtained from the devices to be matched is processed directly.

For single matching problems, Carlin describes the lossless equalizer in terms of its minimum driving point input immittance $F_B(j\omega)=R_B(\omega)+jX_B(\omega)$ at the backend as shown Figure 11.1.

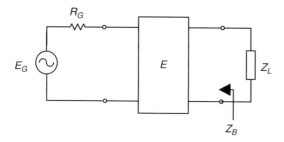

Figure 11.1 Single matching problem: power transfer from a resistive generator to a complex load

In Carlin's approach $F_B(j\omega)=R_B(\omega)+jX_B(\omega)$ is assumed to be a minimum function.[2] This assumption does not disturb the general form of the driving point immittance since we can always add a Foster part to it as mentioned in Chapter 1.

The real part $R_B(\omega)$ is chosen as the unknown of the matching problem. The imaginary part $X_B(\omega)$ is generated from $R_B(\omega)$ using the Hilbert transformation relation.

Furthermore, $R_B(\omega)$ is piecewise linearized as shown in Figure 11.2. In other words, $R_B(\omega)$ is expressed as the linear combination of line segments as follows:

$$R_B(\omega) = \begin{cases} a_j\omega + b_j & \text{for} \quad \omega_j < \omega < \omega_{j+1}, j = 1, 2, \ldots, (N-1) \\ 0 & \omega \geq \omega_N \end{cases}$$

where

$$a_j = \frac{R_j - R_{j+1}}{\omega_j - \omega_{j+1}} = \frac{\Delta R_j}{\omega_{j+1} - \omega_j} \quad \text{and} \quad b_j = \frac{\omega_j(R_{j+1}) - \omega_{j+1}(R_j)}{\omega_j - \omega_{j+1}}$$

$$\Delta R_j = R_j - R_{j+1}$$

(11.1)

[1] H. J. Carlin, 'A New Approach to Gain-Bandwidth Problems', *IEEE Transactions on Circuits and Systems*, Vol. 23, pp. 170–175, April, 1977.
[2] If $F(p)$ is selected as an impedance function, then it is minimum reactance. Otherwise, it is minimum susceptance.

In Equation (11.1) ω_j is called the break frequency, where R_B takes the value $R_j = R_B(\omega_j)$. Therefore, R_j is called the break point, ΔR_j is the excursion of the jth segment and N designates the number of break points.

From the numerical Hilbert transformation relation, $X_B(\omega)$ is given as in Equation (1.39):

$$X_B(\omega) = \sum_{j=1}^{N-1} B_j(\omega) \Delta R_j \qquad (11.2)$$

where

$$B_j(\omega) = \frac{1}{\pi(\omega_j - \omega_{j+1})} [F_{j+1}(\omega) - F_j(\omega)] \qquad (11.3)$$

and

$$F_j(\omega) = (\omega + \omega_j) \, ln(|\omega + \omega_j|) + (\omega - \omega_j) \, ln(|\omega - \omega_j|) \qquad (11.4)$$

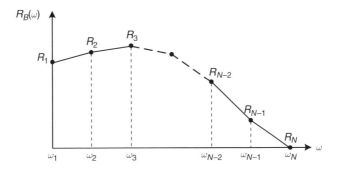

Figure 11.2 Piecewise linear approximation of $R_B(\omega)$

The immittance function $F_B(j\omega)$ can be selected as either minimum reactance or minimum susceptance depending on the variation of the load.

For example, if the load data is capacitive over the band of operation, then, at the backend, the equalizer demands an inductive driving point impedance so that possible reactance cancellation occurs to maximize TPG. In this case, the designer should rather start with a minimum susceptance admittance which avoids capacitive input. On the other hand, if the load is inductive, then the designer may prefer to start with minimum reactance impedance for the above-mentioned reasons.

Let the measured load immittance be $F_L(j\omega) = R_L(\omega) + jX_L(\omega)$; then the transducer gain $T(\omega)$ is given by

$$T(\omega) = \frac{4R_B(\omega)R_L(\omega)}{[R_B(\omega) + R_L(\omega)]^2 + [X_B(\omega) + X_L(\omega)]^2} \qquad (11.5)$$

It should be emphasized that, in Equation (11.5), the form of TPG is generic in terms of F_B and F_L. In other words, the pair $\{F_B, F_L\}$ is either impedance or admittance, preserving the form of Equation (11.5).

The ultimate goal of the matching problem is to maximize TPG as high and as flat as possible. Let the physically attainable flat value of the gain be T_0. This level may be achievable by means of an optimization over the passband $\omega_{c1} \leq \omega \leq \omega_{c2}$.

Then, loosely speaking, we can state that[3]

$$T_0 \approx \frac{4R_B(\omega)R_L(\omega)}{\left[R_B(\omega) + R_L(\omega)^2\right] + [X_B(\omega) + X_L(\omega)]^2} \geq 0, \quad \forall\, \omega$$

or

$$\boxed{\begin{aligned} T_0\Big\{[R_B(\omega) + R_L(\omega)]^2 + [X_B(\omega) + X_L(\omega)]^2\Big\} \\ -4R_B(\omega)R_L(\omega) \approx 0 \end{aligned}}$$

(11.6)

By selecting a flat gain level T_0, taking $R_B(\omega)$ as the unknown of the matching problem as specified by Equation (11.1), and generating $X_B(\omega)$ as in Equation (11.2), one can always determine a reasonable spread for $R_B(\omega)$ by means of a nonlinear curve-fitting algorithm. This is the crux of Carlin's RFLT.

The above can be summarized as follows.

In RFLT, the unknowns of the single matching problem are the break points R_j or equivalently the excursions ΔR_j.

The imaginary part $X_B(\omega)$ is generated in terms of the same break points by employing Equations (11.2) and (11.3).

Thus, the single matching problem is converted to a nonlinear curve-fitting problem with the following objective function which stems from Equation (11.6):

$$\boxed{\begin{aligned} \text{error}(\omega) &= T_0\Big\{[R_B + R_L(\omega)]^2 \\ &\quad + \left[\sum_{j=1}^{N-1} B_j(\omega)\Delta R_j + X_L(\omega)\right]\Big\} \\ &\quad - 4R_B(\omega)R_L(\omega) \\ \text{or, in a direct way,} \\ \text{error}(\omega) &= T(\omega) - T_0 \end{aligned}}$$

(11.7)

where $R_B(\omega)$ is also specified in terms of ΔR_j in a linear manner as expressed by Equations (11.1) and (11.2).

Once the break frequencies are selected by the designer, the objective function error (ω) is minimized to determine the unknown break points R_j or equivalently ΔR_j, $j=1, 2, 3, \ldots, N-1$.[4]

Clearly, the objective function error(ω) is quadratic in the unknown break points. Therefore, it is always convergent.

---

[3] The sign '$\approx$' means 'approximately equal'.

[4] It is customary to distribute the break frequencies evenly within the passband and $R_B(\omega)$ is set zero to at the stop. Therefore, for low-pass problems, the total number of break points runs up to $N-1$ since the last break point $R_N = 0$ at ω_N, for which $T(\omega_N) = 0$.

Solution to Single Matching Problem with Reactance Cancellation: Generation of Initial Values for the Nonlinear Optimization

Close examination of Equation (11.5) reveals that TPG is maximized if reactance cancellation occurs, yielding $X_B = -X_L$.[5]

Under reactance cancellation, let T_0 be the physically attainable flat gain level imposed by the theory of broadband matching. Then, the following equation must be satisfied:

$$T_0 = \frac{4R_B R_L}{(R_B + R_L)^2}$$

or

$$T_0 R_B^2 - 2R_B R_L (1 - T_0) + T_0 R_L^2 = 0$$

(11.8)

In this case, the idealized form of the real part $R_B(\omega)$ is obtained as

$$R_{B0}(\omega) = R_L(\omega) \left[\frac{(2 - T_0) + 2 \text{ sign } \sqrt{1 - T_0}}{T_0} \right] \geq 0, \quad \forall \omega$$

(11.9)

and the corresponding driving point immittance $F_{eq}(j\omega)$ of the equalizer is

$$F_{eq}(j\omega) = R_B + jX_{eq}$$

$$F_{eq}(j\omega) = R_L(\omega) \left[\frac{(2 - T_0) + 2 \text{ sign } \sqrt{1 - T_0}}{T_0} \right] - jX_L(\omega)$$

$$X_{eq} = -X_L$$

(11.10)

In general, over the entire real frequency axis, an arbitrary immittance $F_{eq}(j\omega)$ can be expressed as

$$F_{eq}(j\omega) = R_B(\omega) + jX_{eq}(\omega)$$

or

$$F_{eq}(j\omega) = F_B(j\omega) + jX_F(\omega)$$

(11.11)

where

$$X_B(\omega) = \text{Hilbert transform of} \{R_B(\omega)\}$$

$$X_F(\omega) = X_{eq}(\omega) - X_B(\omega) = -(X_L + X_B)$$

is the Foster part of the immittance such that

$$\frac{dX_F(\omega)}{d\omega} \geq 0, \quad \forall \omega$$

(11.12)

---

[5] Maximization of gain requires that $\partial T / \partial X_B = 0$, which in turn yields $X_B = -X_L$.

From an equalizer design point of view, Equation (11.12) is quite important in understanding if $F_{eq}(j\omega)$ is realizable within acceptable tolerance.

Let us investigate this situation with an example.

Example 11.1: Let the immittance $L + R\|C$ with $L=1.0, C=3, R=1$ be the load network to be matched to a resistive generator over the normalized angular frequency band $0 \leq \omega \leq 1$.[6]

(a) Select $T_0=0.75$ and generate the ideal driving point impedance $Z_{eq}(\omega)=R_B(\omega) + jX_B(\omega) + jX_F(\omega)$ over $N=7$ break points by setting $\omega_N=2$.

Note that, in Equation (11.9), R_B has two values, namely low (sigma $= -1$) and high (sigma $= +1$). Generate both high and low values of R_B and determine the corresponding values of $X_B(\omega)$ and $X_F(\omega)$. Finally, plot $\{R_B(\omega), X_B(\omega)$ and $X_F(\omega)\}$vs $\cdot \omega$.

(b) Comment on the realizability of $Z_{eq}(\omega)$ by examining the plot of the Foster part $X_F(\omega)$.

(c) Plot TPG of Equation (11.5) with the minimum reactance function $F_B(j\omega)=Z_B(j\omega)=R_B(\omega) + jX_B(\omega)$.

(d) Minimize the error function given by Equation (11.7) which results in optimum $F_B(j\omega) = R_B(\omega) + jX_B(\omega)$.

Solution:

(a) This part of the example is handled in two steps.

Step 1: In this step we develop the following MATLAB® function:

```
function WBR=Break_Frequencies(N,ws1,ws2,wc1,wc2)
```

which generates the break frequencies for specified integer N (total number of break points), ws1 (lower edge of the stopband), ws2 (upper edge of the stopband frequency), wc1 (lower edge of the passband frequency), wc2 (upper edge of the passband frequency). The first and the last break frequencies are placed at the frequencies ws1 and ws2 respectively. Then, the rests are evenly distributed over the passband (wc2-wc1). In function WBR, from a programming point of view, for low-pass problems (wc1=0 case), ws1 is chosen to be equal to wc1 to end up with WBR(1)=ws1=wc1=0.

As specified, we select $N=7$ break frequencies to sample the load impedance. The passband starts from $\omega_{c1}=0$ and ends at $\omega_{c2}=1$ Therefore, we select ws1=wc1=0. The first break point is placed at DC. The total number of $N - 1=6$ points are evenly distributed within the passband $0 \leq \omega \leq 1$. WBR(7) is placed at $\omega = 2$ Thus, the break frequencies WBR (or w_j) are given as

$$\text{WBR}= [0\ 0.2\ 0.4\ 0.6\ 0.8\ 1\ 2]$$

and

$$\text{function}[\text{RL}, \text{XL}] = \text{RLC\_Load}(w, R, C, L, \text{KFlag})$$

generates the load data at a given angular frequency ω. Obviously, for the given $L + R\|C$ load, the inputs are L=1.0, C=3, R=1.

The input KFlag controls the type of immittance. If KFlag=1, the user works with impedances. If KFlag=0 then admittances are employed throughout the program.

Then, over the break frequencies, the load impedance is generated as ZLA=RLA+JXLA such that

---

[6]Referring to Example 10.11, note that this load has no monotone ladder Chebyshev solution since $(L = 1) > (L_{max} < 0.89)$.

$$\text{RLA} = [1\ 0.73529\ 0.40984\ 0.23585\ 0.14793\ 0.1\ 0.027027]$$
$$\text{XLA} = [0\ -0.44118\ -0.4918\ -0.42453\ -0.35503\ -0.3\ -0.16216]$$

Step 2: In this step the function called

$$[\text{RBA},\text{XBA},\text{XFA}] = \text{Ideal\_SingleMatching}(\text{T0},\text{sign},\text{WBR},\text{RLA},\text{XLA})$$

generates the ideal driving point immittance as in Equations (11.10–11.12).
In this case, function XB is determined as in Chapter 1 by employing the MATLAB function

$$\text{XB} = \text{num\_hilbert}(w,\text{WBR},\text{RBA})$$

All the above functions are gathered together in the main program `Example11_1.m`. The execution of `Example11_1.m` yields the following results.
In Table 11.1, low and high values of the driving point impedance and the corresponding Foster parts are listed.
Plots of these quantities are depicted in Figures 11.3–11.5.

Table 11.1 Generation of idealized driving point impedance using function `[RBA,XBA,XFA]=` `Ideal_SingleMatching(T0,sign,WBR,RLA,XLA)`

WBR	R_{B-low}	X_{B-low}	R_{B-low}	X_{B-high}	X_{F-low}	X_{F-high}
0	0.333 3	0	3	0	0	0
0.2	0.24 51	−0.130 81	2.205 9	−1.177 3	0.371 98	1.418 4
0.4	0.136 61	−0.160 16	1.229 5	−1.441 4	0.251 96	1.533 2
0.6	0.078 616	−0.139 96	0.707 55	−1.259 7	−0.035 507	1.084 2
0.8	0.049 31	−0.115 94	0.443 79	−1.043 5	−0.329 03	0.598 5
1	0.033 333	−0.095 18	0.3	−0.856 62	−0.604 82	0.156 62
2	0.009 009	−0.051 877	0.081 081	−0.466 89	−1.786	−1.370 9

Thus, part (a) of the example is completed.
(b) Close examination of the $X_F(\omega)$vs · ω plot indicates that, at higher frequencies $(\omega > 0.4)$, $\partial X_F/\partial \omega < 0$. Therefore, the ideal solution does not exist. Nevertheless, at the lower frequencies $(\omega < 0.4)$, $\partial X_F/\partial \omega > 0$. So, it may be possible to obtain a reasonable approximation for the Foster part at lower frequencies.
As we can see from Figure 11.5, lower values of R_B result in a smaller Foster part. In this case, we expect to obtain a reasonable initial guess for the minimization of the error function specified by Equation (11.7).
(c) For this part of the example, the gain function of Equation (11.5) is programmed under the MATLAB function

$$[\text{Gain\_SM}] = \text{gain\_singleMatching}(\text{WBR},\text{RLA},\text{XLA},\text{RBA},\text{XBA})$$

and placed in the main program `Example11_1.m` so that we can plot the gain with high and low values of R_B. The result is depicted in Figure 11.6.
As pointed out in part (b), up to $\omega = 0.8$, lower values of R_B result in better gain approximation. However, for $0.8 \leq \omega \leq 1$, higher values of R_B yield higher gain.

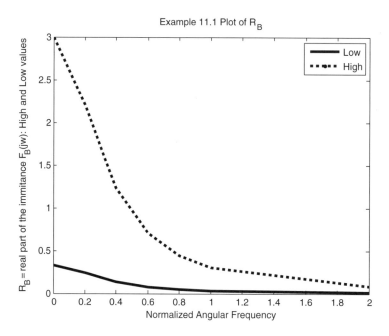

Figure 11.3   High and low values of the real part R_B of Equation (11.9)

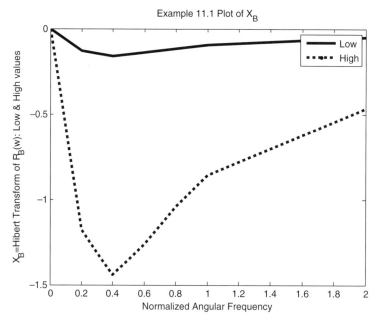

Figure 11.4   High and low values of $X_B = $ Hilbert transform of $\{R_B\}$

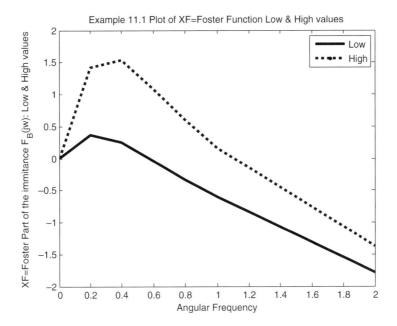

Figure 11.5 High and low values of the Foster part $X_F(\omega) = -(X_L + X_B)$

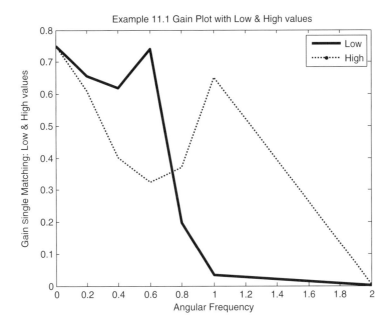

Figure 11.6 Initial gain plot for the single matching problem introduced in Example 11.1

Gain optimization for RFLT:

(d) In this part of the example, the error function given by Equation (11.7) is minimized using the Levenberg–Marquardt algorithm, which is called as a MATLAB function block as

```
OPTIONS = OPTIMSET('MaxFunEvals',20000,'MaxIter',50000);
x = lsqnonlin('error_RFLT',x0[],[],OPTIONS,WBR,wc1,wc2,N,T0,R,C,L,KFlag)
```

there, `'error_RFLT'` is a user-supplied function which programs the error function for RFLT. At this point we should note that, in the main program, the optimization is non linear; it requires an excellent initial guess. For the least-squares nonlinear optimization (`lsqnonlin`), the initial non-negative break points are generated by a MATLAB function called

$$RBA = initials(R,C,L,KFlag,T0,sign,WBR)$$

as in Equation (11.7). Furthermore, all the break points must be non-negative. This simple constraint can easily be achieved by defining the unknown vector x as

$$RBA(j)=x^2(j) \geq 0$$

or the initial vector $x0$ is given by

$$x0(j)=\mp\sqrt{RBA(j)}$$ (11.13)

where index j runs from 1 to $N-1$.

Execution of the program `Example11_1.m` optimizes TPG which in turn results in the optimum minimum reactance driving point impedance $Z_B(j\omega)=R_B(\omega)+jX_B(\omega)$ in the form of break points.

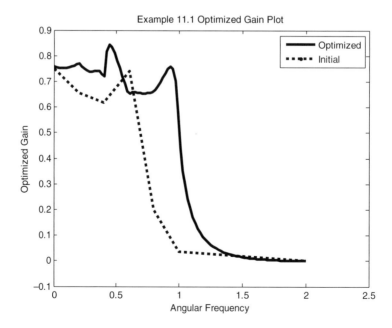

Figure 11.7 Optimized gain performance of Example 11.1

The optimized gain is depicted in Figure 11.7. In Figure (11.8), the optimized break points are given as

$$\text{RBA}=[0.34077\ 0.30775\ 0.12076\ 0.86689\ 0.55783\ 0.1635\ 0]$$

and plotted in Figure 11.8.

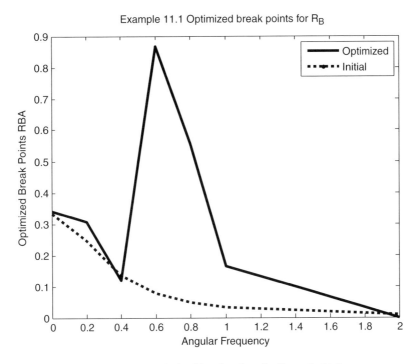

Figure 11.8 Optimized break points for Example 11.1

Recall that, in Chapter 10, for the R‖C + L (with R = 1, C = 3) type of load, we investigated the upper bound imposed on the value of inductor L to obtain a low-pass ladder equalizer. There, it was found that L < 0.89.

In other words, there is no simple analytic solution for $(R = 1)\|(C = 3) + (L = 1)$. However, one may obtain a reasonable ladder solution when the minimum reactance driving impedance data is modeled analytically. In fact, this is the case, and it will be shown in the following section. Furthermore, the RFLT solution can be improved by shifting the location of last break frequency toward wc2 and also by increasing the total number of break points.

The MATLAB main program and related functions developed for Example 11.1 are as follows.

Program Listing 11.1 Main program for Example 11.1

```
% Main Program Example11_1
% Solution to Single Matching Problem
% via Real Frequency Line Segment Technique
% Solution to Example (11.1)
```

Program Listing 11.1   (Continued)

```
% For this problem load is given as R//C+L load.
clear
clc
% Inputs:
N=input('Enter total number of break points N=')%Total Number of
Break Points
R=input('L+R//C Load: Enter resistance of the load R=')
C=input('Enter Capacitor value of the load C=')
L=input('Enter Inductor value of the load L=')
wc1=input('Enter lower edge of the passband wc1=')
wc2=input('Enter upper edge of the passband wc2=')
ws1=input('Enter lower edge of the stop band;(ws1=0 for Lowpass
problems) ws1=')
ws2=input('Enter upper edge of the stop band ws2=')
T0=input('Enter ideal flat gain level T0=')
KFlag=input('Enter KFlag,for Z=>1; for Y=>0; KFlag=')
%
%- - - - - - - - - - - - - - - - - - - - - - - - - - - - - - - - - - - -
%
% Step 1: Generate the load data
WBR=Break_Frequencies(N,ws1,ws2,wc1,wc2);
for i=1:N
    w=WBR(i);
    [RL,XL]=RLC_Load(w,R,C,L,KFlag);
    RLA(i)=RL;XLA(i)=XL;
end
% Step 2: Generate the low
%and the high values of RB,XB and XF
sign=-1;% Low Values
[RBA_Low,XBA_Low,XFA_Low]=Ideal_SingleMatching(T0,sign,WBR,RLA,X
LA);
sign=+1;% High Values
[RBA_High,XBA_High,XFA_High]=Ideal_SingleMatching(T0,sign,WBR,RL
A,XLA);
%
figure(1)
plot(WBR,RBA_Low, WBR,RBA_High)
xlabel('Normalized Angular Frequency')
ylabel('R_B=real part of the immittance F_B(jw): High and Low
values')
title('Example 11.1 Plot of R_B')
legend('Low','High')
%
figure(2)
plot(WBR,XBA_Low, WBR, XBA_High)
xlabel('Normalized Angular Frequency')
ylabel('X_B=Hilbert Transform of R_B(w): Low & High values')
```

```
title('Example 11.1 Plot of X_B')
legend('Low','High')
%
figure(3)
plot(WBR,XFA_Low, WBR,XFA_High)
xlabel('Angular Frequency')
ylabel('XF=Foster Part of the immittance F_B(jw): Low & High
values')
title('Example 11.1 Plot of XF=Foster Function Low & High
values')
legend('Low','High')
%- - - - - - - - - - - - - - - - - - - - - - - - - - - - - - - - - - -
% Part (b) of Example (11.1)
Gain_Low=gain_singleMatching(WBR,RLA,XLA,RBA_Low,XBA_Low)
Gain_High=gain_singleMatching(WBR,RLA,XLA,RBA_High,XBA_High)
%
figure(4)
plot(WBR,Gain_Low, WBR,Gain_High)
xlabel('Angular Frequency')
ylabel('Gain_Single Matching: Low & High values')
title('Example 11.1 Gain Plot with Low & High values')
legend('Low','High')
% Part (c) of Example (11.1):
%    Optimization of the Driving Point Immittance
%
%- - - - - - - - - - - - - - - - - - - - - - - - - - - - - - - - - - -
% Optimization using Levenberg Marquardt algorithm
% Generate Initials
sign=input('=>1 high values; =>-1 low values. Enter sign=')
RB0=initials(R,C,L,KFlag,T0,sign,WBR);
%
% Define unknowns for the optimization:
for j=1:(N-1)
        x0(j)=sqrt(RB0(j));%Initial
end
%
OPTIONS=OPTIMSET('MaxFunEvals',20000,'MaxIter',50000);
x=lsqnonlin('error_RFLT',x0,[],[],OPTIONS,WBR,wc1,wc2,N,T0,R,C,L
,KFlag);
%- - - - - -END OF OPTIMIZATION- - - - - - - - - - - - - - - - - -%
% Generate optimized driving point impedance
for j=1:N-1
   RBA(j)=x(j)^2;
end
   RBA(N)=0.0;
%- - - - - - - - - - - - - - - - - - - - - - - - - - - - - - - - -
N_Print=100;
DW=(ws2-ws1)/(N_Print-1);
w=wc1+1e-11;
```

Program Listing 11.1   (Continued)

```
NBR=length(WBR);
% Generate load and the driving point immittance
for i=1:N_Print
    WA(i)=w;
    [RL,XL]=RLC_Load(w,R,C,L,KFlag);
    RQ=line_seg(NBR,WBR,RBA,w);
    RQA(i)=RQ;
    XQ=num_hilbert(w,WBR,RBA);
    XQA(i)=XQ;
    RTSQ=(RQ+RL)*(RQ+RL);
    XTSQ=(XQ+XL)*(XQ+XL);
    TPG=4*RL*RQ/(RTSQ+XTSQ);
    TA(i)=TPG;
        w=w+DW;
end
%------------------------------------
%
%------------------------------------
figure(5)
plot(WA,TA,WBR,Gain_Low)
xlabel('Angular Frequency')
ylabel('Optimized Gain')
title('Example 11.1 Optimized Gain Plot')
legend('Optimized',' Initial')

figure(6)
plot(WBR,RBA,WBR,RBA_Low)
xlabel('Angular Frequency')
ylabel('Optimized Break Points RBA')
title('Example 11.1 Optimized Break Points for R_B')
legend('Optimized',' Initial')
```

Program Listing 11.2   Generation of break frequencies for specified inputs: N, ws1, ws2, wc1, wc2

```
function
WBR=Break_Frequencies(N,ws1,ws2,wc1,wc2)
% Generate break frequencies
% Inputs:
%    N :Total number of Break Points.
%    ws1: Lower edge of the stop band.
%    ws2: Upper edge of the stop band.
%    wc1: Left edge of the passband.
```

```
%     wc2: Upper edge of the passband.
% Outputs:
% WBR: Break Frequencies
%
% Generate Angular Break Frequencies
WBR(1)=ws1;
DW=(wc2-wc1)/(N-2);
W=WBR(1);
for j=2:(N-1)
    W=W+DW;
    WBR(j)=W;
end
WBR(N)=ws2;
```

Program Listing 11.3   Load function for Example 11.1

```
function [RL,XL]=RLC_Load(w,R,C,L,KFlag)
% This function generates the impedance data for a
given R//C load
% Inputs:
%      w,R,C,L,KFlag
% Generate the load impedance
    p=sqrt(-1)*w;
    YL=(1/R)+p*C;
    ZL=p*L+1/YL;
      if KFlag==1;% Work with impedances
        RL=real(ZL);
        XL=imag(ZL);
      end
        if KFlag==0;%Work with admittances
            RL=real(YL);
            XL=imag(YL);
        End
```

Program Listing 11.4   Ideal solution to single matching problems

```
function
[RBA,XBA,XFA]=Ideal_SingleMatching(T0,sign,WBR,RLA,XLA)
% Computation of idealized immitance for single
% matching problems
% See Chapter 11, section 11.2 Equations (11.6.7.8.9.10,11
% and 12)
% Inputs:
```

Program Listing 11.4   (Continued)

```
%        T0: Ideal flat gain level
%        WBR: Break frequencies of the load
%        RLA: Real Part of the immitance data
N=length(WBR);
for j=1:N
    RL=RLA(j);
    %RB(w)=RL(w)*((2-T0 )+2*sign*sqrt(1-T0 ))/T0
    RB=RL*((2-T0)+2*sign*sqrt(1-T0))/T0;
            %See Equation(11.9)
    RBA(j)=RB;
end
% Computation of Hilbert Transform Loop
    for j=1:N
        w=1e-11+WBR(j);
    %XBA(w)=Hilbert Transform of {RB(w)}
    XB=num_hilbert(w,WBR,RBA);
    XBA(j)=XB;
    XA(j)= - XLA(j);
    XFA(j)=XA(j)-XBA(j);
            %see Equation (11.12)
    End
```

Program Listing 11.5   User-supplied objective function for the nonlinear least-squares optimization

```
function
Func=error_RFLT(x,WBR,wc1,wc2,N,T0,R,C,L,KFlag)
% Objective function to be minimized
% It should be noted that
% number of sampling points N or N-1 must be
% much greater than the unknowns.
% RFLT for single matching problems
% Load: R//C+L
%- - - - - - - - - - - - - - - - - - - - - - - - - -
% Generate RBA from the given initials:
for j=1:N-1
    RBA(j)=x(j)^2;
end
RBA(N)=0.0;
%- - - - - - - - - - - - - - - - - - - - - - - - - -
N_opt=30;
DW=(wc2-wc1)/(N_opt-1);
w=wc1+1e-11;
```

```
NBR=length(WBR);
% Generate load and the driving point
immittance data
for i=1:N_opt
    [RL,XL]=RLC_Load(w,R,C,L,KFlag);
    RQ=line_seg(NBR,WBR,RBA,w);
    XQ=num_hilbert(w,WBR,RBA);
    RTSQ=(RQ+RL)*(RQ+RL);
    XTSQ=(XQ+XL)*(XQ+XL);
    TPG=4*RL*RQ/(RTSQ+XTSQ);
    Func(i)=TPG-T0;
    w=w+DW;
end
```

Program Listing 11.6   Initial break points for the optimization package

```
function RBA=initials(R,C,L,KFlag,T0,sign,WBR)
%function
%[RBA,XBA,XFA]=Ideal_SingleMatching(T0,sign,WBR,RLA,XLA)
% Computation of idealized immittance for single
matching
% problems
% See Chapter 11, section 11.2 Equations (11.6-11.12)

%Inputs:
%       T0: Ideal flat gain level
%       WBR: Break frequencies of the load
%       RLA: Real Part of the immitance data
N=length(WBR);
for j=1:N
    w=WBR(j);
    [RL,XL]=RLC_Load(w,R,C,L,KFlag);
    RB=RL*((2-T0)+2*sign*sqrt(1-T0))/T0;
              %See Equation (11.9)
    RBA(j)=RB;
End
```

Program Listing 11.7 Generation of ideal solution for single matching problems

```
function
[RBA,XBA,XFA]=Ideal_SingleMatching(T0,sign,WBR,RLA,XLA)
% Computation of idealized immitance for single
```

Program Listing 11.7   (Continued)

```
matching problems
% See Chapter 11, section 11.2 Equations (11.6.7.8.9.10,11
% and 12)
% Inputs:
%        T0: Ideal flat gain level
%        WBR: Break frequencies of the load
%        RLA: Real Part of the immitance data
N=length(WBR);
for j=1:N
    RL=RLA(j);
  %RB(w)=RL(w)*((2-T0 )+2*sign*sqrt(1-T0 ))/T0
    RB=RL*((2-T0)+2*sign*sqrt(1-T0))/T0;
%(See Equation (11.9))
    RBA(j)=RB;
end
% Computation of Hilbert Transform Loop
    for j=1:N
        w=1e-11+WBR(j);
  %XBA(w)=Hilbert Transform of {RB(w)}
    XB=num_hilbert(w,WBR,RBA);
    XBA(j)=XB;
    XA(j)= - XLA(j);
    XFA(j)=XA(j)-XBA(j);%see Equation (11.12)
    End
```

Effect of the Last Break Point and Total Number of Unknowns on the Gain Performance

Let us investigate this effect in the following example.

Example 11.2: By choosing ws2=WBR(N)=1.2, solve the above problem in Example 11.1 for $N = 19$ break points.

Solution: In the new case, the upper edge of the stopband frequency ws2 is shifted from ws2=2 to 1.2. In this case, power must be confined under the gain curve spread from w=0 to 1.2, which in turn increases the gain at the cut-off frequency wc2=1. Furthermore, we have a better chance to smooth the wiggles of the gain function in the passband since the total number of break points has been drastically increased.

Performance measure of the matched system: Usually, the performance of the matched system is measured by means of the gain fluctuation ΔT about the average level T_0. In this case, the gain function is expressed by

$$T(\omega) = T_0 \mp \Delta T \qquad (11.14)$$

In addition, the minimum and maximum of the passband gain are given as follows:

$$T_{min} = T_0 - \Delta T$$
$$T_{max} = T_0 + \Delta T$$

or equivalently

$$T_0 = \frac{T_{max} + T_{min}}{2}$$

and

$$\Delta T = T_0 - T_{min}$$

(11.15)

The objective of the broadband matching problem is to minimize the fluctuation while T_0 is maintained as high as possible. Therefore, in the optimization process, T_0 is provided by the user and the mean square error is minimized by employing the MATLAB optimization function `lsqnonlin`.

For the example under consideration, we run the main program `Example11_1.m` with the following inputs:

- Enter total numer of break points `N=19`.
- $L + R\|C$ load: enter resistance of the load `R=1`.
- Enter capacitor value of the load `C=3`.
- Enter inductor value of the load `L=1`.
- Enter lower edge of the passband `wc1=0`.
- Enter upper edge of the passband `wc2=1`.
- Enter lower edge of the stopband (`ws1=0` for low-pass problems) `ws1=0`.
- Enter upper edge of the stopband `ws2=1.2`.
- Enter ideal flat gain level `T0=0.75`.
- Enter `KFlag`, for `Z=1`; for `Y=>0`; `KFlag=1=>1` high values; `=>-1` low values. Enter `sign=1`.

Note that, for this example, we preferred to use initial break points with high values. Therefore, the 'sign' of Equation (11.9) is chosen as `sign=1`.

The results are given as follows.

The break points are as shown in Table 11.2 and plotted in Figure 11.9.

Table 11.2   Break frequencies and optimized break points for Example 11.2

Break frequencies WBR	Optimized break points RBA
0	3.099 1
0.058 824	1.252 2
0.117 65	0.798 5
0.176 47	0.691 04
0.235 29	0.636 39
0.294 12	0.607 63
0.352 94	0.585 88
0.411 76	0.567 66

(continued overleaf)

Table 11.2 (*continued*)

Break frequencies WBR	Optimized break points RBA
0.470 59	0.550 06
0.529 41	0.531 92
0.588 24	0.512 64
0.647 06	0.491 38
0.705 88	0.467 74
0.764 71	0.441 1
0.823 53	0.410 28
0.882 35	0.374 39
0.941 18	0.332 28
1.000 00	0.272 44
1.200 00	0.000 00

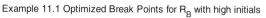

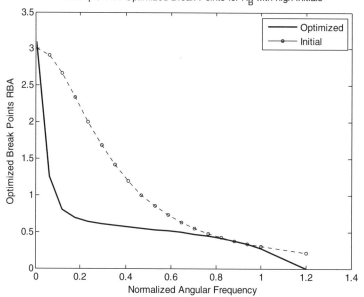

Figure 11.9   Optimized break points. RBA

The optimized gain is depicted in Figure 11.10.

Close examination of Figure 11.10 reveals that moving WBR(N) from 2 to 1.2 improves the passband gain performance of the system and by increasing the number of break points in the passband, fluctuations are drastically reduced. It is found that the average gain level is

$$T_{average} = 0.750\ 83 \mp \Delta T$$

and

$$T_{min} = 0.764\ 53 \quad T_{min} = 0.737\ 12 \quad \Delta T = 0.013\ 704$$

This is a superior result.

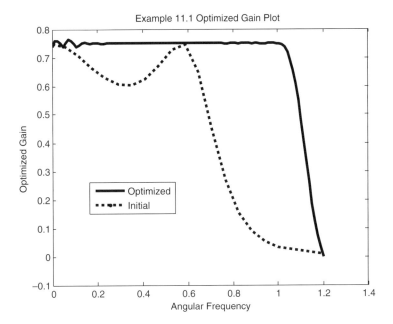

Figure 11.10 Gain variation for Example 11.2

From the theoretical point of view, we should mention that 'any immittance function $F(j\omega)=R_B(\omega)+jX_B(\omega)$ generated by the RFLT is realizable since the real part $R_B(\omega)$ is non-negative for all values of ω'. However, from the practical point of view, one needs to obtain an analytic form of the immittance. This can be easily done with the immittance modeling techniques introduced in Chapter 7 and Chapter 8 employing Bode or Gewertz methods.

Practical aspects of modeling techniques are discussed in the following section.

Practical Models for RFLT Generated Minimum Immittance Data

The first step of the RFLT single matching approach is to determine the minimum immittance data which optimizes TPG as discussed in the previous sections.

In the second step, generated immittance data is modeled as a practical positive real minimum function which can be realized as a lossless matching network. In this step, it is sufficient to build a good analytic fit to the real part and then the analytic form of the immittance is generated by employing either the Bode or Gewertz procedure.

For practical matching problems, it is proper to choose the rational form of the real part as in Equation (7.12):

$$R_B(\omega^2) = \frac{A_0\omega^{2ndc}\prod_{i=1}^{nz}(\omega_i^2-\omega^2)^2}{B_1\omega^{2n}+B_2\omega^{2(n-1)}+\ldots+B_n\omega^2+1}$$

$$= \frac{A(\omega^2)}{B(\omega^2)} \geq 0, \forall \omega$$

where

$$B(\omega^2) = \frac{1}{2}\left[c^2(\omega)+c^2(-\omega)\right] > 0$$

such that

$$c(\omega) = c_1\omega^n + c_2\omega^{(n-1)}+\ldots+c_n\omega+1$$

(11.16)

For simple cases, a simplified version of Equation (11.16) is given by

$$R_B(\omega^2) = \frac{A_0\omega^{2ndc}}{B_1\omega^{2n} + B_2\omega^{2(n-1)} + \ldots + B_n\omega^2 + 1} = \frac{A(\omega^2)}{B(\omega^2)} \geq 0, \forall\omega$$

This means that all the finite transmission zeros beyond DC are skipped.

Hence, for specified break frequencies and break points, a rational form of Equation (11.16) can be obtained as in the main program developed for Example 7.16. Details of the modeling are given in Sections 7.6 and 7.7 and in Section 8.6 for the parametric (or Bode) and Gewertz approaches.

Now, let us revise MATLAB code `Example7_16` for the following example.

Example 11.3: Referring to Example 11.2, model the optimized minimum reactance driving point impedance data which matches a $(R=1)\|(C=3) + (L=1)$ load to a resistive generator.

Solution: For this problem the data modeling program `Example7_16.m` is slightly modified to generate the gain function after data modeling. The new version of the modeling program is called `Example11_3.m`.

This program reads the generated data from the data files called

```
WBR.txt; % the break frequencies
RBA.txt; % Break points for the real part of the immitance data
XBA.txt; % Imaginary part of the immitance data
```

Inputs:

- Enter transmission zeros at DC; `ndc=0` (low-pass case).
- Enter finite transmission zeros at `[W]=[...]=0`.
- Enter `a0=1` (note that $A_0=a_0^2$).
- Enter initial values for `[c0]=[...]=[+1 +1 +1 -1 -1 +1 -1]`.
- (`ntr=1`, yes, model with transformer; `ntr=0`, no transformer): `ntr=1`.

Execution of the program yields the following results.

Coefficients of the polynomial $c(\omega)$ of Equation (11.16):

$$c = [156.55 \ -9.1399 \ -244.66 \ 34.512 \ 111.71 \ -21.195 \ -15.654]$$

which yields

$$R_B(\omega^2) = \frac{a_0^2}{B_B(\omega^2)}$$

with optimized values of

$$a_0 = 1.6299$$
$$BB = [24507 \ -76519 \ 94206 \ -57987 \ 18659 \ -2979.3 \ 202.66 \ 1]$$

Or, equivalently, in the complex p-plane,

$$R_B(p^2)$$
$$= \frac{2.6565}{-24507p^{14} - 76519p^{12} - 94206p^{10} - 57987p^8 - 18659p^6 - 2979.9p^4 - 202.66p^2 + 1}$$

From the Bode method, MATLAB function `[a,b]=RtoZ(A,B)` yields the analytic form of the positive real minimum reactance impedance function $Z_B(p)=a(p)/b(p)$ as in Table 11.3.

Table 11.3 Analytic form of the driving point impedance

$$Z_B(p) = \frac{a(p)}{b(p)}$$

$a(p)$	$b(p)$
0	1
0.377 87	0.614 5
0.232 2	1.75
0.583 6	0.777 59
0.246 09	0.868 65
0.232 53	0.228 64
0.050 719	0.105 79
0.016 969	0.006 387 9

Synthesis of the Equalizer for RFLT

The third step of RFLT is devoted to the synthesis of the driving point immittance.

Once the analytic form of the driving point impedance is generated, it is synthesized as a lossless two-port in resistive termination.

Circuit Schematic

C1= +2.65 F   L4= +2.73 H   C7= +6.126e−001 F

L2= +1.84 H   C5= +1.23 F   R8= +2.66 Ω

C3= +1.74 F   L6= +3.37 H

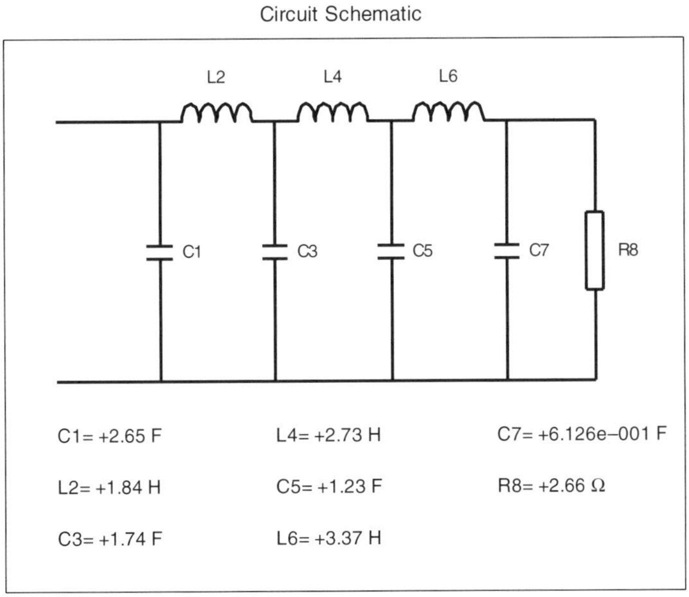

Figure 11.11 Synthesis of the equalizer as an LC ladder in resistive termination `R8=2.66`

For the problem under consideration, synthesis of the immittance can be automatically carried out using our MATLAB function[7]

```
CVal =general_synthesis(a,b)
```

For Example 11.3, the equalizer is a typical LC ladder network as shown in Figure 11.11. In this figure, it assumed that the load is connected to the left of the equalizer and the generator is on the right with termination resistance. `R8=2.66`. Obviously, `R8` can be level to $R=1\Omega$ resistance using an ideal transformer.

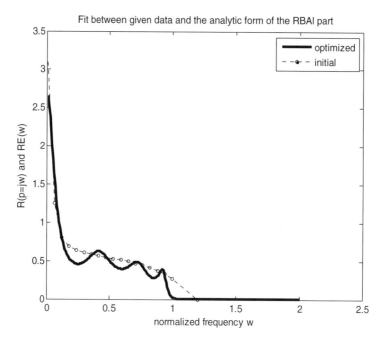

Figure 11.12 Fit between RFLT generated real part data and the rational form of the real part $R_B = (\omega)$ for Example 11.3

The corresponding data fit and gain plots are depicted in Figure 11.12, Figure 11.13 and Figure 11.14 respectively.

The performance parameters of the design are given as

$$\text{average passband gain is } T_0 = 0.785\ 47 \mp \Delta T$$

$$T_{min} = 0.631\ 04 \text{ at } wmin = 0.176$$

$$T_{max} = 0.9399 \text{ at } wmax = 0.976$$

$$\Delta T = 0.154$$

---

[7] The synthesis function was developed by Dr Ali Kilinc of Okan University, Istanbul, Turkey.

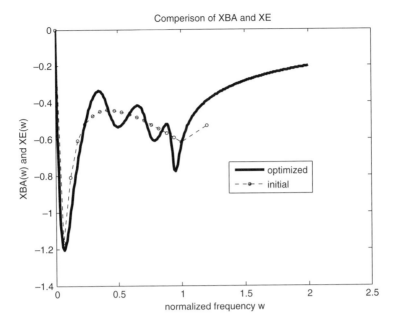

Figure 11.13 Fit between the imaginary parts of the driving point minimum reactance input impedance

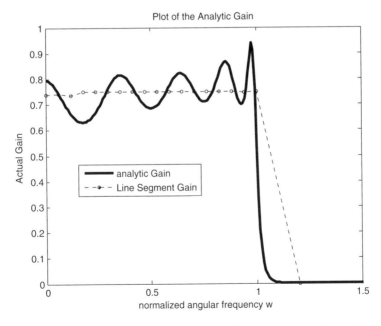

Figure 11.14 After nonlinear curve fitting, actual gain of the matched equalizer of Example 11.14 compared to the line segment gain of Example 11.3

Summary of RFLT Algorithm

RFLT consists of three major steps:

Step 1 (Generation of optimum line segments): In the first step, the lossless equalizer is described in terms of its driving point input immittance $F_B(j\omega)=R_B=JX_B(\omega)$ over the angular real frequency axis ω. Furthermore, it is assumed to be a minimum function.

The real part of this immittance is approximated by means of lines in a piecewise manner. These line segments are expressed in terms of break points $\{RBA=R_1, R_2, \ldots, R_{N-1}, R_N\}$ and user-selected break frequencies $\{\omega_1, \omega_2, \ldots, \omega_{N-1}, \omega_N\}$. Usually, the real part is assumed to be practically zero for $\omega > WBR(N)=\omega_N$. Therefore, it is meaningful to produce the corresponding imaginary part XBA by means of the Hilbert transformation.

Hence, in RFLT, the break points $\{RBA=R_1, R_2, \ldots, R_{N-1}, R_N\}$ are chosen as the unknowns of the **single matching** problem and they are determined via optimization of the TPG function.

Step 2 (Model for the line segments): In the second step, the data obtained for the real part is modeled as a non-negative rational function which in turn yields the minimum driving point immittance of the equalizer by means of the Bode (or parametric) approach or Gewertz procedure.

Step 3 (Synthesis of the single matching equalizer): Eventually, the minimum driving point immittance is synthesized as a lossless network terminated in a resistance. Later this resistance is removed by an ideal transformer to give the original generator internal resistance.

The following remarks are useful for implementation of the RFLT algorithm.

Remarks:

- The first step of the RFLT algorithm can be regarded as the numerical or ideal design of the equalizer. It provides substantial insight into the problem. In this step, the design is virtual, but theoretically it yields a realizable driving point immittance for the matching network. Therefore, this step is excellent for understanding the theoretical limits of the single matching problem by trial and error. Trials can be run for different flat gain levels T_0. In fact, one can generate a sweeping algorithm to obtain the highest level of T_0.

- For RFLT, the designer selects the break frequencies as desired. However, for low-pass designs, the passband starts from DC $(\omega_{c1}=0)$ and runs up to ω_{c2}. In this case, as a programming gimmick, we set $\omega_{s1}=\omega_{c1}=0$. In general, the choice for ω_{s1} and ω_{s2} determines the sharpness of the gain function as it rolls off at both the lower and upper edges of the passband, which in turn yields the degree of the numerator and denominator of the real part, or equivalently the number of reactive elements in the equalizer.

- For bandpass problems, at the band edges, the break frequencies and break points are fixed as follows:

$$
\begin{array}{c}
WBR(1)=\omega_{s1}, R(1)=0 \\
\text{and} \\
WBR(N)=\omega_{s2}, R(N)=0
\end{array}
$$

and the rest of the break frequencies are distributed over the passband as desired. However, we usually prefer an even distribution and they are paired with the unknown break points in the passband. Therefore, in bandpass problems, if the number of break points is N, then the total number of unknowns is specified as $N-2$.

- In the second step, which is devoted to modeling, the designer should select a proper rational form for the real part $R_B(\omega)$. The following issues are quite important in this selection:
  - The selected rational form must provide a good fit to real part data.
  - The resulting equalizer must easily be built using the technology available to the designer. In this regard, one may wish to avoid finite transmission zeros on the real frequencies. If this is the case, one should set the input vector $W=0$, which includes user fixed transmission zeros in the main program.
  - For practical bandpass problems, it may be sufficient to select the numerator polynomial as $A(p^2)=(-1)^{ndc}A_0p^{2ndc}$.
  - Eventually, the degree of the denominator polynomial n is chosen to give an acceptable fit. In the main program, the designer enters the initial coefficient for the nonlinear curve fitting which in turn yields the degree n.
- Selection of the minimum reactance or minimum susceptance driving point immittance function does not violate the generalization of RFLT since one can always add a Foster part to a minimum function, which results in a general form of the driving point immittance. In this regard, when we start with a minimum function to design the equalizer, we can always extract a Foster part from it which is also included among the unknowns in a linear manner.

  For example, let the real part $R_B(\omega)$ be

$$R_B(\omega)=\sum_{i=1}^{N} a_i(\omega)R_i$$

  Then, a practical form of the reactive part of a non-minimum immittance can be given as

$$X_{eq}(\omega)=\sum_{i=1}^{N} b_i(\omega)R_i + \text{Foster}(\omega)$$

  In the above, the simple form of the frequency-dependent Foster function is given by

$$f(\omega)R_{N+1}=\left\{\begin{matrix} \omega \\ \text{or} \\ 1 \\ \dfrac{1}{\omega} \end{matrix}\right\}R_{N+1}$$

  where R_{N+1} is included among the unknowns.
- The beauty of RFLT is the success of the gain optimization. No matter what the total numbers of break points are, one can always define an objective function that is quadratic in terms of the unknowns. Hence, nonlinear curve fitting is always convergent.[8]
- As pointed out in Sections 7.6, 7.8 and 8.8, in the course of optimization one may employ either the Bode or Gewertz procedure to generate the analytic form of the driving point immittance. However, the Bode method provides robust numerical stability. Therefore, it may be preferred over the Gewertz method.
- Our final remark is that, once the analytic form of the driving point immittance is obtained via modeling, the resulting gain performance of the matched system can be improved by re optimization of the gain function directly, on the coefficients c_i of Equation (11.16). In this case, RFLT is used as an initial solution for the direct optimization. This approach is called the 'real frequency direct method to matching problems'. It is detailed in the following section.

The main program and the related functions for Example 11.3 are as follows.

---

[8] In optimization theory, it is well established that a quadratic or a convex objective function always leads to convergent optimization.

Program Listing 11.8   Main program for Example 11.3

```
% Main Program Example11_3
clc
clear
% Nonlinear Immitance Modeling
%                 via Positive Polynomial Approach:
%
%     Direct Modeling Method
%                 via Parametric Approach
%
% Immittance F(p)=a(p)/b(p)
% Real Part of F(p)is expressed as
%                         in direct computational
% techniques R(w^2)=AA(w^2)/BB(w^2)
% Where AA(w^2) if fixed by means of
%                         transmission zeros
% BB(w^2)=(1/2)[C^2(w)+C^2(-w)] where C(w)=C(1)w^n+C(2)w^(n-
% 1)+...+C(n)w+1
% Note that AA(w^2) is computed by means of
%                 function R_num which is
% specified in p-domain as
% A(-p^2)=(a0^2)*R_Num=(ndc,W)=
%             A(1)p^2m+A(2)p^2(m-1)+...+A(m)p^2+A(m+1)
% Inputs:
% Input Set #1:
% WBR.txt; % Sampling Frequency Array to be loaded
% RBA.txt; % Real Part of the Immittance data
%                                     to be loaded
% XBA.txt; % Imaginary part of the immittance data
%                                 to be loaded
%           ndc, W(i), B1: Transmission zeros of
%               the model immittance
%         W=[...] Finite transmission zeros of
%                                     even part
% Input Set #2:
%       a0: Constant multiplier of
%                     the denominator as a0^2
%       c0; Initial coefficients of BB(w^2)
%       ntr: Model with or without transformer/
%       ntr=1> Yes Transformer. ntr=0>No transformer
% Output:
%         Optimized values of a0,
%                 and coefficients of c(p)
% Direct form of Even Part R(p^2)=A(-p^2)/B(-p^2)
% Polynomials a(p) and b(p) of Minimum function F(p)
%------------
% Step 1: Load Data to be modeled
```

```
load WBR.txt; % Sampling Frequency Array
load RBA.txt; % RBAl Part of the Immitance data
load XBA.txt; % Imaginary part of the immitance data


% Step 2: Enter zeros of the RBAl part data
ndc=input('Enter transmission zeros at DC ndc=')
W=input('Enter finite transmission zeros at [W]=[....]=')
%
a0=input('Enter a0=')
[c0]=input('Enter initials for [c0]=[....]=')%sigma(i): RBAl
roots of b(p)
ntr=input('ntr=1, Yes, model with Transformer, ntr=0 No
Transformer ntr=')
%
Nc=length(c0);
c0 =[172.53 -12.545 -266.47 39.425 119.77 -22.863 -16.164];
%
% Step 3: Generate inverse of the immittance data
NA=length(WBR);
cmplx=sqrt(-1);
for i=1:NA
    Z(i)=RBA(i)+cmplx*XBA(i);
    Y(i)=1/Z(i);
    GEA(i)=real(Y(i));
    BEA(i)=imag(Y(i));
end
% Step 4: Call nonlinear data fitting function
%%%%%%%%%% Preparetion for the optimization
OPTIONS=OPTIMSET('MaxFunEvals';20000,'MaxIter',50000);
%%%%%%%%%%%%%%%%%%%%%%%%%%%%%%%%%%%%%%%%%%%%%%%%%%%%%%
if ntr==1; x0=[c0 a0];Nx=length(x0);end;
                    %Yes transformer case
if ntr==0; x0=c0;Nx=length(x0);end;
                    %No transformer case
%
% Call optimization function lsqnonlin:
x=lsqnonlin('direct',x0,[],[],OPTIONS,ntr,ndc,W,a0,WBR,RBA,XBA);
%
% After optimization
%          separate x into sigma, alfa and beta
if ntr==1; %Yes transformer case
    for i=1:Nx-1;
    c(i)=x(i);
    end;
    a0=x(Nx);
end
if ntr==0;% No Transformer case
    for i=1:Nx;
```

Program Listing 11.8   (Continued)

```
            c(i)=x(i);
        end;
end
%
% Step 5: Generate analytic form of Fmin(p)=a(p)/b(p)
% Generate B(-p^2)
        C=[c 1];
        BB=Poly_Positive(C);
% This positive polynomial is in w-domain
        B=polarity(BB);
% Now, it is transferred to p-domain
% Generate A(-p^2) of R(-p^2)=A(-p^2)/B(-p^2)
nB=length(B);
        A=(a0*a0)*R_Num(ndc,W);
% A is specified in p-domain
nA=length(A);
if (abs(nB-nA)>0)
  A=fullvector(nB,A);
% work with equal length vectors
end


% Generation of minimum immitance function using
% Bode or Parametric method
        [a,b]=RtoZ(A,B);
% Here A and B are specified in p-domain
% Generate analytic form of Fmin(p)=a(p)/b(p)
        NA=length(WBR);
        cmplx=sqrt(-1);
% Step 6: Print and Plot results
Nprint=251;
ws1=0;ws2=2;
DW=(ws2-ws1)/(Nprint-1);
w=ws1;
Tmax=0.0;
Tmin=1.;
wc1=0;
wc2=1;
for j=1:Nprint
        WA(j)=w;
        p=cmplx*w;
        aval=polyval(a,p);
        bval=polyval(b,p);
        F=aval/bval;
        %
        RE(j)=real(F);
        XE(j)=imag(F);
```

```
            %
            RQ=RE(j);
            XQ=XE(j);
            %
            R=1;C=3;L=1;KFlag=1;
            [RL,XL]=RLC_Load(w,R,C,L,KFlag)
            %
            RTSQ=(RQ+RL)*(RQ+RL);
XTSQ=(XQ+XL)*(XQ+XL);
TPG=4*RL*RQ/(RTSQ+XTSQ);
TA(j)=TPG;
if max(TA(j))>Tmax
    wmax=WA(j);Tmax=TA(j);
end
if w>=wc1
    if w<=wc2
    if min(TA(j))<Tmin
        wmin=WA(j);Tmin=TA(j);
    end
    end
end
    w=w+DW;
end
%
% Generation of optimized line segment gain
% for comparison purpose
N=length(WBR);
for i=1:N
w=WBR(i);
KFlag=1;
[RL,XL]=RLC_Load(w,R,C,L,KFlag)
RLA(i)=RL;XLA(i)=XL;
end
    Gain_Line=gain_singleMatching(WBR,RLA,XLA,RBA,XBA);


        figure(1)
        plot(WA,RE,WBR,RBA)
        title('Fit between given data and the analytic form of
the RBA1 part')
        xlabel('normalized angular frequency w')
        ylabel('R(p=jw) and RE(w)')
        legend('optimized' ,'initial')
%
        figure(2)
        plot(WA,XE,WBR,XBA)
        title('Comparison of XBA and XE')
        xlabel('normalized frequency w')
        ylabel('XBA(w) and XE(w)')
```

Program Listing 11.8   (Continued)

```
         legend('optimized' ,'initial')
%
         figure(3)
         plot(WA,TA, WBR,Gain_Line)
         title('Plot of the Analytic Gain')
         xlabel('normalized angular frequency w')
         ylabel('Actual Gain')
         legend('analytic Gain', 'Line Segment Gain')
%
CVal=general_synthesis(a,b)
```

Program Listing 11.9   Objective function for the main program `Example11_3`

```
function
fun=direct(x0,ntr,ndc,W,a0,WA,REA,XEA)
% Modeling via Direct - Parametric approach


if ntr==1; %Yes transformer case
   Nx=length(x0);
       for i=1:Nx-1;
           c(i)=x0(i);
       end;
       a0=x0(Nx);
end
if ntr==0;% No Transformer case
    Nx=length(x0);
    for i=1:Nx;
        c(i)=x0(i);
    end;
end
%
% Step 5: Generate analytic form of
Fmin(p)=a(p)/b(p)
% Generate B(-p^2)
        C=[c 1];
        BB=Poly_Positive(C);
% This positive polynomial is in w-domain
        B=polarity(BB);
% Now, it is transferred to p-domain
% Generate A(-p^2) of R(-p^2)=A(-p^2)/B(-
p^2)
```

```
nB=length(B);
        A=(a0*a0)*R_Num(ndc,W);
% A is specified in p-domain
nA=length(A);
if (abs(nB-nA)>0)
    A=fullvector(nB,A);
% Note that function RtoZ requires
%               same length vectors A and B
end
        [a,b]=RtoZ(A,B);
% Here, A and B are specified in p-domain
%
        NA=length(WA);
        cmplx=sqrt(-1);
%
        for j=1:NA
            w=WA(j);
            p=cmplx*w;
            aval=polyval(a,p);
            bval=polyval(b,p);
            Z=aval/bval;
            RE=real(Z);
            XE=imag(Z);
            error1=abs(REA(j)-RE);
            error2=abs(XEA(j)-XE);
            fun(j)=error1+error2;
end
```

11.3 Real Frequency Direct Computational Technique for Double Matching Problems[9]

Referring to Figure 11.15 and Figure 11.16, in the real frequency direct computation technique (RFDT) TPG of the double matched system is described in terms of the driving point immittances of the generator $[Z_G$ or $Y_G]$, equalizer $[Z_B$ or $Y_B]$ and the load $[Z_L$ or $Y_L]$.

In this case, from Equation (6.76) of Chapter 6, TPG of Figure 11.6 is given by

$$T(\omega)=\frac{1-|G_{22}|^2}{|1-G_{22}S_{in}|^2}T_{[EL]} \tag{11.17}$$

[9] B. S. Yarman, "Broadband Matching a Complex Generator to a Complex Load", PhD thesis, Cornell University, 1982.
H. J. Carlin and B. S. Yarman, 'The Double Matching Problem: Analytic and Real Frequency Solutions', *IEEE Transaction on Circuits and Systems*, Vol. 30, pp. 15–28, January, 1983.

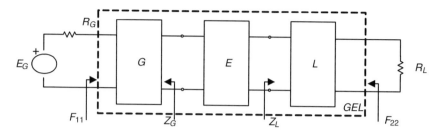

Figure 11.15   Double matching problem

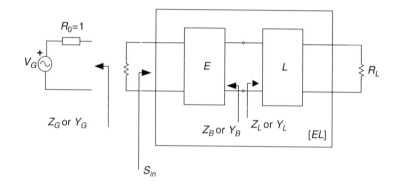

Figure 11.16   Cascaded connection of two lossless two-ports [G] and [EL]

where

$$G_{22} = \frac{Z_G - 1}{Z_G + 1} = \frac{1 - Y_G}{1 + Y_G} \qquad (11.18)$$

is the unit normalized generator reflectance.

As proven by the main theorem of Yarman and Carlin in Chapter 10, S_{in} is given by Equation (10.19) as

$$S_{in} = \eta_B(j\omega) \left[\frac{Z_L - Z_B(-j\omega)}{Z_L(j\omega) + Z_B(j\omega)} \right] \qquad (11.19)$$

which is the unit normalized input reflectance of the lossless two-port [EL].

In Equation (11.19), the all-pass function $\eta_B(p)$ is given by

$$\eta_B(p) = \left[\frac{W_B(-p)}{W_B(p)} \right]$$

Its detailed description is postponed to Equation (11.23).

It is interesting to observe that, by Equation (10.19), TPG T_{EL} of the lossless two-port [EL] is given by

$$T_{EL} = 1 - |S_{in}(j\omega)|^2 = \frac{4R_B R_L}{[R_B + R_L]^2 + [X_B + X_L]^2}$$

which is the immittance-based conventional single matching gain.

Assuming $F_B(j\omega)$ as a minimum function, Equation (11.17) is expressed as a function of the real part R_B as

$$T_G(\omega) = \left[\frac{1 - |G_{22}|^2}{|1 - G_{22}S_{in}|^2} \right] \left\{ \frac{4R_B R_L}{[R_B + R_L]^2 + [X_B + X_L]^2} \right\} \tag{11.20}$$

where $X_B(\omega) = \text{Hilbert}\{R_B(\omega)\}$.

In the above formulation, the complex generator drives the lossless two-port [EL]. Therefore, $T_G(\omega)$ is referred to as the generator-based TPG of the double matching problem. On the other hand, we can turn the matching problem the other way round, feeding the equalizer [E] with a complex generator of internal impedance Z_L while terminating it in Z_G at the front end. This type of formulation of TPG may be referred to as load based. In this case, it may be useful to make the following definitions.

Definition 1 (Generator-based TPG $T_G(\omega)$): $T_G(\omega)$ of Equation (11.20) is called the generator-based TPG of the double matched system.

Definition 2 (Load-based TPG $T_L(\omega)$): Similarly, we can define the load-based TPG of the double matched system as

$$T_L(\omega) = \left[\frac{1 - |L_{11}|^2}{|1 - L_{11}S_{in}|^2} \right] \left\{ \frac{4R_B R_G}{[R_B + R_G]^2 + [X_B + X_G]^2} \right\}$$

where

$$L_{11} = \frac{Z_L - 1}{Z_L + 1} = \frac{1 - Y_L}{1 + Y_L} \tag{11.21}$$

In this case, $F_B = R_B + jX_B$ is the driving point immittance of the resistively terminated equalizer at the frontend (or at the generator [Z_G] end).

Obviously, the generator-and load-based defined TPG must be identical. Thus,

$$T(\omega) = T_G(\omega) \equiv T_L(\omega) \tag{11.22}$$

In Equation (11.19), the fictitious function $W_B(j\omega)$ is defined on the explicit factorization of the even part $R_B(p^2)$ in a similar manner to that of Equation (10.10) or Equation (10.23) such that

$$F_B(p) = \frac{a(p)}{b(p)} R_B(p^2) = \text{even}\{F_B(p)\} = \frac{A(p^2)}{B(p^2)} = \frac{n_B(p)n_B(-p)}{b(p)b(-p)}$$

$$= W_B(p)W_B(-p)$$

where

$$W_B(-p) = \frac{n_B(-p)}{b(p)} \tag{11.23}$$

and $n_B(-p)$ includes the proper RHP and $j\omega$ axis zero of $R_B(p^2)$ as described in Chapter 10.

Thus, for the direct method of broadband matching, the unknown of the problem is the rational form of the real part R_B such that

$$R_B(\omega^2) = \frac{A_1\omega^{2m} + A_2\omega^{2(m-1)} + \ldots + A_m\omega^2 + A_{m+1}}{B_1\omega^{2n} + B_2\omega^{2(n-1)} + \ldots + B_n\omega^2 + B_{n+1}}$$

$$\geq 0 \text{ and bounded for all } \omega$$

$$\text{therefore, } n \geq m \tag{11.24}$$

In this case, the coefficients $\{A_i, B_j, i = 1, 2, \ldots, m,$ and $j = 1, 2, \ldots, n\}$ must be determined in such a way that TPG of Equation (11.20) or, equivalently, Equation (11.21) is optimized as high and as flat as possible as in RFLT of the single matching problems.

However, we should note that the general form of R_B specified by Equation (11.24) is not at all practical. It may result in complicated equalizer structures which cannot be built. Therefore, we prefer to work with a simple form of R_B which has all its zeros on the $j\omega$ axis as in Equation (11.16) such that

$$R_B(\omega^2) = \frac{A_0 \omega^{2ndc} \prod_{i=1}^{nz} (\omega_i^2 - \omega^2)^2}{B_1 \omega^{2n} + B_2 \omega^{2(n-1)} + \ldots + B_n \omega^2 + 1}$$

$$= \frac{A(\omega^2)}{B(\omega^2)} \geq 0, \forall \omega$$

where

$$B(\omega^2) = \frac{1}{2} \left[c^2(\omega) + c^2(-\omega) \right] > 0$$

and

$$c(\omega) = c_1 \omega^n + c_2 \omega^{(n-1)} + \ldots + c_n \omega + 1$$

(11.25)

From Equation (11.25), it is straightforward to see that if all the zeros of $n_B(-p)$ are on the real frequency axis, then the all-pass function $\eta_B(p) = W_B(-p)/W_B(p)$ is simplified as

$$\eta_B(p) = \frac{W_B(-p)}{W_B(p)} = (-1)^{ndc} \frac{b(-p)}{b(p)}$$

(11.26)

Once T_0 is selected, and the coefficients $\{c_j, j = 1, 2, \ldots, n\}$ and $A_0 = a_0^2 \geq 0$ of the real part $R_B(\omega)$ are initialized, we can generate the generator-based error function ε_G or equivalently the load-based error function ε_L as follows:

or

$$\varepsilon_G = \left[\frac{1 - |G_{22}|^2}{|1 - G_{22} S_{in}|^2} \right] \left\{ \frac{4 R_B R_L}{\left[R_B + R_L \right]^2 + \left[X_B + X_L \right]^2} \right\} - T_0$$

$$\varepsilon_L = \left[\frac{1 - |L_{11}|^2}{|1 - L_{11} S_{in}|^2} \right] \left\{ \frac{4 R_B R_G}{\left[R_B + R_G \right]^2 + \left[X_B + X_G \right]^2} \right\} - T_0$$

(11.27)

Then, the error function is minimized, which in turn yields the realizable driving point input immittance $F_B(p) = a(p)/b(p)$ of the lossless equalizer. Eventually, $F_B(p) = a(p)/b(p)$ is synthesized, yielding the desired lossless equalizer in resistive termination R. Finally, the resistive termination is replaced by an ideal transformer with a turns ratio $R = n^2 : 1$ which completes the design.

As is always the case, initial guesses for the nonlinear optimizations are quite important. Therefore, in the following, we will investigate the degree of nonlinearity of the double matching gain function and introduce a method to generate the initial values as in RFLT.

Investigation of the Nonlinearity of the Double Matching Gain

Close examination of Equation (11.20) reveals that, while Z_B converges to Z_L^*, if $S_{in}(j\omega)$ approaches $G_{22}(-j\omega)$, the gain of the double matched system is maximized. However, instead of assuming a perfect match, let us consider reactance cancellation at the load end. In this case, the generator-based maximum flat gain level of the double matched system may be approximated as

$$T_0 = T_{G0} \cong \left[1 - |G_{22}|^2\right]^{-1} \left\{ \frac{4R_B R_L}{[R_B + R_L]^2} \right\}$$

when $X_B = -X_L$ and $S_{in} = G_{22}^*$.

(11.28)

Definition 3 (Generator-based attainable flat gain level T_{G0}): T_{G0} is called the generator-based attainable flat gain level for the double matched system $[G]-[E]-[L]$.

Definition 4 (Load-based attainable flat gain level T_{L0}): Similarly, by Equation (11.21), we can define the load-based attainable flat gain level as

$$T_0 = T_{L0} \cong \left[1 - |L_{11}|^2\right]^{-1} \left\{ \frac{4R_B R_G}{[R_B + R_G]^2} \right\}$$

when $X_B = -X_G$ and $S_{in} = L_{11}^*$, where

$$L_{11} = \frac{Z_L - 1}{Z_L + 1} = \frac{1 - Y_L}{1 + Y_L}$$

(11.29)

Definition 5 (Generator-based real part variation $R_{GB0}(\omega)$): Using Equation (11.28), we can produce the generator-based initial variation for the real part function $R_{GB0}(\omega)$ as in RFLT as

$$R_{GB0}(\omega) = R_L(\omega) \left[\frac{\left(2 - W_{TG} + 2\,\mathrm{sign}\,\sqrt{1 - W_{TG}}\right)}{W_{TG}} \right]$$

$$\geq 0, \forall \omega$$

provided that $1 - W_{TG} \leq 1, \forall \omega$ in the passband.

(11.30)

Definition 6 (Frequency-dependent, generator-based weight factor W_{TG}): The quantity W_{TG} is called the frequency-dependent, generator-based weight factor and is given by

$$W_{TG}(\omega) = \left[1 - |G_{22}(j\omega)|^2\right] T_0 \text{ for the best case}$$

and

$$W_{TG}(\omega) = \frac{T_0}{1 - |G_{22}(j\omega)|^2} = \frac{T_0}{|G_{21}|^2} \text{ for the worst case}$$

provided that $T_0 \leq 1 - |G_{22}(j\omega)|^2, \forall \omega$ in the passband.

(11.31)

Definition 7 (Frequency-dependent, load-based weight factor W_{TL}): Similarly, the frequency-dependent, load-based weight factor can be defined as

$$W_{TL}(\omega) = \left[1 - |L_{11}(j\omega)|^2\right] T_0 \text{ for the best case}$$

and

$$W_{TL}(\omega) = \frac{T_0}{1 - |L_{11}(j\omega)|^2} = \frac{T_0}{|L_{11}|^2} \text{ for the worst case}$$

provided that $T_0 \leq 1 - |L_{11}(j\omega)|^2, \forall \omega$ in the passband

(11.32)

Definition 8 (Load-based real part variation $R_{LB0}(\omega)$): As in Equation (11.28), one can generate the load-based real part variation $R_{LB0}(\omega)$ as

$$
\boxed{
\begin{array}{c}
R_{LB0}(\omega) = R_G(\omega) \left[\dfrac{(2 - W_{TL}) + 2 \text{ sign } \sqrt{1 - W_{TL}}}{W_{TL}} \right] \\[2mm]
\geq 0, \forall \omega \\[1mm]
\text{provided that } 1 - W_{TL} \leq 1, \forall \omega \text{ in the passband}
\end{array}
}
\qquad (11.33)
$$

On the other hand, we see that the double matching gain is bounded by

$$
\boxed{
\begin{array}{c}
\left[1 - |G_{22}|^2 \right] \left\{ \dfrac{4 R_B R_L}{[R_B + R_L]^2} \right\} \leq T_G(\omega) \\[3mm]
\leq \dfrac{1}{1 - |G_{22}|^2} \left\{ \dfrac{4 R_B R_L}{[R_B + R_L]^2} \right\}
\end{array}
}
\qquad (11.34)
$$

In any case, the term $1 - |G_{22}|^2$ or its inverse can be considered as a weight factor on the single matching gain. Therefore, we say that 'the double matching gain function is quasi-quadratic in terms of the real part function R_B'.

The initial frequency variation of the unknown real part R_{B0} can be generated by means of either Equation (11.30) or Equation (11.33). Its rational approximation and corresponding analytic form for $F_{B0}(p)$ can be obtained by employing the Bode or Gewertz methods as in Chapter 7 or Chapter 8 respectively.

In light of the above explanations, we can develop a MATLAB code to solve the double matching problem by revising our previously developed single matching programs, as follows.

Algorithm for RFDT

Actually, this algorithm is straightforward. The main program and the objective function (or error function) developed for Example 11.3 can be revised to handle the double matching problems. Recall that, in Example 11.3, we generated the analytic form of the back end driving point immittance $F_B(p) = a(p)/b(p)$ to fit the given real part data RBA. In the present case, all we have to do is revise the error function to express it as the difference between the double matching gain and the flat gain level as in Equation (11.27). For this purpose, we use the rational form for the real part $R_B(\omega)$ as specified by Equation (11.25). In this case, the designer must select the transmission zeros of the equalizer as in previous examples.

Now we present the formal layout of the algorithm to generate the error function to be minimized.

Inputs:

- x_0: Initial values for the unknowns. Actually, the unknowns of the double matching problem are the coefficients a_0 and $\{c_i, i = 1, 2, \ldots, n\}$ of the rational form of $R_B(\omega)$ as specified by Equation (11.25).
- T_0: Expected flat gain level for the double matching problem. Note that T_0 may be estimated either by means of the analytic theory if the generator and load networks are simplified to a single reactive element, or by Equation (11.34).
- ω_{c1}: Beginning of the optimization frequency (lower edge of the passband).
- ω_{c2}: End of the optimization frequency (upper edge of the passband).
- KFlag: Decide whether to work with impedance or admittance functions. If impedances, set KFlag=1; if admittances, then set KFlag=0.

- `ntr`: If you wish to employ an ideal transformer in the equalizer set `ntr=1`, otherwise `ntr=0`.
- `ndc`: Transmission zeros at DC. If there is no transmission zeros at DC, set `ndc=0`. At this point we should note that if `ndc` is not equal to 0, then `ntr` must be `ntr=1` since the equalizer will be terminated in a non-unity resistance.
- `W`: This vector includes the finite transmission zeros of the real part on the real frequency axis. Usually, we do not wish to employ finite transmission zeros on the real frequency axis, therefore, for many problems, it is set to zero.
- a_0: Coefficient of the numerator of Equation (11.25). For transformerless design it is fixed in advance; otherwise, it is included among the unknowns of the double matching problem.
- Generator immittance $F_{Gen} = R_G(\omega) + jX_G(\omega)$ is supplied as the function of normalized angular frequencies.
- Load immittance $F_{Load} = R_L(\omega) + jX_L(\omega)$ is supplied as the function of the normalized angular frequencies.

Computational Steps:

Step 1: From the given inputs and the initial vector x_0, extract the coefficients c_i and a_0 to generate the realizable rational open coefficient form of the real part $R_B(p) = A(p^2)/B(p^2)$. Actually, at the input of the main program we enter the coefficients a_0 and c_i, so the initial vector x_0 is formed accordingly.

Step 2: From the generated $A(p^2)$ and $B(p^2)$, construct the minimum driving point immittance function $F_B(p)$ by employing the Bode or Gewertz procedure. In this step, we can use the previously developed MATLAB functions `[a,b]=RtoZ(A,B)` or `[a,b,x0,x,y0,y]=Gewertz(A,B)`.

Note that, as mentioned in Chapter 8, the Bode procedure is numerically more stable than the Gewertz method. Therefore, in the present MATLAB code, we prefer to use the Bode approach. However, the reader is encouraged to use our MATLAB codes with the Gewertz procedure.

Step 3: Over the passband frequencies ω_j, generate the following quantities: S_{in} as in Equation (11.19); and the double matching gain as in Equation (11.20).

Step 4: Generate the error function as in Equation (11.27) as

$$\text{Error=Double Matching Gain-T0}$$

Once the error function is developed, it is minimized using the MATLAB mean square function called `zlsqnonlin`, as in previous examples.

Now let us run an example to utilize the above algorithm for the solution of the double matching problem.

Example 11.4: Solve the double matching problem for the following specification.

Generator network: $R_G \| C_G + L_G$ with $R_G = 1, C_G = 0, L_G = 1.2$.

Load network: $R_L \| C_L + L_L$ with $R_L = 1, C_L = 3, L_L = 0.87$.

Passband: $\omega_{c1} = 0, \omega_{c2} = 1$, (low-pass design which requires `ndc=0`).

Total number of elements in the equalizer: $n = 3$.

Transmission zeros at DC: `ndc=0` (no transmission zeros at DC).

Finite transmission zeros vector: `W=[0]` (*no finite $j\omega$ transmission zero*).

Stopband frequency: `ws1=0` (generate the gain plot starting from `ws1`).

Stopband frequency: `ws2=1.5` (generate the gain function up to frequency `ws2`).

Flat Gain Level: `T0=0.85`.

Solution: For this example, we developed an error function called

$$\text{function error = direct\_RFDT}$$

with the following inputs:

`(x0,T0,wc1,wc2,KFlag,ntr,ndc,W,a0,RGen,CGen,LGen,RLoad,CLoad,LLoad)`

as outlined above. The error function is minimized by the main MATLAB program `Example11_4`.

Inputs:

Enter T0=0.85.

Enter KFlag=1 (we work with the impedance function).

Enter RGen=1; enter CGen=0; enter LGen=1.2.

Enter RLoad=1; enter CLoad=3; enter LLoad=.87.

Enter transmission zeros at DC, ndc=0 (no DC zero).

Enter finite transmission zeros at [W]=[. . .]=0 (no finite frequency zeros).

Enter a0=0.5 (first initial value).

Enter initial values for [c0]=[. . .]=[-1 -1 -1]. (This is an ad hoc choice for the initial values; as will be seen later, it works fine.)

ntr=1, yes, work with transformer; ntr=0, no transformer: ntr=1. (Here, we employ an ideal transformer to provide extra freedom for the gain optimization.)

Enter ws1=0; enter ws2=1.5. (These values are used just to plot the gain function at the end of the optimization. Therefore, they have no effect on the optimization.)

Enter wc1=0; enter wc2=1 (passband frequencies).

The results of the optimization are summarized as follows:

$a_0 = 0.647\,41$, which yields $A_0 = a_0^2 = 0.41913$. Optimized coefficients c_j are given as

$$c = [-4.7604 \quad -0.59582 \quad 1.8461]$$

$$B(\omega^2) = \frac{1}{2}\left[c^2(\omega) + c^2(-\omega)\right]$$

which in turn yields the real part as in Equation (11.25) with $p^2 = -\omega^2$ as

$$R_B(p^2) = \frac{A_0}{B(p^2)} = \frac{0.419\,13}{-22.661p^6 - 17.221p^4 - 2.2164^2 + 1}$$

By the Bode method, the backend minimum reactance driving point input impedance of the equalizer is given by

$$Z_B(p) = \frac{a(p)}{b(p)}$$

with

$$a = [0 \quad 0.2635 \quad 0.18861 \quad 0.088047]$$
$$b = [1 \quad 0.70287 \quad 0.62698 \quad 0.2100]$$

The gain performance of the double matched system is depicted in Figure 11.17.

The equalizer is constructed via synthesis of the driving point impedance:

$$Z_B(p) = \frac{a(p)}{b(p)} = \frac{0.268\,35p^2 + 0.188\,61p + 0.088\,047}{p^3 + 0.702\,87p^2 + 0.626\,98p + 0.21}$$

At the generator end, the equalizer must include an ideal transformer with turns ratio

$$n^2 = \frac{0.088}{0.21} = 0.4191$$

The complete matched system is depicted in Figure 11.18.

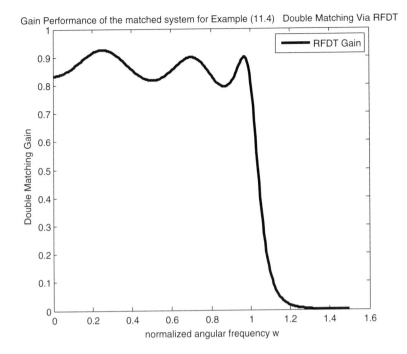

Figure 11.17 Gain plot for the matched generator and load of Example 11.4

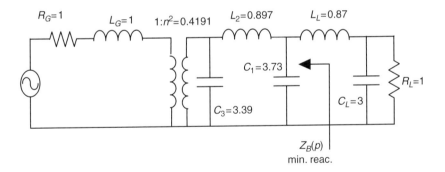

Figure 11.18 Matched generator and load networks for Example 11.4

A summary of the gain performance is as follows:

$$\text{wmax} = 0.252, \text{Tmax} = 0.92588$$
$$\text{wmin} = 0.87, \text{Tmim} = 0.79332$$
$$T_{average} = 0.8596 \mp 0.066\,284$$
$$T_0 = 0.85$$

Remarks:

- The above solution is quite good for optimization of the performance of the matched system. Note that our initial target was $T_0 = 0.85$ and we ended up with

$$T_{average} = 0.8596 \mp 0.066\ 284$$

which is close enough.
- For the sake of orientation of the reader, let us clarify the following.

Our MATLAB function `CVal=general_synthesis(a,b)` generates the circuit shown in Figure 11.19 which corresponds to the actual equalizer design with a transformer as shown in Figure 11.18.

In Figure 11.19, the load is considered on the left and the termination resistance is replaced by an ideal transformer connected to the generator network as shown in Figure 11.18.

RFDT developed for double matching problems can be employed to improve solutions obtained with the real frequency line segment technique. As an exercise, let us use it to improve the results obtained in Example 11.3.

Circuit Schematic

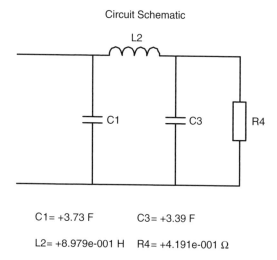

C1= +3.73 F   C3= +3.39 F

L2= +8.979e-001 H   R4= +4.191e-001 Ω

Figure 11.19 Direct synthesis of $Z_B(p) = a(p)/b(p)$ as completed by the function `CVal=general_synthesis(a,b)`

Example 11.5: Solve the single matching problem described in Example 11.3 by employing RFDT. The specifications are:

Load network: R//C + L load with $R = 1$, $CL = 3$, $LL = 1$
Passband: $0 \le \omega \le 1$.
Generator network: $RG = 1$, $CG = 0$, $LG = 0$.
Solution: For this example, we run the main program `Example11_4` with the following inputs.

Inputs:

- Enter `T0=0.9`; enter `KFlag=1`; (impedance-based design choice).
- Enter `RGen=1`; enter `CGen=0`; enter `LGen=0`.

- Enter RLoad=1; enter CLoad=3; enter LLoad=1.
- Enter transmission zeros at DC, ndc=0.
- Enter finite transmission zeros at [W]=[...]=0.
- Enter a0=1.8; enter initial values for [c0]=[...]=[1 1 1 -1 -1 1 -1].
- ntr=1, yes, model with transformer; ntr=0, no transformer: ntr=1.
- Enter ws1=0; enter ws2=1.5; enter wc1=0; enter wc2=1.

The results are given as in Table 11.4

$$a0 = 0.78419$$

Table 11.4

$$R_B = \frac{a_0^2}{\frac{1}{2}\left[c^2(\omega) + c^2(-\omega)\right]} \text{ with } s_0^2 = 0.784\,19$$

$$Z_B(p) = \frac{a(p)}{b(p)}$$

$$T_{average} = 0.847\,28 \mp 0.0958\,72$$

$c(\omega)$	$a(p)$	$b(p)$
31.851	0	1
−8.0279	0.304 86	1.174 2
−50.755	0.357 97	2.251 2
20.376	0.602 95	1.731 3
25.3	0.429 94	1.401 8
−11.388	0.285 54	0.621 25
−5.3639	0.0988 94	0.211 94
	0.0193 07	0.031 396

The gain plot is depicted in Figure 11.20.

The synthesis of $Z_B(p) = a(p)/b(p)$ is shown in Figure 11.21.

Finally, let us run an example to design a double matching equalizer by employing RFDT, which cannot be handled with the analytic theory of broadband matching.

Example 11.6: Referring to Figure 11.23 below, construct an equalizer for the double matching problem which is described as follows.

Generator network: $\{1/pC_G + pL_G\}\|R_G$ with $C_G = 2, L_G = 0.3$.
Load network: $\{R\|[[L_1]]\} + [pL_2]$ with $R = 1, L_1 = 4, L_2 = 0.75$.
Passband: wc1=0.3, wc2=1.

Solution: This double matching problem can be solved using the MATLAB main program developed for Example 11.4. However, for the present case, we should modify both the generator and the load functions to include the given specification. Hence, with the above-mentioned modifications, we run the program Example11_6.m with the following inputs.

Inputs:

- Enter T0=0.85. In this problem, it is our objective to achieve the flat gain level .85. If this does not work, then it can be lowered in the next run.

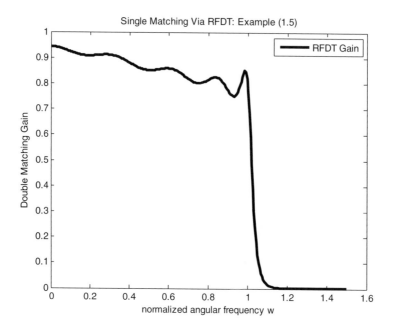

Figure 11.20   Gain performance of the single matched system of Example 11.5 via RFDT

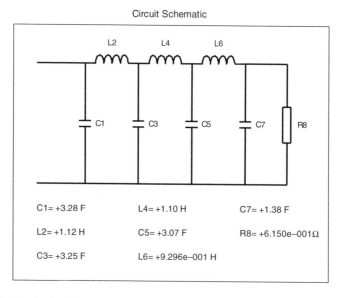

Figure 11.21   Synthesis of minimum reactance driving point impedance $Z_B(p)$ for Example 11.5

- `KFlag=0` > admittance; `KFlag=1` > impedance. Enter `KFlag=0`. In the present design, we preferred to work with admittance functions. Therefore, `KFlag` is set to `KFlag=0`
- Enter `RGen=1`; Enter `CGen=2`; enter `LGen=3`. These are the specified parameters of the generator networks.
- Enter `RLoad=1`; enter `L1=4`; enter `L2=0.75`. These are the specified parameters of the load network.
- Enter transmission zeros at DC, `ndc=2`. Since the matching problem is not low pass, then it may be proper to use DC transmission zeros in the equalizer. For the present example we preferred to employ two DC transmission zeros in [E]. Eventually, these zeros will be realized as series capacitors and/or shunt inductors.
- Enter finite transmission zeros at `[W]=[...]=0`. Here, we do not wish to employ any finite transmission zero on the real frequency axis. Therefore, we set `W=0`.
- Enter `a0=1`. This is an ad hoc initialization for the unknown `a0`.
- Enter initial values for `[c0]=[...]=[1 1 -1 -1 1]`. These are the ad hoc initializations of the unknown coefficients c_i of the real part R_B.
- `ntr=1`, yes, model with transformer; `ntr=0`, no transformer: `ntr=1`. Since we have chosen to work with `ndc=2` then it is essential to employ a transformer in the final equalizer. Therefore, `ntr` must be set to `ntr=1`.
- Enter `ws1=0`; enter `ws2=1.5`. In this program, we use `ws1` and `ws2` as the edge frequencies to generate the gain function after optimization. In the present example, it is proper to plot the TPG function over the frequency range of $0 \leq \omega \leq 1.5$. Therefore, `ws1=0` and `ws2=1.5` are chosen.
- Enter `wc1=0.3`; enter `wc2=1.0`. There are the specified frequency bands of operation.

The results of optimization are given as follows:

$$a0 = 1.8332$$
$$c = [0.55556 \quad 1.102 \quad -3.2435 \quad -6.7647 \quad 0.0044049]$$
$$a = [0 \quad 6.528 \quad 14.24 \quad 13.6597 \quad 3.8283 \quad 0]$$
$$b = [1 \quad 2.1815 \quad 6.2504 \quad 12.999 \quad 1.7207 \quad 1.8]$$

The synthesis of $Z_B(p) = a(p)/b(p)$ is shown in Figure 11.22.

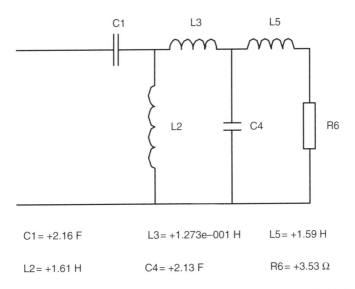

C1 = +2.16 F   L3 = +1.273e–001 H   L5 = +1.59 H

L2 = +1.61 H   C4 = +2.13 F   R6 = +3.53 Ω

Figure 11.22 Double matching problem as described by Example 11.6

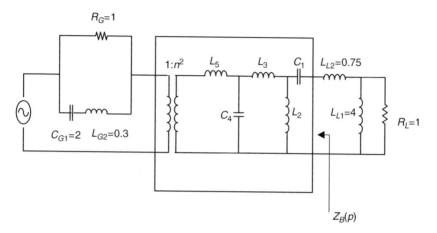

Figure 11.23 Construction of the double matched equalizer with an ideal transformer of turns ratio $1 : n^2 = \text{R6} = 3.53$

As mentioned earlier, $Z_B(p)$ is the driving point impedance of the equalizer, seen from the load end. Therefore, in Figure 11.22, the load network must be connected to the left of Z_B. On the other hand, the generator network must be placed on the resistive termination side. Therefore, by removing the termination resistor R6 by an ideal transformer of turns ratio $n^2 = \text{R6}$, we complete the equalizer design as shown in Figure 11.23.

The performance of the matched system is summarized as follows:

$$\texttt{wmax=0.336, Tmax=0.87}$$
$$\texttt{wmin=0.999, Tmin=0.641}$$
$$T_{average} = 0.75739 \mp 0.11$$
$$T_0 = 0.85$$

The gain plot of the matched system is depicted in Figure 11.24.

Obviously, the performance of the matched system can be improved by reiterating the solution or by increasing the number of elements in the equalizer.

For example, we ran the program `Example11_6` with seven elements and obtained the following solution. The ad hoc initial values were

$$a0 = 1$$
$$c0 = [-1 \ -1 \ 1 \ 1 \ -1 \ -1 \ 1]$$

The results of the optimization were

$$a0 = 1.2113$$
$$c = [-6.0379 \ -11.425 \ 12.706 \ 25.361 \ -5.3055 \ -11.866 \ 0.65382]$$
$$Z_B(p) = \frac{a(p)}{b(p)}$$
$$a = [0 \ \ 3.0465 \ \ 6.1713 \ \ 3.424 \ \ 5.4835 \ \ 0.70472 \ \ 0.73133 \ \ 0]$$
$$b = [1 \ \ 2.0257 \ \ 3.9431 \ \ 7.5108 \ \ 2.8115 \ \ 4.1226 \ \ 0.368 \ 19 \ \ 0.382 \ 09]$$

The performance summary of the equalizer is given as

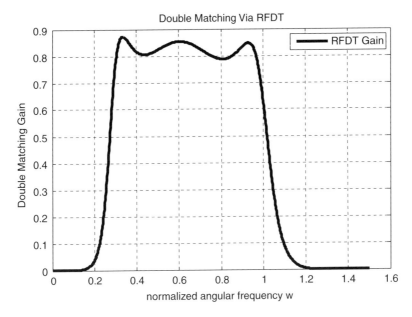

Figure 11.24 Gain performance of the double matched system described by Figure 11.23

$$\begin{array}{l} \text{wmax} = 0.3255, \ \text{Tmax} = 0.90468 \\ \text{wmin} = 0.999, \ \text{Tmin} = 0.73466 \\ T_{average} = 0.81967 \mp 0.085012 \\ T_0 = 0.82 \end{array}$$

The synthesis of the driving point impedance $Z_B(p) = a(p)/b(p)$ is shown in Figure 11.25.
   Finally, the gain performance of the new matched system is depicted in Figure 11.26.
   Clearly, the gain performance of the equalizer has improved.

Remarks:

- Example 11.6 cannot be handled with the analytic theory of broadband matching since there is no function that we can use to imbed the given generator and load networks.
- Assume that we wish to solve this double matching problem by choosing an equalizer topology with unknown element values. Then, we optimize the gain function to determine the unknown element values. First of all, how should we select the topology in advance? This is not clear. Secondly, even if we choose the topology for the equalizer, how should we initiate the element values to optimize gain? This is an ambiguous question.
- Based on our experience, we know that it is not practically possible to determine the element values of the seven element equalizer by optimization of the gain function using any commercially available design package, since the gain function of the double matched system is highly nonlinear in terms of the unknown element values.
- However, RFDT does not need to choose a transfer function in advance or require the selection of equalizer topology. These are natural consequences of RFDT.
- Therefore, we can state that the real frequency solutions to broadband matching problems are superior to other available techniques if they provide solutions; otherwise, RFDT provides a unique solution to the matching problem under consideration.

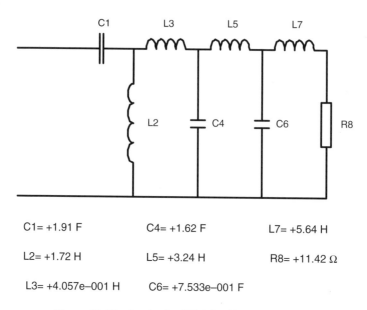

C1= +1.91 F   C4= +1.62 F   L7= +5.64 H

L2= +1.72 H   L5= +3.24 H   R8= +11.42 Ω

L3= +4.057e–001 H   C6= +7.533e–001 F

Figure 11.25 Synthesis of $Z_B(p)$ with seven elements

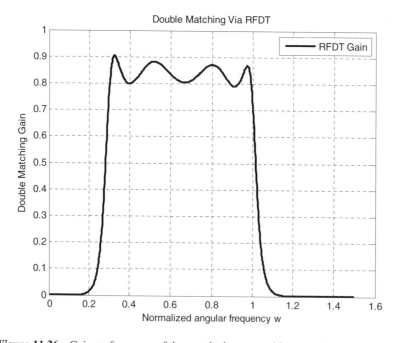

Figure 11.26 Gain performance of the matched system with seven-element equalizer

The MATLAB program lists and related functions are given below.

Program Listing 11.10   Main program for Example 11.4 to design equalizers using RFDT

```
% Main Program Example11_4
clc
clear
% -----------------------------------------------------
% Direct Computational Technique for Double Matching Problems
%
% -----------------------------------------------------
%
% Optimize the back-end Immitance F(p)=a(p)/b(p)
% Real Part of F(p)is expressed as in direct
% computational techniques R(w^2)=AA(w^2)/BB(w^2)
% see Equation (11.25)
% Where AA(w^2) if fixed via jw transmission zeros
% BB(w^2)=(1/2)[C^2(w)+C^2(-w)]
% where C(w)=C(1)w^n+C(2)w^(n-1)+...+C(n)w+1
% Note that AA(w^2) is computed by means of function
%           R_num which is
% specified in p-domain as
% A(-p^2)=(a0^2)*R_Num=(ndc,W)=A(1)p^2m+A(2)p^2(m-
% 1)+...+A(m)p^2+A(m+1)
% Inputs:
% T0=input('Enter T0=')
%         Target value of the flat gain level
% Step 1: Generator and Load Data
%----------------------------
% RGen=input('Enter RGen=')
% CGen=input('Enter CGen')
% Lgen=input('Enter LGen=')
%----------------------------
% RLoad=input('Enter RLoad=')
% CLoad=input('Enter CLoad=')
% LLoad=input('Enter Load=')
%----------------------------
% Step 2: Generic form of RB(w^2):
%         ndc, W(i), B1: Transmission zeros of the
%                      model immitance
%         W=[...] Finite transmission zeros of the
%                      even part
% Input Set #2:
% a0: Constant multiplier of the denominator as a0^2
%         c0; Initial coefficients of BB(w^2)
%         ntr: Model with or without transformer/
%         ntr=1> Yes Transformer. ntr=0>No transformer
% Output:
```

Program Listing 11.10   (Continued)

```
% Optimized values of a0, and coefficients of c(p)
% Direct form of Even Part R(p^2)=A(-p^2)/B(-p^2)
% Polynomials a(p) and b(p) of Minimum function F(p)
% -------------------------------------
% Step 1: Enter T0, Generator and Load Networks
% ----------------------------
T0=input('Enter T0=')
KFlag=input('Enter KFlag=')
%
RGen=input('Enter RGen=')
CGen=input('Enter CGen=')
LGen=input('Enter LGen=')
%
%--------------------------------
RLoad=input('Enter RLoad=')
CLoad=input('Enter CLoad=')
LLoad=input('Enter LLoad=')
%
%--------------------------------
% Step 2: Enter zeros of the RB
ndc=input('Enter transmission zeros at DC ndc=')
W=input('Enter finite transmission zeros at [W]=[....]=')
a0=input('Enter a0=')
[c0]=input('Enter initials for [c0]=[....]=')%sigma(i): RBA1
roots of b(p)
ntr=input('ntr=1, Yes, model with Transformer, ntr=0 No
Transformer ntr=')
%----------------------------
% Step 3 stop band and corner frequencies
ws1=input('Enter ws1=')
ws2=input('Enter ws2=')
wc1=input('Enter wc1=')
wc2=input('Enter wc2=')
%
%c0=[ 0.18045 0.12197 -5.183 -3.4992 3.1246]
%c0=[ -4.6046 -0.60396 1.8004];a0=0.64893;T0=0.9
Nc=length(c0);
%
% Step 4: Call nonlinear data fitting function
%%%%%%%%%% Preparation for the optimization
%
OPTIONS=OPTIMSET('MaxFunEvals',20000,'MaxIter',50000);
%
%%%%%%%%%%%%%%%%%%%%%%%%%%%%%%%%%%%%%%%%%%%%%%%%%%%%%%%

if ntr==1; x0=[c0 a0];Nx=length(x0);end;
%Yes transformer case
```

```
if ntr==0; x0=c0;Nx=length(x0);end;
%No transformer case
%
% Call optimization function lsqnonlin:
x=lsqnonlin('direct_RFDT',x0,[],[],OPTIONS,T0,wc1,wc2,KFlag,ntr,
ndc,W,a0,RGen,CGen,LGen,RLoad,CLoad,LLoad);
%
% After optimization separate x into coefficients
if ntr==1; %Yes transformer case
    for i=1:Nx-1;
    c(i)=x(i);
    end;
    a0=x(Nx);
end
if ntr==0;% No Transformer case
    for i=1:Nx;
        c(i)=x(i);
    end;
end
%
% Step 5: Generate analytic form of Fmin(p)=a(p)/b(p)
% Generate B(-p^2)
        C=[c 1];
        BB=Poly_Positive(C);
% This positive polynomial is in w-domain
        B=polarity(BB);
% Now, it is transferred to p-domain
% Generate A(-p^2) of R(-p^2)=A(-p^2)/B(-p^2)
nB=length(B);
        A=(a0*a0)*R_Num(ndc,W);
% A is specified in p-domain
nA=length(A);
if (abs(nB-nA)>0)
  A=fullvector(nB,A);
% work with equal length vectors
end
% Generation of minimum immitance function
%using Bode or Parametric method
        [a,b]=RtoZ(A,B);
% Here A and B are specified in p-domain
% Generate analytic form of Fmin(p)=a(p)/b(p)
        cmplx=sqrt(-1);
% Step 6: Print and Plot results
Nprint=251;
DW=(ws2-ws1)/(Nprint-1);
w=ws1;
Tmax=0.0;
Tmin=1.;
for j=1:Nprint
```

Program Listing 11.10   (Continued)

```
                    WA(j)=w;
                    p=cmplx*w;
                    [RG,XG]=RLC_Gen(w,RGen,CGen,LGen,KFlag);
                    [RL,XL]=RLC_Load(w,RLoad,CLoad,LLoad,KFlag);
        %
                    aval=polyval(a,p);
                    bval=polyval(b,p);
                    bval_conj=conj(bval);
        %
                    eta=(-1)^ndc*bval_conj/bval;
        %
                    ZB=aval/bval;
                    RB=real(ZB);
                    XB=imag(ZB);
        %
                    Sin=inputref_EL(KFlag,eta,RL,XL,RB,XB);
        %
                    TEL=1-abs(Sin)*abs(Sin);
                    ZG=complex(RG,XG);
                    G22=(ZG-1)/(ZG+1);
                    if KFlag==0
                        G22=-G22;
                    end
                    Weight=(1-abs(G22)*abs(G22))/abs((1-G22*Sin))^2;
                    T_Double=Weight*TEL;
                    TA(j)=T_Double;
%
% Compute the performance parameters:
%Tmax,Tmin,Tave and detT
        if max(TA(j))>Tmax
            wmax=WA(j);Tmax=TA(j);
        end
if w>=wc1
        if w<=wc2
        if min(TA(j))<Tmin
                wmin=WA(j);Tmin=TA(j);
        end
        end
end
        w=w+DW;
end

    %

    %
                figure(1)
                plot(WA,TA)
```

```
            title('Double Matching Via RFDT')
            xlabel('normalized angular frequency w')
            ylabel('Double Matching Gain')
            legend('RFDT Gain')
  %
     CVal=general_synthesis(a,b)
      % Print Performance parameters:
wmax, Tmax,
wmin,Tmin,
Tzero=(Tmax+Tmin)/2
delT=Tzero-Tmin
%
```

Program Listing 11.11   Error function for RFDT

```
function
error=direct_RFDT(x0,T0,wc1,wc2,KFlag,ntr,ndc,W,a0,RGen,CGen,LGe
n,RLoad,CLoad,LLoad)
% Modeling via Direct - Parametric approach

if ntr==1; %Yes transformer case
    Nx=length(x0);
        for i=1:Nx-1;
            c(i)=x0(i);
        end;
        a0=x0(Nx);
end
if ntr==0;% No Transformer case
    Nx=length(x0);
    for i=1:Nx;
        c(i)=x0(i);
    end;
end
%
% Step 5: Generate analytic form of Fmin(p)=a(p)/b(p)
% Generate B(-p^2)
        C=[c 1];
        BB=Poly_Positive(C);
% This positive polynomial is in w-domain
        B=polarity(BB);
% Now, it is transferred to p-domain
% Generate A(-p^2) of R(-p^2)=A(-p^2)/B(-p^2)
nB=length(B);
        A=(a0*a0)*R_Num(ndc,W);
% A is specified in p-domain
```

Program Listing 11.11   (Continued)

```
nA=length(A);
if (abs(nB-nA)>0)
    A=fullvector(nB,A);
% Note that function RtoZ requires same length
% vectors A and B
end
    [a,b]=RtoZ(A,B);
% Here A and B are specified in p-domain
%
    N_opt=21;
    w=wc1;
    del=(wc2-wc1)/(N_opt-1);
    cmplx=sqrt(-1);
%
    for j=1:N_opt
        p=cmplx*w;
        [RG,XG]=RLC_Gen(w,RGen,CGen,LGen,KFlag);
        [RL,XL]=RLC_Load(w,RLoad,CLoad,LLoad,KFlag);
%
        aval=polyval(a,p);
        bval=polyval(b,p);
        bval_conj=conj(bval);
%
        eta=(-1)^ndc*bval_conj/bval;
%
        ZB=aval/bval;
        RB=real(ZB);
        XB=imag(ZB);
%
        Sin=inputref_EL(KFlag,eta,RL,XL,RB,XB);
%
        TEL=1-abs(Sin)*abs(Sin);
        ZG=complex(RG,XG);
        G22=(ZG-1)/(ZG+1);
        if KFlag==0
            G22=-G22;
        end
        Weight=(1-abs(G22)*abs(G22))/abs((1-G22*Sin))^2;
        T_Double=Weight*TEL;
        error(j)=T_Double-T0;
        w=w+del;
    end
```

Program Listing 11.12   Function for input reflectance `Sin`

```
function
Sin=inputref_EL(KFlag,eta,RL,XL,RB,XB)
ZL=complex(RL,XL);
ZB=complex(RB,XB);
ZBC=conj(ZB);
S=(ZL-ZBC)/(ZL+ZB);
Sin=eta*S;
if KFlag==0
    Sin=-Sin;
end
```

Program Listing 11.13   Function to generate generator data

```
function
[RGen,XGen]=RLC_Gen(w,RGen,CGen,LGen,KFlag)
% This function generates the impedance data
%    for a given R//C+L load
% Inputs:
%      w,RGen,CGen,LGen,KFlag
% Generate the load impedance
   p=sqrt(-1)*w;
   YGen=(1/RGen)+p*CGen;
   ZGen=p*LGen+1/YGen;
      if KFlag==1;% Work with impedances
        RGen=real(ZGen);
        XGen=imag(ZGen);
      end
       if KFlag==0;%Work with admittances
          RGen=real(YGen);
          XGen=imag(YGen);
       end
```

Program Listing 11.14   Main Program for Example 11.6

```
% Main Program Example11_6
clc
clear
input('Example (11.6)')
input('Welcome to RFDT Double Matching Program')
input('Generator Network is (1/pC+pL)//R ')
input(' Load Network is R//pL1+pL2')
```

Program Listing 11.14   (Continued)

```
%-----------------------------------------------------------
% Direct Computational Technique for Double Matching Problems
%
% --------------------------------------------------------
%
% Optimize the back-end Immittance F(p)=a(p)/b(p)
% Real Part of F(p)is expressed as in direct
%                                  computational
% techniques R(w^2)=AA(w^2)/BB(w^2) see Equation (11. 25)
% Where AA(w^2) if fixed by means of
%                                 transmission zeros
% BB(w^2)=(1/2)[C^2(w)+C^2(-w)]
% where C(w)=C(1)w^n+C(2)w^(n-1)+...+C(n)w+1
% Note that AA(w^2) is computed by means of function
% R_num which is specified in p-domain as
% A(-p^2)=(a0^2)*R_Num=(ndc,W)=A(1)p^2m+A(2)p^2(m-
% 1)+...+A(m)p^2+A(m+1)
% Inputs:
% T0=input('Enter T0=')Target value of the flat gain
%                                     level
% Step 1: Generator and Load Data
%-------------------------------
% RGen=input('Enter RGen=')
% CGen=input('Enter CGen')
% Lgen=input('Enter LGen=')
%-------------------------------
% RLoad=input('Enter RLoad=')
% LL1=input('Enter CLoad=')
% LL2=input('Enter Load=')
%-------------------------------
% Step 2: Generic form of RB(w^2):
% ndc, W(i), B1:
% Transmission zeros of the model immittance
% W=[...] Finite transmission zeros of even part
% Input Set #2:
%         a0: Constant multiplier of the denominator as a0^2
%         c0; Initial coefficients of BB(w^2)
%         ntr: Model with or without transformer/
%         ntr=1> Yes Transformer. ntr=0>No transformer
% Output:
% Optimized values of a0, and coefficients of c(p)
% Direct form of Even Part R(p^2)=A(-p^2)/B(-p^2)
% Polynomials a(p) and b(p) of Minimum function F(p)
% ---------------------------------------------
% Step 1: Enter T0, Generator and Load Networks
% ----------------------------
T0=input('Enter T0=')
```

```
KFlag=input('KFlag=0>Admittance, KFlag=1>Impedance..Enter
KFlag=')
%
RGen=input('Enter RGen=')
CGen=input('Enter CGen=')
LGen=input('Enter LGen=')
%
%----------------------------
RLoad=input('Enter RLoad=')
LL1=input('Enter the parallel load inductor L1=')
LL2=input('Enter the series load inductor L2=')
%
%----------------------------
% Step 2: Enter zeros of the RB
ndc=input('Enter transmission zeros at DC ndc=')
W=input('Enter finite transmission zeros at [W]=[....]=')
a0=input('Enter a0=')
[c0]=input('Enter initials for [c0]=[....]=')%sigma(i): RBAl
roots of b(p)
ntr=input('ntr=1, Yes, model with Transformer, ntr=0 No
Transformer ntr=')
%----------------------------
% Step 3 stop band and corner frequencies
ws1=input('Enter ws1=')
ws2=input('Enter ws2=')
wc1=input('Enter wc1=')
wc2=input('Enter wc2=')
%
a0 =0.94437;
c0=[-3.5008 -5.8409 11.264 19.16 -5.8967 -10.624 0.9017];
   Nc=length(c0);
%
% Step 4: Call nonlinear data fitting function
%%%%%%%%%%% Preparation for the optimization %%%%%

OPTIONS=OPTIMSET('MaxFunEvals',20000,'MaxIter',50000);
%
%%%%%%%%%%%%%%%%%%%%%%%%%%%%%%%%%%%%%%%%%%%%%%%%%%%%%%%

if ntr==1; x0=[c0 a0];Nx=length(x0);end;
%Yes transformer case
if ntr==0; x0=c0;Nx=length(x0);end;
%No transformer case
%
% Call optimization function lsqnonlin:
x=lsqnonlin('direct_RFDT',x0,[],[],OPTIONS,T0,wc1,wc2,KFlag,ntr,
ndc,W,a0,RGen,CGen,LGen,RLoad,LL1,LL2);
%
% After optimization separate x into coefficients
```

Program Listing 11.14   (Continued)

```
if ntr==1; %Yes transformer case
    for i=1:Nx-1;
    c(i)=x(i);
    end;
    a0=x(Nx);
end
if ntr==0;% No Transformer case
    for i=1:Nx;
        c(i)=x(i);
    end;
end
%
% Step 5: Generate analytic form of Fmin(p)=a(p)/b(p)
% Generate B(-p^2)
        C=[c 1];
        BB=Poly_Positive(C);
% This positive polynomial is in w-domain
        B=polarity(BB);
% Now, it is transferred to p-domain
% Generate A(-p^2) of R(-p^2)=A(-p^2)/B(-p^2)
nB=length(B);
        A=(a0*a0)*R_Num(ndc,W);
% A is specified in p-domain
nA=length(A);
if (abs(nB-nA)>0)
  A=fullvector(nB,A);
% work with equal length vectors
end

% Generation of minimum immittance function using Bode
%                  or Parametric method
        [a,b]=RtoZ(A,B);
% Here A and B are specified in p-domain
% Generate analytic form of Fmin(p)=a(p)/b(p)
        cmplx=sqrt(-1);
% Step 6: Print and Plot results
Nprint=1001;
DW=(ws2-ws1)/(Nprint-1);
w=ws1;
Tmax=0.0;
Tmin=1.;
for j=1:Nprint
        WA(j)=w;
        p=cmplx*w;
        [RG,XG]=RLC_Gen(w,RGen,CGen,LGen,KFlag);
        [RL,XL]=RLC_Load(w,RLoad,LL1,LL2,KFlag);
        %
```

```
          aval=polyval(a,p);
          bval=polyval(b,p);
          bval_conj=conj(bval);
     %
          eta=(-1)^ndc*bval_conj/bval;
     %
          ZB=aval/bval;
          RB=real(ZB);
          XB=imag(ZB);
     %
          Sin=inputref_EL(KFlag,eta,RL,XL,RB,XB);
     %
          TEL=1-abs(Sin)*abs(Sin);
          ZG=complex(RG,XG);
          G22=(ZG-1)/(ZG+1);
          if KFlag==0
              G22=-G22;
          end
          Weight=(1-abs(G22)*abs(G22))/abs((1-G22*Sin))^2;
          T_Double=Weight*TEL;
          TA(j)=T_Double;
%
    % Compute the performance
%       parameters: Tmax, Tmin, Taverage and delT
    if max(TA(j))>Tmax
          wmax=WA(j);Tmax=TA(j);
    end
if w>=wc1
    if w<=wc2
    if min(TA(j))<Tmin
        wmin=WA(j);Tmin=TA(j);
    end
    end
end
    w=w+DW;
end

  %

  %

        figure(1)
        plot(WA,TA)
        title('Double Matching Via RFDT')
        xlabel('normalized angular frequency w')
        ylabel('Double Matching Gain')
        legend('RFDT Gain')
%
if KFlag==1
```

Program Listing 11.14   (Continued)

```
    CVal=general_synthesis(a,b);
end
if KFlag==0
    CVal=general_synthesis(b,a);
end
  % Print Performance parameters:
wmax, Tmax,
wmin,Tmin,
T_Average=(Tmax+Tmin)/2
delT=T_Average-Tmin
%
```

Program Listing 11.15   Generator function for Example 11.6

```
function
[RGen,XGen]=RLC_Gen(w,RGen,CGen,LGen,KFlag)
% This function generates the impedance data
% for a given R//C load
% Inputs:
%       w,RGen,CGen,LGen,KFlag
% Generate the load impedance
    p=sqrt(-1)*w;
    imp=1/p/CGen+p*LGen;
    adm=1/imp+1/RGen;
        YGen=adm;
        ZGen=1/YGen;
     if KFlag==1;% Work with impedances
        RGen=real(ZGen);
        XGen=imag(ZGen);
     end
        if KFlag==0;%Work with admittances
            RGen=real(YGen);
            XGen=imag(YGen);
        end
```

Program Listing 11.16   Load function for Example 11.6

```
function [RL,XL]=RLC_Load(w,R,LL1,LL2,KFlag)
% This function generates the impedance data
% for a given R//C load
% Inputs:
```

```
% w,R,C,L,KFlag
% Generate the load impedance
p=sqrt(-1)*w;
YL1=(1/R)+1/LL1/p;
ZL=p*LL2+1/YL1;
YL=1/ZL;
if KFlag==1;% Work with impedances
RL=real(ZL);
XL=imag(ZL);
end
if KFlag==0;%Work with admittances
RL=real(YL);
XL=imag(YL);
end
```

11.4 Initialization of RFDT Algorithm

Ad Hoc Initialization

As mentioned earlier, for nonlinear optimization problems the initial guess is quite important. As we have practiced so far, initial values were selected on an ad hoc basis for all the matching problems solved using RFDT.

In other words, it was sufficient to initialize the double matching optimization on the coefficients by setting the initial values for $a0$ and $c0$ as $+1$ or -1, such that

$$a0 = +1 \, \text{or} -1$$
$$c0(j) = \mp 1$$

The sign can be determined by trial and error.

Initialization via RFLT

A meaningful initial guess may be generated using the single matching line segment technique. In this case, the double matching problem can be considered as a single matching one by ignoring the reactive elements of the generator. Then, the rational form of the real part is generated as in Equation (11.25). Thus, the initial coefficients are obtained.

Initialization of the Best Case Solution

Equation (11.20) may be utilized to generate the initial values for the RFDT algorithm.

Assume that, in the best case, the phase variation of $G_{22}(j\omega)$ and $S_{in}(j\omega)$ occurs in such a way that the term

$$|1 - G_{22}(j\omega_k)S_{in}(j\omega)|^2$$

becomes

$$(1 - |G_{22}(\omega_k)|^2)^2$$

In this case,

$$\frac{T_G}{1 - |G_{22}|^2}\left[1 - |G_{22}|^2\right]^2 = T_G\left[1 - |G_{22}|^2\right] = \frac{4R_B R_L}{\left[(R_B + R_L)^2 + (X_B + X_L)^2\right]}$$

Let

$$\boxed{0 \leq WG = T_G\left[1 - |G_{22}|^2\right] \leq 1} \tag{11.35}$$

With the present assumption, we can try to achieve the flat gain level T_0 such that perfect reactance cancellation occurs at the load end. In this case, the real part variation is given as in Equation (11.30):

$$\boxed{R_{B0}(\omega) = R_L(\omega)\left[\frac{(2 - WG) + \text{sign}\sqrt{1 - WG}}{WG}\right]} \tag{11.36}$$

Equation (11.36) can be modeled as an initial guess for the main optimization package, as mentioned in RFLT.

We should note that, in working with RFDT, optimization of TPG, for both single and double matching problems, is so well behaved in the coefficients of the real part $R_B(p^2)$ that an ad hoc choice for the initial values provides even better numerical stability than that from forced initial values. Therefore, we always prefer to work with ad hoc initial choices.

For example, if we wish to design an equalizer of five elements with an acceptable gain level, then we start RFDT optimization with

$$T0 = 0.85$$
$$a0 = 1$$

and

$$c0 = [1\ {-}1\ 1\ {-}1\ 1\ {-}1 \ldots 1]$$

Or we can start with

$$c0 = [+1\ +1\ {-}1\ {-}1\ +1]$$

In any case, a proper selection of the 'sign' will make the optimization converge rapidly.

As far as the choice for $T0$ is concerned, the flat gain level can be found by trial and error. For example, when we start with 0.85, if the resulting gain performance is 1 or higher values than 0.85, we can increase the gain, but if the passband gain leads to huge variations below 0.85, then we must reduce the level of flat gain. Let us develop the above proposed approaches in the following actual antenna example.

11.5 Design of a Matching Equalizer for a Short Monopole Antenna

Short monopole antennas over the frequency range $20\,\text{MHz} \leq f \leq 100\,\text{MHz}$ have various commercial and military uses. In this section, we examine the design of an actual matching network for a physical monopole antenna manufactured for a special purpose. The measured impedance data for the antenna is given in Table 11.5 and depicted in Figure 11.27.

Table 11.5 Measured impedance data $Z_L = R_L + jX_L$ for a
short monopole antenna

Actual frequency (MHz)	$R_L(\Omega)$	$X_L(\Omega)$
20	0.6	−6
30	0.8	−2.2
40	0.8	0
45	1	1.4
50	2	2.8
55	3.4	4.6
60	7	7.6
65	15	8.8
70	22.4	−5.4
75	11	−13
80	5	−10.8
90	1.6	−6.8
100	1	−4.4

Before we start with the design, the data must be normalized with respect to both frequency and standard resistive termination $R_0 = 50\,\Omega$.

Let us choose the frequency normalization with respect to $f0 = 100$ MHz.

Close examination of Table 11.5 reveals that, at $f = 20$ MHz, the real part is R_L (20 MHz) $= 0.6\,\Omega$. On the other hand, at $f = 70\,\text{MHz}$, $R_L(70\,\text{MHz}) = 22.4\,\Omega$. Hence, the real part variation is $22.4/0.6 = 37.333$,

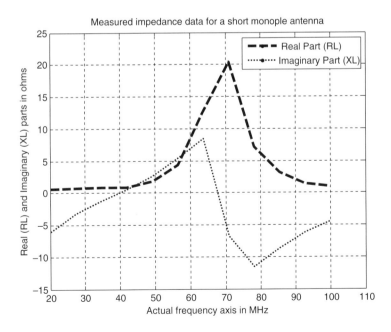

Figure 11.27 Measured impedance data for a short monopole antenna

which is quite high. Therefore, this is a very tough matching problem. At this point we have no idea about the gain–bandwidth limitation of the antenna. Actually, a model for the antenna can be obtained by means of the technique presented in the next chapter. A model is shown in Figure 11.28.

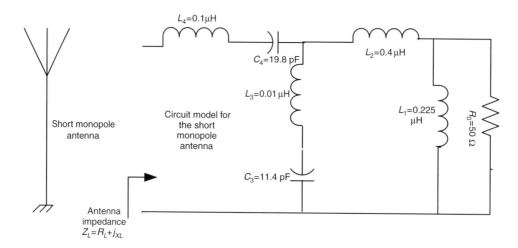

Figure 11.28   Model for the measured impedance data for the short monopole antenna

As can be seen from Figure 11.28, to our knowledge, there is no known transfer function which can imbed the above complicated load network to utilize the analytic theory of broadband matching. The theory is not accessible for the present case.

From a practical point of view, we do not wish to deal with an antenna that delivers less than 50% of the available power of the generator. Therefore, we should seek a useful bandwidth for the antenna such that it yields at least -3 dB or higher gain over the band of operation.

Our first inclination is to use the antenna over the 20 to 100 MHz bandwidth. But, practically, the real part is almost zero (or about 0.6 Ω) at 20 MHz. Therefore, we do not have much hope of getting a gain at the lower edge of the measured frequencies. However, at 30 MHz, the real part increases slightly to 0.8 Ω.

Now, let us try to capture at least -3 dB gain over 30 to 100 MHz.

For this purpose, we developed a MATLAB program to construct an equalizer from the measured data. The name of the program is `Monopole_Antenna.m`. In this program, RFDT is employed with a nonlinear optimization algorithm.

As in previous examples, a lossless equalizer is described in terms of the real part of the driving point input impedance R_B. RB is described as in Equation (11.25) with $a0$, ndc, W, and the denominator coefficients $c_j, j = 1, 2, \ldots, n$.

In other words, these are the unknowns of the matching problem. The integer n refers to the total number of elements in the equalizer. Minimum reactance driving point impedance $Z_B(p)$ is generated from R_B using the Bode method.

For the RFDT algorithm, the load function is written directly on the measured data as follows.

Program Listing 11.17   Load function written for the RFDT algorithm which processes the measured data for the monopole antenna

```
function [RLN,XLN]=Load(w, KFlag)
Antenna_Data =[


20 0.6 -6
30 0.8 -2.2
40 0.8 0
45 1 1.4
50 2 2.8
55 3.4 4.6
60 7 7.6
65 15 8.8
70 22.4 -5.4
75 11 -13
80 5 -10.8
90 1.6 -6.8
100 1 -4.4];
% Normalization Numbers
R0=50; f0=100;
N=length(Antenna_Data);
for j=1:N
WNA(j)=Antenna_Data(j,1)/f0;
RLNA(j)=Antenna_Data(j,2)/R0;
XLNA(j)=Antenna_Data(j,3)/R0;
end
RLN=line_seg(N,WNA,RLNA,w);
%Interpolation of the given real part data
XLN=line_seg(N,WNA,XLNA,w);
%Interpolation of the given imaginary part
data
% ya=line_seg(n,x,y,xa)is a line segment
%
approximation
ZL=complex(RLN,XLN);
YL=1/ZL;
if KFlag==0;
RLN=real(YL);
XLN=imag(YL);
End
```

The error function, which is subject to minimization, is generated under the MATLAB function direct_Monopole, as listed below.

Program Listing 11.18   Error function written to design matching equalizer for the monopole antenna via RFDT

```
function
fun=direct_Monopole(x0,KFlag,T0,wc1,wc2,ntr,ndc,W,a0)
% Modeling via Direct - Parametric approach


if ntr==1; %Yes transformer case
Nx=length(x0);
for i=1:Nx-1;
c(i)=x0(i);
end;
a0=x0(Nx);
end
if ntr==0;% No Transformer case
Nx=length(x0);
for i=1:Nx;
c(i)=x0(i);
end;
end
%
% Generate analytic form of Fmin(p)=a(p)/b(p)
% Generate B(-p^2)
C=[c 1];
BB=Poly_Positive(C);
% This positive polynomial is in w-domain
B=polarity(BB);
% Now, it is transferred to p-domain
% Generate A(-p^2) of R(-p^2)=A(-p^2)/B(-p^2)
nB=length(B);
A=(a0*a0)*R_Num(ndc,W);
% A is specified in p-domain
nA=length(A);
if (abs(nB-nA)>0)
A=fullvector(nB,A);
% Note that function RtoZ requires same length
% vectors A and B
end
[a,b]=RtoZ(A,B);
% Here A and B are specified in p-domain
%
NA=100;
DW=(wc2-wc1)/(NA-1);
w=wc1;
```

```
cmplx=sqrt(-1);
%
for j=1:NA
p=cmplx*w;
aval=polyval(a,p);
bval=polyval(b,p);
Z=aval/bval;
RB=real(Z);
XB=imag(Z);
[RL,XL]=Load(w,KFlag);
T=4*RB*RL/((RL+RB)^2+(XB+XL)^2);
error=T-T0;
fun(j)=error;
w=w+DW;
end
```

Eventually, we run the main program `monopole_Antenna.m` with the following inputs.

Inputs (design choices):

- `T0=0.625`. This is the target values of the flat gain level.
- `KFlag=1`. We should mention that we can start the design either with a minimum reactance impedance (the `KFlag=1` case) or with a minimum susceptance admittance (the `KFlag=0` case) Both choices are acceptable for the gain optimization. Here, we have the liberty to select the minimum reactance driving point impedance for the equalizer which results in quite a good solution.
- `ndc=0`. In this design, we wish to work over a wide frequency of operation. Therefore, there is no need to introduce a zero of transmission at DC. Hence, we set `ndc=0`. This choice will let us end up with a low-pass LC ladder-type equalizer structure which is quite easy to build.
- `W=0`. Here, we need no finite zero of transmission on the real frequency axis. Therefore, we set `W=0`.
- `ntr=1`. We know that use of an ideal transformer will greatly help to obtain a reasonable average gain level with smooth fluctuation. Furthermore, it is not difficult to realize transformers at frequencies over 20 to 100 MHz. Therefore, we decided to employ an ideal transformer in the equalizer.

The initial coefficients for the optimization are based as our prefered ad hoc choice. Therefore, we start with

$$a0 = 1$$
$$c0 = [-1 \ 1 \ 1 \ -1 \ -1 \ 1]$$

which means that the equalizer will include six elements.

Thus, we obtain the following optimized solution:

$$a0 = 0.28103$$
$$A0 = a0 \times a0 = 0.0789$$
$$c = [-1.5256 \quad 1.8821 \quad 2.5705 \quad -4.3191 \quad -1.2094 \quad 2.1366]$$

$$B[2.3274 \quad 4.3004 \; -5.9611 \; -17.429 \; -11.853 \; -2.1464\,1]$$

$$a = [0.1447 \quad 0.35816 \quad 0.42631 \quad 0.40499 \quad 0.19034 \quad 0.051768]$$

$$b = [1\,2.4742 \quad 3.9848 \quad 5.3703 \quad 4.0675 \quad 2.5009 \quad 0.65549]$$

$$\texttt{wmax=0.9596, Tmax=0.63906}$$

$$\texttt{wmin=0.30505, Tmin=0.5212}$$

$$T_{average} = 0.580\,13 \mp 0.589\,32$$

$$\texttt{dbmax=-1.9446, dbmin=-2.83}$$

$$dB_{Average} = -2.3873 \mp 0.4427$$

This means that

$$R_B(p^2) = \frac{0.0789}{2.327p^{12} + 4.3p^{10} - 5.961p^8 - 17.429p^6 - 11.853p^4 - 2.146p^2 + 1}$$

and

$$Z_B(p) = \frac{0.144p^5 + 0.35p^4 + 0.42p^3 + 0.404p^2 + 0.19p + 0.0517}{p^6 + 2.47p^5 + 3.98p^4 + 5.73p^3 + 4.06p^2 + 2.5p + 0.655}$$

$Z_B(p)$ is synthesized as a lossless two-port in resistive termination, which yields the equalizer as shown in Figure 11.29.

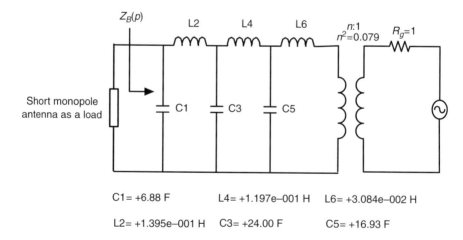

C1= +6.88 F   L4= +1.197e–001 H   L6= +3.084e–002 H

L2= +1.395e–001 H   C3= +24.00 F   C5= +16.93 F

Figure 11.29 Single matching equalizer designed to match short a monopole antenna to a resistive generator

The actual element values of the equalizer can be computed as follows. Actual capacitors are given by

$$C_A = \frac{C_{normalized}}{2\pi f_0 R_0}$$

and actual inductors are given by

$$L_A = \frac{L_{normalized}}{2\pi f_0} R_0$$

Thus, we have

$$C_1 = 2.1906 \times 10^{-10} \text{ F} = 0.219 \text{ nF} \qquad L_2 = 1.1103 \times 10^{-8} = 11.1 \text{ nH}$$
$$C_3 = 7.6404 \times 10^{-10} \text{ F} = 0.746 \text{ nF} \qquad L_4 = 9.5277 \times 10^{-9} = 9.277 \text{ nH}$$
$$c_5 = 5.3894 \times 10^{-10} \text{ F} = 0.5389 \text{ nF} \qquad L_6 - 2.4538 \times 10^{-9} = 2.4538 \text{ nH}$$

$\frac{1}{n^2} = 12.6$ *turns ratio of the ideal transformer*

The gain performance of the matched antenna is depicted in Figure 11.30 in dB. It is clear that more than 50% of the generator power is transferred to the antenna over 30 to 100 MHz. This result is pretty satisfying.

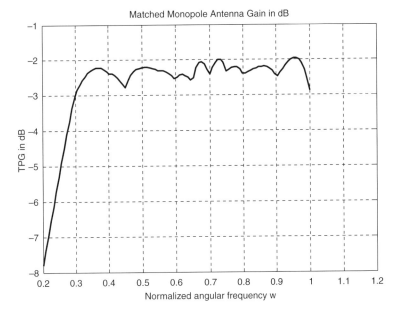

Figure 11.30   Gain performance of the matched monopole antenna

We should mention that this is a unique solution which expands the band of the antenna by more than three octaves.

Remarks:

• We tried to solve the same matching problem with commercially available packages by optimizing the gain function on the element values of the chosen circuit topology. Unfortunately, optimization was not successful. This clearly indicates that the solution obtained via RFDT is unique and superior.

- Various bandpass solutions for the same antenna are given employing the simplified real frequency technique. Therefore, we encourage interested readers to go over those solutions as in Yarman's, book.[10]

We present the main program `Monopole_Antenna.m` as follows.

Program Listing 11.19   Design of an equalizer for a monopole antenna

```
% Monopole Antenna Design
clear
clc
% User Information
input('Example: Monopole Antenna Matching Problem via RFDT
Single
Matching')
input('The measured load Data is directly imbedded into function
Load(w,KFlag)')
%
% Normalization numbers for the measured load
% Measured antenna data is normalized with respect to
%R0=50 ohm, f0=100 MHz
R0=50;
f0=100e6;
%
% Program inputs: Real Frequency Direct Computational
% Technique
ws1=0.2;ws2=1.0; %This is fixed by the measured data
% Step 2: Enter zeros of the RB
T0=input('Enter Flat Gain Level T0=')
KFlag=input('KFlag=0>Admittance, KFlag=1>Impedance..Enter
KFlag=')
ndc=input('Enter transmission zeros at DC ndc=')
W=input('Enter finite transmission zeros at [W]=[....]=')
a0=input('Enter a0=')
[c0]=input('Enter initials for [c0]=[....]=')%sigma(i): Real
roots of b(p)
c0=[-4.1705 0.4066 9.2306 -1.887 -5.464 1.7575 ];
a0=0.19241;
ntr=input('ntr=1, Yes, model with Transformer, ntr=0 No
Transformer ntr=')
%
wc1=input('Enter wc1=')
wc2=input('Enter wc2=')
```

---

[10] See Chapter 10 of the book by B. S. Yarman, *Design of Ultra Wideband Antenna Matching Networks*, Springer-Verlag, 2008.

```
%
Nc=length(c0);


% Optimization
% Step 4: Call nonlinear data fitting function
%%%%%%%%%% Preparation for the optimization

OPTIONS=OPTIMSET('MaxFunEvals',20000,'MaxIter',50000);
%
%%%%%%%%%%%%%%%%%%%%%%%%%%%%%%%%%%%%%%%%%%%%%%%%%%%%%%%%%
%
if ntr==1; x0=[c0 a0];Nx=length(x0);end;
%Yes transformer case
if ntr==0; x0=c0;Nx=length(x0);end;
%No transformer case
%
% Call optimization function lsqnonlin:
x=lsqnonlin('direct_Monopole',x0,[],[],OPTIONS,KFlag,T0,wc1,wc2,
ntr,ndc,W,a0);
% fun= direct_Monopole (x0, KFlag, T0, wc1, wc2,
% ntr,ndc,W,a0)
%
% After optimization separate x into coefficients
%
if ntr==1; %Yes transformer case
Nx=length(x);
for i=1:Nx-1;
c(i)=x(i);
end;
a0=x(Nx);
end
if ntr==0;% No Transformer case
Nx=length(x);
for i=1:Nx;
c(i)=x(i);
end;
end
%
% Step 5: Generate analytic form of Fmin(p)=a(p)/b(p)
% Generate B(-p^2)
C=[c 1];
BB=Poly_Positive(C);
% This positive polynomial is in w-domain
B=polarity(BB);
% Now, it is transferred to p-domain
% Generate A(-p^2) of R(-p^2)=A(-p^2)/B(-p^2)
nB=length(B);
A=(a0*a0)*R_Num(ndc,W);
```

Program Listing 11.19   (Continued)

```
% A is specified in p-domain
nA=length(A);
if (abs(nB-nA)>0)
A=fullvector(nB,A);
% Note that function RtoZ requires same length
% vectors A and B
end
[a,b]=RtoZ(A,B);
% Here A and B are specified in p-domain
%

% Initialize the performance measures.
Tmax=0;
Tmin=1;
NPrint=100;
DW=(ws2-ws1)/(NPrint-1);
w=ws1;
cmplx=sqrt(-1);
%
for j=1:NPrint
WA(j)=w;
p=cmplx*w;
aval=polyval(a,p);
bval=polyval(b,p);
Z=aval/bval;
RB=real(Z);
XB=imag(Z);
[RL,XL]=Load(w, KFlag)
T=4*RB*RL/((RL+RB)^2+(XB+XL)^2);
TA(j)=T;
TdB(j)=10*log10(TA(j));
%
% Compute the performance parameters: Tmax,Tmin,Tave
% and detT
if max(TA(j))>Tmax
wmax=WA(j);Tmax=TA(j);
end
if w>=wc1
if w<=wc2
if min(TA(j))<Tmin
wmin=WA(j);Tmin=TA(j);
end
end
end
% End of Performance measure loop
%
```

```
w=w+DW;
end
%


figure(1)
plot(WA,TA)
xlabel('Normalized angular frequency w')
ylabel('Transducer Power Gain')
title('Matched Monopole Antenna Gain')
%
figure(2)
plot(WA,TdB)
xlabel('Normalized angular frequency w')
ylabel('TPG in dB')
title('Matched Monopole Antenna Gain in dB')
%
%Synthesis of the driving point immittance FB(p)
if KFlag==1
CVal=general_synthesis(a,b);
end
if KFlag==0
CVal=general_synthesis(b,a);
end
% Print Performance parameters:
wmax, Tmax,
wmin,Tmin,
T_Average=(Tmax+Tmin)/2
delT=T_Average-Tmin
dbmax=10*log10(Tmax)
dbmin=10*log10(Tmin)
dB_Average=(dbmax+dbmin)/2
dbdel=dB_Average-dbmin
```

11.6 Design of a Single Matching Equalizer for the Ultrasonic T1350 Transducer

The T1350 transducer is a high-power piezoelectric ultrasonic transducer which resonates in the frequency range of 1300 to 1500 kHz. It is utilized in various commercial and military underwater applications.

For the T1350 transducer, the real and imaginary parts of the measured impedance data are as shown in Figure 11.31 and Figure 11.32 respectively. The frequency range of the measurements is 600 to 2100 kHz. We can normalize the data with respect to the normalization frequency $f_0 = 2100$ kHz and the standard resistance $R_0 = 50\,\Omega$. The normalized data is summarized in Table 11.6.

Close examination of Table 11.6 indicates that the T1350 transducer resonates at $f = 1450$ KHz.

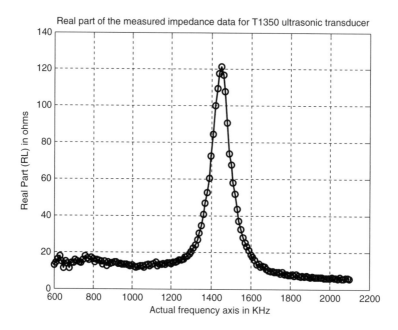

Figure 11.31   Real part variation of the measured impedance data of the T1350

In this section, it is our purpose to match the T1350 to a resistive generator over the frequency band of

$$f_{c1} = 1100 \leq f \leq f_{c2} = 1600 \, \text{KHz}$$

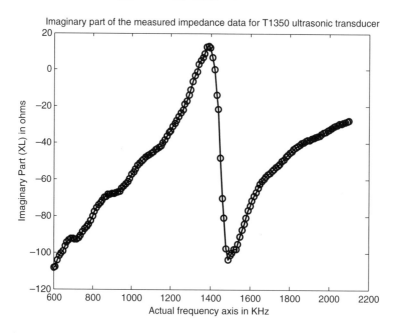

Figure 11.32   Imaginary part variation of the measured impedance for the T1350

Table 11.6 Summary of the impedance measurements for the ultrasonic T1350 transducer. In this table all entries are normalized with respect to $f0 = 2100$ kHz, $R0 = 50$ Ω. The resonance frequency of the transducer is at $f = 1450$ kHz

Normalized angular frequency ω	Normalized real part data $RL(\omega)$	Normalized imaginary part data $RL(\omega)$
0.285 71	0.256 72	−2.169
0.345 24	0.291 64	−1.784 6
0.404 76	0.303 2	−1.478 7
0.464 29	0.263 72	−1.199 9
0.523 81	0.263 56	−0.910 12
0.583 33	0.307 54	−0.553 86
0.642 86	0.928 86	0.008 704
0.690 48 resonance frequency	**2.42 44**	**−0.962 34**
0.695 24	2.337 5	−1.404 5
0.702 38	1.380 1	−1.459 2
0.761 9	0.371 62	−1.485
0.821 43	0.184 46	−1.069 8
0.880 95	0.136 35	−0.836 4
0.940 48	0.118 93	−0.678 56
1	0.110 39	−0.557 52

For this single matching problem, we modified the main program, the error function for the optimization and the load function developed for the monopole antenna to incorporate the measured load data of the T1350. Thus the new names are respectively given as

```
Ultrasonic_Transducer.m
fun=direct_Transducer(x0,KFlag,T0,wc1,wc2,ntr,ndc,W,a0)function
[RLN,XLN]=Load_Transducer(w,KFlag)
```

The main program `Ultrasonic_Transducer.m` directly processes the normalized data with respect to $f0 = 2100$ kHz and $R0 = 50$ Ω.

TPG of the matched system is optimized over the normalized frequency band of

$$f_{c1} = 1100/2100 = 0.523\,81 \text{ to} f_{c2} = 1600/2100 = 0.7619$$

Firstly, we will try to achieve `T0=0.85` flat gain level without using a transformer in the equalizer. In this case, we must choose `ndc=0, ntr=1`. In order to get the final result over the complete measured frequency band we choose

`ws1=600/2100=0.28571` and `ws2=2100/2100=1`

To make the equalizer as simple as possible, we set the finite transmission zero vector `W=0`.

Firstly, let us work with the impedance-based design. Therefore, `KFlag` is set to 1.

Hence, we run the main program `Ultrasonic_Transducer` with the following inputs.

Inputs:

- T0=0.85.
- KFlag=0 > admittance; KFlag=1 > impedance. Enter KFlag=1 (work with impedance functions).
- Enter transmission zeros at DC, ndc=0.
- Enter finite transmission zeros at [W]=[...]=0.
- Initial guess on the coefficients of the real part, RB.
- Enter a0=1; enter initial values for [c0]=[...]=[1 1 -1 -1 1].
- ntr=1, yes, work with transformer; ntr=0, no transformer: ntr=0.
- Enter ws1=600/2100=0.28571.
- Enter ws2=2100/2100=1.
- Enter wc1=1100/2100=0.52381.
- Enter wc2=1600/2100=0.7619.

The result of the optimization is summarized as follows:

$$R_B = \frac{a_0^2}{B(p^2)} \quad Z_B(p) = \frac{a(p)}{b(p)}$$

(minimum reactance driving point impedance)

$$B(\omega^2) = \frac{1}{2}\left[c^2(\omega) + c^2(-\omega)\right]$$

$$a_0 = 1 \,(\text{transformerless design})$$

$$C = [-33.448 \quad -3.8855 \quad 36.433 \quad 1.7357 \quad -9.0796 \quad 1]$$

$$B(p) = [-1118.8 \quad -2422.1 \quad -1921.2 \quad -666.35 \quad -85.911 \quad 1]$$

$$a(p) = [0 \; 0.62956 \; 0.24185 \; 0.40182 \; 0.068798 \; 0.029897]$$

$$b(p) = [1 \; 0.38416 \; 1.1563 \; 0.30828 \; 0.30858 \; 0.029897]$$

$$\text{wmax}=0.5619, \text{Tmax}=0.85822$$

$$\text{wmin}=0.7619, \text{Tmin}=0.72742$$

$$T_{average} = 0.79282 \mp 0.065401$$

$$\text{dbmax}=-0.66401, \text{dbmin}=-1.3822$$

$$dB_{average} = -1.0231 \mp 0.35907$$

The gain performance of the T1350 (in dB) is shown in Figure 11.33. The result is quite good and confirms the laboratory test.

We should mention that the gain performance of the matched transducer may be improved by reiteration.

The synthesis of $Z_B(p)$ which in turn yields the desired equalizer is shown in Figure 11.34.

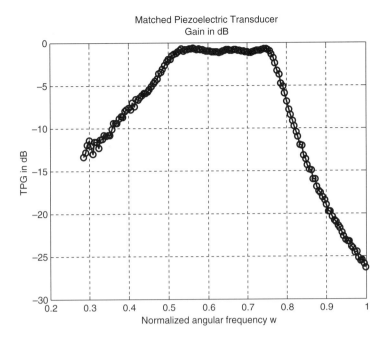

Figure 11.33   TPG of the matched T1350 transducer

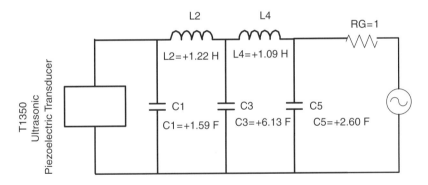

Figure 11.34   Design of single matching equalizer for the T1350 piezoelectric transducer

Finally, by denormalization, the actual element values are given as

$$C_1 = 2.4077 \times 10^{-9} = 2.4 \text{ nF}$$
$$L_2 = 4.6054 \times 10^{-6} = 4.6 \text{ μH}$$
$$C_3 = 9.2917 \times 10^{-9} = 9.29 \text{ nF}$$
$$L_4 = 4.1147 \times 10^{-6} = 4.1 \text{ μH}$$
$$C_5 = 3.9457 \times 10^{-9} = 3.95 \text{ nF}$$

11.7 Simplified Real Frequency Technique (SRFT): 'A Scattering Approach'[11]

In Chapter 5, SRFT was introduced for the equalization of arbitrary gain shapes by employing resistively terminated lossless two-ports consisting of commensurate transmission lines. In Chapter 6, a scattering description of lossless two-ports was given in the complex variable $p = \sigma + j\omega$ which results in lumped element circuit components. In this chapter, we utilize SRFT to construct single and double matching equalizers as well as microwave amplifiers utilizing lumped circuit components.

Let us refresh our minds with the following short description of SRFT.

In this technique, a lossless equalizer is simply described in terms of its unit normalized scattering parameters. Therefore, neither the impedance nor the admittance parameters are employed. All the computations are carried out with the scattering parameters of the blocks to be matched. Therefore, the gain optimization of the matched system is well behaved numerically. It is faster than the other existing CAD algorithms and easier to use. It is also naturally suited for the design of broadband microwave amplifiers.

The basis of SRFT or the scattering approach is to describe the lossless equalizer $[E]$ in terms of its unit normalized input reflection coefficient $E_{11}(p)$ in Belevitch form such that

$$E_{11}(p) = \frac{h(p)}{g(p)} \tag{11.37}$$

For selected transmission zeros, the complete scattering parameters of a lossless reciprocal equalizer can be generated from the numerator polynomial $h(p)$, using the para-unitary condition given by Equations (6.37–6.54).

This idea constitutes the crux of SRFT. In other words, TPG of the system to be matched is expressed as an implicit function of $h(p)$.

By replacing the generator and load impedances by their Darlington equivalent lossless two-ports, $[G]$ and $[L]$, and utilizing their unit normalized scattering descriptions given by Equation (10.10) and Equation (10.23), TPG of the doubly matched system can be generated in two steps as in Chapter 6, Equations (6.74–6.76).

In the first step, by multiplying the transfer scattering parameters of the lossless two-ports $[G]$ and $[E]$, the gain of the cascaded duo $[GE]$ is generated as

$$T_1 = |G_{21}|^2 \frac{|E_{21}|^2}{|1 - E_{11}G_{22}|^2}$$

Then, in the second step, TPG of the cascaded two-ports $[GE]$ and $[L]$ is obtained as

$$T = T_1 \frac{|L_{21}|^2}{|1 - \hat{E}_{22}L_{11}|^2}$$

[11] B. S. Yarman, 'A Simplified Real Frequency Technique for Broadband Matching Complex Generator to Complex Loads, *RCA Review*, Vol. 43, pp. 529–541, 1982. B. S. Yarman and H. J. Carlin, 'A Simplified Real Frequency Technique Applied to Broadband Multi-stage Microwave Amplifiers', *IEEE Transactions on Microwave Theory and Techniques*, Vol. 30, pp. 2216–2222, December, 1982.

Thus, using the open form of T_1 in this equation, the double matching gain of the trio $\{[G] - [E] - [L]\}$ is expressed as follows:

$$T(\omega) = |G_{21}|^2 \frac{|E_{21}|^2}{|1 - E_{11}G_{22}|^2 |1 - \hat{E}_{22}L_{11}|^2} |L_{21}|^2 \tag{11.38}$$

where the scattering parameters of the generator $\{G_{ij}\}$ and the load $\{L_{ij}\}$ networks are either specified by measurements or provided as circuit models. More specifically, in terms of the generator $Z_G(j\omega)$ and the load $Z_L(j\omega)$ impedances,

the generator reflectance is given by
$$G_{22} = \frac{Z_G - 1}{Z_G + 1}; \ |G_{21}|^2 = 1 - |G_{22}|^2$$
the load reflectance is given by
$$L_{11} = \frac{Z_L - 1}{Z_L + 1}; \ |L_{21}|^2 = 1 - |L_{11}|^2 \tag{11.39}$$
and the back end reflectance of the equalizer, while it is terminated in the complex generator, is given by
$$\hat{E}_{22} = E_{22} + \frac{E_{21}^2 G_{22}}{1 - E_{11}G_{22}}$$

Now let us construct TPG, once $h(p)$ is initialized.

For simplicity, we assume that $[E]$ is a minimum phase structure with transmission zeros only at $\omega = 0$ and $\omega = \infty$. This is a convenient assumption because it assures realization of the equalizer without coupled coils, except possibly for an impedance level transformer. Then, the real normalized scattering parameters of $[E]$ are given in Belevitch form as follows:

$$E_{11}(p) = \frac{h(p)}{g(p)} = \frac{h_n p^n + h_{n-1}p^{n-1} + \ldots + h_1 p + h_0}{g_n p^n + g_{n-1}p^{n-1} + \ldots + g_1 p + g_0}$$
$$E_{12}(p) = E_{21}(p) = \mp \frac{p^k}{g(p)} \tag{11.40}$$
$$E_{22}(p) = (-1)^k \frac{h(-p)}{g(p)}$$

where the integer n specifies the number of total reactive elements in $[E]$ and the integer k is the order of the transmission zeros at DC.[12]

Since the matching network is lossless, it follows that

$$G(p^2) = g(p)g(-p) = h(p)h(-p) + (-1)^k p^{2k} > 0, \text{ for all } p \tag{11.41}$$

In the SRFT algorithm, the goal is to optimize TPG as high and as flat as possible over the band of operation as in the other real frequency techniques.

---

[12] Note that, in Equation (11.40), the indexing used for polynomial coefficients $h(p)$ and $g(p)$ is classical rather than MATLAB convention. However, by replacing i with $n - i + 1$ we can always end up with MATLAB indexing.

The coefficients of the numerator polynomial $h(p)$ are selected as the unknowns of the matching problem. To construct the scattering parameters of $[E]$, it is sufficient to generate the Hurwitz denominator polynomial $g(p)$ from $h(p)$.

It can readily be shown that, once the coefficients of $h(p)$ are initialized, and the complexity of the equalizer $[E]$ is specified (i.e. n and k are fixed in advance), then $g(p)$ is generated as a Hurwitz polynomial by explicit factorization of Equation (11.41). Thus, the bounded realness of the scattering parameters $\{E_{ij}, i, j = 1, 2\}$ is already built into the design procedure.

Note that, in order to assure the realizability of the lossless equalizers, we cannot simultaneously allow $h(p)$ and $g(p)$ to become zero. Therefore, the selection of initial coefficients of $h(p)$ must be proper. For example, if $k > 0$, we cannot let $h_0 = 0$ since it makes $g(0) = 0$.

In generating the Hurwitz denominator polynomial $g(p)$ from the initialized coefficients of $h(p)$, we first construct $G(p^2)$ as in Equation (11.41). That is,

$$G(p^2) = g(p)g(-p) = G_n p^{2n} + G_{n-1} p^{2(n-1)} + \ldots + G_1 p^2 + G_0 \qquad (11.42)$$

where the coefficients G_i are given as the convolution of $h(p)$ and $h(-p)$ with the contributing term $(-1)^k p^{2k}$ as follows:

$$
\begin{aligned}
G_0 &= h_0^2 \\
G_1 &= -h_1^2 + 2h_2 h_0 \\
&\;\;\vdots \\
G_i &= (-1)^i h_i^2 + 2\left[h_{2i} h_0 + \sum_{j=2}^{i} (-1)^{j-1} h_{j-1} h_{2j-i+1} \right] \\
&\;\;\vdots \\
G_k &= G_i|_{i=k} + (-1)^k \\
&\;\;\vdots \\
G_n &= (-1)^n h_n^2
\end{aligned}
\qquad (11.43)
$$

Then, explicit factorization of Equation (11.43) follows. At the end of the factorization process, polynomial $g(p)$ is formed on the LHP zeros of $G(p^2)$.

Hence, the scattering parameters of $[E]$ are obtained as in Equation (11.40) and the TPG $T(\omega)$ is generated using Equation (11.38). Then, as in the other real frequency techniques, by selecting flat gain level T_0, the error function $error = T(\omega) - T_0$ is minimized. As a result of optimization, the unknown coefficients h_i are determined.

Details of the numerical work can be found in the Bibliography at the end of the chapter.

In brief, examination of Equation (11.40) together with Equation (11.38) indicates that TPG is almost inversely quadratic in the unknown coefficients h_i. Therefore, the SRFT algorithm is always convergent. Furthermore, the numerical stability of the algorithm is excellent, since all the scattering parameters E_{ij} and reflection coefficients G_{22} and L_{11} are BR, i.e. $\{|E_{ij}|, |G_{22}| \text{ and } |L_{11}|\} \leq 1$.

As is usually the case, an intelligent initial guess is important to run the program efficiently. It is our experience that, for many practical problems, an ad hoc direct choice for the coefficients h_i (e.g. $h_i = \mp 1$) provides satisfactory initialization to start the SRFT algorithm.

As indicated previously, SRFT is naturally suited for the design of microwave amplifiers since it employs scattering parameters for all the units to be matched. For over 30 years, matching networks and amplifiers designed using SRFT showed excellent agreement with laboratory performance measurements.

Now, let us develop the SRFT matching algorithm to design a matching network for a helix antenna. The antenna can be utilized for general purpose GPS applications.

11.8 Antenna Tuning Using SRFT: Design of a Matching Network for a Helix Antenna

For this section, a helix antenna was designed for a GPS receiver.[13]

The antenna is supposed to operate in the neighborhood of the center frequency $f0 = 1.575$ GHz.

The real part RL and the imaginary part XL of the measured antenna impedance are depicted in Figure 11.35 and Figure 11.36 respectively. A summary of the measurements is given in Table 11.7.

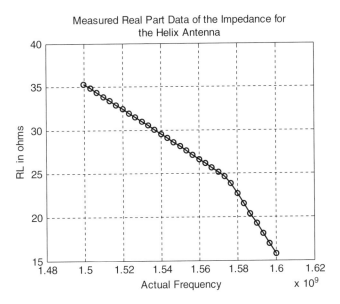

Figure 11.35   Measured real part data of the impedance of a helix antenna

Here, the idea is to tune the antenna at 1.575 GHz by employing SRFT.

For this purpose, we wish to design an equalizer to match the antenna to a 50 Ω generator over the passband $R(1.5 - 1.6\,\text{GHz})$ which also captures the tuning frequency $f_0 = 1.57$ GHz. In this case, we must implement an SRFT algorithm for the given load to minimize the error function

$$
\boxed{
\begin{array}{l}
\text{error} = T(\omega) - T_0 \\
\text{or} \\
\text{error} = |G_{21}|^2 \dfrac{|E_{21}|^2}{|1 - E_{11}G_{22}|^2 |1 - \hat{E}_{22}L_{11}|^2} |L_{22}|^2 - T_0
\end{array}
} \tag{11.44}
$$

where $G_{21} = 1$ since the generator is resistive with $R_G = R_0 = 50\Omega$,

---

[13] The helix antenna was designed, physically realized and measured by Miss Eda Konakyeri of Technical University of Istanbul (ITU) as her BSc thesis under the supervision of Prof. Dr Osman Palamutcuogullari in 2008.

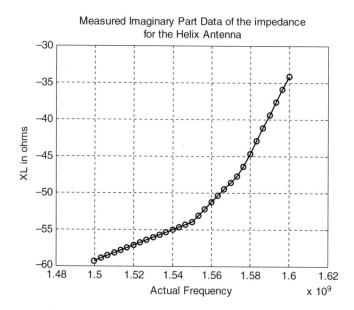

Figure 11.36 Imaginary part of the measured impedance data for a helix antenna

Table 11.7 Summary of the measured impedance data for a helix antenna

Actual frequencies (GHz)	Real part RL (Ω)	Imaginary part XL (Ω)
1.5	35.31	59.3
1.55	28.14	−53.98
1.575	24.44	−47.32
1.6	15.77	−34.2

$$L_{22} = \frac{Z_L - R_0}{Z_L + R_0}$$

and

$$\hat{E}_{22} = E_{22} = -(-1)^k \frac{h(-p)}{g(p)}$$

which is the back end reflection coefficient of the equalizer. Obviously, here, it is presumed that all the transmission zeros of the equalizer are on the $j\omega$ axis. Furthermore, transmission zeros at DC are of order k. In this regard, $f(p)$ is chosen as $f(p) = p^k$.

On the other hand, this is a narrow-bandwidth problem. Therefore, during the optimization process we may be able to achieve the ideal flat gain level $T_0 = 1$.

It is our desire to make the matching network as simple as possible. So, we do not wish to employ an ideal transformer. Therefore, we set `ntr=0`. In this case, we can work with an LC ladder-type structure with $k = 0$.

The MATLAB code developed for this problem is described in the following algorithm.

SRFT Algorithm to Design Matching Networks

This algorithm basically minimizes the error function specified by Equation (11.44) which in turn determines the Belevitch form of the scattering parameters of the lossless equalizer [E] as described above.

Within the algorithm, firstly the complexity of the matching network is selected by the designer. In other words, the total number of elements n and the transmission zeros of the equalizer are set in advance.

For practical matching networks, we prefer to utilize transmission zeros only at DC and infinity. In this case, we set $f(p) = p^k$ so that

$$E_{21} = \frac{f(p)}{g(p)} = \frac{p^k}{g(p)}$$

This choice of equalizer is simple enough to be implemented by employing existing manufacturing technologies either with discrete components or with silicon CMOS technology.

The unknown of the matching problem is the numerator polynomial

$$h(p) = h_n p^n + h_{n-1} p^{n-1} + \ldots + h_1 p + h_0$$

Here, the coefficients $\{h_0, h_1, \ldots, h_n\}$ are determined to minimize the error function. For the optimization, we use the standard MATLAB Levenberg–Marquardt function called `lsqnonlin`, which minimizes the sum of the mean square error.

The main program developed for this section is called `Helix_Antenna.m`. Inputs to the main program are given as follows.

Inputs:

- `FA`: An array, sampling frequencies to measure the load data.
- `RLA`: An array, real part of the antenna impedance.
- `XLA`: An array, imaginary part of the antenna impedance.
- `T0`: Targeted flat gain level.
- `ntr`: Choice for transformer. For an ideal transformer set `ntr=1`, otherwise `ntr=0`.
- `k`: Order of the transmission zeros at DC.
- `wlow`: Low end of the optimization frequency (normalized with respect to selected frequency which is $f_0 = 1.575$ GHz).
- `whigh`: High end of the optimization frequency (normalized as above).
- `h0`: Initial value for the coefficient $h0 = h(0)$.
- `h`: Initial values for the coefficients $h(p)$.

Computational Steps:

Step 1: Using the above initial values, start optimization employing the error function called

```
fun=sopt_Hellix(x,T0,ntr,k,nopt,wlow,whigh)
```

The input integer `nopt` is the total number of sampling points which is used to sample the error function over the normalized angular frequencies between `wlow` and `whigh`. For the present case, it may be sufficient to set `nopt=30`.

The input `x` starts with the initial value `x0` on the coefficients `x0={h0,h1,h2,...,hn}`. Obviously, if there is no transformer used in the circuit, then `h0=0` and it is not part of the unknowns.

Step 2: Call the optimization function

```
x=lsqnonlin('spot_Helix',x0,[],[],OPTIONS,T0,ntr,k,nopt,wlow,whigh)
```

which determines the optimum set of the numerator coefficients `x={h0,h1,...,hn}` minimizing the error function $error = T(\omega) - T0$.

Step 3: Once the optimum $h(p)$ is determined, the complete scattering parameters of the equalizer are constructed as specified by Equation (11.40) and it is synthesized as a lossless two-port.

In this step, it is useful to compare the performance of the antenna as it is matched and unmatched. The unmatched antenna performance can be analyzed from the input reflectance

$$L_{11} = \frac{Z_L - R_0}{Z_L + R_0} = \rho_L e^{j\phi_L}$$

In this regard, the measure of mismatch is given by

$$L(\text{dB}) = 10\log_{10}(\rho_L)$$

This is called the input mismatch. It is always worth while to plot matched and unmatched reflection coefficients on the same sheet to understand the success of the optimization. The matched input reflectance in dB or the matched 'input mismatch' is given by

$$\rho_M(\text{dB}) = 10\log_{10}[1 - T(\omega)]$$

Mismatch is the measure of the power reflection from the load. As it grows in absolute value, power reflection decreases. The ideal value of the mismatch is $-\infty$ dB, which corresponds to full power transfer to the load. In this case, TPG becomes unity. The worst mismatch case is 0 dB, which corresponds to full reflection from the load end. In this case, TPG becomes zero.

From a practical point of view, a mismatch of -20 dB is quite good. On the other hand, a mismatch of -4 or -5 dB is considered quite poor.

The equalizer is synthesized by long division of its driving point impedance function

$$Z_{in}(p) = \frac{a(p)}{b(p)} = \frac{g(p) + h(p)}{g(p) - h(p)}$$

As in the other real frequency techniques, we use the function CVal=general_synthesis(h,g) to synthesize the equalizer.[14] In this function, $g(p)$ and $h(p)$ are defined as conventional MATLAB polynomials. Therefore, we have to shuffle the coefficients of the polynomials $h(p)$ and $g(p)$ as they are obtained from the optimizations.

Now, let us run the main program Helix_Antenna.m with the following inputs.

Inputs:

- f0=1.575 GHz (normalization frequency).
- R0=50 (normalization resistance).
- FA=[1.5 1.55 1.575 1.6] (Actual frequencies in GHz; normalized input frequencies with respect to f0.)
- WBR=[0.95238 0.98413 1 1.0159].
- RLA=[35.31 28.14 24.44 15.77]/R0 (real part of the antenna impedance).
- XLA=[-59.30 -53.98 -47.32 -34.2]/R0 (imaginary part of the antenna impedance).
- T0=1 (We try to achieve the best value over the band of operation.)
- k=0 (Note that $f(p) = p^k = 1$ which is fixed at the input. Here, we wish to work with a low-pass LC ladder to ease the implementation.)
- ntr=0 (In the equalizer design, we avoid using an ideal transformer.)
- wlow=0.953 (corresponds to 1.5 GHz).
- whigh=1 (corresponds to 1.6 GHz).

---

[14] Function CVal=general_synthesis(h,g) was developed by Dr Ali Kiline of Okan University, Istanbul, Turkey.

Initial guess:

- h0=0, h=[1 -1 1] (ad hoc choice for the initial values).

Result of optimization:

- h=[0.64211 0.28508 0] (regular MATLAB polynomial form).
- g=[0.64211 1.1685 1] (regular MATLAB polynomial form).

The optimized driving point input impedance of the equalizer is given as

$$Z_{in}(p) = \frac{a(p)}{b(p)} = \frac{g+h}{g-h} = \frac{1.2842p^2 + 1.4536p + 1}{0.88346p + 1} = 1.4536p + \frac{1}{0.88346p + 1}$$

Here, the first component of the equalizer is a normalized inductor $L_{N1} = 1.4536$. The second component is a normalized capacitor $C_{N2} = 0.88346$ and the termination is a unit resistance as expected.

The actual element values of the matching network are obtained by denormalization with respect to $f_0 = 1.575\,\text{GHz}$ and $R_0 = 50\,\Omega$. Hence, we obtain the following actual elements:

$$L_{A1} = \frac{L_{N1}}{2\pi f_0} R_0 = 7.3445 \times 10^{-9}\text{H} = 7.3445\,\text{nH}$$

Similarly, the actual capacitor is found as

$$C_{A2} = \frac{C_{N2}}{2\pi f_0} R_0 = 1.7855 \times 10^{-12}\text{F} = 1.7855\,\text{pF}$$

The performance of the matched system is summarized in Table 11.8.

The matched and unmatched input reflectance of the antenna are depicted in Figure 11.37 and the matched TPG is depicted in Figure 11.38.

A photograph of the physical helix antenna is shown in Figure 11.39; the measured performance is depicted in Figure 11.40.

Finally, the antenna with matching network is shown in Figure 11.41.

Table 11.8 Optimized performance of the tuned antenna

Actual frequency (GHz)	Matched input reflectance (dB)	Unmatched input reflectance (dB)	Matched Gain (dB)
1.50	−12.394	−4.6122	−0.257 77
1.51	−13.222	−4.5568	−0.211 91
1.52	−14.175	−4.4931	−0.169 32
1.53	−15.277	−4.4207	−0.130 81
1.54	−16.545	−4.339	−0.097 315
1.55	−17.966	−4.2478	−0.069 927
1.56	−21.914	−4.2777	−0.028 039
1.57	**−24.591**	**−4.2947**	**−0.015 117**
1.58	−18.122	−4.2415	−0.067 452
1.59	−11.659	−4.0438	−0.307 04
1.6	−7.6666	−3.7053	−0.815 16

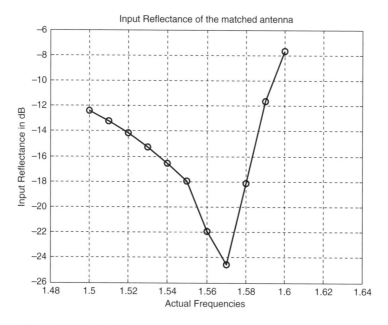

Figure 11.37   Optimized input reflectance of the matched helix antenna

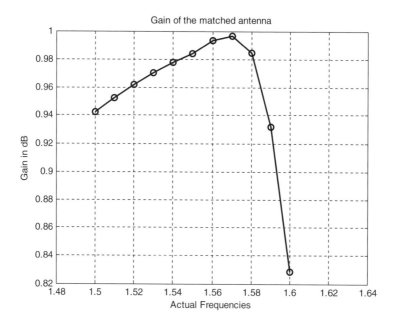

Figure 11.38   Optimized TPG of the matched helix antenna

Figure 11.39 The helix antenna, designed to operate at 1.575 GHz

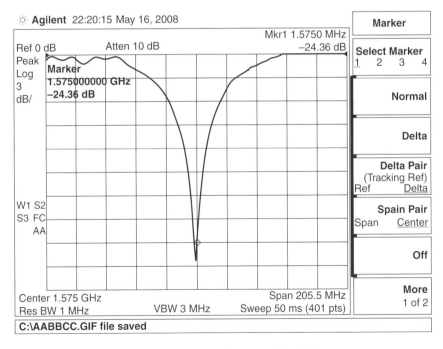

Figure 11.40 Measured performance of the helix antenna

Here, it is clearly shown that a simple LC ladder equalizer design using SRFT significantly improves

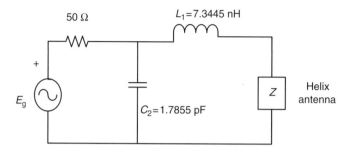

Figure 11.41　Tuned helix antenna employing SRFT

the power transfer from the generator to the load. Especially at $f0 = 1.575$ GHz, without the equalizer, the reflection or mismatch is measured as -4.3 dB, whereas match input reflection is about -25 dB which is almost a perfect match.

We list the main program `Helix_Antenna.m` and related files as follows.

Program Listing 11.20　Main program `Helix_Antenna.m`

```
% PROGRAM Helix_Antenna.m: Design of a Matching
% Network for a helix antenna
% employing Simplified Real Frequency Technique with simple
% optimization
% using MATLAB function lsqnonlin
display('Design of a matching network for a helix antenna using
SRFT')
display('Measured antenna impedance data has been already put in
the arrays FA, RLA and XLA')
% Inputs:
% RLA: Real part of the measured impedance data
%XLA: Imaginary part of the measured impedance data
%FA : Frequency sampling points for measurements
%
%T0: Desired flat gain level
%ntr: Control flag for equalizer design:
%ntr=1; equalizer design with a transformer.
%In this case k may be different than zero.
%ntr=0; equalizer is constructed without transformer.
%In this case k must be zero (k=0)
%and equalizer will be constructed as LC ladders
%
%initial values for h(p), and f(p)=p^k
%nopt: number of sampling points in optimization.
```

```
%wlow: Low frequency end of the optimization.
%whigh: High Frequency end of the optimization
%ccccccccccccccccccccccccccccccccccccccccccc
% Outputs:
%Constructed antenna matching network
%Optimization using Levenberg-Marquardt Technique.
%Optimized Scattering Parameters of the Equalizer.
%Optimized Transducer Power Gain
%Realization of the Equalizer as an LC ladder
% actual element values.
% Plot of the Gain Performance
clear
clc
% ccccccccccccccccccccccccccccccccccccccccccc
% Input Data:
% Measured antenna impedance ZL=RL+jXL
% Normalization numbers:
f0=1.575; %Normalization frequency in GHZ
R0=50; % Normalization resistance
FA=[1.5 1.55 1.575 1.6]; %Actual frequencies in GHZ
RLA=[35.31 28.14 24.44 15.77]/R0;%
XLA=[-59.30 -53.98 -47.32 -34.2]/R0;
%
% Data Normalization
WBR=FA/f0;
% Normalized frequencies
display('Normalized Frequencies over the passband');
WBR'
T0=input('Enter desired flat gain level T0=')
ntr=input('Enter ntr; 0> for no transformer 1> with transformer
ntr=')
k=input('Enter transmission zeros at DC k=')
wlow=input('Enter beginning of normalized optimization frequency
wlow=')
whigh=input('Enter end of normalized optimization frequency
whigh=')
h0=input('Enter initial value for h0=')
h=input('Enter initial values for h=[....] use brackets
[.....]=')


nopt=30;
%Total number of sample points in optimization


na=length(h);
%
%%%%%%%%%%%%%%%%%%%%%%%%%%%%%%%%%%%%%%%%%%%%%%
%
```

Program Listing 11.20   (Continued)

```
% Start Optimization using simplest version of the
% LSQNONLIN
%
% Type design preferences:
%ntr=1: design with transformer
%ntr=0: design without transformer.
%---------------------------------------
%
%%%%%%%%%% Preparation for the optimization %%%%%%


OPTIONS=OPTIMSET('MaxFunEvals',20000,'MaxIter',50000);
%
%%%%%%%%%%%%%%%%%%%%%%%%%%%%%%%%%%%%%%%%%%%%%%%%
%Design with transformer:
if ntr==1,
            x0=h;
% Call optimization function lsqnonlin:

x=lsqnonlin('sopt_Helix',x0,[],[],OPTIONS,T0,ntr,k,nopt,wlow,whi
gh);
%
            h=x;
end
%---------------------------------------
%
% Design without transformer:
      if ntr==0,
            for r=1:na-1
              x0(r)=h(r);
%Call optimization function lsqnonlin:

x=lsqnonlin('sopt_Helix',x0,[],[],OPTIONS,T0,ntr,k,nopt,wlow,whi
gh);
            end
    h=x; %Regular MATLAB polynomial Coefficients
    h(na)=0;
       end
% Generate g(p) from optimized h(p):
    g=Hurwitzpoly_g(h,k);%General MATLAB polynomial
% Compute the optimized transducer power gain
W1=wlow;
W2=whigh;
Nsample=100;
DW=(W2-W1)/Nsample;
W=wlow;
```

```
        for j=1:Nsample+1
            Wa(j)=f0*W;
            E22=Equalizer(W,h,g,k);
            % GENERATION OF THE LOAD DATA
            L11=reflection_Helix(W);
            LdB(j)=10*log10(abs(L11)*abs(L11));
            % (Load Mismatch factor)

%COMPUTATION OF GAIN
            T=gain(W,h,g,k,L11);
            Ta(j)=T;
            TdB(j)=10*log10(Ta(j));
            W=W+DW;
        end
%
        figure(1)
        plot(Wa,Ta)
        title('Gain of the Matched antenna')
        xlabel('Actual Frequencies in GHz')
        ylabel('Gain in dB')
%
Sin_dB=10*log10(1-Ta);%input mismatch factor
figure(2)
plot(Wa,Sin_dB, Wa,LdB)
title('Matched and unmatched Input Reflectance of the helix
antenna')
xlabel('Actual Frequencies in GHz')
ylabel('Matched Reflectance in dB')
legend('matched',' unmatched')
%
figure (3)
plot(Wa,LdB)
title('Input Reflectance of the unmatched antenna')
xlabel('Actual Frequencies in GHz')
ylabel('unmatched Reflectance in dB')


CV=general_synthesis(h,g);
a=g+h;
b=g-h;
if a(1)<1e-7;kimm=0;
display('Start with a capacitor')
%Actual element values starts with an inductor
display('Actual Element Values AEV=')
fnorm=f0*1e9
AEV=Actual(kimm,R0,fnorm,CV)
end
if a(1)>1e-7; kimm=1;
display('start with an inductor')
```

Program Listing 11.20   (Continued)

```
%Actual element values starts with an inductor
display('Actual Element Values AEV=')
fnorm=f0*1e9
AEV=Actual(kimm,R0,fnorm,CV)
End
```

Program Listing 11.21   Error function subject to optimization for helix antenna problem

```
function fun=sopt_Helix(x,T0,ntr,k,nopt,wlow,whigh)
%Inputs:
%  T0: Desired gain level
%  ntr: Control flag for equalizer design:
%  ntr=1; equalizer design with a transformer.
%  In this case k may be different than zero.
%  ntr=0; equalizer is constructed without
%transformer.
%  In this case k must be zero (k=0)
%  and equalizer will be constructed with Low Pass LC
%elements
%x=Row Vector includes the unknown coefficients
%  of the polynomial h(p).
%for ntr=0 and k=0, dimension of x must be n.
%for ntr=1, dimension of x must be n+1.
%nopt: total number of sampling points within
%  optimization.
%  This is the dimension of the objective function
%invT subject to optimization.
%  wlow: Low frequency end of the optimization.
%  whigh: High Frequency end of the optimization
%%%%%%%%%%%%%%%%%%%%%%%%%%%%%%%%%%%%%%%%%%%%%%%%%
%  Output:
%  fun: Objective function Vector subject to
%  optimization.
%  For the design of Antenna Matching Networks
fun=T-T0
%%%%%%%%%%%%%%%%%%%%%%%%%%%%%%%%%%%%%%%%%%%%%%%%%%%
nx=length(x);
w=wlow;
DW=(whigh-wlow)/nopt;
%  Start frequency loop to compute objective function
%  within passband:
for j=1:nopt
%  Prepare to generate transducer power gain:
```

```
% Call load function to compute load reflection
%coefficient.
      L11=reflection_Helix(w);
      % Set h vector for the values of ntr:
      if ntr==1,
         h=x;
      end
      %
      if ntr==0,
         for r=1:nx
            h(r)=x(r);
         end
       na=nx+1;
       h(na)=0;
      end
      %
      % Generate g polynomial from h:
       g=Hurwitzpoly_g(h,k);
       % Generate the Transducer power Gain
       %
         T=gain(w,h,g,k,L11);
         %
         % Generate objective function as
            fun(j)=T-T0;
            w=w+DW;
      end
            return
```

Program Listing 11.22   TPG of helix antenna

```
function T=gain(W,h,g,k,L11)
% Define complex variable p=jW; j=sqrt(-1)
p=sqrt(-1)*W;
%Compute the values of h and g using MATLAB function
%polyval(p,X)

gval=polyval(g,p);
hval=polyval(h,p);
fval=p^k;
% Construct the input reflection coefficient of the
equalizer from the
% back-end.
            E22=hval/gval;
%COMPUTATION OF GAIN
%
Weight=fval*conj(fval)*(1-
```

Program Listing 11.22   (Continued)

```
L11*conj(L11));
%Compute the denominator of the gain function as in
%Equation (21).

D=hval*conj(hval)*(1+L11*conj(L11))+fval*conj(fval)-
2*real(L11*hval*conj(gval));
T=Weight/D;
return
```

Program Listing 11.23   Reflectance of the helix antenna

```
function SL=reflection_Helix(w)
% This function computes the Helix Antenna
%reflection coefficient provided by EDA
% KONAKYERI.
% SL is a complex quantity.
%%%%%%%%%%%%%%%%%%%%%%%%%%%%%%%%%%%%%%%%%
RLA=[35.31 28.14 24.44 15.77]/50;
%Data provided by EDA of ITU
XLA=[-59.30 -53.98 -47.32 -34.2]/50;
%Data provided by EDA of ITU
WBR =[0.95238 0.98413 1 1.0159];
nL=length(WBR);
RL=line_helix(nL,WBR,RLA,w);
XL=line_helix(nL,WBR,XLA,w);
j=sqrt(-1);
ZL=RL+j*XL;
SL=(ZL-1)/(ZL+1);
```

Program Listing 11.24   Input reflectance of the equalizer

```
function SE=Equalizer(W,h,g,k)
p=sqrt(-1)*W;
gval=polyval(g,p);
hval=polyval(h,p);
fval=p^k;
SE=hval/gval;
return
```

Program Listing 11.25   Generation of strictly Hurwitz polynomial from $h(p)$ and $f(p)=p^k$

```
function g=Hurwitzpoly_g(h,k)
%Computation of g(p)=g1p^n+g2p^(n-
%
% 1)+...+g(n+1)*p+g(n+1).
% Inputs: Row vector h which contains
% coefficients of h(p) in descending order.
% k which is the order of
f(p)=p^k.
na=length(h);
%
% Step 1: Generate even polynomial H(-
p^2)=h(p)h(-p);
%
% Computation of even polynomial with
convolution: W=conv(U,V)
U=h;
% Generate row vector V
sign=-1;
for i=1:na
sign=-sign;
V(na-i+1)=sign*U(na-
i+1);
end
W=conv(U,V);
% Select non-zero terms of W and form the
vector H
for i=1:na
H(i)=W(2*i-1);
end
%
% Step 2: Generate G(-p^2)=H(-p^2)+(-
1)^k
for i=1:na
G(i)=H(i);
end
ka=na-k; sign=(-1)^(k);
G(ka)=H(ka)+sign;
%
% Step 3:Generate the row vector X and
Compute
% the roots of G(-p^2)
% Arrange the sign of vector X.
sign=-1;
for i=1:na
sign=-sign;
```

Program Listing 11.25   (Continued)

```
X(na-i+1)=sign*G(na-i+1);
end
%
% Compute the RHP and LHP Roots of G(-p^2)
Xr=roots(X);pr=sqrt(-Xr);prm=-pr;
%
% Step 4: Generate the monic
polynomial g'(p)
C=poly(prm);
%
% Step 5: Generate Scattering Hurwitz
polynomial g(p)
Cof=sqrt(G(1));
for i=1:na
g(i)=Cof*C(i);
end
```

11.9 Performance Assessment of Active and Passive Components by Employing SRFT[15]

In designing wireless communication systems via analog VLSI-silicon CMOS processing technologies, it is essential to know the electrical performance limitations of active and passive components. Especially as the geometries shrink, the statistical variations due to processing start to play a more significant role in the predictability of the metal geometry. For example, during the manufacturing processes of inductors and capacitors, the metal thickness and its width depend on the mask as well as on the proximity metal density and its geometry. On the other hand, geometry variations due to these effects significantly change the target values of the lumped components. In other words, the geometry of the layout deeply affects the resulting parasitic capacitances and loss of the components' self- and mutual inductances. Just to give an idea to the reader, the atomic spacing of approximately 5–6 angstroms (65 nm) becomes roughly on the order of 100 atoms wide and this number shrinks to the order of 50 atoms in finer lithographies. In this case, one atom variation in the width will translate to 2% variation.

As far as practical implementations of the active and passive lumped components are concerned, it would be wise to work with well-established circuit models which precisely imbed the physical and geometrical properties at the atomic level.

The amplification performance of an active component, namely a MOS transistor, may be assessed by designing a linear Class A amplifier at the cutting edge of the useful frequency band. For example, 0.18 μm silicon processing technology has been used to design commercial wireless products up to 10 GHz. The question is whether this technology can be utilized up to 20 GHz or beyond. In this case, one may

---

[15] B. S. Yarman *et al.* 'Performance Assesement of Active and Passive Components Manufactured Employing 0.18 μm Silicon CMOS Processing Technology up to 22GHz', Proceedings of 2008 IEEJ, International Analog VLSI Workshop, Istanbul, Turkey, July 30–August 1, 2008, pp.129–132.

attempt to design a single stage microwave amplifier over a preset frequency band. In the course of amplifier design, frontend and backend matching networks must be constructed, preferably using lumped elements, i.e. inductors and capacitors. Therefore, in this section, by employing SRFT, we investigated the electrical performance of active and passive components using the measured data for the active device and reliable models for the capacitors and inductors by designing a wideband power MOSFET amplifier over the band of $B = 11.23$–22.39 GHz using typical 0.18 μm processing technology.

For this purpose, firstly a single stage Class A power amplifier was designed with ideal circuit elements using SRFT to assess the ideal electrical performance of the MOS transistor. Then, the amplifier performance was reoptimized by varying the actual layout of the inductors, considering the metal and substrate losses. In order to implement this step, we expressed the actual value of an inductor as a function of its loss over the band of operation. Eventually, degraded gain performance of the amplifier is compared to that of its ideal. Finally, it is concluded that 0.18 μm silicon technology can be utilized even beyond 20 GHz to design analog circuits for wireless communications

SRFT to Design Microwave Amplifiers

It is well known that SRFT is well suited to design microwave amplifiers in two steps.

Referring to Figure 11.42, in the first step, the frontend matching network $[E_F]$ is designed while the output port is terminated in 50 Ω (or in the normalization resistance $R_0 = 50\Omega$). In this step, TPG T_1 which is specified by

$$T_1(\omega) = \frac{|S_{21F}|^2 |S_{21}|^2}{|1 - S_{22F}S_{11}|^2} \tag{11.45}$$

is optimized to achieve the ideal stable flat gain level

$$T_{f\,lat1} = \text{minimum of } T_{01}$$
$$= \left\{ \frac{|S_{21}(\omega)|^2}{1 - |S_{11}(\omega)|^2} \right\} \text{ over } B \tag{11.46}$$

In Equation 11.45 and Equation 11.46, $\{S_{ijF}, i,j = 1, 2\}$ and $\{S_{ij}, i,j = 1, 2\}$ are the real normalized (50 Ω) scattering parameters of the frontend matching network $[E_F]$ and the MOS transistor respectively. B is the frequency band which runs from 11.23 to 22.39 GHz (more than 10 GHz bandwidth).

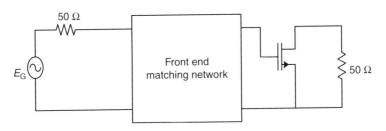

Figure 11.42 Step 1: Design of the frontend matching network

In the second step, the backend equalizer $[E_B]$ is designed to optimize the overall TPG $T_2(\omega)$ as shown by Figure 11.43. In this step, $T_2(\omega)$ is specified by

$$T_2(\omega) = T_1(\omega) \cdot \frac{|S_{21B}|^2}{|1 - S_{2F}S_{11B}|^2} \qquad (11.47)$$

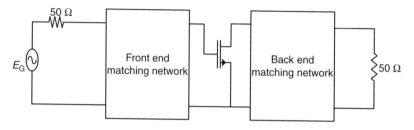

50 Ω

E_G

Front end matching network

Back end matching network

50 Ω

Figure 11.43   Step 2: Design of the backend matching network

In this step, the optimization algorithm targets the flat gain level of

$$T_{f\,lat2} = \text{minimum of } T_{02}(\omega) = \frac{T_1(\omega)}{1 - |S_{22}(\omega)|^2} \qquad (11.48)$$

In Equation (11.47) the backend reflection coefficient is given by

$$S_{2F} = S_{22} + \frac{S_{12}S_{21}}{1 - S_{11}S_{22F}} \qquad (11.49)$$

In SRFT, the frontend $[E_F]$ and backend $[E_B]$ matching networks are described in terms of their unit normalized input reflection coefficients in Belevitch form:

$$\begin{aligned} S_{22F} &= \frac{h_F}{g_F} = \frac{h_{0F} + h_{1F}p + \ldots + h_{n_F F}p^{n_F}}{g_{0F} + g_{1F}p + \ldots + g_{n_F F}p^{n_F}} \\ S_{11B} &= \frac{h_{0B} + h_{1B}p + \ldots + h_{n_B F}p^{n_B}}{g_{0B} + g_{1B}p + \ldots + g_{n_B B}p^{n_B}} \end{aligned} \qquad (11.50)$$

In step 1, selecting the total number of equalizer elements n_F in $[E_F]$ first, the coefficients $\{h_{0F}, h_{1F}, \ldots, h_{nF}\}$ are initialized. Then, the denominator polynomial $g_F(p)$ is generated as a strictly Hurwitz polynomial by spectral factorization of

$$g_F(p)g_F(-p) = h_F(p)h_F(-p) + f_F(p)f_F(-p)$$

In this step, S_{21F} is given as

$$S_{21F} = \frac{f_F(p)}{g_F(p)}$$

and the monic polynomial $f_F(p)$ is selected by the designer to include transmision zeros of the frontend equalizer $[E_F]$. For example, if $[E_F]$ is a simple low-pass LC ladder, then $f_F(p) = 1$ is selected. Eventually, coefficients $\{h_{0F}, h_{1F}, \ldots, h_{nF}\}$ are determined by minimizing the error function

$$\varepsilon_1(\omega) = T_1(\omega) - T_{f\,lat1}$$

over the band of interest $B = f_2 - f_1$ where f_1 and f_2 designate the lower and upper ends of the frequency band respectively.

Similarly, in step 2, the backend equalizer $[E_B]$ is constructed by determining the unknown numerator coefficients $\{h_{0B}, h_{1B}, \ldots, h_{nB}\}$ of S_{11B}. In this case, the error function

$$\varepsilon_2(\omega) = T_2(\omega) - T_{f\,lat2}$$

is minimized over the same frequency band.

Once S_{22F} and S_{11B} are determined, they are synthesized as lossless two-ports as in the Darlington procedure, yielding the circuit topologies for frontend and backend matching networks with element values.

SRFT Algorithm for Single Stage Microwave Amplifier Design

In this subsection, the gain–bandwidth performance of a typical 0.18 μm NMOS FET is assessed by designing an amplifier over 11.23–22.39 GHz.

The 50 Ω based scattering parameters of the device are give in Table 11.9.

Table 11.9 Selected samples from the measured scattering parameters of a 0.18 μm TSMC* MOS transistor (`MSij` are the amplitudes and `PSij` are the phases of the S parameters; the phases are given in radians)

Fr(GHz)	MS11	PS11	MS21	PS21	MS12	PS12	MS22	PS22
10.0	0.90	−2.99	1.11	1.1	0.05	−0.34	0.77	−2.96
11.2	0.90	−3.01	0.97	1.1	0.05	−0.39	0.78	−2.96
12.6	0.91	−3.01	0.84	1.1	0.06	−0.44	0.79	−2.96
14.1	0.91	−3.01	0.73	1.0	0.05	−0.49	0.80	−2.96
15.8	0.92	−3.02	0.63	0.9	0.05	−0.54	0.82	−2.96
17.8	0.92	−3.02	0.55	0.8	0.05	−0.60	0.83	−2.96
20.0	0.93	−3.03	0.46	0.81	0.04	−0.66	0.85	−2.96
22.4	0.94	−3.04	0.39	0.74	0.04	−0.72	0.86	−2.96

* Taiwan Semiconductor Manufacturing Company.

The biasing conditions and design parameters of the device are given as follows:

- $V_{DD} = 1.8$ V; I_D (drain DC current) = 200 mA.
- NMOS size is $W/L = 1000\,\mu m/0.18\,\mu m$.
- Simulation temperature is $50°$C.
- Expected output power is approximately 100 mW up to 23 GHz when operated as a Class A amplifier.

For this purpose, we developed the MATLAB main program called `TSMC_amp.m`.

As explained above, this program constructs a microwave amplifier in two parts. In the first part, the frontend equalizer $[E_F]$ is constructed. Then, in the second part, the backend matching network of the amplifier is completed.

We desire to construct both equalizers by employing LC ladder networks without a transformer. Therefore, in the main program, we set $k = 0$ and we fix $h_{0F} = h_{0B} = 0$.

Thus, in the first part, the objective function

```
function Func=levenberg_TR1(x,k,w,f11,f12,f21,f22,TS1)
```

minimizes $\varepsilon_1(\omega) = T_1(\omega) - T_{flat1}$, which in turn generates the numerator polynomial $h_F(p) = h_{nF}p^n + \ldots + h_{1F}p + 0$.

In this case, the designer provides the following inputs to the main program `TMSC_amp.m`.

Inputs to part I:

`TMSC_Spar.txt`: Scattering parameters of the CMOS transistor loaded from the text file. In this file, S parameters are read as amplitude and phase in degrees in the following order:

> MS11(i)=TMSC_Spar(i,2), PS11(i)=TMSC_Spar(i,3)
> MS21(i)=TMSC_Spar(i,4), PS21(i)=TMSC_Spar(i,5)
> MS12(i)=TMSC_Spar(i,6), PS12(i)=TMSC_Spar(i,7)
> MS22(i)=TMSC_Spar(i,8), PS22(i)=TMSC_Spar(i,9)

($i = 1$ to 48).

The program converts all the phases from degrees to radians and then generates the maximum stable gain of the amplifier at the first step as

$$T_{01} = \left\{ \frac{|S_{21}(\omega)|^2}{1 - |S_{11}(\omega)|^2} \right\}$$

and plots it as figure (1). For the CMOS transistor under consideration this figure is depicted in Figure 11.44.

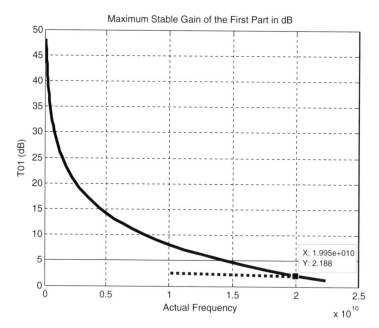

Figure 11.44   Maximum stable gain of part I of the single stage amplifier design

Close examination of Figure 11.44 reveals that one can obtain about 2.1 dB flat gain level even beyond 20 GHz.

At this point, the designer inputs the frequency band of operation. In this case, low (f_1) and the high (f_2) ends of the frequency band must be entered. The low end of the frequency band is denoted by

$$f_1 = \texttt{FR(nd1)}$$

and, similarly, the high end is denoted by

$$f_2 = \texttt{FR(nd2)}$$

However, the user only enters the index of the low and high ends of the frequency band.

The lower end of the optimization starts from the frequency $\texttt{FR(nd1)} = 11.23$ GHz which is placed at the 42nd sampling point of the measured data. Therefore, we set

$$\texttt{nd1=42}$$

The high end of the optimization frequency is $\texttt{FR(nd2)} = 22.39$ GHz which located at the 48th sampling point. Hence, we set

$$\texttt{nd2=48}$$

Now the program asks the designer to enter the desired flat gain level $\texttt{Tflat1}$ of part I.

Here, it may be appropriate to select the minimum or a little higher value of

$$T_{01} = \left\{ \frac{|S_{21}(\omega)|^2}{1 - |S_{11}(\omega)|^2} \right\}$$

in the passband to assure the absolute stability of the design.

Referring to Figure 11.44, it may be proper to select

$$\texttt{TFlat1} = 2.1 \text{ dB}$$

As the last input, we should enter the initial values for the unknown coefficients such that

$$hF(0) = 0 \quad [xF(1) = hF(1), xF(2) = hF(2), \ldots, XF(n) = hF(n)]$$

Here, we use an ad hoc input such that

$$[xF(1) = hF(1), xF(2) = hF(2), \ldots, XF(n) = hF(n)] = [-1 \; -1 \; -1 \; 1 \; 1]$$

This means that $\texttt{nF=length(hF)} = 4$. Therefore, we will employ only four elements in the LC ladder without an ideal transformer, since we set $\texttt{h0F=0}$ in advance.

Result of optimization:

In this part of the main program the optimization function

```
xF = lsqnonlin('levenberg_TR1',xF0,[],[],OPTIONS,k,ww,f11,f12,f21,
                                    f22,Tflat1)
```

is called.

function $\texttt{lsqnonlin}$ is a least mean square minimization algorithm which employs the error function $\texttt{levenber\_TR1}$ such that

```
function Func = levenberg_TR1(x,k,w,f11,f12,f21,f22,Tflat1)
```

This function basically generates the error term

$$\varepsilon_1 = T_1 - \texttt{Tflat1}$$

in dB.

Minimization of $\varepsilon_1 = T_1 - \texttt{Tflat1}$ yields the optimum reflection coefficient

$$S_{22F}(p) = \frac{h_F(p)}{g_F(p)}$$

as shown in Table 11.10.

Table 11.10   Front end reflection coefficient $S_{22F} = h_F/g_F$

h_{0F}	0	g_{0F}	1
h_{1F}	$-4.997\,2$	g_{1F}	$5.890\,5$
h_{2F}	$-1.744\,8$	g_{2F}	$4.863\,1$
h_{3F}	$-4.747\,3$	g_{3F}	$5.940\,1$
h_{4F}	$0.964\,63$	g_{4F}	$0.964\,63$

The above reflection coefficient leads to the frontend driving point impedance

$$Z_F(p) = \frac{g_F + h_F}{g_F - h_F} = \frac{1.9293p^4 + 1.1928p^3 + 3.1183p^2 + 0.893\,32p + 1}{10.687p^3 + 6.6079p^2 + 10.888p + 1}$$

and it is synthesized by long division as

$$Z_F(p) = 0.180\,52p + \cfrac{1}{9.2703p + \cfrac{1}{0.7128p + \cfrac{1}{1.6174p + 1}}}$$

The normalized circuit elements of the frontend matching network are given in Table 11.11.

Table 11.11   Normalized and actual element values of frontend equalizer for the microwave amplifier designed using a TSMC 180μm Si CMOS transistor

	Normalized element values (CVF)	Actual element values (AEVF)
L_1(H)	0.18052	6.4167e-011
C_2(F)	9.2703	1.3181e-012
L_3(H)	0.7128	2.5337e-010
C_4(F)	1.6174	2.2996e-013

The actual circuit element values are generated by denormalization using the function `AEVF=Actual(kimm,R0,fnorm,CVF)` as shown in the same table.

Thus, the first part of the amplifier design procedure is completed as shown in Figure 11.45.

The optimized gain performance of the first part of the design problem is depicted in Figure 11.46.

In the second part of the design process the backend equalizer of the amplifier is constructed by minimizing the error

$$\varepsilon_2 = T_2 - \texttt{Tflat2}$$

At this point, the main program `TSMC_amp.m` calls the optimization routine `lsqnonlin` with error function

```
function Func = levenberg_TR2(x,k,w,T1,SF2,f11,f12,f21,f22,Tflat2)
```

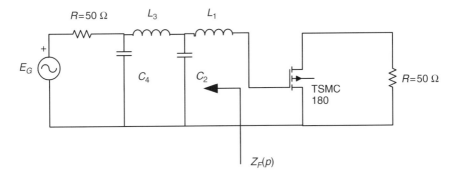

Figure 11.45 Completion of part I of the single stage amplifier design using the TMSC 180 μm Si CMOS transistor

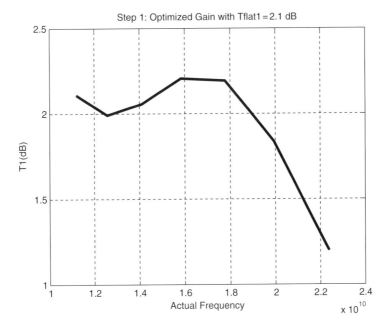

Figure 11.46 Optimized gain performance of part I

The function `levenberg_TR2` generates $\varepsilon_2 = T_2 - $ `Tflat2` in dB and it is minimized by `lsqnonlin`, which in turn yields the reflectance

$$S_{11B} = \frac{h_B(p)}{g_B(p)}$$

of the backend equalizer of the amplifier.

Inputs to part II:

At this point, the main program generates the maximum stable gain at the high end of the passband which may be used as a possible target for the flat gain level of the amplifier.

For the present case, the main program returns to the designer with the following statements:

```
Suggested maximum flat gain level for the second part of the design
                        Tflat2 = 7.2059
    Enter the desired target gain value in dB for Step 2: T flat2 =
```
In this case it may be appropriate to enter
$$Tflat2 = 7.1 \text{ dB}$$
Then, the main program asks for the initial values for the backend equalizer such that
```
enter back end equalizer; [xB(1)=hB(1),xB(2)=hB(2),...,XB(n)=hB(n)]=
```
Here, we use an ad hoc initialization for the coefficient as
$$xB0 = [-1 \ -1 \ -1 \ 1 \ -1 \ 1]$$
meaning that we will have nB=length(hB)=6 elements in the backend matching network. We already fixed the structure as a LC ladder without an ideal transformer by setting k=0 and hB0=0 in advance

Thus the optimization returns with the following results.

Results of optimization:

The backend reflection coefficient is determined as in Table 11.12.

Finally, $S_{11B} = h_B/g_B$ is synthesized as in part I as shown in Figure 11.47.

The element values of the backend equalizer are summarized in Table 11.13.

The gain performance of the amplifier is depicted in Figure 11.48 and listed in Table. 11.14.

Table 11.12 Backend reflection coefficient $S_{11B} = h_B/g_B$ (or, in short, S_B)

h_{0B}	0	g_{0B}	1
h_{1B}	−5.9242	g_{1B}	7.8643
h_{2B}	−1.9747	g_{2B}	13.375
h_{3B}	−15.969	g_{3B}	25.796
h_{4B}	2.8824	g_{4B}	20.767
h_{5B}	−9.2939	g_{5B}	17.764
h_{6B}	−6.4072	g_{6B}	6.4072

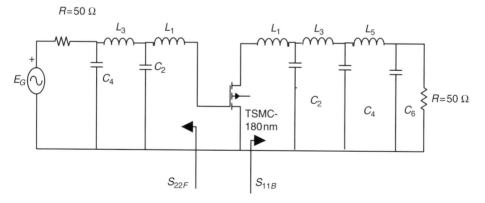

Figure 11.47   A single stage amplifier designed using SRFT for a typical 0.18 μm NMOS transistor over 11.23–22.39 GHz

Table 11.13 Normalized and actual element values of backend equalizer for the microwave amplifier designed using the TSMC 180 μm Si CMOS transistor

	Normalized element values (CVB)	Actual element values (AEVB)
L_1(H)	0.4736	1.6834e-010
C_2(F)	6.9925	9.9422e-013
L_3(H)	0.50192	1.7841e-010
C_4(F)	5.2831	7.5117e-013
L_5(H)	0.96454	3.4285e-010
C_6(F)	1.5129	2.1511e-013

Table 11.14 Gain performance of the amplifier

Actual frequency	Gain (dB)
1.122e+010	6.473
1.2589e+010	6.9404
1.4125e+010	7.0587
1.5849e+010	7.1319
1.7783e+010	7.0882
1.9953e+010	7.098
2.2387e+010	7.0987

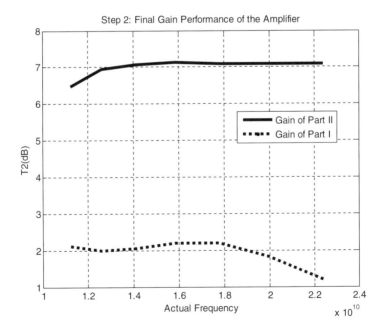

Figure 11.48 Ideal performance of the amplifier over the normalized frequencies from 0.55 (11.22 GHz) to 1 (22.39 GHz)

Stability of the Amplifier

In order to analyze the stability of the amplifier, the input and the output impedances of the matched amplifier must be computed.

Let $Z_{in}(j\omega) = R_{in}(\omega) + jX_{in}(\omega)$ and $Z_{out}(j\omega) = R_{out}(\omega) + jX_{out}(\omega)$ be the input and output impedances of the matched amplifier. Then, absolute stability of the amplifier requires that

$$
\boxed{
\begin{array}{c}
R_{in}(\omega) \geq 0, \forall \omega \\[6pt]
\text{and} \\[6pt]
R_{out}(\omega) \geq 0, \forall \omega
\end{array}
}
\tag{11.51}
$$

Sometimes, it may be difficult to satisfy the above conditions. In this case, we may be content with conditional stability.

In general, let the amplifier be terminated in a complex generator with internal impedance $Z_G = R_G(\omega) + jX_G(\omega)$ and a complex load $Z_L(j\omega) = R_L(\omega) + jX_L(\omega)$. Then, conditional stability requires that

$$
\boxed{
\begin{array}{c}
R_{in}(\omega) + R_{in}(\omega) \geq 0, \forall \omega \\[6pt]
\text{and} \\[6pt]
R_L(\omega) + R_{out}(\omega) \geq 0, \forall \omega
\end{array}
}
\tag{11.52}
$$

For the problem under consideration, $S_G = (Z_G - 1)/(Z_G + 1)$ is the input reflectance of the frontend equalizer which is generated in the first part of the amplifier design process as $S_F = h_F/g_F$. Similarly, complex termination $S_L = (Z_L - 1)/(Z_L + 1)$ of the active device is generated as the input reflectance of the backend equalizer such that $S_L = S_B = h_B/g_B$.

In this regard, we need to generate input and the output impedances to investigate the stability situation of the matched amplifier.

The input reflectance S_{in} is given by

$$
S_{in} = f_{11} + \frac{f_{12}f_{21}}{1 - f_{22}S_B} S_B
$$

where $\{f_{ij}, i, j = 1, 2\}$ are the S parameters of the active device (Previously defined $S_{ij} = f_{ij}$)

and

$$
S_B = \frac{h_B}{g_B}
$$

is the input reflectance of the backend equalizer.

Then, the input impedance is generated from S_{in} as

$$
Z_{in} = \frac{1 + S_{in}}{1 - S_{in}}
$$

such that

$$
Z_{in}(j\omega) = R_{in}(\omega) + jX_{in}(\omega)
$$

$$
\tag{11.53}
$$

Similarly, the output reflectance is given as

$$S_{out} = f_{22} + \frac{f_{12}f_{21}}{1 - f_{11}S_{11F}} S_F$$

where

$$S_F = \frac{h_F}{g_F}$$

is the input reflectance of the frontend equalizer.   (11.54)

Then, the output impedance is generated from S_{out} as

$$Z_{out} = \frac{1 + S_{out}}{1 - S_{out}}$$

such that

$$Z_{out}(j\omega) = R_{out}(\omega) + jX_{out}(\omega)$$

All the above computations are carried out using the following statements in the main program TSMC_amp.m.

Investigation of the stability of the amplifier

Stability check at the output port:

```
SB11=hoverg(hB,k,ww);
```
(input reflectance of the backend equalizer from hB)
```
SL=SB11;
ZL=stoz(SL);
RL=real(ZL);
```
(complex termination of the active device)
```
Sout=SF2;
Zout=stoz(Sout);
Rout=real(Zout);
```
(output reflectance of the active devices when it is terminated in frontend equalizer, i.e. in
```
SG=SF)
StabilityCheck_output=RL+Rout;
```
(must be always non-negative)

Stability check at the input port:

```
SF=hoverg(hF,k,ww);
SG=SF;
```
(complex input termination of the active device)
```
ZG=stoz(SG);
RG=real(ZG);
[Sin]=INPUT_REF(SB11,f11,f12,f21,f22);
Zin=stoz(Sin);
Rin=real(Zin);
StabilityCheck_input=RG+Rin;
```
(must be always non-negative)

Now, let us investigate the stability of the amplifier as designed with the above frontend and backend equalizers.

Table 11.15 indicates that the amplifier is absolutely stable from 12.6 to 22.39 GHz but conditionally stable at 11.22 GHz.

Table 11.15   Stability check of the amplifier (all the impedances given in this table are normalized with respect to $R_0 = 50\ \Omega$)

Frequency (GHz)	Absolute stability check		Conditional stability check	
	Rin	Rout	RG+Rin	RL+Rout
11.22	-0.0067505	0.11592	0.050926	0.22373
12.589	0.0064048	0.092387	0.059	0.22254
14.125	0.017844	0.067221	0.067412	0.23336
15.849	0.023586	0.042898	0.071946	0.24681
17.783	0.025254	0.024308	0.073142	0.24775
19.953	0.023913	0.01286	0.068118	0.2474
22.387	0.016272	0.0033594	0.048029	0.078335

Practical aspects of the design:

- It is our experience that SRFT is naturally suited for the design of microwave amplifiers in a sequential manner. In fact, one can even design a multistage amplifier step by step as described by Yarman and Carlin.[16] Details are omitted here. Interested readers are referred to the Bibliography provided at the end of this chapter. In this regard, the MATLAB main program TSMC_Amplifier.m can easily be modified to design multistage amplifiers.
- The stability of Equations (11.50) and (11.51) can be utilized as constraints in the course of TPG optimization.
- On completion of the equalizer design with ideal elements, lumped circuit components can be realized on a selected substrate as described by the examples in Chapters 2, 3 and 4.
- Physical implementation of lumped circuit elements introduces material loss and some inductive and capacitive parasitics as expected. These additions degrade the final electrical performance of the power transfer networks designed via other real frequency techniques. Nevertheless, once we have an idealized solution, we can always reoptimize it with commercially available software design packages employing the circuit models provided by the foundries. For example, an ideal inductor may be replaced by the practical model shown in Figure 11.49.
- Similarly, an ideal capacitor acts like a passive device and may be described by the model shown in Figure 11.50.
- Parameters of the selected models are specified by geometric layouts and material properties of the components which are provided by the manufacturers.
- In order to achieve the actual electrical performance of the matching networks or microwave amplifiers designed by employing the real frequency techniques, we need to replace the ideal element values with those models which include loss and parasitics of the components from the way they are produced with the technology under consideration. On the other hand, degraded performance of the matched system can be improved by reoptimizing the gain of the system on the element values of the preset models.

---

[16] B. S. Yarman and H. J. Carlin, 'A Simplified Real Frequency Technique Applied to Broad-Band Multistage Microwave Amplifiers', *IEEE Transactions on Microwave Theory and Techniques*, Vol. 30, pp. 2216–2222, 1982.

• It should also be noted that, in practice, there exist neither an ideal lumped component nor an ideal distributed element. To be more realistic, a physical passive layout may be modeled with both lumped and distributed elements which lead the engineer to design lossless two-ports with both lumped and distributed elements. From a theoretical point of view, designs consisting of mixed elements are highly complicated. However, in Chapter 13, we will present an extension of SRFT which handles mixed element designs.

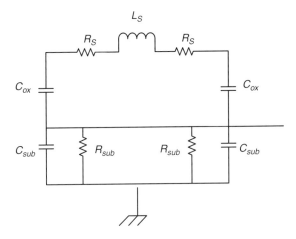

Figure 11.49   Nine-element model of an ideal inductor L_S

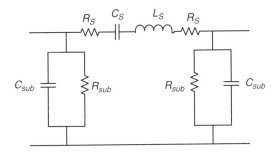

Figure 11.50   Eight-element model of an ideal capacitor C_S

All the SRFT MATLAB codes employed to run the SRFT examples presented in this section are listed at the end of the chapter.

Design of an Ultra Wideband Microwave Amplifier Using Commensurate Transmission Lines

In Chapter 2, transmission lines were introduced in conjunction with electromagnetic wave theory; in Chapter 3, the properties of transmission lines were investigated from a circuit theory point of view; in Chapter 4, classical circuit theory was revisited using the Richard variable, and commensurate transmission lines were employed to construct lossless two-ports and microwave filters; and in Chapter 5, arbitrary gain equalizers were designed utilizing commensurate transmission lines. In all these chapters, it was emphasized that ideal transmission lines are very handy for introducing physical sizes and geometric layouts into microwave circuit design. In fact, lumped circuit components such as resistors, capacitors and inductors are realized in terms of the geometric parameters of the layouts. A simple layout may be a TEM line. Therefore, in this subsection, we will present the design of a wideband microwave amplifier using commensurate transmission lines.

In the present case, the active device used in the microwave amplifier is a 180 nm CMOS transistor which was designed by Dr N. Retdian of Tokyo Institute of Technology using the VDEC foundry tools of Tokyo University.

The original amplifier chip consisted of lumped components.[17] In the present case, however, a similar design is made with commensurate transmission lines. The scattering parameters of the 180 nm silicon CMOS transistor are summarized in Table 11.16 and stored in the MATLAB text file `newcmos.txt`.

As in the previous example, here we also design the microwave amplifier in two steps. In the first step, the

Table 11.16 $R_0 = 50\,\Omega$ normalized scattering parameters of a 180 nm Si CMOS transistor

Freq. (GHz)	S_{11R}	S_{11X}	S_{21R}	S_{21X}	S_{12R}	S_{12X}	S_{22R}	S_{22X}
0.45	0.99	−0.05	−1.71	0.10	0.000	0.010	0.95	−0.04
0.7	1.00	−0.07	−1.71	0.15	0.001	0.015	0.94	−0.07
1	0.99	−0.10	−1.70	0.22	0.002	0.022	0.94	−0.09
2	0.97	−0.20	−1.64	0.43	0.009	0.042	0.92	−0.19
6	0.76	−0.51	−1.14	1.02	0.064	0.096	0.75	−0.49
10	0.51	−0.64	−0.56	1.19	0.127	0.101	0.54	−0.65

frontend and, in the second step, the backend equalizers are constructed. However, in this case, designs are made using commensurate transmission lines. Therefore, the scattering parameters of the equalizers are described in the Richard variable λ such that, on the real frequency axis $j\omega$,

$$\lambda = j\tan(\omega\tau)$$

where the constant delay τ is fixed as

$$\tau = \frac{1}{4(f_{end} \times \text{kLength})}$$

`kLength` is a user-defined parameter which refers to the fraction of the upper edge of the passband frequency f_{end}.

The amplifier design program developed for the previous example is revised to handle the design with commensurate transmission lines. The new program is called `Amplifier_Distributed.m`.

Let us summarize the design steps as follows.

Step I: Description of the frontend matching network:
The frontend equalizer is described in terms of its unit normalized scattering parameters (normalization number is $R_0 = 50\,\Omega$).

In this step, the unknown of the design problem is the numerator polynomial

$$h_F(\lambda) = h_1 \lambda^{nF} + h_2 \lambda^{(nF-1)} + \ldots + h_{nF} + h_{nF} + 1$$

[17] B. S. Yarman, N. Retdian, S. Takagi and N. Fujii, 'Gain-Bandwidth Limitations of 0.18μ Si-CMOS RF Technology', Proceedings of ECCTD 2007, Seville, Spain, August 26–30, 2007.

of the input reflection coefficient

$$S_{22F} = \frac{h_{F(\lambda)}}{g_F(\lambda)}$$

On the other hand, the transfer scattering parameter of the equalizer is specified by

$$S_{11F}(\lambda) = \frac{\lambda^{qF}(1-\lambda^2)^{kF/2}}{g_F}$$

Once the coefficients $h_F(\lambda)$ are initialized, the strictly Hurwitz polynomial $g_F(\lambda)$ is generated by the losslessness condition as in Equation (5.2).

Based on the above nomenclature we set the following inputs to the MATLAB main program `Amplifier_Distributed.m`.

Passband subject to optimization:

$$f_{start} = f_{C1} = 1\,\text{GHz} \quad f_{end} = f_{C2} = 10\,\text{GHz}$$

where

f_{start} is the beginning frequency of the optimization

f_{end} is the high end of the passband where the optimization ends

$q_F = 0$ is the count of transmission zeros at DC (this means that there is no transmission zero at DC)

$k_F = 6$ is the total number of cascaded sections

$n_F = 6$ is the total number of elements in the frontend equalizer.

It should be noted that, by choosing $n_F = k_F$, we avoid using series short or open shunt stubs as described in Chapter 4. This fact makes the implementation easy. Thus, the frontend equalizer will consist of six cascade connections of unit elements.

`ntr=0` means that no ideal transformer is used in the design.

`kLength=0.37`: This is our choice to fix the constant delay of UE.

`Tflat1=6.3` dB: Idealized flat gain level subject to optimization in the first part.

`xF=[-1 -1 -1 1 -1 1]`: Initial values for the optimization.

Here, it should be noted that the unknown of the frontend equalizer is the numerator polynomial $h_F(\lambda) = h_1\lambda^{nF} + h_2\lambda^{(nF-1)} + \ldots + h_{nF} + h_{nF+1}$. In other words, by using MATLAB polynomial convention, at the beginning of the optimization we set the numerator polynomial as

$$h_{F0} = [xF \ h_{F+1}] = [xF \ 0]$$

In this equation, we automatically set $h_{F+1} = 0$ since we desire no ideal transformer in the frontend equalizer. Actually, in order to simplify the main program, we set $h_{nF+1} = 0.0$ in advance. In this case, the simplified version of the transfer scattering parameter is given as

$$S_{21F}(\lambda) = \frac{(1-\lambda^2)^{kF/2}}{g_F}$$

In this step, the error function error $= T_1(\omega) - T_{flat1}$ which is generated by

`function Func=Distributed_TR1(x,kLength,qF,kF,w,f11,f12,f21,f22,Tflat1)`

is minimized. A listing of this function together with the main program `Amplifier_ Distributed.m` are given at the end of this chapter.

Step II: Description of the backend matching network:

In this step, the backend equalizer is described in terms of its input scattering parameters as

$$S_{11B} = \frac{h_{B(\lambda)}}{g_B(\lambda)}$$

where

$$h_B(\lambda) - h_1\lambda^{nB} + h_2\lambda^{(nB-1)} + \ldots + h_{nB} + h_{nB+1}$$

The transfer scattering parameter is specified by

$$S_{21B}(\lambda) = \frac{\lambda^{qB}(1 - \lambda^2)^{kB/2}}{g_B}$$

Similarly, the denominator polynomial $g_B(\lambda)$ is generated via losslessness of the backend equalizer as in Equation (5.2).

In this step, error $= T_2(\omega) - T_{f\,lat2}$ which is generated by

```
function Func=Distributed_TR2(x,kLength,qB,kB,w,T1,SF2,Tflat2)
```

is minimized. A listing of this function together with the main program `Amplifier_Distributed.m` and other related MATLAB functions are given at the end of this chapter.

Now, let us use the following inputs for the backend equalizer design.

Passband subject to optimization:

$$f_{start} = f_{C1} = 1\,\text{GHz} \quad f_{end} = f_{C2} = 10\,\text{GHz}$$

$q_B = 0$: total count of transmission zeros at DC
$k_B = 6$: total number of cascaded sections
$n_B = 6$: total number of elements in the backend matching network
`ntr=0`: no ideal transformer (actually, the main program `Amplifier_Distributed.m` is organized in such a way that it makes designs without a transformer automatically, therefore, a user of the program does not need to bother with `ntr`; in this regard, h_{F+1} is also fixed at zero, i.e. $h_{F+1} = 0$)
`kLength=0.37`
`Tflat2=10.5` dB: idealized flat gain level subject to optimization in the first part
`xB=[-1 -1 -1 1 -1 1]`: initial values for the optimization.

As explained above, initial values for the polynomial for $h_B(\lambda)$ are expressed in the main program as

$$h_{B0} = [xB \; 0]$$

Result of optimization:

In the first step, optimization of the TPG $T_1(\omega)$ yields the results summarized in Table 11.17.
In the second step, optimization of the TPG $T_2(\omega)$ reveals the results given in Table 11.18.
The gain performance of step 1 is depicted in Figure 11.51.
The overall gain performance is depicted in Figure 11.52.
The schematic of the amplifier is shown in Figure 11.53.

Table 11.17   Results of optimization of step I

Tflat1=6.3 dB	kLength=0.37	nd1=51 freq(51)=1 GHz	nd2=101 freq(101)=10 GHz
hF	gF	zF (normalized)	ZF Actual(Ω)
-15.65	15.682	1.8171	Z1F=90.854
42.125	63.873	2.6136	Z2F=130.68
8.5322	64.792	0.5207	Z3F=26.035
27.25	57.638	3.1297	Z4F=156.49
4.1329	26.387	0.52992	Z5F=26.496
2.4754	8.0561	1.9205	Z6F=96.026
0	1	1	R0=50

Table 11.18   Results of optimization of step 2

Tflat2=10.5 dB	kLength=0.37	nd1=51 freq(51)=1 GHz	nd2=101 freq(101)=10 GHz
hB	gB	zB (normalized)	ZB Actual (Ω)
5.0982	51.954	5.7794	Z1B=288.97
81.673	83.839	0.83948	Z2B=41.974
23.821	57.315	4.0919	Z3B=204.60
54.973	70.825	1.2459	Z4B=62.293
7.3537	25.852	1.1117	Z5B=55.587
5.8952	9.6154	2.4422	Z6B=122.11
0	1	1	R0=50

Practical notes:

- Using the same transistor and the same passband, a similar amplifier was designed by employing lumped elements.[18]
- When the performance of the lumped element amplifier is compared to the amplifier designed with UEs, it is observed that the lumped element design offers a much better gain performance over the amplifier design with UEs. This is expected. The reason is become the lumped element design offers a smooth roll-off in the passband. In other words, power delivered to the load is very much confined within the band of operation, whereas the design made with UEs spreads the gain somewhat periodically over the entire frequencies by suppressing it. This is due to the periodicity of the Richard variable $\lambda = j\tan(\omega\tau)$. This fact is clearly shown in Figure 11.52. At 10 GHz (high end of the passband), the gain starts dropping drastically and reaches its minimum value of –55 dB at 19 GHz. Then, it starts increasing up to –15 dB at 38 GHz. Of course, the shape is not exactly periodic as opposed to that in

---

[18] B. S. Yarman, N. Retdian, S. Takagi and N. Fujii, 'Gain-Bandwidth Limitations of 0.18μ Si-CMOS RF Technology', Proceedings of ECCTD 2007, Seville, Spain, August 26–30, 2007.

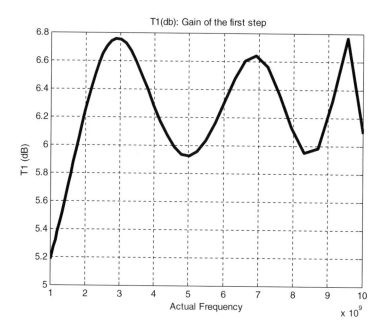

Figure 11.51   Gain performance of step 1

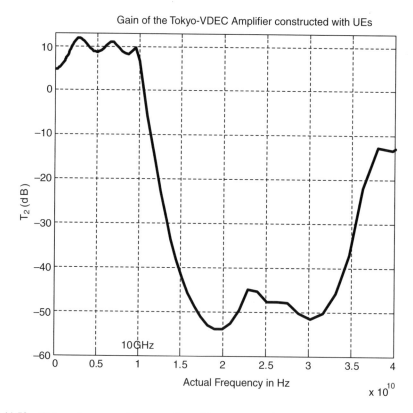

Figure 11.52   Gain performance of the overall amplifier constructed with commensurate transmission lines

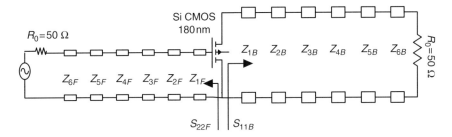

Figure 11.53   Design of a microwave amplifier using 180 nm Si CMOS of the Tokyo University VDEC with TEM commensurate transmission lines

the filter design with a UE. This is due to the presence of the active device. Remember that active devices are modeled with lumped elements which suppress the gain as the frequency increases, maintaining the periodic nature of the Richard frequency $\Omega = \tan(\omega\tau)$.

- As shown by Examples 3.1 and 3.3 of Chapter 3, a TEM line in cascade configuration resembles the operation of lumped elements. For example, if a UE has low characteristic impedance it acts like a shunt capacitor. Similarly, if a UE has high characteristic impedance, it acts like a series inductor.
- In this regard, we might think that the frontend and backend equalizers of the amplifier designed for the above example resemble the operation of a low-pass CLC or LCL section. On the other hand, one should keep in mind that this is a loose statement since we know exactly what the line impedance equations are.
- It is crucial to point out that, in the process of gain optimization, fixed delay length τ may also be included among the unknowns of the design problem. It also affects the characteristic impedance of the UEs.

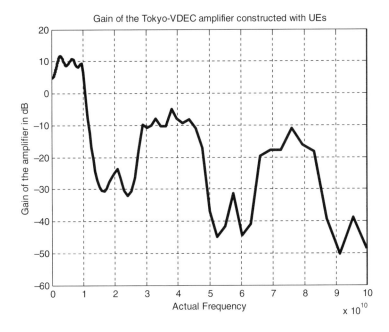

Figure 11.54   Gain performance of the amplifier designed for `Length=0.3`

For example, with the above inputs to the main program `Amplifier_Distributed.m`, if we choose `KLength=0.3`, we obtain a similar gain performance but in different characteristic impedances such that

$$ZFActual=[102.48\ 87.361\ 36.289\ 116.77\ 34.321\ 80.041]$$
$$ZBActual=[229.98\ 54.205\ 157.95\ 65.844\ 69.276\ 90.29]$$

The corresponding gain performance of the amplifier is shown in Figure 11.54.

Clearly, new characteristic impedances are smaller and perhaps easier to realize as microstrip lines. The reader can run the main program `Amplifier_Distributed.m` with different values of `KLength` and investigate the effect of this parameter on the gain performance as well as on the characteristic impedances.

Physical Realization of Characteristic Impedances

As mentioned in Chapters 3 and 4, the physical delay length L_{actual} of the UEs can be determined after selection of the substrate which is described by its thickness h, dielectric constant or permittivity $\varepsilon = \varepsilon_r \varepsilon_0$ and permeability $\mu = \mu_r \mu_0$.

In this case,

$$L_{actual} = v\tau_{actual} = \frac{\tau_{actual}}{\sqrt{\mu\varepsilon}}$$

Based on our description,

$$\tau = \frac{1}{4 \times f_{e-normalized} \times \text{kLength}} = \frac{1}{4 \times \text{kLength}}$$

Then, the actual delay is given by

$$\tau_{actual} = \frac{\tau}{f_e}\ \text{s}$$

For example, for a silicon substrate, $\varepsilon_r = 3.8$ and $\mu_r = 1$. Then, for `kLength=0.3`,

$$\tau = \frac{1}{4\text{kLength}} = \frac{1}{4 \times 0.3} = 0.83333$$

and

$$\tau_{actual} = \frac{\tau}{f_e} = \frac{0.8333}{10 \times 10^9} = 0.0833\,ns$$

Thus, the actual physical length is

$$L_{actual} = \frac{0.0833 \times 10^{-9}}{\sqrt{3.8}} \times 3 \times 10^8 = 0.004\,273\,2\ \text{m} = 4.2732\ \text{mm}$$

If UEs are realized as microstrip TEM lines, then the width of the lines can be determined by means of readily available formulas given for the characteristic impedance calculation as in Bahl's books.[19]

---

[19] Iaen Bahl, *Lumped Elements for RF and Microwave Circuits*, Artech House, 2003; Iner Bahl and Prakash Bhartia, *Microwave Solid State Circuit Design*, Wiley-Interscience, 2003.

For example, the main program `Microstrip_CharateristicImpedance.m` calculates the characteristic impedance Z of a given microstrip line for specified ε_r, width w and substrate thickness h. The listing of this program is as follows.

Program Listing 11.26 `Microstrip_CharatecristicImpedance.m`

```
% Main Program Microstrip_CharacteristicImpedance.m
clear
% Inputs:
% er : relative dielectric constant
% W : width of track
% h : thickness of the substrate
%
er=input('Enter relative permittivity: ');
W=input('Enter width of the substrate (mm): ');
h=input('Enter thickness of the substrate (mm): ');


if (W/h<1)

        e_eff=(er+1)/2+((er-
1)/2)*((1/(sqrt(1+(12*h/W))))+0.04*(1-W/h)^2);
else
    e_eff=((er+1)/2)+((er-1)/2)*(1/sqrt(1+12*h/W));
end
%
if (W/h<1)
    Z_0=60/sqrt(e_eff)*log(8*h/W+W/4*h);
else

Z_0=(120*pi/sqrt(e_eff))*(1/(W/h+1.393+0.677*log((W/h)+1.444)));
end
%
[s,errmsg]=sprintf('Z_0 = %d',Z_0);
disp(s);
%
```

Just to orient the reader, let us consider the following example.

A Hypothetical Example for the Calculation of Characteristic Impedance

Let $\varepsilon_r = 3.8$, $w = 0.02\,\text{mm} = 20\,\mu\text{m}$ and $h = 1\,\text{mm}$. Then, the characteristic impedance Z is found as

$$Z = 226.87\,\Omega$$

which is a high value of characteristic impedance; it acts like a series inductor.

On the other hand, if $w = 1\,\text{mm}$ then the characteristic impedance becomes $Z = 75.3\,\Omega$. If $w = 4\,\text{mm}$ then $Z = 32.7\,\Omega$. Similarly, if $w = 10\,\text{mm}$, then $Z = 15.8\,\Omega$ as shown in Figure 11.55.

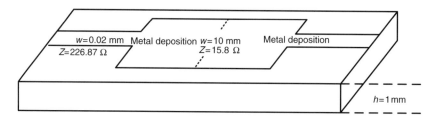

Figure 11.55 Realization of UE microstrip lines on a selected substrate with fixed actual length $L_{actual} = 4273 \, \mu m$

All the above calculations are symbolic and may not necessarily be feasible for printing on a silicon chip.

We should mention that this book is intended to cover the design capability of power transfer networks with ideal circuit elements. However, just to give some insight, we have also included elementary design concepts of circuit components stemming from electromagnetic field theory. Nevertheless, details and the related technologies for manufacturing circuit components are beyand the scope of this book. Interested readers are referred to the Bibliography presented at the end of the chapter.

We will conclude this chapter by providing the listings of the amplifier design programs and related functions developed in MATLAB. These programs and functions have already been utilized to design the single stage amplifiers included in this chapter.

Program Listing 11.27 Main program to design single stage microwave amplifier using 180 nm TSMC device

```
% Program TSMC_amp.m
clc
clear
% DESIGN OF A SINGLE STAGE AMPLIFIER
% USING A TSMC 0.18 micron NMOS SILICON TRANSISTOR
% via SIMPLIFIED REAL FREQUENCY TECHNIQUE
%------------------------------------------------------
% General Information about the device NMOS Transistor, 0.18
% micron gate
%                   Frequency range : 100MHz - 22.4 GHz;
%                   VDD = 1.8V; ID (drain DC current) = 200mA,
%                   NMOS size W/L = 1000u/0.18u;
%                   Simulation temperature = 50C.
%                   Output power= 200mW up to 20 GHz
%                   Operation Class: Class A amplifier.
%------------------------------------------------------
% INPUTS:
%                   FR(i): ACTUAL FREQUENCIES
%                   SIJ: SCATTERING PARAMETERS OF THE TSMC-CMOS FET.
%                   THESE PARAMETERS ARE READ FROM A FILE CALLED
% TMSC_Spar.txt
%                   Tflat1: FLAT GAIN LEVEL OF STEP 1 in dB.
%                   Tflat2: FLAT GAIN LEVEL OF STEP 2 in dB.
%                   FR(nd1); nd1: BEGINNING OF THE OPTIMIZATION
```

```
FREQUENCY
%                     FR(nd2); nd2: END OF OPTIMIZATION FREQUENCY
%
% OUTPUT: FRONT & BACK END EQUALIZERS ARE GENERATED EMPLOYING
% SRFT
%                     Step 1:
%                             hF: Front End Equalizer;
%                             T1: Gain of the First Step
%                             CVF:Normalized Element Values of the
% Front-End
%                             equalizer
%                             AEVF: Actual Element Values of the
% Front-End
%                     Step 2:
%                             hB: Back End Equalizer
%                             T2: Gain of the Second Step
%                             CVB:Normalized Element Values of the
% Back-End
%                             Equalizer
%                             AEVB: Actual Element Values of the
% Back-End
%
%
%                             DESIGN OF A MICROWAVE AMPLIFIER
% READ THE TRANSISTOR DATA ersad_1.txt
%
% -------------------------------------------
% Special Notes: From the given data file one should obtain that
% nd=48 (Total Number of sampling points)
% For front-end Equalizer:
%      freq(48)=22.38 GHz; freq(41)=10 GHz; freq(35)=5.02
% GHz
%      T01(48)=1.317 T01(41)=6.54 T01(35)=26.1
%      (MSG for front-end)
%      T02(48)=5.255 T02(41)=16.53 T02(35)=58.25
%      MSG(48)=7.2dB; MSGdB(41)=12.1dB;
% MSGdB(35)=17.65dB
%      (MSG for the back-end)
% ----------------------------------------------------
% ----
load TMSC_Spar.txt; % Load cmos data to the program
nd=length(TMSC_Spar)
pi=4*atan(1);
rad=pi/180;
%
% EXTRACTION OF SCATTERING PARAMETERS FROM THE GIVEN FILE
% TMSC_Spar.txt
for i=1:nd
    % Read the Actual Frequencies from the file ersad_1.txt
```

Program Listing 11.27   (Continued)

```
                        freq(i)=TMSC_Spar(i,1);
   % Read the magnitude and phase of the scattering parameters
   % from the data file ersad_1.txt
MS11(i)=TMSC_Spar(i,2);PS11(i)=rad*TMSC_Spar(i,3);
MS21(i)=TMSC_Spar(i,4);PS21(i)=rad*TMSC_Spar(i,5);
MS12(i)=TMSC_Spar(i,6);PS12(i)=rad*TMSC_Spar(i,7);
MS22(i)=TMSC_Spar(i,8);PS22(i)=rad*TMSC_Spar(i,9);
% Convert magnitude and phase to real and imaginary parts.
S11R(i)=MS11(i)*cos(PS11(i));S11X(i)=MS11(i)*sin(PS11(i));
S12R(i)=MS12(i)*cos(PS12(i));S12X(i)=MS12(i)*sin(PS12(i));
S21R(i)=MS21(i)*cos(PS21(i));S21X(i)=MS21(i)*sin(PS21(i));
S22R(i)=MS22(i)*cos(PS22(i));S22X(i)=MS22(i)*sin(PS22(i));

% Get rid off the numerical errors due to extraction of the
% scattering
% parameters from cadnace...(Vdec extraction
   if S11R(i)>=1;S11R(i)=0.99999999;S11X(i)=0.0;end
   end
%
% CONSTRUCT THE AMPLITUDE SQUARES OF THE SCATTERING PARAMETERS
 for i =1:nd
SQS21(i)=S21R(i)*S21R(i)+S21X(i)*S21X(i);
SQS12(i)=S12R(i)*S12R(i)+S12X(i)*S12X(i);
SQS11(i)=S11R(i)*S11R(i)+S11X(i)*S11X(i);
SQS22(i)=S22R(i)*S22R(i)+S22X(i)*S22X(i);
S1(i)=1.0/(1-SQS11(i));S2(i)=1.0/(1-SQS22(i));
   end
     % COMPUTATION OF MAXIMUM STABLE GAIN 'MSG'
%
   for i=1:nd;if SQS11(i)>=1;SQS11(i)=0.999999999;end
if SQS22(i)>=1;SQS22(i)=0.999999999;end
%Ideal Flat Gain Levels
      % Step 1:
                   T01(i)=SQS21(i)/(1-SQS11(i));
                   T01dB(i)=10*log10(T01(i));
      % Step 2:
                   T02(i)=T01(i)/(1-SQS22(i));
                   T02dB(i)=10*log10(T02(i));
      % Maximum Stable Gain T02=MSG
                   MSG(i)=SQS21(i)/(1-SQS11(i))/(1-SQS22(i));
                   DB(i)=10*log10(MSG(i));
end
                %
%-------------------------------------------
% Part I: DESIGN OF THE FRONT-END EQUALIZER
%-------------------------------------------
figure (1); plot(freq,T01dB);
```

```
title('Maximum Stable Gain of the First Part in dB')
ylabel('T01 (dB)');xlabel('Actual Frequency')
% Selection of data points from the given data
% between the points nd1 and nd2 out of nd points.
%First Frequency given by data file
FR1=freq(1)
%Last Frequency given by data file
FR2=freq(nd)
nd1=input('10GHz is at nd1=41,11.2GHz is at nd1=42 Enter start
point nd1=')
nd2=input('20GHz is at nd2=47;22.4GHz is at nd2=48;Enter end
point nd2=')
%nd1=42; %This corresponds to 11.23 GHz
%nd2=48;% This corresponds to 22.39 GHz
Fstart=freq(nd1)
Fend=freq(nd2)
%Define complex number j
j=sqrt(-1);
%
for k=1:(nd2-nd1+1)
            i=nd1-1+k;
% ww(i) is the normalized angular frequency indexed from k=1 to
(nd2-nd1)
            ww(k)=freq(i)/Fend;
% FR(k) is the actual frequency
            FR(k)=freq(i);
%
% Scattering Parameters of CMOS Field effect transistor
f11r(k)=S11R(i); f11x(k)=S11X(i); f11(k)=f11r(k)+j*f11x(k);
f12r(k)=S12R(i); f12x(k)=S12X(i); f12(k)=f12r(k)+j*f12x(k);
f21r(k)=S21R(i); f21x(k)=S21X(i); f21(k)=f21r(k)+j*f21x(k);
f22r(k)=S22R(i); f22x(k)=S22X(i); f22(k)=f22r(k)+j*f22x(k);
  end
%
% Step1:
Tflat1=10*log10(T01(nd2))
Tflat1=input('Enter the desired targeted gain value in dB for
Step 1: Tflat1=')
% For low pass case k=0
k=0;
xF0=input('hF(0)=0, enter front end equalizer;
[xF(1)=hF(1),xF(2)=hF(2),...,XF(n)=hF(n)]=')
OPTIONS=OPTIMSET('MaxFunEvals',20000,'MaxIter',50000);
xF=
lsqnonlin('levenberg_TR1',xF0,[],[],OPTIONS,k,ww,f11,f12,f21,f22
,Tflat1);
%
% Here it is assumed that there is no transformer in the circuit.
% That is
```

Program Listing 11.27   (Continued)

```
% h(0)=1. This fact is used in the function xtoh(xF)
hF=xtoh(xF);gF=poly_g(k,hF);
%
k=0;
T1=Gain1(hF,k,ww,f11,f12,f21,f22);
T1dB=10*log10(T1);
%
        figure (2);
        plot(ww*Fend,10*log10(T1))
        title('Step 1: Optimized Gain with Tflat1=2.1 dB')
        xlabel('Actual Frequency')
        ylabel(' T1(dB)')
%
% Computation of the actual element values of the front-end
        a=gF+hF;b=gF-hF;
nF=length(hF);
%Swap a(p) and b(p)to end up with standard MATLAB Polynomials
for i=1:nF
aF(i)=a(nF-i+1);bF(i)=b(nF-i+1);
end
   if aF(1)<1e-7;kimm=0;
display('Start with a capacitor')
%Actual element values starts with an inductor
display('Actual Element Values AEVF=')
fnorm=Fend; R0=50;
CVF=general_synthesis(aF,bF)
AEVF=Actual(kimm,R0,fnorm,CVF)
end
if aF(1)>1e-7; kimm=1;
display('start with an inductor')
%Actual element values starts with an inductor
display('Actual Element Values AEV=')
fnorm=Fend; R0=50;
CVF=general_synthesis(aF,bF)
AEVF=Actual(kimm,R0,fnorm,CVF)
end
%----------------------------------------
% Part II: DESIGN OF THE BACK-END EQUALIZER
%----------------------------------------
    SF2=reflection(hF,ww,f11,f12,f21,f22);
% output reflectance of the active device when it is terminated
in SF=hF/gF
%
% Step1:
figure (3); plot(freq,T02dB);
   title('Maximum Stable Gain of the Second Part in dB')
   ylabel('T02 (dB)');xlabel('Actual Frequency')
```

```
   display('Suggested maximum flat gain level for the second part
of the design')
Tflat2=10*log10(T02(nd2))
Tflat2=input('Enter the desired targeted gain value in dB for
Step 2: Tflat2=')
    k=0;
    xB0=input('hB(0)=0, enter back end equalizer;
[xB(1)=hB(1),xB(2)=hB(2),...,XB(n)=hB(n)]=')
    OPTIONS=OPTIMSET('MaxFunEvals',20000,'MaxIter',50000);
    xB=
lsqnonlin('levenberg_TR2',xB0,[],[],OPTIONS,k,ww,T1,SF2,f11,f12,
f21,f22,Tflat2);
%
%
  hB=xtoh(xB);gB=poly_g(k,hB);
%
T2=Gain2(hB,k,ww,T1,SF2,f11,f12,f21,f22);
T2dB=10*log10(T2);
%
    figure (3);
    plot(ww*Fend,T2dB,ww*Fend,T1dB);
    title('Step 2: Final Gain Performance of the Amplifier')
    ylabel('T2(dB)')
    xlabel('Actual Frequency')
    legend('Gain of Part II','Gain of Part I')
%
    a=gB+hB;b=gB-hB;
    nB=length(hB);
%Swap a(p) and b(p) to endup with standard MATLAB Polynomials
    for i=1:nB
    aB(i)=a(nB-i+1);bB(i)=b(nB-i+1);
    end
if aB(1)<1e-7;kimm=0;
display('Start with a capacitor')
%Actual element values starts with an inductor
display('Actual Element Values AEV=')
fnorm=Fend;R0=50;
CVB=general_synthesis(aB,bB)
AEVB=Actual(kimm,R0,fnorm,CVB)
end
if aB(1)>1e-7; kimm=1;
display('start with an inductor')
%Actual element values starts with an inductor
display('Actual Element Values AEV=')
fnorm=Fend;R0=50;
CVB=general_synthesis(aB,bB)
AEVB=Actual(kimm,R0,fnorm,CVB)
end
%
```

Program Listing 11.27   (Continued)

```
% Investigation on the stability of the amplifier:
%
% Stability Check at the output port:
        SB11=hoverg(hB,k,ww);%input ref. Cof. Back-end from
hB(p)
        SL=SB11;
        ZL=stoz(SL);
        RL=real(ZL);
        Sout=SF2;
        Zout=stoz(Sout);
        Rout=real(Zout);
        StabilityCheck_output=RL+Rout;
%
%Stability Check at the input port:
        SF=hoverg(hF,k,ww);
        SG=SF;
        ZG=stoz(SG);
        RG=real(ZG);
[Sin]=INPUT_REF(SB11,f11,f12,f21,f22);
            Zin=stoz(Sin);
            Rin=real(Zin);
StabilityCheck_input=RG+Rin;
            %
```

Program Listing 11.28 $\varepsilon_1 = T_1 -$ Tflat1

```
function
Func=levenberg_TR1(x,k,w,f11,f12,f21,f22,TS1)
%Design with no transformer. h(0)=0 case.
Na=length(w);

%Evaluate the polynomial values for h and g.
% Here use standard MATLAB function Polyval.
In this case change the order
% of the coefficients h and g
n=length(x);
        n1=n+1;
    % For Lowpass case with no transformer
        h(1)=0;
        for j=2:n1
            h(j)=x(j-1);
        end
        T1=Gain1(h,k,w,f11,f12,f21,f22);
```

```
                              for i=1:Na

Func(i)=10*log10(T1(i))-TS1;
                                        end
```

Program Listing 11.29 $\varepsilon_2 = T_2 - \text{Tflat2}$

```
function
Func=levenberg_TR2(x,k,w,T1,SF2,f11,f12,f21,f22,TS2)
%Design with no transformer. h(0)=0 case.
% Optimization function for the second step
n=length(x);
        n1=n+1;
         h(1)=0;
          for j=2:n1
               h(j)=x(j-1);
          end
%
T2=Gain2(h,k,w,T1,SF2,f11,f12,f21,f22);
Na=length(w);
                     for i=1:Na
                         Func(i)=10*log10(T2(i))-
TS2;
                  End
```

Program Listing 11.30 Generation of reflection coefficient for a given equalizer from $h(p)$ and $g(p)$ provided that $f(p) = p^k$

```
function S=hoverg(h,k,w)
% Computation of Reflection Coefficient from
h and g ; S=h/g
% INPUTS:
%           h: Coefficients of polynomial
h(p)=h(0)+h(1)p+h(2)p**2+...
%           k: power of the numerator
polynomial f(p)=p^2
% OUTPUT:
%           S: Reflection coefficient S=h/g
.
Na=length(w);
g=poly_g(k,h);
```

Program Listing 11.30   (Continued)

```
%Evaluate the polynomial values for h and g.
% Here use standard MATLAB function Polyval.
In this case change the order
% of the coefficients h and g
n1=length(h);
for i=1:n1
    hc(i)=h(n1-i+1);
    gc(i)=g(n1-i+1);
end
%
%Compute the complex variable p=jw
j=sqrt(-1);
for i=1:Na
    p=j*w(i);
    hval=polyval(hc,p);
    gval=polyval(gc,p);
    S(i)=hval/gval;
End
```

Program Listing 11.31   Computation of the backend reflection when the equalizer is terminated in a complex load SF2

```
function SE2=reflection2(h,k,w,SF2)
% Back-end Reflection (SE2) Coefficient of
the Back-end equalizer. Next is
% new FET
Na=length(w);
g=poly_g(k,h);
%Evaluate the polynomial values for h and g.
% Here use standard MATLAB function Polyval.
In this case change the order
% of the coefficients h and g
n1=length(h);
for i=1:n1
    hc(i)=h(n1-i+1);
    gc(i)=g(n1-i+1);
end
%
%Compute the complex variable p=jw
j=sqrt(-1);
for i=1:Na
    p=j*w(i);
    hval=polyval(hc,p);
```

```
gval=polyval(gc,p);
S11=hval/gval;
S22=-conj(hval)/gval;
fval=p^k;
S21=fval/gval;

D(i)=1-S11*SF2(i);

SE2(i)=S22+S21*S21*SF2(i)/D(i);
end
```

Program Listing 11.32 Generation of an input impedance Z from a given real normalized reflection coefficient

```
function Z=stoz(S)
% This function computes the impedance from
the given reflection
% coefficient
% input: Array
%    S: reflection coefficient
% Output: Array
%    Z: Impedance

num=1+S;
denom=1-S;
nz=length(S);
for i=1:nz
    Z(i)=num(i)/denom(i);
End
```

Program Listing 11.33 Computation of the input reflectance of an equalizer from the given LC ladder components

```
function Sin=Circuit(kimm,w,a)
%This function computes the equalizer
driving point reflection coefficient from
the
%given circuit components....
% INPUTS:
%                kimm>Control Integer:
%                            if kimm=1
```

Program Listing 11.33   (Continued)

```
equalizer starts with a inductance,
%                                  if kimm=0
equalizer starts with a
%                                    capacitor.
%                 fnorm: Normalization
frequency
%                 w(i) Normalized angular
frequency for which driving point
%                 immitance is computed...
%                 a(i): Normalized Circuit Component Values
% OUTPUT:
%                 Sin(i): Array, driving point
reflection coefficient
nd=length(a);
%
n=length(w);
%Frequency Normalization: Normalized
frequencies are equal to normalized angular
frequencies w(i)=freq(i)/fnorm
j=sqrt(-1);
for i=1:n
                        p=j*w(i);
            % Set the End Immitance
            A(nd-1)=a(nd-1)*p+a(nd);
                            for r=2:(nd-
1)

A(nd-r)=a(nd-r)*p+1/A(nd-r+1);
                            end
%Driving point immitance of the equalizer
                        Q(i)=A(1);

if kimm==0

Sin(i)=(1-Q(i))/(1+Q(i));

end

if kimm==1

Sin(i)=(Q(i)-1)/(Q(i)+1);

end
                end
```

Program Listing 11.34   Computation of the input reflectance of an active device when it is terminated in a complex load

```
function
[Sin]=INPUT_REF(SB11,f11,f12,f21,f22)
% Sin: input reflectance of an active device
[F] when it is
% terminated in a complex reflectance SB11
% INPUTS:
%                         SCATTERING PARAMETERS OF
THE ACTIVE DEVICE fij
%                         complex termination SB11
% OUTPUT:
%                         Sin: Input Reflectance of the active
%                         device [F]
Na=length(f11);
for i=1:Na
    Sin(i)=f11(i)+f12(i)*f21(i)*SB11(i)/(1-
f22(i)*SB11(i));
    End
```

Program Listing 11.35   Computation of the backend reflectance of an active device

```
function SF2=reflection(h,w,f11,f12,f21,f22)
% Computation of the back-end reflection of
the active when it is
% terminated in S22=h/g
%Design with no transformer. h(0)=1 case.
% INPUTS
%                         h(p): Numerator
polynomial (Coefficients h array)
%                         w: Normalized
angular frequencies; array of size Na
%         fij: Scattering parameters of the
active device
% OUTPUT:
%                         SF2: Back-end
reflection Coefficients looking at
%         the back-end of the active
device.
Na=length(w);
%
n1=length(h);
k=0;
    g=poly_g(k,h);
```

Program Listing 11.35   (Continued)

```
   % Change the order of the coefficients to
% compute
% the values of the polynomials.
    for i=1:n1
    hc(i)=h(n1-i+1);
    gc(i)=g(n1-i+1);
end
%
%Compute the complex variable p=jw
j=sqrt(-1);
for i=1:Na
    p=j*w(i);
    hval=polyval(hc,p);
    gval=polyval(gc,p);
    S22=hval/gval;
        SF2(i)=f22(i)+f12(i)*f21(i)*S22/(1-
f11(i)*S22);
end
```

Program Listing 11.36   Design of a microwave amplifier with commensurate transmission lines

```
% Program Amplifier_Distributed.m
clc
clear
% DESIGN OF A SINGLE STAGE AMPLIFIER USING A 0.18 micron CMOS
% SILICON TRANSISTOR
% with Distributed elements via SIMPLIFIED REAL FREQUENCY
TECHNIQUE
% INPUTS:
%               FR(i): ACTUAL FREQUENCIES
%               SIJ: SCATTERING PARAMETERS OF THE CMOS FET.
%               THESE PARAMETERS ARE READ FROM A FILE CALLED
% newcmos.txt
%               Tflat1: FLAT GAIN LEVEL OF STEP 1 in dB.
%               Tflat2: FLAT GAIN LEVEL OF STEP 2 in dB.
%               nd1: BEGINNING OF THE OPTIMIZATION FREQUENCY
%(Note that f(nd1) is the beginning of the passband;
%nd1 is the index of the frequency f(nd1)
%               nd2: END OF OPTIMIZATION FREQUENCY
%(Note that f(nd2) is the end of the passband;
%nd2 is the index of the frequency f(nd2)
```

```
%               q: number transmission zeros at DC for
%               k: total number of cascaded section
% Notes: f(lmbda)=(lmbda)^q(1-lmbda^2)^k/2
% In the optimization, all the frequencies are normalized with
% respect to
% upper edge of the passband. In other words, wc2=1 is fixed.
% All the polynomials are in MATLAB format.
% No Transformer is employed in the equalizers.
% Therefore, hF(nF+1)=0
% Front end equalizer is described by
%       hF=[hF(1) hF(2)...hF(nF) 0]; Standard MATLAB form
% Back-end equalizer is described by
%       hB=[hB(1) hB(2)...hB(nB) 0]
%
% OUTPUT: FRONT & BACK END EQUALIZERS ARE GENERATED EMPLOYING
% SRFT
%    Step 1:
%    hF: Optimized Front End Equalizer;
%    T1: Gain of the First Step in dB
%    [ZF]:Normalized Cascaded Characteristic impedances of the
% Front-End
%    [ZFActual]:Actual Characteristic impedances


%    Step 2:
%    hB: Optimized Back End Equalizer
%    T2: Gain of the Second Step in dB
% [ZB]:Normalized Cascaded Characteristic impedances of the
% Back-End
% [ZFActual]:Actual Characteristic impedances
%
%DESIGN OF A MICROWAVE AMPLIFIER
%    READ THE TRANSISTOR DATA from newcmos.txt
load newcmos.txt
display('nd=Total number of sampling points for the S-Par
measurements')
nd=length(newcmos)
%
% EXTRACTION OF SCATTERING PARAMETERS FROM THE GIVEN FILE
% newcmos.txt
for i=1:nd
    % Read the Actual Frequencies from the file nico_1.txt
                freq(i)=newcmos(i,1);
                % Read the real and the imaginary parts of
the scattering parameters
% from the data file newcmos.txt
                S11R(i)=newcmos(i,2);
                S11X(i)=newcmos(i,3);
                %
```

Program Listing 11.36   (Continued)

```
                                S12R(i)=newcmos(i,4);
                                S12X(i)=newcmos(i,5);
                                %
                                S21R(i)=newcmos(i,6);
                                S21X(i)=newcmos(i,7);
                                %
                                S22R(i)=newcmos(i,8);
                                S22X(i)=newcmos(i,9);
%
% Get rid of the numerical errors due to extraction of the
scattering
% parameters from cadnace...(Vdec extraction
                                            if S11R(i)>=1;
S11R(i)=0.99999999;
                                            S11X(i)=0.0;
                                            end
end
%
display('Beginning of the frequency measurements freqn(1) for S-
Parameters')
freq(1),
display('End of measurement frequency freq(nd)')
freq(nd),
% CONSTRUCT THE AMPLITUDE SQUARES OF THE SCATTERING PARAMETERS
          for i =1:nd

SQS21(i)=S21R(i)*S21R(i)+S21X(i)*S21X(i);
                                %

SQS12(i)=S12R(i)*S12R(i)+S12X(i)*S12X(i);
                                %

SQS11(i)=S11R(i)*S11R(i)+S11X(i)*S11X(i);
                                %

SQS22(i)=S22R(i)*S22R(i)+S22X(i)*S22X(i);
                %
                                            S1(i)=1.0/(1-SQS11(i));
                                            S2(i)=1.0/(1-SQS22(i));
                end
        % COMPUTATION OF MAXIMUM STABLE GAIN 'MSG'
        %
                        for i=1:nd
                                if SQS11(i)>=1
SQS11(i)=0.999999999;
                                end
```

```
                                    %
                                                    if
SQS22(i)>=1;

SQS22(i)=0.999999999;
                                                            end
%Ideal Flat Gain Levels
      % Step 1:
                      T01(i)=SQS21(i)/(1-SQS11(i));
      % Step 2:
                      T02(i)=T01(i)/(1-SQS22(i));
      % Maximum Stable Gain T02=MSG
                      MSG(i)=SQS21(i)/(1-SQS11(i))/(1-SQS22(i));
                      DB(i)=10*log10(MSG(i));
                        end
                      %
                      figure(1)
                      plot(freq,DB)
                      title('Maximum Stable Gain of the 180 nm CMOS
% of VDEC of Tokyo')
                      xlabel('Actual Frequency')
                      ylabel('MSG(dB)')
                      display('i=51; freq(51)= 1GHz')
                      display('i=102; freq(101)= 10GHz')
                  % COMPUTATION OF NORMALIZED INPUT (zin) OUTPUT
% (zout) IMPEDANCES
                          j=sqrt(-1);
                          for i=1:nd
                                 if S11R(i)>=1;
                                 S11R(i)=0.999999;
                                 end

zin(i)=(1+S11R(i)+j*S11X(i))/(1-S11R(i)-j*S11X(i));

zout(i)=(1+S22R(i)+j*S22X(i))/(1-S22R(i)-j*S22X(i));
                                          rin(i)=real(zin(i));
                                          xin(i)=imag(zin(i));
                                          rout(i)=real(zout(i));
                                          xout(i)=imag(zout(i));
                                   end
      % Selection of data points from the given data between
% the points nd1and nd2 out of nd points.
                      %First Frequency given by data file
                      FR1=freq(1)
                      %Last Frequency given by data file
                      FR2=freq(nd)
  nd1=input('Enter start point nd1=')
  nd2=input('Enter end point nd2=')
  display('Your optimization starts at ')
```

Program Listing 11.36   (Continued)

```
    Fstart=freq(nd1)
    display(' Your optimization ends at')
    Fend=freq(nd2)
 for ka=1:(nd2-nd1+1)
                           i=nd1-1+ka;
% ww(i) is the normalized angular frequency indexed from ka=1 to
(nd2-nd1)
                           ww(ka)=freq(i)/Fend;
                       %
                       % FR(ka) is the actual frequency
                             FR(ka)=freq(i);
                       %
                       j=sqrt(-1);
                       % Scattering Parameters of CMOS
                                  f11r(ka)=S11R(i);
                                  f11x(ka)=S11X(i);
                                  f11(ka)=f11r(ka)+j*f11x(ka);
f12r(ka)=S12R(i);
f12x(ka)=S12X(i);
f12(ka)=f12r(ka)+j*f12x(ka);
f21r(ka)=S21R(i);
f21x(ka)=S21X(i);

f21(ka)=f21r(ka)+j*f21x(ka);
f22r(ka)=S22R(i);
f22x(ka)=S22X(i);
f22(ka)=f22r(ka)+j*f22x(ka);
end
%
display('UE has a delay tou=1/4/Fend/kLength')
kLength=input('Enter kLength=')
% Step1:
display('Ideal flat gain level in dB')
Tflat1=10*log10(T01(nd2))
display(' It is suggested that Tfalt1=6dB')
Tflat1=input('Enter the desired targeted gain value in dB for
Step 1: Tflat1=')
% For low pass case q=0; n=k total number of cascaded elements
qF=0;%Transmission zeros at DC
    xF0=input('hF(0)=0, enter front end equalizer;
[xF(1)=hF(1),xF(2)=hF(2),...,XF(n)=hF(n)]=')
    OPTIONS=OPTIMSET('MaxFunEvals',20000,'MaxIter',50000);
nF=length(xF0);%Total number of UE or cascaded elements
kF=nF;
% Here we assume that n=k;
% in other words, total number of cascaded elements are equal to
% total
```

```
% elements
% number of
    xF=
lsqnonlin('levenberg_TR1',xF0,[],[],OPTIONS,kLength,qF,kF,ww,f11
,f12,f21,f22,Tflat1);
%
%
%        Here it is assumed that there is no transformer in the
circuit.
% That is q=0 and h(n+1)=0. Hence,
                         hF=[xF 0];
              qF=0.0;
              T1=Gain1(hF,kLength,qF,kF,ww,f11,f12,f21,f22);
figure(2)
plot(Fend*ww,10*log10(T1))
title(' T1(db): Gain of the first step')
xlabel(' Actual Frequency')
ylabel('T1 (dB) ')
%Synthesis By Richard extractions
[GF,HF,FF,gF]=SRFT_htoG(hF,qF,kF);
%----------------------------------------
[zF]=UE_sentez(hF,gF)
%----------------------------------------
% Actual impedances
R0=50;
[ZFActual]= R0*[zF]
%----------------------------------------
SF2=reflection(hF,kLength,qF,kF,ww,f11,f12,f21,f22);
% DESIGN OF THE BACK-END EQUALIZER
%
% Step2:
qB=0;
Tflat2=10*log10(T02(nd2))
display(' It is suggested that Tflat2=10.5 dB')
Tflat2=input('Enter the desired targeted gain value in dB for
Step 1: Tflat2=')
xB0=input('hB(0)=0, enter back end equalizer;
[xB(1)=hB(1),xB(2)=hB(2),...,XB(n)=hB(n)]=')
 nB=length(xB0);
 kB=nB;%Total number of cascaded sections
 OPTIONS=OPTIMSET('MaxFunEvals',20000,'MaxIter',50000);
 xB=
lsqnonlin('levenberg_TR2',xB0,[],[],OPTIONS,kLength,qB,kB,ww,T1,
SF2,Tflat2);
%
%
              hB=[xB 0];
              T2=Gain2(hB,kLength,qB,kB,ww,T1,SF2);
%
```

Program Listing 11.36   (Continued)

```
figure (3)
plot(Fend*ww,10*log10(T2))
title('Gain of the amplifier designed with distributed
elements')
xlabel('Actual Frequency')
ylabel('Gain of the amplifier designed with UEs in dB')
[GB,HB,FB,gB]=SRFT_htoG(hB,qB,kB);
%-------------------------------------------
% Synthesis of the back-end equalizer
[zB]=UE_sentez(hB,gB)
%-------------------------------------------
% Actual impedances
R0=50;
[ZBActual]= R0*[zB]
%-------------------------------------------
% Print the result over the measured frequencies of the S-
Parameters
[wprint,f11 f12 f21 f22]=Scattering_Parameters(newcmos,Fend);
SF_Print=reflection(hF,kLength,qF,kF,wprint,f11,f12,f21,f22);
T1_Print=Gain1(hF,kLength,qF,kF,wprint,f11,f12,f21,f22);
T2_Print=Gain2(hB,kLength,qB,kB,wprint,T1_Print,SF_Print);
Gain_dB=10*log10(T2_Print);
figure (4)
plot(Fend*wprint,Gain_dB)
title('Gain of the Tokyo-VDEC amplifier constructed with UEs')
xlabel('Actual Frequency')
ylabel('Gain of the amplifier in dB')
```

Program Listing 11.37   Design of a microwave amplifier using unit elements, main
program `Amplifier_Distributed.m`

```
% Program Amplifier_Distributed.m
clc
clear
% DESIGN OF A SINGLE STAGE AMPLIFIER USING A 0.18 micron CMOS
% SILICON TRANSISTOR
% with Distributed elements via SIMPLIFIED REAL FREQUENCY
% TECHNIQUE
% INPUTS:
%         FR(i): ACTUAL FREQUENCIES
%         SIJ: SCATTERING PARAMETERS OF THE CMOS FET.
```

```
%            THESE PARAMETERS ARE READ FROM A FILE CALLED
% newcmos.txt
%         Tflat1: FLAT GAIN LEVEL OF STEP 1 in dB.
%         Tflat2: FLAT GAIN LEVEL OF STEP 2 in dB.
%         nd1: BEGINNING OF THE OPTIMIZATION FREQUENCY
% (Note that f(nd1) is the beginning of the passband;
% nd1 is the index of the frequency f(nd1)
%         nd2: END OF OPTIMIZATION FREQUENCY
% (Note that f(nd2) is the end of the passband;
% nd2 is the index of the frequency f(nd2)
%         q: number transmission zeros at DC for
%         k: total number of cascaded section
% Notes: f(lmbda)=(lmbda)^q(1-lmbda^2)^k/2
% In the optimization, all the frequencies are normalized with
% respect to
% upper edge of the passband. In other words, wc2=1 is fixed.
% All the polynomials are in MATLAB format.
% No Transformer is employed in the equalizers.
% Therefore, hF(nF+1)=0
% Front end equalizer is described by
%         hF=[hF(1) hF(2)...hF(nF) 0]; Standard MATLAB form
% Back-end equalizer is described by
%         hB=[hB(1) hB(2)...hB(nB) 0]
%
% OUTPUT: FRONT & BACK END EQUALIZERS ARE GENERATED EMPLOYING
% SRFT
%    Step 1:
%    hF: Optimized Front End Equalizer;
%    T1: Gain of the First Step in dB
%    [ZF]:Normalized Cascaded Characteristic impedances of the
% Front-End
%    [ZFActual]:Actual Characteristic impedances

%    Step 2:
%    hB: Optimized Back End Equalizer
%    T2: Gain of the Second Step in dB
%   [ZB]:Normalized Cascaded Characteristic impedances of the
% Back-End
%   [ZFActual]:Actual Characteristic impedances
%
%DESIGN OF A MICROWAVE AMPLIFIER
% READ THE TRANSISTOR DATA from newcmos.txt
load newcmos.txt
display('nd=Total number of sampling points for the S-Par
measurements')
nd=length(newcmos)
%
% EXTRACTION OF SCATTERING PARAMETERS FROM THE GIVEN FILE
% newcmos.txt
```

Program Listing 11.37   (Continued)

```
for i=1:nd
    % Read the Actual Frequencies from the file newcmos.txt
                     freq(i)=newcmos(i,1);
                     % Read the real and the imaginary parts of
% the scattering parameters
    % from the data file newcmos.txt
                     S11R(i)=newcmos(i,2);
                     S11X(i)=newcmos(i,3);
                     %
                     S12R(i)=newcmos(i,4);
                     S12X(i)=newcmos(i,5);
                     %
                     S21R(i)=newcmos(i,6);
                     S21X(i)=newcmos(i,7);
                     %
                     S22R(i)=newcmos(i,8);
                     S22X(i)=newcmos(i,9);
%
% Get rid of the numerical errors due to extraction of the
% scattering
% parameters from cadnace...(Vdec extraction
.
                                       if S11R(i)>=1;
S11R(i)=0.99999999;
                                       S11X(i)=0.0;
                                       end
end
%
display('Beginning of the frequency measurements freqn(1) for S-
Parameters')
freq(1),
display('End of frequency measurements freq(nd)')
freq(nd),
% CONSTRUCT THE AMPLITUDE SQUARES OF THE SCATTERING PARAMETERS
             for i =1:nd

SQS21(i)=S21R(i)*S21R(i)+S21X(i)*S21X(i);
                                 %

SQS12(i)=S12R(i)*S12R(i)+S12X(i)*S12X(i);
                                 %

SQS11(i)=S11R(i)*S11R(i)+S11X(i)*S11X(i);
                                 %

SQS22(i)=S22R(i)*S22R(i)+S22X(i)*S22X(i);
                     %
                                 S1(i)=1.0/(1-SQS11(i));
```

```
                                        S2(i)=1.0/(1-SQS22(i));
                end
        % COMPUTATION OF MAXIMUM STABLE GAIN 'MSG'
        %
                        for i=1:nd
                                if SQS11(i)>=1

SQS11(i)=0.999999999;

                                end
                                %
                                                        if
SQS22(i)>=1;

SQS22(i)=0.999999999;
                                                        end
%Ideal Flat Gain Levels
        % Step 1:
                        T01(i)=SQS21(i)/(1-SQS11(i));
        % Step 2:
                        T02(i)=T01(i)/(1-SQS22(i));
        % Maximum Stable Gain T02=MSG
                        MSG(i)=SQS21(i)/(1-SQS11(i))/(1-SQS22(i));
                        DB(i)=10*log10(MSG(i));
                  end
                %
                figure(1)
                plot(freq,DB)
                title('Maximum Stable Gain of the 180 nm CMOS
% of VDEC of Tokyo')
                xlabel('Actual Frequency')
                ylabel('MSG(dB)')
                display('i=51; freq(51)= 1GHz')
                display('i=101; freq(101)= 10GHz')
                % COMPUTATION OF NORMALIZED INPUT (zin) OUTPUT
% (zout) IMPEDANCES
                                j=sqrt(-1);
                                for i=1:nd
                                        if S11R(i)>=1;
                                        S11R(i)=0.999999;
                                        end

zin(i)=(1+S11R(i)+j*S11X(i))/(1-S11R(i)-j*S11X(i));

zout(i)=(1+S22R(i)+j*S22X(i))/(1-S22R(i)-j*S22X(i));
                                        rin(i)=real(zin(i));
                                        xin(i)=imag(zin(i));
                                        rout(i)=real(zout(i));
                                        xout(i)=imag(zout(i));
```

Program Listing 11.37   (Continued)

```
                                         end
         % Selection of data points from the given data between
    % the points nd1and nd2 out of nd points.
                      %First Frequency given by data file
                      FR1=freq(1)
                      %Last Frequency given by data file
                      FR2=freq(nd)
    nd1=input('Enter start point nd1=')
    nd2=input('Enter end point nd2=')
    display('Your optimization starts at ')
    Fstart=freq(nd1)
    display(' Your optimization ends at')
    Fend=freq(nd2)
                      for ka=1:(nd2-nd1+1)
                            i=nd1-1+ka;
    % ww(i) is the normalized angular frequency indexed from ka=1 to
    % (nd2-nd1)
                         ww(ka)=freq(i)/Fend;
                      %
                      % FR(ka) is the actual frequency
                         FR(ka)=freq(i);
                      %
                      j=sqrt(-1);
                      % Scattering Parameters of CMOS
                                  f11r(ka)=S11R(i);
                                  f11x(ka)=S11X(i);
                                  f11(ka)=f11r(ka)+j*f11x(ka);
                                  %
                                  f12r(ka)=S12R(i);
                                  f12x(ka)=S12X(i);
                                  f12(ka)=f12r(ka)+j*f12x(ka);
                                  %
                                  f21r(ka)=S21R(i);
                                  f21x(ka)=S21X(i);
                                  f21(ka)=f21r(ka)+j*f21x(ka);
                                  %
                                  f22r(ka)=S22R(i);
                                  f22x(ka)=S22X(i);
                                  f22(ka)=f22r(ka)+j*f22x(ka);
                      end
    %
    display('UE has a delay tou=1/4/Fend/kLength')
    kLength=input('Enter kLength=')
    % Step1:
    display('Ideal flat gain level in dB')
```

```
Tflat1=10*log10(T01(nd2))
display(' It is suggested that Tfalt1=6dB')
Tflat1=input('Enter the desired targeted gain value in dB for
Step 1: Tflat1=')
% For low pass case q=0; n=k total number of cascaded elements
qF=0;%Transmission zeros at DC
     xF0=input('hF(0)=0, enter front end equalizer;
[xF(1)=hF(1),xF(2)=hF(2),...,XF(n)=hF(n)]=')
     OPTIONS=OPTIMSET('MaxFunEvals',20000,'MaxIter',50000);
nF=length(xF0);%Total number of UE or cascaded elements
kF=nF;
% Here we assume that n=k;
% in other words, total number of cascaded elements are equal to
% total
% elements
% number of
    xF=
lsqnonlin('Distributed_TR1',xF0,[],[],OPTIONS,kLength,qF,kF,ww,f
11,f12,f21,f22,Tflat1);
%
%
%          Here it assumed that there is no transformer in the
% circuit.
% That is q=0 and h(n+1)=0. Hence,
                         hF=[xF 0];
          qF=0.0;
          T1=Gain1(hF,kLength,qF,kF,ww,f11,f12,f21,f22);
figure(2)
plot(Fend*ww,10*log10(T1))
title(' T1(db): Gain of the first step')
xlabel(' Actual Frequency')
ylabel('T1 (dB) ')
%Synthesis By Richard extractions
[GF,HF,FF,gF]=SRFT_htoG(hF,qF,kF);
%-----------------------------------------
[zF]=UE_sentez(hF,gF)
%-----------------------------------------
% Actual impedances
R0=50;
[ZFActual]= R0*[zF]
%-----------------------------------------
  SF2=reflection(hF,kLength,qF,kF,ww,f11,f12,f21,f22);
% DESIGN OF THE BACK-END EQUALIZER
%
% Step2:
qB=0;
Tflat2=10*log10(T02(nd2))
```

Program Listing 11.37   (Continued)

```
display(' It is suggested that Tflat2=10.5 dB')
Tflat2=input('Enter the desired targeted gain value in dB for
Step 1: Tflat2=')
 xB0=input('hB(0)=0, enter back end equalizer;
[xB(1)=hB(1),xB(2)=hB(2),...,XB(n)=hB(n)]=')
 nB=length(xB0);
 kB=nB;%Total number of cascaded sections
 OPTIONS=OPTIMSET('MaxFunEvals',20000,'MaxIter',50000);
 xB=
lsqnonlin('Distributed_TR2',xB0,[],[],OPTIONS,kLength,qB,kB,ww,T
1,SF2,Tflat2);
 %
 %
          hB=[xB 0];
          T2=Gain2(hB,kLength,qB,kB,ww,T1,SF2);
%
figure (3)
plot(Fend*ww,10*log10(T2))
title('Gain of the amplifier designed with distributed
elements')
xlabel('Actual Frequency')
ylabel('Gain of the amplifier designed with UEs in dB')
[GB,HB,FB,gB]=SRFT_htoG(hB,qB,kB);
%------------------------------------------
% Synthesis of the back-end equalizer
[zB]=UE_sentez(hB,gB)
%------------------------------------------
% Actual impedances
R0=50;
[ZBActual]= R0*[zB]
%------------------------------------------
% Print the result over the measured frequencies of the S-
% Parameters
[wprint,f11 f12 f21 f22]=Scattering_Parameters(newcmos,Fend);
 SF_Print=reflection(hF,kLength,qF,kF,wprint,f11,f12,f21,f22);
 T1_Print=Gain1(hF,kLength,qF,kF,wprint,f11,f12,f21,f22);
 T2_Print=Gain2(hB,kLength,qB,kB,wprint,T1_Print,SF_Print);
 Gain_dB=10*log10(T2_Print);
 figure (4)
plot(Fend*wprint,Gain_dB)
title('Gain of the Tokyo-VDEC amplifier constructed with UEs')
xlabel('Actual Frequency')
ylabel('Gain of the amplifier in dB')
```

Program Listing 11.38 Step I of the amplifier design with UEs, objective function to be minimized

```
function
Func=Distributed_TR1(x,kLength,q,k,w,f11,f12,f21,f22,TS1)
%Design with no transformer. h(0)=0 case.
Na=length(w);

%Evaluate the polynomial values for h and g.
% Here use standard MATLAB function Polyval. In this case
% change the order
% of the coefficients h and g
n=length(x);
      n1=n+1;
% For Lowpass case with no transformer
  h(n1)=0;
    for j=1:n
        h(j)=x(j);
    end
    T1=Gain1(h,kLength,q,k,w,f11,f12,f21,f22);
                      for i=1:Na

Func(i)=10*log10(T1(i))-TS1;
                        end
```

Program Listing 11.39 Step II of the amplifier design with UEs, objective function to be minimized

```
function
Func=Distributed_TR2(x,kLength,q,k,w,T1,SF2,TS2)
%Design with no transformer. h(0)=0 case.
% Optimization function for the second step
n=length(x);
        n1=n+1;
        h(n1)=0;
        for j=1:n
            h(j)=x(j);
        end
%
T2=Gain2(h,kLength,q,k,w,T1,SF2);
Na=length(w);
                    for i=1:Na

Func(i)=10*log10(T2(i))-TS2;
                      end
```

Program Listing 11.40   Generation of backend reflectance in the Richard domain

```
function
SF2=reflection(h,kLength,q,k,w,f11,f12,f21,f22)
% Computation of the back-end reflectance of
% the active in Richard domain
% when it is terminated in S22F=hF/gF
%Design with no transformer. h(0)=1 case.
% INPUTS
%                        h(lmbda): Numerator
% polynomial (Coefficients h array)
%                        w: Normalized angular
% frequencies; array of size Na
%                        fij: Scattering
% parameters of the active device
% OUTPUT:
%                        SF2: Back-end
% reflection Coefficients looking at
%                        the back-end of the
% active device.
Na=length(w);
%
n1=length(h);
%Compute the complex variable lmbda=j*tn(w*tau)
j=sqrt(-1);
Na=length(w);
[G,H,F,g]=SRFT_htoG(h,q,k);%
%Compute the complex variable
lmbda=j*tan(w*tau)
% Note that fe=stop band frequency=1.5*fc2;
wc2=ww(nd2)
tau=1/4/1/kLength;
%
j=sqrt(-1);
for i=1:Na
    teta=w(i)*tau;
    omega=tan(teta);
    lmbda=j*omega;
    %
    hval=polyval(h,lmbda);
    gval=polyval(g,lmbda);
    S22=hval/gval;
        SF2(i)=f22(i)+f12(i)*f21(i)*S22/(1-
f11(i)*S22);
End
```

Program Listing 11.41   Gain of the first step in the Richard domain

```
function
T1=Gain1(h,kLength,q,k,w,f11,f12,f21,f22)
%Gain1 of step 1. With front end equalizer
% in Richard domain
% This gain is computed in lmbda domain.
% lmbda=j*OMEGA
% OMEGA=tan(w*tau)
% tau=1/4/w(na)
%         INPUTS:
                             % h:
% Coefficients of the h(p) polynomial for
                             % the input
% equalizer
                             %k=0 for lowpass
% transformerless design
                             % w(i): Array,
% Normalized angular frequency
                             %fij: Scattering
% parameters of the active
                             %device. In the
% present case 0.18 Micron Si
                             %CMOSFET...
%                    %
%      OUTPUT: T1; gain of the first stage......

Na=length(w);
[G,H,F,g]=SRFT_htoG(h,q,k);%
%Compute the complex variable
lmbda=j*tan(w*tau)
% Note that fe=stop band frequency=1.5*fc2;
wc2=ww(nd2)
tau=1/4/1/kLength;
%
j=sqrt(-1);
for i=1:Na
    teta=w(i)*tau;
    omega=tan(teta);
    lmbda=j*omega;
    %
    fval=(-1)^q*(lmbda)^q*(1-lmbda^2)^(k/2);
  %
    hval=polyval(h,lmbda);
    gval=polyval(g,lmbda);
    S22=hval/gval;
    S21=fval/gval;
```

Program Listing 11.41   (Continued)

```
MS21=abs(S21);
MS21SQ=MS21*MS21;
Mf21(i)=abs(f21(i));
Mf21SQ(i)=Mf21(i)*Mf21(i);
MD(i)=abs(1-S22*f11(i));
MDSQ(i)=MD(i)*MD(i);
T1(i)=MS21SQ*Mf21SQ(i)/MDSQ(i);
end
```

Program Listing 11.42   Gain of the second step in the Richard domain

```
function T2=Gain2(h,kLength,q,k,w,T1,SF2)
%Gain2 of step 2 in Richard domain.
%With back-end equalizer
Na=length(w);
%Na=length(w);
[G,H,F,g]=SRFT_htoG(h,q,k);%
%Compute the complex variable
lmbda=i*tan(w*tau)
tau=1/4/1/kLength;
j=sqrt(-1);
for i=1:Na
    teta=w(i)*tau;
    omega=tan(teta);
    lmbda=j*omega;
    %
    fval=(-1)^q*(lmbda)^q*(1-lmbda^2)^(k/2);
%
    hval=polyval(h,lmbda);
    gval=polyval(g,lmbda);
    S11=hval/gval;
    S21=fval/gval;
    MS21=abs(S21);
    MS21SQ=MS21*MS21;
    MD(i)=abs(1-S11*SF2(i));
    MDSQ(i)=MD(i)*MD(i);
    T2(i)=T1(i)*MS21SQ/MDSQ(i);
    end
```

Program Listing 11.43 Generation of the scattering parameters for a lossless two-port in the Richard domain

```
function [G,H,F,g]=SRFT_htoG(h,q,k)
%This function generates the complete
% scattering parameters
% from the given h and F in lambda domain
% all the functions are given in lmbda
% square.
% Inputs:
%          given h and q,k
%          where F=(lmbd)^2q*(1-lmbd^2)^k
% Outputs:
%          G,H,F and g
% Note that F(lmbda)=(-1)^q*(lmbda^2q)*(1-
% lmbda^2)^k
F=lambda2q_UE2k(k,q);% Generation of
F=f(lambda)*f(-lambda)
h_=paraconj(h);
He=conv(h,h_);%Generation of h(lmbda)*h(-
% lambda)
H=poly_eventerms(He);%select the even terms
% of He to set H
G=vector_sum(H,F);% Generate G=H+F with
% different sizes
% Construction of g(lmbda) from h(lmbda)
r=roots(G);
z=sqrt(r);
g1=sqrt(abs(G(1)));
g=g1*real(poly(-z));
return
```

Program Listing 11.44 Richard extraction of UE lines

```
function [z]=UE_sentez(h,g)
% Richard Extraction written by Dr. Metin Sengul
% of Kadir Has University, Istanbul, Turkey.
% Inputs:
% h(lambda), g(lambda)as MATLAB Row vectors
% Output:
% [z] Characteristic impedances of UE lines
%_____
```

Program Listing 11.44   (Continued)

```
na=length(g);
for i=1:na
     h0i(i)=h(na-i+1);
     g0i(i)=g(na-i+1);
end
m=length(g0i);
for a=1:m
     gt=0;
     ht=0;
     k=length(g0i);
     for b=1:k
          gt=gt+g0i(b);
          ht=ht+h0i(b);
     end
     if a==1;
          zp=1;
     else
          zp=zz(a-1);
     end
     zz(a)=((gt+ht)/(gt-ht))*zp;

     x=0;
     y=0;
     for c=1:k-1
     x(c)=h0i(c)*gt-g0i(c)*ht;
     y(c)=g0i(c)*gt-h0i(c)*ht;
     end

     N=0;
     D=0;
     n=0;
     for e=1:k-1
          N(e)=n+x(e);
          n=N(e);
          d=0;
          for f=1:e
               ara=((-1)^(e+f))*y(f);
               D(e)=ara+d;
               d=D(e);
          end
     end

     h0i=N;
     g0i=D;
end
z=zz;
return
```

Bibliography

Real Frequency Line Segment Technique (LST)

[1] H. J. Carlin, 'A New Approach to Gain-Bandwidth Problems,' *IEEE Transactions on Circuits and Systems*, Vol. **23**, pp. 170–175, April, 1977.

[2] H. J. Carlin and J. J. Komiak, 'A New Method of Broadband Equalization Applied to Microwave Amplifiers,' *IEEE Transactions on Microwave Theory and Techniques*, Vol. **27**, pp. 93–99, February, 1979.

[3] H. J. Carlin and P. Amstutz, 'On Optimum Broadband Matching,' *IEEE Transactions on Circuits and Systems*, Vol. **28**, pp. 401–405, May, 1981.

[4] B. S. Yarman, 'Real Frequency Broadband Matching Using Linear Programming', *RCA Review*, Vol. **43**, pp. 626–654, 1982.

Direct Computational Technique (DCT)

[5] B. S. Yarman, 'Broadband Matching a Complex Generator to a Complex Load', PhD thesis, Cornell University, 1982.

[6] H. J. Carlin and B. S. Yarman, 'The Double Matching Problem: Analytic and Real Frequency Solutions,' *IEEE Transactions on Circuits and Systems*, Vol. **30**, pp. 15–28, January, 1983.

[7] H. J. Carlin and P. P. Civalleri, 'On Flat Gain with Frequency-dependent Terminations,' *IEEE Transactions on Circuits and Systems*, Vol. **32**, pp. 827–839, August, 1985.

Parametric Approach

[8] A. Fettweis, 'Parametric Representation of Brune Functions,' *International Journal of Circuit Theory and Applications*, Vol. **7**, pp. 113–119, 1979.

[9] J. Pandel and A. Fettweis, 'Broadband Matching Using Parametric Representations,' *IEEE International Sympoxium on Circuits and Systems*, Vol. **41**, pp. 143–149, 1985.

[10] J. Pandel and A. Fettweis, 'Numerical Solution to Broadband Matching Based on Parametric Representations,' *Archivdes Elektrischen Übertragung*, Vol. **41**, pp. 202–209, 1987.

[11] B. S. Yarman and A. Fettweis, 'Computer Aided Double Matching via Parametric Representation of Brune Functions', *IEEE Transactions on Circuits and Systems*, Vol. **37**, No. 2, pp. 212–222, February, 1990.

[12] E. G. Cimen, A. Aksen and B. S. Yarman, 'A Numerical Real Frequency Broadband Matching Technique Based on Parametric Representation of Scattering Parameters', Melecon'98, 9th Mediterranean Electrotechnical Conference, Tel Aviv, Israel, May 18–20, 1998.

[13] A. Aksen, E. G. Çimen and B. S.Yarman, 'A Numerical Real Frequency Broadband Matching Technique Based on Parametric Representation of Scattering Parameters', IEEE APCCAS'98, Asia Pasific Conference on Circuits and Systems, Chiangmai, Thailand, November 24–27, 1998.

Simplified Real Frequency Technique (SRFT)

[14] B. S. Yarman, 'A Simplified Real Frequency Technique for Broadband Matching Complex Generator to Complex Loads', *RCA Review*, Vol. **43**, pp. 529–541, 1982.

[15] B. S. Yarman and H. J. Carlin, 'A Simplified Real Frequency Technique Applied to Broadband Multi-stage Microwave Amplifiers', *IEEE Transactions on Microwave Theory and Techniques*, Vol. **30**, pp. 2216–2222, December, 1982.

[16] B. S. Yarman, 'A Dynamic CAD Technique for Designing Broad-band Microwave Amplifiers', *RCA Review*, Vol. **44**, pp. 551–565, December, 1983.

[17] B. S. Yarman, 'Modern Approaches to Broadband Matching Problems', *Proceedings of the IEE*, Vol. **132**, pp. 87–92, April, 1985.

[18] P. Jarry and A. Perennec, 'Optimization of Gain and VSWR in Multistage Microwave Amplifier Using Real Frequency Method', *European Conference on Circuit Theory and Design*, Vol. **23**, pp. 203–208, September, 1987.

[19] L. Zhu, B. Wu and C. Cheng, 'Real Frequency Technique Applied to Synthesis of Broad-band Matching Networks with Arbitrary Nonuniform Losses for MMIC's,' *IEEE Transactions on Microwave Theory and Techniques*, Vol. **36**, pp. 1614–1620, December, 1988.

[20] M. Sengul and B. S. Yarman, 'Real Frequency Technique without Optimization', *ELECO 2005*, 4th International Conference on Electrical and Electronics Engineering, Bursa, Turkey, December 07–11, 2005.

[21] P. Lindberg, M. Şengül, E. Çimen, B. S. Yarman, A. Rydberg, and A. Aksen, 'A Single Matching Network Design for a Dual Band PIFA Antenna via Simplified Real Frequency Technique', *First European Conference on Antennas and Propagation (EuCAP 2006)*, Nice, France, November 6–10, 2006.

[22] B. S. Yarman *et al.*, 'A Single Matching Network Design for a Double Band PIFA Antenna via Simplified Real Frequency Technique', *Asia Pacific Microwave Conference*, Yokohama, Japan, December 13–15, 2006.

[23] B. S. Yarman *et al.*, 'Design of Broadband Matching Networks,' *ECT*, Okinawa, Japan (Invited Talk), January 24–27, 2007, pp. 35–40.

[24] B. S. Yarman, N. Retdian, S. Takagi and N. Fujii, 'Gain-Bandwidth Limitations of 0.18µ Si-CMOS RF Technology', *Proceedings of ECCTD 2007*, Seville, Spain, August 26–30, 2007.

[25] B. S. Yarman, 'Modern Techniques to Design Wide Band Power Transfer Networks and Microwave Amplifiers on Silicon RF Chips', *IEEE International Conference on Recent Advances in Microwave Theory and Applications,* Jaipur, India, November 21–24, 2008.

[26] B. S. Yarman *et al.*, 'Performance Assesement of Active and Passive Components Manufactured Employing 0.18µm Silicon CMOS Processing Technology up to 22GHz', *Proceedings of 2008 IEEJ*, International Analog VLSI Workshop, Istanbul, Turkey, July 30–August 1, 2008, pp. 129–132.

[27] B. S. Yarman, *Design of Ultra Wideband Antenna Matching Networks via Simplified Real Frequency Technique*, Springer-Verlag, 2008.

SRFT with Mixed Lumped and Distributed Elements

[28] B. S. Yarman and A. Aksen, 'An Integrated Design Tool to Construct Lossless Matching Networks with Mixed Lumped and Distributed Elements', *IEEE Transactions on Circuits and Systems*, Vol. **39**, pp. 713–723, March, 1992.

[29] A. Aksen, 'Design of Lossless Two-Ports with Lumped and Distributed Elements for Broadband Matching', *PhD thesis*, Ruhr University at Bochum, 1994.

[30] A. Sertbaş, A. Aksen and S. Yarman, 'Construction of Some Classes of Two-variable Lossless Ladder Networks with Simple Lumped Elements and Uniform Transmission Lines', *IEEE APCCAS'98*, Asia Pacific Conference on Circuits and Systems, Chiangmai, Thailand, November 24–27, 1998.

[31] A. Sertbaş, A. Aksen and B.S. Yarman, 'Explicit Formulas for a Special Class of Two-variable Resonant Ladder Networks with Simple Lumped Elements and Commensurate Stubs', *ELECO'99*, Bursa, Turkey, December 1–5, 1999, pp. 132–135.

[32] B. S. Yarman, A. Sertbaş and A. Aksen, 'Construction of Analog RF Circuits with Lumped and Distributed Components for High Speed/High Frequency Mobile Communication MMICs', *European Conference on Circuit Theory and Design (ECCTD'99)*, Stresa, Italy, August 29–September, 2, 1999.

[33] A. Aksen and B. S. Yarman, 'Cascade Synthesis of Two Variable Lossless Two-port Networks of Mixed, Lumped Elements and Transmission Lines: A Semi-Analytic Procedure', *NDS-98*, The First International Workshop on Multidimensional Systems, Zielona Gora, Poland, July 12–14, 1998.

[34] A. Aksen, A. Sertbaş and B. S. Yarman, 'Explicit Two-Variable Description of a Class of Band-pass Lossless Two-ports with Mixed, Lumped Elements and Transmission Lines', *NDS-98*, The First International Workshop on Multidimensional Systems, Zielona Gora, Poland, July 12–14, 1998. Technical.

[35] A. Aksen and B. S. Yarman, 'A Real Frequency Approach to Describe Lossless Two-ports Formed with Mixed Lumped and Distributed Elements', *International Journal of Electronics and Communications (AEÜ)*, Vol. **6**, pp. 389–396, 2001.

[36] B. S. Yarman and E. G. Cimen, 'Design of a Broadband Microwave Amplifier Constructed with Mixed Lumped and Distributed Elements for Mobile Communication', *ICCSC'02*, 1st International Conference on Circuits and Systems for Communications, St Petersburg, Russia, June 26–28, 2002, pp. 334–337.

[37] B. S. Yarman, A. Aksen and E. G. Cimen, 'Design and Simulation of Miniaturized Communication Systems Employing Symmetrical Lossless Two-ports Constructed with Two Kinds of Elements' *ISCAS'03*, *Proceedings of the IEEE International Symposium on Circuits and Systems*, Thailand, May 25–28, 2003, Vol. **2** pp. 336–339.

[38] A. Aksen and B. S. Yarman, 'A Parametric Approach to Describe Distributed Two-ports with Lump Discontinuities for the Design of Broadband MMIC'S', ISCAS'03, *Proceedings of the IEEE International Symposium on Circuits and Systems*, Thailand, May 25–28, 2003, Vol. **1**, pp. 336–339.

[39] A. Sertbaş and B. S. Yarman, 'A Computer-Aided Design Technique for Lossless Matching Networks with Mixed, Lumped and Distributed Elements', *International Journal of Electronics and Communications (AEÜ)*, Vol. **58**, pp. 424–428, 2004.

Modeling via Real Frequency Techniques

[40] B. S. Yarman, A. Aksen and A. Kilinç, 'An Immitance Based Tool for Modelling Passive One-port Devices by Means of Darlington Equivalents', International Journal of Electronics and Communications (AEÜ*)*, Vol. **55**, pp. 443–451, 2001.

[41] B. S. Yarman, A. Kilinç and A. Aksen, 'Immitance Data Modeling via Linear Interpolation Techniques', *ISCAS 2002*, IEEE International Symposium on Circuits and Systems, Scottsdale, Arizona, May 2002.

[42] A. Kilinç, H. Pinarbaşi, M. Şengül and B. S. Yarman, 'A Broadband Microwave Amplifier Design by Means of Immitance Based Data Modelling Tool', *IEEE Africon 02, 6th Africon Conference in Africa*, George, South Africa, October 2–4, 2002, Vol. **2**, pp. 535–540.

[43] B. S. Yarman, A. Kilinç and A. Aksen, 'Immitance Data Modelling via Linear Interpolation Techniques: A Classical Circuit Theory Approach', *International Journal of Circuit Theory and Applications*, Vol. **32**, pp. 537–563, 2004.

[44] B. S. Yarman, M. Sengul and A. Kilinc, 'Design of Practical Matching Networks with Lumped Elements via Modeling', *IEEE Transactions on Circuits and Systems*, Vol. **54**, No. 8, pp.1829–1837, 2007.

Component Models

[45] O. E. Akcasu, 'Interconnect Parasitic RCL and Delay Variability Below 90nm, Physical Origins and its Impact on the Future Geometry Scaling', *IDV 2007*, (Invited Paper), Bangalore, India, December 13–14, 2007.

[46] O. E. Akcasu, O. Uslu, N.S., Nagaraj, T. Colak, S. Hale and E. Soo, Invited paper, 'A General and Comparative Study of RC(0), RC, RCL and RCLK Modeling of Interconnects and their Impact on the Design of Multi-giga Hertz Processors', Invited paper, ISQED 2002.

[47] O. E. Akcasu, '60nm 90nm Interconnect Modeling Challenges', Invited paper, FSA 2004.

[48] Spiral User Manual and Application Notes, OEA International, Inc.: www.oea.com.

[49] O. E. Akcasu, 'High Capacitance Structure in a Semiconductor Device', US Patent 5,208,725, May 1993.

[50] R. Aparicio and A. Hajimiri, 'Capacity Limits and Matching Properties of Integrated Capacitors', *IEEE Journal of Solid-State Circuits*, Vol. **37**, pp. 384–393, 2002.

[51] Cadence Design System: www.cadence.com.

[52] SmartSpice: Simucad Design Automation, Santa Clara, CA: www.simucad.com.

[53] AWR: Microwave Office of Applied Wave Research Inc.: www.appwave.com.

[54] VLSI Design and Education Center (VDEC) of Tokyo University: www.vdec.u-tokyo.ac.jp/English/index.html.

12

Immittance Data Modeling via Linear Interpolation Techniques: A Classical Circuit Theory Approach[1]

12.1 Introduction

For many communications engineering applications, circuit models for measured data obtained from physical devices or subsystems are essential. For example, models are required to analyze the electrical behavior of physical devices, such as gain–bandwidth limitations, determination of the noise figure merits or power delivering capabilities, etc [1, 2]. Similarly, computer-aided simulation of the communication subsystems printed on VLSI chips employs various kinds of models for nonlinear active and passive devices as well as interconnects [3–12]. Moreover, modern design techniques for constructing matching networks and microwave amplifiers also demand Darlington models for the computed data, which is generated theoretically to optimize the system performance [13–17].

Until recently, there had been no well-established technique in the literature which enables modeling of the given immittance or reflectance data. In practice, a common exercise to model the given data starts with the choice of an appropriate circuit topology. Then, the element values of the chosen topology are determined to best fit the given data by means of a nonlinear optimization. Although this approach is straightforward, it presents serious difficulties. Firstly, the optimization is heavily nonlinear in terms of element values that may result in local minimums or that may not converge at all. Secondly, there is no customary process for initializing the element values of the chosen circuit topology. Worst of all, 'the optimum choice of the circuit topology which best describes the physical device is in question'.

On the other hand, it may be wise to generate circuit models by means of interpolation techniques. As a matter of fact, many mathematicians and circuit theoreticians had studied interpolation with positive real

---

[1] This chapter, co-authored by Ali Kilinc (Okan University, Istanbul, Turkey), is adopted from the following paper: B. S. Yarman, A. Kilinç, and A. Aksen, 'Immitance Data Modelling via Linear Interpolation Techniques: A Classical Circuit Theory Approach', *International Journal of Circuit Theory and Applications*, Vol. 32, pp. 537–563, 2004.

(PR) functions extensively in the literature. In this regard, it may be worth mentioning the contributions of Smilen [18], Zaheb and Lempel [19] and Youla and Saito [20]. Among all these works, the last one presents a rather complete picture of the problem. In this work, PR impedance data $\{Z_i = R_i + jX_i, R_i \geq 0\}$ is specified on the complex s plane such that $\{s_i = \sigma_i + j\omega_i, \sigma_i \geq 0\}$. The data set $\{Z_i, s_i, i = 1, 2, \ldots, n\}$ is called a prepared sequence. Then, the necessary and sufficient conditions, for the so-called given prepared sequences to generate PR functions, are given. Once data is validated, it is interpolated as a PR function step by step, by extracting each pair(Z_i, s_i) as a cascade block. For a prepared sequence, the point i may be specified only on the real axis $\sigma \geq 0$, or only on the $j\omega$ axis, or in the right half plane (RHP) of the complex domain. Furthermore, the real part R_i or the imaginary part X_i may be specified as zero. Depending on the structure of the data, a cascade block is realized in the form of a Foster, Brune C or D-type section, or as a Richard section or Type E section. In this scheme, the real and imaginary parts of the driving point function are simultaneously realized. Regrettably, the use of the above-mentioned sections in a cascade block introduces complex transmission zeros and requires gyrators and perfectly coupled coils that might make the model impractical. Moreover, in [20], inter-polation on the real frequency axis demands special attention and requires derivatives of the measured immittance data, which are practically impossible to obtain. Needless to say, a general formula for the minimum degree of any PR interpolating function is also lacking in [20].

Lately, we have introduced handy tools to model given immittance or reflectance data by means of nonlinear curve-fitting or regression techniques [22, 23]. These tools overcome most of the difficulties of the brute-force and existing interpolation techniques of [18–20]. That is to say, in these tools it is neither necessary to select a fixed circuit topology in advance nor to force any complicated cascade structure into the model. Rather, the modeler is required to choose a certain analytical form for the driving point function with unknown parameters. In this form, transmission zeros are placed as required by the physical nature of the problem. Then, the unknown parameters are determined to best fit the given data by means of a nonlinear curve-fitting algorithm. Eventually, the driving point function is synthe-sized. Hence, the desired circuit topology is obtained as a result of the modeling process. On the other hand, these techniques require an initialization for the determination of the unknown parameters due to the nonlinear optimization. This may be considered a drawback.

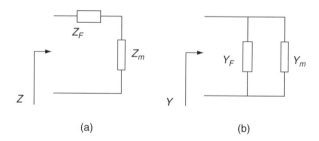

(a)   (b)

Figure 12.1   Representation of an immittance function: (a) $Z(s) = Z_F(s) + Z_m(s)$; (b) $Y(s) = Y_F(s) + Y_m(s)$

In this chapter, we have transformed the previously introduced nonlinear curve-fitting techniques to a linear interpolation scheme to model the given immittance data. In maintaining all the outstanding features of the previously introduced curve-fitting techniques, the proposed methods of modeling via interpolation require neither nonlinear optimization with initial values nor the implementation of virtual transmission zeros with complicated cascade structures. Therefore, models are simplified and algorith-mic computations are drastically reduced.

It is well known that any PR immittance function $F(s)$ can be decomposed into its minimum $F_m(s)$ and the Foster $F_f(s)$ part. In short, $F(s) = F_m(s) + F_f(s)$. This outstanding property constitutes the major philosophy of the new modeling techniques. The proposed modeling methods are put into practice in three major steps. Firstly, it is assumed that the data is specified on the real frequency axis and the given real part data is interpolated with a non-negative, even rational function $R(\omega^2) = N(\omega^2)/D(\omega^2)$. Then, the corresponding PR minimum function (PR-MF) $F_m(s)$ is generated by employing the Gewertz procedure [21]. At this point, the Foster part of the immittance data $F_f(j\omega) = F(j\omega) - F_m(j\omega)$ is extracted. In the second step, the Foster data is interpolated with a Foster function $F_f(s)$. Hence, the PR function (PR-F), which interpolates the given data, is expressed as $F(s) = F_m(s) + F_f(s)$ (Figure 12.1). Finally, the resulting immittance function $F(s)$ is synthesized and the desired model for the one-port device is obtained (third step).

In the following sections, we introduce the new modeling techniques via linear interpolation with the associated algorithms. Modeling examples are included to display the implementation of the computer algorithms. It is stated that the models obtained by employing the linear interpolation methods provide excellent initial circuit topology with element values that would serve as input to commercially available software packages.

In [26], due to space limitations, only the core concepts of the proposed modeling techniques were delivered and many important issues had to be omitted. Therefore, in this extended version, the reader is furnished with the full theoretical and numerical aspects of the new methods. In order to maintain the integrity of the chapter, the major clauses of [26] are repeated herewith.

12.2　Interpolation of the Given Real Part Data Set

Interpolation of the given real part data is not straightforward since it must always be associated with a non-negative, even rational function over the entire frequencies. In the following proposition, it is shown that any non-negative real part data can always be interpolated in terms of a non-negative, even rational function of minimum degree.

Proposition 1 (Interpolation of real data): Let $\{(\omega_i, R_i), R_i > 0, i = 1, 2 \ldots n_d\}$ be the given n_d independent ordered data pairs to be utilized to interpolate the real part $R(\omega)$ of the immittance function. It is asserted that one can always find an even positive polynomial designated by $D(\omega^2)$ of degree $4(n_d - 1)$ which is constructed by means of:

(a)　the sum of squares of the even Lagrange interpolation polynomials (SS-ELIP); or
(b)　the perfect square of an even interpolation polynomial (PS-EIP), to form an even rational real part function such that

$$R(\omega^2) = \frac{1}{D(\omega^2)} \geq 0, \forall \omega \tag{12.1}$$

which interpolates the given data pairs.

12.3　Verification via SS-ELIP

Let $L_i(\omega^2)$ be an even auxiliary Lagrange polynomial (ALP) of degree $2(n_d - 1)$ which exactly passes through the data set $\{(\omega_1, 1), (\omega_2, 1), (\omega_3, 1), \ldots, (\omega_{n_d}, 1)\}$ such that

$$L_i(\omega^2) = \prod_{j=1}^{n_d} \frac{\left(\omega^2 - \omega_j^2\right)}{\left(\omega_i^2 - \omega_j^2\right)}, \{i = 1, 2, \ldots, n_d\} \text{ and } (i \neq j) \tag{12.2}$$

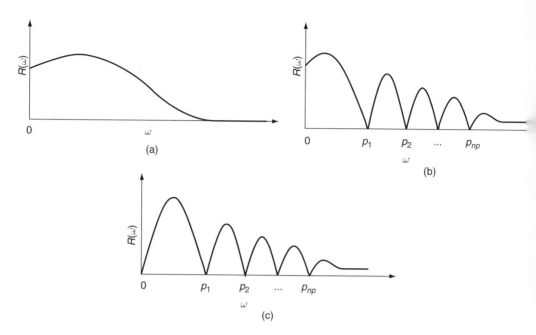

Figure 12.2   Possible shapes of the given real part data: (a) RPF-I; (b) RPF-IIA; (c) RPF-IIB

Clearly, $L_i(\omega^2) = 1$ for $\omega = \omega_i$ and is zero otherwise. Let us consider an even positive polynomial $D_i(\omega^2)$ with a minimum degree of $4(n_d - 1)$ constructed with $L_i(\omega^2)$ as

$$\boxed{D_i(\omega^2) = L_i^2(\omega^2)} \tag{12.3}$$

One can readily see that $D_i(\omega^2)$ is always non-negative.

Let $G_i = 1/R_i$ and the even polynomial $D(\omega^2)$ be

$$\boxed{D(\omega^2) = \sum_{i=1}^{n_d} G_i D_i(\omega^2)} \tag{12.4}$$

Then, the even rational interpolation function $R(\omega^2)$, which is defined in Equation (12.1), is always non-negative over the entire frequency axis and passes through the given data pairs $\{(\omega_1, R_1), (\omega_2, R_2), (\omega_3, R_3), \ldots, (\omega_{n_d}, R_{n_d})\}$. This form of the real part is called the Real Part Form I or RPF-I (Figure 12.2(a)).

Note that, for $i = 1$, the first frequency point ω_1 might be equal to zero. This situation does not affect the validity of the above proposition. If this is the case, then $R(0)$ is given by $R(0) = 1/G_1 = R_1$.

It should also be noted that, in the above proposition, the given real part data was assumed to be free of real frequency zeros. However, this may not be the case. Therefore, in the following lemma the real frequency zeros of $R(\omega^2)$ are considered.

Lemma: Let $\{(\omega_i, R_i), R_i > 0, i = 1, 2, \ldots, n_d\}$ be one set of the given pairs for which the real part is different than zero. Also let $\{(\omega_{zj}, R_{zj}), R_{zj} = 0, zj = 1, 2, \ldots, m_z, m_z < n_d, \omega_{zj} > 0\}$ be the other set with zero real part. Then, one can always find an even rational function $R(\omega^2)$

$$R(\omega^2) = \frac{N(\omega^2)}{D(\omega^2)} \geq 0, \forall \omega \tag{12.5}$$

that interpolates the given data set.

Proof: Let the even numerator polynomial $N(\omega^2)$ be

$$N(\omega^2) = \prod_{zj=1}^{m_z} \left(\omega^2 - \omega_{zj}^2\right)^2 \tag{12.6}$$

$N(\omega^2)$ is always non-negative over the entire frequency axis since it is a perfect square, even polynomial. Furthermore, it is zero at $\omega = \omega_{zj}, zj = 1, 2, \ldots, m_z$, as desired.

Moreover, at discrete frequencies the numerator polynomial is given by

$$N_i = \prod_{zj=1}^{m_z} \left(\omega_i^2 - \omega_{zj}^2\right)^2, \quad \{i = 1, 2, \ldots, n_d\} \tag{12.7}$$

In a similar manner to that in Equation (12.4), the denominator polynomial $D(\omega^2)$ is expressed as

$$D(\omega^2) = \sum_{i=1}^{n_d} N_i G_i D_i (\omega^2) \tag{12.8}$$

Thus, the rational even function Equation (12.5) interpolates the given data pairs as desired. This form of the real part is called the Real Part Form IIA or RPF-IIA (Figure 12.2(b)). If the given real part data also includes a zero at DC (i.e. $\omega = 0$) of order k, then the numerator polynomial $N(\omega^2)$ of Equation (12.5) must take the following form:

$$N(\omega^2) = \omega^{2k} \prod_{zj=1}^{m_z} \left(\omega^2 - \omega_{zj}^2\right)^2, \quad (k/2 + m_z + 1) \leq n_d \tag{12.9}$$

In this case, $R(\omega^2)$ is called the Real Part Form IIB or RPF-IIB (Figure 12.2(c)).

Note that multiple zeros on the real frequency axis might also be considered in the above lemma. This situation can easily be incorporated as the power of the associated zero term in $N(\omega^2)$, without affecting the generalization.

12.4 Verification via PS-EIP

Even though interpolation with Lagrange polynomials presents excellent numerical stability, it may be tedious to carry out numerous summations and multiplications to determine the polynomial coefficients. Instead, what we refer to as the 'via' or 'direct polynomial interpolation' technique may be employed to generate a mathematical model for the real part.

In this method, firstly, a Real Part Form (RPF-I or IIA, B) is chosen. Then, the coefficients of the denominator polynomial are computed by utilizing the real part data point by point as follows.

Let us define $R(\omega^2)$ in terms of the perfect square, even numerator and the denominator polynomials as

$$R(\omega^2) = \frac{N(\omega^2)}{D(\omega^2)} = \frac{N_S^2(\omega^2)}{D_S^2(\omega^2)}, \quad \text{such that } D_S^2(\omega^2) > 0, \forall \omega \tag{12.10}$$

In this representation, $N_S^2(\omega^2)$ is fixed in advance with the chosen real part form and is only a function of the frequency. At any single point (ω_i, R_i), the square root of the denominator polynomial is given by

$$D_s(\omega_i^2) = B_0 + B_1\omega_i^2 + \ldots + B_{n_d-1}\omega_i^{2(n_d-1)} = \alpha_i \neq 0 \tag{12.11}$$

where

$$\alpha_i = \frac{|N_S(\omega_i^2)|}{\sqrt{R_i}} > 0 \tag{12.12}$$

with

$$N_S(\omega_i) = 1 \quad \text{for RPF} - \text{I} \tag{12.13a}$$

$$N_S(\omega_i) = \prod_{zj=1}^{m_z}\left(\omega_i^2 - \omega_{zj}^2\right) \quad \text{for RPF} - \text{IIA} \tag{12.13b}$$

$$N_S(\omega_i) = \omega_i^k \cdot \prod_{zj=1}^{m_z}\left(\omega_i^2 - \omega_{zj}^2\right) \quad \text{for RPF} - \text{IIB} \tag{12.13c}$$

For the given n_d data pairs, Equation (12.11) can be expressed in matrix form as

$$[\Omega][B] = [\alpha] \tag{12.14}$$

In Equation (12.14), the vector $[B]$ includes all the unknown coefficients B_i to be determined as a result of the interpolation process, and is given by $[B]^T = [B_0 \quad B_1 \quad \ldots \quad B_{n_d-1}]$. The matrix

$$[\Omega] = \begin{bmatrix} 1 & \omega_1^2 & \omega_1^4 & \cdots & \omega_1^{2(n_d-1)} \\ 1 & \omega_2^2 & \omega_2^4 & \cdots & \omega_2^{2(n_d-1)} \\ \vdots & \vdots & \vdots & \ddots & \vdots \\ 1 & \omega_{nd}^2 & \omega_{nd}^4 & \cdots & \omega_{nd}^{2(n_d-1)} \end{bmatrix}$$

and the vector

$$[\alpha] = \begin{bmatrix} \alpha_1 \\ \alpha_2 \\ \vdots \\ \alpha_{n_d} \end{bmatrix}$$

are specified with the given data. Obviously, the solution of Equation (12.14) yields the desired coefficients B_i of $D(\omega^2)$. On the other hand, we need to find the open form of the denominator polynomial $D^2(\omega^2)$.

Let $D(\omega^2) = D_S^2(\omega^2)$ be expressed as

$$D_S^2(\omega^2) = C_0 + C_1\omega^2 + \ldots + C_{2(n_d-1)}\omega^{4(n_d-1)} \tag{12.15}$$

By taking the square of Equation (12.11), coefficients C_i can easily be computed in terms of B_i. Hence,

$$C_0 = B_0^2 \tag{12.16a}$$

$$C_i = 2\sum_{j=0}^{(i-1)/2}\left(B_j B_{i-j}\right), \quad i = 1, 3, 5, \ldots < 2(n_d - 1) \tag{12.16b}$$

$$C_i = B_{i/2}^2 + 2\sum_{j=0}^{i/2}\left(B_j B_{i-j}\right), \quad i = 2, 4, 6, \ldots < 2(n_d - 1) \tag{12.16c}$$

$$C_{2(n_d-1)} = B_{nd}^2 - 1 \tag{12.16d}$$

Note that this technique requires the solution of a linear equation set as opposed to the straightforward multiplication and addition of the Lagrange polynomial method. Therefore, in the process of interpolation, one would expect to obtain better numerical stability with the help of the Lagrange method. Nevertheless, in practice we have not yet encountered any serious difficulty regarding the numerical robustness of the direct interpolation technique. Another important issue is that the 'via polynomial' approach yields a different solution than that of the 'Lagrange polynomial method' for the same interpolation problem due to the differences in the numerical casting of the methods (i.e. sum of squares of Lagrange polynomials versus perfect square of a single polynomial). Consequently, the modeler will have the opportunity to select one of these techniques, depending on the final circuit models, which are supposed to reflect the nature of the physical problem under consideration.

Based on the above theoretical derivations, the following algorithm is proposed to interpolate the given real part data.

Algorithm I: Interpolation of the Given Real Part Data

Step 1: Plot and analyze the given data and select the proper analytic form for the real part as follows:

(a) If the real part does not touch the real frequency axis at all, and its DC values are known, then RPF-I is chosen for $R(\omega)$.

(b) If the real part data reaches zero at finite frequencies beyond DC, then we choose RPF-IIA for $R(\omega)$.

(c) If the real part reaches zero both at $\omega = 0$ and at finite frequencies as well, then the RPF-IIB form is chosen for $R(\omega)$.

Step 2: Based on the selected form of $R(\omega^2)$ (or, equivalently, after selection of $N(\omega^2)$), generate the data for the denominator polynomial $D(\omega^2)$ as $\{\alpha_i = N(\omega_i^2)/R_i, i = 1, 2, \ldots, n_d\}$ for the interpolation. Then, construct $D(\omega^2)$ using either the Lagrange interpolation technique or the via polynomial approach. Hence, the desired analytic form of $R(\omega^2)$ in Equation (12.5) or Equation (12.10) is obtained. Here, note that $D(\omega^2)$ should never reach zero over the entire frequencies. In order to avoid this, especially in the implementation of the Lagrange method (SS-ELIP), a numerical trick is utilized as follows. Firstly, the minimum of α_i is detected and set to $D_{min} = min\{\alpha_i, i = 1, 2, \ldots, n_d\}$. Then, it is assumed that $D(\omega^2) = 0.5 \cdot D_{min} + D_x(\omega^2)$ to assure always the positivity of the denominator polynomial. In this case, one generates $\alpha_{xi} = \alpha_i - 0.5 \cdot D_{min}$ for the given sample points to construct $D_x(\omega^2)$ by means of the proposed interpolation techniques. Hence, positivity of $D(\omega^2) = 0.5 \cdot D_{min} + D_x(\omega^2)$ is guaranteed.

Step 3: Using the Gewertz procedure, form the minimum part immittance function

$$F_m(S) = \frac{n(s)}{d(s)} = \frac{a_0 + a_1 s + \ldots + a_{n-1} s^{n-1}}{b_0 + b_1 s + \ldots + b_n s^n}$$

Step 4: Extract the Foster portion $X_f(j\omega_i)$ point by point from the immittance data such that $F(j\omega_i) = R(\omega_i) + jX(\omega_i)$ is specified by the modeler: $F_m(j\omega_i) = R_m(\omega_i) + jX_m(\omega_i)$ is generated in step 3 such that $R_m(\omega_i) = R(\omega_i)$. Then,

$$\boxed{X_f(j\omega_i) = X(j\omega_i) - X_m(j\omega_i), i = 1, 2, \ldots, n_d} \qquad (12.17)$$

In the following section a straightforward procedure is outlined to interpolate the given Foster data set of Equation (12.17).

12.5 Interpolation of a Given Foster Data Set $X_f(\omega)$

The general form of a Foster function on the $j\omega$ axis is given by

$$\boxed{X_f(\omega) = \sum_{r=1}^{n_p}\left(\frac{k_r \omega}{p_r^2 - \omega^2}\right) + k_\infty \omega - \frac{k_0}{\omega}} \qquad (12.18)$$

We can always introduce a pole p_r in Foster form, which is specified by Equation (12.18), that passes through a given point (ω_i, X_{fi}). Selecting the pairs (ω_i, X_{fi}) properly among the given Foster data and the poles p_r in advance, the residues k_r, k_0, and k_∞ can easily be computed by solving Equation (12.18). Hence, we state the following proposition to interpolate the given Foster data.

Proposition 2 (Interpolation of the Foster data): Let $\left\{ (\omega_i, X_{fi}), i = 1, 2, \ldots, n_f \right\}$ represent the selected n_f ordered pairs for the Foster portion of the immittance data. It is asserted that one can always generate a realizable Foster function as in Equation (12.18) which interpolates the selected data. In this case, the residues of Equation (12.18) are given by the vector K such that

$$[K] = [A]^{-1}[X_f]$$

$$[K]^T = \begin{bmatrix} k_1 & k_1 & \cdots & k_{n_p} & k_\infty & k_0 \end{bmatrix}$$

$$[X_f]^T = \begin{bmatrix} X_{f_1} & X_{f_1} & \cdots & X_{f_{n_f}} \end{bmatrix} \tag{12.19a}$$

$$A = \begin{bmatrix} \dfrac{\omega_1}{p_1^2 - \omega_1^2} & \cdots & \dfrac{\omega_1}{P_{n_p}^2 - \omega_1^2} & \omega_1 & -\dfrac{1}{\omega_1} \\[2ex] \dfrac{\omega_2}{p_1^2 - \omega_2^2} & \cdots & \dfrac{\omega_2}{P_{n_p}^2 - \omega_2^2} & \omega_2 & -\dfrac{1}{\omega_2} \\[2ex] \vdots & \vdots & \vdots & \vdots & \vdots \\[2ex] \dfrac{\omega_{n_f}}{p_1^2 - \omega_{n_f}^2} & \cdots & \dfrac{\omega_{n_f}}{P_{n_p}^2 - \omega_{n_f}^2} & \omega_{n_f} & -\dfrac{1}{\omega_{n_f}} \end{bmatrix} \tag{12.19b}$$

with

(i) $n_f = n_p + 2$, for the full form of Equation (12.18);
(ii) $n_f = n_p + 1$, if one of the poles at DC or at infinity is not required in Equation (12.18);
(iii) $n_f = n_p$, if there is no need to employ poles at DC and at infinity.

Verification: Once the poles are selected properly as described below, the given data is then simply plugged into Equation (12.18) and the unknown residues that are included in $[K]$ are solved. Hence, Equation (12.19) follows.

Note that the total number of selected points n_f must be equal to the total number of residues to be determined.

Depending on the distribution of the data, the full form or partial forms of Equation (12.18) may be utilized. *Foster forms*, to be employed for the interpolation process, may be classified as follows:

(1) Having selected all the finite poles beyond zero, if a DC pole together with a pole at infinity are required, then we set $n_f = n_p + 2$. This form of Equation (12.18) is called the full form or the Foster form I (or FF-I); it is depicted in Figure 12.3(a).
(2) Having selected all the finite poles beyond zero:

(a) If one does not wish to employ a pole at infinity but the interpolation problem requires a pole at DC, then the term $k_\infty \omega$ is dropped in Equation (12.18) and the last column of the coefficient matrix $[A]$ is deleted. In this case, $n_f = n_p + 1$ is chosen. This form of the Foster function is called foster form IIA as shown in Figure 12.3(b).
(b) In a similar manner, if there is no need to utilize a DC pole in Equation (12.18) but the pole at infinity is inserted, then the column which appears on the one before the last in $[A]$ will be deleted. In this case, we still set $n_f = n_p + 1$. This form of Equation (12.18) is called Foster form IIB or FF-IIB (Figure 12.3(c)).

(3) If having neither a DC pole nor a pole at infinity is required, then $n_f = n_p$ is selected. In this case, the last two terms of Equation (12.18) are dropped and the last two columns of $[A]$ are deleted. This form of Equation (12.18) is called Foster form III (or FF-III); it is illustrated in Figure 12.3(d).
(4) If the Foster data is always positive tracing a linear line, then $X_f(\omega) = k_\infty \omega$. This form of X_f is called Foster form IVA or FF-IVA. If the Foster data is always negative tracing the shape of a hyperbola,

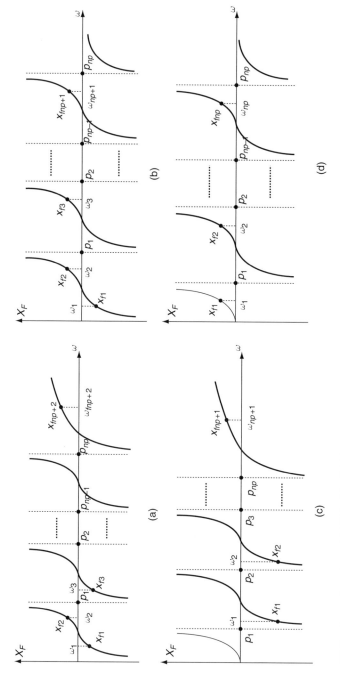

Figure 12.3 Foster functions: (a) FF-I; (b) FF-IIA; (c) FF-IIB; (d) FF-III

then $X_f(\omega) = -k_0/\omega$. This form of X_f is called Foster form IVB or FF-IVB. Furthermore, one may also describe a case for the form of $X_f(\omega) = k_\infty\omega - k_0/\omega$, for which the Foster data starts from negative values (or $-\infty$) and increases monotonically toward positive values (or $+\infty$). This form is referred to as Foster form IVC or FF-IVC.

(5) To ensure the realizability of the Foster form of Equation (12.18) and the existence of the solution of Equation (12.19), the following remarks are essential:

    (a) First of all, the spread of the Foster data must be carefully examined to develop insight into the shape of the Foster form to be utilized. Then, proper points and associated poles are selected in such a way that Equation (12.19) yields a realizable Foster function.

    (b) The solution of Equation (12.19) is not unique since it depends on the location of the poles.

    (c) The solution exists if det $(A) \neq 0$. Clearly, in Equation (12.19b), ω_i must be distinct and should not coincide with the poles p_r (i.e. $p_r \neq \omega_i, \forall i, r$). Furthermore, elements of the last column of $[A]$ must be finite, which requires $\omega_i \neq 0$.

    (d) If one ends up with negative residue values, then this means that the pole locations are improper and they must be corrected accordingly.

    (e) Location of the poles: Referring to the full form of the Foster function in Equation (12.18), the general rule is to 'locate the poles in an interlacing manner with selected points'. However, there may be exceptional cases for which two adjacent points can be placed on the ω axis without interlacing the poles, but yielding a realizable Foster function.

For practical applications, interpolation of the Foster data can be accomplished by employing the following algorithm.

Algorithm II: Interpolation of the Foster Data

Step 1: Plot the given ordered data pairs point by point and choose the desired form of the Foster function of Equation (12.18) to interpolate the given data.

Step 2: To remove ambiguity in the selection of the pole locations, a finite pole may be inserted on the arithmetic or geometric mean of the end points of the interval under consideration. If suitable, one may also utilize a pole location estimation procedure, which is described in [16].

Step 3: On selecting the poles, generate the coefficient matrix $[A]$ of Equation (12.19b) properly.

Step 4: Once the coefficient matrix $[A]$ has been generated, solve Equation (12.19a) for the unknown residues. Hence, the Foster data interpolation is complete.

Note that, as an alternative method, one might consider employing the Youla and Saito method to interpolate the Foster data introduced in [20]. However, as indicated earlier, implementation of this method on the real frequency axis demands derivatives at each point of interpolation, which is not at all practical.

12.6 Practical and Numerical Aspects

In this section, we wish to consider some practical and numerical aspects of the modeling techniques presented in this chapter:

(1) No matter what the practical case is, the resulting model must reflect the physical nature of the problem with properly inserted transmission zeros.

(2) In the existing techniques of [18–20], each separate point $\{Z_i(s_i) = R_i(s_i) + jX_i(s_i), s_i = \sigma_i + j\omega_i,$ with $R_i(s_i) \geq 0$ and $\sigma_i \geq 0\}$ is realized in the form of a cascade block by employing the Brune C, D, E, Richard-type sections or as resonance circuits. These sections introduce artificial complex RHP zeros

into the immittance function, which in turn unnecessarily complicates the resulting model with an increased number of circuit elements. In this case, if one is dealing with a design problem, then the electrical performance of the system is penalized.

(3) It is well known that realization of complex transmission zeros requires perfectly coupled coils and gyrators. At microwave frequencies, fabrication of these elements is very difficult.

(4) For almost all engineering problems, immittance data is specified over the real frequency axis. In this context, the elegant results of [20], which are basically given in the complex s plane, are not applicable to practical problems. Those results of [20] given on the real frequency axis require derivatives of the immittance data, which are practically impossible to obtain.

(5) As opposed to [18–20], in our proposed methods it is sufficient to specify the given data over the real frequencies. In the implementation phase, firstly, the real part of the immittance data is modeled. Then, the resulting Foster part is constructed.

Thus, the transmission zeros of the models are perfectly controlled on both the real and the Foster parts to yield practical circuits.

In other words, locations of the transmission zeros are selected by the modeler and are fixed to reflect the nature of the problem. This constitutes the major difference between the methods presented in this chapter and other methods of interpolation.

(6) At this point, let us comment on the maximum number of sections required to construct the model utilizing the interpolation techniques proposed in this chapter. As a hypothetical case, consider the given data set generated by employing a PR function specified in the complex s plane rather than on the $j\omega$ axis. Then, the degree of the PR function or total number of non-redundant points n can be estimated as described in [20]. However, if the data is specified over the real frequencies, then the numerical procedure proposed in the following (Section 12.8) may be employed.

(7) Regardless of what the modeling case is, it is always possible to generate many points, perhaps infinitely many points, over the real frequencies for the actual physical devices to be modeled. Moreover, these points may not be redundant. If this were the case, one would expect to obtain infinitely many elements in the model. This situation presents an impossible case. Therefore, the modeler must carefully select a few points from the given data to build an acceptable model.

In this manner, the interpolation problem may be understood as an approximation problem with a properly selected and reduced number of points. This understanding provides substantial simplicity in the resulting model that can be clearly demonstrated as in the following example.

(8) The modeling process may start with impedance or admittance data. While working with a reduced number of data points for the interpolation, the choice of immittance depends on the quality of the fit.

(9) Finally, one should not exclude selection of the sample points by trial and error for the interpolation problem under consideration to get the best fit. In this regard, 'the best fit' may be achieved in the least mean square (LMS) sense.

12.7 Estimation of the Minimum Degree n of the Denominator Polynomial $D(\omega^2)$

As an alternative method to [20], degree n may be estimated by taking the sequential numerical derivative of the denominator polynomial $D(\omega^2)$ of the real part $R(\omega^2) = N(\omega^2)/D(\omega^2)$. Obviously, this degree is associated with the total number of elements of the model. In this approach, we assume that data is provided over equally sampled frequency intervals $\{\Delta\omega_i = \omega_{i+1} - \omega_i, i = 1, 2, \ldots, (n_d - 1)\}$, which are set to unity (i.e. $\Delta\omega_i = 1$). Once the numerator polynomial $N(\omega^2)$ is selected, data for the denominator polynomial is generated from the given or measured real part as $\{D_i = N(\omega_i^2)/R_i, i = 1, 2, \ldots, n_d\}$. Then, the first-order derivatives over the unity intervals are approximated by $\{\Delta D_j = D_{j+1} - D_j, j = 1, 2, \ldots, (n_d - 1)\}$. The second-order derivatives are approximated by

$\{\Delta^2 D_j = \Delta D_{j+1} - \Delta D_j, j = 1, 2, \ldots, (n_d - 2)\}$. Similarly, the kth-order derivatives are approximated by $\{\Delta^k D_j = \Delta^{k-1} D_{j+1} - \Delta^{k-1} D_j, j = 1, 2, \ldots, (n_d - k)\}$. This process continues until we end up with zero differences. At this point, let us designate the step index k by k_{max}. Then, the minimum degree n will be $n = k_{max}$. Obviously, k_{max} cannot exceed 'the total number of given data points $n_d - 1$' (i.e. $k_{max} \leq 4(n_d - 1)$).

12.8 Comments on the Error in the Interpolation Process and Proper Selection of Sample Points

The methods introduced in this chapter for the interpolation of the real part data rely on the interpolation of the denominator polynomial $D(\omega^2)$. In this regard, the error due to the selection of sample points can be explored by means of the error function given for polynomial interpolation. Once the numerator $N(\omega^2)$ of the real part $R(\omega^2) = N(\omega^2)/D(\omega^2)$ is chosen, the denominator polynomial $D(\omega^2)$ of degree n_D is constructed by means of interpolation techniques introduced in this chapter, utilizing the selected sample points $\alpha_i = N(\omega_i^2)/R_i$ generated over the given frequencies. In this representation, the subscript i runs from 1 to n_k and the real part, $R_i = R(\omega_i)$, is specified by properly selecting n_k ordered pairs $\{(R_1, \omega_1), (R_2, \omega_2), \ldots, (R_{n_k}, \omega_{n_k})\}$ among the given data points. The polynomial degree n_D (or the minimum degree n) is $n_D = 4(n_k - 1)$ and it is estimated as discussed in the above sections. Let us now introduce the error expression in polynomial interpolation.

Let $\alpha(\omega)$ be a hypothetical continuous function, which is differentiable at least n_D times, associated with the specified data points α_i. Then, the interpolation error $\varepsilon(\omega)$ is given by

$$\boxed{\begin{aligned} \varepsilon(\omega) &= \alpha(\omega) - D(\omega^2) \\ &= \psi(\omega)\left\{\frac{1}{n_D!}\left[\frac{d^{(n_D)}}{d\omega^{(n_D)}}\alpha(\omega)\right]_{\text{at } (\omega = c_x)}\right\} \end{aligned}} \qquad (12.20)$$

for $(\omega_{min} = \omega_1) \leq \omega \leq (\omega_{max} = \omega_{n_k})$.

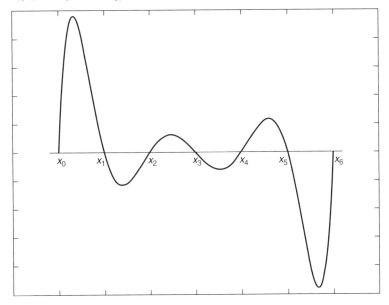

Figure 12.4   Error in polynomial interpolation

In (20), $\psi(\omega) = (\omega - \omega_1)(\omega - \omega_2) \ldots (\omega - \omega_{n_k})$ is called the 'monic error polynomial' and C_x is an unknown point between the minimum and maximum of $\omega_1, \omega_2, \ldots, \omega_{n_k}$ and ω. In determining the behavior of the error, the quantity $\psi(\omega) = (\omega - \omega_1)(\omega - \omega_2) \ldots (\omega - \omega_{n_k})$ is the most important. Obviously, the error is zero at the sample points $\omega_1, \omega_2, \ldots, \omega_{n_k}$ as expected. For the time being, assume that the node points are evenly spaced. In this case, it is known that for large values of n_k ($n_k \geq 5$), values of $\psi(\omega)$ and therefore $\varepsilon(\omega)$ change greatly over the interval $\omega_1 \leq \omega \leq \omega_{n_k}$ as shown in Figure 12.4. Values of the error function in the first interval $[\omega_1, \omega_2]$ and in the last interval $\left[\omega_{(n_k-1)}, \omega_{n_k}\right]$ become much larger than the values in the middle of $[\omega_1, \omega_{n_k}]$. Theoretical aspects of this issue can be found in any numerical analysis textbook such as [27]. On the other hand, the behavior of the error function may be controlled by properly distributing node points $\omega_1, \omega_2, \ldots, \omega_{n_k}$ of $\psi(\omega) = (\omega - \omega_1)$ $(\omega - \omega_2) \ldots (\omega - \omega_{n_k})$. First of all, in order to capture the shape of $D(\omega^2)$ to restrict the error fluctuations between two sample points (or between two zeros of $\varepsilon(\omega)$), it may be wise to choose $\omega_1, \omega_2, \ldots, \omega_{n_k}$ as the roots of a Chebyshev polynomial so that $\varepsilon(\omega)$ is forced to be a 'near mini–max or almost-equal ripple function' [27]. This means that it is appropriate to designate the error polynomial $\psi(\omega)$ as a 'symmetrical, even, monic Chebyshev polynomial of degree $m = 2n_k$'. Hence, $\psi(\omega) = \psi_{2n_k}(\omega^2) = \left(\omega^2 - \omega_1^2\right)\left(\omega^2 - \omega_2^2\right) \ldots \left(\omega^2 - \omega_{n_k}^2\right)$ such that, for any $m = 2n_k$, $\psi(\omega) = \left(1/2^{m-1}\right)T_m(\omega)$ where $T_m(\omega) = cos[m \cdot cos^{-1}(\omega)]$. In this regard, the following algorithm is proposed to select the proper data set for interpolation of the denominator polynomial $D(\omega^2)$ to control the error fluctuations.

As far as interpolation of the Foster part is concerned, the sample points are selected in the neighborhood of the poles, which are determined as described in Section 12.5.

Algorithm III: Selection of Proper Data Set for the Interpolation of $D(\omega^2)$ to Control the Ripples of the Error in the Chebyshev (or in the Near Mini–Max) Sense

Inputs:

- n_d: Total number of given data pairs for the immittance function to be modeled.
- Real part data set $\{(R_1, \omega_1), (R_2, \omega_2), \ldots, (R_{n_d}, \omega_{n_d})\}$.
- n_D: Estimated least degree of $D(\omega^2)$.

Step 1: Compute the data, which is associated with the denominator as $\alpha_i = N(\omega_i^2)/R_i$ for $i = 1, 2, \ldots, n_d$.

Step 2: Plot α_i vs. ω_i and detect its extremums as $\{\alpha_{mi}, \omega_{mi}, mi = 1, 2, \ldots, M\}$ where M designates the total number of extremums. It is expected that $M + 1 \approx n_D$. However, if $M + 1$ is greater than n_D, then set $n_D = M + 1$. Then the required sample count is $n_k = (n_D/4) + 1$. Obviously, in this step, $n_D/4$ must be rounded as an integer if necessary. Note that, in this step, the angular frequency ω_i is normalized with respect to the highest actual angular frequency of the given data. Therefore, the normalized frequency set $\{\omega_i, i = 1, 2, \ldots, n_d\}$ is placed within $0 \leq \omega_i \leq 1$.

Step 3: Find the positive roots $\omega_{r1}, \omega_{r2}, \ldots \omega_{rm}$ of the Chebyshev polynomial of degree $m = 2n_k$.

Step 4: Select n_k pairs from $\left\{(R_1, \omega_1), (R_2, \omega_2), \ldots, (R_{n_k}, \omega_{n_k})\right\}$ in such a way that the normalized frequencies $\{\omega_i\}$ either coincide or get close enough to positive Chebyshev roots $\omega_{r1}, \omega_{r2}, \ldots, \omega_{rn_k}$ of $\psi(\omega)$.

12.9 Examples

In this section two examples are presented to demonstrate the utilization of the proposed modeling algorithms introduced in this chapter.

Example 12.1 *(Model for a monopole antenna)*: In this example, we will construct circuit models for the data that belongs to a short monopole antenna, employing the techniques introduced in this chapter. The measured impedance data for the antenna is given over 13 points, covering the frequency band of 20 to 100 MHz as specified in Table 12.1.

Firstly, we carried out a redundancy check for the measured points by employing the techniques given in the previous section and also by taking sequential derivatives with several preselected $N(\omega^2)$ forms. As expected, it was found that all the measured points were independent and could be modeled with lumped circuit elements. Therefore, it was thought wise to select a few proper points from the table to give a reasonable model for the antenna. In fact, this example had already been worked out utilizing the non-linear curve-fitting methods presented in [16] and [22]; it was found that four-element low-pass LC ladders were sufficient to model the minimum part of the given driving point impedance. For the Foster part, it was adequate to utilize a series capacitor together with an inductance. For the interpolation techniques proposed in this chapter, if we were to utilize all the measured data, at least $2(n_d - 1) = 24$ sections or simple circuit elements (capacitors, inductors) would be required just for the realization of the minimum immittance function. Furthermore, we could employ 13 sections in the realization of the Foster part. This would mean that interpolation of the Foster part would require at least 24 elements. In this case, the total number of circuit elements would be $24 + 24 = 48$. On the other hand, in this modeling problem, our ultimate goal is to design a matching network between the antenna and a source (single matching [1, 2]). The model may be required to determine the theoretical gain–bandwidth limitations of the antenna over a wide band. In this regard, the circuit model with 48 elements would unnecessarily complicate the problem. Moreover, each Foster section would also contribute to the transmission zeros of the matched system [1, 2]. Therefore, the transmission zeros, which are introduced for the sake of interpolation of the Foster part, may be superfluous and impose artificial restrictions on the bandwidth of the matched system. However, we can get round all the above-mentioned difficulties by selecting the proper samples as desired from the given measured data set. In order to compare the results obtained via the curve-fitting and interpolation methods, we properly select only three points to model the real part and only two points to model the Foster part from the given impedance data. Regarding the above comments, we will model the antenna with both the direct and Lagrange interpolation methods.

Table 12.1 Impedance measurement of the monopole antenna problem [16,20] for Example 12.1

Sample index	Normalized frequency ω	Real part of the input impedance $R(\omega)$	Imaginary part of the input impedance $X(\omega)$
1	0.2	0.6	−6.0
2	0.3	0.8	−2.2
3	0.4	0.8	0
4	0.45	1.0	1.4
5	0.5	2.0	2.8
6	0.55	3.4	4.6
7	0.60	7.0	7.6
8	0.65	15.0	8.8
9	0.70	22.4	−5.4
10	0.75	11.0	−13.0
11	0.80	5.0	−10.8
12	0.9	1.6	−6.8
13	1.0	1.0	−4.4

Direct interpolation technique:

Following the implementation steps of the direct interpolation algorithm, first the real part data is interpolated by using sample numbers 7, 9 and 10 of Table 12.1 to give the best fit in a LMS sense. In this case, RPF-I is chosen. For this example, it is not necessary to employ an auxiliary denominator polynomial $D_x(\omega^2)$, which is described in step 2 of Algorithm I. Thus, the real part of Equation (12.1) is selected with $D(\omega^2) = D_s^2(\omega^2)$ and $D_s(\omega^2) = B_0 + B_1\omega^2 + B_2\omega^4$. Coefficients B_0, B_1 and B_2 are computed as

$$B_2 = 12.476\,91 \quad B_1 = -11.887\,50 \quad B_0 = 3.040\,456$$

On the other hand, by Equation (12.15),

$$D(\omega^2) = D_s^2(\omega^2) = C_0 + C_1\omega^2 + C_2\omega^4 + C_3\omega^6 + C_4\omega^8$$

By employing Equation (12.16), coefficients C_i are computed:

$$C_4 = 155.6733 \quad C_3 = -296.6385 \quad C_2 = 217.1836 \quad C_1 = -72.286\,82 \quad C_0 = 9.244\,371$$

From the analytic form of the real part, the minimum reactance function

$$Z_m = \frac{a_0 + a_1 s + a_2 s^2 + a_3 s^3}{b_0 + b_1 s + b_2 s^2 + b_3 s^3 + b_4 s^4}$$

is determined using the Gewertz procedure:

$$a_3 = 27.41884 \quad a_2 = 10.19049 \quad a_1 = 15.367 \quad a_0 = 0.3288981$$
$$b_4 = 12.47691 \quad b_3 = 4.637169 \quad b_2 = 12.74922 \quad b_1 = 2.289121 \quad b_0 = 3.040456$$

Once the analytic form of $Z_m(s)$ is computed, then the imaginary part X_m is computed and the Foster portion, Equation (12.17), is extracted from the original input impedance $Z(j\omega) = R(\omega) + jX(\omega)$ point by point as shown in Table 12.2. Then, the Foster data is interpolated as described in Section 12.5.

After carefully analyzing the Foster data, sample points 1 and 13 are selected (Table 12.2) and FF-I is chosen for the interpolation. Hence, $n_f = 2, n_p = 0$.

Table 12.2 Extraction of the Foster portion and selection of samples for interpolation (verification A via SS-ELIP) for Example 12.1

Sample index	Normalized frequency ω	$X_M = \text{Imag}\{Z_M\}$	Foster data $X_f = X - X_M$
1	0.2	1.0943	−7.0943
2	0.3	1.875	−4.075
3	0.4	3.1067	−3.1067
4	0.45	4.0738	−2.6738
5	0.5	5.4842	−2.6842
6	0.55	7.5918	−2.9918
7	0.60	10.257	−2.6567
8	0.65	9.3224	−0.52241
9	0.70	−5.1533	−0.24672
10	0.75	−14.148	1.148
11	0.80	−11.353	0.55345
12	0.9	−6.2697	−0.53026
13	1.0	−4.2894	−0.11062

Based on the above form, the residues are computed by solving Equation (12.21): $k_\infty = 1.3628$; $k_0 = 1.4734$

Eventually, the impedance $Z(s) = Z_m(s) + Z_f(s)$ is synthesized to give the desired network topology. Synthesis of minimum reactance $Z_m(s)$ yields a low-pass LC ladder network with resistive termination R_0. Starting with a shunt capacitor C_1, we have the following normalized element values:

$$C_1 = 0.455 \text{ F} \quad L_2 = 4.76 \text{ H} \quad C_3 = 6.5 \text{ F} \quad L_4 = 0.291 \text{ H} \quad R_0 = 0.1082 \text{ } \Omega$$

Synthesis of the Foster part is trivial. Element values are given by $C_0 = 1/k_0 = 0.679$ F for the capacitor and $L_\infty = k_\infty = 1.36$ H for the inductor.

The interpolation results are shown in Figure 12.5(a–c) and the resulting circuit model is depicted in Figure 12.5(d).

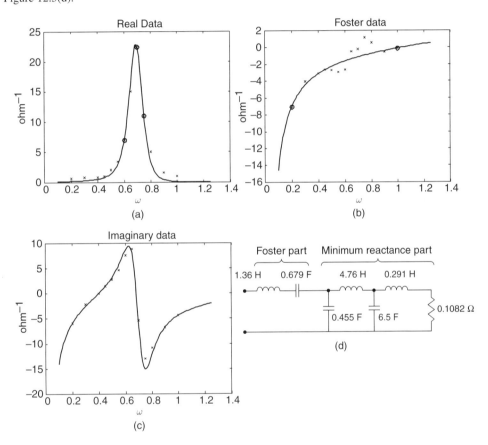

Figure 12.5 Interpolation of impedance data with PS-EIP for Example 12.1: (a) interpolation of real part impedance data with PS-EIP (x, real data; o, interpolation points; —, interpolated function); (b) plot of the Foster part data after interpolation; (c) plot of the imaginary part after interpolation; (d) modeled circuit

Lagrange interpolation technique:

In this part, we present the implementation of the Lagrange interpolation technique by employing the admittance data. Therefore, the measured impedance data is inverted as $Y = 1/z(j\omega) = G(\omega) + jB(\omega)$ and it is listed in Table 12.3. For the real part interpolation, sample points 1, 3 and 13 of Table 12.3

are employed. RPF-IIB is selected to model the given real part data $G(\omega)$. By employing Equation (12.2–12.4) and rearranging the coefficients of the even Lagrange interpolation polynomial, the analytic form of the real part $G(\omega^2) = N(\omega^2)/D(\omega^2)$ is constructed. Using similar notation as above, the numerator polynomial is selected as $N(\omega^2) = \omega^2$ and the denominator polynomial is given by $D(\omega^2) = C_0 + C_1\omega^2 + C_2\omega^4 + C_3\omega^6 + C_4\omega^8$ with the computed coefficients

Table 12.3 Admittance measurement of the monopole antenna, extraction of the Foster portion (verification B via PS-EIP) for Example 12.1

Sample index	Normalized frequency ω	Real part of the input admittance $G(\omega)$	Imaginary part of the input admittance $B(\omega)$	Foster data $B_m = \text{Imag}\{Y_m\}$	$B_f = B - B_m$
1	0.2	0.016 502	0.165 02	0.165 33	−0.000 314 4
2	0.3	0.145 99	0.401 46	0.401 36	0.000 103 17
3	0.4	1.25	0	−0.110 11	0.110 11
4	0.45	0.337 84	−0.472 97	−0.559 37	0.086 398
5	0.5	0.168 92	−0.236 49	−0.412 13	0.175 64
6	0.55	0.103 91	−0.140 59	−0.313 64	0.173 05
7	0.60	0.065 568	−0.071 188	−0.252 81	0.181 62
8	0.65	0.049 597	−0.029 097	−0.212 64	0.183 54
9	0.70	0.042 191	0.010 171	−0.184 93	0.195 1
10	0.75	0.037 931	0.044 828	−0.166	0.210 83
11	0.80	0.035 301	0.076 25	−0.154 63	0.230 88
12	0.9	0.032 787	0.139 34	−0.154 23	0.293 57
13	1.0	0.049 116	0.216 11	−0.161 37	0.377 48

$$C_4 = 226.560 \quad C_3 = -462.4827 \quad C_2 = 320.514 \quad C_1 = -68.9292 \quad C_0 = 4.697\ 365$$

From the analytic form of the real part, the minimum susceptance function Y_m is determined by employing the Gewertz procedure. Thus,

$$Y_m = \frac{a_0 + a_1 s + a_2 s^2 + a_3 s^3}{b_0 + b_1 s + b_2 s^2 + b_3 s^3 + b_4 s^4}$$

with computed coefficients

$$b_4 = 15.051\ 93 \quad b_3 = 6.339\ 863 \quad b_2 = 16.698\ 08 \quad b_1 = 1.857\ 859 \quad b_0 = 2.167\ 340$$
$$a_3 = 1.656\ 38 \quad a_2 = 0.697\ 666\ 2 \quad a_1 = 1.352\ 137 \quad a_0 = 0$$

The synthesis of $Y_m(S)$ is depicted in Figure 12.6(d) below with the following element values:

$$C_1 = 0.62\ \text{F} \quad L_2 = 9.09\ \text{H} \quad C_3 = 0.943\ \text{F} \quad L_4 = 1.3\ \text{H} \quad R_0 = 0.547\ \Omega$$

Once the analytic form of $Y_m(s)$ is found, then the Foster part $B_f = B - B_m$ is extracted from the original input admittance point by point as shown in Table 12.3.

After carefully analyzing the Foster data and selecting the sample points 1 and 12, the analytic expression for B_f is found by employing Foster form I with poles only at DC and infinity. Hence, the residues are $k_0 = 0.013\ 791\ 63$ and $k_\infty = 0.343\ 218\ 4$, or the corresponding element values of the parallel resonance circuit are given by $L_0 = 1/k_0 = 72$ H and $C_\infty = k_\infty = 0.34$ F

The results of the interpolation and the admittance-based circuit model for the antenna are depicted in Figure 12.6.

Example 12.2 *(Proper selection of sample points in the 'Chebyshev sense')*: In this example, the measured impedance data of Table 12.4, which is given over 21 sample points (SPs), is modeled by selecting proper points in the Chebyshev sense. The mathematical form of the impedance is built step by step by employing the various modeling procedures proposed in this chapter, as follows.

Step 1 (Search for the transmission zeros of the real part data): Examination of the real part data of Table 12.4 reveals that there is a transmission zero at SP 1 (i.e. at $\omega = 0$, $R(0) = 0$). The order of this transmission zero is simple. This fact can easily be verified by taking the numerical derivative in the neighborhood of $\omega = 0$. Therefore, the numerator polynomial is set to $N(\omega^2) = \omega^2$.

Step 2 (Generation of the denominator polynomial of the real part in Equation (12.5)): Using Algorithm III, denominator data $\alpha_i = N(\omega_i^2)/R_i$ is generated as shown in Table 12.5. Note that at $\omega = 0$, due to the zero division, data for the denominator is not available. Therefore, SP 1 must be skipped in the interpolation process. Here, in this step, it is desired to employ the Lagrange interpolation method. Therefore, to guarantee the absolute positivity of $D(\omega^2)$ numerically, an auxiliary polynomial $D_x(\omega)$ is utilized such that $D(\omega^2) = 0.5D_{min} + D_x(\omega^2)$ as explained in step 2 of Algorithm I. In this case, the minimum of the denominator is searched for. This minimum is detected at $\omega = 0.25$ (SP 6) and it is found as $D_{min} = 0.1064$. Then, one sets $\alpha_{xi} = \alpha_i - 0.5D_{min} = \alpha_i - 0.0532$. The result is presented in Table 12.5. Thus, auxiliary polynomial $D_x(\omega^2)$ can be constructed by means of the Lagrange method. However, prior to interpolation, the degree of $D_x(\omega^2)$ must be estimated.

Table 12.4 Normalized impedance data given for Example 12.2

Index no.	Normalized frequency ω	Real part of the input impedance $R(\omega)$	Imaginary part of the input impedance $X(\omega)$
1	0	0	0
2	0.0500	0.0105	0.1140
3	0.1000	0.0483	0.2361
4	0.1500	0.1365	0.3676
5	0.2000	0.3174	0.4750
6	0.2500	0.5872	0.4386
7	0.3000	0.7425	0.1966
8	0.3500	0.6755	−0.0197
9	0.4000	0.5503	−0.0991
10	0.4500	0.4562	−0.1014
11	0.5000	0.4000	−0.0750
12	0.5500	0.3737	−0.0405
13	0.6000	0.3719	−0.0073
14	0.6500	0.3930	0.0175
15	0.7000	0.4367	0.0218
16	0.7500	0.4944	−0.0159
17	0.8000	0.5274	−0.1174
18	0.8500	0.4687	−0.2477
19	0.9000	0.3295	−0.3088
20	0.9500	0.2003	−0.2825
21	1.0000	0.1176	−0.2206

Step 3 (Estimation of the least degree n_D of the denominator polynomial $D(\omega^2)$): Degree n_D of $D_x(\omega^2)$ is estimated as described in Section 12.8, by the taking the sequential numerical derivative as presented in Table 12.6. Clearly, in the sequential differentiation process shown by Table 12.6, numerical derivatives disappear at the ninth step (i.e. $\Delta^9 D(\omega_i) \approx 0$ for $i = 1, 2, \ldots, 21$). Therefore, we say that the degree n_D of $D_x(\omega)$ or $D(\omega)$ is '$9 - 1 = 8$'. Hence, $n_D = 8$.

Step 4 (Selection of the proper sample points in the Chebyshev sense): Since the degree of $D_x(\omega^2)$ is $n_D = 8$, then one needs to select $n_k = (n_p/4) + 1 = 3$ proper sample points in the Chebyshev sense to construct the polynomial $D_x(\omega^2) = C_0 + C_1\omega^2 + C_2\omega^4 + C_3\omega^6 + C_4\omega^8$ by employing the SS-ELIP technique. In this case, one finds the roots of the Chebyshev polynomial of degree $m = 2n_k = 6$. These roots are located at the following frequencies: $+0.9659$, $+0.7071$, $+0.2588$, -0.9659, -0.7071, -0.2588. Then, the positive roots $+0.2588$, $+0.7071$, $+0.9659$ are selected as the proper sample points on the frequency axis. For this example, it is found that positive roots of the sixth-order Chebyshev polynomial nearly coincide with the sample points SP 6, SP 15 and SP 20 of Table 12.5.

Step 5 (Construction of $D_x(\omega^2)$ and $D(\omega^2)$ by employing the SS-ELIP technique): Selecting the sample points SP 6, SP 15 and SP 20 of Table 12.5, and utilizing SS-ELIP, $D_x(\omega^2)$ is constructed as $D_x(\omega^2) = 71.8743\omega^8 - 108.473\omega^6 + 50.6462\omega^4 - 5.5055\omega^2 + 0.2249$.

Table 12.5 Data generated for the interpolation of the denominator polynomial and the auxiliary polynomial for Example 12.2

Index no.	Normalized frequency ω	Denominator data α_i	Auxiliary denominator data α_{xi}
1	0	Not available	Not available
2	0.0500	0.2384	0.1852
3	0.1000	0.2069	0.1537
4	0.1500	0.1648	0.1116
5	0.2000	0.1260	0.0728
6	0.2500	0.1064	0.0532
7	0.3000	0.1212	0.0680
8	0.3500	0.1813	0.1281
9	0.4000	0.2907	0.2375
10	0.4500	0.4439	0.3906
11	0.5000	0.6250	0.5718
12	0.5500	0.8095	0.7562
13	0.6000	0.9680	0.9148
14	0.6500	1.0750	1.0218
15	0.7000	1.1222	1.0689
16	0.7500	1.1377	1.0845
17	0.8000	1.2136	1.1604
18	0.8500	1.5414	1.4882
19	0.9000	2.4586	2.4053
20	0.9500	4.5060	4.4528
21	1.0000	8.5000	8.4468

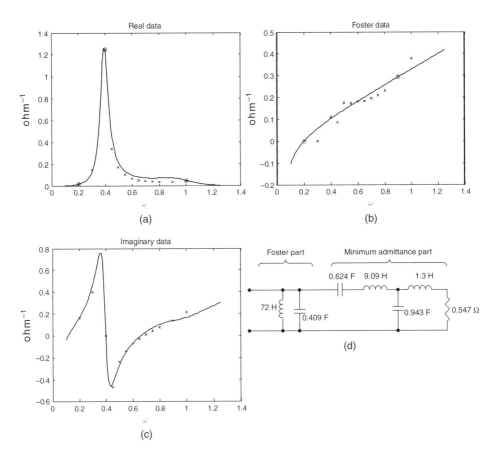

Figure 12.6 Interpolation of admittance data with SS-ELIP for Example 12.1: (a) interpolation of real part admittance data with SS-ELIP (x, real data; o, interpolation points; —, interpolated function); (b) plot of the Foster part data after interpolation; (c) plot of the imaginary part after interpolation; (d) modeled circuit

On the other hand,

$$D(\omega^2) = 0.5D_{min} + D_x(\omega^2) \quad \text{or} \quad D(\omega^2) = 0.0532 + D_x(\omega^2)$$

Thence,

$$D(\omega^2) = 71.8743\omega^8 - 108.4734\omega^6 + 50.6462\omega^4 - 5.5055\omega^2 + 0.2781$$

Step 6 (Generation of the minimum reactance impedance function $Z_m(s)$ using the Gewertz procedure): From step 5, one can easily construct the analytic form of the real part as $R(\omega^2) = \omega^2/D(\omega^2)$. Then, the minimum reactance impedance $Z_m(s)$ is generated using the Gewertz procedure. Thus, we have

$$Z_m = \frac{4.0224s^3 + 1.8566s^2 + 1.9411s}{16.0766s^4 + 7.4205s^3 + 13.8440s^2 + 2.8090s + 1}$$

Table 12.6 Sequential numerical derivatives of the auxiliary polynomial data given by Table 12.5

	1	2	3	4	5	6	7	8	9	10	11	12	13	14	15	16	17	18	19	20	21
α_{xi}	NA	1.85e-1	1.54e-1	1.12e-1	7.28e-2	5.32e-2	6.80e-2	1.28e-1	2.38e-1	3.91e-1	5.72e-1	7.56e-1	9.15e-1	1.02	1.07	1.08	1.16	1.49	2.41	4.45	8.45
$\Delta^{(1)}D_x$	—	-3.15e-2	-4.21e-2	-3.88e-2	-1.96e-2	1.48e-2	6.01e-2	1.09e-1	1.53e-1	1.81e-1	1.84e-1	1.59e-1	1.07e-1	4.71e-2	1.55e-2	7.59e-2	0.328	0.917	2.05	3.99	0
$\Delta^{(2)}D_x$	—	-1.06e-2	3.27e-3	1.92e-2	3.43e-2	4.54e-2	4.93e-2	4.38e-2	2.80e-2	3.32e-3	-2.59e-2	-5.15e-2	-5.99e-2	-3.16e-2	6.04e-2	2.52e-1	0.589	1.13	1.95	0	0
$\Delta^{(3)}D_x$	—	1.38e-2	1.60e-2	1.51e-2	1.10e-2	3.89e-3	-5.50e-3	-1.57e-2	-2.47e-2	-2.93e-2	-2.55e-2	-8.42e-3	2.83e-2	9.20e-2	1.92e-1	3.37e-1	0.541	0.816	0	0	0
$\Delta^{(4)}D_x$	—	2.13e-3	-8.59e-4	-4.09e-3	-7.12e-3	-9.39e-3	-1.02e-2	-8.93e-3	-4.59e-3	3.74e-3	1.71e-2	3.67e-2	6.37e-2	9.96e-2	1.46e-1	2.04e-1	0.275	0	0	0	0
$\Delta^{(5)}D_x$	—	-2.99e-3	-3.23e-3	-3.03e-3	-2.27e-3	-8.55e-4	1.32e-3	4.34e-3	8.33e-3	1.34e-2	1.96e-2	2.70e-2	3.59e-2	4.61e-2	5.80e-2	7.15e-2	0	0	0	0	0
$\Delta^{(6)}D_x$	—	-2.48e-4	2.05e-4	7.60e-4	1.41e-3	2.17e-3	3.03e-3	3.99e-3	5.04e-3	6.20e-3	7.46e-3	8.82e-3	1.03e-2	1.18e-2	1.35e-2	0	0	0	0	0	0
$\Delta^{(7)}D_x$	—	4.54e-4	5.54e-4	6.55e-4	7.56e-4	8.57e-4	9.58e-4	1.06e-3	1.16e-3	1.26e-3	1.36e-3	1.46e-3	1.56e-3	1.66e-3	0	0	0	0	0	0	0
$\Delta^{(8)}D_x$	—	1.01e-4	1.01e-4	1.01e-4	1.01e-4	1.01e-4	1.01e-4	1.01e-4	1.01e-4	1.01e-4	1.01e-4	1.01e-4	1.01e-4	0	0	0	0	0	0	0	0
$\Delta^{(9)}D_x$	—	5.2e-15	-6.4e-15	6.1e-15	-7.5e-16	-5.1e-15	4.4e-15	-4.1e-15	1.6e-14	-3.2e-14	2.6e-14	1.2e-14	0	0	0	0	0	0	0	0	0

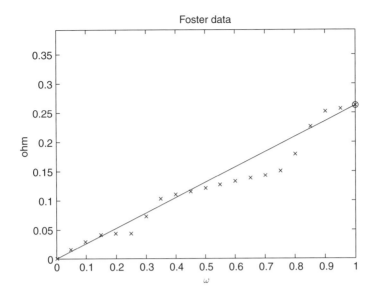

Figure 12.7 Plot of Foster data $X_f(\omega)$ for Example 12.2.

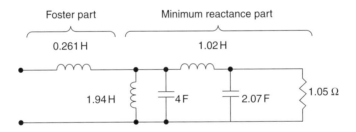

Figure 12.8 Synthesis of the impedance model for Example 12.2

Step 7 (Generation of the Foster part): In this step, firstly the real and imaginary parts of Z_m are generated on the $j\omega$ axis as $Z_m(j\omega) = R(\omega) - jX_m(\omega)$. Then, point by point the Foster part data $X_f(\omega_i) = X(\omega_i) - X_m(\omega_i)$ is generated as presented in Table 12.7 and depicted in Figure 12.7. Close examination of Table 12.7 and Figure 12.7 reveals that $X_f(\omega) \geq 0, \forall \omega$. This situation is described as FF-IVA. Therefore, there is no need to select finite poles to interpolate the Foster part. In this case, 21 sample points can easily be approximated by a simple line. Thus, $X_f(\omega) = L_\infty \omega$. For interpolation purposes, it is sufficient to pick only one sample point to determine L_∞. Let this be SP 21. Then, $L_\infty = 0.261\ 49$ H.

Step 8 (Complete model for the impedance data and the final results): From the last two steps, one ends up with the following mathematical form for the impedance data:

$$Z_m = \frac{4.0224s^3 + 1.8566s^2 + 1.9411s}{16.0766s^4 + 7.4205s^3 + 13.8440s^2 + 2.8090s + 1} + 0.261\ 49s$$

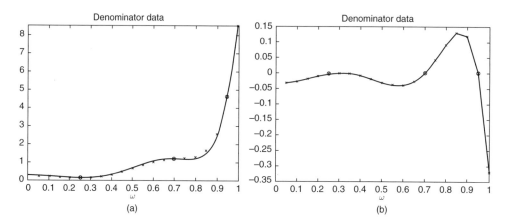

Figure 12.9 The denominator function and its error for Example 12.2: (a) plot of the denominator data; (b) plot of the near mini–max error

Table 12.7   Foster part of the impedance data for Example 12.2.

Index no.	Normalized frequency ω	Foster data X_f
1	0	0
2	0.0500	0.0153
3	0.1000	0.0298
4	0.1500	0.0407
5	0.2000	0.0428
6	0.2500	0.0429
7	0.3000	0.0734
8	0.3500	0.1024
9	0.4000	0.1112
10	0.4500	0.1156
11	0.5000	0.1208
12	0.5500	0.1269
13	0.6000	0.1329
14	0.6500	0.1379
15	0.7000	0.1424
16	0.7500	0.1514
17	0.8000	0.1796
18	0.8500	0.2267
19	0.9000	0.2532
20	0.9500	0.2579
21	1.0000	0.2615

Eventually, it is synthesized as shown in Figure 12.8.

For comparison purposes, interpolated and computed data for the denominator are depicted in Figure 12.9(a) and the near mini–max error is shown in Figure 12.9(b). The error fluctuation is almost-equal ripple over the frequencies $0 \leq \omega \leq 0.75$ with the ripple factor $\varepsilon \leq 0.025$; it jumps at $\omega \approx 0.85$ with

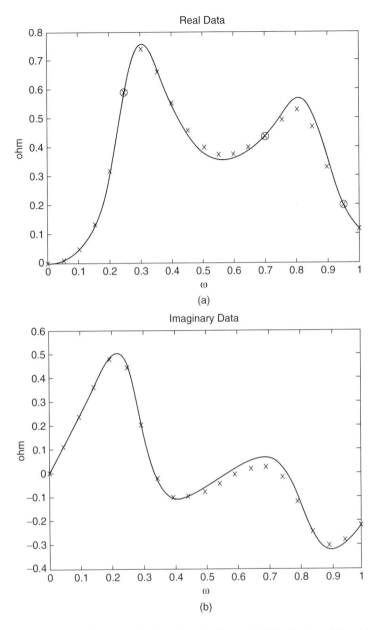

Figure 12.10   Given data and the modeled function for Example 12.2: (a) plot of the real part data; (b) plot of the imaginary part data

$\varepsilon \approx 0.125$ and has its maximum value at $\omega = 1$ with $\varepsilon \approx 0.32$. Finally, as a function of normalized angular frequency ω, the real and imaginary parts of the generated model are shown in Figure 12.10(a) and Figure 12.10(b) respectively together with the measured data. It is observed that the error in the real part model exhibits quasi-mini–max behavior. This error behavior also propagates to the imaginary part of the impedance due to the generation of the imaginary part by means of the Gewertz procedure and Foster part interpolation.

12.10 Conclusion

In this chapter, techniques have been presented to model the given immittance data by utilizing linear interpolation methods. The crux of the idea for the new modeling tools is the 'point-by-point decomposition of the immittance data into its minimum and Foster parts'.

Firstly, the real part of the given immittance data is interpolated with a non-negative, even rational function, which in turn yields the realizable positive real minimum immittance. Then, the Foster data is interpolated. Finally, the minimum and Foster functions are synthesized, preferably as a lossless two-port in resistive termination. Hence, the desired circuit model is obtained as a natural consequence of the interpolation process. Two major approaches were introduced to model the real part data. The first technique resides on the 'summation of the perfect squares of even Lagrange interpolation polynomials'. The second one utilizes the 'perfect square of an even interpolation polynomial'. Related interpolation algorithms are introduced to ease the implementation of the modeling process. In the course of interpolation, the total number of residues of the Foster function is set equal to the total number of data points to be interpolated. The modeler selects the locations of the finite poles for the construction of the Foster part. Obviously, this process affects the resulting element values of the Foster section. In addition, the modeler is free to employ either impedance or admittance data to start with. Therefore, the models generated by employing the new interpolation techniques are by no means unique. A numerical method was introduced to estimate the least degree of the immittance function by means of 'sequential numerical differentiation' which in turn determines the minimum number of circuit elements to be used in the model. Furthermore, a 'proper data selection process' is presented to minimize the interpolation error in the near mini–max sense or in the 'Chebyshev sense'.

As a result of this research work, a complete software package called 'ALIYAR' has been developed to construct circuit models for the given immittance data. *ALIYAR* estimates the least degree of the immittance model from the given data, selects the proper sample points either to minimize the error fluctuations in the 'near mini–max sense' or to minimize the sum of squares error by forming possible combinations of sample points for the selected least degree of the immittance function. Then, it generates the realizable mathematical model of the immittance data. Finally, it synthesizes the constructed model as a lossless circuit in unit termination. Obviously, *ALIYAR* can be utilized as the front end to any commercially available microwave circuit design/analysis software package like ANSOFT or EAGLE-WARE to design and to practically realize passive and active microwave systems, such as matching networks, amplifiers, etc., from scratch without selecting the circuit topology. Rather, the optimum circuit topology is obtained as an output of *ALIYAR* from the given data with the least number of elements. Examples were presented to portray implementation of the modeling processes via the interpolation techniques introduced in this chapter. The proposed modeling techniques via interpolation can easily be extended to handle models with distributed or with mixed lumped and distributed elements, by employing the description of the two-variable immittance functions given in [23–25].

References

[1] B. S. Yarman, 'Broadband Networks', in *Wiley Encyclopedia of Electrical and Electronics Engineering*, Vol. **II**, pp. 589–604, 1999.

[2] H. J. Carlin and B. S. Yarman, 'The Double Matching Problem: Analytic and Real Frequency Solutions', *IEEE Transactions on Circuit and Systems*, Vol. **30**, pp. 15–28, 1983.

[3] F. Maloberti, 'Design of High-speed Analog Circuits for Mobile Communications', *Proceedings of International Conference on Micro Electronics*, Vol. **2**, pp. 673–680, September, 1997.

[4] F.Y. Chang, 'Waveform Relaxation Analysis of RLCG Transmission Lines', *IEEE Transactions on Circuits and Systems*, Vol. **37**, pp. 1394–1415, 1990.

[5] R. Wang and O. Wing, 'A Circuit Model of a System of VLSI Interconnects for Time Response Computation', *IEEE Transactions on Microwave Theory and Techniques*, Vol. **39**, pp. 688–693, 1991.

[6] L.T. Pillage and R.A. Rohrer, 'Asymptotic Waveform Evaluation for Timing Analysis', *IEEE Transactions on Computer Aided Design*, Vol. **9**, pp. 352–366, 1990.

[7] D.H. Xie and M. Nakhla, 'Delay and Cross-talk Simulation of High-speed VLSI Interconnects with Non-linear Terminations', *IEEE Transactions on Computer Aided Design*, Vol. **12**, pp. 1798–1811, 1993.

[8] Q. Yu, J. Wang and E. Kuh, 'Passive Multipoint Moment Matching Model Order Reduction Algorithm on Multiport Distributed Interconnect Networks', *IEEE Transactions on Circuits and Systems*, Vol. **46**, pp. 140–160, 1999.

[9] E. McShane, M. Trivedi, Y. Xu, P. Khandewal, A. Mulay and K. Shenai, 'One Chip Wonders', *IEEE Circuits and Devices Magazine*, Vol. **14**, pp. 35–42, 1998.

[10] S. Sercu and L. Martens, 'High-frequency Circuit Modeling of Large Pin Count Packages', *IEEE Transactions on Microwave Theory and Techniques*, Vol. **45**, pp. 1897–1904, 1997.

[11] J.R. Long and A. Miles, 'Modeling Characterization and Design of Monolithic Inductors for Silicon RFICs', *IEEE Custom Integrated Circuits Conference (CICC)*, May 1996, pp. 185–188.

[12] C. O'Connor, 'RFIC Receiver Technology for Digital Mobile Phones', *Microwave Journal*, Vol. **40**, pp. 64–75, 1997.

[13] B.S. Yarman, 'Modern Approaches to Broadband Matching Problems', *Proceedings of the IEE*, Vol. **132**, pp. 87–92, 1985.

[14] H.J. Carlin, 'New Approach to Gain Bandwidth Problems', *IEEE Transactions on Circuits and Systems* Vol. **23**, pp. 170–175, 1977.

[15] F. Güneş, A. Cetiner, 'A Novel Smith Chart Formulation of Performance Characterization for a Microwave Transistor', *IEE Proceedings – Circuit Devices and Systems*, Vol. **145**, pp. 419–429, 1998.

[16] B.S. Yarman, A. Aksen and A. Kilinç, 'An Immitance Based Tool for Modeling Passive One-port Devices by Means of Darlington Equivalents', International Journal of Electronics and Communications (AEÜ), Vol. **55**, pp. 443–451, 2001.

[17] B.S. Yarman and A. Aksen, 'A Reflectance-based Computer Aided Modeling Tool for High Speed/High Frequency Communication Systems', *Proceedings of IEEE-ISCAS*, Sydney, Australia, Vol. **4**, pp. 270–273, 2001.

[18] L.I. Smilen, "Interpolation on the Real Frequency Axis", *IEEE International Convention Record*, Vol. **13**, pp. 42–50, 1965.

[19] E. Zaheb and A. Lempel, "Interpolation in the Network Sense", *IEEE Transactions on Circuits and Systems*, Vol. **CT-13**, No. 1, pp. 118–119, 1966.

[20] D.C. Youla and M. Saito, "Interpolation with Positive Real Functions", *Journal of the Franklin Institute*, Vol. **284**, No. 2, pp. 77–108, 1967.

[21] C.M. Gewertz, 'Synthesis of a Finite, Four-terminal Network from its Prescribed Driving Point Functions and Transfer Functions', *Journal of Mathematics and Physics*, Vol. **12**, pp. 1–257, 1933.

[22] A. Kilinç, 'Novel Data Modeling Procedures: Impedance and Scattering Approaches', PhD thesis, Istanbul University, 1995.

[23] A. Aksen and B.S. Yarman, 'Cascade Synthesis of Two-variable Lossless Two-port Networks with Lumped Elements and Transmission Lines: A Semi-analytic Procedure', in *Multidimensional Signals, Circuits and Systems*, ed. K. Galkowski and J.Wood, Taylor & Francis, 2001, pp. 219–231.

[24] B.S. Yarman and A. Aksen, 'An Integrated Design Tool to Construct Lossless Matching Networks with Mixed Lumped and Distributed Elements', *IEEE Transactions on Circuits and Systems*, Vol. **39**, pp. 713–723, 1992.

[25] A. Aksen and B.S. Yarman, 'A Real Frequency Approach to Describe Lossless Two-ports Formed with Mixed Lumped and Distributed Elements', International Journal of Electronics and Communications (AEÜ), Vol. **55**, pp. 389–396, 2001.

[26] B.S. Yarman, A. Aksen and A. Kilinç, 'Immitance Data Modelling via Linear Interpolation Techniques', *Proceedings of the IEEE International Symposium on Circuits and Systems, ISCAS' 2002*, Phoenix, Arizona, Vol. **2**, pp. 527–530, May, 2002.

[27] K. Atkinson, *Elementary Numerical Analysis*, 3rd edition, John Willy & Sons, Inc., 2005, Chapter 4.

13

Lossless Two-ports Formed with Mixed Lumped and Distributed Elements: Design of Matching Networks with Mixed Elements[1]

13.1 Introduction

For almost half a century, one of the major concerns of the circuit and microwave community has been the insertion loss synthesis with mixed lumped and distributed elements. In earlier works [1–4], network functions such as the driving point impedance of a lossless two-port Z_{in} was expressed in terms of a rational function of a complex exponential $e^{p\tau}$ with polynomial coefficients P_{Di} and P_{Ni} in the complex variable $p = \sigma + j\omega$ such that

$$Z_{in} = \frac{\sum_{i=0}^{N} P_{Ni}(p)e^{-2p\tau}}{\sum_{i=0}^{M} P_{Di}(p)e^{-2p\tau}} \tag{13.1}$$

In this representation, the polynomial coefficients are related to lumped circuit elements and $e^{p\tau}$ refers to equal length transmission lines with electrical delay τ. As in the case of networks with single type elements, it was claimed that positive realness of the driving point impedance is a necessary and sufficient condition for realizability. Unfortunately, this did not turn out to be the case. Right after this claim, Rhodes gave his famous counterexample [6]. Later, Youla, Rhodes and Marston introduced necessary and sufficient conditions imposed on the driving point impedance that is constructed with mixed lumped and distributed elements for a special class of networks [7]. Results of this paper did not

[1] This chapter, co-authorized by A. Aksen (Isik University, Istanbul, Turkey) and M. Sengul (Kadir Mas University, Istanbul, Turkey), is adopted from the following paper: A. Aksen and B. S. Yarman, 'A Real Frequency Approach to Describe Lossless Two-ports Formed with Mixed Lumped and Distributed Elements', *International Journal of Electronics and Communications (AEÜ)*, Vol. 6, pp. 389–396, November, 2001.

prove to possess practicality. Following this work, Scanlan and Baher introduced their classical paper to describe mixed element lossless networks with commensurate stubs [8]. In the follow-up work, Fettweis introduced the new concept of scattering Hurwitz parameters to describe passive two-ports in two variables; namely, the $p = \sigma + j\omega$ variable and the Richard variable $\lambda = \tanh(p\tau) = \Sigma + j\Omega$ [9–12]. All these papers point out the fact that positive realness of driving point impedance $Z_{in}(p, \lambda)$ or bounded realness of the corresponding input reflectance $S_{in} = (Z_{in} - 1)/(Z_{in} + 1)$ is necessary but not sufficient for the synthesis. Rather, sufficient conditions are imposed by the topological restrictions of the two-port under consideration. In other words, restrictions imposed on the scattering parameters via the connection order of circuit elements yield the sufficiency conditions. In this regard, sufficiency conditions turn out to be the connectivity information of the two kinds of elements which form the lossless network.

In this chapter, lossless two-ports are described in terms of Fettweis's scattering Hurwitz bounded real parameters [9–12]. Furthermore, as in the case of simplified real frequency techniques [13–14], it is shown that the scattering parameters can be completely described in terms of some selected independent coefficients of the two-variable numerator polynomial of the input reflectance. This, in turn, allows the practical design of microwave/millimeter wave broadband networks. In the course of synthesis, Fettweis's decomposition technique is employed [15]. Explicit descriptive equations are given for some selected topologies. Finally, a double matching design example with lumped and distributed elements is presented.

Construction of Two-Variable Scattering Functions for Lossless Two-ports Formed with Cascaded Lumped and Distributed Elements

Consider the generic form of a lossless two-port formed with cascade connections of lumped and distributed two-ports as shown in Figure 13.1.

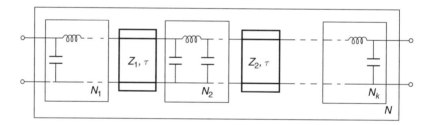

Figure 13.1   Generic form of cascaded lumped and distributed two-ports

The scattering matrix describing the mixed element two-port can be expressed in the Belevitch canonical form as [9, 10]

$$S(p, \lambda) = [1/g(p, \lambda)] \begin{bmatrix} h(p, \lambda) & \sigma f(-p, -\lambda) \\ f(p, \lambda) & -\sigma h(-p, -\lambda) \end{bmatrix} \tag{13.2a}$$

$$S(p, \lambda)S(-p, \lambda) = I \tag{13.2b}$$

where:

$f(p, \lambda)$ is a monic real polynomial that consists of transmission zeros;

$h(p, \lambda)$ and $g(p, \lambda)$ are real polynomials in the complex variables $p = \sigma + j\omega$ and $\lambda = \Sigma + j\Omega (\lambda = \tanh(p\tau), \tau$ being the delay length of unit elements);

$g(p, \lambda)$ is a scattering Hurwitz polynomial;

σ is a unimodular constant, $|\sigma| = 1$.

Losslessness of the mixed element two-port requires that Equation (13.2b), where I is the identity matrix. The open form of the above equations yields the following:

$$g(p, \lambda)g(-p, -\lambda) = h(p, \lambda)h(-p, -\lambda) + f(p, \lambda)f(-p, -\lambda)$$

Furthermore, it can be shown that

$$g(p, \lambda) = \sum_{i=0}^{n_\lambda} g_i(p)\lambda^i$$

$$h(p, \lambda) = \sum_{i=0}^{n_\lambda} h_i(p)\lambda^i$$

(13.3)

with

$$g_i(p) = \sum_{j=0}^{n_p} g_{ji}p^j$$

$$h_i(p) = \sum_{j=0}^{n_p} h_{ji}p^j$$

(13.4)

where n_p and n_λ are the total number of lumped and distributed elements respectively.

Alternatively, Equation (13.3) can be written as

$$g(p, \lambda) = \sum_{i=0}^{n_p} g'_i(\lambda)p^i$$

$$h(p, \lambda) = \sum_{i=0}^{n_p} h'_i(\lambda)p^i$$

(13.5)

with

$$g'_i(\lambda) = \sum_{j=0}^{n_\lambda} g_{ij}\lambda^j$$

$$h'_i(\lambda) = \sum_{j=0}^{n_\lambda} h_{ij}\lambda^j$$

(13.6)

The real polynomials $g(p, \lambda)$ and $h(p, \lambda)$ with partial degrees n_p and n_λ can also be expressed in matrix form as follows:

$$g(p, \lambda) = \bar{p}^T \Lambda_g \bar{\lambda}$$

$$h(p, \lambda) = \bar{p}^T \Lambda_h \bar{\lambda}$$

(13.7)

where

$$\bar{p}^T = \begin{bmatrix} 1 \ p \ p^2 \ \cdots \ p^{n_p} \end{bmatrix} \qquad \bar{\lambda}^T = \begin{bmatrix} 1 \ \lambda \ \lambda^2 \ \cdots \ \lambda^{n_\lambda} \end{bmatrix}$$

$$\Lambda_g = \begin{bmatrix} g_{00} & g_{01} & \cdots & g_{0n_\lambda} \\ g_{10} & g_{11} & \cdots & g_{1n_\lambda} \\ \vdots & \vdots & & \vdots \\ g_{n_p0} & g_{n_p1} & \cdots & g_{n_pn_\lambda} \end{bmatrix}$$
$$\Lambda_h = \begin{bmatrix} h_{00} & h_{01} & \cdots & h_{0n_\lambda} \\ h_{10} & h_{11} & \cdots & h_{1n_\lambda} \\ \vdots & \vdots & & \vdots \\ h_{n_p0} & h_{n_p1} & \cdots & h_{n_pn_\lambda} \end{bmatrix}$$
(13.8)

The scattering parameters and hence the canonical polynomials $f(p, \lambda), g(p, \lambda)$ and $h(p, \lambda)$ must satisfy some additional conditions to ensure the realizability as a passive lossless cascade structure. In this regard, the following properties may be stated.

Property 1: For the mixed element lossless two-ports, the polynomial $f(p, \lambda)$ defines the transmission zeros of the cascade and is given by

$$f(p, \lambda) = f_0(p) f_1(\lambda)$$
(13.9)

where $f_0(p)$ and $f_1(\lambda)$ contain the transmission zeros due to the lumped sections and unit elements in the cascade, respectively.

If n_λ unit elements (or commensurate transmission lines) are considered in the cascade topology, then $f_1(\lambda) = \left(1 - \lambda^2\right)^{n_\lambda/2}$.

When the transmission lines are removed from the cascade structure, one ends up with a lumped network whose transmission zeros are defined by $f_0(p)$. In most practical cases, it is appropriate to choose $f_0(p)$ as an even/odd real polynomial, which corresponds to reciprocal lumped structures. A particularly useful case is obtained for $f_0(p) = 1$, which corresponds to a typical low-pass structure having transmission zeros only at infinity. In this case, the polynomial $f(p, \lambda)$ takes the simple form

$$f(p, \lambda) = \left(1 - \lambda^2\right)^{n_\lambda/2}$$
(13.10)

Another practical case is to choose $f_0(p) = p^{n_p}$, which corresponds to a typical high-pass structure having transmission zeros at the origin of the complex p plane (i.e. at DC or $\omega = 0$).

Property 2: Let us consider a low-pass-based structure having all transmission zeros at infinity in the p domain. When the transmission lines are removed from the mixed cascade structure, one ends up with a lumped network whose transmission zeros are defined by $f_0(p)$. In this case, the scattering matrix of the resulting lumped prototype can be fully described in terms of the canonical real polynomials $f_0(p)$, $g_0(p)$ and $h_0(p)$ as in the Belevitch representation specified by Equation (13.2). This would correspond to the boundary case where we set $\lambda = 0$ in the scattering description given by Equations (13.1–13.8). In other words, the boundary polynomials $h_0(p) = h(p, 0)$, $g_0(p) = g(p, 0)$ and $f_0(p) = f(p, 0)$ define the cascade of lumped sections which take place in the composite structure, where $g_0(p)$ is strictly Hurwitz and satisfies the losslessness condition:

$$g_0(p)g_0(-p) = h_0(p)h_0(-p) + f_0(p)f_0(-p) \qquad (13.11)$$

Property 3: In a similar manner to that of Property 2, when the lumped elements are removed from the cascade structure, one obtains a lossless two-port formed with a typical cascade connection of transmission lines. In this case, the resulting distributed prototype whose transmission zeros are defined by $f_1(\lambda) = \left(1 - \lambda^2\right)^{n_\lambda/2}$ can be fully described by means of the real polynomials $f_1(\lambda)$ of Equations (13.9–13.10), $g'_0(\lambda)$ and $h'_0(\lambda)$ of Equation (13.6). Setting $p = 0$, the scattering description given by Equation (13.5) results in the boundary polynomials $f_1(\lambda) = f(0, \lambda)$, $g'_0(\lambda) = g(0, \lambda)$ and $h'_0(\lambda) = h(0, h)$. Here, the boundary polynomials are given by

$$h'_0(\lambda) = \sum_{j=0}^{n_\lambda} h_{0j}\lambda^j \quad \text{and} \quad g'_0(\lambda) = \sum_{j=0}^{n_\lambda} g_{0j}\lambda^j$$

In this case, the boundary polynomials $h(0, \lambda)$, $g(0, \lambda)$ and $f(0, \lambda)$ define the cascade of UEs (Unit Elements) which appear in the composite structure, where $g(0, \lambda)$ is strictly Hurwitz and is given by

$$g(0, \lambda)g(0, -\lambda) = h(0, \lambda)h(0, -\lambda) + \left(1 - \lambda^2\right)^{n_\lambda} \qquad (13.12)$$

Transmission zeros at finite complex or real frequencies beyond DC and at infinity may also be placed in the lumped sections. For example, if the lumped sections in the cascade are assumed to consist of solely high-pass types of elements with transmission zeros at zero ($p = 0$) in the p domain only, then, as p approaches infinity, all the lumped elements in the two-port are removed and we obtain the corresponding boundary polynomials

$$h'_{n_p}(\lambda) = \sum_{j=0}^{n_\lambda} h_{n_p j}\lambda^j \quad \text{and} \quad g'_{n_p}(\lambda) = \sum_{j=0}^{n_\lambda} g_{n_p j}\lambda^j$$

In this case, the para-unitary relation Equation (13.12) is modified as

$$g'_{n_p}(\lambda)g'_{n_p}(-\lambda) = h'_{n_p}(\lambda)h'_{n_p}(-\lambda) + \left(1 - \lambda^2\right)^{n_\lambda} \qquad (13.13)$$

where $g'_{n_p}(\lambda)$ is strictly Hurwitz.

Property 4: For the topologies where all the transmission zeros are at infinity in the p domain, the single variable boundary polynomial pairs $\{h(p, 0), g(p, 0)\}$ and $\{h(0, \lambda), g(0, \lambda)\}$, related by Equation (13.11) and Equation (13.12) respectively, refer to the first row and the first column entries of the Λ_h and Λ_g matrices given by Equations (13.7–13.8). In this case, having chosen the first row and first column entries of the Λ_h and Λ_g matrices as independent real parameters to describe the mixed element structure, the remaining matrix elements can be generated by employing these independent entries and using the cascade connectivity information and the para-unitary condition of Equation (13.2b). Hence, the following definitions are made.

Definition 1: Referring to a low-pass-based mixed element structure with transmission zeros at infinity only in the p domain, the submatrix obtained by deleting the first row and first column of the numerator coefficient matrix Λ_h, is called the connectivity matrix of the numerator.

Definition 2: Similarly, the submatrix obtained by deleting the first row and first column of the denominator coefficient matrix Λ_g is called the connectivity matrix of the denominator.

Property 5: Cascade connection information can easily be derived via direct analysis of the selected topology. Given that complete topologic connectivity information is provided, construction of the corresponding two-variable scattering parameters of the mixed element structure and, hence, the connectivity submatrices of Λ_h and Λ_g, can be generated using the following procedure.

Procedure for generation of connectivity matrices: The above-mentioned boundary polynomial sets $h(p,0), g(p,0), f(p,0)$, and $h(0,\lambda), g(0,\lambda), f(0,\lambda)$ describe two independent lumped and distributed network prototypes respectively. These prototypes can be decomposed algebraically into their subsections by use of the transfer matrix factorization technique [14,16]. The subsections can thus be connected sequentially as desired, to form a cascade. As a result of this process, the entries of the connectivity matrices are generated by means of the independently selected entries of the first row and first column of Λ_h and Λ_g.

Property 6: For the generation of the connectivity matrices Λ_h and Λ_g, which satisfy the boundary conditions given by Properties 1 and 2, it is essential to establish the para-unitary relation of Equation (13.2a). By equating the coefficients of the same powers of the complex frequency variables in Equation (13.2b), we obtain the following equation set:

$$
\begin{aligned}
&g_{0,k}^2 + 2\sum_{i=0}^{k-1}(-1)^{k-1}g_{0,l}g_{0,2k-l} \\
&= h_{0,k}^0 + f_{0,k}^2 \\
&\quad + 2\sum_{l=0}^{k-1}(-1)^{k-l}\left(h_{0,l}h_{0,2k-l} + f_{0,l}f_{0,2k-1}\right), (k=0,1,\ldots,n_\lambda)
\end{aligned}
\tag{13.14a}
$$

$$
\begin{aligned}
&\sum_{j=0}^{i}\sum_{l=0}^{k}(-1)^{i-j-l}g_{j,l}g_{i-j,2k-1-l} \\
&= \sum_{j=0}^{i}\sum_{l=0}^{k}(-1)^{i-j-l}\left[h_{j,i}h_{i-j,2k-1-l} + f_{j,l}f_{i-j,2k-1-l}\right], \\
&\left(i=1,3,\ldots,2n_p-1, \quad k=0,1,\ldots,n_\lambda\right)
\end{aligned}
\tag{13.14b}
$$

$$
\begin{aligned}
&\sum_{j=0}^{i}(-1)^{i-j}g_{j,k}g_{i-j,k} + 2\sum_{l=0}^{k-1}(-1)^{k-l}g_{j,l}g_{i-j,2k-l} \\
&= \sum_{j=0}^{i}(-1)^{i-j}\left(h_{j,k}h_{i-j,k} + f_{j,k}f_{i-j,k} + 2\sum_{l=0}^{k-1}(-1)^{k-1}\left[h_{j,l}h_{i-j,2k-1} + f_{j,l}f_{i-j,2k-1}\right]\right), \\
&\left(i=2,4,\ldots,2n_p-2, \quad k=0,1,\ldots,n_\lambda\right)
\end{aligned}
$$

$$
\tag{13.14c}
$$

$$g_{n_p,k}^2 + 2 \sum_{l=0}^{k-1} (-1)^{k-l} g_{n_p,l} g_{n_p,2k-l}$$

$$= h_{n_p,k}^2 + f_{n_p,k}^2$$

$$+ 2 \sum_{l=0}^{k-1} (-1)^{k-l} \left(h_{n_p,l} h_{n_p,2k-l} + f_{n_p,l} f_{n_p,2k-l} \right), (k=0,1,\ldots,n_\lambda)$$

(13.14d)

Definition 3 (Fundamental equation set): Usually, the construction of mixed element lossless two-ports is initiated by selecting transmission zeros of the topology. That is, the form of $f(p,\lambda)$ is chosen in advance. Then, some entries of the matrix Λ_h are selected as the independent parameters to describe the system. Evidently, the total number of independent parameters must be equal to the total number of circuit elements to be placed in the mixed structure. The rest of the entries of the matrices Λ_h and Λ_g are computed subsequently; or, if possible, explicitly derived in terms of the selected independent parameters.

At this point, it is worthwhile mentioning that the solution of the above equation set for the coefficients g_{ij} of $g(p,\lambda)$ is equivalent to factorization of the two-variable polynomial

$$g(p,\lambda)g(-p,-\lambda) = h(p,\lambda)h(-p,-\lambda) + f(p,\lambda)f(-p,-\lambda)$$

Therefore, the equation set Equation (13.13) will be referred to as the fundamental equation set (FES) [16].

Since the fundamental equation set is nonlinear in the coefficients g_{ij}, the solution for the scattering Hurwitz polynomial $g(p,\lambda)$ is not unique. Therefore, in order to obtain a realizable system, further properties of FES must be investigated and the necessary constraints leading to an acceptable solution must be established. In this case, in addition to the boundary conditions given by Equation (13.11) and Equation (13.12), we need to consider additional coefficient constraints, which contain the cascade connectivity information.

Based on the above discussion, we developed the following semi-analytical procedure to construct two-variable canonical polynomials, which in turn define the scattering matrices for cascaded lumped and distributed structures.

A semi-analytical procedure to construct two-variable canonical polynomials:

- Assuming a regular cascaded structure as in Figure 13.1, select the total number of lumped and distributed elements (n_p, n_λ) in the circuit topology and fix the form of $f(p,\lambda)$.
- Select the coefficients of the polynomials $h(p,0)$ and $h(0,\lambda)$ as the independent parameters and generate the strictly Hurwitz polynomials $g(p,0)$ and $g(0,\lambda)$ (or $g'_{n_p}(\lambda)$) by employing Equation (13.11) and Equation (13.12) (or Equation (13.13)) respectively.
- In addition to the boundary conditions stated above, if further topological constraints on the coefficients are imposed, then FES given by Equation (13.14) is solved for the unknown coefficients. This requires the establishment of coefficient constraints reflecting the connectivity information for each class of cascade topology. In the case where the derived connectivity constraints are sufficient, we obtain a unique solution of FES. In this context, by straightforward analysis, the coefficient constraints leading to an explicit solution of FES can easily be generated for symmetric structures or those regular structures in which the lumped sections are confined to be simple low-pass, high-pass or band pass configurations which consist of inductors and capacitors [17–19]. In the following section, a FES solution is considered for a given specific class of topology.

13.2 Construction of Low-Pass Ladders with UEs

From a physical implementation point of view, one practical circuit configuration is that of simple low-pass ladder sections connected with UEs as depicted in Figure 13.2.

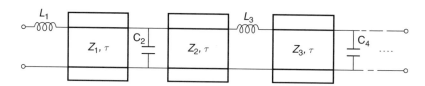

Figure 13.2 Low-pass ladder with UEs

The semi-analytical procedure discussed in the previous section can easily be applied for the construction of canonical polynomials describing the generic low-pass ladder structures with UEs (LPLUE) as follows [17–19].

Procedure: Construction of LPLUE Structures

Step 1: An LPLUE structure includes n_p simple lumped sections with transmission zeros at ∞ and n_λ UEs which introduce transmission zeros at $\lambda = \pm 1$. Thus, the polynomial $f(p, \lambda)$, corresponding to an LPLUE structure has the form $f(p, \lambda) = \left(1 - \lambda^2\right)^{n_\lambda/2}$.
Therefore, choose the form $f(p, \lambda) = \left(1 - \lambda^2\right)^{n_\lambda/2}$.

Step 2: By setting $\lambda = 0$, UEs are removed from the mixed structure. In this case, the polynomials $h(p, 0), g(p, 0)$ and $f(p, 0)$ completely characterize the resulting lumped ladder and the following condition holds:

$$g(p,0)g(-p,0) = h(p,0)h(-p,0) + 1 \tag{13.15}$$

where $g(p,0)$ is strictly Hurwitz.

Therefore, select the coefficients of the $h(p,0)$ polynomial as one group of independent parameters to describe the lumped sections of the LPLUE structure.

Step 3: By setting $p = 0$, lumped elements are removed. In this case, the polynomials $h(0, \lambda), g(0, \lambda)$ and $f(0, \lambda)$ completely characterize the resulting UE cascade and the following condition holds:

$$g(0, \lambda)g(0, -\lambda) = h(0, \lambda)h(0, -\lambda) + \left(1 - \lambda^2\right)^{n_\lambda} \tag{13.16}$$

where $g(0, \lambda)$ is strictly Hurwitz.

Therefore, select the coefficients of $h(0, \lambda)$ as another group of independent parameters to describe the distributed sections of the LPLUE structure.

Step 4: As a result of the straightforward cascade analysis, we can readily obtain the generic forms for the coefficient matrices Λ_h and Λ_g:

$$\Lambda_h = \begin{bmatrix} 0 & h_{01} & h_{02} & \cdots & h_{0n_\lambda} \\ h_{10} & h_{11} & h_{12} & \cdots & h_{1n_\lambda} \\ h_{20} & h_{21} & h_{22} & 0 & 0 \\ \vdots & \vdots & 0 & \ddots & \vdots \\ h_{n_p0} & 0 & 0 & \cdots & 0 \end{bmatrix}$$

$$\Lambda_g = \begin{bmatrix} 1 & g_{01} & g_{02} & \cdots & g_{0n_\lambda} \\ g_{10} & g_{11} & g_{12} & \cdots & g_{1n_\lambda} \\ g_{20} & g_{21} & g_{22} & 0 & 0 \\ \vdots & \vdots & 0 & \ddots & \vdots \\ g_{n_p0} & 0 & 0 & \cdots & 0 \end{bmatrix} \tag{13.17}$$

where:

the non-zero entries of Λ_g are non-negative real numbers;
$g_{11} = g_{01}g_{10} - h_{01}h_{10}$;
$g_{kl} = h_{kl} = 0$ for $k + l > n_\lambda + 1, k, l, = 0, 1, \ldots, n_\lambda$;
$h_{kl} = \mu g_{kl}$ for $k + l = n_\lambda + 1, k, l = 0, 1, \ldots, n_\lambda$; and

$$\boxed{\mu = \pm 1} \qquad\qquad (13.18)$$

$\mu = h_{n_p 0}/g_{n_p 0} = \pm 1$, if $n_p = n_\lambda + 1$.

The above relations describe the connectivity information for the selected LPLUE topology.

Therefore, set the above LPLUE connectivity relations between the independently chosen parameters and the rest of the entries of the connectivity matrices.

Note that for any other regular mixed cascade topology, different sets of coefficient relations and hence different generic matrix forms will be obtained.

Step 5: Utilizing the boundary conditions given by steps 1, 2 and 3, together with the coefficient properties of step 4, FES can be solved algebraically for the unknown coefficients.

Here, it is important to note that we were able to come up with explicit relations for the entries of the Λ_h and Λ_g matrices up to total degree $n_p + n_\lambda = 5$. The coefficient properties given above constitute the sufficiency conditions that lead to a unique explicit solution. For selected LPLUE topologies presented in Table 13.1, corresponding FES solutions and the explicit coefficient relations are given in Table 13.2.

For more complicated LPLUE topologies, explicit solutions are unfortunately not available. In this case, having imposed the connectivity information given by Equation (13.18) in step 4, FES must be solved numerically.

The results presented in this chapter may be employed to design various kinds of practical microwave passive and active systems with two kinds of elements. In the following section, we introduce a typical example to design a double matching network by employing the procedure discussed to construct LPLUE structures.

13.3 Application

A major potential area for application of the results obtained in this chapter is computer-aided real frequency broadband matching. In the real frequency techniques, the lossless matching network is described in terms of an independent set of real parameters. These parameters are determined to optimize the system performance, usually in the form of transducer power gain over the prescribed frequency band [20,21].

In the mixed element design, the matching network is described by means of the partially selected polynomial coefficients of $h(p, \lambda)$ by utilizing the procedure discussed above. For example, if the LPLUE structure is employed, firstly the complexity of the matching network is fixed by setting the $f(p, 0)$ and $f(p, \lambda)$ polynomials. These polynomials identify the transmission zeros of the selected topology. In this case, $f(p, 0) = 1$ and $f(0, \lambda) = (1 - \lambda^2)^{n_\lambda/2}$ are chosen. Subsequently, the coefficients of the polynomials $h(p, 0)$ and $h(p, \lambda)$ are selected as the unknown independent parameters of the problem and the rest of the entries of the connectivity matrices are determined in terms of the unknowns of $g(p, \lambda)$. Eventually, transducer power gain of the doubly terminated system is optimized by means of a nonlinear optimization algorithm, which in turn yields the unknown coefficients.

Table 13.1   Low-order LPLUE structures

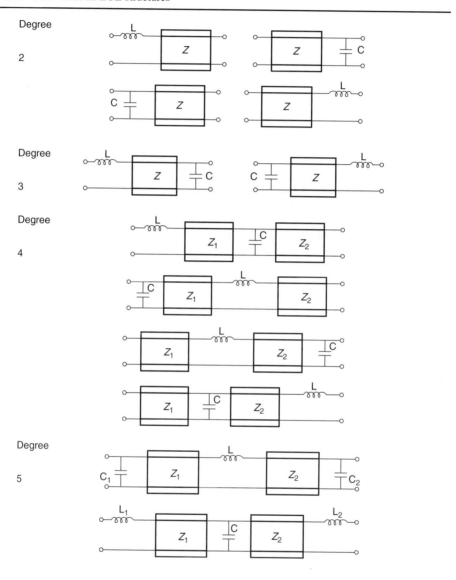

Table 13.2   Explicit formulas for low-order LPLUE

Degree 2	(h_{01}, h_{10}) independent coefficients

$$g_{01} = \left(1 + h_{01}^2\right)^{1/2} \; g_{11} = g_{01}g_{10} - h_{01}h_{10}$$
$$g_{10} = |h_{10}| \; ; \; |h_{11}| = \mu g_{11} \; ; \; \mu = \pm 1\,(= h_{10}/g_{10})$$

Degree 3	(h_{01}, h_{10}, h_{20}) independent coefficients

$$g_{01} = \left(1 + h_{01}^2\right)^{1/2} \; ; \; g_{20} = |h_{20}|$$
$$g_{10} = \left(h_{10}^2 + 2g_{20}\right)^2 \; ; \; g_{11} = g_{01}g_{10} - h_{01}h_{10}$$
$$h_{11} = g_{11}g_{20}/h_{10}$$

Degree 4	$(h_{01}, h_{02}, h_{10}, h_{20})$ independent coefficients

$$g_{01} = \left(g_{01}^2 + 2g_{02} + 2\right)^{1/2} \; ; \; g_{11} = g_{01}g_{10} - h_{01}h_{10}$$
$$h_{12} = \mu g_{12} \; ; \; g_{02} = \left(1 + h_{02}^2\right)^{1/2}$$
$$h_{11} = h_{02}\beta/\alpha + h_{20}\alpha/\beta \; ; \; h_{21} = \mu g_{21}$$
$$g_{10} = \left(h_{10}^2 + 2g_{20}\right)^{1/2} \; ; \; g_{12} = (g_{11}g_{02} - h_{11}h_{02})/\alpha$$
$$\mu = \pm 1\,(= h_{20}/g_{20}) \; ; \; \alpha = g_{01} - \mu h_{01} \; ; \; \beta = g_{10} - \mu h_{10}$$
$$g_{20} = |h_{20}| \; ; \; g_{21} = (g_{11}g_{20} - h_{11}h_{20})/\beta$$

Degree 5	$(h_{01}, h_{02}, h_{10}, h_{20}, h_{30})$ independent coefficients

$$g_{01} = \left(h_{01}^2 + 2g_{02} + 2\right)^{1/2}$$
$$g_{02} = \left(1 + h_{02}^2\right)^{1/2} \; ; \; g_{11} = g_{01}g_{10} - h_{01}h_{10}$$
$$h_{12} = \mu g_{12} \; ; \; g_{10} = \left(h_{10}^2 + 2g_{20}\right)^{1/2}$$
$$h_{11} = h_{02}\beta/\alpha + h_{20}\alpha/\beta \; ; \; h_{21} = \mu g_{21}$$
$$g_{20} = \left(h_{20}^2 + 2g_{30}g_{10} - 2h_{30}h_{10}\right)^{1/2}$$
$$g_{12} = (g_{11}g_{02} - h_{11}h_{02})/\alpha$$
$$\mu = \pm 1\,(= g_{30}/h_{30}) \; ; \; \alpha = g_{01} - \mu h_{01}$$
$$\beta = g_{10} - \mu h_{10} \; ; \; g_{30} = |h_{30}|$$
$$g_{21} = (g_{11}g_{20} - h_{11}h_{20} - g_{01}g_{30} + h_{01}h_{30})/\beta$$

Transducer power gain of the mixed structure is given as follows [16]:

$$T(\omega) = \frac{\left(1 - |S_G|^2\right)\left(1 - |S_L|^2\right)|f(j\omega, j\Omega)|^2}{|g(j\omega, j\Omega) - h(j\omega, j\Omega)S_G + \sigma S_L h(-j\omega, -j\Omega) - \sigma S_G g(-j\omega, -j\Omega)|^2} \tag{13.19}$$

where $\Omega = \tan(\omega\tau)$, τ being the delay length of the UEs, S_G and S_L denote the reflectance of the generator and the load networks to be matched, respectively; h, g and f are the two-variable polynomials describing the mixed element matching network.

Close examination of Equation (13.19) reveals that $T(\omega)$ assumes an almost inverse quadratic form in the unknown coefficients h_{ij} of the polynomials $h(p, 0)$ and $h(0, \lambda)$. Therefore, a least-squares optimization algorithm yields satisfactory results, especially if a flat gain characteristic T_0 is to be approximated. In this case, the objective function to be minimized may be defined as

$$\delta = \sum_{k=0}^{N_\omega} \left[T\left(\omega_k, \{ h_{i0}, h_{0j} \} \right) - T_0 \right]^2$$

$$i = 1, 2, \ldots, n_p, \ j = 1, 2, \ldots, n_\lambda$$

(13.20)

where N_ω denotes the number of sampling frequencies over the band of operation.

Once the final forms of $g(p, \lambda)$ and $h(p, \lambda)$ are generated, the mixed element realization is obtained by employing the algebraic decomposition technique of Fettweis [15, 16] on the polynomial sets describing the lumped and distributed prototypes independently. Let us trace the above-described process in the following classical example.

Example 13.1:[2] In this example, we solve the classical double matching problem, which is depicted in Figure 13.3, to demonstrate the application of the design procedure with mixed lumped and distributed elements [23]. Here, an LPLUE network of degree 4 $\left(n_p = 2 + 2 \right)$ will be constructed over the normalized frequency band of $0 \leq \omega \leq 1$.

In the course of the optimization process, the coefficients $\{ h_{01}, h_{02}, h_{10}, h_{20} \}$, the sign constant μ and the delay length τ of the UEs are chosen as the unknown independent parameters; the other coefficients are derived by means of the explicit expressions given in Table 13.2. As a result of optimization, we obtain the following coefficient matrices that completely describe the scattering parameters of the matching network under consideration:

$$\Lambda_h = \begin{bmatrix} 0 & -4.325 & -0.824 \\ 0.687 & -3.303 & -20.188 \\ 0.798 & -4.965 & 0 \end{bmatrix} \quad \Lambda_g = \begin{bmatrix} 1 & 4.827 & 1.296 \\ 1.439 & 9.917 & 20.188 \\ 0.798 & 4.965 & 0 \end{bmatrix}$$

The final network with the component values and the corresponding gain performance of the system are shown in Figure 13.3 and Figure 13.4, respectively.

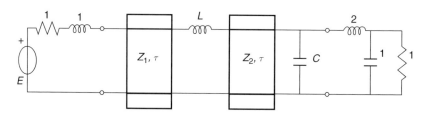

Figure 13.3 Double matching example with lumped and distributed elements $(L = 2.126, C = 0.751, Z_1 = 0.161, Z_2 = 0.341, \tau = 0.21)$

The gain performance of the matched system is in good agreement with the available lumped element designs given in [20] and [21]. Solutions employing solely lumped sections may create implementation problems due to the physical length of interconnections [22]. In the present case, however, the UEs

---

[2] The MATLAB® codes in this example are given on the companion website as well as at the end of this chapter. The main program is called `Mixed_matching.m` Interested readers can reproduce the result of the example using these codes. The codes were developed by Dr Metin Sengul of Kadir Has University, Istanbul.

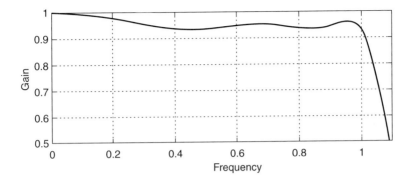

Figure 13.4 Gain performance of double matching example

provide the physical connections between lumped elements and also contribute to the gain performance since they are part of the matching network. The capacitor C may also regarded as part of the discontinuity of the UE.

13.4 Conclusion

In this chapter, the simplified real frequency technique has been extended to describe lossless two-ports in two kinds of elements, namely lumped and distributed elements. In the above-mentioned description, Fettweis's 'bounded real scattering Hurwitz' parameters have been employed. It was shown that connectivity information is crucial in order to obtain realizable network structures. In this regard, connectivity matrices, which constitute the realizability information, were introduced. A fundamental equation set (FES) that completely describes the mixed element lossless two-ports was presented. Explicit equations were given to describe lumped ladders connected with commensurate transmission lines (UE) of up to five elements. The construction of the lossless two-ports with mixed lumped and distributed elements was demonstrated via a double matching example. It has been shown that the new approach makes implementation of microwave circuits practical by providing physical length connections between the lumped elements. It has also been exhibited that parasitic effects and discontinuities are naturally embedded in the design process by utilizing the mixed element structures for matching networks. It is expected that the proposed approach will facilitate the design and implementation of analog microwave monolithic integrated circuits for modern communication systems.

Program Listing 13.1 `Mixed_matching.m`

```
%*****Mixed matching main program*****
%*****Sub-programs:Error_find, g0ifind, gj0find, hg11, hg12,
hg21, hg22, hg23, paraconjugate*****
%*Sub-programs:sentez11, sentez12, sentez21,
% sentez22, sentez23*****

clear
clc
close all
```

Program Listing 13.1   (Continued)

```
b1=0;
while b1<1

%*****source and load impedance, source and load
% reflection coefficient calculation*****
w=[0.1 0.2 0.3 0.4 0.5 0.6 0.7 0.8 0.9 1];
ind1=1;
ind2=2;
cap=1;
ZL=i.*w.*ind2+1./(1+i.*w.*cap);
ZS=1+i.*w.*ind1;
sg=(ZS-1)./(ZS+1);
sl=(ZL-1)./(ZL+1);
%*****************************************************************

%*****Initial
values*****************************************
numdist=3;
while numdist~=1 & numdist~=2
    h0i=input('Enter initial h0i coefficients [h0,n h0,n-1 ...
h0,1 h0,0]:');
    numdist=length(h0i)-1;
    if numdist>2
        disp('There must be no more than 2 unit elements, enter
again.');
    end
end

diff1=2;
while diff1>1
    hj0=input('Enter initial hj0 coefficients [hm,0 hm-1,0 ...
h1,0 h0,0]:');
    numlump=length(hj0)-1;
    diff1=abs(numdist-numlump);
    if diff1>1
        disp('The difference between number of lumped elements and
unit elements cannot exceed one.');
    end
end

thau=2;
while (thau)<0 | (thau)>1
    thau=input('Enter initial delay length (0<thau<1):');
    if (thau)<0 | (thau)>1
        disp('thau must be between 0 and 1.');
    end
end
```

```
if hj0(1)>0
    m=1;
else
    m=-1;
end

T0=2;
while (T0)<0 | (T0)>1
    T0=input('Enter desired tpg level (0<tpg<1):');
    if (T0)<0 | (T0)>1
        disp('TPG must be between 0 and 1.');
    end
end
%*****************************************************************


%*****Optimisation vector
construction***************************
dist=length(h0i)-1;
lump=length(hj0)-1;
dimension=dist+lump;

for a=1:dist;
    v(a)=h0i(a);
end;
for a=1:lump;
    v(dist+a)=hj0(a);
end;

v(dimension+1)=thau;
%*****************************************************************


%*****optimisation
part***************************************
const=length(v);
LB=zeros(1,const);
UB=ones(1,const);
for a=1:const-1
    LB(a)=-Inf;
    UB(a)=Inf;
end
OPTIONS=OPTIMSET('MaxFunEvals',10000,'MaxIter',40000);
v_new =
lsqnonlin('error_find',v,LB,UB,OPTIONS,dist,lump,w,sg,sl,T0);
%*****************************************************************


%*****Getting h0i, hj0 and thau after
```

Program Listing 13.1   (Continued)

```
optimisation*****************
for a=1:dist;
    h0i(a)=v_new(a);
end;
h0i(dist+1)=0;
for a=1:lump;
    hj0(a)=v_new(dist+a);
end;
hj0(lump+1)=0;
if hj0(1)>0
    m=1;
else
    m=-1;
end
thau=v_new(dist+lump+1);
%*******************************************************************


%*****Calculation of optimised h and g
% matrices*******************
if dist==1 & lump==1;
    [Ah,Ag]=hg11(h0i,hj0,m);
elseif dist==1 & lump==2;
    [Ah,Ag]=hg12(h0i,hj0,m);
elseif dist==2 & lump==1;
    [Ah,Ag]=hg21(h0i,hj0,m);
elseif dist==2 & lump==2;
    [Ah,Ag]=hg22(h0i,hj0,m);
elseif dist==2 & lump==3;
    [Ah,Ag]=hg23(h0i,hj0,m);
end;

Ah
Ag
thau
%*******************************************************************


%****Calculation of load and source impedances, source and load
reflection coefficients over new frequency range*****
w=[0.1 0.2 0.3 0.4 0.5 0.6 0.7 0.8 0.9 1 1.1 1.2 1.3 1.4 1.5 1.6
1.7 1.8 1.9 2];
ind1=1;
ind2=2;
cap=1;
ZL=i.*w.*ind2+1./(1+i.*w.*cap);
ZS=1+i.*w.*ind1;
```

```
sg=(ZS-1)./(ZS+1);
sl=(ZL-1)./(ZL+1);
%***************
%**According to optimized h and g matrices, getting
% the values of h, hpara, g, gpara and f*****
k=length(w);
for a=1:k
    d=i*tan(w(a)*thau);
    hv(a)=0;
    for b=1:lump+1;
        for c=1:dist+1;
            hv(a)=hv(a)+Ah(b,c)*((i*w(a))^(b-1))*(d)^(c-1);
        end;
    end;
end;

for a=1:k;
    d=-i*tan(w(a)*thau);
    hpv(a)=0;
    for b=1:lump+1;
        for c=1:dist+1;
            hpv(a)=hpv(a)+Ah(b,c)*((-i*w(a))^(b-1))*(d)^(c-1);
        end;
    end;
end;

for a=1:k;
    d=i*tan(w(a)*thau);
    gv(a)=0;
    for b=1:lump+1;
        for c=1:dist+1;
            gv(a)=gv(a)+Ag(b,c)*((i*w(a))^(b-1))*(d)^(c-1);
        end;
    end;
end;

for a=1:k;
    d=-i*tan(w(a)*thau);
    gpv(a)=0;
    for b=1:lump+1;
        for c=1:dist+1;
            gpv(a)=gpv(a)+Ag(b,c)*((-i*w(a))^(b-1))*(d)^(c-1);
        end;
    end;
end;

for a=1:k;
    f(a)=(1+((tan(w(a)*thau))*(tan(w(a)*thau))))^(dist/2);
end;
```

Program Listing 13.1   (Continued)

```
%********
%*****Calculation of tpg over the frequencies
for a=1:k;
    tpg(a)=[(1-(abs(sg(a)))^2)*(1-(abs(sl(a)))^2)*(abs(f(a)))^2]/
(abs([gv(a)-hv(a)*sg(a)+sl(a)*(hpv(a)-sg(a)*gpv(a))]))^2;
end;
%***************
%*****Plotting the result*****
hold on
plot(w,tpg,'k-')
xlabel('Frequency')
ylabel('Gain')
grid on
axis([0 2 0 1])
%**********************
%*****synthesize******
if dist==1 & lump==1;
    [LE1,UE1]=sentez11(Ah,Ag,m)
    if m>0
        if Ah(2,2)>0
            disp('First element is an inductor, second is a unit
element.')
        else
            disp('First element is a unit element, second is an
inductor.')
        end
    else
        if Ah(2,2)<0
            disp('First element is a capacitor, second is a unit
element.')
        else
            disp('First element is a unit element, second is a
capacitor.')
        end
    end
elseif dist==1 & lump==2;
    [LE1,LE2,UE1]=sentez12(Ah,Ag,m)
    if m>0
        disp('First element is an inductor, second is a unit
element, third is a capacitor.')
    else
        disp('First element is a capacitor, second is a unit
element, third is an inductor.')
    end
elseif dist==2 & lump==1;
    [LE1,UE1,UE2]=sentez21(Ah,Ag,m)
    if m>0
```

```
        disp('First element is a unit element, second is an
inductor, third is a unit element.')
    else
        disp('First element is a unit element, second is a
capacitor, third is a unit element.')
    end
elseif dist==2 & lump==2;
    [LE1,LE2,UE1,UE2]=sentez22(Ah,Ag,m)
    if m>0
        if Ah(3,2)<0
            disp('First element is a unit element, second is an
inductor.')
            disp('Third element is a unit element, fourth is a
capacitor.')
        else
            disp('First element is an inductor, second is a unit
element.')
            disp('Third element is a capacitor, fourth is a unit
element.')
        end
    else
        if Ah(3,2)<0
            disp('First element is a capacitor, second is a unit
element.')
            disp('Third element is an inductor, fourth is a unit
element.')
        else
            disp('First element is a unit element, second is a
capacitor.')
            disp('Third element is a unit element, fourth is an
indictor.')
        end
    end
elseif dist==2 & lump==3;
    [LE1,LE2,LE3,UE1,UE2]=sentez23(Ah,Ag,m)
    if m>0
        disp('First element is an inductor, second is a unit
element.')
        disp('Third element is a capacitor, fourth is a unit
element.')
        disp('Fifth element is an inductor.')
    else
        disp('First element is a capacitor, second is a unit
element.')
        disp('Third element is an inductor, fourth is a unit
element.')
        disp('Fifth element is a capacitor.')
    end
end;
```

Program Listing 13.1   (Continued)

```
%***********************
%*another design ****
a1=0;
while a1<1
select10=7;
while select10~=1 & select10~=2 & select10~=3 & select10~=4 &
select10~=5 & select10~=6
    select10=input('Another mixed matching network design (1-
Yes 2-No) : ');
    if (select10==1)
        a1=1;
    elseif (select10==2)
      a1=1;
       b1=1;
     end
     end
   end
end
```

Program Listing 13.2

```
function eps=error(v,dist,lump,w,sg,sl,T0)
%*****error sub-program*****

%*****calculation of h0i, hj0 and thau from optimisation
vector*****
k=length(w);
dimension=dist+lump;

for a=1:dist;
    h0i(a)=v(a);
end
h0i(dist+1)=0;

for a=1:lump;
    hj0(a)=v(dist+a);
end
hj0(lump+1)=0;

if hj0(1)>0
    m=1;
else
    m=-1;
end
```

```
thau=v(dist+lump+1);
%***********************
%*****Calculation of h and g
% matrices**************************
if dist==1 & lump==1;
    [Ah,Ag]=hg11(h0i,hj0,m);
elseif dist==1 & lump==2;
    [Ah,Ag]=hg12(h0i,hj0,m);
elseif dist==2 & lump==1;
    [Ah,Ag]=hg21(h0i,hj0,m);
elseif dist==2 & lump==2;
    [Ah,Ag]=hg22(h0i,hj0,m);
elseif dist==2 & lump==3;
    [Ah,Ag]=hg23(h0i,hj0,m);
end
%calculation of h, hpara, g, gpara and
for a=1:k;
    d=i*tan(w(a)*thau);
    hv(a)=0;
    for b=1:lump+1;
        for c=1:dist+1;
            hv(a)=hv(a)+Ah(b,c)*((i*w(a))^(b-1))*(d)^(c-1);
        end
    end
end

for a=1:k;
    d=-i*tan(w(a)*thau);
    hpv(a)=0;
    for b=1:lump+1;
        for c=1:dist+1;
            hpv(a)=hpv(a)+Ah(b,c)*((-i*w(a))^(b-1))*(d)^(c-1);
        end
    end
end

for a=1:k;
    d=i*tan(w(a)*thau);
    gv(a)=0;
    for b=1:lump+1;
        for c=1:dist+1;
            gv(a)=gv(a)+Ag(b,c)*((i*w(a))^(b-1))*(d)^(c-1);
        end
    end
end

for a=1:k;
    d=-i*tan(w(a)*thau);
    gpv(a)=0;
```

Program Listing 13.2   (Continued)

```
    for b=1:lump+1;
        for c=1:dist+1;
            gpv(a)=gpv(a)+Ag(b,c)*((-i*w(a))^(b-1))*(d)^(c-1);
        end
    end
end

for a=1:k;
f(a)=(1+((tan(w(a)*thau))*(tan(w(a)*thau))))^(dist/2);
end
%*********************

%calculation of tpg******
for a=1:k;
    tpg(a)=[(1-(abs(sg(a)))^2)*(1-(abs(sl(a)))^2)*(abs(f(a)))^2]/
(abs([gv(a)-hv(a)*sg(a)+sl(a)*(hpv(a)-sg(a)*gpv(a))]))^2;
end
%**************************
%*****calculation of error**
for a=1:k;
    eps(a)=((tpg(a)-T0)/tpg(a))^2;
end
%*******
return
```

Program Listing 13.3

```
function g0i=g0ifind(h0i)
h_Lambda=h0i;
n_Lambda=length(h0i)-1;
F_temp=[-1 0 1];
F_Lambda=[-1 0 1];
for i=1:(n_Lambda - 1)
    F_Lambda = conv(F_Lambda,F_temp);
end
HP_Lambda = paraconjugate(h_Lambda);

% Constructing the even polynomial
G = conv(h_Lambda , HP_Lambda);
GP_Lambda = G + F_Lambda;
RO = roots(GP_Lambda);
s = length(RO);
for h = 1 : s
    if real (RO(h)) < 0
```

```
           rootlhp(h)= RO(h);
      else
           rootlhp(h)= 0;
      end
end
[k,l,m] = find(rootlhp);
g_Lambda = poly (m);
g_Lambda = g_Lambda./g_Lambda(length(g_Lambda));
g0i=g_Lambda;
return
```

Program Listing 13.4

```
function gj0=gj0find(hj0);
H_S=hj0;
boy=length(H_S);
Fara=zeros(1,(boy-1));
Fara(1,boy)=1;
F_S=Fara;
HS=paraconjugate(H_S);
FS=paraconjugate(F_S);
% Constructing the even polynomial
G = conv(H_S , HS);
F = conv(F_S,FS);
G_S = G + F;
     RO = roots(G_S);
   s = length(RO);
   for h = 1 : s
     if real (RO(h)) < 0
         rootlhp(h)= RO(h);
      else
         rootlhp(h)= 0;
      end
end
[k,l,m] = find(rootlhp);
g_s = poly (m);
g0iara=g_s;
b=length(g_s);
for a=1:b;
   g_s(a)=g_s(a)/g0iara(b);
end
gj0=g_s;
return
```

Program Listing 13.5

```
function [Ah,Ag]=hg11(h0i,hj0,m);
%*****hg11 sub-program*****

%*****calculation of h and g matrices of the networks contains
% one distributed and one lumped elements*****
g11=1;
g12=(1+(h0i(1))^2)^(1/2);
g21=abs(hj0(1));
g22=g21*g12-hj0(1)*h0i(1);
h22=m*g22;
h11=h0i(2);
h12=h0i(1);
h21=hj0(1);
Ah=[h11 h12;h21 h22];
Ag=[g11 g12;g21 g22];
%************************
Return
```

Program Listing 13.6

```
function [Ah,Ag]=hg12(h0i,hj0,m)
%*****hg12 sub-program*****

%*****calculation of h and g matrices of the networks
%contains
% one distributed and two lumped elements*****
g11=1;
h11=0;
h12=h0i(1);
h21=hj0(2);
h31=hj0(1);
g12=(1+h0i(1)^2)^(1/2);
g31=abs(hj0(1));
g21=(hj0(2)^2+2*g31)^(1/2);
g22=g12*g21-h0i(1)*hj0(2);
a=g12-m*h0i(1);
b=g21-m*hj0(2);
g32=(2/b)*(g31*g22-(a/b)*(hj0(1)^2));
if g32==0
    h22=(g31*g22)/hj0(1);
else
    h22=(2*a*hj0(1))/b-(g31*g22)/hj0(1);
end
h32=m*g32;
```

```
Ah=[h11 h12;h21 h22;h31 h32];
Ag=[g11 g12;g21 g22;g31 g32];
%***************************
return
```

Program Listing 13.7

```
function [Ah,Ag]=hg21(h0i,hj0,m)
%*****hg21 sub-program*****

%*****calculation of h and g matrices of the networks
% contains
% two distributed and one lumped elements*****
m1=-m;
g11=1;
h11=0;
h12=h0i(2);
h13=h0i(1);
h21=hj0(1);
g13=(1+h0i(1)^2)^(1/2);
g21=abs(hj0(1));
g12=(h0i(2)^2+2*(g13+1))^(1/2);
a=g12-m1*h0i(2);
b=g21-m1*hj0(1);
g22=g12*g21-h0i(2)*hj0(1);
h22=h13*b/a;
g23=(g22*g13-h22*h13)/a;
h23=m1*g23;
Ah=[h11 h12 h13;h21 h22 h23];
Ag=[g11 g12 g13;g21 g22 g23];
Ah=numeric(Ah);
Ag=numeric(Ag);
%*************************************
return
```

Program Listing 13.8

```
function [Ah,Ag]=hg22(h0i,hj0,m)
%*****hg22 sub-program*****
%*****calculation of h and g matrices of the networks
% contains
% two distributed and two lumped elements*****
```

Program Listing 13.8   (Continued)

```
gj0=gj0find(hj0);
g0i=g0ifind(h0i);

g1=g0i(1)+g0i(2)+g0i(3);
h1=h0i(1)+h0i(2)+h0i(3);

x1=h0i(3)*g1-g0i(3)*h1;
x2=h0i(2)*g1-g0i(2)*h1;
y1=g0i(3)*g1-h0i(3)*h1;
y2=g0i(2)*g1-h0i(2)*h1;

Z1=(g1+h1)/(g1-h1);
Z2=((y2+2*x1+x2)/(y2-2*x1-x2))*Z1;

G1=(gj0(1))/(gj0(2)-m*hj0(2));
G2=gj0(2)-G1;

h13=(1/2)*(Z1/Z2-Z2/Z1);
h12=(1/2)*(Z1+Z2-1/Z1-1/Z2);
h11=0;

h23=(1/2)*(Z2*Z1*G2+Z2*G1/Z1+Z2*m*G2*Z1+m*G2/Z2*Z1+m*G1*Z1/
Z2+Z2*m*G1/Z1-G1*Z1/Z2-G2/Z2*Z1);
h22=(1/2)*(m*G1/Z2-m*G2*Z2-m*G2/Z2+m*G1*Z2+m*G1/Z1+m*G1*Z1-G2/
Z1+m*G2/Z1+m*G2*Z1-Z2*G2-Z2*G1+G1/Z2-G1*Z1+G1/Z1+Z1*G2+G2/Z2);
h21=m*G1-m*G2;

h33=0;
h32=(1/2)*(-G1*G2/Z2+G2*G1*Z1+Z2*G1*G2-G2*G1/Z1+G1*G2/
Z1+Z2*G2*G1-G1*Z1*G2-G2*G1/Z2)+(m*G2*G1*Z2+m*G2*G1/Z2);
h31=2*m*G2*G1;

g13=(1/2)*(Z1/Z2+Z2/Z1);
g12=(1/2)*(Z1+Z2+1/Z1+1/Z2);
g11=1;

g23=(1/2)*(Z2*Z1*G2+Z2*G1/Z1+Z2*m*G2*Z1-m*G2/Z2*Z1-m*G1*Z1/
Z2+Z2*m*G1/Z1+G1*Z1/Z2+G2/Z2*Z1);;
g22=(1/2)*(m*G1/Z2+m*G2*Z2-m*G2/Z2-m*G1*Z2+m*G1/Z1-m*G1*Z1+G2/
Z1-m*G2/Z1+m*G2*Z1+Z2*G2+Z2*G1+G1/Z2+G1*Z1+G1/Z1+Z1*G2+G2/Z2);
g21=G1+G2;

g33=0;
g32=(1/2)*(G1*G2/Z2-G2*G1/Z1+Z2*G1*G2+G2*G1/Z2+G1*Z1*G2+G1*G2/
Z1+Z2*G2*G1-G2*G1*Z1)+(-m*G2*G1/Z2+m*G2*G1*Z2);
g31=2*G1*G2;
```

```
Ah=[h11 h12 h13;h21 h22 h23;h31 h32 h33];
Ag=[g11 g12 g13;g21 g22 g23;g31 g32 g33];
return
```

Program Listing 13.9

```
function [Ah,Ag]=hg23(h0i,hj0,m)
%*****hg23 sub-program*****

%calculation of h and g matrices
gj0=gj0find(hj0);
g0i=g0ifind(h0i);

h33=0;
h42=0;
h43=0;
g33=0;
g42=0;
g43=0;
h11=h0i(3);
g11=g0i(3);
h12=h0i(2);
h13=h0i(1);
h21=hj0(3);
h31=hj0(2);
h41=hj0(1);
g13=g0i(1);
g12=g0i(2);
g31=gj0(2);
g21=gj0(3);
g41=gj0(1);
g22=g12*g21-h12*h21;
a=g12-m*h12;
b=g21-m*h21;
h22=(b*h13)/a+(a*h31)/b;
g23=(1/a)*(g22*g31-h22*h31-g12*g41+h12*h41);
h23=m*g23;
g32=(1/b)*(g22*g31-h22*h31-g12*g41+h12*h41);
h32=m*g32;
Ah=[h11 h12 h13;h21 h22 h23;h31 h32 h33;h41 h42 h43];
Ag=[g11 g12 g13;g21 g22 g23;g31 g32 g33;g41 g42 g43];
%*************************
return
```

Program Listing 13.10

```
function P_S=paraconjugate (PS)
c = length (PS)-1;
for k = 0:c
   if rem(k , 2)== 0
      P_S(c-k+1) = (+1)*PS(c-k+1);
   else
      P_S(c-k+1) = (-1)*PS(c-k+1);
   end
end
return
```

Program Listing 13.11

```
function [LE1,UE1]=sentez11(Ah,Ag,m)

g0i=fliplr(Ag(1,:));
h0i=fliplr(Ah(1,:));

G1=g0i(1);
LE1=2*G1;
g1=g0i(1)+g0i(2);
h1=h0i(1)+h0i(2);
UE1=(g1+h1)/(g1-h1);

return
```

Program Listing 13.12

```
function [LE1,LE2,UE1]=sentez12(Ah,Ag,m)

g0i=fliplr(Ag(1,:));
h0i=fliplr(Ah(1,:));
G1=(Ag(3,1))/(Ag(2,1)-m*Ah(2,1));
G2=Ag(2,1)-G1;
LE1=2*G1;
LE2=2*G2;
g1=g0i(1)+g0i(2);
h1=h0i(1)+h0i(2);
UE1=(g1+h1)/(g1-h1);
Return
```

Program Listing 13.13

```
function [LE1,UE1,UE2]=sentez21(Ah,Ag,m)

g0i=fliplr(Ag(1,:));
h0i=fliplr(Ah(1,:));

G1=g0i(1);
LE1=2*G1;
g1=g0i(1)+g0i(2)+g0i(3);
h1=h0i(1)+h0i(2)+h0i(3);
x1=h0i(3)*g1-g0i(3)*h1;
x2=h0i(2)*g1-g0i(2)*h1;
y1=g0i(3)*g1-h0i(3)*h1;
y2=g0i(2)*g1-h0i(2)*h1;
UE1=(g1+h1)/(g1-h1);
UE2=((y2+2*x1+x2)/(y2-2*x1-x2))*UE1;

return
```

Program Listing 13.14

```
function [LE1,LE2,UE1,UE2]=sentez22(Ah,Ag,m)

g0i=fliplr(Ag(1,:));
h0i=fliplr(Ah(1,:));

G1=(Ag(3,1))/(Ag(2,1)-m*Ah(2,1));
G2=Ag(2,1)-G1;
LE1=2*G1;
LE2=2*G2;
g1=g0i(1)+g0i(2)+g0i(3);
h1=h0i(1)+h0i(2)+h0i(3);
x1=h0i(3)*g1-g0i(3)*h1;
x2=h0i(2)*g1-g0i(2)*h1;
y1=g0i(3)*g1-h0i(3)*h1;
y2=g0i(2)*g1-h0i(2)*h1;
UE1=(g1+h1)/(g1-h1);
UE2=((y2+2*x1+x2)/(y2-2*x1-x2))*UE1;

return
```

Program Listing 13.15

```
function [LE1,LE2,LE3,UE1,UE2]=sentez23(Ah,Ag,m)

g0i=fliplr(Ag(1,:));
h0i=fliplr(Ah(1,:));
gj0=fliplr(Ag(:,1)');
hj0=fliplr(Ah(:,1)');

G1=[gj0(1)/(gj0(2)-(m*hj0(2)))];
G2=[((gj0(2)-(G1*(gj0(3)-(m*hj0(3)))))/(gj0(3)-G1+(m*((hj0(3))-
(m*G1)))))];
G3=[gj0(3)-G1-G2];

LE1=2*G1;
LE2=2*G2;
LE3=2*G3;

g1=g0i(1)+g0i(2)+g0i(3);
h1=h0i(1)+h0i(2)+h0i(3);
x1=h0i(3)*g1-g0i(3)*h1;
x2=h0i(2)*g1-g0i(2)*h1;
y1=g0i(3)*g1-h0i(3)*h1;
y2=g0i(2)*g1-h0i(2)*h1;

UE1=(g1+h1)/(g1-h1);
UE2=((y2+2*x1+x2)/(y2-2*x1-x2))*UE1;

return
```

References

[1] H. Ozaki and T. Kasami, 'Positive Real Functions of Several Variables and Their Applications to Variable Networks', *IEEE Transactions on Circuit Theory*, Vol. 7, pp. 251–260, 1960.

[2] H. G. Ansell, 'On Certain Two-variable Generalization of Circuit Theory with Applications to Networks of Transmission Lines and Lumped Reactances', *IEEE Transactions on Circuit Theory*, Vol. 11, pp. 214–223, 1964.

[3] B. K. Kinariwala, 'Theory of Cascaded Structures: Lossless Transmission Lines', *Bell Systems Technical Journal*, Vol. 45, pp. 631–649, 1966.

[4] M. Saito, 'Generalized Networks', in *Synthesis of Transmission Line Networks by Multivariable Techniques*, ed P. I. Brooklyn pp. 353–392, Polytechnic Press, 1966.

[5] T. Koga, 'Synthesis of a Resistively Terminated Cascade of Uniform Lossless Transmission Lines and Lumped Passive Lossless Two-ports', *IEEE Transactions on Circuit Theory*, Vol. 18, pp. 444–455, 1971.

[6] J. D. Rhodes and P. C. Marston, 'Cascade Synthesis of Transmission Lines and Lossless Lumped Networks', *Electronics Letters*, Vol. 7, pp. 621–623, 1971.

[7] D. C. Youla, J. D. Rhodes and P. C. Marston, 'Driving Point Synthesis of Resistor Terminated Cascades Composed of Lumped Lossless Passive 2-ports and Commensurate TEM Lines', *IEEE Transactions on Circuit Theory*, Vol. 19, pp. 648–664, 1972.

[8] S. O. Scanlan and H. Baher, 'Driving Point Synthesis of a Resistor Terminated Cascade Composed of Lumped Lossless 2-ports and Commensurate Stubs', *IEEE Transactions on Circuits and Systems*, Vol. 26, pp. 947–955, 1979.

[9] A. Fettweis, 'Scattering Properties of Real and Complex Lossless 2-ports', *Proceedings of the IEE*, Vol. 128, pp. 147–148, 1981.

[10] A. Fettweis, 'On the Scattering Matrix and the Scattering Transfer Matrix of Multidimensional Lossless Two-ports', *Archiv Elektrischen Übertragung*, Vol. 36, pp. 374–381, 1982.

[11] A. Fettweis, 'Multidimensional Circuit and Systems Theory', *IEEE International Symposium on Circuits and Systems*, Vol. 2, pp. 951–957, May, 1984.

[12] A. Fettweis, 'A New Approach to Hurwitz Polynomials in Several Variables', *Circuits, Systems and Signal Processing*, Vol. 5, pp. 405–417, 1986.

[13] B. S. Yarman, 'Modern Approaches to Broadband Matching Problems', *Proceedings of the IEE*, Vol. 132, pp. 87–92, 1985.

[14] B. S. Yarman, and H. J. Carlin, 'A Simplified Real Frequency Technique Applied to Broadband Multistage Microwave Amplifiers', *IEEE Transactions on Microwave Theory and Techniques*, Vol. 30, pp. 15–28, 1983.

[15] A. Fettweis, 'Factorization of Transfer Matrices of Lossless Two-ports', *IEEE Transactions Circuit Theory*, Vol. 17, pp. 86–94, 1970.

[16] A. Aksen, 'Design of Lossless Two-ports with Mixed, Lumped and Distributed Elements for Broadband Matching', *PhD dissertation, Ruhr Universitaet Bochum, 1994.*

[17] B. S. Yarman and A. Aksen, 'An Integrated Design Tool to Construct Lossless Matching Networks with Mixed Lumped and Distributed Elements for Matching Problems', *IEEE Transactions on Circuits and Systems*, Vol. 39, pp. 713–723, 1992.

[18] A. Aksen and B. S. Yarman, 'A Semi-analytical Procedure to Describe Lossless Two-ports with Mixed Lumped and Distributed Elements', *IEEE International Symposium on Circuits and Systems*, Vol. 5–6, pp. 205–208, 1994.

[19] A. Sertbas, A. Aksen and B. S. Yarman, 'Construction of Some Classes of Two-variable Lossless Ladder Networks with Simple Lumped Elements and Uniform Transmission Lines', *IEEE APCCAS'98, Asia Pacific Conference on Circuits and Systems*, Thailand, November 1998.

[20] B. S. Yarman and A. Fettweis, 'Computer Aided Double Matching via Parametric Representation of Brune Functions', *IEEE Transaction on Circuits and Systems*, Vol. 37, pp. 212–222, 1990.

[21] W. K. Chen, *Broadband Matching: Theory and Implementations*, 2nd edition World Scientific, 1988.

[22] P. L. D. Abrie, *Design of RF and Microwave Amplifiers and Oscillators*, Artech House, 1999.

[23] H. J. Carlin and B. S. Yarman, 'The Double Matching Problem: Analytic and Real Frequency Solutions', *IEEE Transactions on Circuits and Systems*, Vol. 30, pp. 15–28, 1983.

Index